A Systematic Vademecum *to the* Vascular Plants *of* Puerto Rico

Franklin S. Axelrod

A Systematic Vademecum *to the* Vascular Plants *of* Puerto Rico

Franklin S. Axelrod

BRIT
PRESS
34
2011

Number 34
SBM: Sida, Botanical Miscellany
ISBN 13-978-1889878-33-1
ISSN 0883-1475

Editor: Barney Lipscomb
Botanical Research Institute of Texas
1700 University Dr.
Fort Worth, Texas 76107-3400, USA
Contributing Spanish Editor: Félix Llamas
Dpto. de Biodiversidad y Gestión Ambiental, Universidad de León
Campus de Vegazana s/n, 24071 León, Spain
Production: Rebecca Horn, rhorngraphics, rlrchorn@verizon.net
Cover Design: Christian Franklin (BRIT)

A Systematic Vademecum to the Vascular Plants of Puerto Rico
By Franklin S. Axelrod
Front cover photograph: Photo by Amelia Merced Alejandro
Vicinity of Dos Picachos, El Yunque National Forest
Inside back cover maps: Physiographic maps adapted from Gould et al. (2009) by Olga Ramos of the GIS and Remote Sensing Lab, USDA Forest Service, International Institute of Tropical Forestry.

A Systematic Vademecum to the Vascular Plants of Puerto Rico
is published with the support of:
Center for Applied Tropical Ecology and Conservation of the University of Puerto Rico – Río Piedras
The National Science Foundation
Botanical Research Institute of Texas

Distribution of copies by:
Botanical Research Institute of Texas
1700 University Dr.
Fort Worth, Texas 76107-3400, USA
Telephone: 1 817 332 4441
Fax: 1 817 332 4112
Website: http://www.brit.org/brit-press/
Email: jbrit@brit.org

BRIT
PRESS

18 Apr 2011

Printed in the United States of America

A Systematic Vademecum *to the* Vascular Plants *of* Puerto Rico

Dedicated

To my wife, Jean Turnquist,
and
children Ysaaca, Amittai, and Barzilai,
all quondam companions willy nilly of my
collecting and sufferers from a surfeit of my specimens,
I offer this work to make amends.

A Systematic Vademecum *to the* Vascular Plants *of* Puerto Rico

A Book to Promote Botanical and Environmental Education, Scientific Exploration, and Conservation in Puerto Rico

The following institutions have generously provided institutional and financial support for the publication of this book. The author would like to thank them for their support.

Center for **A**pplied **T**ropical **E**cology
and **C**onservation
University of Puerto Rico – Río Piedras

The Center for Applied Tropical Ecology and Conservation (CATEC) of the University of Puerto Rico-Río Piedras (UPR-RP) is an agency that seeks to promote the development of education and research in tropical ecology in Puerto Rico. Its mission is fourfold: **1)** to provide training to students (at undergraduate and graduate levels) and postdoctoral researchers, **2)** to conduct and support research in conservation biology and other environmental issues, **3)** to improve the infrastructure of research facilities and **4)** to contribute relevant data to policy and conservation management of the island's natural resources.

The University of Puerto Rico, Río Piedras Campus (UPRRP)—or Universidad de Puerto Rico, Recinto de Río Piedras in Spanish—is a state university and located in the city of San Juan, Puerto Rico. UPRRP is the largest campus of the University of Puerto Rico System, and the first public university in the history of Puerto Rico.

University of **P**uerto **R**ico, **R**ío **P**iedras
(UPRRP)

National Science Foundation

The National Science Foundation (NSF) is an independent federal agency created by Congress in 1950 "to promote the progress of science; to advance the national health, prosperity, and welfare; to secure the national defense..."

TABLE OF CONTENTS

NEW COMBINATIONS

The following three new combinations are made in this work:

A Systematic Vademecum *to the* Vascular Plants *of* Puerto Rico

Franklin S. Axelrod

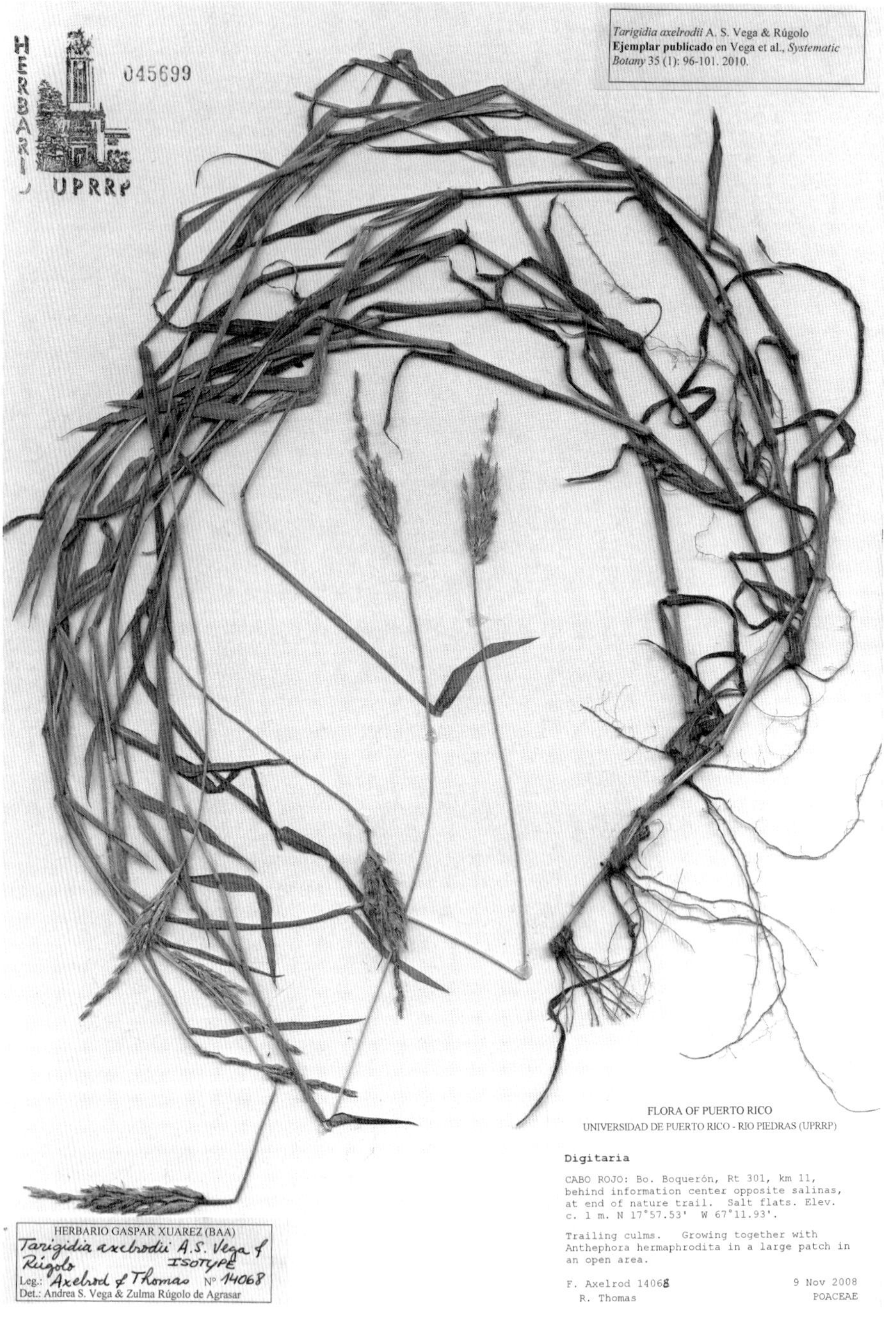

Isotype of *Tarigidia axelrodii* A.S. Vega & Rúgolo

A grass serendipitously discovered by the author and his wife on a field trip to the southwestern tip of Puerto Rico in 2007. As can been seen from the scan, he thought for a time that it belonged in the genus *Digitaria*. It was subsequently determined to belong to a rare genus and named after the author by the two gracious Argentinian agrostologists. This scan is used as a frontispiece both from pride of discovery and in homage to the largest, yet most unfavored family, in the Puerto Rican flora.

A SYSTEMATIC VADEMECUM TO THE VASCULAR PLANTS OF PUERTO RICO

Franklin S. Axelrod

Herbarium, Department of Biology
University of Puerto Rico-Río Piedras
P.O. Box 70377
San Juan, Puerto Rico 00936–8377, U.S.A.

ABSTRACT

This book is a guide to the vascular plants presently found growing in the wild on the island of Puerto Rico and on its dependent adjacent islands. It lists 2912 taxa that are included in 1057 genera and 210 families. Of these taxa 2337 (80.3%) are native and 575 (19.7%) are exotic; of the native ones, 243 (10.4%) are endemic.

Many taxa given in previous flora have been excluded due to their dubious or unsupported presence, their misidentification, or their confinement to cultivation or ornament. There is a special section at the end of each family devoted to accepted taxa which have not been collected for at least 50 years.

For each entry, online nomenclators such as GRIN, IPNI and TROPICOS and/or original literature and recent journal articles have been consulted in order to provide a correct name and an accurate literature citation. The Angiosperm Phylogeny Website was consulted for the most recent alignment in genera and families. Taxonomic controversy and embellishment, where appropriate, are aired out in a 'Notes' section.

The distribution of the vascular plants within Puerto Rico, based solely on verified specimens, is provided according to a scheme that incorporates both broad physiographic areas and specific geographical sites. The latter are described in the text and illustrated by maps.

Plant elevation and times of reproductive phenology are also provided. When the distribution is based on just one or two specimens, these specimens are referenced.

The distribution of these plants outside of Puerto Rico is also given. Where possible, references are made to modern floras of Puerto Rico where may be found a more detailed description of individual taxa. Where known, common Spanish names are provided.

Also, four new combinations have been made in the names of taxa.

RESUMEN

Este libro es una guía de las plantas vasculares que se encuentran actualmente creciendo de modo natural en la isla de Puerto Rico y en las islas adyacentes bajo su dependencia. Se listan 2912 taxa, que se incluyen en 1057 géneros y 210 familias. De estos taxa 2337 (80.3%) son nativos y 575 (19.7%) son exóticos; de los nativos, 243 (10.4%) son endémicos.

Muchos taxa, dados previamente, se han excluido debido a su presencia dudosa o no documentada, a identificaciones erróneas, o su confinamiento a cultivo u ornamentación. Hay una sección especial al final de cada familia dedicada a taxa aceptados que no han sido colectados desde hace más de 50 años.

Para cada entrada, se han consultado nomenclátores online tales como GRIN, IPNI y TROPICOS y/o bibliografía original y artículos recientes de publicaciones para ofrecer un nombre correcto y una citación bibliográfica precisa. El Angiosperm Phylogeny Website se consultó para la más reciente colocación en géneros y familias. La controversia taxonómica y embellecimiento, cuando son apropiados, son aireados en la sección 'Notes'.

Se aporta la distribución de las plantas vasculares en Puerto Rico, basada únicamente en especimenes verificados, de acuerdo a un esquema que incorpora tanto áreas fisiográficas amplias como lugares geográficos específicos. Estos últimos se describen en el texto y se ilustran mediante mapas.

Se aporta también la elevación de la planta y las épocas de su fenología reproductiva. Cuando la distribución se basa en uno o dos especimenes, estos especimenes están referenciados.

Se da también la distribución de estas plantas fuera de Puerto Rico. Cuando es posible, se hacen referencias a floras modernas de Puerto Rico en las que puede encontrarse una descripción más detallada de taxa individuales. Se aportan, cuando son conocidos, los nombres comunes en español.

Finalmente, son efectuadas cuatro nuevas combinaciones en los nombres de los taxa.

INTRODUCTION

This book is a guide to the vascular plants presently found growing in the wild on the island of Puerto Rico and on its dependent adjacent islands. It is intended to meet the needs of botanists who wish to find and study specific plants now present on the island. Because the book is meant it to accompany them in the field, I have called it a 'vademecum.' To compile it, I have relied on practical knowledge gained from twenty-five years experience in collecting almost 14000 specimens of Puerto Rico plants and on taxonomic knowledge gleaned both from my own study of taxonomic literature and from the tutelage of experts. The practical knowledge has been used to vet at least 50000 specimens contained in Puerto Rican and United States herbaria; the taxonomic knowledge has been used to verify the names, authors and places of publication of all relevant Puerto Rican taxa. 'Taxa' (singular 'taxon') are basic taxonomical entities such as species, subspecies and varieties; it is a word I prefer over 'species' since it is less fraught with theoretical and forensic baggage.

This book treads on the heels of a similar work, the second edition (2000) of *Flora of Puerto Rico and Adjacent Islands: a Systematic Synopsis* by H.A. Liogier and L.F. Martorell. Though it covers the same subject matter as the latter, it presents more focused and accurate information. It is grounded upon a thorough search and study of existing specimens and of taxonomic literature, the end result of which is the addition and subtraction of many taxa in the flora, a whole raft of new names for the taxa retained and many shifts in the genera and families to which they belong. Furthermore, there is not only a more extensive and accurate description of the distribution of these taxa in Puerto Rico and elsewhere, but there is also a chronology of their reproductive phenology. The nomenclatural changes, while at first disconcerting, better reflect the evolutionary relationships of the plants that are to be classified and are largely the product of new and powerful research tools, the apotheosis of which has been reached in molecular systematics. A further consequence of such studies has been the blurring of the boundary within the flowering plants between monocots and dicots. To co-opt further changes in this alignment, I have thought it best to list the families of both putative groups together under Magnoliophyta. Cycads, conifers, ferns and fern allies thus far remain inviolately separate.

In sum, this work presents **1)** what are for the moment the taxonomically accepted names of Puerto Rican vascular plants, together with the names employed for them in previous accounts of the Puerto Rican flora; **2)** the localities and elevation at which these plants are to be found and the times at which they flower and fruit or produce spores, all such data vouchered by herbarium specimens; **3)** either, in the case of presumed native plants, their general distribution outside of Puerto Rico or, in the case of exotics, their place of origin; **4)** notes that discuss problematic issues relating to their taxonomy or ecology; and **5)** wherever known, Spanish common names for them.

Organization

The scientific names of Puerto Rican plants, in the form of Latin binomials, are herein organized alphabetically, first by family and then by genus and specific epithet, within five separate sections: Lycopodiophyta, Pteridophyta, Cycadophyta, Pinophyta and Magnoliophyta. The currently accepted name has been determined by consulting the latest taxonomy agreed upon by experts in the group to which it belongs. This information has been obtained from either recently published works and online databases or by communication with experts. If there is no consensus in the use of a new name, then the old accepted name has been retained. Where there are conflicts in the taxonomy and decisions have to be made of a nontrivial nature, such decisions are explained in the notes after each taxon. The literature citation for all accepted names has been given, since I have found many errors in previous citations. Such errors were resolved through consultation of original sources, either directly or through query of experts with access to such sources. With rare exception, authorities are abbreviated according to Brummitt & Powell (1992); sources of publication, given only for the accepted name, are abbreviated according to the usage in Stafleu & Cowan's *Taxonomic Literature*. 2nd ed. (TL-2) and in the two volumes of the *Botanico-Periodicum-Huntianum* (BPH), sometimes modified by the online version (BPH-2).

The **currently accepted name of a taxon** is followed by its various **synonyms,** that is, those validly published names that were formerly applied to the plant but have been superseded for reasons either of priority, legitimacy or revision, such as to cause placement in a new or different genus. The synonym first given is often a special one called a **basionym,** which is the name that was given in the first published description of the plant, but later superseded. It is preceded by '**Bas.**' and followed by its date of publication in parentheses. The basionym is followed by other synonyms, whose number I have restricted to only those that were used as accepted names by Britton & Wilson (1923–1930) and later authors of floras dealing with Puerto Rico. These synonyms are preceded by '**Syn.**' and those belonging to the same type specimen are grouped together within a separate indentation; again, each citation is followed by its date of publication in parentheses. In the section of excluded taxa, I do not treat basionyms apart, but lump all the synonyms together and give no dates. When a name has been misapplied to a taxon or has not been validly published, it is indicated in the following notes rather than kept with the true synonyms.

A brief description of the **plant growth habit** is given and is usually confined to simple terms such as tree, shrub, herb, terrestrial, epiphyte, lithophyte, with an occasional embellishment, such as 'robust herb.' While I would have liked to have been able to state whether a plant was common, occasional or rare, I found in practice that this circumscription was no easy matter; for example, what to do, on the one hand, with an uncommon plant that is widely distributed or, on the other hand, with one that is very locally abundant but not found elsewhere on the island. I decided, in the interest of keeping my text a reasonable length, not to tackle the issue.

A perennial issue is what plants to include in a flora. **Endemic, native and invasive plants** all have their place; **cultivated and ornamental plants** are the problem. In general, unlike Liogier & Martorell (1982, 2000), I have excluded strictly ornamental plants from the flora. By 'strictly ornamental' I mean plants that either never stray from cultivation or, if they do, do so rarely and transiently. As for accepted ornamentals, I describe them as 'planted and persistent,' which indicates that they are to be encountered, at least occasionally, in areas outside of cultivation, even though their presence is commonly due to human intervention. I have avoided the word 'naturalized,' since I am not quite sure how to apply it to an island flora where practically all the plants are in some sense 'naturalized'—the question being not whether, but when.

Localities are listed in which these plants may presently be found. To this end only specimens recently collected—by my fiat, from the 1960s onwards, when extensive collecting was begun in Puerto Rico by Henri Alain Liogier and Roy Woodbury—are used. Undetermined sightings and checklists without substantiating vouchers are for the most part not used, though they may be mentioned. My reason for restricting this work to collections made after 1960 is that I want it to reflect the present state of the vegetation of Puerto Rico. The habitats of many plants collected prior to that year are no longer with us, though at present those plants are usually to be found elsewhere on the island.

For purposes of describing locality in a succinct, but useful fashion, I have divided Puerto Rico into not only **topographically distinct regions,** roughly following Picó (1974), but also **botanically unique areas,** which have been the foci of collectors. I have also included various **Commonwealth Forest Reserves,** since some are the ultimate repository for many of the rarer taxa (bear in mind in mind that to collect plants in these reserves a permit must be obtained from the Puerto Rico Department of Natural Resources and the Environment, as one must also be obtained from the U.S. Forest Service for collecting in El Yunque National Forest). The result of this locality schema is a pastiche, but one which, in my opinion, most succinctly represents where the plants are. Purely political localities, such as municipios (municipalities) encompass too much ecological diversity; purely geographical localities, though somewhat better, need to be better circumscribed. Some of the areas noted below, though often having no salient botanical or ecological boundaries, are noted because they are well known to collectors and contain plants not found elsewhere in the larger geographical area to which they belong.

The following, in alphabetical order, are the **geographical terms** which I use to categorize the distribution of plants within Puerto Rico:

Caguas Valley: a valley extending east-west from Las Piedras to Caguas and also including Gurabo, Juncos, and the northern part of San Lorenzo

Caja de Muertos: an island situated off the southern coast at Ponce; it is included within the Southern Coastal Lowlands (the subject of special collection)

Cambalache Forest Reserve: low haystack hills (*mogotes* in Spanish) along the coast at Arecibo; it is included within the Northern Limestone Hills

Cañón San Cristóbal: a steep-sided and deep ravine in Aibonito/Barranquitas; it is included within the Eastern Cordillera (the subject of special collection)

Cayo Santiago: an island with an introduced monkey colony situated off the eastern coast at Playa Santiago in Humacao; it is included within the Eastern Coast (the subject of special collection)

Central Cordillera: the part of the central mountains stretching from Orocovis/Villalba west to Yauco and including the Toro Negro and Monte Guilarte Forest Reserves and Los Tres Picachos; I have here deviated from standard geographical usage, which has the Central Cordillera running from Cayey to Maricao

Cerro de Las Cuevas: an isolated limestone ridge in Juana Díaz south of the Central Cordillera (the subject of special collection)

Cerro La Peña: an isolated peak in Bayamón (about 400); is included within the Eastern Cordillera (the subject of special collection)

Cerro Las Mesas: a ridge east of Mayagüez that is a continuation of the serpentine outcrop found in the Maricao and Susúa Forest Reserves (the subject of special collection)

Cuchilla de Panduras: a ridge running from the Sierra de Cayey to the eastern coast of the island at Yabucoa/Maunabo; it is included within the Eastern Coast (the subject of special collection)

Culebra: an island off the eastern coast

Desecheo: an island off the western coast

Eastern Coast: the coastal strip from Ceiba to Maunabo, including hills and valleys adjacent to the coastal plain; it includes Cayo Santiago, Punta Guayanés and Cuchilla de Panduras

Eastern Cordillera: the lower eastern part of the central mountains, stretching westward from Cidra and western Cayey, through Aibonito and Comerío to Barranquitas, including also the upper hills of Guaynabo, Bayamón, Aguas Buenas and Naranjito; it includes Cerro de Peña, Cañón San Cristóbal and La Torrecilla

Guajataca Forest Reserve: haystack hills (*mogotes*) in southern Isabela; it is included within the Northern Limestone Hills

Guánica Forest Reserve: an extension of the Southern Coastal Limestone which contains a unique dry forest flora

Islands Adjacent to Fajardo: small islands scattered off the northeastern coast from Fajardo to Ceiba; they are included within the Northern Coastal Lowlands

Laguna Tortuguero: a natural freshwater lagoon on the north central coast, extending from western Vega Baja to eastern Manatí, which contains a unique flora on its shores of white sand; it is included within the Northern Coastal Lowlands (the subject of special collection)

Las Piedras Chiquitas: razorback hills on the border between Coamo and Salinas (the subject of special collection)

Las Tetas de Cayey: isolated set of peaks (about 800 m) in Cayey/Aibonito just west of the Sierra de Cayey (the subject of special collection)

La Torrecilla: isolated peak (just over 900 m) in Barranquitas, with a relict cloud forest; it is included within the Eastern Cordillera (the subject of special collection)

Los Tres Picachos: an isolated ridge of the central mountains in Jayuya; it is included within the central mountains (the subject of special collection)

Maricao Forest Reserve: a mostly serpentine outcrop occupying the border region of Maricao, Sabana Grande and San Germán; it is included within the Western Cordillera

Mona: an island off the western coast, also a nature reserve

Monte Guilarte Forest Reserve: an isolated peak (about 1200 m) in Adjuntas at the western end of the Central mountains; it is included within the Central Cordillera

Northern Coastal Lowlands: mostly coastal plain extending from Fajardo to Aguadilla; much of it long ago cleared and planted, but subsequently abandoned; it includes the Islands Adjacent to Fajardo, Pinoñes Forest Reserve and Laguna Tortuguero

Northern Limestone Hills: a band of haystack hills (*mogotes*) just south of, but sometimes extending into, the Northern Coastal Lowlands; it is at first narrow and sporadic in the east from Loíza to Bayamón and thereafter broad and continuous up to Moca on the Western Coast; it includes the Cambalache and the Guajataca Forest Reserves

Piñones Forest Reserve: mostly mangrove and marsh in Carolina at the northeast coast; it is included within the Northern Coastal Lowlands

Punta Guayanés: a headland and slopes in Yabucoa, just south of Humacao and north of Cuchilla de Panduras, which contains a unique flora; it is included within the Eastern Coast (the subject of special collection)

Quebrada de Los Cedros: a ravine running northwards from Guaynilla up into Peñuelas; it is included within the Southern Coastal Limestone (the subject of special collection)

Río Abajo Forest Reserve: haystack hills (*mogotes*) south of Arecibo, about 200 meters higher than those of the Northern Limestone Hills; which contain a unique flora

Sierra Bermeja: a range of hills in the southwest corner of the island in Cabo Rojo/Lajas, unusual in that the soil substrate is chert (the subject of special collection)

Sierra de Cayey: a mountain region which contains parts of the municipalities of Caguas, Cayey, Guayama, Patillas, and San Lorenzo, southwest of the Sierra de Luquillo and east of the Eastern Cordillera, but separated from them both by valleys; part of it consists of the Carite Forest Reserve

Sierra de Luquillo: a mountain region containing El Yunque (formerly Caribbean) National Forest and the higher parts of the municipalities of Luquillo, Río Grande, Canóvanas, Naguabo and Ceiba

Southeastern Foothills: a semi-arid region, between the Sierra de Cayey and the Southern Coastal Lowlands, which contains parts of the municipalities of Cayey, Coamo, Guayama, Salinas and Juana Díaz

Southern Coastal Lowlands: mostly coastal plain extending from Patillas west to Yauco, up to the Guánica Forest Reserve, now mostly agricultural land or pasture; Quebrada de Los Cedros and Caja de Muertos and various other small islands are included within it, while the coastal band of limestone is not

Southern Coastal Limestone: coastal band of limestone, not karsted, extending from Ponce to Guayanilla, which consists primarily of scrub forest

Southwestern Coastal Lowlands: an area extending from Guánica, excluding the Forest Reserve, to Cabo Rojo and including parts of Sabana Grande and San Germán

Susúa Forest Reserve: lower foothills of the Western Cordillera, northwest of Guánica Forest Reserve and southeast of the Maricao Forest Reserve, which consist mostly of serpentine, but with some limestone

Toro Negro Forest Reserve: the highest part of the island (up to about 1340 m), extending from Orocovis to Ponce; it is included within the Central Cordillera

Vieques: an island off the eastern coast

Western Coast: an area extending southward from the limestone hills (*mogotes*) of Aguada to Hormigueros and part of San Germán; it also includes hills and valleys adjacent to the coastal plain

Western Cordillera: the western end of the central mountains, treated separately because much of it consists of serpentine with a concomitant unique flora; it includes the Maricao Forest Reserve

When the range of a taxon within one of the above areas is limited, the municipalities to which it is restricted will be put in parentheses, as in 'Northern Limestone Hills (Bayamón, Toa Baja).

The geographical locality is followed by an **altitude range** in meters ('m') and then by the phenology of the plant. Altitude is determined directly or indirectly from information on specimen labels; it is either based on altimeter readings or on topographic maps and is subject, even in cases of accurate reporting of locality, to the vagaries of the instruments or maps so that it is to be taken only as a rough estimate. By my convention, an altitude of '0 m' signifies that a given plant actually grows in the tidal zone, whereas one of '1 m' signifies that it is always above that mark, if only barely.

Phenology here concerns only the times of reproductive events, namely the times of **flowering** ('*fl*'), **fruiting** ('*fr*') and **cone or spore formation** ('*spr*'). These times are determined from herbarium specimens and are given as a range made by ordering the consecutive months of activity. A gap of a single month will not affect this ordering, that is, the range of a plant which flowers in April, May and July will be described as Apr–Jul. When a month falls out of that range, it is listed separately, for example, 'Apr–Jul, Dec.' When months are known from only one or two specimens, that fact will be indicated, for example, *fl* Jun (one specimen); when they are not known, they will given as '*fl* (?)' or '*fr* (?).' Note that because of the existence of sterile specimens, the number of specimens used to determine phenology is often fewer than those determining locality, with the result that the one or two specimens determining phenology are not cited as vouchers (see below).

Many recent floristic studies cite representative herbarium specimens for all taxa. Numerous citations, though desirable, would have made this work much too ponderous and cluttered. The purpose of citation, after all, is to attest to the accuracy of a work. Sticking to this purpose, I came up with the following compromise: since the accuracy of the work is most likely to be flawed in the identification of the least common specimens, those that have been collected only rarely, I have cited such specimens in brackets at the end of the above section so that they might be checked by the inquisitive. For all other specimens, I refer the reader to the plant databases of the herbaria of the Departments of Biology of the University of Puerto Rico at Río Piedras (UPRRP) and at Mayagüez (MAPR), both to be encountered on the Botanical Research and Herbarium Management System (BRAHMS) website (cited in the Selected References Section) and on the websites of the individual herbaria. The **citation of specimens** is given by *collector, collection number*—when that number is not given, by the *herbarium accession* number preceded by '*a.n.*' or, when that is not known, by the *date of collection* or, when even that is not known, simply by '*s.n.*' (*sine numero*)—and finally the herbarium where the specimen is located is given by the herbarium acronym in parentheses, such as '(UPRRP).'

Finally, the broader **geographical distribution** in which the plant is native is given. When native only to Puerto Rico, it is listed as 'endemic.' When native to Puerto Rico and other areas, only those others are listed; when not native to Puerto Rico, the list of areas to which it is native is prefaced by 'Native to.' Certain conventions will apply here: when more three or more countries in a region contain the taxon in question, only the whole region will usually be listed; when fewer than three contain it, the countries in question will be put in parenthesis after the region, e.g., Lesser Antilles (Guadeloupe, Martinique). The following convention is used for the area in and around the Caribbean: 'Greater Antilles,' besides Puerto Rico, denotes the islands of Jamaica, the Cayman Islands, Cuba, Hispaniola (the Dominican Republic and Haiti); 'Lesser Antilles' denotes the island chain from Grenada to Anguilla, the Southern Caribbean Islands (Aruba, Curaçao, Bonaire, Isla Margarita) and Trinidad and Tobago being excluded; 'Antilles' denotes both the Greater and Lesser Antilles; 'West Indies' denotes the Antilles together with the Bahamas. The Virgin Islands, though part of the Puerto Rican Bank, and thus part of the Greater Antilles, are treated as a separate entity when vagaries of distribution merit a separate listing. This information on plant nativity, of which I have little direct knowledge, has been taken from GRIN and from various floras, judiciously supplemented from lists of specimens in TROPICOS. My feeling upon retailing this information is that much of the attribution of nativity is guesswork and that it is far easier to determine where a plant is not native than where it is. Much work needs to be done with formal studies of genetic markers, pollinators, pests and such before attribution of plant distribution in the West Indies is reliable.

For the convenience of locating descriptions of the individual taxa cited, I have appended **a page reference to the following flora dealing with Puerto Rico** (see the Bibliography for the complete references), using the abbreviation given in brackets:

Acevedo-Rodríguez (2005) – [AR]
Acevedo-Rodríguez & Strong (2005) – [AR&S]
Ackerman (1995) – [A]
Britton & Wilson (1923–1930) – [B&W5], [B&W6]
Liogier (1985–1997) – [L1], [L2]…[L5]
Proctor (1989) – [P]

Where commentary is appropriate, **Notes** are appended to the above information.

The last line is the **Spanish name** or names, where known, for the taxon. When more than one name is given, they are listed in alphabetical order, with no preference given to any particular given name. For these names, I have relied entirely on secondary sources, primarily Martorell, Liogier and Woodbury (1981), with additions and subtractions from works by Little and Wadsworth (1964), Little, Woodbury and Wadsworth (1974), Liogier and Acevedo-Rodríguez (2005), Acevedo-Rodríguez and Strong (2005). Since I am not directly conversant with most of the names, I can give no guarantees that many of them are actually used by, or even known to, native speakers; nor do I consider this selection exhaustive. I set these names down therefore not as a guide to usage but rather as a reference for those that might be heard or read.

At the end of the list of recently collected taxa of a given family, where applicable, I have provided two separate sections. The first section is a Limbo, that is, a place for **Taxa not recently collected,** by which I mean ones collected before 1960, but not collected since. I have given them the same treatment as that of accepted taxa and, where possible, have included the historical information for their collection, namely the collector, locality, collection number and date of collection. Perhaps someone using this book will again find a specimen of *Rorippa portoricensis* in Puerto Rico. The second section is a list of taxa under the heading **Taxa cited by previous authors, but not included.** This list is composed of four subsections: (**1**) 'No specimens seen' refers to records based on the citation of other authors whose accuracy I cannot judge because I have been unable to examine the material, (**2**) 'Doubtful occurrence' refers to records based on the citation of taxa without corroborating vouchers, such as specimens or photographs deposited in herbaria, or based on sterile collections of dubious identification, (**3**) 'Cited due to misidentification' refers to records based on specimens wrongly cited as the given taxon and (**4**) 'Only under cultivation' refers to records of ornamentals which have been collected only under conditions of cultivation and trees and shrubs that are 'planted and persistent' but of very limited occurrence.

ANNOTATED CHECKLIST OF THE VASCULAR PLANTS

LYCOPODIOPHYTA

LYCOPODIACEAE

Huperzia acerosa (Sw.) Holub, Folia Geobot. Phytotax. 20(1):70. 1985.
Bas.: *Lycopodium acerosum* Sw. (1806).
Syn.: *Lycopodium portoricense* Underw. & F.E. Lloyd (1906).

Epiphyte in Sierra de Luquillo, Sierra de Cayey, Monte Guilarte Forest Reserve, Maricao Forest Reserve; 450–1150 m; *spr* Aug–Apr.

West Indies, Trinidad, southern Central America, northern South America. [P: 17].

Notes: Øllgaard (1983) applies the above name to only Neotropical populations of what was formerly considered a broader taxon. The previously used names, *Huperzia verticillata* (L.f.) Trevis., *Lycopodium verticillatum* L.f. and *Lycopodium setaceum* Lam., now apply to only Paleotropical populations.

Colchón de pobre.

Huperzia aqualupiana (Spring) Rothm., Feddes Repert. Spec. Nov. Regni Veg. 54:62. 1944.
Bas.: *Lycopodium aqualupianum* Spring (1841).

Epiphyte in Sierra de Luquillo, Sierra de Cayey, Central Cordillera, Maricao Forest Reserve; 600–1250 m; *spr* throughout the year.

Greater Antilles (Cuba, Hispaniola), Lesser Antilles, Trinidad, northern South America (Venezuela, Colombia). [P: 21].

Huperzia dichotoma (Jacq.) Trevis., Atti Soc. Ital. Sci. Nat. 17(3):248, no. 22. 1875.
Bas.: *Lycopodium dichotomum* Jacq. (1762).

Epiphyte in Northern Limestone Hills (Quebradillas), Sierra de Luquillo, Sierra de Cayey, Maricao Forest Reserve; 200–900 m; *spr* Feb–Nov.

Antilles, Trinidad, southeastern United States (Florida), Mexico, Central America, northern South America to Bolivia, Paraguay. [P: 20].

Huperzia funiformis (Cham. ex Spring) Trevis., Atti Soc. Ital. Sci. Nat. 17(3):248, no. 30. 1875.
Bas.: *Lycopodium funiforme* Cham. ex Spring (1841).

Epiphyte in Sierra de Luquillo, Sierra de Cayey, Toro Negro Forest Reserve, Maricao Forest Reserve; 275–925 m; *spr* Nov–Apr.

Antilles, southern Central America, northern South America to Peru. [P: 19].

Huperzia linifolia (L.) Trevis., Atti Soc. Ital. Sci. Nat. 17(3):248, no. 15. 1875.
Bas.: *Lycopodium linifolium* L. (1753).

Epiphyte in Sierra de Luquillo, Sierra de Cayey, Eastern and Central Cordillera, Maricao Forest Reserve; 300–925 m; *spr* throughout the year.

Antilles, Trinidad, Mexico, Central America, northern South America to Bolivia. [P: 20].

Huperzia reflexa (Lam.) Trevis., Atti Soc. Ital. Sci. Nat. 17(3):248, no. 41. 1875.
Bas.: *Lycopodium reflexum* Lam. (1792).
Epiphyte in Sierra de Luquillo, Eastern Coast (Cuchilla de Panduras), Sierra de Cayey, Central Cordillera, Maricao Forest Reserve; 600–1325 m; *spr* throughout the year.
Antilles, Trinidad, Mexico, Central America, Venezuela and northwestern South America to Bolivia, Paraguay. [P: 16].

Huperzia taxifolia (Sw.) Trevis., Atti Soc. Ital. Sci. Nat. 17(3):248, no. 17. 1875.
Bas.: *Lycopodium taxifolium* Sw. (1788).
Epiphyte in Sierra de Luquillo, Sierra de Cayey, Las Tetas de Cayey, Eastern Cordillera (La Torrecilla), Central Cordillera, Maricao Forest Reserve; 375–1300 m; *spr* throughout the year.
Antilles, Trinidad, Mexico, Central America, northern South America to Bolivia. [P: 20].

Huperzia tenuicaulis (Underw. & F.E. Lloyd) B. Øllg., Opera Bot. 92:169. 1987.
Bas.: *Lycopodium tenuicaule* Underw. & F.E. Lloyd (1906).
Epiphyte in Sierra de Luquillo, Sierra de Cayey; 800–1050 m; *spr* Oct–Jan, May.
Lesser Antilles (Guadeloupe, Dominica). [P: 19].

Huperzia wilsonii (Underw. & F.E. Lloyd) B. Øllg., Opera Bot. 92:170. 1987.
Bas.: *Lycopodium wilsonii* Underw. & F.E. Lloyd (1906).
Epiphyte in Sierra de Luquillo; 600–900 m; *spr* Sep, Dec (two specimens). [*Sotomayor 8b, 16c* (MAPR)].
Greater Antilles (Jamaica), Lesser Antilles, southern Mexico, Central America, northern South America to Peru. [P: 17].

Lycopodiella caroliniana (L.) Pic. Serm. var. **meridionalis** (Underw. & F.E. Lloyd) B. Øllg. & P.G. Windisch, Bradea 5:27. 1987.
Bas.: *Lycopodium meridionale* Underw. & F.E. Lloyd (1906).
Syn.: *Lycopodium carolinianum* L. var. *meridionale* (Underw. & F.E. Lloyd) Nessel & Hoehne (1927).
Terrestrial in Northern Coastal Lowlands (Laguna Tortuguero), Sierra de Luquillo; 1–1025 m; *spr* Dec–Sep.
Greater Antilles, Lesser Antilles (Guadeloupe, Dominica), Trinidad, southern Mexico, Central America, South America (except Chile). [P: 22].

Lycopodiella cernua (L.) Pic. Serm. var. **cernua**, Webbia 23(1):166. 1968.
Bas.: *Lycopodium cernuum* L. (1753).
Terrestrial in Northern Coastal Lowlands, Northern Limestone Hills (Moca), Sierra de Luquillo, Eastern Coast (Cuchilla de Panduras), Caguas Valley (Las Piedras), Sierra de Cayey, Eastern Cordillera (Cerro La Peña), Central Cordillera, Western Cordillera, Cerro Las Mesas, Western Coast (Mayagüez); 40–1300 m; *spr* throughout the year.
Pantropical. [P: 23].
Azufre vegetal, Colchón de pobre, Pata de gato.

Lycopodiella cernua (L.) Pic. Serm. var. **curvata** (Sw.) Kartesz & Gandhi, Phytologia 70(3):202. 1991.
Bas.: *Lycopodium curvatum* Sw. (1801).
Syn.: *Lycopodium cernuum* L. var. *curvatum* (Sw.) Hook. & Grev. (1831).
Terrestrial in Sierra de Luquillo, Toro Negro Forest Reserve; 775–1250 m; *spr* Feb–Mar, Aug–Sep (two specimens). [*Proctor 39452, 41132* (SJ)].
Greater Antilles (Jamaica). [P: 23].

Lycopodium clavatum L., Sp. Pl. 2:1101. 1753.
Viny terrestrial in Toro Negro Forest Reserve; 1100–1300 m; *spr* throughout the year.
Nearly cosmopolitan. [P: 24].

TAXA NOT RECENTLY COLLECTED

Huperzia sintenisii (Herter) Holub, Folia Geobot. Phytotax. 20:76. 1985.
Bas.: *Urostachys sintenisii* Herter (1923).
Syn.: *Lycopodium sintenisii* (Herter) Maxon ex C. Chr. (1937).
Lithophyte collected in the Sierra de Naguabo on 27 Oct 1886 [*Sintenis 5429* (B), scan seen].
Greater Antilles (Hispaniola). [P: 17].

SELAGINELLACEAE

Selaginella armata Baker, J. Bot. 22:90. 1884.
Terrestrial in Northern Limestone Hills (Quebradillas); 225 m; *spr* (?). [*Proctor 39563, 39564* (SJ)].
Greater Antilles. [P: 31].
Notes: Maxon (1926) misapplied the name *Selaginella plagiochila* Baker to this taxon.

Selaginella cordifolia (Desv. ex Poir.) Spring, Bull. Acad. Roy. Sci. Bruxelles 10(1):228. 1843.
Bas.: *Lycopodium cordifolium* Desv. ex Poir. (1814).

Terrestrial in Northern Limestone Hills, Río Abajo Forest Reserve, Sierra de Luquillo, Eastern Coast (Yabucoa), Sierra de Cayey, Las Tetas de Cayey, Eastern Cordillera (Corozal, Cañón San Cristóbal); 50–825 m; *spr* Jan–Apr, Aug.

Greater Antilles (Cuba, Hispaniola). [P: 26].

Selaginella krugii Hieron. in Urban, Symb. Antill. 3(3):526. 1903.

Terrestrial in Sierra de Luquillo, Sierra de Cayey; 475–1000 m; *spr* throughout the year.

Endemic. [P: 29].

Flor de piedra, Mano de gato silvestre.

Selaginella laxifolia Baker in Urban, Bot. Jahrb. Syst. 24(1):151. 1897.

Terrestrial in Northern Limestone Hills, Río Abajo Forest Reserve, Sierra de Cayey, Toro Negro Forest Reserve, Susúa Forest Reserve, Maricao Forest Reserve; 175–1200 m; *spr* throughout the year.

Endemic. [P: 26].

Selaginella ovalifolia Baker, J. Bot. 22:90. 1884.

Terrestrial in Northern Limestone Hills (Arecibo), Sierra de Luquillo, Central Cordillera; 50–1200 m; *spr* Oct–Apr, Jul.

Greater Antilles (Jamaica, Hispaniola), northern Central America (Belize), northern South America (Venezuela). [P: 32].

Flor de piedra, Mano de gato silvestre.

Selaginella plana (Desv. ex Poir.) Hieron. in Engler & Prantl, Nat. Pflanzenfam. 1(4):703. 1901.
Bas.: *Lycopodium planum* Desv. ex Poir. (1814).

Terrestrial in Northern Limestone Hills (Arecibo, Quebradillas), Sierra de Luquillo, Central Cordillera, Western Cordillera (Maricao), Western Coast (Mayagüez); 1–1050 m; *spr* throughout the year.

Native to Indo-China, Indonesia. [P: 26].

Notes: Liogier & Martorell (1982) misapplied the name *Selaginella flabellata* (L.) Spring to this taxon.

Selaginella substipitata Spring, Bull. Acad. Roy. Sci. Bruxelles 10(1):227. 1843.
Syn.: *Selaginella portoricensis* A. Braun (1865).

Terrestrial in Sierra de Luquillo, Central Cordillera; 425–1250 m; *spr* Jul–Apr.

Greater Antilles (Cuba, Hispaniola), Lesser Antilles, Trinidad, northern South America to Bolivia. [P: 28].

Selaginella tenella (P. Beauv.) Spring, Bull. Acad. Roy. Sci. Bruxelles 10(1):234. 1843.
Bas.: *Diplostachyum tenellum* P. Beauv. (1804).
Syn.: *Selaginella albonitens* Spring (1843).
Selaginella subcaulescens Baker (1884).
Selaginella sintenisii Hieron. (1917).

Terrestrial in Northern Limestone Hills, Río Abajo Forest Reserve, Sierra de Luquillo, Caguas Valley (Gurabo), Sierra de Cayey (Patillas), Eastern Cordillera (Guaynabo), Central Cordillera, Maricao Forest Reserve; 100–750 m; *spr* throughout the year.

West Indies, Trinidad and Tobago, Mexico, Central America, northern South America (Venezuela, Colombia) and Brazil. [P: 28, 31].

Notes: Smith & Mickel (2004), having examined specimens and types of both *Selaginella tenella* and *Selaginella subcaulescens*, concluded that there is no significant character that separates them throughout their ranges.

Flor de piedra, Selaginela.

Selaginella willdenowii (Desv. ex Poir.) Baker, Gard. Chron. 1867: 783, 950. 1867 "willdenovii."
Bas.: *Lycopodium willdenowii* Desv. ex Poir. (1814).

Terrestrial in Sierra de Luquillo, Sierra de Cayey, Western Coast (Mayagüez); 75–700 m; *spr* Feb (one specimen).

Native to southeastern tropical Asia. [P: 25; AR: 41].

Mano de gato, Sombra.

TAXA CITED BY PREVIOUS AUTHORS, BUT NOT INCLUDED

ONLY UNDER CULTIVATION

Selaginella plumosa (L.) C. Presl. Syn.: *Selaginella stolonifera* (Sw.) Spring.

Terrestrial collected in Naguabo on 16 Nov 1886 [*Sintenis 5441* (US)], but cited as cultivated; also collected in Río Piedras on 29 Jul 1916 [*Stevenson 5651* (US)] and cited by Maxon (1926) as occurring on rocky banks in partial shade, though this information is not on the label of the specimen that I saw. [P: 29].

PTERIDOPHYTA

ANEMIACEAE

Anemia adiantifolia (L.) Sw., Syn. Fil. 157. 1806.
Bas.: *Osmunda adiantifolia* L. (1753).

Terrestrial in Northern Coastal Lowlands (Laguna Tortuguero), Northern Limestone Hills, Río Abajo Forest Reserve, Central Cordillera (Utuado), Maricao Forest Reserve; 1–900 m; *spr* throughout the year.

West Indies, Trinidad, southeastern United States (Florida), Mexico, Central America, northwestern South America and Venezuela, Brazil. [P: 46].

Anemia hirsuta (L.) Sw., Syn. Fil. 156. 1806.
Bas.: *Osmunda hirsuta* L. (1753).
Terrestrial in Maricao Forest Reserve, Western Coast (Añasco); 95–420 m; *spr* Mar, Jul, Oct.
Greater Antilles, Mexico, Central America, northwestern, west-central South America and Venezuela, Brazil. [P: 47].

Anemia hirta (L.) Sw., Syn. Fil. 155. 1806.
Bas.: *Osmunda hirta* L. (1753).
Terrestrial in Eastern Cordillera (Cayey, La Torrecilla); 350–925 m; *spr* Apr, Jul–Sep.
Greater Antilles (Hispaniola), Lesser Antilles, northern South America. [P: 47].

Anemia portoricensis Maxon, N. Amer. Fl. 16:48. 1909.
Terrestrial in Northern Limestone Hills, Río Abajo Forest Reserve; 50–450 m; *spr* Nov–Jul.
Endemic. [P: 46].

ASPLENIACEAE

Asplenium abscissum Willd., Sp. Pl. 5(1):321. 1810.
Epiphyte/terrestrial in Northern Limestone Hills, Río Abajo Forest Reserve, Sierra de Luquillo, Eastern Cordillera (Aguas Buenas), Central Cordillera (Ponce), Western Cordillera; 5–950 m; *spr* throughout the year.
Antilles, Trinidad, southeastern United States (Florida), Mexico, Central America, South America (scattered). [P: 226].

Asplenium auriculatum Sw., Kongl. Svenska Vetensk. Acad. Handl. 1817(1):68. 1817.
Epiphyte/terrestrial in Sierra de Luquillo, Sierra de Cayey, Eastern Cordillera (La Torrecilla), Central Cordillera, Maricao Forest Reserve; 475–1260 m; *spr* throughout the year.
Greater Antilles, Lesser Antilles (Guadeloupe), Trinidad, Mexico, northern South America to Bolivia. [P: 227].

Asplenium cirrhatum Rich. ex Willd., Sp. Pl. 5(1):321. 1810.
Syn.: *Asplenium radicans* L. var. *cirrhatum* (Rich. ex Willd.) Rosenst. (1906).
Epiphyte/terrestrial in Sierra de Luquillo, Sierra de Cayey, Central Cordillera (Utuado), Maricao Forest Reserve; 475–900 m; *spr* Jan–Oct.
Greater Antilles (Jamaica, Hispaniola), Lesser Antilles (Guadeloupe), southern Mexico, Central America, northern South America to Bolivia. [P: 223].

Asplenium corderoanum Proctor, Mem. New York Bot. Gard. 53:234. 1989.
Terrestrial in Northern Limestone Hills, Río Abajo Forest Reserve; 150–350 m; *spr* throughout the year.
Greater Antilles (Cuba). [P: 234].
Notes: Maxon (1926) and Liogier & Martorell (1982) misapplied the name *Asplenium myriophyllum* (Sw.) C. Presl to this taxon.

Asplenium cristatum Lam., Encycl. 2(1):310. 1786.
Terrestrial in Northern Coastal Lowlands (Guaynabo), Northern Limestone Hills, Río Abajo Forest Reserve, Caguas Valley (Gurabo), Sierra de Cayey, Las Tetas de Cayey, Eastern and Central Cordillera, Quebrada de Los Cedros, Maricao Forest Reserve; 65–1050 m; *spr* throughout the year.
Antilles, Trinidad and Tobago, southeastern United States (Florida), Mexico, Central America, northern South America to Bolivia. [P: 231].

Asplenium cuneatum Lam., Encycl. 2(1):309. 1786.
Epiphyte/terrestrial in Northern Limestone Hills (Ciales, Camuy), Río Abajo Forest Reserve, Sierra de Luquillo, Caguas Valley (Caguas), Sierra de Cayey, Eastern Cordillera (Guaynabo, La Torrecilla), Central Cordillera, Western Cordillera; 100–975 m; *spr* throughout the year.
Antilles, Trinidad and Tobago, Central America, northern South America to Bolivia, Polynesia, Africa. [P: 231].

Asplenium cuspidatum Lam., Encycl. 2(1):310. 1786.
Syn.: *Asplenium lacerum* Schltdl. & Cham. (1830).
Asplenium auritum Sw. var. *bipinnatifidum* Kunze (1845).
Epiphyte/terrestrial in Central Cordillera, Maricao Forest Reserve; 550–1100 m; *spr* throughout the year.
Greater Antilles, Trinidad, Mexico, Central America, northwestern, west-central South America and Venezuela. [P: 230].

Asplenium dentatum L., Sp. Pl. 2:1080. 1753 "Trich. dentatum."
Lithophyte/terrestrial in Northern Limestone Hills, Río Abajo Forest Reserve, Eastern Cordillera (Corozal, Aguas Buenas), Southwestern Coastal

Lowlands (Cabo Rojo); 10–350 m; *spr* throughout the year.

Bahamas, Greater Antilles, Trinidad, southeastern United States (Florida), southern Mexico, northern Central America, northern South America (Venezuela, Colombia). [P: 226].

Notes: Liogier & Martorell (1982, 2000) and Proctor (1989) used *Asplenium trichomanes-dentatum* L. for the name of this taxon; according to the ICBN (Vienna Code 23.8, ex. 19), it is to be corrected to the above.

Culantrillo.

Asplenium erosum L., Syst. Nat. (ed. 10) 2:1324. 1759.

Syn: *Asplenium auritum* Sw. (1801).

Epiphyte/terrestrial in Río Abajo Forest Reserve, Central Cordillera, Western Cordillera; 250–1050 m; *spr* Aug–May.

Greater Antilles, Trinidad, southeastern United States (Florida), southern Mexico, Central America, South America. [P: 230].

Asplenium feei Kunze ex Fée, Mém. Foug. 5:194. 1852.

Syn.: *Asplenium sanguinolentum* Kunze ex Mett. (1859).

Epiphyte/terrestrial in Sierra de Cayey, Eastern Cordillera (La Torrecilla), Central Cordillera, Maricao Forest Reserve; 500–1300 m; *spr* throughout the year.

Greater Antilles (Cuba, Hispaniola), southern Mexico, Central America, northern South America to northern Argentina and Paraguay. [P: 220].

Asplenium flabellulatum Kunze, Linnaea 9(1):71. 1834.

Syn.: *Asplenium radicans* L. var. *partitum* (Klotzsch) Hieron. (1904) "partita."

Epiphyte/terrestrial in Eastern Cordillera (La Torrecilla), Central Cordillera; 875–1225 m; *spr* Mar, Jul–Nov.

Greater Antilles, southern Mexico, Central America, northwestern, west-central South America and Venezuela. [P: 224].

Asplenium formosum Willd., Sp. Pl. 5(1):329. 1810.

Terrestrial in Eastern Cordillera (Cañón San Cristóbal), Central Cordillera, Western Cordillera (Maricao); 700–850 m; *spr* Sep–Dec, Mar–May.

West Indies, Mexico, Central America, northern South America to northern Argentina and Paraguay. [P: 224].

Asplenium heterochroum Kunze, Linnaea 9(1):67. 1834.

Lithophyte/terrestrial in Northern Coastal Lowlands (Dorado), Northern Limestone Hills, Guánica Forest Reserve; 10–225 m; *spr* Aug–Sep, Dec–Apr.

Greater Antilles (Cuba, Hispaniola), southeastern United States (South Carolina, Georgia, Florida), southern Mexico, northern Central America (Belize), northwestern South America and Venezuela. [P: 224].

Asplenium juglandifolium Lam., Encycl. 2(1):307. 1786.

Syn.: *Asplenium integerrimum* Spreng. (1821).

Epiphyte in Río Abajo Forest Reserve, Sierra de Luquillo, Sierra de Cayey, Central Cordillera; 300–1100 m; *spr* Jul–Feb.

Greater Antilles, Trinidad, southern Mexico, Central America, northern South America to Bolivia. [P: 220].

Asplenium laetum Sw., Syn. Fil. 79, 271. 1806.

Terrestrial in Northern Limestone Hills, Río Abajo Forest Reserve, Sierra de Luquillo, Central Cordillera (Adjuntas), Western Cordillera (Las Marías); 150–350 m; *spr* Oct–Jul.

Antilles, Trinidad and Tobago, southern Mexico, Central America, South America. [P: 225].

Asplenium obtusifolium L., Sp. Pl. 2:1080. 1753.

Epiphyte/terrestrial in Central Cordillera; 500–1250 m; *spr* Jan–Oct.

Greater Antilles (Jamaica), Lesser Antilles, Trinidad and Tobago, northern South America (Venezuela, Colombia) and Brazil. [P: 225].

Asplenium ocoense C. Chr., Kongl. Svenska Vetensk. Acad. Handl., ser. 3 16(2):53, *t. 13, ff.* 6–8. 1937.

Terrestrial in Northern Limestone Hills, Río Abajo Forest Reserve, Sierra de Luquillo, Caguas Valley (Gurabo), Eastern Cordillera (Aguas Buenas, Cañón San Cristóbal), Central Cordillera, Western Cordillera; 10–1100 m; *spr* throughout the year.

Greater Antilles (Hispaniola). [P: 232].

Asplenium pseuderectum Hieron., Hedwigia 60:239. 1919.

Epiphyte/terrestrial in Sierra de Luquillo, Sierra de Cayey, Central Cordillera (Yauco); 475–950 m; *spr* Jul–Apr.

Greater Antilles (Jamaica, Hispaniola), southern Mexico, Central America. [P: 228].

Asplenium pteropus Kaulf., Enum. Filic. 170. 1824.

Epiphyte/terrestrial in Central Cordillera; 850–1200 m; *spr* Aug–Nov, Feb–Apr.

Antilles, Trinidad, southern Mexico, Central America, northern South America to Bolivia. [P: 228].

Asplenium pumilum Sw., Prodr. 129. 1788.

Terrestrial in Las Piedras Chiquitas, Eastern Cordillera (Cayey, Cañón San Cristóbal), Central Cordillera, Vieques; 175–800 m; *spr* Jun–Dec.

Antilles, Trinidad and Tobago, southeastern United States (Florida), Mexico, Central America, South America (scattered). [P: 221].

Asplenium radicans L., Syst. Nat. (ed. 10) 2:1323. 1759.

Terrestrial in Central Cordillera; 875–1240 m; *spr* Jul–Mar.

Greater Antilles, southern Mexico, Central America, northern South America to Bolivia. [P: 224].

Asplenium rhomboidale Desv., Mém. Soc. Linn. Paris 6(3):272. 1827.

Syn.: *Asplenium sintenisii* Hieron. (1919).

Terrestrial in Eastern Cordillera (La Torrecilla), Central Cordillera, Maricao Forest Reserve; 500–1100 m; *spr* Dec–Mar, Jun–Sep.

Greater Antilles. [P: 228].

Asplenium salicifolium L., Sp. Pl. 2:1080. 1753.

Epiphyte/terrestrial in Sierra de Luquillo, Eastern Coast (Cuchilla de Panduras), Sierra de Cayey, Las Tetas de Cayey, Central Cordillera; 350–1250 m; *spr* Sep–Jun.

Antilles, southern Mexico, Central America, northern South America to Bolivia. [P: 227].

Asplenium serra Langsd. & Fisch., Pl. Voy. Russes Monde 1:16. 1810 (*tabula* published in 1818).

Terrestrial in Central Cordillera; 630–950 m; *spr* Jul–Dec.

Antilles, Trinidad, Mexico, Central America, South America (scattered). [P: 219].

Asplenium serratum L., Sp. Pl. 2:1079. 1753.

Epiphyte/terrestrial in Northern Coastal Lowlands (Loíza), Sierra de Luquillo, Eastern Coast (Cuchilla de Panduras), Sierra de Cayey, Las Tetas de Cayey, Eastern Cordillera (Bayamón), Western Cordillera; 1–750 m; *spr* throughout the year.

Antilles, Trinidad and Tobago, southeastern United States (Florida), southern Mexico, Central America, South America. [P: 219].

Culantrillo.

Asplenium sessilifolium Desv., Mag. Neuesten Entdeck. Gesammten Naturk. Ges. Naturf. Freunde Berlin 5:322. 1811.

Terrestrial in Maricao Forest Reserve; 650–800 m; *spr* Jul (one specimen). [*Proctor 41899* (SJ)].

Greater Antilles (Hispaniola), Central America, northern South America to Bolivia. [P: 232].

TAXA CITED BY PREVIOUS AUTHORS, BUT NOT INCLUDED

DOUBTFUL OCCURRENCE

Asplenium rutaceum (Willd.) Mett.

Terrestrial ascribed by Maxon (1926) to eastern Puerto Rico on the basis of a collection by Schwanecke; subsequently cited by Proctor (1989); also cited by Liogier & Martorell (1982) as only occurring in mountain forests and by Liogier & Martorell (2000) as occurring in shaded places, at upper to middle elevations, though collected only once by Schwanecke. [P: 221].

Asplenium uniseriale Raddi. Syn.: *Asplenium radicans* L. var. *uniseriale* (Raddi) L.D. Gómez.

Terrestrial cited by Maxon (1926) as occurring in woods near Adjuntas and Utuado; subsequently cited by Liogier & Martorell (1982); similarly cited by Proctor (1989), who also adds Orocovis; and cited by Liogier & Martorell (2000) as occurring in moist montane forests, at upper middle to rather high elevations in the Central Cordillera. [P: 224].

Notes: It is unlikely that this taxon occurs in Puerto Rico. One of the two specimens upon which Maxon based his citation [*Britton 5221* (NY)], the only one I could obtain, is better referred to *Asplenium flabellulatum* Kunze (determined by Alan R. Smith from a scan).

AZOLLACEAE

Azolla caroliniana Willd., Sp. Pl. 5(1):541. 1810.

Aquatic in Northern Coastal Lowlands; 1–15 m; *spr* Jan (one specimen).

Greater Antilles, Trinidad and Tobago, eastern and central United States, Mexico, Central America, South America. [P: 348].

Notes: Mickel & Smith (2004) placed this taxon in synonymy with *Azolla filiculoides* Lam.

Alfombra de agua.

BLECHNACEAE

Blechnum divergens (Kunze) Mett., Ann. Sci. Nat., Bot., sér. 5 2:225. 1864.
Bas.: *Lomaria divergens* Kunze (1834).
Syn.: *Struthiopteris exaltata* (Fée) Broadh. (1912).

Terrestrial in Sierra de Luquillo, Toro Negro Forest Reserve; 600–1260 m; *spr* Jul–Aug, Jan.

Greater Antilles (Cuba, Hispaniola), Lesser Antilles, southern Mexico, Central America, northwestern, west-central South America and Venezuela. [P: 173].

Blechnum fragile (Liebm.) C.V. Morton & Lellinger, Amer. Fern J. 57(2):68. 1967.
Bas.: *Lomaria fragilis* Liebm. (1849).
Syn.: *Struthiopteris polypodioides* (Sw.) Trevis. (1869).

Epiphyte/terrestrial in Sierra de Luquillo, Sierra de Cayey, Eastern Cordillera (La Torrecilla), Central Cordillera, Maricao Forest Reserve; 500–1325 m; *spr* throughout the year.

Greater Antilles, Trinidad, southern Mexico, Central America, western South America and Venezuela. [P: 173; AR: 22].

Blechnum insularum C.V. Morton & Lellinger, Amer. Fern J. 57(2):70. 1967.

Epiphyte/terrestrial in Sierra de Luquillo, Toro Negro Forest Reserve; 950–1300 m; *spr* Dec, Mar, Jun.

Lesser Antilles. [P: 174].

Notes: Maxon (1926) misapplied the name *Struthiopteris underwoodiana* Broadh. to this taxon.

Blechnum lherminieri (Bory) Mett., Fil. Hort. Bot. Lips. 64, *t. 4, ff. 14–15*. 1856.
Bas.: *Lomaria lherminieri* Bory (1845) "l'hermi-nieri."

Terrestrial in Sierra de Luquillo, Toro Negro Forest Reserve; 825–1240 m; *spr* Apr–Jan.

Antilles, northern South America (Venezuela, Colombia) and Ecuador.

Blechnum lineatum (Sw.) Hieron., Bot. Jahrb. Syst. 34:473. 1904.
Bas.: *Osmunda lineata* Sw. (1788).
Syn.: *Struthiopteris lineata* (Sw.) Broadh. (1912).

Terrestrial in Sierra de Luquillo, Sierra de Cayey, Central Cordillera, Maricao Forest Reserve; 775–1325 m; *spr* Aug–May.

Greater Antilles, western South America (Ecuador, Bolivia). [P: 174].

Blechnum occidentale L., Sp. Pl. 2:1077. 1753 "orientale"; Sp. Pl. (ed. 2) 2:1534. 1763.

Terrestrial in Northern Coastal Lowlands, Northern Limestone Hills, Río Abajo Forest Reserve, Sierra de Luquillo, Eastern Coast (Cuchilla de Panduras), Caguas Valley (Gurabo), Sierra de Cayey, Southeastern Foothills (Coamo), Eastern and Central Cordillera, Susúa Forest Reserve, Western Cordillera, Western Coast, Vieques; 10–950 m; *spr* throughout the year.

Antilles, Trinidad, southeastern United States (Florida), Mexico, Central America, South America. [P: 172].

Blechnum occidentale ×polypodioides.

Terrestrial in Northern Limestone Hills (Lares), Western Cordillera (Maricao); 150–450 m; *spr* (?).

Western South America (Ecuador, Peru). [P: 172].

Blechnum polypodioides Raddi, Opusc. Sci. 3:294. 1819.
Syn.: *Blechnum unilaterale* Sw. (1810), *nom. illeg. superfl.*

Terrestrial in Northern Limestone Hills (Lares), Eastern Cordillera (Cerro La Peña), Central Cordillera, Western Cordillera, Western Coast (Mayagüez); 250–1075 m; *spr* Jun–Mar.

Greater Antilles, Lesser Antilles (Dominica, Martinique), Trinidad, Mexico, Central America, South America. [P: 172].

Blechnum serrulatum Rich., Actes Soc. Hist. Nat. Paris 1:114. 1792.

Terrestrial in Northern Coastal Lowlands, Western Coast (Añasco); 1–2 m; *spr* throughout the year.

West Indies, Trinidad, southeastern United States (Florida), Mexico, Central America, northern and eastern South America. [P: 170].

Notes: Maxon (1926) misapplied the name *Blechnum indicum* Burm. f. to this taxon.

CYATHEACEAE

Alsophila amintae D.S. Conant, Rhodora 83(833): 149. 1981 "aminta."
Syn.: *Cyathea dryopteroides* Maxon (1925).

Tree in Central Cordillera; 1000–1250 m; *spr* Aug–May.

Endemic. [P: 84].

Notes: This taxon is on the U.S. Fish & Wildlife Service list of threatened and endangered plants.

Alsophila amintae ×bryophila.
Syn.: *Cyathea bryophila ×dryopteroides*.

Tree in Central Cordillera; 1150–1225 m; *spr* (?). [P: 86].

Alsophila amintae ×portoricensis.
Syn.: *Cyathea dryopteroides ×portoricensis*.
Tree in Central Cordillera; 1100–1300 m; *spr* Nov–Apr, Jul–Aug. [P: 86].

Alsophila brooksii (Maxon) R.M. Tryon, Contr. Gray Herb. 200:29. 1970.
Bas.: *Cyathea brooksii* Maxon (1909).
Tree in Maricao Forest Reserve; 450–850 m; *spr* Nov–Jun.
Greater Antilles (Cuba, Hispaniola). [P: 85].

Alsophila brooksii ×portoricensis.
Syn.: *Cyathea brooksii ×portoricensis*.
Tree in Maricao Forest Reserve; 600–760 m; *spr* Jul–Sep (two specimens). [*Conant 1677, 1952* (GH); *Lehnert 1739* (UPRRP)]. [P: 86].

Alsophila bryophila R.M. Tryon, Rhodora 74:443, *ff. 9, 10.* 1972.
Syn.: *Cyathea bryophila* (R.M. Tryon) Proctor (1989).
Tree in Sierra de Luquillo, Sierra de Cayey, Toro Negro Forest Reserve; 700–1325 m; *spr* throughout the year.
Endemic. [P: 85].
Notes: Maxon (1926) misapplied the name *Cyathea pubescens* Mett. ex Kuhn to this taxon.
Helecho gigante de la sierra.

Alsophila bryophila ×portoricensis.
Syn.: *Cyathea bryophila ×portoricensis*.
Tree in Sierra de Luquillo, Central Cordillera; 725–1300 m; *spr* Jan–Sep. [P: 86].

Alsophila portoricensis (Spreng. ex Kuhn) D.S. Conant, J. Arnold Arbor. 64:370. 1983.
Bas.: *Cyathea portoricensis* Spreng. ex Kuhn (1869).
Syn.: *Nephelea portoricensis* (Spreng. ex Kuhn) R.M. Tryon (1970).
Tree in Northern Limestone Hills (Quebradillas, Isabela), Río Abajo Forest Reserve, Sierra de Luquillo, Sierra de Cayey, Las Tetas de Cayey, Eastern Cordillera (La Torrecilla), Central Cordillera, Maricao Forest Reserve, Western Coast (Mayagüez); 5–1225 m; *spr* throughout the year.
Endemic. [P: 86].
Helecho gigante espinoso, Palma de camarón.

Cyathea andina (H. Karst.) Domin, Pterid. Isl. Dominica 263. 1929.
Bas.: *Hemitelia andina* H. Karst. (1856).
Syn.: *Hemitelia escuquensis* H. Karst. (1869).
Cyathea escuquensis (H. Karst.) Domin (1929).
Tree in Río Abajo Forest Reserve, Sierra de Luquillo, Central Cordillera, Maricao Forest Reserve; 325–1000 m; *spr* Dec–Apr, Jul–Sep.
Greater Antilles (Hispaniola), southern Central America (Panama), northern South America to Bolivia. [P: 88].
Helecho gigante.

Cyathea arborea (L.) Sm., Mém. Acad. Roy. Sci. (Turin) 5(1790–1791):417. 1793.
Bas.: *Polypodium arboreum* L. (1753).
Tree in Northern Limestone Hills (Quebradillas), Sierra de Luquillo, Eastern Coast (Cuchilla de Panduras), Caguas Valley (Las Piedras), Sierra de Cayey, Eastern Cordillera (La Torrecilla), Central Cordillera, Maricao Forest Reserve; 150–1100 m; *spr* throughout the year.
Antilles. [P: 90].
Camarón, Camaroncillo, Helecho arbóreo, Helecho gigante, Palmilla.

Cyathea armata (Sw.) Domin, Acta Bot. Bohem. 9:93. 1930.
Bas.: *Polypodium armatum* Sw. (1788).
Tree in Toro Negro Forest Reserve; 1150–1300 m; *spr* Jan–May, Aug–Sep.
Greater Antilles. [P: 88].

Cyathea borinquena (Maxon) Domin, Acta Bot. Bohem. 9:99. 1930.
Bas.: *Alsophila borinquena* Maxon (1925).
Syn.: *Trichipteris borinquena* (Maxon) R.M. Tryon (1970).
Tree in Sierra de Luquillo, Sierra de Cayey, Eastern Cordillera (Cerro La Peña), Toro Negro Forest Reserve, Maricao Forest Reserve; 200–1225 m; *spr* throughout the year.
Endemic. [P: 87].

Cyathea borinquena ×horrida.
Tree in Central Cordillera; 970–980 m; *spr* Sep, Jan (two specimens).

Cyathea brittoniana Maxon, J. Wash. Acad. Sci. 14:139. 1924.
Tree in Central Cordillera, Maricao Forest Reserve; 800–1325 m; *spr* Jul–Jan.
Greater Antilles (Cuba, Hispaniola). [P: 92].
Notes: Liogier & Martorell (1982) treated the above taxon as conspecific with *Cyathea furfuracea*; however, *Cyathea brittoniana* and *Cyathea furfuracea* belong to different types. Proctor (1989) treated it as conspecific with *Cyathea tenera* (J. Sm. ex Hook.) T.

Moore; however, according to Marcus Lehnert (pers. comm.), the type of *Cyathea tenera* from St. Vincent, is alignable with other material of that taxon in the Lesser Antilles, but not with what has been called *Cyathea 'tenera'* in the Greater Antilles.

Helecho gigante.

Cyathea caracasana (Klotzsch) Domin, Pterid. Isl. Dominica 262. 1929.
Bas.: *Alsophila caracasana* Klotzsch (1844).

Tree in Central Cordillera; 1100–1200 m; *spr* Jan (one specimen). [*Lehnert 1692, 1713* (UPRRP)].

Greater Antilles (Cuba, Hispaniola), northern South America (Venezuela, Colombia).

Cyathea furfuracea Baker in Hooker & Baker, Syn. Fil. (ed. 2) 450. 1874.

Tree in Eastern Cordillera (La Torrecilla), Central Cordillera, Maricao Forest Reserve; 950–1325 m; *spr* Jan–Sep.

Greater Antilles. [P: 92].

Notes: Puerto Rican specimens identified as the above taxon are, according to Marcus Lehnert (pers. comm.), divergent from the Jamaican type and may represent a different taxon. On the other hand, such specimens are similar to those of *Cyathea tenera* from Dominica, so that the former taxon may just be an aberrant form of the latter.

Cyathea horrida (L.) Sm., Mém. Acad. Roy. Sci. (Turin) 5(1790–1791):416. 1793.
Bas.: *Polypodium horridum* L. (1753).
Syn.: *Hemitelia horrida* (L.) R. Br. ex Spreng. (1827). *Cnemidaria horrida* (L.) C. Presl (1836).

Tree in Northern Limestone Hills (Quebradillas), Río Abajo Forest Reserve, Sierra de Luquillo, Sierra de Cayey, Eastern Cordillera (Cerro La Peña), Central Cordillera, Western Cordillera; 150–1100 m; *spr* throughout the year.

Greater Antilles, southern Central America, northwestern, west-central South America and Venezuela. [P: 94].

Cyathea horrida ×parvula.

Tree in Toro Negro Forest Reserve; 1000 m; *spr* Jun (one specimen). [*Conant 612* (GH)]. [P: 96].

Cyathea parvula (Jenman) Domin, Pterid. Isl. Dominica 264. 1929.
Bas.: *Alsophila parvula* Jenman (1879).
Syn.: *Alsophila aquilina* H. Christ (1897). *Cyathea aquilina* (H. Christ) Domin (1929).

Tree in Sierra de Luquillo, Central Cordillera, Western Cordillera; 775–1100 m; *spr* Jan–Mar, Jul–Oct.

Greater Antilles. [P: 89].

Helecho gigante.

Cyathea pungens (Willd.) Domin, Pterid. Isl. Dominica 263. 1929.
Bas.: *Polypodium pungens* Willd. (1810).
Syn.: *Trichipteris procera* (Willd.) R.M. Tryon (1970).

Tree in Río Abajo Forest Reserve, Sierra de Luquillo; 100–450 m; *spr* Jul–Jan.

Greater Antilles (Hispaniola), Trinidad and Tobago, southern Central America (Panama), northern South America to Bolivia. [P: 87].

Notes: I have maintained the above name for a Puerto Rican population of tree ferns, many of which, according to Marcus Lehnert (pers. comm.), show some characters more typical of *Cyathea aspera* (L.) Sw. Although the latter taxon has not thus far been collected in Puerto Rico, it does occur in the rest of the Antilles.

Cyathea ×wilsonii (Hook.) Domin (pro sp.), Pterid. Isl. Dominica 264. 1929.
Bas.: *Hemitelia wilsonii* Hook. (1865).

Tree in Sierra de Luquillo, Central Cordillera; 200–1225 m; *spr* Dec–Feb, May–Sep.

Greater Antilles (Jamaica, Hispaniola). [P: 90].

Notes: The name of this taxon represents the hybrid *Cyathea arborea ×horrida*.

Helecho gigante.

DENNSTAEDTIACEAE

Dennstaedtia bipinnata (Cav.) Maxon, Proc. Biol. Soc. Wash. 51(8):39. 1938.
Bas.: *Dicksonia bipinnata* Cav. (1802).
Syn.: *Dennstaedtia adiantoides* (Humb. & Bonpl. ex Willd.) T. Moore. (1857).

Terrestrial in Northern Limestone Hills, Río Abajo Forest Reserve, Sierra de Cayey, Las Tetas de Cayey, Central Cordillera, Quebrada de Los Cedros, Western Cordillera; 225–1000 m; *spr* throughout the year.

Greater Antilles, Trinidad, southeastern United States (Florida), Mexico, Central America, northwestern, west-central South America and Venezuela. [P: 100].

Dennstaedtia cicutaria (Sw.) T. Moore, Index Fil. xcvii. 1857.
Bas.: *Dicksonia cicutaria* Sw. (1801).
Syn.: *Dennstaedtia rubiginosa* (Kaulf.) T. Moore (1857).

Terrestrial in Northern Limestone Hills, Sierra de Luquillo, Central Cordillera, Western Cordillera (Maricao); 150–975 m; *spr* Jan–Mar, Jun–Sep.

Greater Antilles, Mexico, Central America, South America (scattered). [P: 102].

Dennstaedtia dissecta (Sw.) T. Moore, Index Fil. 305. 1861.
Bas.: *Dicksonia dissecta* Sw. (1801).
Syn.: *Dennstaedtia ordinata* (Kaulf.) T. Moore (1861).

Terrestrial in Northern Limestone Hills (Arecibo, Hatillo), Río Abajo Forest Reserve, Sierra de Luquillo, Sierra de Cayey, Central Cordillera, Western Cordillera; 200–850 m; *spr* throughout the year.

Antilles, southern Central America, northwestern South America (Ecuador). [P: 102].

Notes: Liogier & Martorell (1982, 2000) and Proctor (1989) misapplied the name *Dennstaedtia obtusifolia* (Willd.) T. Moore to this taxon.

Dennstaedtia globulifera (Poir.) Hieron., Bot. Jahrb. Syst. 34:455. 1904.
Bas.: *Polypodium globuliferum* Poir. (1804).

Terrestrial in Northern Limestone Hills (Florida), Río Abajo Forest Reserve; 150–325 m; *spr* Jun–Jul.

Greater Antilles, Mexico, Central America, South America (scattered). [P: 101].

Histiopteris incisa (Thunb.) J. Sm., Hist. Fil. 295. 1875.
Bas.: *Pteris incisa* Thunb. (1800).

Viny terrestrial in Sierra de Luquillo, Sierra de Cayey, Central Cordillera; 200–1325 m; *spr* Dec–Aug.

Pantropical. [P: 110].

Hypolepis nigrescens Hook., Sp. Fil. 2:66, *t. 90, f. C.* 1852 (*tabula* published in 1851), *nom. cons. prop.*

Viny terrestrial in Central Cordillera; 950–1300 m; *spr* Jan–Mar, Jun–Oct.

Greater Antilles, Trinidad, southern Mexico, Central America, northwestern, west-central South America and Venezuela. [P: 104; AR: 25].

Hypolepis repens (L.) C. Presl, Tent. Pterid. 162. 1836.
Bas.: *Lonchitis repens* L. (1753).

Viny terrestrial in Sierra de Luquillo, Sierra de Cayey, Central Cordillera, Maricao Forest Reserve; 475–1100 m; *spr* Jan–Oct.

Bahamas, Greater Antilles (Cuba, Hispaniola), Lesser Antilles, Trinidad and Tobago, southeastern United States (Florida), Mexico, Central America, South America (scattered). [P: 103; AR: 25].

Hypolepis tenerrima Maxon, J. Washington Acad. Sci. 14:196. 1924.

Viny terrestrial in Eastern Coast (Cuchilla de Panduras); 300–500 m; *spr* Jun (one specimen). [*Woodbury Mar 1970* (UPR); *Proctor 46998* (SJ)].

Greater Antilles (Hispaniola). [P: 104; AR: 25].

Hypolepis urbanii Brause in Urban, Symb. Antill. 7(4):486. 1913.

Viny terrestrial in Toro Negro Forest Reserve; 1300 m; *spr* Feb (one specimen). [*Proctor 40130* (SJ)].

Greater Antilles (Hispaniola). [P: 106; AR: 25].

Pteridium arachnoideum (Kaulf.) Maxon, J. Washington Acad. Sci. 14:89. 1924.
Bas.: *Pteris arachnoidea* Kaulf. (1824).
Syn.: *Pteridium aquilinum* (L.) Kuhn var. *arachnoideum* (Kaulf.) Brade (1920).

Viny terrestrial in Northern Limestone Hills (Barceloneta), Eastern Cordillera (Barranquitas), Central Cordillera; 50–1335 m; *spr* (?).

Greater Antilles, Lesser Antilles (St. Lucia), Trinidad and Tobago, Mexico, Central America, South America (scattered). [P: 108].

Helecho hembra.

Pteridium caudatum (L.) Maxon, Proc. U.S. Natl. Mus. 23(1226):631. 1901.
Bas.: *Pteris caudata* L. (1753).
Syn.: *Pteridium aquilinum* (L.) Kuhn var. *caudatum* (L.) Sadeb. (1896).

Viny terrestrial in Northern Coastal Lowlands (Laguna Tortuguero), Río Abajo Forest Reserve, Sierra de Luquillo, Eastern Cordillera (Barranquitas), Susúa Forest Reserve, Western Cordillera, Cerro Las Mesas; 1–650 m; *spr* Mar (two specimens).

Bahamas, Greater Antilles, Lesser Antilles (Antigua, Montserrat), Trinidad, southeastern United States (Florida), Mexico, Central America, northern South America to Bolivia. [P: 107].

Felpa, Helecho marrano, Helecho de alambre.

DICKSONIACEAE

Lophosoria quadripinnata (J.F. Gmel.) C. Chr. in Skottsberg, Nat. Hist. Juan Fernandez 2:16. 1920.
Bas.: *Polypodium quadripinnatum* J.F. Gmel. (1792).
Syn.: *Alsophila quadripinnata* (J.F. Gmel.) C. Chr. (1905).

Tree in Sierra de Luquillo, Sierra de Cayey, Central Cordillera; 850–1250 m; *spr* throughout the year.

Greater Antilles, Trinidad, Mexico, Central America, South America (scattered). [P: 83].

DRYOPTERIDACEAE

Bolbitis aliena (Sw.) Alston, Bull. Misc. Inform. Kew 1932(7):310. 1932.
Bas.: *Acrostichum alienum* Sw. (1801).
Syn.: *Leptochilus alienus* (Sw.) C. Chr. (1904).

Terrestrial in Northern Coastal Lowlands (Carolina), Sierra de Luquillo, Sierra de Cayey, Central Cordillera (Adjuntas); 50–800 m; *spr* Feb–Jul, Oct.

Greater Antilles, Lesser Antilles (St. Eustatius, Guadeloupe), Trinidad, southern Mexico, Central America, northwestern, west-central South America and Venezuela. [P: 285].

Bolbitis nicotianifolia (Sw.) Alston, Bull. Misc. Inform. Kew 1932(7):310. 1932.
Bas.: *Acrostichum nicotianifolium* Sw. (1806).
Syn.: *Leptochilus nicotianifolius* (Sw.) C. Chr. (1904).

Terrestrial in Northern Coastal Lowlands (San Juan), Sierra de Luquillo, Sierra de Cayey, Central Cordillera (Adjuntas); 20–630 m; *spr* Jan–Jul, Oct.

Greater Antilles (Jamaica, Hispaniola), Lesser Antilles, Central America, northern South America to Bolivia. [P: 286].

Bolbitis pergamentacea (Maxon) Ching in Christensen, Index Filic., Suppl. 3, 3:49. 1934.
Bas.: *Leptochilus pergamentaceus* Maxon (1924).

Terrestrial in Northern Limestone Hills (Ciales, Quebradillas), Río Abajo Forest Reserve, Sierra de Luquillo, Eastern Coast (Humacao), Caguas Valley (San Lorenzo), Eastern Cordillera (Aguas Buenas, Cañón San Cristóbal), Central Cordillera, vicinity of Susúa Forest Reserve (Yauco), Western Cordillera; 100–700 m; *spr* throughout the year.

Greater Antilles, southern Mexico, Central America, northwestern South America and Venezuela. [P: 288].

Bolbitis portoricensis (Spreng.) Hennipman, Amer. Fern J. 65(1):30. 1975.
Bas.: *Acrostichum portoricense* Spreng. (1821).
Syn.: *Leptochilus cladorrhizans* (Spreng.) Maxon (1926).

Epiphyte in Sierra de Luquillo, Sierra de Cayey, Eastern Cordillera (Corozal), Central Cordillera (Ponce, Utuado); 350–1100 m; *spr* Nov–Feb, May–Jun.

Antilles, Trinidad and Tobago, Mexico, Central America, northwestern South America and Venezuela. [P: 286].

Ctenitis hirta (Sw.) Ching var. **portoricana** (C. Chr.) Proctor, Mem. New York Bot. Gard. 53: 248. 1989.
Bas.: *Dryopteris hirta* (Sw.) Kunze f. *portoricana* C. Chr. (1920).

Terrestrial in Northern Limestone Hills, Río Abajo Forest Reserve, Sierra de Luquillo, Central Cordillera (Utuado); 150–450 m; *spr* Jun–Feb.

Endemic. [P: 248].

Notes: Maxon (1926) used *Dryopteris hirta* (Sw.) Kuntze, *sensu lato*, and Liogier & Martorell (1982) used *Ctenitis hirta* (Sw.) Ching, *sensu lato*, for the name of this taxon.

Ctenitis nemorosa (Willd.) Ching, Sunyatsenia 5(4):250. 1940.
Bas.: *Aspidium nemorosum* Willd. (1810).
Syn.: *Dryopteris nemorosa* (Willd.) Urb. (1903).

Terrestrial in Northern Limestone Hills, Río Abajo Forest Reserve, Central Cordillera (Yauco), Quebrada de Los Cedros; 10–875 m; *spr* throughout the year.

Greater Antilles (Hispaniola). [P: 248].

Elaphoglossum apodum (Kaulf.) Schott ex J. Sm., J. Bot. (Hooker) 4:148. 1841.
Bas.: *Acrostichum apodum* Kaulf. (1824).

Epiphyte/lithophyte in Sierra de Luquillo, Eastern Cordillera (Guaynabo), Central Cordillera (Peñuelas), Maricao Forest Reserve; 300–1025 m; *spr* Sep (one specimen).

Antilles, Central America, northern and west-central South America. [P: 299].

Elaphoglossum chartaceum (Baker ex Jenman) C. Chr., Index Filic. 304. 1905.
Bas.: *Acrostichum chartaceum* Baker ex Jenman (1882).

Epiphyte in Sierra de Luquillo, Sierra de Cayey, Central Cordillera; 450–1325 m; *spr* throughout the year.

Greater Antilles. [P: 294].

Elaphoglossum crinitum (L.) H. Christ, Neue Denkschr. Allg. Schweiz. Ges. Gesammten Naturwiss. 36(1):102, *t.* 53 (Monogr. Elaphoglossum). 1899.
Bas.: *Acrostichum crinitum* L. (1753).
Syn.: *Hymenodium crinitum* (L.) Fée (1845).

Epiphyte in Sierra de Luquillo, Sierra de Cayey, Maricao Forest Reserve; 300–900 m; *spr* Jan, Jun–Aug.

Antilles, Trinidad and Tobago, Central America, northern and northwestern South America. [P: 300].

Elaphoglossum eggersii (Baker) H. Christ, Neue Denkschr. Allg. Schweiz. Ges. Gesammten Naturwiss. 36(1):71 (Monogr. Elaphoglossum). 1899.

Bas.: *Acrostichum eggersii* Baker (1888).

Epiphyte/terrestrial in Central Cordillera; 1000–1150 m; *spr* (?). [*Proctor 41311, 43253* (SJ)].

Greater Antilles. [P: 296].

Elaphoglossum erinaceum (Fée) T. Moore, Index Fil. 9. 1857.

Bas.: *Acrostichum erinaceum* Fée (1845).

Epiphyte/terrestrial in Central Cordillera, Maricao Forest Reserve; 480–1225 m; *spr* Jul (one specimen).

Antilles, Mexico, Central America, northwestern, west-central South America and Venezuela, Brazil. [P: 299].

Elaphoglossum glabellum J. Sm., London J. Bot. 1:197. 1842.

Epiphyte in Sierra de Luquillo, Sierra de Cayey, Toro Negro Forest Reserve, Maricao Forest Reserve; 500–1250 m; *spr* May, Aug–Oct.

Greater Antilles, Lesser Antilles (Dominica, Martinique), Trinidad, southern Mexico, Central America (scattered), northern South America to Bolivia. [P: 291].

Elaphoglossum herminieri (Bory ex Fée) T. Moore, Index Fil. 16. 1857.

Bas.: *Acrostichum herminieri* Bory ex Fée (1845).

Epiphyte/terrestrial in Sierra de Luquillo, Maricao Forest Reserve; 250–550 m; *spr* Jul (one specimen).

Antilles, Trinidad, southern Mexico, Central America, northern South America to Ecuador. [P: 292].

Elaphoglossum latifolium (Sw.) J. Sm., London J. Bot. 1:197. 1842.

Bas.: *Acrostichum latifolium* Sw. (1788).

Syn.: *Elaphoglossum longifolium* J. Sm. (1846).

Epiphyte in Sierra de Luquillo, Sierra de Cayey, Eastern Cordillera (La Torrecilla), Central Cordillera, Maricao Forest Reserve; 350–1325 m; *spr* Mar–Nov.

Antilles, Trinidad, Central America, northern South America to Bolivia. [P: 293].

Notes: *Elaphoglossum longifolium*, which cannot be clearly and consistently differentiated from this taxon by either Alan R. Smith (pers. comm.), J. Mickel (pers. comm.) or me, has been placed in synonymy with it.

Elaphoglossum lonchophyllum (Fée) T. Moore, Index Fil. 11. 1857.

Bas.: *Acrostichum lonchophyllum* Fée (1857).

Epiphyte in Central Cordillera (Utuado, Adjuntas); 910–1150 m; *spr* Jul–Sep (two specimens). [*Proctor 47121, 48874* (SJ)].

Antilles, southern Mexico, Central America.

Elaphoglossum martinicense (Desv.) T. Moore, Index Fil. 11. 1857.

Bas.: *Acrostichum martinicense* Desv. (1811).

Syn.: *Elaphoglossum underwoodianum* Maxon (1926).

Epiphyte in Sierra de Luquillo, Maricao Forest Reserve; 350–875 m; *spr* Jun–Sep.

Greater Antilles (Hispaniola), Lesser Antilles. [P: 295].

Elaphoglossum maxonii Underw. ex C.V. Morton, Amer. Fern J. 29(1):12. 1939.

Epiphyte/terrestrial in Sierra de Luquillo, Sierra de Cayey, Central Cordillera, Maricao Forest Reserve; 425–1025 m; *spr* May–Nov.

Greater Antilles. [P: 295].

Notes: Maxon (1926) misapplied the name *Elaphoglossum pteropus* C. Chr. to this taxon.

Elaphoglossum peltatum (Sw.) Urb., Symb. Antill. 4(1):60. 1903.

Bas.: *Osmunda peltata* Sw. (1788).

Syn.: *Rhipidopteris peltata* (Sw.) Schott ex Fée (1845). *Peltapteris peltata* (Sw.) C.V. Morton (1955).

Epiphyte in Sierra de Luquillo, Sierra de Cayey, Toro Negro Forest Reserve, Maricao Forest Reserve; 500–1225 m; *spr* throughout the year.

Antilles, Mexico, Central America, northern South America to Bolivia. [P: 289].

Elaphoglossum petiolatum (Sw.) Urb., Symb. Antill. 4(1):61. 1903.

Bas.: *Acrostichum petiolatum* Sw. (1788).

Syn.: *Elaphoglossum dussii* Underw. ex Maxon (1926). *Elaphoglossum petiolatum* (Sw.) Urb. var. *dussii* (Underw. ex Maxon) Proctor (1966).

Lithophyte in Sierra de Luquillo, Eastern Cordillera (Cañón San Cristóbal); 300–650 m; *spr* Jun–Jul.

Greater Antilles (Hispaniola), Lesser Antilles, northern, northwestern South America and Brazil. [P: 296].

Elaphoglossum piloselloides (C. Presl) T. Moore, Index Fil. 13. 1857.

Bas.: *Acrostichum piloselloides* C. Presl (1825).

Lithophyte in Sierra de Luquillo; 435 m; *spr* Mar (one specimen). [*Areces 261* (UPRRP)].

Antilles, Mexico, Central America, northern and western South America. [P: 297].

Notes: The above name applies only to Neotropical

populations of what was formerly considered a broader taxon; the previously used name, *Elaphoglossum spatulatum* (Bory) T. Moore, now applies only to Paleotropical populations.

Elaphoglossum rigidum (Aubl.) Urb., Symb. Antill. 9(3):374. 1925.
Bas.: *Polypodium rigidum* Aubl. (1775).
Syn.: *Elaphoglossum flaccidum* (Fée) T. Moore (1862).

Epiphyte in Northern Limestone Hills (Arecibo), Sierra de Luquillo, Sierra de Cayey, Eastern Cordillera (Guaynabo), Maricao Forest Reserve; 110–860 m; *spr* Jul–Nov, Feb–Mar.

Greater Antilles (Cuba, Hispaniola), Lesser Antilles (Guadeloupe, Dominica, Martinique), Trinidad, South America (Suriname, Ecuador). [P: 292].

Notes: Maxon (1926) misapplied the above name to *Elaphoglossum latifolium*.

Elaphoglossum serpens Maxon & C.V. Morton, Proc. Biol. Soc. Wash. 60:124. 1947.

Epiphyte in Toro Negro Forest Reserve; 1300 m; *spr* Jan–Feb, Aug.

Endemic. [P: 291].

Notes: This taxon is on the U.S. Fish & Wildlife Service list of threatened and endangered plants.

Elaphoglossum simplex (Sw.) Schott ex J. Sm., J. Bot. (Hooker) 4:148. 1841.
Bas.: *Acrostichum simplex* Sw. (1788).
Syn.: *Elaphoglossum firmum* (Mett. ex Kuhn) Urb. (1903).

Epiphyte in Sierra de Luquillo, Sierra de Cayey, Central Cordillera, Maricao Forest Reserve; 175–1200 m; *spr* Feb–Aug.

Greater Antilles. [P: 294].

Elaphoglossum smithii (Baker) H. Christ, Neue Denkschr. Allg. Schweiz. Ges. Gesammten Naturwiss. 36(1):106 (Monogr. Elaphoglossum). 1899.
Bas.: *Acrostichum smithii* Baker (1891).

Epiphyte in Central Cordillera; 900–1250 m; *spr* Jul (one specimen). [*Acevedo-Rodríguez 390* (SJ); *Axelrod 13589* (UPRRP)].

Greater Antilles (Hispaniola), Lesser Antilles (St. Vincent, Grenada), southern Central America. [P: 297].

Lastreopsis effusa (Sw.) Tindale, Victoria Naturalist 73:184. 1957.
Bas.: *Polypodium effusum* Sw. (1788).
Syn.: *Dryopteris effusa* (Sw.) Urb. (1903).

Terrestrial in Northern Limestone Hills, Sierra de Luquillo, Sierra de Cayey, Las Tetas de Cayey, Eastern Cordillera (La Torrecilla), Central Cordillera, Western Cordillera; 175–1100 m; *spr* throughout the year.

Antilles, southern Mexico, Central America, northern South America to Bolivia, Paraguay. [P: 246].

Lomagramma guianensis (Aubl.) Ching, Amer. Fern J. 22(1):17. 1932.
Bas.: *Polypodium guianense* Aubl. (1775).
Syn.: *Leptochilus guianensis* (Aubl.) C. Chr. (1904).

Viny epiphyte in Sierra de Cayey, Central Cordillera (Utuado), Western Cordillera (Las Marías), Western Coast (Mayagüez); 1–750 m; *spr* Oct, Feb.

Greater Antilles (Cuba, Hispaniola), South America. [P: 283; AR: 33].

Megalastrum gilbertii (Clute) R.C. Moran, J. Prado & Labiak, Brittonia 61(3):281. 2009.
Bas.: *Dryopteris gilbertii* Clute (1900).

Terrestrial in Río Abajo Forest Reserve, Sierra de Luquillo, Western Cordillera (Maricao); 300–500 m; *spr* Oct–Jan.

Greater Antilles. [P: 249].

Notes: Both this taxon and *Megalastrum martinicense* were included by Maxon (1926) under the name *Dryopteris subincisa* (Willd.) Urb. and by Liogier & Martorell (1982, 2000) and Proctor (1989) under the name *Ctenitis subincisa* (Willd.) Ching, a taxon which is no longer thought to occur in Puerto Rico.

Megalastrum martinicense (Spreng.) R.C. Moran, J. Prado & Labiak, Brittonia 61(3):284. 2009.
Bas.: *Phegopteris martinicensis* Spreng. (1872).

Terrestrial in Northern Limestone Hills (Hatillo), Sierra de Luquillo, Sierra de Cayey, Central Cordillera; 200–700 m; *spr* throughout the year.

Antilles, Trinidad. [P: 249].

Olfersia cervina (L.) Kunze, Flora 7(1):312. 1824.
Bas.: *Osmunda cervina* L. (1753).
Syn.: *Polybotrya cervina* (L.) Kaulf. (1824).

Terrestrial in Northern Limestone Hills (Arecibo), Río Abajo Forest Reserve, Sierra de Luquillo, Sierra de Cayey, Eastern Cordillera (Guaynabo, Cerro La Peña), Central Cordillera, Maricao Forest Reserve; 250–1200 m; *spr* throughout the year.

Antilles, Trinidad, southern Mexico, Central America, northern South America to Bolivia. [P: 278].

Polystichopsis chaerophylloides (Poir.) C.V. Morton, Amer. Fern J. 50(1):155. 1960.
Bas.: *Polypodium chaerophylloides* Poir. (1804).
Syn.: *Dryopteris chaerophylloides* (Poir.) C. Chr. (1920). *Arachniodes chaerophylloides* (Poir.) Proctor (1982).

Terrestrial in Northern Coastal Lowlands (San

Juan), Northern Limestone Hills, Río Abajo Forest Reserve, Sierra de Luquillo, Eastern Coast (Cuchilla de Panduras), Caguas Valley (Caguas), Sierra de Cayey, Las Tetas de Cayey, Eastern and Central Cordillera, Southern Coastal Limestone (Guayanilla), Western Cordillera; 25–1050 m; *spr* throughout the year.

Greater Antilles. [P: 276].

Polystichum calderonense Proctor, Mem. New York Bot. Gard. 53:273. 1989.

Terrestrial in Central Cordillera (Adjuntas, Peñuelas); 1000–1150 m; *spr* Feb–Apr, Aug.

Endemic. [P: 273].

Notes: This taxon is on the U.S. Fish & Wildlife Service list of threatened and endangered plants.

Polystichum echinatum (J.F. Gmel.) C. Chr., Index Filic. 581. 1906.

Bas.: *Polypodium echinatum* J.F. Gmel. (1792).

Terrestrial in Northern Limestone Hills, Central Cordillera (Yauco); 200–1100 m; *spr* Jan–Feb, Sep.

Greater Antilles. [P: 272].

Notes: Maxon (1926) misapplied the name *Polystichum triangulum* (L.) Fée to this taxon.

Polystichum platyphyllum (Willd.) C. Presl, Tent. Pterid. 84. 1836.

Bas.: *Aspidium platyphyllum* Willd. (1810).

Terrestrial in Central Cordillera, Western Cordillera (Maricao); 425–1100 m; *spr* Dec, Mar (two specimens).

Greater Antilles (Jamaica), southern Mexico, Central America, South America. [P: 275].

Polystichum polystichiforme (Fée) Maxon, Contr. U.S. Natl. Herb. 13(1):35. 1909 "polystichiformis."

Bas.: *Phegopteris polystichiformis* Fée (1852).

Terrestrial in Central Cordillera; 250–1100 m; *spr* Jan–Sep.

Greater Antilles. [P: 273].

Polystichum rhizophyllum (Sw.) C. Presl, Tent. Pterid. 82. 1836.

Bas.: *Polypodium rhizophyllum* Sw. (1788).

Syn.: *Polystichum adiantiforme* (G. Forst.) J. Sm. (1875).

Terrestrial in Northern Limestone Hills (Ciales), Caguas Valley (Gurabo), Eastern Cordillera (Bayamón), Central Cordillera, Maricao Forest Reserve; 65–800 m; *spr* Dec-Jan, Apr–Jun, Sep.

Greater Antilles, Lesser Antilles (Guadeloupe). [P: 272].

Rumohra adiantiformis (G. Forst.) Ching, Sin. Contr. 5:70. 1934.

Bas.: *Polypodium adiantiforme* G. Forst. (1786).

Epiphytic/terrestrial in Sierra de Luquillo, Eastern Cordillera (La Torrecilla), Central Cordillera, Maricao Forest Reserve; 400–1000 m; *spr* throughout the year.

Greater Antilles, Bermuda, southern, west-central South America and Venezuela, Africa (Zimbabwe, South Africa), islands of the western Indian Ocean, Papua New Guinea, Australia, New Zealand. [P: 269].

TAXA NOT RECENTLY COLLECTED

Elaphoglossum decoratum (Kunze) T. Moore, Index Fil. 8. 1857.

Bas.: *Acrostichum decoratum* Kunze (1834).

Terrestrial collected in El Duque in the Sierra de Naguabo on 13 Aug 1914 [*Shafer 3653* (US)].

Greater Antilles (Jamaica, Hispaniola), Lesser Antilles (Guadeloupe), Central America, northern South America to Bolivia. [P: 296].

GLEICHENIACEAE

Dicranopteris flexuosa (Schrad.) Underw., Bull. Torrey Bot. Club 34(5):254. 1907.

Bas.: *Mertensia flexuosa* Schrad. (1824).

Viny terrestrial in Northern Coastal Lowlands, Northern Limestone Hills (Moca), Sierra de Luquillo, Caguas Valley (Las Piedras), Sierra de Cayey, Central Cordillera, Western Cordillera, Cerro Las Mesas; 5–1330 m; *spr* Feb–May, Aug–Sep.

Antilles, Trinidad, southeastern United States (Florida, Alabama), southern Mexico, Central America, northern South America to Bolivia, Paraguay. [P: 52; AR: 29].

Helecho bifurcado.

Gleichenella pectinata (Willd.) Ching, Sunyatsenia 5(4):276. 1940.

Bas.: *Mertensia pectinata* Willd. (1804).

Syn.: *Dicranopteris pectinata* (Willd.) Underw. (1907).

Viny terrestrial in Northern Limestone Hills (Lares), Río Abajo Forest Reserve, Sierra de Luquillo, Caguas Valley (Las Piedras), Sierra de Cayey, Central Cordillera, Western Cordillera; 225–850 m; *spr* throughout the year.

Antilles, Trinidad and Tobago, Mexico, Central America, northern South America to Bolivia. [P: 52; AR: 29].

Sticherus bifidus (Willd.) Ching, Sunyatsenia 5(4):282. 1940.

Bas.: *Mertensia bifida* Willd. (1804).

Syn.: *Gleichenia bifida* (Willd.) Spreng. (1827). *Dicranopteris bifida* (Willd.) Maxon (1909).

Viny terrestrial in Northern Limestone Hills (Lares), Río Abajo Forest Reserve, Sierra de Luquillo, Eastern Coast (Cuchilla de Panduras), Caguas Valley (Caguas), Sierra de Cayey, Central Cordillera, Western Cordillera, Cerro Las Mesas; 225–1300 m; *spr* throughout the year.

Antilles, Trinidad, Mexico, Central America, northern South America to Bolivia. [P: 56; AR: 31].

Sticherus fulvus (Desv.) Ching, Sunyatsenia 5(4):283. 1940.
Bas.: *Mertensia fulva* Desv. (1827).
Syn.: *Gleichenia brevipubis* H. Christ (1906).

Viny terrestrial in Central Cordillera (Jayuya); 1050–1225 m; *spr* Aug–Sep (two specimens). [*Proctor 39420, 41419* (SJ)].

Greater Antilles, southern Mexico, Central America, northwestern South America (Colombia, Ecuador). [P: 56; AR: 31].

Sticherus rubiginosus (Mett.) Nakai, Bull. Natl. Sci. Mus. Tokyo 29:28. 1950.
Bas.: *Gleichenia rubiginosa* Mett. (1864).

Viny terrestrial in Toro Negro Forest Reserve; 1250–1300 m; *spr* Dec–Aug.

Southern Central America, northwestern and west-central South America. [P: 54].

HYMENOPHYLLACEAE

Abrodictyum rigidum (Sw.) Ebihara & Dubuisson, Blumea 51(2):243. 2006.
Bas.: *Trichomanes rigidum* Sw. (1788).

Epiphyte in Sierra de Luquillo, Sierra de Cayey, Eastern Cordillera (Corozal), Central Cordillera; 475–1225 m; *spr* throughout the year.

Pantropical. [P: 70].

Didymoglossum angustifrons Fée, Mém. Foug. 11:113, *t. 28, f.* 5. 1866.
Syn.: *Trichomanes angustifrons* (Fée) Wess. Boer (1962).

Epiphyte in Northern Limestone Hills (Hatillo, Camuy), Río Abajo Forest Reserve, Sierra de Luquillo, Sierra de Cayey, Maricao Forest Reserve; 175–625 m; *spr* Sep–Jun.

Antilles, Trinidad and Tobago, southern Mexico, southern Central America, northern South America to Bolivia, Paraguay. [P: 74].

Didymoglossum hymenoides (Hedw.) Copel., Philipp. J. Sci. 67:77. 1938.
Bas.: *Trichomanes hymenoides* Hedw. (1799).

Epiphyte in Central Cordillera; 850–1200 m; *spr* Feb, Aug–Sep.

Antilles, southern Mexico, Central America, South America (scattered). [P: 73].

Didymoglossum kapplerianum (J.W. Sturm) Ebihara & Dubuisson, Blumea 51(2):236. 2006.
Bas.: *Trichomanes kapplerianum* J.W. Sturm (1859).

Epiphyte in Río Abajo Forest Reserve; 275 m; *spr* Jul (one specimen). [*Axelrod 7952* (UPRRP)].

Lesser Antilles, Trinidad, southern Central America, northern South America to Bolivia. [P: 75].

Didymoglossum krausii (Hook. & Grev.) C. Presl, Hymenophyllaceae 23. 1843 (preprint of Abh. Königl. Böhm. Ges. Wiss., ser. 5 3:115. 1843).
Bas.: *Trichomanes krausii* Hook. & Grev. (1830).

Epiphyte in Northern Coastal Lowlands (Río Grande), Northern Limestone Hills, Río Abajo Forest Reserve, Sierra de Luquillo, Eastern Coast, Caguas Valley (Gurabo), Eastern Cordillera (Cerro La Peña, Aguas Buenas); 1–550 m; *spr* throughout the year.

Antilles, southeastern United States (Florida), Mexico, Central America, South America. [P: 75].

Didymoglossum lineolatum Bosch, Ned. Kruidk. Arch. 5(2):136. 1861.
Syn.: *Trichomanes lineolatum* (Bosch) Hook. (1867).

Epiphyte in Northern Coastal Lowlands (San Juan), Northern Limestone Hills, Río Abajo Forest Reserve, Eastern Cordillera (Aguas Buenas), Central Cordillera (Utuado), Maricao Forest Reserve; 10–600 m; *spr* throughout the year.

Greater Antilles, Lesser Antilles (St. Vincent), southeastern United States (Florida), northern South America (Venezuela) and Brazil. [P: 73].

Didymoglossum membranaceum (L.) Vareschi, Fl. Venezuela 1(1):222. 1969.
Bas.: *Trichomanes membranaceum* L. (1753).

Epiphyte in Sierra de Luquillo, Sierra de Cayey, Central Cordillera (Los Tres Picachos); 450–850 m; *spr* Mar–Nov.

Antilles, Trinidad and Tobago, southern Mexico, Central America, northern South America to Bolivia. [P: 71].

Didymoglossum ovale E. Fourn., Bull. Soc. Bot. France 19:240. 1873.
Syn.: *Trichomanes ovale* (E. Fourn.) Wess. Boer. (1962).

Epiphyte in Río Abajo Forest Reserve, Sierra de Luquillo, Maricao Forest Reserve; 110–600 m; *spr* Jan–Jul, Oct.

Greater Antilles, southern Mexico, Central America, northwestern, west-central South America and Venezuela. [P: 72].

Didymoglossum punctatum (Poir.) Desv., Mém. Soc. Linn. Paris 6(3):330. 1827.
Bas.: *Trichomanes punctatum* Poir. (1808).
Syn.: *Trichomanes sphenoides* Kunze (1834). *Trichomanes punctatum* Poir. subsp. *sphenoides* (Kunze) Wess. Boer (1962).

Epiphyte in Sierra de Luquillo, Eastern Cordillera (Aguas Buenas), Southern Coastal Limestone (Peñuelas), Southwestern Coastal Lowlands (Cabo Rojo); 50–550 m; *spr* Nov–Apr, Aug.

Antilles, Trinidad and Tobago, Central America, northwestern, west-central South America and Venezuela. [P: 72, 73].

Didymoglossum pusillum (Sw.) Desv., Mém. Soc. Linn. Paris 6(3):330. 1827.
Bas.: *Trichomanes pusillum* Sw. (1788).

Epiphyte in Río Abajo Forest Reserve, Sierra de Cayey; 275–775 m; *spr* Jul–Aug (two specimens). [*Proctor 39406* (SJ); *Axelrod 7981* (UPRRP)].

Greater Antilles (Jamaica, Hispaniola), southern Central America (Costa Rica), South America (Venezuela, Brazil). [P: 74].

Didymoglossum reptans (Sw.) C. Presl, Hymenophyllaceae 23. 1843 (preprint of Abh. Königl. Böhm. Ges. Wiss., ser. 5, 3:115. 1843).
Bas.: *Trichomanes reptans* Sw. (1788).

Herb in Central Cordillera (Yauco); 850–950 m; *spr* Mar–Apr (one specimen). [*Axelrod 7666* (UPRRP)].

Greater Antilles (Jamaica, Hispaniola), Trinidad, Mexico, Central America, South America.

Didymoglossum wesselsboeri Ebihara & Dubuisson, Blumea 51(2):237. 2006.
Syn.: *Trichomanes hookeri* C. Presl (1843).

Epiphyte in Río Abajo Forest Reserve, Sierra de Luquillo, Central Cordillera (Adjuntas); 100–630 m; *spr* Dec–Aug.

Greater Antilles. [P: 75].

Hymenophyllum asplenioides (Sw.) Sw., J. Bot. (Schrader) 1800(2):98. 1801.
Bas.: *Trichomanes asplenioides* Sw. (1788).

Epiphyte in Maricao Forest Reserve; 850 m; *spr* Sep (one specimen). [*Cedeño 1024* (MAPR)].

Greater Antilles, southern Mexico, Central America, northern South America to Ecuador, Brazil. [P: 63].

Hymenophyllum axillare Sw., J. Bot. (Schrader) 1800(2):101. 1801.

Epiphyte in Toro Negro Forest Reserve; 1300 m; *spr* Aug (one specimen). [*Proctor 41293* (SJ)].

Greater Antilles, South America (Venezuela, Bolivia). [P: 62].

Hymenophyllum brevifrons Kunze, Bot. Zeitung (Berlin) 5:185. 1847.

Epiphyte in Sierra de Luquillo, Toro Negro Forest Reserve; 850–1300 m; *spr* Sep, Jan–Feb (two specimens). [*Woodbury a.n. 17833* (SJ); *Axelrod 5167* (UPRRP)].

Greater Antilles (Cuba, Hispaniola), Central America, northern South America. [P: 59].

Hymenophyllum decurrens (Jacq.) Sw., J. Bot. (Schrader) 1800(2):99. 1801.
Bas.: *Adiantum decurrens* Jacq. (1789).
Syn.: *Hymenophyllum polyanthos* (Sw.) Sw. var. *protrusum* (Hook.) Farw. (1931).

Epiphyte in Sierra de Luquillo, Sierra de Cayey, Maricao Forest Reserve; 250–800 m; *spr* throughout the year.

Greater Antilles (Jamaica, Hispaniola), Lesser Antilles (Martinique, St. Vincent), Mexico, Central America, northern South America. [P: 61].

Hymenophyllum fragile (Hedw.) C.V. Morton, Contr. U.S. Natl. Herb. 29(3):172. 1947.
Bas.: *Trichomanes fragile* Hedw. (1802).

Epiphyte in Toro Negro Forest Reserve; 1125–1300 m; *spr* Dec–Apr.

Greater Antilles, southern Mexico, Central America, northwestern, west-central South America and Venezuela, Brazil. [P: 65].

Hymenophyllum fucoides (Sw.) Sw., J. Bot. (Schrader) 1800(2):99. 1801.
Bas.: *Trichomanes fucoides* Sw. (1788).

Epiphyte in Central Cordillera; 1200–1300 m; *spr* Oct–Mar.

Antilles, Trinidad, southern Mexico, Central America, northern South America to Bolivia. [P: 58].

Hymenophyllum hirsutum (L.) Sw., J. Bot. (Schrader) 1800(2):99. 1801.
Bas.: *Trichomanes hirsutum* L. (1753).
Syn.: *Hymenophyllum ciliatum* (Sw.) Sw. (1801).

Epiphyte in Sierra de Luquillo, Sierra de Cayey, Central Cordillera, Maricao Forest Reserve; 250–1300 m; *spr* throughout the year.

Antilles, Trinidad and Tobago, southern Mexico, Central America, northern South America to Bolivia. [P: 63].

Hymenophyllum lineare (Sw.) Sw., J. Bot. (Schrader) 1800(2):100. 1801.

Bas.: *Trichomanes lineare* Sw. (1788).

Epiphyte in Sierra de Luquillo, Sierra de Cayey, Toro Negro Forest Reserve; 825–1300 m; *spr* Jan–Aug.

Antilles, Central America. [P: 64].

Hymenophyllum macrothecium Fée, Mém. Foug. 11:115, *t. 31*, *f.* 2. 1866.

Epiphyte in Sierra de Luquillo, Sierra de Cayey; 700–1050 m; *spr* Dec–Jan, May–Sep.

Lesser Antilles. [P: 59].

Notes: Maxon (1926), Liogier & Martorell (1982, 2000) and Proctor (1989) used the orthographic variant *Hymenophyllum macrothecum* for the name of this taxon; the protologue has the above spelling.

Hymenophyllum microcarpum Desv., Mém. Soc. Linn. Paris 6(3):333. 1827.

Epiphyte in Sierra de Luquillo, Central Cordillera; 725–1300 m; *spr* Jun–Jul, Oct–Feb.

Greater Antilles, Lesser Antilles (Dominica), southern Mexico, Central America, northern South America to Bolivia. [P: 64].

Hymenophyllum paucicarpum Jenman, Bull. Bot. Dept. Jamaica 18:4. 1890.

Epiphyte in Sierra de Luquillo; 750–900 m; *spr* Apr (two specimens). [*Proctor 40454* (SJ); *Axelrod 12511A* (UPRRP)].

Greater Antilles (Jamaica), southern Mexico, Central America (scattered). [P: 59].

Hymenophyllum polyanthos (Sw.) Sw., J. Bot. (Schrader) 1800(2):102. 1801.

Bas.: *Trichomanes polyanthos* Sw. (1788).

Epiphyte in Sierra de Luquillo, Sierra de Cayey, Eastern Cordillera (Cerro La Peña), Central Cordillera (Jayuya), Maricao Forest Reserve; 175–1300 m; *spr* throughout the year.

Pantropical. [P: 61].

Hymenophyllum sieberi (C. Presl) Bosch, Ned. Kruidk. Arch. 4:414. 1859.

Bas.: *Sphaerocionium sieberi* C. Presl (1843).

Epiphyte in Sierra de Cayey; 850–900 m; *spr* (?). [*Axelrod 10738* (UPRRP)].

Lesser Antilles, southern Mexico, Central America. [P: 66].

Hymenophyllum tegularis (Desv.) Proctor & Lourteig, Bradea 5(40):385. 1990.

Bas.: *Davallia tegularis* Desv. (1827).

Syn.: *Hymenophyllum elegantulum* Bosch. (1859). *Hymenophyllum elegantulum* Bosch var. *petiolulatum* C.V. Morton. (1947).

Epiphyte in Sierra de Luquillo, Toro Negro Forest Reserve; 850–1300 m; *spr* May–Jul, Dec.

Greater Antilles (Cuba, Hispaniola), Mexico, Central America, northwestern, west-central South America and Venezuela. [P: 64].

Hymenophyllum undulatum (Sw.) Sw., J. Bot. (Schrader) 1800(2):101. 1801.

Bas.: *Trichomanes undulatum* Sw. (1788).

Epiphyte in Sierra de Luquillo, Toro Negro Forest Reserve; 850–1300 m; *spr* Nov–Dec, Mar–Aug.

Greater Antilles, Lesser Antilles (Guadeloupe), southern Mexico, Central America, northern South America to Bolivia. [P: 62].

Notes: Maxon (1926) misapplied the name *Hymenophyllum crispum* Kunth, and Liogier & Martorell (1982) misapplied *Hymenophyllum contortum* Bosch, to this taxon.

Polyphlebium capillaceum (L.) Ebihara & Dubuisson, Blumea 51(2):240. 2006.

Bas.: *Trichomanes capillaceum* L. (1753).

Epiphyte in Sierra de Cayey, Central Cordillera; 625–1200 m; *spr* Jul–Mar.

Greater Antilles, southern Mexico, Central America, northwestern and west-central South America and Venezuela. [P: 69].

Polyphlebium hymenophylloides (Bosch) Ebihara & Dubuisson, Blumea 51(2):240. 2006.

Bas.: *Trichomanes hymenophylloides* Bosch (1863).

Epiphyte in Sierra de Luquillo, Sierra de Cayey, Central Cordillera, Maricao Forest Reserve; 625–1250 m; *spr* Oct–Apr.

Antilles, Trinidad, southern Mexico, Central America, northern South America to Ecuador, Brazil. [P: 69].

Trichomanes alatum Sw., J. Bot. (Schrader) 1800(2):97. 1801.

Epiphyte in Sierra de Luquillo, Central Cordillera; 600–1300 m; *spr* throughout the year.

Antilles, Southern Caribbean Islands (Isla Margarita), Trinidad. [P: 80].

Trichomanes crispum L., Sp. Pl. 2:1097. 1753.

Epiphyte in Northern Limestone Hills, Río Abajo Forest Reserve, Sierra de Luquillo, Sierra de Cayey, Eastern Cordillera (Cerro La Peña), Central Cordillera, Maricao Forest Reserve; 200–1000 m; *spr* throughout the year.

Antilles, Trinidad, southern Mexico, Central America, northern South America to Bolivia, Paraguay. [P: 79].

Trichomanes holopterum Kunze, Farrnkräuter 1:185, *t. 77*, *f.* 2. 1845.

Epiphyte in Northern Limestone Hills, Maricao Forest Reserve; 200–875 m; *spr* Feb–Oct.

Antilles, southeastern United States (Florida), southern Mexico. [P: 80].

Trichomanes padronii Proctor, Amer. Fern J. 74(1):7, *t. 1*. 1984.

Epiphyte in Sierra de Cayey, Maricao Forest Reserve; 450–800 m; *spr* Mar, Oct–Nov.

Greater Antilles (Cuba, Hispaniola). [P: 77].

Trichomanes pinnatum Hedw., Fil. Gen. Sp. *t. 4, f. 1*. 1799.

Epiphyte in Sierra de Luquillo; 100 m; *spr* Feb (two specimens). [*Proctor 44485, 44500* (SJ)].

Antilles, Trinidad and Tobago, southern Mexico, Central America, northern South America to Bolivia. [P: 76].

Trichomanes polypodioides L., Sp. Pl. 2:1098. 1753.

Epiphyte in Sierra de Luquillo, Eastern Coast (Cuchilla de Panduras), Sierra de Cayey, Eastern and Central Cordillera, Maricao Forest Reserve; 100–1050 m; *spr* throughout the year.

Antilles, Trinidad, southern Mexico, Central America, northern South America to Paraguay, Uruguay. [P: 77].

Trichomanes robustum E. Fourn., Bull. Soc. Bot. France 15:147. 1869.

Epiphyte in Sierra de Luquillo, Toro Negro Forest Reserve; 850–1300 m; *spr* Dec–Sep.

Antilles, northern South America and Ecuador. [P: 79].

Trichomanes scandens L., Sp. Pl. 2:1098. 1753.

Epiphyte in Sierra de Cayey, Eastern Cordillera (La Torrecilla), Central Cordillera, Maricao Forest Reserve; 400–1225 m; *spr* throughout the year.

Greater Antilles. [P: 67].

Vandenboschia radicans (Sw.) Copel., Philipp. J. Sci. 67(1):54. 1938.

Bas.: *Trichomanes radicans* Sw. (1801).

Epiphyte in Sierra de Luquillo, Central Cordillera, Western Cordillera (Las Marías); 250–1125 m; *spr* Feb–Apr, Jul–Nov.

Pantropical and -subtropical. [P: 69].

TAXA CITED BY PREVIOUS AUTHORS, BUT NOT INCLUDED

NO SPECIMENS SEEN:

Hymenophyllum lanatum Fée.

Epiphyte cited by Liogier & Martorell (1982) as recorded by Maxon for Puerto Rico; cited by Proctor (1989) as known from a single collection by Schwanecke, as reported by Maxon, whose exact locality is unknown; and similarly cited by Liogier & Martorell (2000). [P: 65].

LINDSAEACEAE

Lindsaea lancea (L.) Bedd., Suppl. Ferns S. Ind. 6. 1876.

Bas.: *Adiantum lanceum* L. (1763).

Terrestrial in Northern Coastal Lowlands (San Juan), Sierra de Luquillo, Sierra de Cayey, Eastern Cordillera (Corozal), Central Cordillera (Adjuntas); 20–700 m; *spr* throughout the year.

Antilles, Trinidad and Tobago, southern Mexico, Central America, northern South America to Bolivia, Paraguay. [P: 114].

Lindsaea quadrangularis Raddi subsp. **subalata** K.U. Kramer, Acta Bot. Neerl. 6(2):190, *t. 50*. 1957.

Syn.: *Lindsaea quadrangularis* Raddi subsp. *antillensis* K.U. Kramer (1957).

Lindsaea montana Fée (1866).

Terrestrial in Sierra de Luquillo; 300–750 m; *spr* Dec–Apr, Jul.

Lesser Antilles, southern Mexico, Central America, northern South America (Venezuela, Colombia). [P: 114].

Lindsaea stricta (Sw.) Dryand. var. **stricta**, Trans. Linn. Soc. London 3:42. 1797.

Bas.: *Adiantum strictum* Sw. (1788).

Syn.: *Lindsaea portoricensis* Desv. (1811).

Terrestrial in Northern Coastal Lowlands (Dorado to Laguna Tortuguero), Eastern Cordillera (Guaynabo); 1–500 m; *spr* Dec–Jun, Sep.

Greater Antilles, Trinidad, southern Mexico, Central America, northern South America to Bolivia. [P: 112].

Lindsaea stricta (Sw.) Dryand. var. **jamesoniiformis** K.U. Kramer, Acta Bot. Neerl. 6(2):231, *t. 76*. 1957.

Terrestrial in Sierra de Luquillo; 800–850 m; *spr* Dec–Jan, Apr–May.

Northern South America. [P: 112].

Lonchitis hirsuta L., Sp. Pl. 2:1078. 1753.

Syn.: *Anisosorus hirsutus* (L.) Underw. & Maxon (1926).

Terrestrial in Northern Limestone Hills (Camuy), Río Abajo Forest Reserve, Sierra de Luquillo, Eastern Coast (Cuchilla de Panduras), Sierra de Cayey, Las Tetas de Cayey, Eastern Cordillera (Corozal), Central Cordillera, Western Cordillera; 100–1200 m; *spr* throughout the year.

Antilles, Trinidad, southern Mexico, Central America, northern South America to Bolivia. [P: 108].

Odontosoria aculeata (L.) J. Sm., Cult. Ferns 67. 1867.
Bas.: *Adiantum aculeatum* L. (1753).
Viny terrestrial in Northern Coastal Lowlands (San Juan, Laguna Tortuguero), Sierra de Luquillo, Eastern Coast, Caguas Valley (Caguas), Sierra de Cayey, Las Tetas de Cayey, Eastern Cordillera (Cerro La Peña, La Torrecilla), Central Cordillera, Western Cordillera; 5–1300 m; *spr* throughout the year.
Greater Antilles (Cuba, Hispaniola). [P: 116; AR: 27].
Helecho espinoso.

Odontosoria scandens (Desv.) C. Chr., Index Filic. 354. 1905; et Index Filic. 465. 1906.
Bas.: *Humata scandens* Desv. (1827).
Syn.: *Odontosoria uncinella* (Kunze) Fée (1852).
Viny terrestrial in Sierra de Luquillo, Eastern Cordillera (La Torrecilla), Central Cordillera, Maricao Forest Reserve; 550–1300 m; *spr* throughout the year.
Greater Antilles (Cuba, Hispaniola). [P: 116; AR: 27].
Tembladora.

Sphenomeris clavata (L.) Maxon, J. Washington Acad. Sci. 3:144. 1913.
Bas.: *Adiantum clavatum* L. (1753).
Terrestrial in Northern Limestone Hills, Río Abajo Forest Reserve; 50–350 m; *spr* throughout the year.
Bahamas, Greater Antilles, southeastern United States (Florida), southern Mexico. [P: 115].

LOMARIOPSIDACEAE

Cyclopeltis semicordata (Sw.) J. Sm., Bot. Mag. 72(Comp.):36. 1846.
Bas.: *Polypodium semicordatum* Sw. (1788).
Terrestrial in Northern Coastal Lowlands (Trujillo Alto), Northern Limestone Hills, Río Abajo Forest Reserve, Eastern Coast (Cuchilla de Panduras), Eastern Cordillera (Corozal), Central Cordillera (Ponce), Quebrada de Los Cedros, Western Coast (Hormigueros), Mona; 10–350 m; *spr* throughout the year.
Antilles, Trinidad and Tobago, southern Mexico, Central America, northern South America to Bolivia. [P: 259].

Lomariopsis amydrophlebia (Sloss. ex Maxon) Holttum, Bull. Misc. Inform. Kew 1939:617. 1940.
Bas.: *Stenochlaena amydrophlebia* Sloss. ex Maxon (1924).
Viny epiphyte in Sierra de Luquillo, Sierra de Cayey, Central Cordillera; 250–1150 m; *spr* Aug, Jan.
Greater Antilles (Jamaica, Hispaniola). [P: 281; AR: 35].

Lomariopsis sorbifolia (L.) Fée, Mém. Foug. 2:69. 1845.
Bas.: *Acrostichum sorbifolium* L. (1753).
Syn.: *Stenochlaena sorbifolia* (L.) J. Sm. (1841).
Viny epiphyte in Northern Limestone Hills, Río Abajo Forest Reserve, Sierra de Luquillo, Eastern Coast (Cuchilla de Panduras), Central Cordillera (Adjuntas), Maricao Forest Reserve; 175–800 m; *spr* May–Sep, Dec.
Greater Antilles (Hispaniola), Lesser Antilles. [P: 281; AR: 36].

Nephrolepis biserrata (Sw.) Schott, Gen. Fil. *ad t.* 3. 1834.
Bas.: *Aspidium biserratum* Sw. (1801).
Terrestrial in Northern Coastal Lowlands, Northern Limestone Hills (Camuy), Río Abajo Forest Reserve, Sierra de Luquillo, Sierra de Cayey, Eastern Cordillera (Comerío, La Torrecilla), Maricao Forest Reserve; 1–800 m; *spr* throughout the year.
Antilles, southern United States (Florida, Louisiana), Trinidad, Mexico, Central America, northern South America to Bolivia, tropical Africa, tropical Asia, Polynesia. [P: 265].

Nephrolepis brownii (Desv.) Hovenkamp & Miyam., Blumea 50(2):293. 2005.
Bas.: *Nephrodium brownii* Desv. (1827).
Syn.: *Nephrolepis multiflora* (Roxb.) F.M. Jarrett ex C.V. Morton (1974).
Terrestrial in Northern Coastal Lowlands, Northern Limestone Hills, Río Abajo Forest Reserve, Sierra de Luquillo, Eastern Coast, Caguas Valley, Sierra de Cayey, Las Tetas de Cayey, Eastern Cordillera (Cerro La Peña, Cañón San Cristóbal), Central Cordillera, Maricao Forest Reserve, Sierra Bermeja, Cerro Las Mesas, Vieques, Culebra, Mona, Desecheo; 1–1325 m; *spr* throughout the year.
Native to tropical Africa, tropical Asia. [P: 265].

Nephrolepis cordifolia (L.) C. Presl, Tent. Pterid. 79. 1836.
Bas.: *Polypodium cordifolium* L. (1753), *nom. cons. prop.*
Terrestrial in Northern Coastal Lowlands (Dorado), Eastern Coast (Naguabo), Sierra de Cayey,

Central Cordillera (Ciales, Peñuelas), Maricao Forest Reserve; 20–1025 m; *spr* Feb–Oct.

Pantropical; widely cultivated. [P: 262].

Helecho rizado gigante.

Nephrolepis exaltata (L.) Schott, Gen. Fil. *ad t. 3.* 1834.

Bas.: *Polypodium exaltatum* L. (1759).

Terrestrial in Northern Coastal Lowlands (Laguna Tortuguero), Río Abajo Forest Reserve, Sierra de Luquillo, Las Tetas de Cayey, Central Cordillera, Maricao Forest Reserve; 1–1200 m; *spr* throughout the year.

Bahamas, Greater Antilles, southeastern United States (Florida), southern Mexico, possibly Central America, northern South America possibly to Bolivia. [P: 263].

Helecho Boston, Helecho espada.

Nephrolepis falcata (Cav.) C. Chr. f. **furcans** (T. Moore) Proctor, Rhodora 63:32. 1961.

Bas.: *Nephrolepis davallioides* var. *furcans* T. Moore (1885).

Terrestrial planted and persistent in Northern Coastal Lowlands (San Juan), Sierra de Luquillo, Central Cordillera (Ponce), Western Cordillera (Maricao), Vieques; 175–450 m; *spr* Feb, May–Jul, Nov.

Widely cultivated; origin obscure. [P: 261, key].

Notes: Proctor (1989) used *Nephrolepis falcata* (Cav.) C. Chr. '*Furcans*' for the name of this taxon.

Nephrolepis rivularis (Vahl) Mett. ex Krug in Urban, Bot. Jahrb. Syst. 24(1):122. 1897.

Bas.: *Polypodium rivulare* Vahl (1807).

Terrestrial in Northern Limestone Hills (Quebradillas), Río Abajo Forest Reserve, Sierra de Luquillo, Eastern Coast (Cuchilla de Panduras), Sierra de Cayey, Eastern Cordillera (Guaynabo, Cerro La Peña), Central Cordillera (Adjuntas), Maricao Forest Reserve; 225–950 m; *spr* throughout the year.

West Indies, Trinidad and Tobago, southern Mexico, Central America, northern South America to Bolivia. [P: 263].

TAXA NOT RECENTLY COLLECTED

Lomariopsis kunzeana (Underw.) Holttum, Bull. Misc. Inform. Kew 1939:617. 1940.

Bas.: *Stenochlaena kunzeana* Underw. (1906).

Viny epiphyte collected in Lares on 6 Feb 1887 [*Sintenis 6134* (US)]; between Utuado and Arecibo on 13 Jul 1901 [*Underwood & Griggs 843, 843a* (US)], on 14 Mar 1914 [*Britton & Cowell 2047*(NY)] and on 12 Sep 1943 [*Sargent 3107* (US)]; in the vicinity of Utuado between 7–10 Mar 1915 [*Britton 5215* (NY)] and in Bayamón on 7 Feb 1912 [*Hioram a.n. 566781* (US)].

Greater Antilles (Cuba, Hispaniola), southeastern United States (Florida). [P: 281; AR: 35].

TAXA CITED BY PREVIOUS AUTHORS, BUT NOT INCLUDED

DOUBTFUL OCCURRENCE

Nephrolepis undulata (Afzel. ex Sw.) J. Sm. Syn.: *Nephrolepis occidentalis* Kunze.

Epiphyte cited by Maxon (1926) as collected long ago by Bertero (according to Urban); subsequently cited by Liogier & Martorell (1982) and by Proctor (1989); cited by Liogier & Martorell (2000) as occurring in dry, savanna-like areas toward the southwestern part of the island. [P: 261].

LYGODIACEAE

Lygodium japonicum (Thunb.) Sw., J. Bot. (Schrader) 1800(2):106. 1801.

Bas.: *Ophioglossum japonicum* Thunb. (1784).

Terrestrial in Northern Coastal Lowlands, Northern Limestone Hills, Río Abajo Forest Reserve, Western Coast (Mayagüez); 10–300 m; *spr* Mar–Dec.

Native to Indian Subcontinent, southeastern tropical Asia, Japan, Korea. [P: 44; AR: 39].

Notes: Liogier & Martorell (1982) misapplied the name *Lygodium venustum* Sw. to this taxon.

MARATTIACEAE

Danaea geniculata Raddi, Opusc. Sci. 3:281; et Syn. Fil. Bras. 1:75, *t. 5, f. 1* (reprint). 1819.

Terrestrial in Northern Coastal Lowlands (San Juan, Guaynabo), Sierra de Luquillo, Sierra de Cayey, Eastern Cordillera (Cerro La Peña, Barranquitas), Central Cordillera, Western Cordillera; 25–1050 m; *spr* throughout the year.

Antilles, Mexico, Central America, northern South America to Bolivia. [P: 39].

Danaea grandifolia Underw., N. Amer. Fl. 16:18. 1909.

Terrestrial in Sierra de Luquillo, Sierra de Cayey, Eastern Cordillera (La Torrecilla), Central Cordillera, Maricao Forest Reserve; 365–1075 m; *spr* Jan–Oct.

Greater Antilles (Hispaniola), northern South America (Venezuela, Colombia).

Notes: This taxon has previously been treated as conspecific with *Danaea nodosa*.

Danaea jenmanii Underw., Bull. Torrey Bot. Club 29(12):677. 1902 "Jenmani."

Terrestrial in Sierra de Cayey, Central Cordillera; 850–1220 m; *spr* Mar, Jul–Oct.

Greater Antilles. [P: 41].

Danaea mazeana Underw., Bull. Torrey Bot. Club 29(12):676. 1902.

Syn.: *Danaea jamaicensis* Underw. (1902).

Terrestrial in Toro Negro Forest Reserve; 1000–1250 m; *spr* Jan, Jul (two specimens). [*Christenhusz 3515* (MAPR)].

Greater Antilles, Lesser Antilles (Guadeloupe). [P: 42].

Danaea nodosa (L.) Sm., Mém. Acad. Roy. Sci. (Turin) 5(1790–1791):420, *t. 9, f. 11*. 1793.

Bas.: *Acrostichum nodosum* L. (1753).

Syn.: *Danaea elliptica* Sm. (1808).

Terrestrial in Northern Coastal Lowlands (San Juan), Northern Limestone Hills, Río Abajo Forest Reserve, Sierra de Luquillo, Eastern Coast (Maunabo), Central Cordillera (Los Tres Picachos), Western Cordillera (Las Marías), Western Coast (Mayagüez); 20–940 m; *spr* Apr, Jul–Jan.

Greater Antilles. [P: 39].

Notes: Maxon (1926), Liogier & Martorell (1982, 2000) and Proctor (1989) misapplied the name *Danaea elliptica* Sm., whose type is conspecific with the above taxon, to specimens of *Danaea geniculata*.

Danaea urbanii Maxon, J. Washington Acad. Sci. 14:195. 1924.

Terrestrial in Sierra de Luquillo, Toro Negro Forest Reserve; 650–1250 m; *spr* Aug, Nov–Jan.

Greater Antilles (Hispaniola). [P: 41].

Eupodium laeve (Sm.) Murdock, Taxon 57(3):744. 2008 "laevis."

Bas.: *Marattia laevis* Sm. (1790).

Syn.: *Marattia kaulfussii* J. Sm. (1839).

Terrestrial in Central Cordillera; 950–1250 m; *spr* Jan–Feb, May, Aug–Sep.

Greater Antilles (Cuba, Hispaniola), southern Central America, South America (scattered). [P: 37].

MARSILEACEAE

Marsilea ancylopoda A. Braun, Monatsber. Königl. Preuss. Akad. Wiss. Berlin 1863:434. 1864.

Syn.: *Marsilea berteroi* A. Braun (1871).

Aquatic in Southwestern Coastal Lowlands (Lajas); 10–15 m; *spr* (?). [*Proctor 40575* (SJ); *Axelrod 12478* (UPRRP)].

Greater Antilles (Jamaica, Hispaniola), Mexico, Central America (Nicaragua), South America (scattered). [P: 343].

Marsilea polycarpa Hook. & Grev., Icon. Filic. 2: *t. 160*. 1830.

Aquatic in Northern Coastal Lowlands; 1–2 m; *spr* Mar–Apr, Sep.

Antilles, west-central Mexico, Central America, South America (scattered). [P: 345].

OLEANDRACEAE

Oleandra articulata (Sw.) C. Presl, Tent. Pterid. 78, *t. 2, f. 12*. 1836.

Bas.: *Aspidium articulatum* Sw. (1801).

Syn.: *Oleandra nodosa* (Willd.) C. Presl (1836).

Viny epiphyte in Northern Limestone Hills, Río Abajo Forest Reserve, Sierra de Luquillo, Sierra de Cayey, Eastern Cordillera (Cerro La Peña), Central Cordillera, Maricao Forest Reserve; 200–1050 m; *spr* throughout the year.

Antilles, Trinidad, southern Mexico, Central America, northern South America to Bolivia. [P: 266; AR: 22].

OPHIOGLOSSACEAE

Ophioglossum palmatum L., Sp. Pl. 2:1063. 1753.

Syn.: *Cheiroglossa palmata* (L.) C. Presl (1846).

Terrestrial in Toro Negro Forest Reserve, Maricao Forest Reserve; 450–1335 m; *spr* Jan–Apr.

Greater Antilles, Lesser Antilles (Montserrat, Guadeloupe), southeastern United States (Florida), southern Mexico, Central America, northern South America to Bolivia. [P: 35].

Ophioglossum reticulatum L., Sp. Pl. 2:1063. 1753.

Terrestrial in Northern Coastal Lowlands, Northern Limestone Hills (Ciales), Sierra de Luquillo, Sierra de Cayey, Toro Negro Forest Reserve, Maricao Forest Reserve, Western Coast (Mayagüez); 1–900 m; *spr* throughout the year.

Pantropical. [P: 35].

Lengua de culebra.

TAXA NOT RECENTLY COLLECTED

Botrychium jenmanii Underw., Fern Bull. 8:59. 1900.

Terrestrial collected in San Germán between 20–23 Oct 1913 [*Chase 6231* (US)] and in Maricao on 18 Nov 1913 [*Stevens & Hess 4703* (NY)] and on 3 Aug 1914 [*Hess 346* (US)].

Greater Antilles, southeastern United States. [P: 33].

OSMUNDACEAE

Osmundastrum cinnamomeum (L.) C. Presl, Gefässbündel Farrn 18. 1847 (preprint of Abh. Königl. Böhm. Ges. Wiss. ser. 5 5:326. 1848).
Bas.: *Osmunda cinnamomea* L. (1753).

Terrestrial in Northern Coastal Lowlands (Vega Alta to Laguna Tortuguero); 1–2 m; *spr* Jan–Sep.

Greater Antilles, eastern half of North America, southern Mexico, Central America, South America (scattered), Indo-China, China, Japan, Korea. [P: 43].

Helecho acuático, Helecho de espiga, Marunguey falso.

POLYPODIACEAE

Campyloneurum angustifolium (Sw.) Fée, Mém. Foug. 5:257. 1852.
Bas.: *Polypodium angustifolium* Sw. (1788).

Epiphyte in Northern Coastal Lowlands (Fajardo), Northern Limestone Hills (Quebradillas), Río Abajo Forest Reserve, Sierra de Luquillo, Sierra de Cayey, Eastern Cordillera (La Torrecilla), Central Cordillera, Cerro de Las Cuevas, Western Cordillera; 225–1300 m; *spr* throughout the year.

Greater Antilles, Lesser Antilles (Guadeloupe), southeastern United States (Florida), Mexico, Central America, South America (scattered). [P: 319].

Campyloneurum brevifolium (Lodd. ex Link) Link, Fil. Spec. 124. 1841.
Bas.: *Polypodium brevifolium* Lodd. ex Link (1833).
Syn.: *Polypodium latum* (T. Moore) Sodiro (1893).

Terrestrial in Northern Coastal Lowlands (Laguna Tortuguero), Northern Limestone Hills, Río Abajo Forest Reserve, Sierra de Luquillo, Eastern Coast (Cuchilla de Panduras), Sierra de Cayey, Eastern Cordillera (Naranjito, Cayey), Central Cordillera (Ponce), Southern Coastal Lowlands (Juana Díaz); 10–550 m; *spr* throughout the year.

Antilles, Trinidad, southeastern United States (Florida), Central America, west-central South America and Venezuela. [P: 320].

Campyloneurum costatum (Kunze) C. Presl, Tent. Pterid. 190, *t.* 7. 1836.
Bas.: *Polypodium costatum* Kunze (1834).

Epiphyte in Northern Limestone Hills (Quebradillas, San Sebastián), Río Abajo Forest Reserve; 200–300 m; *spr* Jun–Oct.

Greater Antilles, Trinidad, southern Mexico, Central America, northern South America to Ecuador. [P: 319].

Campyloneurum phyllitidis (L.) C. Presl, Tent. Pterid 190, *t.* 7, *ff. 18–20.* 1836.
Bas.: *Polypodium phyllitidis* L. (1753).

Terrestrial in Northern Coastal Lowlands, Northern Limestone Hills, Río Abajo Forest Reserve, Eastern Coast (Cuchilla de Panduras), Sierra de Cayey (San Lorenzo), Las Tetas de Cayey, Eastern Cordillera, Cerro de Las Cuevas, Western Cordillera (Maricao), Sierra Bermeja, Vieques, Mona; 1–825 m; *spr* throughout the year.

West Indies, Trinidad and Tobago, southeastern United States (Florida), Mexico, Central America, South America. [P: 320].

Campyloneurum repens (Aubl.) C. Presl, Tent. Pterid. 190. 1836.
Bas.: *Polypodium repens* Aubl. (1775).

Terrestrial in Sierra de Luquillo; 500–550 m; *spr* Aug (one specimen). [*Woodbury a.n. 6749* (SJ)].

Greater Antilles (Cuba, Jamaica), Lesser Antilles, southern Mexico, Central America, northern South America to Bolivia. [P: 319].

Cochlidium jungens L.E. Bishop, Amer. Fern J. 68(3):84. 1978.
Syn.: *Grammitis jungens* (L.E. Bishop) Proctor (1989).

Epiphyte in Sierra de Luquillo, Sierra de Cayey; 175–1025 m; *spr* Jul–Mar.

Lesser Antilles (Dominica), southern Central America (Panama). [P: 327].

Notes: Maxon (1926) misapplied the name *Cochlidium minus* (Jenman) Maxon to this taxon.

Cochlidium rostratum (Hook.) Maxon ex C. Chr., Dansk Bot. Ark. 6(3):23. 1929.
Bas.: *Monogramme rostrata* Hook. (1864).
Syn.: *Grammitis rostrata* (Hook.) R.M. Tryon & A.F. Tryon (1982).

Epiphyte in Toro Negro Forest Reserve; 1200–1300 m; *spr* Mar (two specimens). [*Proctor 40358, 44599* (SJ)].

Greater Antilles, Lesser Antilles (Guadeloupe), southern Mexico, Central America, northern South America (Venezuela, Colombia). [P: 327].

Cochlidium seminudum (Willd.) Maxon, Sci. Surv. Porto Rico & Virgin Islands 6(3):407. 1926.
Bas.: *Blechnum seminudum* Willd. (1794).
Syn.: *Grammitis seminuda* (Willd.) Willd. (1810).

Epiphyte in Sierra de Luquillo, Eastern Cordillera (Cerro La Peña); 500–1050 m; *spr* Nov–Aug.

Greater Antilles (Jamaica, Hispaniola), Lesser Antilles, Trinidad, northern South America (Venezuela). [P: 327].

Cochlidium serrulatum (Sw.) L.E. Bishop, Amer. Fern J. 68(3):80. 1978.
Bas.: *Acrostichum serrulatum* Sw. (1788).
Syn.: *Grammitis serrulata* (Sw.) Sw. (1801). *Polypodium duale* Maxon (1912).

Epiphyte in Sierra de Luquillo, Sierra de Cayey, Eastern Cordillera (Cerro La Peña, La Torrecilla), Toro Negro Forest Reserve, Maricao Forest Reserve; 175–1300 m; *spr* throughout the year.

Antilles, Trinidad and Tobago, southern Mexico, Central America, northern South America to Bolivia, tropical Africa, Madagascar. [P: 326].

Enterosora trifurcata (L.) L.E. Bishop, Syst. Bot. 17(3):353, *t. 2A-C.* 1992.
Bas.: *Polypodium trifurcatum* L. (1753).
Syn.: *Grammitis trifurcata* (L.) Copel. (1947).

Epiphyte in Sierra de Luquillo, Toro Negro Forest Reserve; 700–1250 m; *spr* throughout the year.

Antilles, Central America, northern South America to Bolivia. [P: 332].

Helecho peine.

Grammitis limbata Fée, Mém. Foug. 5:233. 1852.
Syn.: *Polypodium hessii* Maxon (1915). *Grammitis hessii* (Maxon) Alain (1965).

Epiphyte in Sierra de Luquillo, Sierra de Cayey, Toro Negro Forest Reserve; 700–1300 m; *spr* throughout the year.

Greater Antilles (Jamaica, Hispaniola), Lesser Antilles (Guadeloupe, St. Vincent), southern Central America (Panama), northwestern, west-central South America and Venezuela. [P: 330].

Lellingeria hartii (Jenman) A.R. Sm. & R.C. Moran, Amer. Fern J. 81(3):84. 1991.
Bas.: *Polypodium hartii* Jenman (1886).
Syn.: *Grammitis hartii* (Jenman) Proctor (1953).

Epiphyte in Sierra de Luquillo, Toro Negro Forest Reserve; 925–1300 m; *spr* Feb–Mar (two specimens). [*Proctor 40355, 41481* (SJ)].

Greater Antilles (Jamaica), Lesser Antilles. [P: 334].

Notes: According to Alan R. Smith (pers. comm.), this taxon is soon to be placed in an as yet unnamed genus by Paolo Labiak.

Lellingeria suspensa (L.) A.R. Sm. & R.C. Moran, Amer. Fern J. 81(3):87. 1991.
Bas.: *Polypodium suspensum* L. (1753).
Syn.: *Grammitis suspensa* (L.) Proctor (1962).

Epiphyte in Sierra de Luquillo, Sierra de Cayey; 450–1050 m; *spr* throughout the year.

Antilles, Trinidad, southern Central America, northern South America to Bolivia. [P: 337].

Microgramma heterophylla (L.) Wherry, Amer. Fern J. 54(3):145; et S. Fern Guide 346. 1964.
Bas.: *Polypodium heterophyllum* L. (1753).

Epiphyte in Northern Coastal Lowlands (San Juan to Laguna Tortuguero), Northern Limestone Hills, Southeastern Foothills (Salinas), Southern Coastal Limestone (Peñuelas), Sierra Bermeja, Cerro Las Mesas, Vieques; 10–300 m; *spr* throughout the year.

Bahamas, Greater Antilles, Virgin Islands, northern Lesser Antilles, southeastern United States (Florida). [P: 316; AR: 37].

Microgramma lycopodioides (L.) Copel., Gen. Fil. (Ann. Cryptog. Phytopathol. 5:) 185. 1947.
Bas.: *Polypodium lycopodioides* L. (1753).

Epiphyte in Northern Coastal Lowlands, Northern Coastal Limestone (Hatillo), Río Abajo Forest Reserve, Sierra de Luquillo, Eastern Coast, Caguas Valley (Gurabo), Sierra de Cayey, Las Tetas de Cayey, Eastern Cordillera, Central Cordillera (Ponce), Susúa Forest Reserve, Maricao Forest Reserve, Cerro Las Mesas; 1–1075 m; *spr* throughout the year.

Antilles, Trinidad and Tobago, southern Mexico, Central America, northern South America to Bolivia, tropical and southern Africa, Madagascar. [P: 316; AR: 37].

Microgramma piloselloides (L.) Copel., Gen. Fil. (Ann. Cryptog. Phytopathol. 5:) 185. 1947.
Bas.: *Polypodium piloselloides* L. (1753).

Epiphyte in Northern Limestone Hills (Quebradillas), Río Abajo Forest Reserve, Sierra de Luquillo, Eastern Coast (Cuchilla de Panduras), Sierra de Cayey, Eastern and Central Cordillera, Maricao Forest Reserve; 125–1100 m; *spr* throughout the year.

Antilles, Trinidad and Tobago, southern Mexico, northern Central America (Guatemala), northern South America (Venezuela). [P: 315; AR: 37].

Micropolypodium sherringii (Baker) A.R. Sm., Novon 2(4):423. 1992.
Bas.: *Polypodium sherringii* Baker (1882).
Syn.: *Grammitis sherringii* (Baker) Proctor (1953).

Epiphyte in Toro Negro Forest Reserve; 1300 m; *spr* Jul (one specimen). [*Liogier 9535* (IJ), not seen].

Greater Antilles (Cuba, Jamaica). [P: 335].

Micropolypodium taenifolium (Jenman) A.R. Sm., Novon 2(4):423. 1992.
Bas.: *Polypodium taenifolium* Jenman (1897).
Syn.: *Grammitis taenifolia* (Jenman) Proctor (1953).

Epiphyte in Sierra de Luquillo, Sierra de Cayey, Toro Negro Forest Reserve; 700–1250 m; *spr* throughout the year.

Greater Antilles (Jamaica, Hispaniola), Lesser Antilles, Trinidad, southern Mexico, Central America, northern South America to Bolivia. [P: 335].

Notes: Proctor (1989), and subsequently Liogier & Martorell (2000), misapplied the name *Grammitis flabelliformis* (Poir.) C.V. Morton to this taxon.

Micropolypodium trichomanoides (Sw.) A.R. Sm., Novon 2(4):423. 1992.
Bas.: *Polypodium trichomanoides* Sw. (1788).
Syn.: *Grammitis trichomanoides* (Sw.) Ching (1940).

Epiphyte in Toro Negro Forest Reserve; 1175 m; *spr* Feb (one specimen). [*Proctor 44539* (SJ)].

Greater Antilles (Cuba, Jamaica), southern Mexico, northern Central America, northern South America (Guyana, Venezuela). [P: 335].

Neurodium lanceolatum (L.) Fée, Mém. Foug. 3:28. 1852.
Bas.: *Pteris lanceolata* L. (1753).
Syn.: *Paltonium lanceolatum* (L.) C. Presl (1849).

Epiphyte in Northern Limestone Hills, Río Abajo Forest Reserve, Sierra de Luquillo, Eastern Coast (Humacao), Las Tetas de Cayey, Central Cordillera (Yauco), Maricao Forest Reserve; 150–1050 m; *spr* Jun–Mar.

West Indies, southeastern United States (Florida), southern Mexico, Central America, northern South America (French Guiana, Guyana). [P: 324].

Notes: According to Alan R. Smith (pers. comm.), this genus is soon to be placed in *Pleopeltis*.

Niphidium crassifolium (L.) Lellinger, Amer. Fern J. 62(4):106. 1972.
Bas.: *Polypodium crassifolium* L. (1753).

Terrestrial in Río Abajo Forest Reserve, Sierra de Luquillo, Sierra de Cayey, Eastern Cordillera (La Torrecilla), Central Cordillera, Western Cordillera; 250–1225 m; *spr* throughout the year.

Antilles, Trinidad, Mexico, Central America, South America. [P: 322].

Calaguala, Polipodio.

Pecluma camptophyllaria (Fée) M.G. Price var. **camptophyllaria**, Amer. Fern J. 73(3):113. 1983.
Bas.: *Polypodium camptophyllarium* Fée (1857).

Terrestrial in Sierra de Luquillo, Eastern Cordillera (La Torrecilla); 400–930 m; *spr* Nov–Jan (two specimens). [*Muratti 131* (MAPR)].

Greater Antilles, southern Central America, northwestern and west-central South America. [P: 307].

Pecluma consimilis (Mett.) M.G. Price, Amer. Fern J. 73(3):113. 1983.
Bas.: *Polypodium consimile* Mett. (1864).

Epiphyte in Central Cordillera, Western Cordillera (Maricao); 425–1120 m; *spr* Aug, Nov, Feb–Mar.

Greater Antilles (Jamaica, Hispaniola), southern Mexico, Central America, northwestern, west-central South America and Venezuela. [P: 304].

Pecluma dispersa (A.M. Evans) M.G. Price, Amer. Fern J. 73(3):114. 1983.
Bas.: *Polypodium dispersum* A.M. Evans (1968).

Terrestrial in Northern Coastal Lowlands (Laguna Tortuguero), Northern Limestone Hills, Río Abajo Forest Reserve, Sierra de Luquillo, Sierra de Cayey, Las Tetas de Cayey, Southeastern Foothills (Coamo), Eastern Cordillera (Aguas Buenas, Cañón San Cristóbal), Central Cordillera, Cerro de Las Cuevas, Susúa Forest Reserve, Maricao Forest Reserve; 1–850 m; *spr* May–Feb.

Greater Antilles, southeastern United States (Florida), Mexico, Central America, northwestern and west-central South America and Venezuela, Brazil. [P: 303].

Pecluma pectinata (L.) M.G. Price, Amer. Fern J. 73(3):115, *t.* 2. 1983.
Bas.: *Polypodium pectinatum* L. (1753).

Terrestrial in Northern Limestone Hills (Arecibo), Río Abajo Forest Reserve, Sierra de Luquillo, Eastern Coast (Cuchilla de Panduras), Caguas Valley (Caguas), Sierra de Cayey, Las Tetas de Cayey, Eastern and Central Cordillera, Western Cordillera; 100–1050 m; *spr* throughout the year.

Antilles, southern Central America (Panama), northern South America to Bolivia. [P: 306].

Pecluma plumula (Humb. & Bonpl. ex Willd.) M.G. Price, Amer. Fern J. 73(3):115. 1983.
Bas.: *Polypodium plumula* Humb. & Bonpl. ex Willd. (1810).

Terrestrial in Northern Limestone Hills (Quebradillas), Las Tetas de Cayey, Eastern Cordillera (Cayey), Central Cordillera; 100–825 m; *spr* Sep–Jan.

Greater Antilles, Lesser Antilles (Saba, Guadeloupe, Martinique), Trinidad, southeastern United States (Florida), Mexico, Central America, northern South America to Bolivia. [P: 302].

Pecluma ptilodon (Kunze) M.G. Price var. **bourgaeana** (E. Fourn.) A.R. Sm. in Mickel & Smith, Mem. New York Bot. Gard. 88:442, *t. 214G–H.* 2004 "bourgeauana."

Bas.: *Polypodium consimile* Mett. var. *bourgaeanum* E. Fourn. (1872).
Syn.: *Polypodium ptilodon* Kunze var. *caespitosum* (Jenman) A.M. Evans (1968).

Terrestrial in Central Cordillera; 950–1100 m; *spr* Mar, Sep (two specimens). [*Proctor 41398, 43264* (SJ)].

Greater Antilles, Trinidad, southeastern United States (Florida), Mexico, Central America (scattered). [P: 306].

Phlebodium aureum (L.) J. Sm., J. Bot. (Hooker) 4:59. 1841.
Bas.: *Polypodium aureum* L. (1753).

Epiphytic/terrestrial in Northern Coastal Lowlands (Dorado to Laguna Tortuguero), Northern Limestone Hills, Sierra de Luquillo, Eastern Coast (Cuchilla de Panduras), Sierra de Cayey, Eastern and Central Cordillera, Cerro de Las Cuevas, Western Cordillera, Cerro Las Mesas, Vieques, Culebra; 1–925 m; *spr* May–Feb.

Bahamas, Greater Antilles (Cuba, Hispaniola), Lesser Antilles, Trinidad, southeastern United States (Georgia, Florida, Alabama), northern South America to Bolivia, Paraguay. [P: 314].

Calaguala, Polipodio.

Phlebodium decumanum (Willd.) J. Sm., J. Bot. (Hooker) 4:59. 1841.
Bas.: *Polypodium decumanum* Willd. (1810).

Epiphyte in Northern Coastal Lowlands, Eastern Coast (Naguabo), Caguas Valley (Gurabo); 1–120 m; *spr* Nov–Aug.

Greater Antilles (Jamaica, Hispaniola), Lesser Antilles (Guadeloupe, Martinique), Trinidad, southern Mexico, Central America, South America. [P: 314].

Phlebodium pseudoaureum (Cav.) Lellinger, Amer. Fern J. 77(3):101. 1987.
Bas.: *Polypodium pseudoaureum* Cav. (1802).

Epiphyte in Northern Limestone Hills (Isabela), Sierra de Luquillo, Sierra de Cayey, Las Tetas de Cayey, Eastern Cordillera (Cerro La Peña), Central Cordillera, Maricao Forest Reserve; 200–1100 m; *spr* Jun–Feb.

Greater Antilles, Lesser Antilles (Guadeloupe), Mexico, Central America, South America. [P: 312].

Phymatosorus grossus (Langsd. & Fisch.) Brownlie, Beih. Nova Hedwigia 55:385. 1977.
Bas.: *Polypodium grossum* Langsd. & Fisch. (1810).

Terrestrial in Northern Coastal Lowlands (Dorado); 20 m; *spr* Mar, Jul–Nov.

Native to tropical Africa, Sri Lanka, southeastern Asia, northern Australia. [P: 310].

Notes: Proctor (1989), and subsequently Liogier & Martorell (2000), misapplied the name *Polypodium scolopendria* Burm. f. to this taxon.

Pleopeltis astrolepis (Liebm.) E. Fourn., Mexic. Pl. 1:87. 1872.
Bas.: *Polypodium astrolepis* Liebm. (1849).

Epiphyte in Northern Coastal Lowlands (Loíza), Northern Limestone Hills, Río Abajo Forest Reserve, Sierra de Luquillo, Eastern Coast (Cuchilla de Panduras), Caguas Valley, Sierra de Cayey, Las Tetas de Cayey, Eastern Cordillera (La Torrecilla), Central Cordillera, Western Cordillera, Cerro Las Mesas, Western Coast (Mayagüez); 1–1100 m; *spr* throughout the year.

Antilles, Trinidad, Mexico, Central America, northern South America to Bolivia. [P: 317].

Pleopeltis macrocarpa (Bory ex Willd.) Kaulf., Berlin. Jahrb. Pharm. Verbundenen Wiss. 21:41. 1820.
Bas.: *Polypodium macrocarpum* Bory ex Willd. (1810).
Syn.: *Polypodium lanceolatum* L. (1753).

Epiphyte in Toro Negro Forest Reserve; 950–1100 m; *spr* Nov, Feb (two specimens). [*Proctor 41022* (SJ); *Breckon 4417* (MAPR)].

Antilles, Central America, South America, tropical and southern Africa to India. [P: 318].

Pleopeltis polypodioides (L.) E.G. Andrews & Windham, Contr. Univ. Michigan Herb. 19:46. 1993.
Bas.: *Acrostichum polypodioides* L. (1753).
Syn.: *Polypodium polypodioides* (L.) Watt (1867).

Epiphyte in Northern Coastal Lowlands (San Juan), Northern Limestone Hills, Río Abajo Forest Reserve, Eastern Coast, Sierra de Cayey, Las Piedras Chiquitas, Eastern Cordillera (Cerro La Peña), Central Cordillera, Cerro de Las Cuevas, Maricao Forest Reserve, Sierra Bermeja, Western Coast (Mayagüez); 1–1050 m; *spr* throughout the year.

West Indies, Trinidad, southern Mexico, Central America, northern South America (Venezuela, Colombia). [P: 308].

Doradilla.

Pleopeltis squamata (L.) J. Sm., Hist. Fil. 114. 1875.
Bas.: *Polypodium squamatum* L. (1753).

Epiphyte in Northern Limestone Hills (Camuy), Las Tetas de Cayey, Eastern Cordillera (La Torrecilla),

Central Cordillera, Maricao Forest Reserve; 300–1200 m; *spr* Jul–Apr.

Bahamas, Greater Antilles. [P: 309].

Polypodium dulce Poir., Encycl. 5:523. 1804.
Syn.: *Polypodium sororium* Humb. & Bonpl. ex Willd. (1810).

Epiphyte in Sierra de Luquillo, Caguas Valley (Caguas), Sierra de Cayey, Eastern Cordillera, Central Cordillera (Peñuelas, Yauco), Western Cordillera; 350–1000 m; *spr* throughout the year.

Antilles, Trinidad, southern Mexico, Central America, northern South America to Bolivia. [P: 308].

Serpocaulon dissimile (L.) A.R. Sm., Taxon 55(4): 928, *t. 3B, H, f. 4A-C.* 2006.
Bas.: *Polypodium dissimile* L. (1759).
Syn.: *Polypodium chnoodes* Spreng. (1822).

Epiphyte in Sierra de Luquillo, Sierra de Cayey, Eastern Cordillera (Cerro La Peña, La Torrecilla), Toro Negro Forest Reserve, Western Cordillera; 200–950 m; *spr* throughout the year.

Antilles, Trinidad, southern Mexico, Central America, northwestern South America and Venezuela. [P: 311].

Serpocaulon levigatum (Cav.) A.R. Sm., Taxon 55(4):928, *t. 3E.* 2006.
Bas.: *Polypodium levigatum* Cav. (1801).

Terrestrial in Monte Guilarte Forest Reserve; 1050 m; *spr* Feb (one specimen). [*Vélez-Gavilán 527* (MAPR)].

Lesser Antilles (Guadeloupe), southern Central America, northern and northwestern South America.

Serpocaulon loriceum (L.) A.R. Sm., Taxon 55(4):928. 2006.
Bas.: *Polypodium loriceum* L. (1753).

Epiphyte in Northern Limestone Hills (Quebradillas), Sierra de Luquillo, Sierra de Cayey, Eastern Cordillera (La Torrecilla), Central Cordillera, Maricao Forest Reserve; 500–1325 m; *spr* throughout the year.

Antilles, Trinidad and Tobago, southern Mexico, Central America, South America (scattered). [P: 309; AR: 39].

Terpsichore anfractuosa (Kunze ex Klotzsch) B. León & A.R. Sm., Amer. Fern J. 93(2):86. 2003.
Bas.: *Polypodium anfractuosum* Kunze ex Klotzsch (1847).
Syn.: *Grammitis anfractuosa* (Kunze ex Klotzsch) Proctor (1961).

Terrestrial in Central Cordillera (Ponce); 1125 m; *spr* Jan (one specimen). [*Proctor 40084* (SJ)].

Greater Antilles, Lesser Antilles (Guadeloupe), southern Mexico, southern Central America, northwestern, west-central South America and Venezuela. [P: 337].

Notes: According to Alan R. Smith (pers. comm.), this taxon is soon to be placed in a new genus, *Ascogrammitis*, by Michael Sundue.

Terpsichore aspleniifolia (L.) A.R. Sm., Novon 3(4):479, 485. 1993 "asplenifolia."
Bas.: *Polypodium aspleniifolium* L. (1753) "asplenifolium."
Syn.: *Grammitis aspleniifolia* (L.) Proctor (1962) "asplenifolia."

Epiphyte in Sierra de Luquillo, Sierra de Cayey, Central Cordillera, Maricao Forest Reserve; 500–1300 m; *spr* throughout the year.

Antilles, Trinidad and Tobago, southern Mexico, Central America, northwestern South America and Venezuela. [P: 340].

Polipodio colgante.

Terpsichore cultrata (Willd.) A.R. Sm., Novon 3(4):486. 1993.
Bas.: *Polypodium cultratum* Willd. (1810).
Syn.: *Grammitis cultrata* (Willd.) Proctor (1961).

Epiphyte in Toro Negro Forest Reserve; 1225–1260 m; *spr* Aug, Jan (two specimens). [*Proctor 39446* (SJ); *Breckon 4682* (MAPR)].

Greater Antilles, southern Mexico, Central America, northwestern, west-central South America and Venezuela, Brazil. [P: 338].

Notes: According to Alan R. Smith (pers. comm.), this taxon is soon to be placed in a new genus, *Alansmia*, by Michael Kessler and Anna Moguel.

Terpsichore hanekeana (Proctor) A.R. Sm., Novon 3(4):486. 1993.
Bas.: *Grammitis hanekeana* Proctor (1989).

Epiphyte in Sierra de Luquillo, Sierra de Cayey; 400–1050 m; *spr* Apr, Jul–Jan.

Endemic. [P: 341].

Terpsichore liogieri (Proctor) A.R. Sm., Novon 3(4):487. 1993.
Bas.: *Grammitis liogieri* Proctor (1989).

Epiphyte in Sierra de Cayey, Central Cordillera, Maricao Forest Reserve; 650–1330 m; *spr* throughout the year.

Endemic. [P: 341].

Terpsichore mollissima (Fée) A.R. Sm., Novon 3(4):487. 1993.
Bas.: *Polypodium mollissimum* Fée (1866).
Syn.: *Grammitis mollissima* (Fée) Proctor (1961).

Epiphyte in Sierra de Luquillo, Sierra de Cayey; 475–700 m; *spr* throughout the year.

Antilles, southern Mexico, Central America, northern South America. [P: 340].

Notes: According to Alan R. Smith (pers. comm.), is soon to be placed in a new genus, *Alansmia*, by Michael Kessler and Anna Moguel.

Terpsichore taxifolia (L.) A.R. Sm., Novon 3(4):488. 1993.
Bas.: *Polypodium taxifolium* L. (1753).
Syn.: *Grammitis taxifolia* (L.) Proctor (1961).

Epiphyte in Sierra de Luquillo, Central Cordillera, Maricao Forest Reserve; 480–1225 m; *spr* Aug–May.

Greater Antilles (Hispaniola), Lesser Antilles, Trinidad, southern Central America, northwestern, west-central South America and Venezuela. [P: 337].

Notes: According to Alan R. Smith (pers. comm.), this taxon is soon to be placed in a new genus, *Mycopteris*, by Michael Sundue.

Zygophlebia sectifrons (Kunze ex Mett.) L.E. Bishop, Amer. Fern J. 79(3):110. 1989.
Bas.: *Polypodium sectifrons* Kunze ex Mett. (1856).
Syn.: *Grammitis sectifrons* (Kunze ex Mett.) F. Seym. (1975).

Epiphyte in Sierra de Cayey, Toro Negro Forest Reserve; 800–1300 m; *spr* Mar–Aug.

Greater Antilles (Jamaica, Hispaniola), Lesser Antilles (St. Kitts, Guadeloupe), southern Central America, South America (Venezuela, Ecuador). [P: 338].

TAXA NOT RECENTLY COLLECTED

Lellingeria myosuroides (Sw.) A.R. Sm. & R.C. Moran, Amer. Fern J. 81(3):85. 1991.
Bas.: *Polypodium myosuroides* Sw. (1788).
Syn.: *Grammitis myosuroides* (Sw.) Sw. (1801).

Epiphyte collected in Maricao on 2 Apr 1913 [*E.G. Britton 2670a* (NY)].

Greater Antilles (Jamaica). [P: 334].

Notes: According to Alan R. Smith (pers. comm.), this taxon is soon to be placed in an as yet unnamed genus by Paolo Labiak.

Serpocaulon triseriale (Sw.) A.R. Sm., Taxon 55(4):929, *t. 3F, f. 4L-O.* 2006.
Bas.: *Polypodium triseriale* Sw. (1801).
Syn.: *Polypodium attenuatum* Humb. & Bonpl. ex Willd. (1810).

Lithophyte collected in the Sierra de Luquillo on 28 Sep 1941 [*Blomquist 11882* (US)].

Greater Antilles (Jamaica), Virgin Islands (Tortola), Lesser Antilles, Trinidad, southeastern United States (Florida), southern Mexico, Central America, northern South America to Bolivia, Paraguay. [P: 311].

TAXA CITED BY PREVIOUS AUTHORS, BUT NOT INCLUDED

CITED DUE TO MISIDENTIFICATION

Pecluma camptophyllaria (Fée) M.G. Price var. ***abbreviata*** (A.M. Evans) Stolze. Syn.: *Polypodium camptophyllarium* Fée var. *abbreviatum* A.M. Evans.

Terrestrial cited by Proctor (1989) based upon a single collection of his in Maricao that was determined by A.M. Evans; subsequently cited by Liogier & Martorell (2000). According to Alan R. Smith, the specimen is a better match to *Pecluma consimilis.* [P: 307].

PSILOTACEAE

Psilotum nudum (L.) P. Beauv., Prodr. Aethéogam. 112. 1805.
Bas.: *Lycopodium nudum* L. (1753).

Terrestrial in Northern Coastal Lowlands, Northern Limestone Hills, Río Abajo Forest Reserve, Sierra de Luquillo, Central Cordillera, Maricao Forest Reserve, Cerro Las Mesas; 1–850 m; *spr* throughout the year.

Pantropical and -subtropical. [P: 14].

PTERIDACEAE

Acrostichum aureum L., Sp. Pl. 2:1069. 1753.

Terrestrial in Northern Coastal Lowlands, Eastern Coast (Humacao), Southwestern Coastal Lowlands (Cabo Rojo), Mona; 0–2 m; *spr* Mar–Jun, Sep–Dec.

Pantropical. [P: 165].

Palmita del río.

Acrostichum danaeifolium Langsd. & Fisch., Pl. Voy. Russes Monde 1:5. 1810 (*tabula* published in 1818) "daneaefolium."

Terrestrial in Northern Coastal Lowlands, Río Abajo Forest Reserve, Eastern Coast (Humacao), Southern Coastal Lowlands (Coamo), Sierra Bermeja, Vieques, Mona; 0–100 m; *spr* throughout the year.

West Indies, southeastern United States (Florida), Mexico, Central America, South America. [P: 165].

Helecho de pantano.

Adiantopsis paupercula (Kunze) Fée, Mém. Foug. 5:145. 1852.

Bas.: *Adiantum pauperculum* Kunze (1850).

Epiphyte in Maricao Forest Reserve; 600–640 m; *spr* Jan–Feb, May (two specimens). [*Proctor 41649* (SJ); *Sundue 2088* (NY)].

Greater Antilles (Cuba, Jamaica). [P: 131].

Adiantum capillus-veneris L., Sp. Pl. 2:1096. 1753.

Lithophyte in Northern Coastal Lowlands (San Juan), Southeastern Foothills (Salinas), Southern Coastal Lowlands (Coamo), Quebrada de Los Cedros; 10–400 m; *spr* Sep–Feb.

Bahamas, Greater Antilles (Cuba), western Canada, southern half of United States, Mexico, northern Central America, South America (Venezuela, Peru), southern Europe, eastern tropical and southern Africa, Madagascar. [P: 141].

Culantrillo de pozo, Helecho culantrillo.

Adiantum concinnum Humb. & Bonpl. ex Willd., Sp. Pl. 5(1):451. 1810.

Terrestrial in Central Cordillera, Western Cordillera, Western Coast; 100–850 m; *spr* May–Feb.

Antilles, Trinidad, Mexico, Central America, northwestern, west-central South America and Venezuela, Brazil. [P: 140].

Adiantum deltoideum Sw., Prodr. 134. 1788.

Lithophyte in Northern Coastal Lowlands (San Juan); 10 m; *spr* Jan–Feb (one specimen). [*Sundue 2106* (NY)].

Native to Greater Antilles (except Puerto Rico).

Adiantum fragile Sw., Prodr. 135. 1788.

Syn.: *Adiantum rigidulum* Mett. ex Kuhn (1869). *Adiantum fragile* Sw. var. *rigidulum* (Mett. ex Kuhn) Proctor (1989).

Terrestrial in Northern Limestone Hills, Río Abajo Forest Reserve, Las Piedras Chiquitas, Southeastern Foothills, Eastern Cordillera (Cayey), Central Cordillera, Cerro de Las Cuevas, Southern Coastal Lowlands (Patillas), Western Cordillera (Maricao), Guánica Forest Reserve, Vieques, Mona; 10–890 m; *spr* throughout the year.

Antilles. [P: 141, 143].

Adiantum latifolium Lam., Encycl. 1(1):43. 1783.

Terrestrial in Northern Coastal Lowlands, Northern Limestone Hills, Río Abajo Forest Reserve, Sierra de Luquillo, Eastern Coast, Caguas Valley (Gurabo), Sierra de Cayey, Southeastern Foothills (Coamo), Central Cordillera (Ponce), Western Cordillera, Western Coast; 1–600 m; *spr* throughout the year.

Antilles, Trinidad, southern Mexico, Central America, South America. [P: 139].

Adiantum macrophyllum Sw., Prodr. 135. 1788.

Terrestrial in Northern Limestone Hills, Western Cordillera; 200–375 m; *spr* Aug–Nov, Feb–Mar.

Greater Antilles, Lesser Antilles (Guadeloupe, Martinique), Trinidad and Tobago, southern Mexico, Central America, northern South America to Bolivia. [P: 135].

Culantrillo.

Adiantum melanoleucum Willd., Sp. Pl. 5(1):443. 1810.

Terrestrial in Northern Limestone Hills, Río Abajo Forest Reserve; 10–350 m; *spr* throughout the year.

Bahamas, Greater Antilles, southeastern United States (Florida). [P: 136].

Cucaracha.

Adiantum obliquum Willd., Sp. Pl. 5(1):429. 1810.

Terrestrial in Northern Coastal Lowlands (San Juan, Guaynabo), Sierra de Luquillo, Caguas Valley (Caguas), Central Cordillera (Utuado); 20–750 m; *spr* Dec–Apr, Jul–Sep.

Antilles, Trinidad and Tobago, southern Mexico, Central America, northern South America to Bolivia. [P: 136].

Adiantum petiolatum Desv., Mag. Neuesten Entdeck. Gesammten Naturk. Ges. Naturf. Freunde Berlin 5:326. 1811.

Terrestrial in Northern Coastal Lowlands, Río Abajo Forest Reserve, Sierra de Luquillo, Caguas Valley, Western Cordillera (Las Marías); 1–100 m; *spr* Jan–Apr, Jul–Oct.

Antilles, Trinidad and Tobago, southern Mexico, Central America, northern South America to Bolivia. [P: 135].

Adiantum pulverulentum L., Sp. Pl. 2:1096. 1753.

Terrestrial in Northern Coastal Lowlands, Northern Limestone Hills, Río Abajo Forest Reserve, Sierra de Luquillo, Eastern Coast (Cuchilla de Panduras), Caguas Valley (Gurabo), Sierra de Cayey, Eastern Cordillera (Aguas Buenas), Central Cordillera (Ponce), Western Cordillera (Maricao); 1–700 m; *spr* throughout the year.

Greater Antilles, Lesser Antilles (Guadeloupe, Martinique), Trinidad, Mexico, Central America, northern South America to Bolivia, Paraguay. [P: 138].

Adiantum pyramidale (L.) Willd., Sp. Pl. 5(1):442. 1810.
Bas.: *Polypodium pyramidale* L. (1753).
Syn.: *Adiantum cristatum* L. (1759).

Terrestrial in Northern Coastal Lowlands, Northern Limestone Hills, Río Abajo Forest Reserve, Sierra de Luquillo, Eastern Coast (Cuchilla de Panduras), Caguas Valley (Gurabo), Sierra de Cayey, Las Tetas de Cayey, Las Piedras Chiquitas, Southeastern Foothills, Eastern and Central Cordillera, Southern Coastal Lowlands (Patillas), Southern Coastal Limestone (Sabana Grande), Susúa Forest Reserve, Western Cordillera, Cerro Las Mesas, Western Coast (Hormigueros); 1–1075 m; *spr* throughout the year.

Antilles. [P: 137].

Helecho de pozo, Pintador.

Adiantum tenerum Sw., Prodr. 135. 1788.

Terrestrial in Northern Coastal Lowlands (Laguna Tortuguero), Northern Limestone Hills, Río Abajo Forest Reserve, Eastern Cordillera (Cayey), Central Cordillera, Southern Coastal Lowlands (Coamo), Southern Coastal Limestone (Peñuelas), Guánica Forest Reserve, Western Coast; 10–350 m; *spr* throughout the year.

West Indies, Trinidad and Tobago, southeastern United States (Florida), Mexico, Central America, northern South America (Venezuela). [P: 143].

Culantrillo de pozo, Helecho culantrillo.

Adiantum tetraphyllum Humb. & Bonpl. ex Willd., Sp. Pl. 5(1):441. 1810.

Terrestrial in Northern Coastal Lowlands (San Juan, Guaynabo), Northern Limestone Hills, Sierra de Luquillo, Central Cordillera (Utuado), Western Cordillera; 20–750 m; *spr* Jan–Oct.

Antilles, Trinidad and Tobago, southern Mexico, Central America, South America (scattered). [P: 139].

Adiantum villosum L., Syst. Nat. (ed. 10) 2:1328. 1759.

Terrestrial in Northern Coastal Lowlands (Río Grande), Sierra de Luquillo, Vieques; 1–300 m; *spr* Apr, Jul–Nov.

Antilles, Trinidad and Tobago, southeastern United States (Florida), Mexico, Central America, northern South America to Peru. [P: 138].

Adiantum ×vivesii Proctor (pro sp.), Mem. New York Bot. Gard. 53:140, *tt. 41–42*. 1989.

Terrestrial in Northern Limestone Hills (Quebradillas); 200–250 m; *spr* (?).

Endemic. [P: 140].

Notes: According to Marian Sepúlveda-Orengo (unpublished Master's thesis), the only known population of this taxon consists of a single individual that has spread by vegetative reproduction due to rhizome proliferation and appears to be incapable of sexual reproduction. The taxon most closely resembles *Adiantum tetraphyllum* and is probably a hybrid with the latter as one of its parents.

This hybrid is on the U.S. Fish & Wildlife Service list of threatened and endangered plants.

Adiantum wilsonii Hook., Syn. Fil. 2:6, *t. 72a*. 1851 "Wilsoni."

Terrestrial in Northern Limestone Hills (Quebradillas, Isabela), Central Cordillera (Adjuntas), Susúa Forest Reserve, Western Cordillera (Las Marías); 200–275 m; *spr* Aug–Nov, Mar.

Greater Antilles, southern Mexico, Central America, northwestern South America. [P: 135].

Ananthacorus angustifolius (Sw.) Underw. & Maxon, Contr. U.S. Natl. Herb. 10:487. 1908.
Bas.: *Pteris angustifolia* Sw. (1788).
Syn.: *Vittaria costata* Kunze (1834).

Epiphyte in Northern Limestone Hills (Arecibo), Río Abajo Forest Reserve, Eastern Cordillera (Cañón San Cristóbal), Western Cordillera; 175–600 m; *spr* Dec–Aug.

Antilles, Trinidad, Mexico, Central America, northern South America to Bolivia. [P: 154].

Anetium citrifolium (L.) Splitg., Tijdschr. Natuurl. Gesch. Physiol. 7:395. 1840.
Bas.: *Acrostichum citrifolium* L. (1753).

Epiphyte in Sierra de Luquillo, Eastern Coast (Cuchilla de Panduras); 250–600 m; *spr* Jun–Mar.

Greater Antilles (Jamaica, Hispaniola), Lesser Antilles, Trinidad and Tobago, southern Mexico, Central America, northern South America to Bolivia. [P: 146].

Astrolepis sinuata (Lag. ex Sw.) D.M. Benham & Windham, Amer. Fern J. 82(2):56. 1992.
Bas.: *Acrostichum sinuatum* Lag. ex Sw. (1806).
Syn.: *Cheilanthes sinuata* (Lag. ex Sw.) Domin (1913).

Terrestrial in Las Piedras Chiquitas, Southeastern Foothills (Coamo); 350–475 m; *spr* Jun–Jul (two specimens). [*Proctor 40548* (SJ); *Axelrod 2508* (UPRRP)].

Greater Antilles (Hispaniola), south-central and southwestern United States, Mexico, Central America (scattered), northwestern South America to Bolivia and Argentina. [P: 130].

Ceratopteris thalictroides (L.) Brogn., Bull. Sci. Soc. Philom. Paris, sér 3 8:186. 1822.

Bas.: *Acrostichum thalictroides* L. (1753).
Syn.: *Ceratopteris richardii* Brogn. (1823).

Terrestrial in Northern Coastal Lowlands, Eastern Coast (Humacao), Caguas Valley (Gurabo); 1–100 m; *spr* Dec–Apr.

Pantropical. [P: 167, 168].

Cheilanthes microphylla (Sw.) Sw., Syn. Fil. 127. 1806.
Bas.: *Adiantum microphyllum* Sw. (1788).

Terrestrial in Northern Limestone Hills (Quebradillas, Isabela), Las Piedras Chiquitas, Southeastern Foothills (Coamo), Central Cordillera (Yauco), Cerro de Las Cuevas, Southern Coastal Limestone (Peñuelas), Susúa Forest Reserve, Guánica Forest Reserve, Southwestern Coastal Lowlands (Cabo Rojo), Mona, Desecheo; 50–700 m; *spr* Jun–Mar.

West Indies, southeastern United States (Florida, Mississippi), Mexico, northern Central America, northwestern, west-central South America and Venezuela. [P: 130].

Doryopteris pedata (L.) Fée, Mém. Foug. 5:133. 1852.
Bas.: *Pteris pedata* L. (1753).

Terrestrial in Northern Limestone Hills (Ciales), Eastern Coast (Yabucoa), Las Tetas de Cayey, Las Piedras Chiquitas, Southeastern Foothills (Coamo), Eastern and Central Cordillera, Southern Coastal Lowlands (Patillas), Maricao Forest Reserve, Sierra Bermeja, Vieques; 50–825 m; *spr* throughout the year.

Antilles. [P: 127].

Eriosorus hispidulus (Kunze) Vareschi, Fl. Venez. 1(2):640. 1969.
Bas.: *Jamesonia hispidula* Kunze (1844).
Syn.: *Psilogramme portoricensis* Maxon (1914).

Lithophyte in Sierra de Luquillo; 975 m; *spr* Dec–Jun.

Lesser Antilles (Dominica), northern Central America (Guatemala), northern South America. [P: 121].

Hemionitis palmata L., Sp. Pl. 2:1077. 1753.

Terrestrial in Northern Coastal Lowlands, Eastern Cordillera (Cañón San Cristóbal), Central Cordillera (Ponce, Utuado), Southern Coastal Lowlands (Coamo), Sierra Bermeja, Vieques; 10–550 m; *spr* throughout the year.

Antilles, Trinidad, Mexico, Central America, northern South America to Bolivia. [P: 126].

Notholaena trichomanoides (L.) Desv., J. Bot. Agric. 1:92. 1813.
Bas.: *Pteris trichomanoides* L. (1753).
Syn.: *Cheilanthes trichomanoides* (L.) Mett. (1859).

Terrestrial in Las Piedras Chiquitas, Southeastern Foothills (Coamo), Central Cordillera, Susúa Forest Reserve, Sierra Bermeja; 100–825 m; *spr* throughout the year.

Greater Antilles. [P: 129].

Pityrogramma austroamericana Domin, Spisy Přír. Fak. Karlovy Univ. 88:7. 1928.
Syn.: *Pityrogramma calomelanos* (L.) Link var. *austroamericana* (Domin) Farw. (1931).

Terrestrial in Susúa Forest Reserve, Western Cordillera, Sierra Bermeja; 100–800 m; *spr* Mar–Nov.

Central America, South America (scatterted). [P: 125].

Notes: Proctor (1989), and subsequently Liogier & Martorell (2000), misapplied the name *Pityrogramma calomelanos* (L.) Link var. *aureoflava* (Hook.) Weath. ex L.H. Bailey to this taxon.

Pityrogramma calomelanos (L.) Link var. **calomelanos**, Handbuch 3:20. 1833.
Bas.: *Acrostichum calomelanos* L. (1753).

Terrestrial in Northern Coastal Lowlands, Northern Limestone Hills, Río Abajo Forest Reserve, Sierra de Luquillo, Eastern Coast, Caguas Valley (Gurabo), Sierra de Cayey, Eastern Cordillera (Aguas Buenas), Central Cordillera (Adjuntas), Southern Coastal Lowlands (Coamo), Western Cordillera, Susúa Forest Reserve, Western Coast (Mayagüez), Vieques, Mona; 1–800 m; *spr* Mar–Nov.

West Indies, Trinidad, Mexico, Central America, South America. [P: 124].

Helechito blanco.

Pityrogramma chrysophylla (Sw.) Link var. **gabrielae** Domin, Pterid. Isl. Dominica 151.1929.

Lithophyte in Northern Coastal Lowlands; 3–30 m; *spr* Dec–Mar.

Virgin Islands, Lesser Antilles. [P: 124].

Pityrogramma sulphurea (Sw.) Maxon, Contr. U.S. Natl. Herb. 17(2):173. 1913.
Bas.: *Acrostichum sulphureum* Sw. (1788).

Terrestrial in Central Cordillera (Ciales), Western Cordillera (Maricao); 325–450 m; *spr* Oct, Feb, May.

Greater Antilles, Virgin Islands (St. Croix). [P: 123].

Helecho amarillo, Helecho dorado.

Pityrogramma tartarea (Cav.) Maxon, Contr. U.S. Natl. Herb. 17(2):173. 1913.
Bas.: *Acrostichum tartareum* Cav. (1802).

Syn.: *Pityrogramma ebenea* (L.) Proctor (1965).

Terrestrial in Northern Limestone Hills (Florida), Central Cordillera, Western Cordillera (Maricao); 350–1300 m; *spr* throughout the year.

Greater Antilles, Mexico, Central America, South America (scattered). [P: 125].

Pityrogramma trifoliata (L.) R.M. Tryon, Contr. Gray Herb. 189:68. 1962.

Bas.: *Acrostichum trifoliatum* L. (1753).

Syn.: *Trismeria trifoliata* (L.) Diels (1899).

Terrestrial in Northern Limestone Hills (Lares, San Sebastián), Río Abajo Forest Reserve, Toro Negro Forest Reserve; 200–1180 m; *spr* Oct–Mar.

Bahamas, Greater Antilles, southeastern United States (Florida), Mexico, Central America, South America. [P: 123].

Polytaenium feei (W. Schaffn. ex Fée) Maxon, Sci. Surv. Porto Rico & Virgin Islands 6(3):405. 1926.

Bas.: *Antrophyum feei* W. Schaffn. ex Fée (1857).

Epiphyte/terrestrial in Northern Limestone Hills (Arecibo), Río Abajo Forest Reserve, Sierra de Luquillo, Eastern Coast (Cuchilla de Panduras), Caguas Valley (Las Piedras), Sierra de Cayey, Central Cordillera (Ponce); 150–625 m; *spr* throughout the year.

Antilles, Trinidad, southern Mexico, Central America, northern South America. [P: 148].

Polytaenium intramarginale (Baker ex Jenman) Alston, Mutisia 7:9. 1952.

Bas.: *Vittaria intramarginalis* Baker ex Jenman (1877).

Epiphyte in Northern Limestone Hills, Río Abajo Forest Reserve, Eastern Coast (Yabucoa), Caguas Valley (Las Piedras), Western Cordillera (Las Marías); 100–425 m; *spr* throughout the year.

Greater Antilles, southern Central America. [P: 148].

Polytaenium lineatum (Sw.) J. Sm., J. Bot. (Hooker) 4:68. 1841.

Bas.: *Hemionitis lineata* Sw. (1788).

Epiphyte in Central Cordillera; 850–1200 m; *spr* Mar–Apr, Jul–Sep.

Greater Antilles, Trinidad, southern Mexico, Central America, northern South America to Bolivia. [P: 148].

Polytaenium urbanii (Brause) Alain, Phytologia 47(3):168. 1980.

Bas.: *Antrophyum urbanii* Brause (1913).

Epiphyte in Toro Negro Forest Reserve; 1050–1100 m; *spr* Sep (two specimens). [*Proctor 41422, 43983* (SJ)].

Greater Antilles (Jamaica, Hispaniola), west-central South America (Bolivia). [P: 150].

Pteris altissima Poir., Encycl. 5:722. 1804.

Terrestrial in Río Abajo Forest Reserve, Sierra de Luquillo, Central Cordillera, Maricao Forest Reserve; 225–1200 m; *spr* throughout the year.

Antilles, Trinidad, Mexico, Central America, northern South America to Bolivia, Paraguay. [P: 163].

Pteris arborea L., Sp. Pl. 2:1073. 1753.

Syn.: *Pteris crassipes* J. Agardh (1839).

Terrestrial in Western Cordillera; 425–600 m; *spr* Apr (two specimens). [*Proctor 40479* (SJ), *Cedeño 548* (MAPR)].

Greater Antilles (Jamaica, Hispaniola), Lesser Antilles, Trinidad. [P: 164].

Pteris biaurita L., Sp. Pl. 2:1076. 1753.

Terrestrial in Central Cordillera (Ponce), Quebrada de Los Cedros, Western Cordillera; 65–550 m; *spr* Apr–Jul.

Pantropical. [P: 161].

Pteris deflexa Link, Hort. Berol. 2:30. 1833.

Terrestrial in Eastern Cordillera (La Torrecilla), Central Cordillera, Maricao Forest Reserve; 650–1200 m; *spr* Jul–Mar.

South America (scattered). [P: 160].

Pteris ensiformis Burm. f., Fl. Indica 230. 1768.

Terrestrial planted and escaped in Northern Coastal Lowlands (Río Grande, San Juan); 10 m; *spr* Mar (one specimen). [*Liogier 32066* (UPR); *Proctor 40162* (SJ)].

Native to tropical Asia, China, Japan, Australia. [P: 159].

Pteris grandifolia L., Sp. Pl. 2:1073. 1753.

Terrestrial in Northern Coastal Lowlands (Toa Baja), Northern Limestone Hills; 25–200 m; *spr* Aug–Sep, Jan–Mar.

Greater Antilles, Lesser Antilles (Guadeloupe, Grenada), Trinidad and Tobago, Mexico, Central America, northwestern, west-central South America and Venezuela, Brazil. [P: 158].

Pteris hexagona (L.) Proctor, Bull. Inst. Jamaica, Sci. Ser. 5:53. 1953.

Bas.: *Adiantum hexagonum* L. (1753).

Syn.: *Anopteris hexagona* (L.) C. Chr. subsp. *intermedia* C.V. Morton (1957).

Terrestrial in Northern Limestone Hills, Río

Abajo Forest Reserve; 150–350 m; *spr* throughout the year.

Greater Antilles (Hispaniola). [P: 155].

Notes: Maxon (1926) misapplied the name *Onychium strictum* Kunze to this taxon.

Pteris longifolia L., Sp. Pl. 2:1074. 1753.

Terrestrial in Northern Coastal Lowlands (islands adjacent to Fajardo), Northern Limestone Hills, Río Abajo Forest Reserve, Monte Guilarte Forest Reserve, Maricao Forest Reserve; 10–600 m; *spr* throughout the year.

Bahamas, Greater Antilles (Jamaica, Hispaniola), Lesser Antilles (Saba, St. Kitts, Guadeloupe), Trinidad, Mexico, northern Central America (Belize, Guatemala, Honduras), northern South America (Venezuela). [P: 157].

Pteris multifida Poir., Encycl. 5:714. 1804.

Terrestrial in Northern Coastal Lowlands (Río Grande, San Juan); 10 m; *spr* Mar (two specimens). [*Liogier 32043* (UPR); *Axelrod 12838* (UPRRP)].

Native to Indo-China, Indonesia, China, Japan.

Pteris mutilata L., Sp. Pl. 2:1076. 1753.

Terrestrial in Northern Limestone Hills, Río Abajo Forest Reserve; 200–350 m; *spr* May–Feb.

Greater Antilles. [P: 159].

Pteris quadriaurita Retz., Observ. Bot. 6:38. 1791.

Syn.: *Pteris plumula* Desv. (1827).

Terrestrial in Northern Limestone Hills, Las Tetas de Cayey, Eastern Cordillera (Cayey), Central Cordillera, Western Cordillera; 10–1025 m; *spr* Sep–Oct, Feb–Apr.

Greater Antilles, Lesser Antilles (Montserrat, Guadeloupe, Martinique), Mexico, Central America, South America (scattered), Africa, Indian Subcontinent. [P: 160].

Pteris tripartita Sw., J. Bot. (Schrader) 1800(2):67. 1801.

Terrestrial in Northern Coastal Lowlands (San Juan), Northern Limestone Hills (Arecibo, Hatillo), Río Abajo Forest Reserve; 25–350 m; *spr* Nov–May.

Native to tropical Africa, tropical Asia, China, Australia, Polynesia (Fiji). [P: 164].

Pteris vittata L., Sp. Pl. 2:1074. 1753.

Terrestrial in Northern Coastal Lowlands, Northern Limestone Hills, Río Abajo Forest Reserve, Las Piedras Chiquitas, Maricao Forest Reserve, Vieques; 5–800 m; *spr* Aug–Apr.

Native to tropical Africa, Thailand, China, Japan, Australia. [P: 158].

Radiovittaria remota (Fée) E.H. Crane, Syst. Bot. 22(3):515. 1998.

Bas.: *Vittaria remota* Fée (1857).

Epiphyte in Sierra de Luquillo, Sierra de Cayey; 650–1050 m; *spr* throughout the year.

Greater Antilles (Jamaica, Hispaniola), southern Central America, northern South America to Bolivia. [P: 152].

Vittaria graminifolia Kaulf., Enum. Filic. 192. 1824.

Syn.: *Vittaria filifolia* Fée (1851).

Epiphyte in Northern Coastal Lowlands (Guaynabo), Northern Limestone Hills, Río Abajo Forest Reserve, Sierra de Luquillo, Sierra de Cayey, Eastern Cordillera (Cerro La Peña), Central Cordillera, Maricao Forest Reserve; 175–1230 m; *spr* throughout the year.

Antilles, Trinidad, Mexico, Central America, northern South America to Bolivia, Uruguay. [P: 152].

Vittaria lineata (L.) Sm., Mém. Acad. Roy. Sci. (Turin) 5(1790–1791):421, *t. 9*, *f. 5*. 1793.

Bas.: *Pteris lineata* L. (1753).

Epiphyte in Sierra de Luquillo, Eastern Coast (Cuchilla de Panduras), Toro Negro Forest Reserve, Maricao Forest Reserve; 100–1200 m; *spr* Oct–Apr.

West Indies, Trinidad, southeastern United States (Georgia, Florida), Mexico, Central America, northern South America to northeastern Argentina, Uruguay. [P: 152].

TAXA NOT RECENTLY COLLECTED

Hecistopteris pumila (Spreng.) J. Sm., London J. Bot. 1:193. 1842.

Bas.: *Gymnogramma pumila* Spreng. (1828).

Epiphyte collected in Indiera Fria, near Maricao, between 19–22 Feb 1915 [*Britton et al. 4444* (NY)].

Greater Antilles, Lesser Antilles (Guadeloupe, Martinique), Trinidad, southern Mexico, Central America, northern South America to Bolivia. [P: 145].

Polytaenium cajenense (Desv.) Benedict, Bull. Torrey Bot. Club 38(4):169. 1911.

Bas.: *Hemionitis cajenensis* Desv. (1811).

Syn.: *Antrophyum cajenense* (Desv.) Spreng. (1827).

Terrestrial collected near Arecibo [*Sintenis 6534* (purportedly P), not seen].

Greater Antilles (Cuba, Hispaniola), southern Mexico, Central America, northern South America to Bolivia. [P: 150].

Pteris pungens Willd., Sp. Pl. 5(1):387. 1810.

Terrestrial collected in the Sierra de Yabucoa on 4 Sep 1885 [*Sintenis 2589* (NY)] and near Yabucoa on 8 Oct 1886 [*Sintenis 5255* (NY)], in Río Piedras in Feb 1911 [*Crisostomo 121* (US)] and near Pueblo Viejo on 11 Mar 1924 [*Britton & Britton 8098* (NY)].

Greater Antilles (Jamaica, Hispaniola), Lesser Antilles (Guadeloupe, Martinique), Trinidad, Mexico, Central America, northern South America to Bolivia. [P: 159].

TAXA CITED BY PREVIOUS AUTHORS, BUT NOT INCLUDED

CITED DUE TO MISIDENTIFICATION

Pteris cretica L.

Terrestrial cited by Liogier & Martorell (1982) as cultivated and escaped, but not cited in the later edition (2000). The specimen upon which this citation was based is referable to *Pteris multifida*. [LM1:23].

ONLY UNDER CULTIVATION

Adiantum hirsutum Bory.

Terrestrial cited by Proctor (1989) as cultivated, but not yet known to have escaped here; subsequently cited by Liogier & Martorell (2000). [P: 134].

Adiantum raddianum C. Presl.

Terrestrial cited by Proctor (1989) as cultivated, but not observed escaping here; subsequently cited by Liogier & Martorell (2000). [P: 141].

SACCOLOMATACEAE

Saccoloma domingense (Spreng.) C. Chr., Index Filic. 612. 1906.

Bas.: *Davallia domingensis* Spreng. (1804).

Terrestrial in Central Cordillera; 900–1100 m; *spr* Mar, Jun–Jul.

Antilles, Trinidad, southern Central America (Costa Rica), northwestern South America and Venezuela. [P: 118].

Saccoloma inaequale (Kunze) Mett., Ann. Sci. Nat., Bot., sér 4 15:80. 1861.

Bas.: *Davallia inaequalis* Kunze (1834).

Terrestrial in Sierra de Luquillo, Sierra de Cayey, Eastern Cordillera (Barranquitas), Central Cordillera (Utuado); 250–975 m; *spr* throughout the year.

Antilles, Trinidad, southern Mexico, Central America, northern South America to Bolivia. [P: 120].

SALVINIACEAE

Salvinia auriculata Aubl., Hist. Pl. Guiane 2:969, *t. 367*. 1775.

Aquatic introduced and persistent in Northern Coastal Lowlands (San Juan); 25 m; *spr* (?). [*Axelrod 12040* (UPRRP)].

Greater Antilles (Cuba, Cayman Islands, Hispaniola), Mexico, Central America, South America. [P: 347].

Salvinia minima Baker, J. Bot. 24:98. 1886.

Aquatic in Northern Coastal Lowlands (Barceloneta, Arecibo), Western Coast (Mayagüez); 2 m; *spr* May (one specimen).

Greater Antilles (Cuba), Mexico, Central America, South America (scattered). [P: 346].

SCHIZAEACEAE

Actinostachys pennula (Sw.) Hook., Gen. Fil. *t. 111A*. 1842.

Bas.: *Schizaea pennula* Sw. (1806).

Terrestrial in Northern Coastal Lowlands (Dorado to Laguna Tortuguero), Guajataca Forest Reserve; 1–300 m; *spr* Oct, Jan–May.

Greater Antilles (Cuba), Lesser Antilles (Guadeloupe), Trinidad, southeastern United States (Florida), Mexico, Central America, northern South America to Uruguay. [P: 49].

Schizaea fluminensis Miers ex J.W. Sturm in Martius, Fl. Bras. 1(2):184, *t. 15, f. 2*. 1859.

Terrestrial in Sierra de Luquillo; 500–550 m; *spr* Mar, Nov (two specimens). [*Woodbury a.n. 51* (UPR); *Proctor 40316* (SJ)].

Greater Antilles (Cuba, Jamaica), Lesser Antilles (Grenada), northern South America to Bolivia. [P: 49].

Schizaea poeppigiana J.W. Sturm in Martius, Fl. Bras. 1(2):181. 1859.

Syn.: *Lophidium poeppigianum* (J.W. Sturm) Underw. ex Maxon (1909).

Terrestrial in Río Abajo Forest Reserve, Maricao Forest Reserve; 375–700 m; *spr* throughout the year.

Bahamas, Greater Antilles, southern Mexico, Central America, northern South America to Bolivia. [P: 51].

TECTARIACEAE

Hypoderris brownii J. Sm. in Hooker & Bauer, Gen. Fil. *t. 1*. 1838.

Terrestrial in Northern Limestone Hills, Río Abajo Forest Reserve, Central Cordillera (Adjuntas); 200–350 m; *spr* Jun–Mar.

Greater Antilles (Hispaniola), Lesser Antilles (Grenada), Trinidad and Tobago, northern South America (Venezuela). [P: 250].

Tectaria cicutaria (L.) Copel., Philipp. J. Sci., C 2(6):410. 1907.
Bas.: *Polypodium cicutarium* L. (1759).

Terrestrial in Northern Limestone Hills, Río Abajo Forest Reserve, Central Cordillera (Adjuntas); 50–400 m; *spr* throughout the year.

Greater Antilles. [P: 252].

Tectaria estremerana Proctor & A.M. Evans (pro sp.), Mem. New York Bot. Gard. 53:253, *t. 73*. 1989.

Terrestrial in Northern Limestone Hills, Río Abajo Forest Reserve; 150–350 m; *spr* Jun-Jul, Oct–Mar.

Endemic. [P: 253].

Notes: This taxon is possibly a fertile tetraploid hybrid between *Tectaria cicutaria* and *Tectaria incisa*. This taxon is on the U.S. Fish & Wildlife Service list of threatened and endangered plants.

Tectaria fimbriata (Willd.) Proctor & Lourteig, Bradea 5(40):386. 1990.
Bas.: *Aspidium fimbriatum* Willd. (1810).
Syn.: *Tectaria lobata* (C. Presl) C.V. Morton (1966).

Terrestrial in Northern Limestone Hills (Quebradillas); 100 m; *spr* Sep (one specimen). [*Estremera a.n. 7146* (SJ)].

Bahamas, Greater Antilles (Cuba, Hispaniola), southeastern United States (Florida), southern Mexico, northern Central America (Guatemala). [P: 252].

Tectaria heracleifolia (Willd.) Underw., Bull. Torrey Bot. Club 33(3):200. 1906.
Bas.: *Aspidium heracleifolium* Willd. (1810).

Terrestrial in Northern Coastal Lowlands, Northern Limestone Hills, Río Abajo Forest Reserve, Southeastern Foothills (Cayey), Eastern Cordillera (Aguas Buenas, Cayey), Central Cordillera, Quebrada de Los Cedros, Western Cordillera (Maricao), Guánica Forest Reserve; 1–600 m; *spr* throughout the year.

West Indies, Trinidad, southern United States (Florida, Texas), Mexico, Central America, northern South America to Peru. [P: 253].

Tectaria incisa Cav., Descr. Pl. 249. 1802.
Syn.: *Tectaria martinicensis* (Spreng.) Copel. (1907).

Terrestrial in Northern Coastal Lowlands (San Juan, Guaynabo), Northern Limestone Hills, Río Abajo Forest Reserve, Sierra de Luquillo, Eastern Coast (Cuchilla de Panduras), Sierra de Cayey, Eastern Cordillera (Aguas Buenas), Central Cordillera, Quebrada de Los Cedros, Western Cordillera, Western Coast; 25–450 m; *spr* throughout the year.

West Indies, Trinidad, southern Mexico, Central America, South America. [P: 256].

Tectaria plantaginea (Jacq.) Maxon, Contr. U.S. Natl. Herb. 10:494. 1908.
Bas.: *Polypodium plantagineum* Jacq. (1789).
Syn.: *Tectaria plantaginea* (Jacq.) Maxon var. *confluens* C.V. Morton (1966).

Terrestrial in Sierra de Luquillo, Sierra de Cayey; 100–675 m; *spr* Feb–Aug, Nov.

Lesser Antilles, Trinidad and Tobago, Central America, northern South America to Bolivia. [P: 251, 252].

Notes: *Tectaria plantaginea* var. *confluens* is characterized, according to Proctor (1989), by having sori irregularly elongate rather than round. I studied the two specimens from the Sierra de Luquillo upon which his citation was based and found the shape of their sori not markedly different from the known variation of those in the typical variety within Puerto Rico.

Tectaria prolifera (Hook.) R.M. Tryon & A.F. Tryon, Rhodora 83(833):136. 1981.
Bas.: *Fadyenia prolifera* Hook. (1840).
Syn.: *Fadyenia hookeri* (Sweet) Maxon (1908).

Epiphyte in Northern Limestone Hills (Quebradillas, Isabela), Sierra de Luquillo; 15–250 m; *spr* Nov–Dec (two specimens). [*Proctor 39768, 39967* (SJ)].

Greater Antilles. [P: 258].

Tectaria trifoliata (L.) Cav., Descr. Pl. 249. 1802.
Bas.: *Polypodium trifoliatum* L. (1753).

Terrestrial in Northern Limestone Hills (Quebradillas), Sierra de Luquillo, Sierra de Cayey, Eastern Cordillera (La Torrecilla), Central Cordillera, Western Cordillera; 250–1100 m; *spr* Jan–Oct.

Greater Antilles (Cuba, Hispaniola), Lesser Antilles, Trinidad and Tobago, northern South America. [P: 255].

THELYPTERIDACEAE

Macrothelypteris torresiana (Gaudich.) Ching, Acta Phytotax. Sin. 8(4):310. 1963.
Bas.: *Polystichum torresianum* Gaudich. (1828).

Terrestrial in Northern Limestone Hills, Río Abajo Forest Reserve, Sierra de Luquillo, Sierra de Cayey (San Lorenzo), Las Tetas de Cayey, Central Cordillera (Utuado, Ciales), Western Cordillera (Las Marías); 175–900 m; *spr* throughout the year.

Native to South Africa, Madagascar, southeastern Asia, Japan, Australia, New Zealand, Pacific islands (various). [P: 175].

Thelypteris abdita Proctor, Amer. Fern J. 75(2):63, *t. 4.* 1985.

Terrestrial in Northern Limestone Hills, Río Abajo Forest Reserve; 200–350 m; *spr* throughout the year.

Bahamas. [P: 203].

Thelypteris abrupta (Desv.) Proctor, Rhodora 61 (732):305. 1960.

Bas.: *Polypodium abruptum* Desv. (1827).

Terrestrial in Sierra de Luquillo; 120 m; *spr* Aug (one specimen). [*Proctor 40762* (SJ)].

Greater Antilles, Lesser Antilles, Trinidad and Tobago, northern South America to Bolivia. [P: 210].

Thelypteris angustifolia (Willd.) Proctor, Bull. Inst. Jamaica, Sci. Ser. 5:57. 1953.

Bas.: *Meniscium angustifolium* Willd. (1810).

Syn.: *Dryopteris angustifolia* (Willd.) Urb. (1903).

Terrestrial in Northern Limestone Hills, Río Abajo Forest Reserve, Sierra de Luquillo, Eastern Cordillera (Corozal), Central Cordillera, Western Cordillera (Maricao); 100–800 m; *spr* throughout the year.

Greater Antilles, Virgin Islands (St. Thomas), Lesser Antilles (Montserrat), southern Mexico, Central America, northern South America to Bolivia, Paraguay. [P: 215].

Thelypteris asplenioides (Sw.) Proctor, Bull. Inst. Jamaica, Sci. Ser. 5:57. 1953.

Bas.: *Polypodium asplenioides* Sw. (1801).

Terrestrial in Central Cordillera; 850–1125 m; *spr* Jan–Mar, Jul–Aug.

Greater Antilles (Jamaica, Hispaniola), northern South America (Venezuela). [P: 207].

Thelypteris balbisii (Spreng.) Ching, Bull. Fan Mem. Inst. Biol., Bot. 10:250. 1941.

Bas.: *Polypodium balbisii* Spreng. (1821).

Syn.: *Dryopteris sprengelii* (Kaulf.) Kuntze (1891).

Terrestrial in Northern Limestone Hills (Arecibo, Lares), Río Abajo Forest Reserve, Sierra de Luquillo, Sierra de Cayey, Las Tetas de Cayey, Eastern Cordillera (Cañón San Cristóbal), Central Cordillera, Maricao Forest Reserve; 150–1300 m; *spr* throughout the year.

Antilles, Trinidad and Tobago, southern Mexico, Central America, northern and west-central South America. [P: 187].

Helecho de lana.

Thelypteris brittoniae (Sloss.) Alain, Bull. Torrey Bot. Club 92(4):304. 1965 "brittonae."

Bas.: *Dryopteris brittoniae* Sloss. (1926) "brittonae."

Terrestrial in Central Cordillera (Los Tres Picachos), Western Cordillera; 425–1200 m; *spr* throughout the year.

Greater Antilles (Hispaniola). [P: 209].

Thelypteris cheilanthoides (Kunze) Proctor, Bull. Inst. Jamaica, Sci. Ser. 5:58. 1953.

Bas.: *Aspidium cheilanthoides* Kunze (1849).

Terrestrial in Toro Negro Forest Reserve; 1300–1325 m; *spr* Feb, Aug (two specimens). [*Proctor 40129, 40764* (SJ)].

Greater Antilles, Mexico, Central America, northwestern, west-central South America and Venezuela, Brazil. [P: 188].

Thelypteris decussata (L.) Proctor, Bull. Inst. Jamaica, Sci. Ser. 5:59. 1953.

Bas.: *Polypodium decussatum* L. (1753).

Syn.: *Dryopteris decussata* (L.) Urb. (1903).

Terrestrial in Sierra de Luquillo, Sierra de Cayey, Central Cordillera; 600–1225 m; *spr* Jan–Oct.

Antilles, Trinidad, Central America, northern South America to Bolivia. [P: 190].

Thelypteris deltoidea (Sw.) Proctor, Bull. Inst. Jamaica, Sci. Ser. 5:59. 1953.

Bas.: *Polypodium deltoideum* Sw. (1788).

Syn.: *Dryopteris deltoidea* (Sw.) Kuntze (1891).

Terrestrial in Northern Coastal Lowlands (Laguna Tortuguero), Sierra de Luquillo, Eastern Coast (Cuchilla de Panduras), Caguas Valley (Gurabo), Sierra de Cayey, Eastern and Central Cordillera, Maricao Forest Reserve; 1–1225 m; *spr* throughout the year.

Greater Antilles, Virgin Islands (St. Thomas), Lesser Antilles (Martinique). [P: 192].

Thelypteris dentata (Forssk.) E.P. St. John, Amer. Fern J. 26(2):44. 1936.

Bas.: *Polpodium dentatum* Forssk. (1775).

Syn.: *Dryopteris dentata* (Forssk.) C. Chr. (1920).

Terrestrial in Northern Coastal Lowlands, Northern Limestone Hills, Río Abajo Forest Reserve, Sierra de Luquillo, Eastern Coast, Caguas Valley (Juncos), Sierra de Cayey, Eastern Cordillera (Cañón San Cristóbal), Central Cordillera, Western Cordillera; 1–1100 m; *spr* throughout the year.

Probably native to the Paleotropics. [P: 194].

Thelypteris germaniana (Fée) Proctor, Rhodora 61(732):306. 1960.
Bas.: *Phegopteris germaniana* Fée (1866).
Syn.: *Dryopteris germaniana* (Fée) C. Chr. (1905).

Terrestrial in Sierra de Luquillo, Central Cordillera, Maricao Forest Reserve, Western Coast (Mayagüez); 300–1225 m; *spr* throughout the year.

Antilles, southern Central America, northwestern South America (Ecuador). [P: 180].

Thelypteris grandis A.R. Sm. var. **grandis**, Univ. Calif. Publ. Bot. 59:96. 1971.

Terrestrial in Northern Limestone Hills, Río Abajo Forest Reserve, Sierra de Luquillo, Caguas Valley (Caguas), Sierra de Cayey, Central Cordillera, Maricao Forest Reserve; 200–1200 m; *spr* Jan–Oct.

Greater Antilles, Lesser Antilles (St. Kitts). [P: 199].

Thelypteris guadalupensis (Wikstr.) Proctor, Bull. Inst. Jamaica, Sci. Ser. 5:60. 1953.
Bas.: *Polypodium guadalupense* Wikstr. (1826).
Syn.: *Dryopteris domingensis* (Spreng.) Maxon (1926).

Terrestrial in Northern Coastal Lowlands (Laguna Tortuguero), Northern Limestone Hills, Río Abajo Forest Reserve, Quebrada de Los Cedros, Guánica Forest Reserve; 1–350 m; *spr* throughout the year.

Greater Antilles, Lesser Antilles (Guadeloupe, Martinique), southern Mexico. [P: 203].

Thelypteris hastata (Fée) Proctor, Bull. Inst. Jamaica, Sci. Ser. 5:60. 1953.
Bas.: *Goniopteris hastata* Fée (1866).
Syn.: *Dryopteris hastata* (Fée) Urb. (1903).

Terrestrial in Northern Limestone Hills, Río Abajo Forest Reserve; 25–350 m; *spr* throughout the year.

Greater Antilles (Jamaica, Hispaniola), Lesser Antilles (Guadeloupe, Martinique), Trinidad and Tobago. [P: 210].

Thelypteris hildae Proctor, Amer. Fern J. 75(2):67, *t. 6*. 1985.

Terrestrial in Northern Limestone Hills, Río Abajo Forest Reserve, Sierra de Luquillo, Central Cordillera (Ponce); 150–350 m; *spr* throughout the year.

Greater Antilles (Hispaniola). [P: 201].

Thelypteris hispidula (Decne.) Reed var. **inconstans** (C. Chr.) Proctor, Amer. Fern J. 70(3):89. 1980.
Bas.: *Dryopteris dentata* (Forssk.) C. Chr. var. *inconstans* C. Chr. (1937).
Syn.: *Thelypteris quadrangularis* (Fée) Schelpe var. *inconstans* (C. Chr.) A.R. Sm. (1971).

Terrestrial in Northern Coastal Lowlands (Canóvanas to San Juan), Northern Limestone Hills, Sierra de Luquillo, Eastern Coast, Sierra de Cayey, Las Tetas de Cayey, Eastern Cordillera (Cañón San Cristóbal), Central Cordillera, Western Cordillera, Western Coast (Mayagüez); 1–1075 m; *spr* throughout the year.

Antilles. [P: 195].

Thelypteris ×imitata (C. Chr.) Alain (pro sp.), Moscosoa 3:46. 1978.
Bas.: *Dryopteris imitata* C. Chr. (1937).
Syn.: *Thelypteris cordata* (Fée) Proctor var. *imitata* (C. Chr.) Proctor (1985).

Terrestrial in Northern Limestone Hills, Río Abajo Forest Reserve, Western Cordillera (Maricao); 225–350 m; *spr* Aug–May.

Bahamas, Greater Antilles (Hispaniola). [P: 204].

Notes: According to Alan R. Smith (pers. comm.), since this taxon produces, for the most part, deformed sporangia, it is most likely a hybrid with *Thelpteris abdita* as one of the parents.

Thelypteris inabonensis Proctor, Amer. Fern J. 75(2):61, *t. 3*. 1985.

Terrestrial in Toro Negro Forest Reserve; 1125–1250 m; *spr* Jan–Mar.

Endemic. [P: 181].

Notes: This taxon is on the U.S. Fish & Wildlife Service list of threatened and endangered plants.

Thelypteris interrupta (Willd.) K. Iwats., J. Jap. Bot. 38(10):314. 1963.
Bas.: *Pteris interrupta* Willd. (1794).
Syn.: *Dryopteris gongylodes* (Schkuhr) Kuntze (1891).

Terrestrial in Northern Coastal Lowlands (Toa Baja to Laguna Tortuguero), Northern Limestone Hills (Florida), Río Abajo Forest Reserve, Eastern Coast (Humacao), Western Coast (Añasco); 1–325 m; *spr* throughout the year.

Pantropical. [P: 193].

Thelypteris ×invisa (Sw.) Proctor (pro sp.), Rhodora 61(732):306. 1960.
Bas.: *Aspidium invisum* Sw. (1801).
Syn.: *Dryopteris oligophylla* Maxon (1908). *Thelypteris oligophylla* (Maxon) Proctor (1953).

Terrestrial in Northern Limestone Hills, Central Cordillera (Orocovis); 15–900 m; *spr* (?).

Greater Antilles. [P: 199].

Notes: The name of this taxon represents the hybrid *Thelpteris grandis ×serra*.

Thelypteris kunthii (Desv.) C.V. Morton, Contr. U.S. Natl. Herb. 38:53. 1967.
Bas.: *Nephrodium kunthii* Desv. (1827).
Syn.: *Dryopteris normalis* C. Chr. (1910).

Terrestrial in Northern Coastal Lowlands (Laguna Tortuguero), Northern Limestone Hills, Río Abajo Forest Reserve, Sierra de Luquillo, Sierra de Cayey, Las Tetas de Cayey, Southeastern Foothills (Cayey), Central Cordillera (Ponce), Quebrada de Los Cedros, Western Cordillera, Cerro Las Mesas, Western Coast; 1–800 m; *spr* throughout the year.

West Indies, southeastern and south-central United States, Mexico, Central America, northern South America. [P: 197].

Thelypteris leptocladia (Fée) Proctor, Bull. Inst. Jamaica, Sci. Ser. 5:61. 1953.
Bas.: *Goniopteris leptocladia* Fée (1866).
Syn.: *Dryopteris leptocladia* (Fée) Maxon (1926).

Terrestrial in Northern Limestone Hills; 10–200 m; *spr* Dec–Feb, Jun–Aug.

Greater Antilles, Lesser Antilles (Guadeloupe, Martinique), southern Mexico. [P: 210].

Thelypteris namaphila Proctor, Amer. Fern J. 75(2):56, *t. 1.* 1985.

Terrestrial in Maricao Forest Reserve; 450–850 m; *spr* throughout the year.

Endemic. [P: 185].

Thelypteris nephrodioides (Klotzsch) Proctor, Bull. Inst. Jamaica, Sci. Ser. 5:61. 1953.
Bas.: *Aspidium nephrodioides* Klotzsch (1847).
Syn.: *Dryopteris guadalupensis* (Fée) Kuntze (1891).

Terrestrial in Sierra de Luquillo, Quebrada de Los Cedros; 5–80 m; *spr* Dec–Jan (two specimens). [*Woodbury a.n. 7508* (SJ); *Proctor 40026* (SJ)]

Antilles, Trinidad, southern Central America, northern South America (Venezuela, Colombia). [P: 212].

Thelypteris oligocarpa (Humb. & Bonpl. ex Willd.) Ching, Bull. Fan Mem. Inst. Biol., Bot. 10:253. 1941.
Bas.: *Polypodium oligocarpum* Humb. & Bonpl. ex Willd. (1810).

Terrestrial in Northern Limestone Hills (Quebradillas), Central Cordillera, Western Cordillera (Maricao); 150–1300 m; *spr* throughout the year.

Greater Antilles, Mexico, Central America, northern South America to Bolivia, northwest Argentina. [P: 179].

Thelypteris opposita (Vahl) Ching, Bull. Fan Mem. Inst. Biol., Bot. 10:253. 1941.
Bas.: *Polypodium oppositum* Vahl (1807).

Terrestrial in Western Cordillera (Maricao); 400 m; *spr* Feb (one specimen). [*Liogier 32730* (UPR)].

Lesser Antilles, Trinidad and Tobago, southern Central America, northern South America to Bolivia. [P: 187].

Thelypteris opulenta (Kaulf.) Fosberg, Smithsonian Contr. Bot. 8:3. 1972.
Bas.: *Aspidium opulentum* Kaulf. (1824).

Terrestrial in Sierra de Luquillo; 150–250 m; *spr* Aug (one specimen). [*Proctor 48899* (SJ)].

Native to Paleotropics and -subtropics.

Thelypteris patens (Sw.) Small var. **patens**, Ferns S.E. States 243. 1938.
Bas.: *Polypodium patens* Sw. (1788).

Terrestrial in Northern Limestone Hills, Río Abajo Forest Reserve, Sierra de Luquillo, Sierra de Cayey, Maricao Forest Reserve; 150–770 m; *spr* Feb–Nov.

Antilles, Trinidad, southeastern United States (Florida), Mexico, Central America, South America (scattered). [P: 197].

Thelypteris patens (Sw.) Small var. **smithiana** Ponce, Darwiniana 28(1–4):373. 1988.

Terrestrial in Northern Coastal Lowlands (Loíza), Northern Limestone Hills (Arecibo), Río Abajo Forest Reserve, Sierra de Luquillo, Maricao Forest Reserve; 1–500 m; *spr* Dec–Feb, May.

Antilles, Trinidad, Central America, northwestern and west-central South America and Argentina. [P: 197].

Notes: Proctor (1989), and subsequently Liogier & Martorell (2000), misapplied to this taxon the name *Thelypteris patens* (Sw.) Small var. *scabriuscula* (C. Presl) A.R. Sm., the type of which belongs to *Thelypteris patens* var. *patens*.

Thelypteris pennata (Poir.) C.V. Morton, Contr. U.S. Natl. Herb. 38:64. 1967.
Bas.: *Polypodium pennatum* Poir. (1804).
Syn.: *Dryopteris megalodus* (Schkuhr) Urb. (1903).

Terrestrial in Northern Coastal Lowlands (Carolina, San Juan), Northern Limestone Hills (Toa Baja, Vega Alta), Eastern Coast, Las Tetas de Cayey, Eastern Cordillera (Aguas Buenas), Quebrada de Los Cedros, Maricao Forest Reserve; 40–800 m; *spr* Nov–Aug.

Antilles, Trinidad and Tobago, northern South America to Peru. [P: 213].

Thelypteris piedrensis (C. Chr.) C.V. Morton, Amer. Fern J. 53(2):69. 1963.

Bas.: *Dryopteris piedrensis* C. Chr. (1909).

Terrestrial in Las Tetas de Cayey, Central Cordillera, Western Cordillera; 400–1300 m; *spr* Jun–Mar.

Greater Antilles (Cuba, Hispaniola). [P: 185].

Thelypteris poiteana (Bory) Proctor, Bull. Inst. Jamaica, Sci. Ser. 5:63. 1953.

Bas.: *Lastrea poiteana* Bory (1825).

Syn.: *Dryopteris poiteana* (Bory) Urb. (1903).

Terrestrial in Northern Coastal Lowlands, Northern Limestone Hills, Río Abajo Forest Reserve, Sierra de Luquillo, Eastern Coast, Sierra de Cayey, Las Tetas de Cayey, Southeastern Foothills (Cayey), Eastern Cordillera, Central Cordillera (Ponce, Guayanilla), Western Cordillera, Vieques; 1–830 m; *spr* throughout the year.

Antilles, Trinidad and Tobago, southern Mexico, Central America, northern South America to Bolivia. [P: 215].

Thelypteris reptans (J.F. Gmel.) C.V. Morton var. **reptans**, Fieldiana, Bot. 28(1):12. 1951 May 18; et Amer. Fern J. 41(3):87. 1951 Oct 3.

Bas.: *Polypodium reptans* J.F. Gmel. (1792).

Syn.: *Dryopteris reptans* (J.F. Gmel.) C. Chr. (1905).

Terrestrial in Northern Limestone Hills; 150–300 m; *spr* Jul–Apr.

Bahamas, Greater Antilles, Trinidad, southeastern United States (Florida), southern Mexico, Central America, northern South America (Venezuela). [P: 205].

Thelypteris resinifera (Desv.) Proctor, Bull. Inst. Jamaica, Sci. Ser. 5:63. 1953.

Bas.: *Polypodium resiniferum* Desv. (1811).

Terrestrial in Northern Limestone Hills (Arecibo, Lares), Sierra de Luquillo, Eastern Coast (Naguabo), Eastern Cordillera (Bayamón, Cañón San Cristóbal), Central Cordillera; 40–1050 m; *spr* Jul–Mar.

Greater Antilles, southeastern United States (Florida), Mexico, Central America, northwestern and west-central South America. [P: 186].

Thelypteris reticulata (L.) Proctor, Bull. Inst. Jamaica, Sci. Ser. 5:63. 1953.

Bas.: *Polypodium reticulatum* L. (1759).

Syn.: *Dryopteris reticulata* (L.) Urb. (1903).

Terrestrial in Northern Coastal Lowlands (Loíza), Northern Limestone Hills, Río Abajo Forest Reserve, Sierra de Luquillo, Eastern Coast (Humacao), Sierra de Cayey, Central Cordillera (Ponce); 1–950 m; *spr* throughout the year.

Antilles, Trinidad and Tobago, southeastern United States (Florida), southern Mexico, Central America, northern South America (Venezuela, Colombia). [P: 216].

Thelypteris rheophyta Proctor, Amer. Fern J. 75(2): 58, *t.* 2. 1985.

Terrestrial in Northern Limestone Hills (Arecibo), Río Abajo Forest Reserve, Sierra de Luquillo, Central Cordillera, Western Cordillera (Maricao); 100–1100 m; *spr* Jun–Mar.

Endemic. [P: 186].

Thelypteris ×rolandii (C. Chr.) R.M. Tryon (pro sp.), Rhodora 69(777):8. 1967.

Bas.: *Dryopteris rolandii* C. Chr. (1913).

Terrestrial in Northern Limestone Hills; 10–220 m; *spr* Oct–Jan.

Greater Antilles (Jamaica, Hispaniola), Lesser Antilles, Trinidad, Central America (Nicaragua), South America (Venezuela, Ecuador). [P: 212].

Notes: The name of this taxon represents the hybrid *Thelpteris poiteana ×tetragona*.

Thelypteris rudis (Kunze) Proctor, Bull. Inst. Jamaica, Sci. Ser. 5:64. 1953.

Bas.: *Polypodium rude* Kunze (1839).

Terrestrial in Central Cordillera; 900–1300 m; *spr* Dec–Sep.

Greater Antilles, Mexico, Central America, northern and west-central South America. [P: 190].

Thelypteris sancta (L.) Ching, Bull. Fan Mem. Inst. Biol., Bot. 10:254. 1941.

Bas.: *Acrostichum sanctum* L. (1759).

Syn.: *Dryopteris sancta* (L.) Kuntze (1891). *Thelypteris sancta* (L.) Ching var. *portoricensis* (C. Chr.) C.V. Morton (1973).

Terrestrial in Northern Limestone Hills, Río Abajo Forest Reserve, Sierra de Luquillo, Sierra de Cayey, Central Cordillera; 100–800 m; *spr* throughout the year.

Antilles, southern Mexico, northern Central America. [P: 183].

Thelypteris sclerophylla (Poepp. ex Spreng.) C.V. Morton, Fieldiana, Bot. 28(1):12. 1951 May 18; et Amer. Fern J. 41(3):87. 1951 Oct 3.

Bas.: *Aspidium sclerophyllum* Poepp. ex Spreng. (1827).

Syn.: *Dryopteris sclerophylla* (Poepp. ex Spreng.) C. Chr. (1911).

Terrestrial in Eastern Cordillera (Cañón San Cristóbal), Central Cordillera; 350–1100 m; *spr* Mar, Aug.

Greater Antilles, southeastern United States (Florida). [P: 207].

Thelypteris serra (Sw.) Small, Ferns S.E. States 241. 1938.
Bas.: *Polypodium serra* Sw. (1788).
Syn.: *Dryopteris serra* (Sw.) Kuntze (1891).
Terrestrial in Northern Limestone Hills; 15–400 m; *spr* Aug–Nov, Feb–Apr.
Greater Antilles. [P: 199].

Thelypteris tetragona (Sw.) Small, Ferns S.E. States 256. 1938.
Bas.: *Polypodium tetragonum* Sw. (1788).
Syn.: *Dryopteris subtetragona* (Link) Maxon (1926).
Terrestrial in Northern Coastal Lowlands, Northern Limestone Hills, Eastern Coast (Humacao), Sierra de Cayey, Las Tetas de Cayey, Las Piedras Chiquitas, Southeastern Foothills (Cayey), Eastern Cordillera (Naranjito), Central Cordillera (Ponce, Adjuntas), Quebrada de Los Cedros, Vieques; 5–825 m; *spr* throughout the year.
Antilles, Trinidad, southeastern United States (Florida), Mexico, Central America, northern South America to Bolivia. [P: 211].
Gina.

Thelypteris ×verecunda Proctor (pro sp.), Amer. Fern J. 75(2):65, *t. 5.* 1985.
Terrestrial in Northern Limestone Hills; 250–325 m; *spr* Jun, Oct–Nov.
Endemic. [P: 204].
Notes: According to Alan R. Smith (pers. comm.), the name of this taxon represents a hybrid, possibly *Thelypteris abdita ×reptans*.
This taxon is on the U.S. Fish & Wildlife Service list of threatened and endangered plants.

Thelypteris yaucoensis Proctor, Mem. New York Bot. Gard. 53:208. 1989.
Terrestrial in Central Cordillera; 840–890 m; *spr* May–Jul (two specimens). [*Proctor 48010, 43584* (SJ)].
Endemic. [P: 208].
Notes: This taxon is on the U.S. Fish & Wildlife Service list of threatened and endangered plants.

TAXA NOT RECENTLY COLLECTED

Thelypteris linkiana (C. Presl) R.M. Tryon, Rhodora 69(777):6. 1967.
Bas.: *Grammitis linkiana* C. Presl (1836).
Syn.: *Dryopteris linkiana* (C. Presl) Maxon (1924).
Terrestrial collected in Utuado on 12 Mar 1887 [*Sintenis 6445*, not seen], which, according to Proctor (1989), was cited by Maxon (1926), but not seen by him.
Lesser Antilles, Mexico, Central America, northern South America to Bolivia. [P: 180].

Thelypteris reptans (J.F. Gmel.) C.V. Morton var. **tenera** (Fée) Proctor, Rhodora 61(732):306. 1960.
Bas.: *Goniopteris tenera* Fée (1866).
Terrestrial collected in the Sierra de Luquillo on 17 Jul 1885 [*Sintenis 1770* (US)].
Lesser Antilles (St. Kitts, Nevis, Guadeloupe). [P: 207].

Thelypteris serrata (Cav.) Alston, Bull. Misc. Inform. Kew 1932(7):309. 1932.
Bas.: *Meniscium serratum* Cav. (1802).
Syn.: *Dryopteris serrata* (Cav.) C. Chr. (1905).
Terrestrial collected in Utuado on 9 Mar 1887 [*Sintenis 6378* (US)].
Greater Antilles (Cuba, Jamaica), Lesser Antilles, Trinidad, southeastern United States (Florida), Mexico, Central America, South America. [P: 217].

TAXA CITED BY PREVIOUS AUTHORS, BUT NOT INCLUDED

DOUBTFUL OCCURRENCE

Thelypteris hastata (Fée) Proctor var. ***heterodoxa*** Proctor.
According to Alan R. Smith (pers. comm.), who saw an isotype, it is doubtful that this entity is a simple variant of *Thelypteris hastata*; it is more likely a hybrid of *Thelypteris hastata* with possibly *Thelypteris poiteana* or *Thelypteris tetragona*. [P: 210].

CITED DUE TO MISIDENTIFICATION

Thelypteris hispidula (Decne.) Reed var. ***hispidula***.
Terrestrial cited by Proctor (1989) based upon the misidentification of specimens of *Thelypteris hispidula* var. *inconstans*; subsequently cited by Liogier & Martorell (2000). [P: 195].

WOODSIACEAE

Diplazium centripetale (Baker) Maxon, Sci. Surv. Porto Rico & Virgin Islands 6(3):441. 1926.
Bas.: *Asplenium centripetale* Baker (1874).
Terrestrial in Sierra de Luquillo, Sierra de Cayey, Central Cordillera; 600–1325 m; *spr* Jul–Apr.
Greater Antilles (Jamaica, Hispaniola), Lesser Antilles (Guadeloupe, Martinique), Trinidad, northern South America (Venezuela). [P: 239].

Diplazium cristatum (Desr.) Alston, J. Bot. 74:173. 1936.
Bas.: *Meniscium cristatum* Desr. (1797).
Terrestrial in Northern Limestone Hills (Quebradillas), Río Abajo Forest Reserve, Sierra de Luquillo, Sierra de Cayey, Las Tetas de Cayey, Central Cordillera, Quebrada de Los Cedros; 65–1100 m; *spr* Jul–Mar.
Antilles, Trinidad and Tobago, southern Mexico, Central America, South America (scattered). [P: 241].
Notes: Maxon (1926) misapplied the name *Diplazium arboreum* (Willd.) C. Presl to this taxon.

Diplazium expansum Willd., Sp. Pl. 5(1):354. 1810.
Terrestrial in Central Cordillera, Maricao Forest Reserve; 650–1100 m; *spr* Dec–Apr, Jul–Sep.
Greater Antilles, Lesser Antilles (St. Vincent, Grenadines), Trinidad, Mexico, Central America (scattered), northern South America to Bolivia. [P: 243].

Diplazium grandifolium (Sw.) Sw., J. Bot. (Schrader) 1800(2):62. 1801.
Bas.: *Asplenium grandifolium* Sw. (1788).
Terrestrial in Sierra de Luquillo; 475–500 m; *spr* Feb (one specimen). [*Proctor 44489* (SJ)].
Greater Antilles, Lesser Antilles (Guadeloupe, Grenada), Trinidad and Tobago, southern Mexico, Central America, South America (scattered). [P: 239].

Diplazium hymenodes (Mett.) Á. Löve & D. Löve, Taxon 26 (2–3):326. 1977.
Bas.: *Asplenium hymenodes* Mett. (1859).
Terrestrial in Northern Limestone Hills, Quebrada de Los Cedros; 65–275 m; *spr* Aug–Mar.
Greater Antilles (Cuba, Jamaica). [P: 239].

Diplazium lherminieri Hieron. in Urban, Symb. Antill. 4(4):648. 1911 "l'herminieri."
Terrestrial in Sierra de Luquillo, Sierra de Cayey, Toro Negro Forest Reserve; 700–1300 m; *spr* Dec–Aug.
Greater Antilles (Cuba, Hispaniola), Lesser Antilles (Guadeloupe). [P: 241].

Diplazium plantaginifolium (L.) Urb., Symb. Antill. 4(1):31. 1903.
Bas.: *Asplenium plantaginifolium* L. (1759).
Terrestrial in Sierra de Luquillo, Central Cordillera, Maricao Forest Reserve; 500–1100 m; *spr* Jan–Oct.
Antilles, Trinidad, southern Mexico, Central America, South America (scattered). [P: 238].

Diplazium riedelianum (Bong. ex Kuhn) C. Chr., Index Filic. 238. 1905.
Bas.: *Asplenium riedelianum* Bong. ex Kuhn (1869).
Terrestrial in Río Abajo Forest Reserve; 300–350 m; *spr* Jun (one specimen). [*Acevedo-Rodríguez 325* (SJ)].
Central America, South America (Brazil, Paraguay, Argentina). [P: 238].

Diplazium striatum (L.) C. Presl, Tent. Pterid. 114. 1836.
Bas.: *Asplenium striatum* L. (1753).
Terrestrial in Northern Limestone Hills (Arecibo, Isabela), Río Abajo Forest Reserve, Sierra de Luquillo, Eastern Coast (Cuchilla de Panduras), Sierra de Cayey, Central Cordillera, Maricao Forest Reserve; 200–600 m; *spr* throughout the year.
Antilles, Trinidad, southern Mexico, Central America, northern South America to Bolivia. [P: 243].

Diplazium unilobum (Poir.) Hieron., Hedwigia 59:332. 1918.
Bas.: *Asplenium unilobum* Poir. (1812).
Terrestrial in Northern Limestone Hills, Río Abajo Forest Reserve, Sierra de Luquillo, Sierra de Cayey, Eastern Cordillera (Aguas Buenas), Central Cordillera, Maricao Forest Reserve; 100–900 m; *spr* throughout the year.
Greater Antilles, Lesser Antilles (Guadeloupe), northern South America (Colombia). [P: 240].

Hemidictyum marginatum (L.) C. Presl, Tent. Pterid. 111, *t. 3, f. 24.* 1836.
Bas.: *Asplenium marginatum* L. (1753).
Terrestrial in Río Abajo Forest Reserve, Sierra de Luquillo, Eastern Coast (Cuchilla de Panduras), Central Cordillera, Western Cordillera; 250–1100 m; *spr* throughout the year.
Antilles, Trinidad and Tobago, southern Mexico, Central America, northern South America to Bolivia. [P: 235].

CYCADOPHYTA

CYCADACEAE

TAXA CITED BY PREVIOUS AUTHORS, BUT NOT INCLUDED

ONLY UNDER CULTIVATION
Cycas circinalis L.
Suffrutescent cited by Liogier & Martorell (1982, 2000) as much planted and persistent as an ornamental.

Cycas revoluta Thunb.
Suffrutescent cited by Liogier & Martorell (1982, 2000) as much planted as an ornamental and persistent.

ZAMIACEAE

Zamia erosa O.F. Cook & G.N. Collins, Contr. U.S. Natl. Herb. 8:267. 1903.
Syn.: *Zamia amblyphyllidia* D.W. Stev. (1987).
Suffrutescent in Northern Limestone Hills, Río Abajo Forest Reserve; 1–350 m; *spr* Sep–Mar.
Greater Antilles (Cuba, Jamaica). [AR&S: 16].
Notes: Liogier & Martorell (2000) misapplied the name *Zamia pygmaea* Sims to this taxon.
Marunguey, Palmita de jardín, Zamia.

Zamia portoricensis Urb., Symb. Antill. 1(2):291. 1899.
Suffrutescent in Susúa Forest Reserve, Guánica Forest Reserve, Sierra Bermeja; 50–300 m; *spr* Sep–Feb.
Endemic. [AR&S: 16].
Notes: Liogier & Martorell (2000) misapplied the homonym *Zamia pumila* Jacq. to this taxon.
Marunguey, Palmita de jardín.

Zamia pumila L., Sp. Pl. (ed. 2) 2:1659. 1763.
Syn.: *Zamia debilis* L.f. (1789).
Zamia media Jacq. (1798).
Zamia latifoliolata Prenleloup (1872).
Suffrutescent in Southeastern Foothills (Salinas), Cerro de Las Cuevas, Southern Coastal Lowlands (Santa Isabel), Southern Coastal Limestone; 10–575 m; *spr* Nov–Mar, Jul.
Greater Antilles (Cuba, Hispaniola). [AR&S: 17].
Notes: Britton & Wilson (1926) misapplied the name *Zamia latifoliolata*, and Liogier & Martorell (1982) misapplied *Zamia debilis*, to *Zamia erosa*.
Marunguey, Zamia.

TAXA CITED BY PREVIOUS AUTHORS, BUT NOT INCLUDED

CITED DUE TO MISIDENTIFICATION

Zamia angustifolia Jacq.
Suffrutescent cited by Liogier & Martorell (2000) as occurring on serpentine soil, Susúa. This citation is based upon the misidentification of a specimen of *Zamia portoricensis*.

PINOPHYTA

ARAUCARIACEAE

TAXA CITED BY PREVIOUS AUTHORS, BUT NOT INCLUDED

ONLY UNDER CULTIVATION

Agathis robusta (C. Moore ex F. Muell.) C. Moore ex F.M. Bailey.
Tree cited by Liogier & Martorell (1982, 2000) as tested as a forest tree and becoming naturalized in the West Indies.

Araucaria angustifolia (Bertol.) Kuntze.
Tree cited by Liogier & Martorell (1982, 2000) as planted as an ornamental.

Araucaria bidwillii Hook.
Tree cited by Liogier & Martorell (1982, 2000) as planted as an ornamental.

Araucaria heterophylla (Salisb.) Franco.
Tree cited by Liogier & Martorell (1982, 2000) as planted as an ornamental in the west districts.

CUPRESSACEAE

TAXA CITED BY PREVIOUS AUTHORS, BUT NOT INCLUDED

ONLY UNDER CULTIVATION

Chamaecyparis funebris (Endl.) Franco. Syn.: *Cupressus funebris* Endl.
Tree cited by Liogier & Martorell (1982, 2000) as planted as an ornamental.

Cupressus arizonica Greene.
Tree cited by Liogier & Martorell (1982, 2000) as introduced from Arizona and planted.

Cupressus lusitanica Mill. var. ***benthamii*** (Endl.) Carrière.
Tree cited by Liogier & Martorell (1982, 2000) in the form *Cupressus lusitanica* Mill. '*Benthamii*' as occasionally planted as an ornamental and in reforestation.

Cupressus macrocarpa Hartw. ex Gordon.
Tree cited by Liogier & Martorell (1982, 2000) as planted as an ornamental.

Cupressus sempervirens L.
Tree cited by Liogier & Martorell (1982, 2000) as planted as an ornamental.

Platycladus orientalis (L.) Franco. Syn.: *Thuja orientalis* L.
Tree cited by Liogier & Martorell (1982, 2000) as planted as an ornamental.

PINACEAE

Pinus caribaea Morelet, Rev. Hort. Côte d'Or 1:107. 1851.

Tree planted and persistent in Sierra de Luquillo, Eastern Coast (Ceiba, Naguabo), Sierra de Cayey, Susúa Forest Reserve, Maricao Forest Reserve, 50–760 m; *spr* Dec–Mar, Jun–Jul.

Native to Bahamas, Greater Antilles (Cuba), Mexico, northern Central America to Nicaragua. [AR&S: 14].

Pino antillano, Pino caribea, Pino hondureño.

Pinus occidentalis Sw., Prodr. 103. 1788.

Tree planted and persistent in Sierra de Luquillo; 500 m; *spr* Jan (one specimen).

Native to Greater Antilles (Hispaniola).

Pino.

PODOCARPACEAE

Podocarpus coriaceus Rich. & A. Rich., Comm. Bot. Conif. Cycad. 14, *t. 1, f. 3.* 1826.
Syn.: *Nageia coriacea* (Rich. & A. Rich.) Kuntze (1891).

Tree in Sierra de Luquillo, Monte Guilarte Forest Reserve, Maricao Forest Reserve; 500–1050 m; *spr* throughout the year.

Greater Antilles (Hispaniola), Lesser Antilles, Trinidad, South America (Brazil, Peru). [AR&S: 12].

Caoba del país, Caobilla.

TAXA CITED BY PREVIOUS AUTHORS, BUT NOT INCLUDED

ONLY UNDER CULTIVATION

Podocarpus macrophyllus (Thunb.) D. Don var. ***maki*** Siebold & Zucc.
Tree cited by Liogier & Martorell (1982, 2000) as cultivated as an ornamental and persistent.

MAGNOLIOPHYTA

ACANTHACEAE (including AVICENNIACEAE)

Asystasia gangetica (L.) T. Anderson in Thwaites & Hooker, Enum. Pl. Zeyl. 235. 1860.
Bas.: *Justicia gangetica* L. (1756).

Trailing herb in Northern Coastal Lowlands, Northern Limestone Hills, Southeastern Foothills (Salinas), Eastern Cordillera (Comerío), Southern Coastal Lowlands (Ponce), Susúa Forest Reserve, Vieques; 1–400 m; *fl* throughout the year, *fr* Aug–Apr.

Native to the tropical Africa, India, Sri Lanka. [L5: 17; AR: 47].

Avicennia germinans (L.) L., Sp. Pl. (ed. 3) 2:891. 1764.
Bas.: *Bontia germinans* L. (1759).
Syn.: *Avicennia nitida* Jacq. (1760).

Tree in coastal swamps in Northern Coastal Lowlands, Eastern Coast (Cayo Santiago), Southern Coastal Lowlands, Guánica Forest Reserve, Southwestern Coastal Lowlands, Vieques, Culebra; 0 m; *fl* Jan–Jul, *fr* Jan–Oct.

Coastal swamps of West Indies, Trinidad, southeastern United States (Florida), Mexico, Central America, tropical South America, eastern Africa. [L4: 390].

Chifle de vaca, Mangle bobo, Mangle negro, Mangle prieto, Prieto, Salado, Siete cueros.

Barleria cristata L., Sp. Pl. 2:636. 1753.

Suffrutescent herb/shrub planted and persistent in Southwestern Coastal Lowlands; 60 m; *fl* Mar–Jun, Nov–Dec, *fr* (?).

Native to eastern India, Myanmar. [L5: 20].

Enana.

Barleria prionitis L., Sp. Pl. 2:636. 1753.

Suffrutescent herb/shrub in Southeastern Foothills (Salinas), Southern Coastal Lowlands, Southwestern Coastal Lowlands, Vieques; 5–60 m; *fl* & *fr* Nov–May.

Native to tropical Africa, Indian Subcontinent, Malesia. [L5: 20].

Espinosa amarilla.

Barleriola solanifolia (L.) Oerst. ex Lindau, Bot. Jahrb. Syst. 18(1–2):63. 1893.
Bas.: *Barleria solanifolia* L. (1753).

Shrub in Northern Coastal Lowlands (Manatí, Aguadilla); 30–50 m; *fl* Apr–Jun, *fr* Jun (one specimen). [*Woodbury a.n. 3466, 5156* (UPR)].

Greater Antilles (Cuba, Hispaniola). [L5: 22].

Blechum pyramidatum (Lam.) Urb., Repert. Spec. Nov. Regni Veg. 15:323. 1918.
Bas.: *Barleria pyramidata* Lam. (1785).
Syn.: *Blechum brownei* Juss. (1807).
Blechum blechum (L.) Millsp. (1900).

Trailing herb in Northern Coastal Lowlands, Northern Limestone Hills (Utuado, Isabela), Sierra de Luquillo, Eastern Coast, Las Piedras Chiquitas, Eastern Cordillera, Western Cordillera, Southwestern Coastal Lowlands, Western Coast, Vieques, Mona; 1–600 m; *fl* & *fr* throughout the year.

West Indies, Mexico, Central America, northern South America to Bolivia. [L5: 23].

Yerba de papagayo.

Dicliptera krugii Urb., Symb. Antill. 4(4):577. 1911.

Syn.: *Diapedium krugii* (Urb.) Britton (1925).

Herb in Las Tetas de Cayey, Southeastern Foothills, Eastern Cordillera (Cañón San Cristóbal), Western Cordillera (Maricao); 300–825 m; *fl* Nov–Feb, May, *fr* Jan (one specimen).

Endemic. [L5: 25].

Dicliptera sexangularis (L.) Juss., Ann. Mus. Nat. Hist. 9:269. 1807.

Bas.: *Justicia sexangularis* L. (1753).

Syn.: *Justicia assurgens* L. (1759). *Dicliptera assurgens* (L.) Juss. (1807). *Diapedium assurgens* (L.) Kuntze (1891).

Herb in Vieques; 10 m; *fl* & *fr* Jan (one specimen). [*Breckon 7661* (MAPR)].

Bahamas, Greater Antilles, Virgin Islands, southern United States (Florida, Texas), Mexico, northern Central America (Belize, Honduras), northern South America to Bolivia. [L5: 26].

Hemigraphis reptans (G. Forst.) T. Anderson ex Hemsl., Rep. Challenger, Bot. 1(3):173. 1885.

Bas.: *Ruellia reptans* G. Forst. (1786).

Herb in Northern Coastal Lowlands (San Juan); 10 m; *fl* Apr, Aug, *fr* Apr (two specimens). [*Liogier 28658, 36507* (UPR)].

Native to eastern Pacific Islands. [L5: 412].

Hygrophila costata Nees, Pl. Hort. Bonn. Icon. 2:7–8, *t*. 3. 1824.

Syn.: *Hygrophila brasiliensis* (Spreng.) Lindau (1900).

Herb in Northern Coastal Lowlands, Eastern Coast, Caguas Valley (Caguas), Sierra de Cayey; 1–400 m; *fl* Jan–Feb, May–Jul, *fr* Jan, May–Jul.

Native to Mexico, Central America, South America. [L5: 30].

Yerba de hicotea, Yerba de jicotea.

Justicia comata (L.) Lam., Encycl. 1(2):632. 1785.

Bas.: *Dianthera comata* L. (1759).

Syn.: *Stethoma comata* (L.) Britton (1925).

Herb in Northern Coastal Lowlands (Canóvanas, San Juan), Northern Limestone Hills (Arecibo), Eastern Coast (Humacao), Eastern Cordillera (Cañón San Cristóbal), Southwestern Coastal Lowlands (Lajas); 1–400 m; *fl* & *fr* Nov–Jul.

Greater Antilles, Trinidad and Tobago, Mexico, Central America, South America. [L5: 34].

Justicia martinsoniana R.A. Howard, J. Arnold Arbor. 49:400. 1968.

Syn.: *Justicia verticillaris* (Nees) Urb. (1911). *Stethoma verticillaris* (Nees) Britton (1925).

Herb in Sierra de Luquillo, Sierra de Cayey, Eastern Coast (Cuchilla de Panduras), Central Cordillera, Maricao Forest Reserve; 250–1300 m; *fl* & *fr* throughout the year.

Endemic. [L5: 34].

Justicia mirabiloides Lam., Tabl. Encycl. 1:39. 1791.

Syn.: *Drejerella mirabiloides* (Lam.) Lindau (1900) "mirabilioides."

Suffrutescent herb in Northern Coastal Lowlands (Quebradillas, Isabela), Northern Limestone Hills, Río Abajo Forest Reserve; 1–350 m; *fl* Oct–May, *fr* Jan–Apr.

Greater Antilles, Virgin Islands. [L5: 36].

Justicia pectoralis Jacq., Enum. Syst. Pl. 11. 1760.

Syn.: *Stethoma pectoralis* (Jacq.) Raf. (1838).

Herb in Northern Limestone Hills (Florida), Mona; 10–150 m; *fl* & *fr* Feb–May.

Antilles, Mexico, Central America, South America. [L5: 36].

Curía.

Justicia periplocifolia Jacq., Collectanea 5:7. 1797.

Syn.: *Justicia culubritae* Urb. (1919).

Suffrutescent herb/shrub in Northern Coastal Lowlands, Eastern Coast, Las Tetas de Cayey, Eastern Cordillera (Aibonito), Southern Coastal Lowlands (Salinas, Santa Isabel), Southern Coastal Limestone (Guayanilla), Guánica Forest Reserve, Sierra Bermeja, Vieques, Culebra, Mona; 1–825 m; *fl* & *fr* throughout the year.

Greater Antilles (Cuba, Hispaniola), Virgin Islands, Lesser Antilles. [L5: 34, 37].

Notes: In place of the above name, Liogier (1997) & Liogier & Martorell (2000) wrongly used *Justicia sphaerosperma* Vahl, a taxon which, according to Dieter Wasshausen (pers. comm.), does not occur in Puerto Rico.

Since *Justicia culubritae* differs from *Justicia periplocifolia* only in having narrower leaves, I do not consider that it merits separate taxonomic status and have placed it in synonymy with the latter. Britton & Wilson (1925), Liogier & Martorell (1982, 2000) and Liogier (1997) used the orthographic variant *Justicia culebritae* for this synonym.

Justicia sessilis Jacq., Enum. Syst. Pl. 11. 1760.

Syn.: *Justicia borinquensis* Britton (1925). *Siphonoglossa sessilis* (Jacq.) D.N. Gibson (1972).

Suffrutescent herb in Northern Coastal Lowlands (Fajardo), Northern Limestone Hills (coastal), Río Abajo Forest Reserve, Las Piedras Chiquitas, Southeastern Foothills (Salinas), Southern Coastal Lowlands, Southern Coastal Limestone, Guánica

Forest Reserve, Southwestern Coastal Lowlands, Sierra Bermeja, Vieques; 1–400 m; *fl* & *fr* throughout the year.

Antilles, Central America. [L5: 52].

Lepidagathis alopecuroidea (Vahl) R. Br. ex Griseb., Fl. Brit. W.I. 453. 1862.
Bas.: *Ruellia alopecuroidea* Vahl (1798).
Syn.: *Teliostachya alopecuroidea* (Vahl) Nees (1847).

Herb in Northern Coastal Lowlands, Northern Limestone Hills (Ciales), Río Abajo Forest Reserve, Sierra de Luquillo, Eastern Coast, Sierra de Cayey, Las Tetas de Cayey, Eastern Cordillera (Corozal, La Torrecilla), Central Cordillera, Western Cordillera (Maricao), Western Coast (Añasco); 1–900 m; *fl* & *fr* throughout the year.

West Indies, Trinidad and Tobago, central Mexico, Central America, northern South America and Brazil, western tropical Africa. [L5: 55].

Pata de gallina.

Nelsonia canescens (Lam.) Spreng., Syst. Veg. 1:42. 1824.
Bas.: *Justicia canescens* Lam. (1791).
Syn.: *Nelsonia brunelloides* (Lam.) Kuntze (1891).

Low herb in Northern Coastal Lowlands, Southern Coastal Lowlands (Guayama); 1–10 m; *fl* & *fr* Mar–Nov.

Native to tropical Africa. [L5: 38].

Odontonema cuspidatum (Nees) Kuntze, Revis. Gen. Pl. 2:494. 1891.
Bas.: *Thyrsacanthus cuspidatus* Nees (1847).

Suffrutescent herb in Northern Limestone Hills, Río Abajo Forest Reserve, Sierra de Luquillo, Eastern Coast (Ceiba), Sierra de Cayey, Southeastern Foothills (Coamo), Eastern and Central Cordillera, Susúa Forest Reserve, Maricao Forest Reserve, Western Coast (Mayagüez); 1–800 m; *fl* throughout the year, *fr* (?).

Native to Mexico. [L5: 40].

Notes: Britton & Wilson (1925) and Liogier & Martorell (1982) misapplied the name *Odontonema strictum* (Nees) Kuntze to this taxon.

Coral de jardín.

Oplonia microphylla (Lam.) Stearn, Bull. Brit. Mus. (Nat. Hist.), Bot. 4(7):307. 1971.
Bas.: *Justicia microphylla* Lam. (1791).
Syn.: *Anthacanthus microphyllus* (Lam.) Nees (1847).

Shrub in Northern Coastal Lowlands (islands adjacent to Fajardo), Las Piedras Chiquitas, Culebra; 1–550 m; *fl* Mar (one specimen), *fr* (?). [*Woodbury I-45* & *a.n. 26666* (UPR); *Axelrod 2156* (UPRRP)].

Bahamas, Greater Antilles (Cuba, Hispaniola), Lesser Antilles. [L5: 41].

Oplonia spinosa (Jacq.) Raf., Fl. Tellur. 4:65. 1838.
Bas.: *Justicia spinosa* Jacq. (1760).
Syn.: *Anthacanthus spinosus* (Jacq.) Nees (1847).

Shrub in Northern Coastal Lowlands, Northern Limestone Hills, Río Abajo Forest Reserve, Eastern Coast (Punta Guayanés), Las Tetas de Cayey, Southern Coastal Limestone (Peñuelas), Susúa Forest Reserve, Maricao Forest Reserve, Guánica Forest Reserve, Southwestern Coastal Lowlands (Guánica), Vieques; 1–800 m; *fl* & *fr* throughout the year.

Bahamas, Greater Antilles (Cuba, Hispaniola), Virgin Islands. [L5: 42; AR: 47].

Espinosa.

Ruellia coccinea (L.) Vahl, Symb. Bot. 3:83. 1794.
Bas.: *Barleria coccinea* L. (1753).

Suffrutescent herb in Northern Coastal Lowlands, Northern Limestone Hills, Río Abajo Forest Reserve, Sierra de Luquillo, Eastern Coast (Cuchilla de Panduras), Caguas Valley (Caguas), Sierra de Cayey, Las Tetas de Cayey, Eastern and Central Cordillera, Southern Coastal Lowlands, Western Cordillera, Vieques; 1–950 m; *fl* & *fr* throughout the year.

Greater Antilles (Hispaniola), Virgin Islands. [L5: 48].

Maravilla, Yerba maravilla.

Ruellia simplex C. Wright in Sauvalle, Anales Acad. Ci. Méd. Habana 6:321. 1870.
Syn.: *Ruellia tweediana* Griseb.(1879).
Ruellia coerulea Morong (1893).
Ruellia spectabilis Britton (1893).
Ruellia brittoniana Leonard (1941).

Herb in Northern Coastal Lowlands, Northern Limestone Hills, Southeastern Foothills (Salinas), Eastern Cordillera, Central Cordillera (Ponce, Utuado), Susúa Forest Reserve, Southwestern Coastal Lowlands, Western Coast; 10–550 m; *fl* & *fr* throughout the year.

Native to Mexico, southern South America to Brazil. [L5: 47].

A-las-doce-me-voy.

Ruellia tuberosa L., Sp. Pl. 2:635. 1753.

Suffrutescent herb in Northern Coastal Lowlands, Eastern Coast (Ceiba), Southeastern Foothills, Southern Coastal Lowlands, Guánica Forest Reserve, Southwestern Coastal Lowlands, Sierra Bermeja, Vieques, Culebra; 1–150 m; *fl* Mar–Oct, *fr* Feb–Oct.

West Indies, northern South America to Peru. [L5: 49].

Sanchezia speciosa Leonard, J. Washington Acad. Sci. 16:490. 1926.

Shrub planted and persistent in Northern Coastal Lowlands (Río Grande), Sierra de Luquillo, Sierra de Cayey, Maricao Forest Reserve; 10–700 m; *fl* Sep–May, *fr* (?).

Native to South America. [L5: 50].

Notes: Britton & Wilson (1925) misapplied the name *Sanchezia nobilis* Hook. f. to this taxon.

Stenandrium tuberosum (L.) Urb., Symb. Antill. 4(4):576. 1911.

Bas.: *Gerardia tuberosa* L. (1753).

Syn.: *Gerardia portoricensis* Britton & P. Wilson (1925).

Low herb in Northern Limestone Hills, Río Abajo Forest Reserve, Eastern Cordillera (Cañón San Cristóbal), Southern Coastal Limestone (Ponce), Guánica Forest Reserve, Vieques; 1–300 m; *fl* May–Jun, Oct, *fr* May, Dec.

Greater Antilles (Hispaniola), Virgin Islands (St. Thomas, St. John), Lesser Antilles (St. Martin, Guadeloupe). [L5: 53].

Mata espiritista, Yerba maravilla blanca.

Thunbergia alata Bojer ex Sims, Bot. Mag. 52: *t. 2591*. 1825.

Vine in Northern Coastal Lowlands, Northern Limestone Hills, Caguas Valley (Caguas), Sierra de Cayey (San Lorenzo), Eastern and Central Cordillera, Maricao Forest Reserve, Cerro Las Mesas, Western Coast; 1–1000 m; *fl* & *fr* throughout the year.

Native to tropical Africa. [L5: 57; AR: 49].

Culo de poeta, Ojo de poeta, Susana, Viuda.

Thunbergia erecta (Benth.) T. Anderson, J. Proc. Linn. Soc., Bot. 7:18. 1863.

Bas.: *Meyenia erecta* Benth. (1849).

Shrub planted and persistent in Northern Coastal Lowlands (Río Grande), Río Abajo Forest Reserve, Eastern Coast (Cuchilla de Panduras), Eastern Cordillera (Cayey), Southern Coastal Lowlands (Ponce), Western Cordillera (Maricao); 150–500 m; *fl* Jan–Jun, Sep–Oct, *fr* (?).

Native to western tropical Africa. [L5: 57].

Angelina, Boca de vieja, Violeta, Viuda.

Thunbergia fragrans Roxb., Pl. Coromandel 1:47, *t. 67*. 1796.

Vine in Northern Coastal Lowlands, Northern Limestone Hills (Arecibo), Río Abajo Forest Reserve, Eastern Coast (Naguabo), Caguas Valley (Caguas), Eastern Cordillera, Susúa Forest Reserve, Western Cordillera, Western Coast, Vieques; 1–750 m; *fl* & *fr* throughout the year.

Native to Indian Subcontinent, southeastern Asia. [L5: 58; AR: 49].

Flor de nieve, Susana blanca.

Thunbergia grandiflora Roxb., Bot. Reg. 6: *t. 495*. 1820.

Woody vine in Northern Limestone Hills (Utuado), Río Abajo Forest Reserve, Sierra de Luquillo, Eastern Cordillera (Cayey, Aibonito), Vieques; 50–600 m; *fl* Oct–Nov, Feb, Jun, *fr* (?).

Native to Indian Subcontinent, Myanmar. [L5: 58; AR: 51].

Fausto, Pompeya.

TAXA CITED BY PREVIOUS AUTHORS, BUT NOT INCLUDED

NO SPECIMENS SEEN

Barleria lupulina Lindl.

Suffrutescent herb/shrub cited by Liogier & Martorell (1982, 2000) and Liogier (1997) as occurring on hillsides in waste grounds, Vieques. [L5: 20].

CITED DUE TO MISIDENTIFICATION

Dicliptera martinicensis (Jacq.) Juss.

Herb mentioned by Howard (1989) as occurring in Puerto Rico; subsequently cited by Liogier (1997) and Liogier & Martorell (2000). This citation was probably based on a specimen of *Dicliptera krugii* [*Sintenis 6350* (GH)] that had been misidentified as this taxon. [L5: 26].

ONLY UNDER CULTIVATION

Hemigraphis alternata (Burm. f.) T. Anderson. Syn.: *Ruellia colorata* Blume. *Hemigraphis colorata* (Blume) Hallier.

Herb cited by Liogier & Martorell (1982, 2000) and Liogier (1997) as cultivated and escaped in moist situations. [L5: 28].

Hypoestes phyllostachya Baker.

Herb cited by Liogier (1997) and Liogier & Martorell (2000) as cultivated, naturalized in Maricao. [L5: 31].

Pachystachys spicata (Ruiz & Pav.) Wassh.

Suffrutescent herb cited by Liogier (1997) and Liogier & Martorell (2000) as escaped from cultivation. [L5: 43].

Pseuderanthemum carruthersii (Seem.) Guillaumin.

Shrub cited by Liogier (1997) and Liogier & Martorell (2000) as widely cultivated and persistent after cultivation. [L5: 45].

Pseuderanthemum variabile (R. Br.) Radlk.

Herb, to which Liogier (1997) and Liogier & Martorell

(2000) misapplied the name *Pseuderanthemum fasciculatum* (Oerst.) Leonard, cited by them as a weed, escaped in orchid nurseries. [L5: 45].

Ruellia chartacea (T. Anderson) Wassh.
Suffrutescent herb cited by Liogier (1997) and Liogier & Martorell (2000) as cultivated and naturalized. [L5: 48].

ADOXACEAE

Sambucus canadensis L., Sp. Pl. 1:269. 1753.
Syn.: *Sambucus nigra* L. subsp. *canadensis* (L.) Bolli (1994).
Sambucus mexicana C. Presl ex DC. (1830).
Sambucus simpsonii Rehder (1911).

Tree planted and persistent in Northern Coastal Lowlands (San Juan), Northern Limestone Hills, Río Abajo Forest Reserve, Central Cordillera (Villalba, Ponce); 10–500 m; *fl* Apr–Sep, *fr* Jul–Sep (two specimens).

Native to eastern Canada, eastern and central United States, Mexico, Central America. [L5: 178].

Saúco.

AGAVACEAE

Agave minor Proctor in Acevedo-Rodríguez & Strong, Contr. U.S. Natl. Herb. 52:118. 2005.

Robust acaulescent herb in Sierra Bermeja; 100–200 m; *fl* (?), *fr* Jun (one specimen). [*Proctor 46978* (SJ)].

Endemic. [AR&S: 118].

Agave missionum Trel., Mem. Natl. Acad. Sci. 11:37, *t. B, f. 10, tt. 72–75.* 1913.
Syn.: *Agave portoricensis* Trel. (1913).

Robust acaulescent herb in Las Tetas de Cayey, Las Piedras Chiquitas, Southeastern Foothills (Salinas), Eastern Cordillera (Aibonito), Susúa Forest Reserve, Guánica Forest Reserve, Southwestern Coastal Lowlands, Culebra; 10–825 m; *fl* Mar–Jun, *fr* Jun (one specimen).

Virgin Islands. [AR&S: 118].

Cocuisa, Corita, Maguey.

Agave sisalana Perrine, U.S. Congr. 25, Sess. 2, House of Represent. Rep. 564: (8, 9, 16, 47, 60, 86) 87. 8 Mar 1838 (Senate Rep. 300:36, 105, *tt. 1, 2, 4.* 12 Mar 1838).

Robust acaulescent herb planted and persistent in Guánica Forest Reserve, Mona; 5–50 m; *fl* Oct–Dec, May, *fr* (?).

Native to Mexico. [AR&S: 120].

Sisal.

Agave vivipara L., Sp. Pl. 1:323. 1753.

Robust acaulescent herb planted and persistent in Northern Coastal Lowlands (Laguna Tortuguero), Culebra; 1 m; *fl* Oct (one specimen), *fr* (?). [*Liogier 33548* (UPR)].

Native to Southern Caribbean Islands. [AR&S: 120].

Furcraea tuberosa (Mill.) W.T. Aiton, Hort. Kew. (ed. 2) 2:303. 1811.
Bas.: *Agave tuberosa* Mill. (1768).

Rosette herb in Eastern Cordillera (Cerro La Peña, Cidra), Southern Coastal Lowlands (Ponce), Maricao Forest Reserve, Southwestern Coastal Lowlands (Cabo Rojo), Western Coast (Rincón), Vieques; 5–750 m; *fl* Jul–Oct, Feb, *fr* Mar (one specimen).

Greater Antilles (Cuba, Hispaniola), Lesser Antilles. [AR&S: 122].

Cocuisa, Maguey, Maguey criollo.

Yucca aloifolia L., Sp. Pl. 1:319. 1753.

Shrub planted and persistent in Northern Coastal Lowlands (Carolina); 3 m; *fl* (?), *fr* (?). [*Axelrod 12689* (UPRRP)].

Native to West Indies, southeastern, south-central United States, Mexico, Central America; exact native range uncertain. [AR&S: 124].

Aguja de Adán, Bayoneta, Bayoneta española, Mata de huevos.

Yucca guatemalensis Baker, Refug. Bot. 5: *t. 313.* 1872.

Shrub planted and persistent in Guánica Forest Reserve; 2 m; *fl* May (one specimen), *fr* (?). [*Monsegur 729* (MAPR)].

Native to Mexico, Central America. [AR&S: 125].

Bayoneta.

TAXA CITED BY PREVIOUS AUTHORS, BUT NOT INCLUDED

ONLY UNDER CULTIVATION

Yucca gloriosa L.
Shrub cited by Liogier & Martorell (1982, 2000) as cultivated and persistent.

CITED DUE TO MISIDENTIFICATION

Furcraea antillana A. Álvarez.
Rosette herb cited in Acevedo-Rodríguez & Strong (2005) on the basis of two specimens which were subsequently determined by Abisai García-Mendoza (MEXU) as *Furcraea tuberosa*. [AR&S: 121].

AIZOACEAE

Cypselea humifusa Turpin, Ann. Mus. Nat. Hist. 7:219, *t. 121*. 1806.

Herb in Northern Coastal Lowlands (Camuy), Mona; 1 m; *fl* & *fr* Jan (one specimen). [*Woodbury M-321* (UPR); *Proctor 42900* (SJ)].

Bahamas, Greater Antilles, Virgin Islands, Lesser Antilles (Guadeloupe), southwestern United States. [L1: 195].

Panal.

Sesuvium maritimum (Walter) Britton, Sterns & Poggenb., Prelim. Cat. 20. 1888.

Bas.: *Pharnaceum maritimum* Walter (1788).

Fleshy herb in Northern Coastal Lowlands (Carolina), Mona; 1–60 m; *fl* Mar, Jun (two specimens), *fr* Jun (one specimen). [*Woodbury a.n. 27714* (UPR); *Breckon 5592* (MAPR)].

Bahamas, Greater Antilles (Cuba, Jamaica), Lesser Antilles (St. Barts), United States (New York to Florida, Kansas to Texas), Mexico, northern Central America (Belize). [L1: 198].

Sesuvium microphyllum Willd., Enum. Pl. 521. 1809.

Fleshy prostrate herb in Eastern Coast (Ceiba), Vieques; 1–75 m; *fl* & *fr* Jun–Aug (two specimens). [*Proctor 48095, 50801* (SJ)].

Bahamas, Greater Antilles (Cuba), Lesser Antilles (Anguilla). [L5: 395].

Sesuvium portulacastrum (L.) L., Syst. Nat. (ed. 10) 2:1058. 1759.

Bas.: *Portulaca portulacastrum* L. (1753).

Fleshy herb in Northern Coastal Lowlands, Southern Coastal Lowlands, Guánica Forest Reserve, Southwestern Coastal Lowlands (Cabo Rojo), Vieques, Culebra, Mona, Desecheo; 0–5 m; *fl* & *fr* throughout the year.

Warm-temperate and tropical seacoasts. [L1: 198].

Verdolaga rosada, Yerba de vidrio.

Trianthema portulacastrum L., Sp. Pl. 1:223. 1753.

Herb in Northern Coastal Lowlands, Eastern Coast, Southern Coastal Lowlands, Guánica Forest Reserve, Southwestern Coastal Lowlands, Vieques, Culebra, Mona, Desecheo; 1–10 m; *fl* & *fr* throughout the year.

Warm-temperate and tropical seacoasts. [L1: 201].

Verdolaga de hoja ancha.

TAXA CITED BY PREVIOUS AUTHORS, BUT NOT INCLUDED

ONLY UNDER CULTIVATION

Tetragonia tetragonoides (Pall.) Kuntze. Syn.: *Tetragonia expansa* Murray.

Herb cited by Britton & Wilson (1924) as grown in gardens; subsequently cited by Liogier & Martorell (1982, 2000) and Liogier (1985). [L1: 199].

ALISMATACEAE

Echinodorus berteroi (Spreng.) Fassett, Rhodora 57(677):139. 1955.

Bas.: *Alisma berteroi* Spreng. (1825) "Berterii."

Herb in Northern Coastal Lowlands (Vega Baja), Southern Coastal Lowlands, Guánica Forest Reserve, Southwestern Coastal Lowlands (Cabo Rojo), Western Coast (Añasco); 1–35 m; *fl* & *fr* Jan–Sep.

West Indies, southern and central United States, Mexico, South America (scattered). [AR&S: 59].

Notes: Britton & Wilson (1923) misapplied the name *Echinodorus cordifolius* (L.) Griseb. to this taxon.

Llantén de agua.

Echinodorus bolivianus (Rusby) Holm-Niels., Brittonia 31(2):276. 1979.

Bas.: *Alisma bolivianum* Rusby (1927).

Herb in Northern Coastal Lowlands (Laguna Tortuguero), Southern Coastal Lowlands (Salinas, Caja de Muertos); 1–10 m; *fl* Mar–Jun, *fr* Jun (one specimen).

Greater Antilles, Southern Caribbean Islands (Bonaire), Mexico, Central America, South America. [AR&S: 59].

Notes: Britton & Wilson (1923) misapplied the name *Helianthium tenellum* (Mart. ex Schult. & Schult. f.) Britton, and Liogier & Martorell (1982) misapplied *Echinodorus tenellus* (Mart. ex Schult. & Schult. f.) Buchenau var. *latifolius* (Seub.) Fassett, to this taxon.

Sagittaria intermedia Micheli in A. & C. de Candolle, Monogr. Phan. 3:80. 1881.

Herb in Northern Coastal Lowlands, Eastern Coast (Naguabo); 1–15 m; *fl* Dec–Feb, May–Jul, *fr* Dec–Jun.

Greater Antilles. [AR&S: 61].

Sagittaria lancifolia L., Syst. Nat. (ed. 10) 2:1270. 1759.

Herb in Northern Coastal Lowlands, Northern Limestone Hills (Ciales, Florida), Río Abajo Forest Reserve, Western Coast (Aguada); 1–350 m; *fl* & *fr* throughout the year.

West Indies, southeastern and south-central United States, Mexico, Central America, northern South America to Brazil, Ecuador. [AR&S: 61].

Flecha de agua, Saeta de agua.

TAXA CITED BY PREVIOUS AUTHORS, BUT NOT INCLUDED

DOUBTFUL OCCURRENCE

Sagittaria latifolia Willd.

Shrub cited by Liogier & Martorell (2000) as occurring in shallow water in Laguna San José, Santurce, perhaps not persistent, and in Acevedo-Rodríguez & Strong (2005) as a chance introduction and very likely extirpated. [AR&S: 61].

AMARANTHACEAE (including CHENOPODIACEAE)

Achyranthes aspera L. var. **aspera**, Sp. Pl. 1:204. 1753.

Syn.: *Centrostachys aspera* (L.) Standl. (1915). *Achyranthes aspera* var. *indica* L. (1753). *Achyranthes indica* (L.) Millsp. (1768). *Centrostachys indica* (L.) Standl. (1915).

Herb in Northern Coastal Lowlands, Eastern Coast, Caguas Valley (Juncos), Las Piedras Chiquitas, Eastern Cordillera, Southern Coastal Lowlands, Guánica Forest Reserve, Southwestern Coastal Lowlands, Sierra Bermeja, Western Coast, Vieques, Culebra, Mona, Desescheo; 1–300 m; *fl* & *fr* throughout the year.

Probably native to Paleotropics and -subtropics. [L1: 142].

Rabo de gato.

Achyranthes aspera L. var. **pubescens** (Moq.) C.C. Towns., Kew Bull. 29(3):473. 1974.

Bas.: *Achyranthes fruticosa* Lam. var. *pubescens* Moq. (1849).

Herb in Eastern Cordillera (Cayey), Guánica Forest Reserve, Western Coast (Hormigueros); 5–550 m; *fl* & *fr* Jan–Apr.

Probably native to Paleotropics and -subtropics. [L1: 142].

Alternanthera brasiliana (L.) Kuntze, Revis. Gen. Pl. 2:537. 1891.

Bas.: *Gomphrena brasiliana* L. (1756).

Syn.: *Alternanthera dentata* (Moench) Stuchlik ex R.E. Fr. (1921).

Herb planted and persistent in Northern Coastal Lowlands (Fajardo), Eastern Coast (Punta Guayanés), Southeastern Foothills (Guayama), Eastern Cordillera (Aibonito); 25–500 m; *fl* & *fr* Nov–Feb.

Bahamas, Greater Antilles (Hispaniola), Virgin Islands (St. John), Lesser Antilles, southern Mexico, northern Central America, northern South America to Bolivia, Paraguay. [L1: 146].

Alternanthera caracasana Kunth in Humboldt, Bonpland & Kunth, Nov. Gen. Sp. 2:165 [folio]; 2:205 [quarto]. 1818.

Syn.: *Telanthera caracasana* (Kunth) Moq. (1849).

Illecebrum peploides Humb. & Bonpl. ex Schult. (1819). *Alternanthera peploides* (Humb. & Bonpl. ex Schult.) Urb. (1918). *Achyranthes peploides* (Humb. & Bonpl. ex Schult.) Britton & P. Wilson (1924).

Herb in Southeastern Foothills (Salinas), Vieques; 10–620 m; *fl* & *fr* Mar, Jun (two specimens). [*Woodbury a.n. 7014* (UPR); *Liogier 28431* (UPR)].

West Indies, Mexico, Central America, northern South America to Peru. [L1: 145].

Adorna jardín.

Alternanthera crucis (Moq.) Bold., Fl. Dutch W. Ind. Is. 1:58. 1909.

Bas.: *Telanthera crucis* Moq. (1849).

Syn.: *Achyranthes portoricensis* (Kuntze) Standl. (1915).

Prostrate herb in Northern Coastal Lowlands (San Juan), Las Piedras Chiquitas, Southeastern Foothills, Southern Coastal Lowlands (Salinas, Santa Isabel), Southwestern Coastal Lowlands (Spiny Butte Key), Culebra; 1–500 m; *fl* & *fr* throughout the year.

Virgin Islands, Lesser Antilles. [L1: 145].

Alternanthera ficoidea (L.) P. Beauv., Fl. Oware 2:66, *t. 99, f. 1.* 1818 "ficoides."

Bas.: *Gomphrena ficoidea* L. (1753), *nom. cons.*

Syn.: *Alternanthera tenella* Colla (1828).

Spreading herb in Northern Coastal Lowlands (Río Grande, Isabela), Northern Limestone Hills, Sierra de Luquillo, Eastern Coast (Naguabo), Sierra de Cayey, Eastern Cordillera (Cayey, Aibonito), Central Cordillera (Adjuntas), Maricao Forest Reserve, Western Coast; 10–1000 m; *fl* & *fr* Dec–Jun, Sep.

West Indies, Trinidad and Tobago, southern Central America (Panama), northern South America to Brazil, Paraguay. [L1: 148].

Notes: Britton & Wilson (1924) misapplied the homonym *Achyranthes ficoidea* (L.) Lam. to this taxon.

Sanguinaria.

Alternanthera paronychioides A. St.-Hil., Voy. Distr. Diam. 2:439. 1833.

Syn.: *Achyranthes polygonoides* (L.) Lam. (1785).

Prostrate herb in Southwestern Coastal Lowlands (Lajas); 10 m; *fl* & *fr* May (two specimens). [*Woodbury a.n. 28141* (UPR)].

Bahamas, Antilles, Virgin Islands (St. Croix), southeastern United States, Mexico, Central America (Belize, Costa Rica), South America (scattered). [L1: 146].

Alternanthera philoxeroides (Mart.) Griseb., Abh. Königl. Ges. Wiss. Göttingen 24:36. 1879.
Bas.: *Bucholzia philoxeroides* Mart. (1825).

Herb in Northern Coastal Lowlands, Eastern Cordillera (Aibonito), Southern Coastal Lowlands (Coamo), Western Coast (Mayagüez); 1–400 m; *fl* & *fr* Jan, Apr–Jun.

Native to South America. [L1: 146].

Alternanthera pungens Kunth in Humboldt, Bonpland & Kunth, Nov. Gen. Sp. 2:165 [folio]; 2:206 [quarto]. 1818.
Syn.: *Telanthera pungens* (Kunth) Moq. (1849).

Herb in Southeastern Foothills, Southern Coastal Lowlands (Peñuelas, Sabana Grande), Southwestern Coastal Lowlands, Sierra Bermeja; 1–200 m; *fl* & *fr* Sep–Dec, Mar–May.

Native to South America. [L1: 147].

Alternanthera sessilis (L.) R. Br. ex DC., Cat. Pl. Horti Monsp. 77. 1813.
Bas.: *Gomphrena sessilis* L. (1753).
Syn.: *Achyranthes sessilis* (L.) Steud. ex Standl. (1915).

Trailing herb in Northern Coastal Lowlands, Northern Limestone Hills (San Sebastián), Río Abajo Forest Reserve, Caguas Valley (Caguas), Eastern Cordillera (Comerío), Central Cordillera (Ciales), Susúa Forest Reserve, Southwestern Coastal Lowlands, Vieques; 1–800 m; *fl* & *fr* throughout the year.

Native to tropical Asia, China. [L1: 148].

Sanguinaria.

Amaranthus australis (A. Gray) J.D. Sauer, Madroño 13:15. 1955.
Bas.: *Acnida australis* A. Gray (1876).

Herb in Northern Coastal Lowlands (Laguna Tortuguero); 1–2 m; *fl* & *fr* Feb, Aug. [*Woodbury a.n. 882175* (NY), scan seen; *Woodbury a.n. 5146, 5437* (UPR)].

Greater Antilles, Trinidad, southeastern and south-central United States, Mexico, northern South America and Brazil. [L1: 151].

Blero de agua.

Amaranthus blitum L., Sp. Pl. 2:990. 1753.

Herb in Northern Coastal Lowlands, Eastern Cordillera (Barranquitas), Southern Coastal Lowlands (Ponce), Western Cordillera (Maricao); 3–950 m; *fl* & *fr* Nov–May.

Native to southern Europe, northern and eastern Africa, temperate and tropical Asia. [L5: 395].

Amaranthus crassipes Schltdl., Linnaea 6(5):757. 1831.

Prostrate herb in Southeastern Foothills, Southwestern Coastal Lowlands, Vieques, Culebra; 1–200 m; *fl* & *fr* May–Jan.

Bahamas, Greater Antilles, Virgin Islands, southern United States, Mexico, South America (Venezuela). [L1: 151].

Amaranthus dubius Mart. ex Thell., Fl. Adv. Montpellier 203. 1912.

Herb in Northern Coastal Lowlands, Northern Limestone Hills, Río Abajo Forest Reserve, Eastern Coast, Southeastern Foothills (Coamo), Eastern Cordillera, Central Cordillera (Utuado, Yauco), Southern Coastal Lowlands, Southwestern Coastal Lowlands, Western Coast, Vieques, Mona; 1–600 m; *fl* & *fr* throughout the year.

West Indies, Trinidad, Mexico, Central America, northern South America to Bolivia. [L1: 152].

Bledo, Blero, Blero blanco.

Amaranthus spinosus L., Sp. Pl. 2:991. 1753.

Herb in Northern Coastal Lowlands, Eastern Coast, Caguas Valley, Las Piedras Chiquitas, Southeastern Foothills (Coamo), Eastern Cordillera, Central Cordillera (Utuado), Southern Coastal Lowlands (Santa Isabel), Southwestern Coastal Lowlands (Lajas), Sierra Bermeja, Vieques; 1–500 m; *fl* & *fr* throughout the year.

Probably native to the Neotropics. [L1: 152].

Blero espinoso.

Amaranthus viridis L., Sp. Pl. (ed. 2) 2:1405. 1763.

Herb in Northern Coastal Lowlands, Northern Limestone Hills (Florida, Quebradillas), Eastern Coast (Humacao), Caguas Valley (Caguas), Eastern Cordillera (Cayey), Central Cordillera, Southern Coastal Lowlands, Southern Coastal Limestone (Peñuelas), Western Cordillera (Maricao), Southwestern Coastal Lowlands, Sierra Bermeja, Western Coast (Mayagüez), Vieques, Culebra, Mona, Desecheo; 1–600 m; *fl* & *fr* throughout the year.

Possibly native to South America. [L1: 153].

Notes: Britton & Wilson (1924) used *Amaranthus gracilis* Desf., a *nom. nud.*, for the name of this taxon.

Blero blanco, Blero manso.

Atriplex cristata Humb. & Bonpl. ex Willd., Sp. Pl. 4(2):959. 1806.
Syn.: *Atriplex pentandra* (Jacq.) Standl. (1916).
Fleshy herb in Northern Coastal Lowlands (Fajardo and adjacent islands), Southern Coastal Lowlands (Caja de Muertos, Ponce), Guánica Forest Reserve, Southwestern Coastal Lowlands, Vieques, Culebra; 1–25 m; *fl* & *fr* throughout the year.
Bahamas, Greater Antilles, Virgin Islands, Southern Caribbean Islands (Curaçao), eastern and south-central United States, Mexico, northwestern South America and Venezuela. [L1: 136].
Garbancillo.

Blutaparon vermiculare (L.) Mears, Taxon 31(1): 113. 1982.
Bas.: *Gomphrena vermicularis* L. (1753).
Syn.: *Philoxerus vermicularis* (L.) R. Br. ex Sm. (1814). *Caraxeron vermicularis* (L.) Raf. (1837).
Fleshy herb in Northern Coastal Lowlands, Southwestern Coastal Lowlands (Guánica), Western Coast, Vieques, Culebra, Mona; 1–10 m; *fl* & *fr* throughout the year.
West Indies, southeastern United States (Florida), Mexico, Central America (Belize, Panama), northwestern South America (Ecuador), west Africa. [L1: 155].
Yerba de sal.

Celosia argentea L. var. **argentea**, Sp. Pl. 1:205. 1753.
Herb in Northern Coastal Lowlands, Río Abajo Forest Reserve, Eastern Coast (Ceiba), Southern Coastal Lowlands, Southwestern Coastal Lowlands (Cabo Rojo); 1–300 m; *fl* & *fr* throughout the year.
Pantropical; exact native range obscure. [L1: 156].
Albahaca plateada.

Celosia nitida Vahl, Symb. Bot. 2:44. 1791.
Scrambling herb/shrub in Northern Coastal Lowlands, Southern Coastal Lowlands (Caja de Muertos), Southern Coastal Limestone (Guayanilla), Guánica Forest Reserve, Southwestern Coastal Lowlands, Vieques, Culebra, Mona, Desecheo; 1–75 m; *fl* & *fr* throughout the year.
West Indies, southern United States, Mexico, northern Central America (Guatemala), northern South America. [L1: 157; AR: 53].

Chamissoa altissima (Jacq.) Kunth in Humboldt, Bonpland & Kunth, Nov. Gen. Sp. 2:158 [folio]; 2:197 [quarto], *t. 125*. 1818.
Bas.: *Achyranthes altissima* Jacq. (1760).
Shrub in Northern Limestone Hills (Ciales), Río Abajo Forest Reserve, Eastern Coast (Cuchilla de Panduras), Sierra de Cayey, Las Tetas de Cayey, Eastern Cordillera (Cayey), Central Cordillera, Western Cordillera, Western Coast (Mayagüez); 150–1250 m; *fl* & *fr* Sep–Apr.
Antilles, Trinidad and Tobago, Mexico, Central America, South America. [L1: 158; AR: 53].

Chenopodium murale L., Sp. Pl. 1:219. 1753.
Herb in Sierra de Cayey, Eastern Cordillera (Aibonito), Guánica Forest Reserve, Southwestern Coastal Lowlands (Lajas); 1–700 m; *fl* & *fr* Nov, Feb–Mar, Jul.
Native to Europe, northern Africa, eastern Asia. [L1: 139].

Cyathula prostrata (L.) Blume, Bijdr. Fl. Ned. Ind. 549. 1826.
Bas.: *Achyranthes prostrata* L. (1762).
Syn.: *Desmochaeta prostrata* (L.) DC. (1813). *Pupalia prostrata* (L.) Mart. (1826).
Herb in Northern Coastal Lowlands (San Juan), Sierra de Luquillo, Eastern Coast (Cuchilla de Panduras), Western Cordillera; 10–650 m; *fl* & *fr* Nov–May.
Native to tropical Africa, tropical Asia, China. [L1: 160].

Dysphania ambrosioides (L.) Mosyakin & Clemants, Ukrayins'k. Bot. Zhurn. 59(4):382. 2002.
Bas.: *Chenopodium ambrosioides* L. (1753).
Herb in Northern Coastal Lowlands, Caguas Valley (Caguas), Las Tetas de Cayey, Southeastern Foothills, Eastern Cordillera, Southern Coastal Lowlands (Coamo, Santa Isabel), Southern Coastal Limestone (Sabana Grande); 1–700 m; *fl* & *fr* throughout the year.
Possibly native to Mexico and Central America. [L1: 137].
Apazote, Epazote, Pasote, Pazote.

Gomphrena serrata L., Sp. Pl. 1:224. 1753.
Syn.: *Gomphrena decumbens* Jacq. (1804). *Gomphrena dispersa* Standl. (1916).
Herb in Northern Coastal Lowlands, Eastern Coast (Naguabo), Caguas Valley (Gurabo), Southeastern Foothills (Coamo), Eastern Cordillera, Southern Coastal Lowlands (Caja de Muertos), Southern Coastal Limestone, Susúa Forest Reserve, Southwestern Coastal Lowlands, Sierra Bermeja, Western Coast, Vieques, Culebra; 1–600 m; *fl* & *fr* throughout the year.
Antilles, southeastern United States (Georgia,

Florida), Mexico, Central America, west-central South America. [L1: 162].

Arrasa con todo, Siempreviva silvestre.

Iresine angustifolia Euphrasén, Beskr. Svenska Vestindiska 165. 1795.

Erect herb in Northern Coastal Lowlands, Río Abajo Forest Reserve, Eastern Coast (Cayo Santiago), Las Piedras Chiquitas, Southeastern Foothills, Southern Coastal Lowlands (Caja de Muertos), Southwestern Coastal Lowlands (Lajas), Sierra Bermeja, Vieques, Culebra, Mona, Desecheo; 1–500 m; *fl* & *fr* Nov–Aug.

Antilles, Central America, west-central South America. [L1: 164; AR: 55].

Iresine diffusa Humb. & Bonpl. ex Willd., Sp. Pl. 4(2):765. 1806.

Syn.: *Iresine celosia* L. (1759).

Scrambling suffrutescent herb in Northern Coastal Lowlands, Northern Limestone Hills, Río Abajo Forest Reserve, Sierra de Luquillo, Sierra de Cayey, Las Tetas de Cayey, Las Piedras Chiquitas, Southeastern Foothills, Eastern and Central Cordillera, Southern Coastal Lowlands (Caja de Muertos), Maricao Forest Reserve, Western Coast (Aguada), Vieques, Culebra, Mona, Desecheo; 1–1250 m; *fl* & *fr* Dec–Jun.

West Indies, southeastern and south-central United States, Mexico, Central America, South America. [L1: 165; AR: 55].

Felicidad, Velo de novia del monte.

Lithophila muscoides Sw., Prodr. 14. 1788.

Herb on coastal rocks in Northern Coastal Lowlands, Guánica Forest Reserve, Mona; 1–60 m; *fl* & *fr* throughout the year.

West Indies, northern South America. [L1: 166].

Pfaffia aurata (Mart.) Borsch, Novon 5(3):230. 1995.

Bas.: *Trommsdorffia aurata* Mart. (1826).

Syn.: *Iresine argentata* (Mart.) D. Dietr. (1839).

Woody vine in Río Abajo Forest Reserve, Eastern Coast (Yabucoa), Eastern Cordillera (Cañón San Cristóbal), Central Cordillera (Ciales), Maricao Forest Reserve; 260–600 m; *fl* & *fr* Jan–Mar.

Lesser Antilles, Central America (Nicaragua, Panama), northwestern South America to Bolivia. [L1: 165, 167; AR: 56].

Notes: Liogier & Martorell (1982, 2000) and Liogier (1985) misapplied the name *Pfaffia grandiflora* (Hook.) R.E. Fr. to this taxon.

Salicornia bigelovii Torr. in Emory, Rep. U.S. Mex. Bound. 2(1):184. 1859.

Herb in Southwestern Coastal Lowlands (Cabo Rojo); 0 m; *fl* & *fr* Jul, Dec (two specimens). [*Breckon 6358, 8642* (MAPR); *Axelrod 14269* (UPRRP)].

Bahamas, Greater Antilles (Cuba, Hispaniola), Virgin Islands (Anegada), Lesser Antilles (Anguilla, St. Martin, Guadeloupe), Canada (Nova Scotia), eastern and southern United States, Mexico. [L1: 139].

Chifle.

TAXA NOT RECENTLY COLLECTED

Celosia virgata Jacq., Collectanea 2:279. 1789.

Herb collected in Vieques on 14 Feb 1914 [*Shafer 2864* (NY)].

Greater Antilles (Cuba, Hispaniola), Mexico, northern South America. [L1: 157].

TAXA CITED BY PREVIOUS AUTHORS, BUT NOT INCLUDED

DOUBTFUL OCCURRENCE

Alternanthera ramosissima (Mart.) Chodat. Syn.: *Achyranthes ramosissima* (Mart.) Standl.

Herb cited in Britton & Wilson (1924) for coastal sands in San Juan and Bayamón; subsequently cited by Liogier & Martorell (1982), by Liogier (1985), who added that it was possibly confused by Britton & Wilson with *Gomphrena serrata*, and again by Liogier & Martorell (2000). [L1: 147].

ONLY UNDER CULTIVATION

Alternanthera bettzickiana (Regel) G. Nicholson. Syn.: *Achyranthes bettzickiana* (Regel) Standl.

Herb cited as an ornamental, as *Alternanthera tenella* Colla '*Bettzickiana*' by Liogier & Martorell (1982) and Liogier (1985). [L1: 148].

Notes: Liogier & Martorell (2000) used the homonym *Alternanthera bettzickiana* (Regel) Voss for the name of this taxon.

Amaranthus caudatus L.

Herb cited by Liogier & Martorell (1982, 2000) and Liogier (1985) as grown for ornament and naturalized. [L1: 151].

Amaranthus cruentus L. Syn.: *Amaranthus hybridus* L. subsp. *cruentus* (L.) Thell.

Herb cited by Liogier & Martorell (1982, 2000) and Liogier (1985) as grown for ornament and naturalized. [L1: 152].

Amaranthus tricolor L. Syn.: *Amaranthus gangeticus* L.

Herb cited by Liogier & Martorell (1982, 2000) and

Liogier (1985) as cultivated as an ornamental and as an edible green pot herb. [L1: 154].

Celosia argentea L. var. ***cristata*** (L.) Kunze. Syn.: *Celosia cristata* L.
Herb cited in Britton & Wilson (1924) as planted for ornament; subsequently cited by Liogier & Martorell (1982, 2000) and Liogier (1985) as grown as an ornamental. [L1: 156].

Gomphrena globosa L.
Herb cited in Britton & Wilson (1924) as spontaneous after cultivation; subsequently cited by Liogier & Martorell (1982, 2000) and Liogier (1985) as occasionally escaping. [L1: 162].

AMARYLLIDACEAE

Crinum americanum L., Sp. Pl. 1:292. 1753.

Herb planted and persistent in Northern Coastal Lowlands (Loíza, Bayamón); 1–25 m; *fl* Mar–Sep, *fr* (?).

Native to southeastern and south-central United States.

Crinum asiaticum L., Sp. Pl. 1:292. 1753.

Herb planted and persistent in Eastern Cordillera (Barranquitas), Vieques, Mona; 50–600 m; *fl* Jul–Nov, Feb–Apr, *fr* (?).

Native to western Indian Ocean (Mauritius), tropical Asia, Australia, Pacific islands. [AR&S: 107].

Notes: Britton & Wilson (1924) used *Crinum amabile* Donn, a *nom. nud.*, for the name of this taxon.

Crinum zeylanicum (L.) L., Syst. Nat. (ed. 12) 2:236. 1767.
Bas.: *Amaryllis zeylanica* L. (1753).

Herb planted and persistent in Northern Coastal Lowlands, Southwestern Coastal Lowlands (Cabo Rojo), Mona, Desecheo; 1–50 m; *fl* May, Aug–Nov, *fr* (?).

Native to tropical Africa, Indian Subcontinent. [AR&S: 108].

Lirio.

Eucharis amazonica Linden ex Planch., J. Gén. Hort. 12: *tt. 1216–1217*. 1857.

Herb planted and persistent in Northern Limestone Hills (Ciales, Utuado), Río Abajo Forest Reserve, Central Cordillera (Adjuntas), Maricao Forest Reserve; 250–1100 m; *fl* Nov–Mar, *fr* (?).

Native to South America (Peru). [AR&S: 109].

Notes: Liogier & Martorell (1982, 2000) misapplied the name *Eucharis grandiflora* Planch. & Linden to this taxon.

Anunciación, Lirio turco.

Hippeastrum puniceum (Lam.) Kuntze, Revis. Gen. Pl. 2:703. 1891 "purpureum."
Bas.: *Amaryllis punicea* Lam. (1783).
Syn: *Hippeastrum puniceum* (Lam.) Urb. (1903).

Herb in Northern Coastal Lowlands (islands adjacent to Fajardo, Vega Alta), Río Abajo Forest Reserve, Eastern Coast (Naguabo), Sierra de Cayey, Eastern Cordillera (Aguas Buenas), Central Cordillera, Western Coast (Rincón); 5–750 m; *fl* Feb–May, *fr* (?).

Greater Antilles, northern South America to Peru. [AR&S: 109].

Amapola, Lirio rojo.

Hymenocallis caribaea (L.) Herb., Bot. Reg. 7(App.):44. 1821.
Bas.: *Pancratium caribaeum* L. (1753).
Syn.: *Hymenocallis declinata* (Jacq.) M. Roem. (1847).

Herb in Northern Coastal Lowlands (Fajardo), Southern Coastal Lowlands (Arroyo), Western Coast (Rincón); 1–15 m; *fl* Jan, Apr, *fr* (?). [*González Más 2260, 2262* (MAPR); *Axelrod 3889, 7645* (UPRRP)].

Antilles. [AR&S: 110].

Lirio blanco, Lirio de playa.

Hymenocallis latifolia (Mill.) M. Roem., Fam. Nat. Syn. Monogr. 4:168. 1847.
Bas.: *Pancratium latifolium* Mill. (1768).
Syn.: *Hymenocallis expansa* (Herb.) Herb. (1821).

Herb in Northern Coastal Lowlands, Eastern Coast (Cayo Santiago), Sierra de Cayey (San Lorenzo), Vieques, Culebra, Mona; 1–300 m; *fl* Mar–Sep, Dec, *fr* Jul (one specimen).

Bahamas, Greater Antilles, Lesser Antilles (Dominica, Martinique), southern United States (Florida, Texas). [AR&S: 110].

Lirio, Lirio blanco, Lirio de playa.

Zephyranthes carinata Herb., Bot. Mag. 52: *t. 2594*. 1825.
Syn.: *Atamosco carinata* (Herb.) P. Wilson (1924).

Herb in Eastern Cordillera (Aibonito), Toro Negro Forest Reserve; 500–825 m; *fl* Feb–Mar (two specimens), *fr* (?). [*Liogier 28841* (UPR); *Axelrod 2200* (UPRRP)].

Native to Greater Antilles (Cuba), Lesser Antilles (Guadeloupe, Martinique), Mexico, northern Central America to Costa Rica. [AR&S: 113].

Notes: Liogier & Martorell (1982, 2000) and Acevedo-Rodríguez & Strong (2005) misapplied the name *Zephyranthes grandiflora* Lindl. to this taxon.

Adelfa, Duende rosado.

Zephyranthes citrina Baker, Bot. Mag. 108: *t. 6605*. 1882.

Herb in Northern Coastal Lowlands (San Juan), Northern Limestone Hills (Vega Alta), Eastern Cordillera (Cidra), Western Coast (Añasco); 1–400 m; *fl* May–Jul, Oct, *fr* (?).

Native to Greater Antilles (Cuba, Hispaniola), Lesser Antilles, Trinidad & Tobago. [AR&S: 113].

Duende amarillo.

Zephyranthes proctorii Acev.-Rodr. & M.T. Strong, Contr. U.S. Natl. Herb. 52:115, *t. 26A-D*. 2005.

Herb in Guánica Forest Reserve; 1–5 m; *fl* Apr–May, Oct, *fr* Oct (one specimen).

Endemic. [AR&S: 115].

Zephyranthes puertoricensis Traub, Plant Life 7:37, *t. 4a*. 1951.

Herb in Northern Coastal Lowlands (San Juan); 10 m; *fl* May (one specimen), *fr* (?). [*Proctor 47875* (US)].

West Indies, Tobago, northern South America (Venezuela, Colombia). [AR&S: 115].

Notes: Britton & Wilson (1923) misapplied the name *Atamosco tubispatha* (L'Hér.) M. Gómez to this taxon.

Duende blanco.

Zephyranthes rosea Lindl., Bot. Reg. 10: *t. 821*. 1824.

Syn.: *Atamosco rosea* (Lindl.) Greene (1897).

Herb planted and persistent in Northern Coastal Lowlands; 1 m; *fl* Apr–Nov, *fr* Oct–Nov (one specimen).

Native to Greater Antilles (Cuba), Lesser Antilles (Guadeloupe, Martinique). [AR&S: 115].

Duende rojo.

TAXA CITED BY PREVIOUS AUTHORS, BUT NOT INCLUDED

ONLY UNDER CULTIVATION

Hymenocallis speciosa (L.f. ex Salisb.) Salisb.

Herb cited in Acevedo-Rodríguez & Strong (2005) as known only in cultivation. [AR&S: 112].

ANACARDIACEAE

Anacardium occidentale L., Sp. Pl. 1:383. 1753.

Tree in Northern Coastal Lowlands, Eastern Coast, Southern Coastal Lowlands (Santa Isabel), Vieques; 1–50 m; *fl* throughout the year, *fr* Mar–Jul.

Native to northern South America and Brazil. [L2: 438].

Cajuil, Marañón, Pajuil.

Comocladia dodonaea (L.) Urb., Symb. Antill. 4(3):360. 1910.

Bas.: *Ilex dodonaea* L. (1753).

Shrub/small tree in Northern Coastal Lowlands (Fajardo), Northern Limestone Hills, Eastern Coast, Las Piedras Chiquitas, Southern Coastal Lowlands, Susúa Forest Reserve, Guánica Forest Reserve, Southwestern Coastal Lowlands, Sierra Bermeja, Western Coast, Vieques, Culebra, Mona; 1–500 m; *fl* Jan–Jul, *fr* Mar–Jul.

Greater Antilles (Hispaniola), Virgin Islands, Lesser Antilles. [L2: 440].

Carrasco, Chicharrón, Maíz pelado, Maíz tostado, Prapra, Próspera.

Comocladia glabra Spreng., Syst. Veg. 1:176. 1824.

Shrub/small tree in Northern Limestone Hills, Río Abajo Forest Reserve, Sierra de Luquillo, Sierra de Cayey, Eastern Cordillera (Corozal), Susúa Forest Reserve, Maricao Forest Reserve; 25–850 m; *fl* Nov–Jun, *fr* Nov–May.

Greater Antilles (Hispaniola). [L2: 441].

Carrasco, Guao.

Mangifera indica L., Sp. Pl. 1:200. 1753.

Tree planted and persistent in Northern Coastal Lowlands (Carolina, San Juan), Sierra de Luquillo, Southeastern Foothills (Coamo), Susúa Forest Reserve, Western Cordillera, Western Coast (Rincón), Vieques; 1–650 m; *fl* Dec–May, *fr* Feb–Jul.

Native to India, Myanmar. [L2: 442].

Mango, Mangó.

Metopium toxiferum (L.) Krug & Urb., Bot. Jahrb. Syst. 21:612. 1896.

Bas.: *Amyris toxifera* L. (1759).

Tree in Northern Coastal Lowlands (Arecibo), Northern Limestone Hills, Southern Coastal Limestone (Peñuelas), Southwestern Coastal Lowlands (Cabo Rojo), Mona; 1–200 m; *fl* May–Jan, *fr* Feb–Oct.

Bahamas, Greater Antilles (Cuba, Hispaniola), Lesser Antilles (Anguilla, Antigua, Dominica), southeastern United States (Florida). [L2: 443].

Almendrón, Cedro prieto, Papayo.

Schinus terebinthifolia Raddi, Mem. Mat. Fis. Soc. Ital. Sci. Modena, Pt. Mem. Fis. 18(2):399. 1820 "terebinthifolius."

Shrub/small tree in Northern Coastal Lowlands, Sierra de Cayey, Eastern Cordillera (Comerío), Central Cordillera (Utuado), Southern Coastal Lowlands (Ponce); 5–700 m; *fl* Oct–Jan, *fr* Oct–Feb, May.

Native to southern South America and Brazil. [L2: 445].

Pimienta del Brasil.

Spondias dulcis Sol. ex Parkinson, J. Voy. South Seas 39. 1773.
Syn.: *Spondias cytherea* Sonn. (1782).
Tree planted and persistent in Northern Limestone Hills (Barceloneta, Isabela); 50–150 m; *fl* Mar–Apr, *fr* Nov (one specimen).
Probably native to tropical Asia or Oceania. [L2: 447].
Cítara, Jobo de la India.

Spondias mombin L., Sp. Pl. 1:371. 1753.
Tree in Northern Coastal Lowlands, Northern Limestone Hills, Eastern Coast (Humacao), Las Piedras Chiquitas, Southeastern Foothills, Sierra Bermeja, Vieques; 1–500 m; *fl* Feb–Jun, Sep, *fr* Jun–Sep, Dec.
Native to Mexico, Central America, South America; possibly native to some parts of the West Indies. [L2: 448].
Jobillo, Jobo, Jobo de perro, Jobo gusanero, Jobo vano.

Spondias purpurea L., Sp. Pl. (ed. 2) 1:613. 1762.
Tree planted and persistent in Río Abajo Forest Reserve, Eastern Coast (Humacao), Eastern Cordillera (Aibonito), Central Cordillera (Ponce, Peñuelas), Culebra; 10–300 m; *fl* Feb–Apr (two specimens), *fr* Feb–Aug.
Native to southern Mexico, Central America, northwestern and west-central South America. [L2: 449].
Ciruela, Ciruela del país, Jobillo, Jobo francés.

TAXA CITED BY PREVIOUS AUTHORS, BUT NOT INCLUDED

CITED DUE TO MISIDENTIFICATION

Comocladia cuneata Britton.
Shrub cited by Liogier & Martorell (1982) as occurring on limestone in wet, northern districts; Liogier (1988) stated that this report was in error.

Comocladia dentata Jacq.
Shrub cited by Liogier & Martorell (1982) as occurring on limestone in wet, northern wet forests; Liogier (1988) stated that this report was in error.

ONLY UNDER CULTIVATION

Schinus molle L.
Tree cited by Liogier & Martorell (1982, 2000) and Liogier (1988) as being grown experimentally. [L2: 445].

ANNONACEAE

Annona glabra L., Sp. Pl. 1:537. 1753.
Tree in Northern Coastal Lowlands, Eastern Coast (Humacao), Southeastern Foothills (Coamo), Southern Coastal Lowlands (Guayama), Susúa Forest Reserve, Western Coast (Aguada); 1–150 m; *fl* Jan–Aug, *fr* Jan–Oct.
West Indies, southeastern United States (Florida), Mexico, Central America, northwestern South America to Brazil, Ecuador, western tropical Africa. [L1: 237].
Cayur, Cayure, Corazón cimarrón, Corcho, Coyur.

Annona montana Macfad., Fl. Jamaica 1:7. 1837.
Tree in Northern Limestone Hills, Southeastern Foothills (Coamo), Eastern Cordillera (Cañón San Cristóbal), Quebrada de Los Cedros, Western Cordillera (Maricao), Western Coast (Añasco), Vieques; 1–400 m; *fl* Jul–Nov, *fr* May–Dec.
Antilles, southern Central America, South America. [L1: 238].
Guanábana cimarrona.

Annona mucosa Jacq., Observ. Bot. 1:16. 1764.
Syn.: *Rollinia mucosa* (Jacq.) Baill. (1868).
Tree in Northern Limestone Hills (Florida), Eastern Cordillera (Aibonito), Maricao Forest Reserve; 150–425 m; *fl* May–Sep, *fr* (?).
Greater Antilles (Hispaniola), Lesser Antilles, Trinidad, Mexico, Central America, northern South America to Brazil, Bolivia. [L1: 245].
Anón cimarrón, Cachimán.

Annona muricata L., Sp. Pl. 1:536. 1753.
Tree planted and persistent in Eastern Coast, Las Piedras Chiquitas, Southeastern Foothills, Central Cordillera (Ponce), Vieques; 10–450 m; *fl* & *fr* throughout the year.
Commonly cultivated, though probably native to the Antilles. [L1: 238].
Guanábana.

Annona reticulata L., Sp. Pl. 1:537. 1753.
Tree planted and persistent in Northern Limestone Hills, Río Abajo Forest Reserve, Sierra de Cayey (San Lorenzo), Las Piedras Chiquitas, Southeastern Foothills, Eastern Cordillera (Cayey), Central Cordillera (Ponce), Southern Coastal Lowlands, Guánica Forest Reserve, Vieques; 1–500 m; *fl* & *fr* throughout the year.
Commonly cultivated; probably native to Central America. [L1: 238].
Corazón.

Annona squamosa L., Sp. Pl. 1:537. 1753.

Tree planted and persistent in Southeastern Foothills, Southwestern Coastal Lowlands (Cabo Rojo), Vieques; 10–200 m; *fl* May–Jun (two specimens), *fr* Mar, Jun, Nov.

Commonly cultivated, though probably native to the Antilles. [L1: 239].

Anón.

Cananga odorata (Lam.) Hook. f. & Thomson, Fl. Ind. 130. 1855.

Bas.: *Uvaria odorata* Lam. (1785).

Tree planted and persistent in Northern Coastal Lowlands (Río Grande, San Juan), Southern Coastal Lowlands (Ponce), Southwestern Coastal Lowlands (Lajas), Western Coast (Mayagüez); 5–150 m; *fl* Jun–Dec, *fr* Aug–Dec.

Native to Southeastern Asia, Malesia, Australia. [L1: 240].

Ilán-Ilán, Ilang-ilang, Ilanilán.

Guatteria blainii (Griseb.) Urb., Symb. Antill. 4(2):239. 1905.

Bas.: *Asimina blainii* Griseb. (1866).

Syn.: *Cananga blainii* (Griseb.) Britton (1924).

Tree in Northern Limestone Hills (Arecibo), Río Abajo Forest Reserve, Central Cordillera, Maricao Forest Reserve; 275–900 m; *fl* Sep–Jun, *fr* Nov, Feb–May.

Greater Antilles (Cuba, Hispaniola). [L1: 241].

Haya, Haya minga, Negra lora.

Guatteria caribaea Urb., Symb. Antill. 4(2):240. 1905.

Syn.: *Cananga caribaea* (Urb.) Britton (1924).

Tree in Sierra de Luquillo, Sierra de Cayey, Eastern Cordillera (Cerro La Peña); 250–600 m; *fl* Jun–Aug, *fr* Jul, Nov (two specimens).

Lesser Antilles. [L1: 242].

Haya blanca, Yaya.

Oxandra lanceolata (Sw.) Baill., Adansonia 8:168. 1868.

Bas.: *Uvaria lanceolata* Sw. (1788).

Tree in Northern Limestone Hills (San Sebastián), Guánica Forest Reserve, Southwestern Coastal Lowlands; 15–200 m; *fl* Aug (one specimen), *fr* Aug–Oct.

Greater Antilles, Lesser Antilles (Dominica). [L1: 243].

Haya blanca, Haya prieta, Yaya.

Oxandra laurifolia (Sw.) A. Rich. in Sagra, Hist. Fis. Cuba, Bot. 10:20. 1845.

Bas.: *Uvaria laurifolia* Sw. (1800).

Tree in Northern Limestone Hills (Florida, Arecibo), Río Abajo Forest Reserve, Sierra de Cayey, Eastern Cordillera (La Torrecilla), Toro Negro Forest Reserve, Maricao Forest Reserve; 300–900 m; *fl* (?), *fr* Jan–Feb, Aug.

Antilles. [L1: 243].

Haya, Haya mala, Yaya.

TAXA CITED BY PREVIOUS AUTHORS, BUT NOT INCLUDED

ONLY UNDER CULTIVATION

Annona cherimoya Mill.

Tree cited by Liogier & Martorell (1982, 2000) and by Liogier (1985) as occasionally planted and persistent in the mountains. [L1: 237].

APIACEAE (UMBELLIFERAE)

Anethum graveolens L., Sp. Pl. 1:263. 1753.

Herb in Toro Negro Forest Reserve; 1000 m; *fl* & *fr* Jun (one specimen). [*Liogier 34390* (UPR)].

Probably native to southern Asia. [L4: 87].

Hinojo.

Centella asiatica (L.) Urb. in Martius, Fl. Bras. 11(1):287, *t. 78.* 1879.

Bas.: *Hydrocotyle asiatica* L. (1753).

Syn.: *Centella erecta* (L.f.) Fernald (1940).

Herb in Northern Coastal Lowlands, Sierra de Luquillo, Sierra de Cayey, Southeastern Foothills (Salinas); 1–1000 m; *fl* & *fr* Feb–Nov.

Widespread in tropic, subtropic, and warm temperate regions of the Old and New World. [L4: 90].

Yerba de clavo.

Cyclospermum leptophyllum (Pers.) Sprague ex Britton & P. Wilson, Sci. Surv. Porto Rico & Virgin Islands 6(1):52. 1925.

Bas.: *Pimpinella leptophylla* Pers. (1805).

Syn.: *Apium leptophyllum* (Pers.) F. Muell. ex Benth. (1867).

Herb in Northern Limestone Hills (Vega Alta), Sierra de Luquillo, Sierra de Cayey, Las Tetas de Cayey, Las Piedras Chiquitas, Southeastern Foothills, Eastern and Central Cordillera, Maricao Forest Reserve; 100–1200 m; *fl* & *fr* throughout the year.

Probably native to the Neotropics. [L4: 92].

Notes: Liogier (1995) used the orthographic variant '*Ciclospermum*' for the name of the genus; the above spelling has been conserved.

Perejil de pantano.

Eryngium foetidum L., Sp. Pl. 1:232. 1753.

Herb in Northern Coastal Lowlands (Carolina),

Sierra de Luquillo, Southeastern Foothills (Coamo), Eastern Cordillera (Aguas Buenas); 1–800 m; *fl* & *fr* Oct–May.

Antilles, Trinidad and Tobago, Mexico, Central America, northern South America to Peru. [L4: 96].

Culantro, Culantro de monte, Recao.

TAXA CITED BY PREVIOUS AUTHORS, BUT NOT INCLUDED

NO SPECIMENS SEEN

Ptilimnium capillaceum (Michx.) Raf.
Herb cited by Liogier & Martorell (1982, 2000) and Liogier (1995) as recorded from Cayey by Roy Woodbury. [L4: 104].

ONLY UNDER CULTIVATION

Arracacia xanthorrhiza Bancr.
Herb cited by Liogier & Martorell (1982, 2000) and Liogier (1995) as grown at middle elevations, often persistent. [L4: 89].

Coriandrum sativum L.
Herb cited by Liogier & Martorell (1982, 2000) and Liogier (1995) as grown and escaped. [L4: 92].

Daucus carota L.
Herb cited by Liogier & Martorell (1982, 2000) and Liogier (1995) as grown for food, sometimes escaped. [L4: 95].

Foeniculum vulgare Mill. Syn.: *Anethum foeniculum* L.
Herb cited by Liogier & Martorell (1982, 2000) and Liogier (1995) as cultivated and escaped. [L4: 97].

Petroselinum crispum (Mill.) A.W. Hill. Syn.: *Apium petroselinum* L.
Herb cited by Liogier & Martorell (1982, 2000) and Liogier (1995) as cultivated, sometimes escaped. [L4: 102].

APOCYNACEAE (including ASCLEPIADACEAE)

Allamanda cathartica L., Mant. Pl. 2:214. 1771.
Syn.: *Allamanda hendersonii* W. Bull ex Dombrain (1866). *Allamanda cathartica* var. *hendersonii* (W. Bull ex Dombrain) L.H. Bailey & Raffill (1914).

Shrub planted and persistent in Northern Coastal Lowlands, Sierra de Luquillo, Sierra de Cayey, Southwestern Coastal Lowlands (San Germán), Western Coast (Mayagüez); 1–800 m; *fl* throughout the year, *fr* (?).

Native to northeastern South America and Brazil. [L4: 202, 203; AR: 58].

Alamanda, Allamanda, Canario, Cautiva.

Allamanda schottii Pohl, Pl. Bras. Icon. Descr. 1:73, *t. 58.* 1827.
Syn.: *Allamanda neriifolia* Hook. (1851).

Shrub planted and persistent in Central Cordillera; 850–1000 m; *fl* Sep, Jan, *fr* Oct, Jan–Mar, Jun.

Native to Brazil. [L4: 203].

Cautiva.

Allotoonia agglutinata (Jacq.) J.F. Morales & J.K. Williams, Sida 21(1):139, *tt. 8–9.* 2004.
Bas.: *Echites agglutinatus* Jacq. (1760) "agglutinata."
Syn.: *Prestonia agglutinata* (Jacq.) Woodson (1931).

Vine in Northern Coastal Lowlands, Northern Limestone Hills, Río Abajo Forest Reserve, Eastern Coast (Punta Guayanés), Las Piedras Chiquitas, Southern Coastal Lowlands, Southern Coastal Limestone (Guayanilla), Susúa Forest Reserve, Maricao Forest Reserve, Guánica Forest Reserve, Cerro Las Mesas, Western Coast (Rincón), Vieques, Culebra, Mona; 1–650 m; *fl* May–Jan, *fr* May–Feb.

Greater Antilles (Hispaniola), Virgin Islands. [L4: 218; AR: 61].

Azufaifo, Babeiro.

Anechites nerium (Aubl.) Urb., Repert. Spec. Nov. Regni Veg. 16:150. 1919.
Bas.: *Apocynum nerium* Aubl. (1775).

Vine in Northern Limestone Hills (Arecibo); 80–200 m; *fl* Jul–Oct, *fr* Sep–Oct. [*Ackerman 3401* (UPRRP); *Acevedo-Rodríguez 13408* (MAPR)].

Greater Antilles (Cuba, Hispaniola), Central America (Nicaragua to Panama), northwestern South America. [L4: 204; AR: 59].

Asclepias curassavica L., Sp. Pl. 1:215. 1753.

Herb in Northern Coastal Lowlands, Northern Limestone Hills, Río Abajo Forest Reserve, Sierra de Luquillo, Eastern Coast (Humacao), Sierra de Cayey, Las Tetas de Cayey, Las Piedras Chiquitas, Southeastern Foothills (Guayama), Eastern Cordillera, Maricao Forest Reserve, Southwestern Coastal Lowlands (Cabo Rojo), Western Coast (Aguada), Vieques, Culebra; 1–950 m; *fl* & *fr* throughout the year.

Native to the Neotropics and -subtropics. [L4: 229].

Algodoncillo, Platanillo, Platanillo matizado, Yerba de mariposa.

Asclepias nivea L., Sp. Pl. 1:215. 1753.

Herb in Northern Limestone Hills, Río Abajo Forest Reserve, Sierra de Cayey, Las Tetas de Cayey, Eastern Cordillera (Cerro La Peña), Central

Cordillera, Western Cordillera; 200–1300 m; *fl* & *fr* throughout the year.

Greater Antilles. [L4: 231].

Algodoncillo blanco, Platanillo, Platanillo blanco.

Calotropis procera (Aiton) W.T. Aiton, Hort. Kew. (ed. 2) 2:78. 1811.

Bas.: *Asclepias procera* Aiton (1789).

Suffrutescent herb/shrub in Northern Coastal Lowlands, Eastern Coast (Punta Guayanés), Southeastern Foothills (Coamo), Southern Coastal Lowlands, Southern Coastal Limestone (Peñuelas), Guánica Forest Reserve, Southwestern Coastal Lowlands, Sierra Bermeja, Western Coast, Vieques, Culebra, Desecheo; 1–150 m; *fl* & *fr* throughout the year.

Native to northern and tropical Africa, temperate western Asia, Indian Subcontinent, southeastern Asia. [L4: 231].

Algodón de seda, Bomba, Mata de seda, Mudar, Tula.

Catharanthus roseus (L.) G. Don, Gen. Hist. 4:95. 1837.

Bas.: *Vinca rosea* L. (1759).

Herb in Northern Coastal Lowlands, Sierra de Cayey, Southeastern Foothills (Cayey), Southern Coastal Lowlands, Guánica Forest Reserve, Southwestern Coastal Lowlands, Western Coast; 1–300 m; *fl* throughout the year, *fr* Oct–Dec, Mar.

Native to Madagascar. [L4: 207].

Cangrejera, Desbarata casamiento, Flor de todo el año, Jazmín del mar, Playera.

Cryptostegia madagascariensis Bojer ex Decne. in A. de Candolle, Prodr. 8:492. 1844.

Shrub in Northern Coastal Lowlands, Eastern Coast (Ceiba), Southern Coastal Lowlands, Guánica Forest Reserve, Southwestern Coastal Lowlands, Western Coast (Añasco), Vieques; 1–225 m; *fl* throughout the year, *fr* Apr–Jan.

Native to Madagascar. [L4: 235; AR: 74].

Alamanda morada falsa, Bejuco de cueros, Bejuco de goma, Canario morado falso.

Gomphocarpus physocarpus E. Mey. in Meyer & Drège, Comm. Pl. Afr. Austr. 202. 1838.

Erect herb in Central Cordillera, 650–1000 m; *fl* Apr, Jul, Nov, *fr* Apr, Jul, Nov–Dec.

Native to southeastern tropical Africa and South Africa.

Gonolobus stephanotrichus Griseb., Cat. Pl. Cub. 177. 1866.

Syn.: *Vincetoxicum stephanotrichum* (Griseb.) Britton (1925).

Vine in Río Abajo Forest Reserve, Sierra de Luquillo, Sierra de Cayey, Las Tetas de Cayey, Eastern Cordillera (La Torrecilla), Central Cordillera; 250–1100 m; *fl* Jan–Apr, *fr* Sep–Nov (two specimens).

Greater Antilles (Cuba, Hispaniola). [L4: 235; AR: 76].

Marsdenia elliptica Decne. in A. de Candolle, Prodr. 8:616. 1844.

Vine in Northern Limestone Hills (Arecibo), Río Abajo Forest Reserve, Eastern Cordillera (La Torrecilla), Central Cordillera (Utuado); 200–920 m; *fl* Jun–Jul, *fr* Sep, Nov, Feb.

Endemic. [L4: 238; AR: 79].

Marsdenia woodburyana Acev.-Rodr., Brittonia 51(2):167. 1999.

Vine in Southern Coastal Limestone (Peñuelas), Guánica Forest Reserve, Mona; 10–150 m; *fl* Apr–May (one specimen), *fr* Dec–Jan, Jul.

Endemic. [AR: 81].

Matelea maritima (Jacq.) Woodson, Ann. Missouri Bot. Gard. 28:222. 1941.

Bas.: *Asclepias maritima* Jacq. (1760).

Syn.: *Ibatia maritima* (Jacq.) Decne. (1844).

Vine in Northern Coastal Lowlands, Eastern Coast (Ceiba), Las Piedras Chiquitas, Southeastern Foothills, Southern Coastal Lowlands, Susúa Forest Reserve, Guánica Forest Reserve, Southwestern Coastal Lowlands, Vieques, Culebra, Mona; 1–500 m; *fl* & *fr* throughout the year.

Greater Antilles (Cuba, Hispaniola), Virgin Islands, Lesser Antilles, southern Central America (Panama), northern South America to Brazil. [L4: 240; AR: 82].

Guanabanilla cimarrona, Pompón.

Matelea sintenisii (Schltr.) Woodson, Ann. Missouri Bot. Gard. 28:226. 1941.

Bas.: *Gonolobus sintenisii* Schltr. (1899).

Syn.: *Vincetoxicum sintenisii* (Schltr.) Britton (1925).

Vine in Sierra de Luquillo, Sierra de Cayey, Toro Negro Forest Reserve; 800–1300 m; *fl* Mar–May, Sep, *fr* Mar (one specimen).

Endemic. [L4: 240; AR: 83].

Matelea variifolia (Schltr.) Woodson, Ann. Missouri Bot. Gard. 28:226. 1941.

Bas.: *Gonolobus variifolius* Schltr. (1899).

Syn.: *Vincetoxicum variifolium* (Schltr.) Britton (1925).

Matelea borinquensis Alain (1986).

Vine in Northern Limestone Hills (Dorado),

Río Abajo Forest Reserve, Sierra de Luquillo, Sierra de Cayey, Eastern Cordillera (La Torrecilla), Central Cordillera, Maricao Forest Reserve, Mona; 10–1100 m; *fl* Feb–Oct, *fr* Apr (one specimen).

Endemic. [L4: 240; AR: 83].

Notes: Acevedo-Rodríguez (2005) found that the type specimen of *Matelea borinquensis* Alain is "too poor to characterize its morphology correctly" and concluded that "it represents *Matelea variifolia*."

Metastelma decipiens Schltr. in Urban, Symb. Antill. 1(2):249. 1899.
Syn.: *Metastelma fallax* Schltr. (1899).
Cynanchum cheesmanii Woodson (1947).

Vine in Northern Coastal Lowlands, Northern Limestone Hills, Río Abajo Forest Reserve, Eastern Coast, Eastern Cordillera (Cerro La Peña), Southern Coastal Lowlands, Southwestern Coastal Lowlands, Vieques, Culebra, Desecheo; 1–550 m; *fl* & *fr* throughout the year.

Virgin Islands (St. Thomas), Tobago. [L4: 243; AR: 86].

Metastelma grisebachianum Schltr. in Urban, Symb. Antill. 5(3):469. 1908.
Syn.: *Cynanchum grisebachianum* (Schltr.) Alain (1965).
Metastelma decaisneanum Schltr. (1899).

Vine in Northern Coastal Lowlands (islands adjacent to Fajardo: Isla Palominos), Desecheo; 10–15 m; *fl* Jul, Oct–Dec, *fr* Dec (one specimen). [*Breckon 4809, 4830* (UPRRP)]

Virgin Islands, Lesser Antilles. [L4: 243; AR: 87].

Notes: I do not agree with the decision of Acevedo-Rodríguez (2005) to place this taxon in synonymy with *Metastema decipiens*. The shape of the corona lobes is a distinct character which does not intergrade between the two taxa in the specimens that I have seen.

Metastelma leptocladum (Decne.) Schltr. in Urban, Symb. Antill. 1(2):261. 1899.
Bas.: *Vincetoxicum leptocladum* Decne. (1844) "leptocladon."
Syn.: *Cynanchum leptocladum* (Decne.) J. Jiménez Alm. (1960).
Metastelma ephedroides (Griseb.) Schltr. (1899). *Cynanchum ephedroides* (Griseb.) Alain (1955).

Vine in Northern Limestone Hills (Dorado), Río Abajo Forest Reserve, Las Tetas de Cayey, Central Cordillera, Maricao Forest Reserve; 225–1150 m; *fl* throughout the year, *fr* Dec–Jan (two specimens).

Greater Antilles. [L4: 244; AR: 87].

Metastelma lineare Bello, Anales Soc. Esp. Hist. Nat. 10:292. 1881.
Syn.: *Cynanchum lineare* (Bello) Alain (1963).

Vine in Northern Limestone Hills (Arecibo), Río Abajo Forest Reserve, Southern Coastal Limestone (Peñuelas), Susúa Forest Reserve, Maricao Forest Reserve, Guánica Forest Reserve, Sierra Bermeja, Mona; 50–850 m; *fl* & *fr* throughout the year.

Greater Antilles (Cuba), Virgin Islands (St. Thomas). [L4: 244; AR: 87].

Metastelma monense Britton in Britton & Wilson, Sci. Surv. Porto Rico & Virgin Islands 6(1):97. 1925.
Syn.: *Cynanchum monense* (Britton) Alain (1963).

Vine in Guánica Forest Reserve, Mona; 10–50 m; *fl* May–Feb, *fr* Feb (one specimen).

Endemic. [L4: 245; AR: 89].

Metastelma parviflorum (Sw.) R. Br. ex Schult., Syst. Veg. 6:120. 1820.
Bas.: *Cynanchum parviflorum* Sw. (1788).
Syn.: *Cynanchum parviflorum* (R. Br. ex Schult.) Alain (1963), *nom. illeg.*

Vine in the Northern Limestone Hills (Rincón, Moca), Río Abajo Forest Reserve, Las Tetas de Cayey, Las Piedras Chiquitas, Southeastern Foothills, Eastern Cordillera (Cañón San Cristóbal), Central Cordillera (Ponce), Cerro de Las Cuevas, Southern Coastal Lowlands, Maricao Forest Reserve, Guánica Forest Reserve, Southwestern Coastal Lowlands, Western Coast (Hormigueros); 1–800 m; *fl* Oct–Jul, *fr* Nov–Feb.

Greater Antilles (Hispaniola), Virgin Islands (St. Croix), Lesser Antilles, Trinidad, northern South America. [L4: 245; AR: 89].

Nerium oleander L., Sp. Pl. 1:209. 1753.

Shrub planted and persistent in Northern Coastal Lowlands (San Juan), Toro Negro Forest Reserve; 1–1000 m; *fl* May, Aug, Oct, *fr* Dec (one specimen).

Native to southern Europe, northern Africa, temperate western Asia, Indian Subcontinent, China (Yunnan). [L4: 212].

Adelfa, Alhelí, Alhelí extranjero, Laurel rosado, Oleander.

Pentalinon luteum (L.) B.F. Hansen & Wunderlin, Taxon 35(1):167. 1986.
Bas.: *Vinca lutea* L. (1756).
Syn.: *Urechites luteus* (L.) Britton (1907) "lutea."

Vine in Northern Coastal Lowlands, Northern Limestone Hills, Río Abajo Forest Reserve, Southern

Coastal Lowlands (Caja de Muertos), Southern Coastal Limestone (Guayanilla), Guánica Forest Reserve, Southwestern Coastal Lowlands (Cabo Rojo), Western Coast (Rincón), Vieques, Culebra, Mona, Desecheo; 1–350 m; *fl* Mar–Nov, *fr* throughout the year.

West Indies, southeastern United States (Florida), southern Mexico (Chiapas), northern Central America (Honduras). [L4: 214; AR: 62].

Babeiro amarillo.

Pinochia corymbosa (Jacq.) M.E. Endress & B.F. Hansen subsp. **portoricensis** (Woodson) M.E. Endress & B.F. Hansen, Edinburgh J. Bot. 64(2):272. 2007.

Bas.: *Forsteronia portoricensis* Woodson (1934).

Woody vine in Northern Limestone Hills, Río Abajo Forest Reserve, Sierra de Luquillo, Eastern Coast (Cuchilla de Panduras), Sierra de Cayey, Eastern Cordillera (Cerro La Peña), Central Cordillera, Susúa Forest Reserve, Western Cordillera, Cerro Las Mesas; 1–975 m; *fl* Feb–Nov, *fr* May, Sep–Nov.

Endemic. [L4: 209; AR: 61].

Notes: Britton & Wilson (1925) misapplied the name *Forsteronia corymbosa* (Jacq.) G. Mey., which belongs to a similar taxon occurring in Cuba and Hispaniola, to this taxon.

Bejuco lechoso, Bejuco de San Juan, Leche de perra, Sanjuanera.

Plumeria alba L., Sp. Pl. 1:210. 1753.

Tree in Northern Coastal Lowlands, Eastern Coast, Southern Coastal Lowlands (Caja de Muertos), Southern Coastal Limestone (Guayanilla), Susúa Forest Reserve, Guánica Forest Reserve, Southwestern Coastal Lowlands, Sierra Bermeja, Vieques, Culebra, Mona; 1–300 m; *fl* May–Dec, *fr* Mar–Dec.

Virgin Islands, Lesser Antilles. [L4: 215].

Alhelí, Alhelí blanco, Alhelí cimarrón, Tabaiba, Tabeiba, Tapaiba.

Plumeria krugii Urb., Symb. Antill. 1(3):387. 1899.

Tree in Northern Limestone Hills, Río Abajo Forest Reserve, Central Cordillera (Yauco), Susúa Forest Reserve, Western Cordillera, Cerro Las Mesas; 200–850 m; *fl* Feb–Jun, Sep–Nov, *fr* Feb–Jun, Nov.

Endemic. [L4: 216].

Alhelí cimarrón, Alhelí montuno.

Plumeria obtusa L., Sp. Pl. 1:210. 1753.

Syn.: *Plumeria portoricensis* Urb. (1899).

Tree in Cerro Las Mesas, Mona; 5–400 m; *fl* Mar–Nov, *fr* Mar, Nov–Dec.

Bahamas, Greater Antilles, Mexico, northern Central America (Guatemala, Belize, Honduras). [L4: 217].

Alhelí, Alhelí cimarrón, Alhelí montuno, Alhelí de Mona, Tabeiba.

Rauvolfia nitida Jacq., Enum. Syst. Pl. 14. 1760.

Shrub/small tree in Northern Coastal Lowlands, Northern Limestone Hills, Río Abajo Forest Reserve, Eastern Coast, Sierra de Cayey, Southeastern Foothills, Eastern Cordillera (Aibonito), Central Cordillera (Yauco), Southern Coastal Lowlands (Patillas), Susúa Forest Reserve, Southwestern Coastal Lowlands, Sierra Bermeja, Western Coast, Vieques, Culebra, Mona; 1–875 m; *fl* & *fr* throughout the year.

West Indies, Trinidad. [L4: 220].

Notes: Britton & Wilson (1925) misapplied the name *Rauvolfia tetraphylla* L. to this taxon.

Cachimbo, Muñeco, Palo amargo, Palo de muñeco.

Rauvolfia viridis Roem. & Schult., Syst. Veg. 4:805. 1819.

Syn.: *Rauvolfia lamarckii* A. DC. (1844).

Shrub in Northern Coastal Lowlands, Eastern Coast (Maunabo), Las Tetas de Cayey, Southeastern Foothills, Southern Coastal Lowlands, Guánica Forest Reserve, Southwestern Coastal Lowlands (Cabo Rojo), Sierra Bermeja, Vieques, Culebra, Mona; 1–825 m; *fl* & *fr* throughout the year.

Virgin Islands, Lesser Antilles, Trinidad, northwestern South America. [L4: 220].

Rhabdadenia biflora (Jacq.) Müll. Arg. in Martius, Fl. Bras. 6(1):175. 1860.

Bas.: *Echites biflorus* Jacq. (1760).

Woody vine in Northern Coastal Lowlands; 1–5 m; *fl* Feb–Sep, *fr* (?).

West Indies, Trinidad, southeastern United States (Florida), Mexico, Central America, northern South America and Brazil. [L4: 222; AR: 65].

Bejuco negro.

Tabernaemontana citrifolia L., Sp. Pl. 1:210. 1753.

Tree in Northern Coastal Lowlands (Río Grande), Northern Limestone Hills (Bayamón), Río Abajo Forest Reserve, Sierra de Luquillo, Sierra de Cayey, Las Tetas de Cayey, Central Cordillera, Maricao Forest Reserve; 200–1100 m; *fl* throughout the year, *fr* Aug (two specimens).

Bahamas, Greater Antilles (Cuba, Hispaniola), Virgin Islands (St. Thomas), Lesser Antilles, Tobago. [L4: 223].

Notes: Britton & Wilson (1925) misapplied the

name *Tabernaemontana oppositifolia* (Spreng.) Urb. to this taxon.

Palo lechoso, Pegoge.

Tabernaemontana divaricata (L.) R. Br. ex Roem. & Schult., Syst. Veg. 4:427. 1819.
Bas.: *Nerium divaricatum* L. (1753).
Syn.: *Tabernaemontana coronaria* (Jacq.) Willd. (1809).

Shrub planted and persistent in Eastern Coast, Western Coast (Aguada); 1–250 m; *fl* Mar–May, Sep, Dec, *fr* (?).

Native to Indian Subcontinent, southeastern Asia, China (Yunnan). [L4: 225].

Jazmín doble.

TAXA CITED BY PREVIOUS AUTHORS, BUT NOT INCLUDED

DOUBTFUL OCCURRENCE

Oxypetalum cordifolium (Vent.) Schltr.
Vine cited by Britton & Wilson (1925) as having been collected by Read (according to Urban); subsequently cited by Liogier & Martorell (1982, 2000), Liogier (1995) and Acevedo-Rodríguez (2005). [L4: 247; AR: 91].

CITED DUE TO MISIDENTIFICATION

Cryptostegia grandiflora R. Br.
Shrub cited by Liogier & Martorell (1982, 2000) and Liogier (1995) as occasionally planted for ornament, naturalized in borders of mangrove swamps, and by Acevedo-Rodríguez (2005) as along the coast and also on Vieques. All specimens that I have seen identified as this taxon have so far been referable to *Cryptostegia madagascariensis*. [L4: 233; AR: 74].

Metastelma anagadense Britton.
Vine cited by Liogier & Martorell (1982, 2000) and by Liogier (1995) as occurring on a limestone plateau on Mona Island. According to Acevedo-Rodríguez (2005), this citation was based on the misidentification of a specimen of *Metastelma monense*. [L4: 243].

ONLY UNDER CULTIVATION

Allamanda blanchetii A. DC. Syn.: *Allamanda violacea* Gardner & Fielding.
Shrub cited by Liogier & Martorell (1982, 2000) and Liogier (1995) as cultivated and persistent in moist and wet districts. [L4: 202].

Carissa macrocarpa (Eckl.) A. DC. Syn.: *Arduina grandiflora* E. Mey. *Carissa grandiflora* (E. Mey.) A. DC.
Shrub/small tree cited by Liogier & Martorell (1982, 2000) and Liogier (1995) as cultivated as an ornamental and persistent. [L4: 206].

Funtumia elastica (P. Preuss) Stapf.
Tree cited by Liogier & Martorell (2000) and Liogier (1995) as planted and naturalized in experimental plantings. [L4: 211].

Hoya australis R. Br. ex J. Traill.
Vine cited by Britton & Wilson (1925) as naturalized in thickets; subsequently cited by Liogier & Martorell (1982, 2000) and Liogier (1995) as cultivated and escaped and by Acevedo-Rodríguez (2005) as commonly cultivated. Britton & Wilson (1925), Liogier & Martorell (1982, 2000) and Liogier (1995) misapplied the name *Hoya carnosa* (L.f.) R. Br. to this taxon. [L4: 237; AR: 77].

Plumeria rubra L.
Tree cited by Britton & Wilson (1925) as commonly planted for ornament, locally spontaneous after cultivation; subsequently cited by Liogier & Martorell (2000) and Liogier (1995). [L4: 217].

Rauvolfia vomitoria Afzel.
Shrub/small tree cited by Liogier (1997) as reported by Miguel Vives as naturalized; however, the only collection made was on the site of an abandoned nursery. [L5: 412].

Thevetia peruviana (Pers.) K. Schum. Syn.: *Cerbera thevetia* L.
Tree cited by Britton & Wilson (1925) as occurring in coastal thickets, and commonly planted for ornament and interest; subsequently cited by Liogier & Martorell (1982, 2000) and Liogier (1995) as planted as an ornamental and escaped and naturalized. [L4: 226].

AQUIFOLIACEAE

Ilex cookii Britton & P. Wilson, Sci. Surv. Porto Rico & Virgin Islands 6(3):357. 1926.

Shrub in Toro Negro Forest Reserve; 1300–1335 m; *fl* Feb, Aug (two specimens), *fr* Jan (one specimen). [*Liogier 9473* (NY), scan seen].

Endemic. [L3: 15].

Notes: This taxon is on the U.S. Fish & Wildlife Service list of threatened and endangered plants.

Ilex guianensis (Aubl.) Kuntze, Revis. Gen. Pl. 1:113. 1891.
Bas.: *Macoucoua guianensis* Aubl. (1775).

Tree in Northern Coastal Lowlands (Dorado); 1–5 m; *fl* May (one specimen), *fr* Mar (one specimen). [*Woodbury a.n. 1750, 1770, 26948* (UPR)].

Greater Antilles (Hispaniola), southern Mexico,

Central America, northern South America to Bolivia. [L3: 15].

Macoucoua.

Ilex krugiana Loes. in Urban, Bot. Jahrb. Syst. 15:317. 1892.

Tree in Western Coast (Rincón); 25–80 m; *fl* (?), *fr* (?). [*Woodbury a.n. 1760* (UPR)].

Bahamas, Greater Antilles (Cuba, Hispaniola), southeastern United States (Florida). [L3: 16].

Ilex macfadyenii (Walp.) Rehder, J. Arnold Arbor. 3:215. 1922.

Bas.: *Prinos macfadyenii* Walp. (1842).

Shrub/small tree in Sierra de Luquillo, Central Cordillera, Maricao Forest Reserve; 850–1200 m; *fl* Mar–Apr, Jul–Sep, *fr* Jun–Oct, Jan–Feb.

Antilles. [L3: 16].

Acebo de sierra.

Ilex nitida (Vahl) Maxim. var. **repanda** (Griseb.) P.A. González, Harvard Pap. Bot. 12(2):449. 2007.

Bas.: *Ilex repanda* Griseb. (1860).

Tree in Northern Limestone Hills, Río Abajo Forest Reserve, Sierra de Luquillo, Sierra de Cayey, Las Tetas de Cayey, Central Cordillera, Susúa Forest Reserve, Maricao Forest Reserve; 150–1100 m; *fl* Jan–May, *fr* Apr–Oct.

Bahamas, Greater Antilles. [L3: 17].

Notes: Britton & Wilson (1924), Liogier & Martorell (1982, 2000) and Liogier (1994) used *Ilex nitida* (Vahl) Maxim., *sensu lato*, for the name of this taxon.

Aceituna, Aceitunilla, Brigueta, Brigueta naranjo, Cuero de sapo, Hueso prieto, Palo de murta.

Ilex obcordata Sw., Fl. Ind. Occid. 1:338. 1797.

Syn.: *Ilex sintenisii* (Urb.) Britton (1924).

Shrub/small tree in Sierra de Luquillo; 900–1000 m; *fl* Mar–Oct, *fr* Dec–Jan, May–Aug.

Greater Antilles. [L3: 19].

Notes: According to Pedro González Gutiérrez (2007), *Ilex sintenisii* is conspecific with this taxon.

Ilex sintenisii is on the U.S. Fish & Wildlife Service list of threatened and endangered plants.

Guayabota pequeña.

Ilex sideroxyloides (Sw.) Griseb. subsp. **occidentalis** (Macfad.) P.A. González, Harvard Pap. Bot. 12(2):461. 2007.

Bas.: *Ilex occidentalis* Macfad. (1837).

Syn.: *Ilex sideroxyloides* (Sw.) Griseb. var. *occidentalis* Loes. (1899).

Tree in Sierra de Luquillo, Central Cordillera; 300–1335 m; *fl* Feb–Apr, Jul, *fr* Jan–Aug.

Greater Antilles (Cuba, Jamaica), Lesser Antilles, Trinidad and Tobago. [L3: 19].

Gongolí, Gongolín, Jaboncillo.

Ilex urbaniana Loes. in Urban, Bot. Jahrb. Syst. 15:316. 1892.

Shrub/small tree in Northern Coastal Lowlands (Dorado), Northern Limestone Hills (Arecibo), Río Abajo Forest Reserve, Central Cordillera, Maricao Forest Reserve, Cerro Las Mesas; 1–1275 m; *fl* Jan–May, *fr* Mar–Jul.

Virgin Islands. [L3: 17, 19].

Notes: According to Pedro González Gutiérrez (2007), the extant isotype (located at P) of *Ilex riedlaei* Loes. belongs to *Ilex nitida* var. *repanda* and the former name has been misapplied in Puerto Rico to small-leaved specimens of *Ilex urbaniana* that grow on serpentine soils.

Cuero de sapo.

TAXA CITED BY PREVIOUS AUTHORS, BUT NOT INCLUDED

CITED DUE TO MISIDENTIFICATION

Ilex cassine L.

Tree cited by Liogier & Martorell (2000) and Liogier (1994) as collected in a swamp, Dorado Forest. The citation is based on the misidentification of specimens of *Ilex guianensis*. [L3: 15].

Ilex hypaneura Loes.

Tree cited by Liogier & Martorell (1982, 2000) and Liogier (1994) as collected in El Verde, Luquillo Mountains. The citation is based on the misidentification of specimens of *Ilex macfadyenii*. [L3: 16].

Ilex nitida (Vahl) Maxim. var. ***nitida***.

Tree cited by González-Gutiérrez (2007), based on specimens collected in the Maricao and Susúa Forest Reserves. I consider these specimens to be better referable to var. *repanda*.

Ilex sideroxyloides (Sw.) Griseb. var. ***sideroxyloides***.

Tree cited by Liogier (1994) and by Liogier & Martorell (2000) as rare in moist forest at higher elevations, Central Cordillera and Luquillo forests. The citation was probably based on the misidentification of specimens of *Ilex sideroxyloides* var. *occidentalis*. [L3: 18].

ARACEAE (including LEMNACEAE)

Alocasia macrorrhizos (L.) G. Don in Sweet, Hort. Brit. (ed. 3) 631. 1839 "macrorhizon."

Bas.: *Arum macrorrhizon* L. (1753).

Syn.: *Alocasia macrorrhiza* (L.) Schott (1852).

Robust herb in Northern Coastal Lowlands, Río Abajo Forest Reserve, Sierra de Luquillo, Central Cordillera (Ponce); 25–600 m; *fl* Apr (one specimen), *fr* (?).

Possibly native to Indian Subcontinent, Malaysia. [AR&S: 23].

Yautía cimarrona.

Alocasia plumbea Van Houtte, Ann. Gén. Hort. 21:93, *t. 2206.* 1875.

Herb in Northern Limestone Hills (Toa Baja), Sierra de Luquillo, Central Cordillera (Adjuntas); 200–1000 m; *fl* Jul (one specimen), *fr* (?).

Native to Malaysia. [AR&S: 23].

Malanga morada, Yautía ornamental.

Anthurium crenatum (L.) Kunth, Enum. Pl. 3:75. 1841.

Bas.: *Pothos crenatus* L. (1763) "crenata."

Herb in Northern Coastal Lowlands (Vega Alta to Manatí), Northern Limestone Hills, Río Abajo Forest Reserve, Sierra de Luquillo, Eastern Coast, Sierra de Cayey, Las Piedras Chiquitas, Southeastern Foothills, Eastern Cordillera (Cayey), Central Cordillera (Ponce, Ciales), Cerro de Las Cuevas, Susúa Forest Reserve, Maricao Forest Reserve, Southwestern Coastal Lowlands, Sierra Bermeja, Vieques, Culebra; 10–800 m; *fl* throughout the year, *fr* Apr–Sep.

Greater Antilles (Hispaniola), Virgin Islands. [AR&S: 25].

Notes: Britton and Wilson (1923) misapplied the name *Anthurium acaule* (Jacq.) Schott to this taxon.

Flor de culebra, Hoja de costado, Moco de pavo, Rabo de rata.

Anthurium dominicense Schott, Oesterr. Bot. Z. 8:350. 1858.

Herb in Sierra de Luquillo, Sierra de Cayey, Central Cordillera (Adjuntas), Maricao Forest Reserve; 450–850 m; *fl* throughout the year, *fr* Aug, Nov–Mar.

Lesser Antilles (Dominica, Martinique). [AR&S: 26].

Lengua de vaca.

Anthurium scandens (Aubl.) Engl. in Martius, Fl. Bras. 3(2):78. 1878.

Bas.: *Dracontium scandens* Aubl. (1775).

Epiphytic herb in Northern Limestone Hills, Río Abajo Forest Reserve, Sierra de Luquillo, Sierra de Cayey, Las Tetas de Cayey, Eastern Cordillera (Barranquitas), Central Cordillera, Western Cordillera (Maricao); 225–950 m; *fl* Jan–Apr, Jul–Oct, *fr* Feb–Apr, Jul–Nov.

Greater Antilles, Lesser Antilles (Antigua, Barbuda, Guadeloupe), Trinidad and Tobago, Mexico, Central America, northern South America to Bolivia. [AR&S: 26; AR: 404].

Guinda.

Caladium bicolor (Aiton) Vent., Mag. Encycl. 4(16):464. [22 Dec 1800–21 Jan] 1801 (Descr. Pl. Nouv. *ad t. 30.* [Mar] 1801).

Bas.: *Arum bicolor* Aiton (1789).

Syn.: *Cyrtospadix bicolor* (Aiton) Britton & P. Wilson (1923).

Herb in Northern Limestone Hills, Río Abajo Forest Reserve, Sierra de Cayey, Central Cordillera (Ponce), Western Coast (Mayagüez); 100–600 m; *fl* Jan–May, *fr* (?).

Native to northern South America and Brazil. [AR&S: 29].

Paleta de pintor, Malanga cimarrona, Venas de Jesús, Yautía de jardín, Yautía cimarrona.

Colocasia esculenta (L.) Schott in Schott & Endlicher, Melet. Bot. 18. 1832.

Bas.: *Arum esculentum* L. (1753).

Syn.: *Caladium colocasia* (L.) W. Wight (1905).

Herb in Northern Coastal Lowlands, Río Abajo Forest Reserve, Eastern Coast (Humacao), Caguas Valley (Juncos), Toro Negro Forest Reserve, Western Coast; 1–1330 m; *fl* Dec (one specimen), *fr* (?).

Probably native to tropical Asia. [AR&S: 30].

Angustia, Malanga, Yautía malanga.

Dieffenbachia seguine (Jacq.) Schott, Wiener Z. Kunst 1829(3):803. 1829.

Bas.: *Arum seguine* Jacq. (1760).

Syn.: *Dieffenbachia maculata* (Lodd.) G. Don (1839).

Robust herb in Northern Coastal Lowlands, Northern Limestone Hills (Ciales), Río Abajo Forest Reserve, Sierra de Luquillo, Eastern Coast (Cuchilla de Panduras), Sierra de Cayey, Southeastern Foothills (Coamo), Central Cordillera (Utuado), Western Cordillera (Maricao), Western Coast (Añasco); 10–550 m; *fl* Jun–Feb, *fr* (?).

Antilles, Trinidad and Tobago, Mexico, Central America, northern South America to Bolivia. [AR&S: 31].

Rábano, Rábano cimarrón.

Dracontium asperum K. Koch, Wochenschr. Gärtnerei Pflanzenk. 2:259. 1859.

Herb in Northern Limestone Hills, Río Abajo Forest Reserve, Sierra de Luquillo, Western Coast (Mayagüez); 50–325 m; *fl* Feb–Jun, *fr* Jul (one specimen).

Native to northeastern South America and Brazil. [AR&S: 33].

Notes: Britton & Wilson (1923) and Liogier & Martorell (1982, 2000) misapplied the name *Dracontium polyphyllum* L. to this taxon.

Guapa.

Epipremnum pinnatum 'Aureum' Nicolson, Allertonia 1(6):347. 1978.
Syn.: *Epipremnum aureum* (Linden & André) G.S. Bunting (1964).

Vine in Sierra de Luquillo, Northern Limestone Hills (Utuado); 250–350 m; *fl* Jan–Mar (two specimens), *fr* Feb (one specimen).

Only cultivated. [AR&S: 34; AR: 406].

Amapolo amarillo, Bejuco de agua, Trepapalo amarillo.

Landoltia punctata (G. Mey.) Les & D.J. Crawford, Novon 9(4):532. 1999.
Bas.: *Lemna punctata* G. Mey. (1818).
Syn.: *Spirodela punctata* (G. Mey.) C.H. Thomps. (1898).

Aquatic herb in Northern Coastal Lowlands, Northern Limestone Hills (Florida); 1–150 m; *fl* (?), *fr* (?).

Native to South America (scattered), Africa (scattered), temperate and tropical Asia (scattered), Australia, Pacific islands. [AR&S: 55].

Lemna aequinoctialis Welw., Ann. Cons. Ultramar. (Portugal), Parte Não Off., ser. 1 55:578. 1859 (Apont.).

Aquatic herb in Northern Coastal Lowlands, Northern Limestone Hills (Quebradillas), Eastern Coast (Humacao), Caguas Valley (Caguas), Southern Coastal Lowlands (Salinas), Southwestern Coastal Lowlands (Lajas), Sierra Bermeja; 1–120 m; *fl* (?), *fr* Apr (one specimen).

Cosmopolitan. [AR&S: 53].

Notes: Britton & Wilson (1923) and Liogier & Martorell (1982, 2000) misapplied the name *Lemna perpusilla* Torr. to this taxon.

Lentejilla de agua, Yerba de pato.

Lemna minuta Kunth in Humboldt, Bonpland & Kunth, Nov. Gen. Sp. 1:298 [folio]; 1:372 [quarto]. 1816.
Syn.: *Lemna minuscula* Herter (1954).

Aquatic herb in Northern Coastal Lowlands (Barceloneta), Eastern Cordillera (Cayey); 1–300 m; *fl* (?), *fr* (?). [*Liogier 33219* (UPR); *Proctor 46146* (SJ)].

Greater Antilles (Jamaica), Lesser Antilles (Guadeloupe), southern and western United States, Mexico, northern Central America (Guatemala), South America. [AR&S: 53].

Lemna valdiviana Phil., Linnaea 33(2):239. 1864.

Aquatic herb in Northern Coastal Lowlands (Carolina, Barceloneta); 1 m; *fl* (?), *fr* (?).

Greater Antilles (Jamaica, Hispaniola), Lesser Antilles (Guadeloupe, Martinique), Trinidad, United States, Mexico, southern Central America, South America. [AR&S: 54].

Monstera adansonii Schott, Wiener Z. Kunst 1830(4):1028. 1830.

Vine planted and persistent in Northern Coastal Lowlands (Río Grande), Sierra de Luquillo; 65–300 m; *fl* Feb (one specimen), *fr* (?). [*Acevedo-Rodríguez 9313* (UPR); *Proctor 50421* (SJ)].

Native to Lesser Antilles, Trinidad and Tobago, Central America, northern South America to Bolivia. [AR&S: 36; AR: 408].

Monstera, Piñanona.

Philodendron consanguineum Schott, Syn. Aroid. 88. 1856.
Syn.: *Philodendron angustatum* Schott (1856).
Philodendron krebsii Schott (1859).
Philodendron marginatum Urb. (1903).

Vine in Northern Coastal Lowlands, Northern Limestone Hills (Quebradillas), Río Abajo Forest Reserve, Sierra de Luquillo, Sierra de Cayey (San Lorenzo), Central Cordillera, Susúa Forest Reserve, Maricao Forest Reserve; 1–1100 m; *fl* Jun (one specimen), *fr* (?).

Greater Antilles (Cuba, Hispaniola), Virgin Islands (St. Thomas). [AR&S: 38; AR: 410].

Bejuco de calabazón, Calabazón, Rascagarganta.

Philodendron giganteum Schott, Syn. Aroid. 89. 1856.

Erect or climbing robust herb in Northern Limestone Hills (Hatillo), Río Abajo Forest Reserve, Sierra de Luquillo, Sierra de Cayey; 10–750 m; *fl* Oct–Nov, Feb–Apr, *fr* (?).

Greater Antilles (Hispaniola), Virgin Islands, Lesser Antilles, Trinidad, northern South America (Venezuela). [AR&S: 40; AR: 410].

Yautía cimarrona.

Philodendron hederaceum (Jacq.) Schott, Wiener Z. Kunst 1829(3):780. 1829 "Philodendrum."
Bas.: *Arum hederaceum* Jacq. (1760).
Syn.: *Philodendron scandens* K. Koch & Sello (1853).
Philodendron scandens K. Koch & Sello subsp. *scandens* f. *micans* (K. Koch) G.S. Bunting (1968).
Philodendron scandens K. Koch & Sello subsp. *insertianum* (Schott) G.S. Bunting (1968).

Vine in Northern Limestone Hills, Sierra de Luquillo, Western Cordillera, Southwestern Coastal Lowlands (Cabo Rojo); 25–300 m; *fl* May–Jun (two specimens), *fr* (?).

Antilles, Trinidad, Mexico, Central America, northern South America to Bolivia. [AR&S: 40; AR: 410].

Notes: Britton and Wilson (1923) misapplied the name *Philodendron oxycardium* Schott to this taxon.

Bejuco de calabazón, Paisaje.

Philodendron lingulatum (L.) K. Koch in Braun & Klotzsch, Append. Gen. Sp. Hort. Berol. 1855:2. 1855–1856.
Bas.: *Arum lingulatum* L. (1759).

Vine in Northern Limestone Hills (Ciales), Río Abajo Forest Reserve, Sierra de Luquillo, Sierra de Cayey (San Lorenzo); 150–500 m; *fl* (?), *fr* (?).

Antilles. [AR&S: 42; AR: 413].

Bejuco de calabaza, Calabazón cimarrón.

Philodendron ornatum Schott, Oesterr. Bot. Wochenbl. 3(48):378. 1853.

Vine planted and persistent in Northern Limestone Hills (Quebradillas, San Sebastián); 50–200 m; *fl* Mar (one specimen), *fr* (?). [*Woodbury s.n.* (SJ), not seen; *Proctor 50206* (SJ), not seen; *Liogier 31745* (UPR)].

Native to Trinidad and Tobago, northern South America and Brazil. [AR&S: 44; AR: 413].

Pistia stratiotes L., Sp. Pl. 2:963. 1753.

Aquatic herb in Northern Coastal Lowlands (San Juan), Northern Limestone Hills (Florida), Caguas Valley (Gurabo), Guánica Forest Reserve, Southwestern Coastal Lowlands (Lajas); 10–150 m; *fl* Sep–Nov (two specimens), *fr* Sep (one specimen).

Pantropical. [AR&S: 44].

Lamparilla, Lechuguilla de agua, Lechuga de río, Lechuguilla de río.

Spirodela polyrhiza (L.) Schleid., Linnaea 13(4):392. 1839.
Bas.: *Lemna polyrhiza* L. (1753).

Aquatic herb in Northern Coastal Lowlands (Vega Baja, Manatí), Western Coast (Aguada); 1–5 m; *fl* (?), *fr* (?).

Cosmopolitan. [AR&S: 55].

Syngonium podophyllum Schott, Bot. Zeitung (Berlin) 9(5):85. 1851.

Vine in Northern Coastal Lowlands, Northern Limestone Hills (Hatillo), Sierra de Luquillo, Eastern Coast (Naguabo), Sierra de Cayey, Central Cordillera (Utuado), Maricao Forest Reserve, Vieques; 5–800 m; *fl* Jan–Mar, Jun–Jul, Oct, *fr* (?).

Native to Mexico, Central America, northwestern, west-central South America and Venezuela. [AR&S: 46; AR: 415].

Malanga trepadora.

Wolffia brasiliensis Wedd., Ann. Sci. Nat., Bot., sér. 3 12:170. 1849.

Aquatic herb in Southwestern Coastal Lowlands (Cabo Rojo); 35 m; *fl* (?), *fr* (?). [*Proctor 43935* (SJ)].

Greater Antilles, Virgin Islands (St. Thomas), Lesser Antilles (Martinique), United States, Mexico, South America. [AR&S: 56].

Wolffiella welwitschii (Hegelm.) Monod, Mém. Soc. Hist. Nat. Afrique N., Hors Sér. 2:229. 1949.
Bas.: *Wolffia welwitschii* Hegelm. (1865).

Aquatic herb in Northern Coastal Lowlands; 1–50 m; *fl* (?), *fr* (?).

Greater Antilles, Central America (Nicaragua, Costa Rica), western South America (Ecuador, Bolivia), Africa. [AR&S: 57].

Notes: Liogier & Martorell (2000) used *Wolffia welwitschii* (Hegelm.) Daubs, possibly a confusion with the homonym *Wolffiella welwitschii* (Hegelm.) Daubs, for the name of this taxon.

Xanthosoma brasiliense (Desf.) Engl., Pflanzenr. IV, 23E(Heft 71):58. 1920.
Bas.: *Caladium brasiliense* Desf. (1829).

Robust herb planted and persistent in Northern Limestone Hills (Toa Baja), Río Abajo Forest Reserve; 40–300 m; *fl* May (one specimen), *fr* (?). [*Proctor 45337* (SJ); *Proctor 45337a, Acevedo-Rodríguez 10685* (US)].

Greater Antilles (Hispaniola), Lesser Antilles, South America (Brazil); exact native range uncertain. [AR&S: 49].

Belembé, Calalú, Yautía belembé.

Xanthosoma helleborifolium (Jacq.) Schott, Oesterr. Bot. Z. 15:33. 1865.
Bas.: *Arum helleborifolium* Jacq. (1791).

Herb planted and persistent in Río Abajo Forest Reserve; 450 m; *fl* May, *fr* (?). [*Monsegur 982* (MAPR)].

Native to Lesser Antilles, Trinidad, Central America, northern, northwestern South America and Brazil. [ARS: 49].

Xanthosoma sagittifolium (L.) Schott in Schott & Endlicher, Melet. Bot. 19. 1832 "sagittaefolium."
Bas.: *Arum sagittifolium* L. (1753) "sagittaefolium."
Syn.: *Xanthosoma atrovirens* K. Koch & C.D. Bouché (1854–1855).

Robust herb planted and persistent in Northern Limestone Hills (Arecibo), Río Abajo Forest Reserve, Western Coast (Aguada); 10–350 m; *fl* Aug–Oct (two specimens), *fr* Dec (one specimen).

Widely cultivated in the Neotropics; native range uncertain. [AR&S: 50].

Yautía, Yautía amarilla, Yautía blanca.

Xanthosoma undipes (K. Koch & C.D. Bouché) K. Koch, Bonplandia 4(1–2):4. 1856.
Bas.: *Alocasia undipes* K. Koch & C.D. Bouché (1854–1855).

Robust herb in Northern Limestone Hills (Florida, Arecibo), Eastern Cordillera (Corozal), Central Cordillera (Ciales), Vieques; 45–600 m; *fl* Mar–Apr, Aug, *fr* (?).

Lesser Antilles (Guadeloupe, Martinique, St. Vincent), Trinidad, Central America, northwestern, west-central South America and Brazil. [AR&S: 50].

Notes: Britton & Wilson (1923) misapplied the name *Xanthosoma jacquinii* Schott (1832, not 1856), which is a synonym of *Xanthosoma sagittifolium*, to this taxon.

Yautía de palma , Yautía rascana, Yautía rascosa.

Xanthosoma violaceum Schott, Oesterr. Bot. Wochenbl. 3(47):370. 1853.

Robust herb planted and persistent in Northern Limestone Hills (Arecibo), Sierra de Luquillo, Central Cordillera (Ciales, Utuado), Southern Coastal Lowlands (Patillas); 50–850 m; *fl* Oct–Dec, Mar, *fr* (?).

Native to Lesser Antilles (Guadeloupe, St. Lucia), Mexico, Central America, northern South America (Guyana). [AR&S: 51].

Yautía guayamera, Yautía lila, Yautía morada, Yautía prieta.

TAXA CITED BY PREVIOUS AUTHORS, BUT NOT INCLUDED

NO SPECIMENS SEEN

Syngonium auritum (L.) Schott.
Vine cited by Liogier & Martorell (2000) as occurring in shady forests at lower elevations. [AR&S: 48].

DOUBTFUL OCCURRENCE

Montrichardia arborescens (L.) Schott.
Vine cited by Britton & Wilson (1923) as occurring in Puerto Rico, according to Bello; subsequently cited by Liogier & Martorell (1982, 2000) as occurring in wet or moist places, borders of Río Blanco, Naguabo. [AR&S: 52].

CITED DUE TO MISIDENTIFICATION

Wolffiella lingulata (Hegelm.) Hegelm.
Aquatic herb cited by Liogier & Martorell (1982, 2000) as occurring in Florida and Aguada. Specimens identified as this taxon are referable to *Wolffiella welwitschii*.

ONLY UNDER CULTIVATION

Aglaonema commutatum Schott.
Herb cited by Liogier & Martorell (1982, 2000) as cultivated and escaped. [AR&S: 52].

Aglaonema pictum Kunth.
Herb cited by Liogier & Martorell (1982, 2000) as grown as an ornamental and escaped from cultivation. [AR&S: 52].

Epipremnum pinnatum (L.) Engl. Syn.: *Philodendron nechodomae* Britton.
Vine cited in Acevedo-Rodríguez & Strong (2005) as cultivated and perhaps naturalized. [AR: 406; AR&S: 34].

Monstera deliciosa Liebm.
Vine cited by Liogier & Martorell (1982, 2000) as occasionally planted in gardens and persistent after cultivation. [AR&S: 36].

Spathiphyllum wallisii Regel.
Herb cited in Acevedo-Rodríguez & Strong (2005) as cultivated. [AR&S: 46].

Xanthosoma caracu K. Koch & C.D. Bouché.
Robust herb cited by Liogier & Martorell (1982, 2000) as occasionally cultivated and probably naturalized. [AR&S: 51].

Zantedeschia aethiopica (L.) Spreng. Syn.: *Calla aethiopica* L.
Robust herb cited by Liogier & Martorell (1982, 2000) as occasionally cultivated in gardens and escaped in the mountains. [AR&S: 52].

ARALIACEAE

Dendropanax arboreus (L.) Decne. & Planch., Rev. Hort. (Paris), sér. 4 3:107. 1854 "arboreum."
Bas.: *Aralia arborea* L. (1759).

Tree in Northern Coastal Lowlands (Dorado), Northern Limestone Hills, Río Abajo Forest Reserve, Sierra de Luquillo, Sierra de Cayey, Central Cordillera, Western Cordillera, Western Coast (Mayagüez); 20–1000 m; *fl* & *fr* throughout the year.

Greater Antilles, Mexico, Central America, northwestern, west-central South America and Venezuela. [L4: 78].

Muñeca, Palo de cachimba, Palo de cachumba, Palo de pollo, Pollo, Víbona, Víbora.

Dendropanax laurifolius (Marchal ex Urb.) Decne.

& Planch. ex R.C. Schneid., Bull. Torrey Bot. Club 36(12):644. 1909 "laurifolium."
Bas.: *Gilibertia laurifolia* Marchal ex Urb. (1899).

Tree in Sierra de Luquillo, Sierra de Cayey, Las Tetas de Cayey, Eastern Cordillera (La Torrecilla), Central Cordillera, Maricao Forest Reserve; 450–1200 m; *fl* & *fr* Jan–Sep.

Endemic. [L4: 79].

Gongolí, Palo de cachumba, Palo de gangulín, Palo de pollo negro, Palo de vaca, Víbona, Víbora.

Hydrocotyle bonariensis Lam., Encycl. 3(1):153. 1789.

Herb in Northern Coastal Lowlands (San Juan), Eastern Coast (Humacao); 1–10 m; *fl* & *fr* Feb, Aug (two specimens). [*Woodbury a.n. 28817* (UPR); *Axelrod 2064* (UPRRP)].

Greater Antilles (Cuba), southeastern United States, Mexico, Central America, South America. [L4: 99].

Hydrocotyle hirsuta Sw., Prodr. 54. 1788.

Herb in Northern Limestone Hills, Río Abajo Forest Reserve, Sierra de Luquillo, Las Tetas de Cayey, Toro Negro Forest Reserve, Maricao Forest Reserve; 150–875 m; *fl* & *fr* throughout the year.

Bahamas, Greater Antilles (Cuba, Hispaniola), southern South America (Paraguay). [L4: 99].

Yerba de clavo.

Hydrocotyle pusilla A. Rich., Ann. Gén. Sci. Phys. 4:167, *t. 52*, *f. 2*. 1820.

Low herb in Central Cordillera; 1000–1250 m; *fl* & *fr* throughout the year.

Greater Antilles, Mexico, Central America (Nicaragua to Panama), southern South America to Brazil. [L4: 100].

Hydrocotyle umbellata L., Sp. Pl. 1:234. 1753.

Herb in Northern Coastal Lowlands (Toa Baja), Río Abajo Forest Reserve, Southeastern Foothills (Coamo), Southwestern Coastal Lowlands (Lajas); 1–350 m; *fl* & *fr* Mar, Jul.

Bahamas, Greater Antilles, Lesser Antilles (Guadeloupe, Dominica), eastern Canada and United States, Central America, northern and western South America. [L4: 100].

Ombligo de Venus, Sombrerillo de agua, Yerba de cuatro.

Hydrocotyle verticillata Thunb., Hydrocotyle 2, 5, *t. s.n. [2]*. 1798.

Herb in Northern Coastal Lowlands, Las Piedras Chiquitas, Southeastern Foothills (Coamo), Eastern Cordillera (Comerío), Maricao Forest Reserve; 1–500 m; *fl* & *fr* Jan–Oct.

West Indies, southern United States (Florida to Texas), Central America (Nicaragua), southern South America to Brazil. [L4: 100].

Polyscias guilfoylei (W. Bull) L.H. Bailey, Rhodora 18:153. 1916.
Bas.: *Aralia guilfoylei* W. Bull (1873).

Shrub in Northern Coastal Lowlands (Manatí); 20 m; *fl* (?), *fr* (?). [*Axelrod 11547* (UPRRP)].

Cultivated in Malesia and tropical Oceania, but of unknown origin. [L4: 82].

Gallego.

Schefflera gleasonii (Britton & P. Wilson) Alain, Phytologia 58(5):328. 1985.
Bas.: *Didymopanax gleasonii* Britton & P. Wilson (1926).

Shrub/small tree in Sierra de Luquillo, Central Cordillera, Maricao Forest Reserve; 725–1300 m; *fl* & *fr* Dec–Sep.

Endemic. [L4: 83].

Yagrumillo, Yagrumo macho pequeño, Yuquilla.

Schefflera morototoni (Aubl.) Maguire, Steyerm. & Frodin, Mem. New York Bot. Gard. 38:51, *f. 5E-G*. 1984.
Bas.: *Panax morototoni* Aubl. (1775).
Syn.: *Didymopanax morototoni* (Aubl.) Decne. & Planch. (1854).

Tree in Northern Coastal Lowlands (Guaynabo, Laguna Tortuguero), Guajataca Forest Reserve, Río Abajo Forest Reserve, Sierra de Luquillo, Sierra de Cayey, Eastern Cordillera (Cerro La Peña), Maricao Forest Reserve; 1–600 m; *fl* Oct–May, *fr* Aug–Apr.

Greater Antilles (Cuba, Hispaniola), Virgin Islands, Lesser Antilles (Guadeloupe), Trinidad, Mexico, Central America, South America. [L4: 83].

Grayumo macho, Llagrumo macho, Pana cimarrona, Yagrumo macho.

TAXA CITED BY PREVIOUS AUTHORS, BUT NOT INCLUDED

ONLY UNDER CULTIVATION

Brassaia actinophylla Endl.

Tree cited by Liogier & Martorell (1982, 2000) and Liogier (1995) as planted as an ornamental and persistent. [L4: 76].

Polyscias cumingiana (C. Presl) Fern.-Vill. Syn.: *Polyscias filicifolia* (C. Moore ex E. Fourn.) L.H. Bailey.

Tree cited by Liogier & Martorell (2000) and Liogier (1995) as cultivated and persistent. [L4: 81].

Polyscias fruticosa (L.) Harms.
Tree cited by Liogier & Martorell (1982, 2000) and Liogier (1995) as cultivated and persistent. [L4: 81].

Polyscias scutellaria (Burm. f.) Fosberg. Syn.: *Polyscias pinnata* J.R. Forst. & G. Forst. *Polyscias balfouriana* (hort. ex André) L.H. Bailey.
Tree cited by Liogier & Martorell (1982, 2000) and Liogier (1995) as cultivated and persistent. [L4: 82].

ARECACEAE (PALMAE)

Acrocomia aculeata (Jacq.) Lodd. ex Mart., Hist. Nat. Palm. 3(8):286. 1845.
Bas.: *Cocos aculeatus* Jacq. (1763).
Syn.: *Acrocomia media* O.F. Cook (1901).

Tree in Northern Coastal Lowlands (Laguna Tortuguero, Manatí), Las Piedras Chiquitas, Vieques; 5–500 m; *fl* Jun (one specimen), *fr* Oct, Mar (two specimens).

Lesser Antilles, Mexico, Central America, South America (except Ecuador, Peru). [AR&S: 137].

Notes: The above distribution and phenology largely reflect the inconvenience of collection of this palm.

Corozo, Palma de corozo.

Aiphanes minima (Gaertn.) Burret, Notizbl. Bot. Gart. Berlin-Dahlem 11(107):558. 1932.
Bas.: *Bactris minima* Gaertn. (1791).
Syn.: *Bactris acanthophylla* Mart. (1844). *Aiphanes acanthophylla* (Mart.) Burret (1932).

Tree in Northern Limestone Hills (Toa Baja, Dorado), Río Abajo Forest Reserve; 15–350 m; *fl* Apr–May (two specimens), *fr* May–Jun (one specimen).

Greater Antilles (Hispaniola), Lesser Antilles. [AR&S: 138].

Coyor, Coyura, Coyore, Coyure, Palma de coyor.

Calyptronoma rivalis (O.F. Cook) L.H. Bailey, Gentes Herb. 4:171. 1938.
Bas.: *Cocops rivalis* O.F. Cook (1901).

Tree in Northern Limestone Hills (Hatillo, Camuy); 175–200 m; *fl* Apr, Oct, *fr* Apr (one specimen).

Greater Antilles (Hispaniola). [AR&S: 140].

Notes: Britton & Wilson (1923) misapplied the name *Calyptrogyne occidentalis* (Sw.) M. Gómez to this taxon; that name is a synonym of a similar taxon, *Calyptronoma occidentalis* (Sw.) H.E. Moore, found only in Jamaica.

This taxon is on the U.S. Fish & Wildlife Service list of threatened and endangered plants.

Manaca, Palma manaca, Palmilla.

Coccothrinax barbadensis (Lodd. ex Mart.) Becc., Webbia 2:328. 1907.
Bas.: *Thrinax barbadensis* Lodd. ex Mart. (1853).
Syn.: *Coccothrinax alta* (O.F. Cook) Becc. (1907).

Tree in Northern Coastal Lowlands (Florida), Northern Limestone Hills (Florida, Quebradillas), Río Abajo Forest Reserve, Southern Coastal Limestone (Guayanilla), Southwestern Coastal Lowlands (Cabo Rojo), Vieques; 1–350 m; *fl* Dec (one specimen), *fr* Jun, Sep–Dec.

Virgin Islands, Lesser Antilles, Trinidad and Tobago. [AR&S: 141].

Notes: Britton & Wilson (1923) misapplied the name *Coccothrinax argentea* (Lodd.) Sarg. to this taxon.

Palma de abanico, Palma plateada.

Cocos nucifera L., Sp. Pl. 2:1188. 1753.

Tree in Northern Coastal Lowlands (islands adjacent to Fajardo, Río Grande), Western Coast (Mayagüez), Vieques, Mona; 1 m; *fl* Aug (two specimens), *fr* Feb (one specimen).

Native to the shores of the Indian and Pacific Oceans. [AR&S: 142].

Notes: The above distribution and phenology largely reflect the inconvenience of collection of this palm.

Coco, Cocotero, Palma de coco.

Gaussia attenuata (O.F. Cook) Becc., Pomona Coll. J. Econ. Bot. 2:275. 1912.
Bas.: *Aeria attenuata* O.F. Cook (1901).

Tree in Northern Limestone Hills (Manatí), Southwestern Coastal Lowlands (Cabo Rojo); 15 m; *fl* Aug (one specimen), *fr* Aug–Sep (two specimens).

Greater Antilles (Hispaniola). [AR&S: 143].

Llume, Palma de lluvia.

Prestoea acuminata (Willd.) H.E. Moore var. **montana** (Graham) A.J. Hend. & Galeano, Fl. Neotrop. 72:53. 1996.
Bas.: *Euterpe montana* Graham (1841).
Syn.: *Prestoea montana* (Graham) G. Nicholson (1886).

Tree in Sierra de Luquillo, Sierra de Cayey, Eastern Cordillera (Guaynabo), Central Cordillera (Utuado), Maricao Forest Reserve; 475–800 m; *fl* Feb, Jun, Sep, *fr* Sep–Mar, Jun.

Greater Antilles (Cuba, Hispaniola), Lesser Antilles. [AR&S: 144].

Notes: Britton & Wilson (1923) misapplied the name *Euterpe globosa* Gaertn. to this taxon.

Palma de sierra.

Pseudophoenix sargentii H. Wendl. ex Sarg., Bot. Gaz. 11(11):314. 1886.

Tree in Mona; 5 m; *fl* Sep (one specimen), *fr* Sep–Oct (one specimen). [*Proctor 45905* (SJ)].

Bahamas, Greater Antilles (Cuba, Hispaniola), Lesser Antilles (Dominica), southeastern United States (Florida), Mexico, northern Central America (Belize). [AR&S: 145].

Roystonea borinquena O.F. Cook, Bull. Torrey Bot. Club 28(10):552. 1901.

Tree in Guajataca Forest Reserve, Vieques; 200–250 m; *fl* Aug (one specimen), *fr* (?).

Greater Antilles (Hispaniola), Virgin Islands (St. Croix, St. John). [AR&S: 147].

Notes: The above distribution and phenology largely reflect the inconvenience of collection of this palm.

Palma real, Palma real puertorriqueña, Palma de costa, Palma de yaguas.

Sabal causiarum (O.F. Cook) Becc., Webbia 2:71. 1907.

Bas.: *Inodes causiarum* O.F. Cook (1901).

Tree in Northern Coastal Lowlands (Isabela), Western Coast (Cabo Rojo), Mona (planted and persisent); 1–5 m; *fl* Apr–Jul, *fr* (?).

Greater Antilles (Hispaniola), Virgin Islands (St. Croix). [AR&S: 149].

Notes: I have seen a population of this palm in the Western Coast at the southern end of Laguna Joyuda, Cabo Rojo, but no collection has been made from there since 1913 [*Hess 3061* (MAPR)].

Palma de abanico, Palma de cogollo, Palma de escoba, Palma de sombrero, Yarey.

Thrinax morrisii H. Wendl., Gard. Chron., ser. 3 11:104, *tt. 20–21*. 1892.

Syn.: *Thrinax microcarpa* Sarg. (1896).

Tree in Northern Limestone Hills (Vega Alta, Isabela), Río Abajo Forest Reserve, Southern Coastal Limestone, Susúa Forest Reserve, Guánica Forest Reserve, Southwestern Coastal Lowlands (Cabo Rojo), Sierra Bermeja, Vieques, Mona; 10–350 m; *fl* Sep–Jun, *fr* throughout the year.

Bahamas, Greater Antilles (Cuba, Hispaniola), Virgin Islands (Anegada), Lesser Antilles (Anguilla, Barbuda), southeastern United States (Florida Keys). [AR&S: 151].

Palma de cojollo, Palma de escoba, Pandereta, Yaray.

TAXA CITED BY PREVIOUS AUTHORS, BUT NOT INCLUDED

ONLY UNDER CULTIVATION

Dypsis lutescens (H. Wendl.) Beentje & J. Dransf.

Syn.: *Chrysalidocarpus lutescens* H. Wendl.

Tree cited by Liogier & Martorell (1982, 2000) as grown as an ornamental and naturalized.

Roystonea regia (Kunth) O.F. Cook.

Tree cited by Liogier & Martorell (1982, 2000) as planted as an ornamental and locally naturalized.

ARISTOLOCHIACEAE

Aristolochia elegans Mast., Gard. Chron., n.s. 24(610):301, *t. 64*. 1885.

Vine in Vieques; 5–20 m; *fl* Oct (one specimen), *fr* (?). [*Breckon 6697* (MAPR)].

Native to South America. [L1: 112; AR: 66].

Notes: Liogier & Martorell (1982) and Liogier (1985) misapplied the name *Aristolochia littoralis* D. Parodi to this taxon.

Aristolochia oblongata Jacq. subsp. **calceiformis** (Urb.) R. Rankin & Acev.-Rodr., Contr. U.S. Natl. Herb. 51:70. 2005.

Bas.: *Aristolochia calceiformis* Urb. (1899).

Vine in Río Abajo Forest Reserve, Sierra de Cayey, Las Tetas de Cayey, Southeastern Foothills (Salinas), Central Cordillera (Ciales, Utuado); 250–825 m; *fl* Feb–Sep, *fr* Feb, May–Sep.

Greater Antilles (Cuba, Hispaniola). [L1: 112; AR: 70].

Notes: Liogier & Martorell (1982, 2000) and Liogier (1985) misapplied the name *Aristolochia bilabiata* L. to this taxon.

Aristolochia ringens Vahl, Symb. Bot. 3:99. 1794.

Vine in Western Cordillera (Maricao), Cerro Las Mesas; 500 m; *fl* Jul (one specimen), *fr* (?). [*Breckon 7048* (MAPR)].

Native to northwestern, west-central South America and Venezuela. [L1: 113; AR: 71].

Gallito, Panitos, Pelicanos.

Aristolochia trilobata L., Sp. Pl. 2:960. 1753.

Vine in Northern Coastal Lowlands, Northern Limestone Hills (Quebradillas), Eastern Coast (Cayo Santiago), Southern Coastal Lowlands (Caja de Muertos), Culebra; 1–10 m; *fl* May–Aug, *fr* May, Oct (two specimens).

Antilles, Bermuda, Central America, northern South America. [L1: 114; AR: 71].

Bejuco de Santiago, Cachimbo, Panitos.

ASPARAGACEAE

Asparagus aethiopicus L., Mant. Pl. 1:63. 1767.

Vine in Northern Limestone Hills (Manatí), Central Cordillera (Utuado); 75–900 m; *fl* Apr–Jun, *fr* Feb–Aug.

Native to South Africa. [AR&S: 126].

Cabellera de la reina, Canastillo, Espárrago de jardín.

Asparagus setaceus (Kunth) Jessop, Bothalia 9:51. 1966.

Bas.: *Asparagopsis setacea* Kunth (1850).

Syn.: *Protasapargus setaceus* (Kunth) Oberm. (1983).

Vine planted and persistent in Northern Coastal Lowlands (Guaynabo), Northern Limestone Hills (Ciales); 30–300 m; *fl* (?), *fr* Dec–Feb.

Native to South Africa. [AR&S: 127; AR: 418].

Abeto, Ala de pájaro, Helecho plumoso.

ASPHODELACEAE

Aloe vera (L.) Burm. f., Fl. Indica 83. 1768.

Bas.: *Aloe perfoliata* L. var. *vera* L. (1753).

Syn.: *Aloe vulgaris* Lam. (1783).

Robust herb in Southern Coastal Lowlands (Caja de Muertos), Guánica Forest Reserve, Southwestern Coastal Lowlands, Mona; 1–25 m; *fl* Jan–Apr, *fr* (?).

Native to Spain (Canary Islands). [AR&S: 134].

Sábila, Zábila.

ASTERACEAE (COMPOSITAE)

Acanthospermum hispidum DC., Prodr. 5:522. 1836.

Herb in Northern Coastal Lowlands (Laguna Tortuguero, Manatí), Sierra de Cayey, Southeastern Foothills, Southern Coastal Lowlands (Coamo), Southwestern Coastal Lowlands, Vieques; 1–700 m; *fl* & *fr* Aug–May.

Antilles, Tobago, Central America, South America. [L5: 231].

Acmella iodiscaea (A.H. Moore) R.K. Jansen, Syst. Bot. Monogr. 8:53. 1985.

Bas.: *Spilanthes iodiscaea* A.H. Moore (1907).

Herb in Río Abajo Forest Reserve, Sierra de Luquillo; 325–1050 m; *fl* & *fr* throughout the year.

Bahamas, Greater Antilles, Lesser Antilles (Barbuda). [L5: 233].

Maluco.

Acmella pusilla (Hook. & Arn.) R.K. Jansen, Syst. Bot. Monogr. 8:46. 1985.

Bas.: *Spilanthes pusilla* Hook. & Arn. (1841).

Herb in Northern Limestone Hills (Quebradillas); 10–90 m; *fl* & *fr* Oct–Dec, Apr.

Native to southern South America. [L5: 235].

Adenostemma verbesina (L.) Sch. Bip., J. Bot. 1: 235. 1863.

Bas.: *Cotula verbesina* L. (1760).

Herb in Sierra de Luquillo; 500–900 m; *fl* & *fr* throughout the year.

Greater Antilles, Lesser Antilles (St. Kitts, St. Vincent), South America. [L5: 235].

Notes: Britton & Wilson (1925) used the homonym *Adenostemma verbesina* (L.) Kuntze for the name of this taxon.

Ageratina resiniflua (Urb.) R.M. King & H. Rob., Phytologia 19(4):225. 1970.

Bas.: *Eupatorium resinifluum* Urb. (1899).

Shrub in Northern Limestone Hills (Arecibo), Central Cordillera, Maricao Forest Reserve, Cerro Las Mesas; 250–1250 m; *fl* & *fr* Sep–May.

Endemic. [L5: 293].

Ageratum conyzoides L., Sp. Pl. 2:839. 1753.

Suffrutescent herb in Northern Coastal Lowlands (Laguna Tortuguero), Northern Limestone Hills (Florida), Sierra de Luquillo, Eastern Coast (Cuchilla de Panduras), Caguas Valley, Sierra de Cayey, Las Tetas de Cayey, Las Piedras Chiquitas, Eastern Cordillera (Cañón San Cristóbal), Central Cordillera, Vieques; 1–1200 m; *fl* & *fr* Sep–Apr.

Probably native to the Neotropics. [L5: 237].

Mentastro, Yerba de cabrío.

Ambrosia artemisiifolia L., Sp. Pl. 2:988. 1753.

Herb in Northern Coastal Lowlands (Laguna Tortuguero), Southwestern Coastal Lowlands (Guánica); 1–30 m; *fl* & *fr* Sep–Oct (two specimens). [*Woodbury a.n. 3817* (UPR); *Liogier 29499* (UPR)].

Native to North America.

Ambrosia confertiflora DC., Prodr. 5:526. 1836.

Herb in Northern Coastal Lowlands (Fajardo), Eastern Coast (Cuchilla de Panduras), Southern Coastal Lowlands (Salinas), Southern Coastal Limestone (Peñuelas), Southwestern Coastal Lowlands; 20–400 m; *fl* & *fr* Apr, Aug–Sep.

Native to southwestern and central United States, Mexico. [L5: 239].

Ambrosia hispida Pursh, Fl. Amer. Sept. 2:743. 1813.

Herb in Northern Coastal Lowlands, Guánica Forest Reserve; 1–10 m; *fl* & *fr* May–Jun (two specimens).

West Indies, Mexico, northern Central America (Belize). [L5: 239].

Altamisa de playa.

Ambrosia peruviana Willd., Sp. Pl. 4(1):377. 1805.
Suffrutescent herb in Northern Coastal Lowlands, Northern Limestone Hills, Eastern Coast (Cuchilla de Panduras), Central Cordillera (Utuado), Southern Coastal Lowlands (Juana Díaz), Southwestern Coastal Lowlands (Guánica); 1–900 m; *fl* & *fr* Apr–Dec.
West Indies, Mexico, Central America, western South America and Brazil. [L5: 241].
Altamisa, Artemisa.

Ambrosia tenuifolia Spreng., Syst. Veg. 3:851. 1826.
Herb in Northern Coastal Lowlands (Laguna Tortuguero); 1–5 m; *fl* & *fr* Sep–Oct (one specimen). [*Woodbury a.n. 3817* (UPR)].
Native to southeastern and south-central South America. [L5: 241].

Aphanostephus ramosissimus DC., Prodr. 5:310. 1836.
Herb in Vieques; 10 m; *fl* & *fr* Feb (one specimen). [*Breckon 9224* (MAPR)].
Native to southwestern and south-central United States, Mexico.

Baccharis myrsinites (Lam.) Pers., Syn. Pl. 2(2):424. 1807.
Bas.: *Conyza myrsinites* Lam. (1786).
Shrub in Sierra de Luquillo; 1000 m; *fl* & *fr* Oct (two specimens). [*Liogier 33310, 36159* (UPR); *Vives a.n. 6755* (MAPR)].
Greater Antilles (Hispaniola). [L5: 245].

Berylsimpsonia vanillosma (C. Wright) B.L. Turner, Phytologia 74(5):352. 1993.
Bas.: *Proustia vanillosma* C. Wright (1870).
Syn.: *Proustia krugiana* Urb. (1899).
Woody vine/shrub in Northern Limestone Hills (Ciales), Las Tetas de Cayey, Las Piedras Chiquitas, Southeastern Foothills, Eastern Cordillera (Cayey, Cañón San Cristóbal), Central Cordillera, Western Cordillera (Maricao); 200–975 m; *fl* & *fr* Oct–Apr.
Greater Antilles (Cuba, Hispaniola). [L5: 339; AR: 93].

Bidens alba (L.) DC. var. **radiata** (Sch. Bip.) R.E. Ballard ex Melchert, Phytologia 32(4):295. 1975.
Bas.: *Bidens pilosa* L. var. *radiata* Sch. Bip. (1844).
Herb in Northern Coastal Lowlands, Northern Limestone Hills, Río Abajo Forest Reserve, Sierra de Luquillo, Eastern Coast, Sierra de Cayey, Las Tetas de Cayey, Eastern and Central Cordillera, Susúa Forest Reserve, Maricao Forest Reserve, Southwestern Coastal Lowlands (Lajas), Cerro Las Mesas, Western Coast (Mayagüez), Vieques, Mona; 1–1200 m; *fl* & *fr* throughout the year.
West Indies, southeastern United States (Florida), Mexico, Central America (Guatemala, Honduras, Costa Rica), northern South America and Peru. [L5: 246].
Notes: Britton & Wilson (1925) applied, in part, the name *Bidens pilosa* to this taxon.
Margarita, Romerillo.

Bidens cynapiifolia Kunth in Humboldt, Bonpland & Kunth, Nov. Gen. Sp. 4:185 [folio]. 1818; 4:235 [quarto]. 1820.
Syn.: *Bidens bipinnata* L. var. *cynapiifolia* (Kunth) M. Gómez (1890).
Herb in Northern Coastal Lowlands (islands adjacent to Fajardo), Northern Limestone Hills (Florida, Lares), Eastern Coast (Cuchilla de Panduras), Las Piedras Chiquitas, Southeastern Foothills, Central Cordillera, Cerro de Las Cuevas, Southern Coastal Lowlands, Susúa Forest Reserve, Southwestern Coastal Lowlands (Cabo Rojo), Sierra Bermeja, Vieques, Culebra, Mona; 75–800 m; *fl* & *fr* throughout the year.
West Indies, southern Mexico (Yucatan), southern Central America, South America. [L5: 247].
Alfilerillo, Manzanilla negra, Margarita blanca.

Bidens pilosa L., Sp. Pl. 2:832. 1753.
Herb in Northern Limestone Hills (Quebradillas), Southeastern Foothills, Central Cordillera; 100–1050 m; *fl* & *fr* Aug–May.
Pantropical and -subtropical. [L5: 247].

Bidens reptans (L.) G. Don in Sweet, Hort. Brit. (ed. 3) 360. 1839.
Bas.: *Coreopsis reptans* L. (1759).
Vine in Northern Limestone Hills, Río Abajo Forest Reserve, Sierra de Cayey, Las Tetas de Cayey, Las Piedras Chiquitas, Southeastern Foothills, Central Cordillera, Susúa Forest Reserve, Western Cordillera; 35–925 m; *fl* & *fr* Oct–Jun.
Antilles, Central America, South America (Venezuela, Brazil, Bolivia). [L5: 249; AR: 94].
Manzanilla trepadora.

Bidens urbanii Greenm., Publ. Field Columb. Mus., Bot. Ser. 2(6):271. 1907.
Syn.: *Bidens reptans* (L.) G. Don var. *urbanii* (Greenm.) O.E. Schulz (1911).
Vine in Guajataca Forest Reserve, Maricao Forest Reserve; 250–950 m; *fl* & *fr* Nov–Jan.
Greater Antilles (Cuba), Mexico. [L5: 249; AR: 94].

Borrichia arborescens (L.) DC., Prodr. 5:489. 1836.
Bas.: *Buphthalmum arborescens* L. (1759).

Shrub in Northern Coastal Lowlands, Southwestern Coastal Lowlands (Cabo Rojo), Vieques, Culebra, Mona; 1–20 m; *fl* & *fr* throughout the year.

West Indies, southeastern United States (Florida), Mexico, northern Central America (Belize, Honduras). [L5: 250].

Clavelón de playa.

Calyptocarpus vialis Less., Syn. Gen. Compos. 221. 1832.

Spreading herb in Northern Coastal Lowlands (San Juan), Northern Limestone Hills (Hatillo, Quebradillas); 5–80 m; *fl* & *fr* Feb–Mar, Jun–Aug.

Native to Mexico, northern Central America (Honduras, El Salvador).

Centratherum punctatum Cass. in Cuvier, Dict. Sci. Nat. (ed. 2) 7:384. 1817.

Suffrutescent herb in Northern Coastal Lowlands (Manatí), Northern Limestone Hills, Eastern Coast (Cuchilla de Panduras), Sierra de Cayey, Eastern and Central Cordillera, Western Cordillera (Maricao), Vieques; 10–925 m; *fl* & *fr* throughout the year.

Native to Trinidad, southern Central America (Panama), South America. [L5: 252].

Chaptalia albicans (Sw.) Vent. ex B.D. Jacks., Index Kew. 1(1):506. 1893.
Bas.: *Tussilago albicans* Sw. (1788).

Herb in Northern Limestone Hills; 20–300 m; *fl* & *fr* Mar–Nov.

Bahamas, Greater Antilles (Cuba, Hispaniola), southeastern United States (Florida), Mexico, Central America. [L5: 253].

Notes: Liogier (1997) and Liogier & Martorell (2000) used *Chaptalia albicans* (Sw.) Vent. ex Steud., a *nom. inval.*, for the name of this taxon.

Chaptalia dentata (L.) Cass. in Cuvier, Dict. Sci. Nat. (ed. 2) 26:104. 1823.
Bas.: *Tussilago dentata* L. (1763).

Herb in Northern Limestone Hills (San Sebastián, Rincón), Las Piedras Chiquitas, Southern Coastal Limestone (Peñuelas), Guánica Forest Reserve; 100–500 m; *fl* & *fr* Sep–Apr.

Bahamas, Greater Antilles (Cuba, Hispaniola). [L5: 254].

Chaptalia nutans (L.) Pol., Linnaea 41(7):582. 1878.
Bas.: *Tussilago nutans* L. (1759).

Herb in Northern Coastal Lowlands (San Juan), Northern Limestone Hills, Sierra de Luquillo, Eastern Coast, Sierra de Cayey, Las Tetas de Cayey, Southeastern Foothills (Arroyo), Eastern and Central Cordillera, Maricao Forest Reserve; 25–1250 m; *fl* & *fr* throughout the year.

Antilles, Trinidad and Tobago, southern United States, Mexico, Central America, South America. [L5: 256].

Bretónica.

Chromolaena borinquensis (Britton) R.M. King & H. Rob., Phytologia 20(3):199. 1970.
Bas.: *Osmia borinquensis* Britton (1925).
Syn.: *Eupatorium borinquense* (Britton) B.L. Rob. (1928).

Woody vine in Northern Coastal Lowlands, Northern Limestone Hills, Río Abajo Forest Reserve; 1–400 m; *fl* & *fr* Dec–Jun.

Endemic. [L5: 287; AR: 96].

Chromolaena corymbosa (Aubl.) R.M. King & H. Rob., Phytologia 20(3):200. 1970.
Bas.: *Eupatorium corymbosum* Aubl. (1775).
Syn.: *Osmia corymbosa* (Aubl.) Britton & P. Wilson (1925).
Eupatorium oteroi Monach. (1948). *Chromolaena oteroi* (Monach.) R.M. King & H. Rob. (1970).

Shrub in Northern Coastal Lowlands (Isabela), Eastern Coast, Mona; 2–60 m; *fl* & *fr* May–Feb.

Bahamas, Greater Antilles (Hispaniola), Virgin Islands, Lesser Antilles. [L5: 287, 292].

Notes: There are populations of this taxon on Mona Island with non-punctate, serrate leaves that were described as *Eupatorium oteroi*; however, the inflorescences of these populations are indistinguishible from those of *Chromolaena corymbosa*. Moreover, this taxon is highly variable on Mona Island: it also includes populations with punctate, serrate leaves, as well as with non-punctate, crenate leaves.

Chromolaena geraniifolia (Urb.) R.M. King & H. Rob., Phytologia 20(3):201. 1970 "geranifolia."
Bas.: *Eupatorium geraniifolium* Urb. (1899) "geranifolium."
Syn.: *Osmia geraniifolia* (Urb.) Britton (1925).

Shrub in Sierra de Cayey, Las Tetas de Cayey, Eastern Cordillera (Aibonito), Central Cordillera, Maricao Forest Reserve; 500–1000 m; *fl* & *fr* Nov–Jul.

Endemic. [L5: 288].

Chromolaena ivifolia (L.) R.M. King & H. Rob., Phytologia 20(3):202. 1970 "ivaefolia."
Bas.: *Eupatorium ivifolium* L. (1759) "ivaefolium."
Syn.: *Osmia ivifolia* (L.) Sch. Bip. (1866).

Shrub in Cerro Las Mesas; 300 m; *fl* & *fr* Sep (one specimen). [*Proctor 42181* (SJ)].

Greater Antilles, Lesser Antilles (Guadeloupe, Martinique), southern United States, Mexico, Central America, South America. [L5: 290].

Chromolaena odorata (L.) R.M. King & H. Rob., Phytologia 20(3):204. 1970.
Bas.: *Eupatorium odoratum* L. (1759).
Syn.: *Osmia odorata* (L.) Sch. Bip. (1866).

Shrub in Northern Coastal Lowlands, Northern Limestone Hills, Río Abajo Forest Reserve, Sierra de Luquillo, Eastern Coast, Caguas Valley (Caguas), Sierra de Cayey, Las Tetas de Cayey, Southeastern Foothills, Eastern Cordillera (Cerro La Peña, La Torrecilla), Central Cordillera, Southern Coastal Lowlands, Western Cordillera, Guánica Forest Reserve, Southwestern Coastal Lowlands, Sierra Bermeja, Western Coast (Mayagüez), Vieques, Culebra, Mona, Desecheo; 10–1300 m; *fl* & *fr* Nov–Jun.

West Indies, southern United States (Florida, Texas), Mexico, Central America, South America. [L5: 291].

Cariaquillo Santa María, Santa María.

Chromolaena sinuata (Lam.) R.M. King & H. Rob., Phytologia 32(3):283. 1975.
Bas.: *Eupatorium sinuatum* Lam. (1788).
Syn.: *Osmia sinuata* (Lam.) Britton & P. Wilson (1925).

Shrub in Southern Coastal Limestone, Susúa Forest Reserve, Maricao Forest Reserve, Guánica Forest Reserve, Sierra Bermeja; 40–725 m; *fl* & *fr* Nov–Jun.

Antilles. [L5: 294].

Cirsium mexicanum DC., Prodr. 6:636. 1838.

Herb in Northern Coastal Lowlands, Northern Limestone Hills, Río Abajo Forest Reserve, Caguas Valley (Caguas), Sierra de Cayey, Southeastern Foothills, Eastern Cordillera (Aibonito), Central Cordillera, Susúa Forest Reserve, Maricao Forest Reserve, Southwestern Coastal Lowlands (Cabo Rojo), Western Coast (Mayagüez); 1–1300 m; *fl* & *fr* Jan–Jun.

Greater Antilles (Cuba, Hispaniola), Mexico, Central America. [L5: 258].

Cardo.

Clibadium erosum (Sw.) DC., Prodr. 5:506. 1836.
Bas.: *Trixis erosa* Sw. (1788).

Shrub/small tree in Sierra de Luquillo, Sierra de Cayey, Central Cordillera; 250–1335 m; *fl* & *fr* throughout the year.

Lesser Antilles, southern Central America. [L5: 260].

Cachimbo, Carruzo, Tuchima, Turma de toro.

Condylidium iresinoides (Kunth) R.M. King & H. Rob., Phytologia 24(5):381. 1972.
Bas.: *Eupatorium iresinoides* Kunth (1818).

Shrub in Las Tetas de Cayey, Las Piedras Chiquitas, Southeastern Foothills, Central Cordillera (Villalba); 225–550 m; *fl* & *fr* Nov–Jun.

Greater Antilles (Cuba), Lesser Antilles, Central America, northern and western South America to Bolivia. [L5: 289].

Conyza bonariensis (L.) Cronquist, Bull. Torrey Bot. Club 70(6):632. 1943.
Bas.: *Erigeron bonariensis* L. (1753).
Syn.: *Leptilon bonariense* (L.) Small (1903).

Herb in Las Tetas de Cayey, Las Piedras Chiquitas, Southeastern Foothills, Eastern Cordillera (Aibonito), Central Cordillera, Desecheo; 25–900 m; *fl* & *fr* Jun–Dec.

Probably native to the Neotropics. [L5: 262].

Conyza canadensis (L.) Cronquist, Bull. Torrey Bot. Club 70(6):632. 1943.
Bas.: *Erigeron canadensis* L. (1753).
Syn.: *Leptilon pusillum* (Nutt.) Britton (1914).

Herb in Northern Coastal Lowlands, Northern Limestone Hills, Río Abajo Forest Reserve, Sierra de Luquillo, Eastern Coast, Sierra de Cayey, Las Piedras Chiquitas, Eastern and Central Cordillera, Maricao Forest Reserve, Western Coast (Aguada), Mona, Desecheo; 1–1250 m; *fl* & *fr* Jul–Apr.

Native to Canada, United States, Mexico, Central America. [L5: 264].

Mil hojas, Orozús, Pascueta, Yerba de mil hojas.

Conyza laevigata (Rich.) Pruski, Brittonia 50(4):475. 1998.
Bas.: *Erigeron laevigatus* Rich. (1792).
Syn.: *Conyza apurensis* Kunth (1818).
Leptilon chinense (Jacq.) Britton (1925).

Herb in Northern Limestone Hills (Florida, Arecibo), Sierra de Luquillo, Eastern Coast (Cuchilla de Panduras), Caguas Valley (Caguas), Eastern and Central Cordillera, Susúa Forest Reserve, Western Cordillera, Southwestern Coastal Lowlands (Lajas), Western Coast (Mayagüez); 150–1250 m; *fl* & *fr* Jan–Oct.

Greater Antilles (Cuba, Hispaniola), Lesser Antilles, Mexico, Central America, South America. [L5: 262].

Chiriota, Manzanilla silvestre.

Conyza primulifolia (Lam.) Cuatrec. & Lourteig, Phytologia 58(7):475. 1985 "primulaefolia."

Bas.: *Inula primulifolia* Lam. (1789) "primulaefolia."
Syn.: *Conyza chilensis* Spreng. (1818).

Herb in Central Cordillera, Maricao Forest Reserve; 800–950 m; *fl* & *fr* Jan, Apr–May.

Greater Antilles (Hispaniola), Mexico, Central America, South America. [L5: 264].

Cosmos caudatus Kunth in Humboldt, Bonpland & Kunth, Nov. Gen. Sp. 4:188 [folio]. 1818; 4:240 [quarto]. 1820.

Herb in Northern Coastal Lowlands, Eastern Cordillera (Barranquitas), Central Cordillera (Peñuelas); 50–600 m; *fl* & *fr* Nov–Feb, Aug.

Neotropical. [L5: 265].

Clavelillo, Margarita, Piquete.

Cosmos sulphureus Cav., Icon. 1(3):56, *t. 79*. 1791.

Herb in Northern Coastal Lowlands (Toa Baja), Caguas Valley (Juncos), Sierra de Cayey, Central Cordillera (Peñuelas), Southern Coastal Limestone (Sabana Grande), Vieques; 25–450 m; *fl* & *fr* Jan, Sep.

Native to Mexico. [L5: 267].

Cosmos, Panchita.

Crassocephalum crepidioides (Benth.) S. Moore, J. Bot. 50:211. 1912.
Bas.: *Gynura crepidioides* Benth. (1849).

Herb in Northern Coastal Lowlands, Northern Limestone Hills, Río Abajo Forest Reserve, Sierra de Luquillo, Caguas Valley (Caguas), Sierra de Cayey, Las Piedras Chiquitas, Eastern and Central Cordillera, Maricao Forest Reserve, Western Coast (Aguada); 1–1300 m; *fl* & *fr* throughout the year.

Native to tropical Africa. [L5: 267].

Critonia portoricensis (Urb.) Britton & P. Wilson, Sci. Surv. Porto Rico & Virgin Islands 6(3):291. 1925.
Bas.: *Eupatorium portoricense* Urb. (1899).

Shrub/small tree in Northern Limestone Hills, Río Abajo Forest Reserve, Sierra de Cayey, Las Piedras Chiquitas, Southeastern Foothills, Eastern Cordillera (Aibonito, Barranquitas), Central Cordillera (Coamo, Villalba), Susúa Forest Reserve, Maricao Forest Reserve; 10–800 m; *fl* & *fr* Oct–Apr.

Endemic. [L5: 293].

Guerrero.

Cyanthillium cinereum (L.) H. Rob., Proc. Biol. Soc. Wash. 103(1):252. 1990.
Bas.: *Conyza cinerea* L. (1753).
Syn.: *Vernonia cinerea* (L.) Less. (1829).

Herb in Northern Coastal Lowlands, Northern Limestone Hills (San Sebastián), Sierra de Luquillo, Eastern Coast, Caguas Valley (Caguas), Las Tetas de Cayey, Eastern Cordillera (Cayey), Southern Coastal Lowlands, Southern Coastal Limestone, Susúa Forest Reserve, Maricao Forest Reserve, Guánica Forest Reserve, Southwestern Coastal Lowlands, Western Coast, Vieques, Culebra, Mona, Desecheo; 1–825 m; *fl* & *fr* throughout the year.

Native to tropical Africa, eastern temperate Asia, tropical Asia, Australia. [L5: 372].

Rabo de buey, Yerba socialista.

Eclipta prostrata (L.) L., Mant. Pl. 2:286. 1771.
Bas.: *Verbesina prostrata* L. (1753).
Syn.: *Eclipta alba* (L.) Hassk. (1848).

Herb in Northern Coastal Lowlands, Río Abajo Forest Reserve, Sierra de Luquillo, Eastern Coast, Caguas Valley, Las Piedras Chiquitas, Eastern Cordillera (Aibonito), Central Cordillera (Adjuntas), Southwestern Coastal Lowlands, Sierra Bermeja, Western Coast, Vieques, Culebra; 1–1000 m; *fl* & *fr* throughout the year.

Native to the Neotropics and -subtropics. [L5: 269].

Eclipta blanca.

Elephantopus mollis Kunth in Humboldt, Bonpland & Kunth, Nov. Gen. Sp. 4:20 [folio]. 1818; 4:26 [quarto]. 1820.

Herb in Northern Coastal Lowlands, Río Abajo Forest Reserve, Sierra de Luquillo, Eastern Coast, Sierra de Cayey, Las Piedras Chiquitas, Southeastern Foothills, Eastern Cordillera (Cerro La Peña, La Torrecilla), Central Cordillera, Western Cordillera; 1–1000 m; *fl* & *fr* throughout the year.

Greater Antilles (Cuba), Lesser Antilles, Mexico, Central America, South America. [L5: 271].

Lengua de vaca.

Eleutheranthera ruderalis (Sw.) Sch. Bip., Bot. Zeitung (Berlin) 24(21):165. 1866.
Bas.: *Melampodium ruderale* Sw. (1806).

Herb in Northern Coastal Lowlands, Northern Limestone Hills, Río Abajo Forest Reserve, Eastern Coast (Ceiba), Mona; 1–300 m; *fl* & *fr* throughout the year.

Bahamas, Greater Antilles (Jamaica), Lesser Antilles, Central America, South America. [L5: 272].

Ogiera.

Emilia fosbergii Nicolson, Phytologia 32(1):34. 1975.

Herb in Northern Coastal Lowlands, Río Abajo Forest Reserve, Sierra de Luquillo, Eastern Coast, Las Tetas de Cayey, Las Piedras Chiquitas, Southeastern Foothills, Eastern and Central Cordillera, Southern Coastal Lowlands (Ponce), Susúa Forest Reserve,

Maricao Forest Reserve, Southwestern Coastal Lowlands (San Germán), Sierra Bermeja, Western Coast, Vieques; 1–1200 m; *fl* & *fr* throughout the year.

Widely naturalized in the tropics; origin uncertain. [L5: 275].

Clavelillo del cafetal, Clavelito colorado.

Emilia sonchifolia (L.) DC. in Wight, Contr. Bot. India 24. 1834.
Bas.: *Cacalia sonchifolia* L. (1753).

Herb in Northern Coastal Lowlands, Northern Limestone Hills (San Sebastián), Eastern Coast, Las Piedras Chiquitas, Central Cordillera (Utuado, Adjuntas), Southwestern Coastal Lowlands (Cabo Rojo), Vieques; 1–900 m; *fl* & *fr* throughout the year.

Probably native to southern Asia. [L5: 277].

Huye que te cojo, Yerba socialista.

Enydra sessilis (Sw.) DC., Prodr. 5:637. 1836.
Bas.: *Eclipta sessilis* Sw. (1788).

Trailing herb in Northern Coastal Lowlands, Río Abajo Forest Reserve; 1–350 m; *fl* & *fr* Jan–Feb, May, Oct.

Greater Antilles. [L5: 277].

Erechtites hieraciifolius (L.) Raf. ex DC., Prodr. 6:294. 1838 "hieracifolia."
Bas.: *Senecio hieraciifolius* L. (1753) "hieracifolius."
Syn.: *Erechtites hieraciifolius* (L.) Raf. ex DC. var. *cacalioides* (Fisch. ex Spreng.) Griseb. (1861).

Herb in Northern Coastal Lowlands (Toa Alta), Río Abajo Forest Reserve, Sierra de Cayey, Eastern Cordillera (Barranquitas), Western Cordillera (Maricao); 10–650 m; *fl* & *fr* Jan–Apr, Aug.

Bahamas, Greater Antilles (Jamaica, Hispaniola), Lesser Antilles (Dominica), Canada, eastern and central United States, Central America, South America. [L5: 279].

Achicoria de cabra.

Erechtites valerianifolius (Link ex Spreng.) DC., Prodr. 6:295. 1838 "valerianaefolia."
Bas.: *Senecio valerianifolius* Link ex Spreng. (1826) "valerianaefolius."

Herb in Northern Coastal Lowlands (Dorado), Río Abajo Forest Reserve, Sierra de Luquillo, Caguas Valley (Caguas), Sierra de Cayey, Eastern Cordillera (Corozal, Cayey), Central Cordillera, Maricao Forest Reserve; 25–1250 m; *fl* & *fr* throughout the year.

Greater Antilles (Hispaniola), Lesser Antilles, Mexico, Central America, South America. [L5: 281].

Erigeron bellioides DC., Prodr. 5:288. 1836.

Delicate herb in Northern Coastal Lowlands, Sierra de Luquillo, Sierra de Cayey, Eastern Cordillera (Aibonito), Central Cordillera, Maricao Forest Reserve, Vieques; 1–1000 m; *fl* & *fr* Jan–Oct.

Bahamas, Greater Antilles (Cuba, Hispaniola). [L5: 282].

Bellorita.

Erigeron cuneifolius DC., Prodr. 5:288. 1836 "cuneifolium."

Delicate herb in Northern Coastal Lowlands (Laguna Tortuguero), Northern Limestone Hills (Ciales, Florida), Sierra de Luquillo, Sierra de Cayey, Las Tetas de Cayey, Eastern and Central Cordillera, Southwestern Coastal Lowlands (San Germán); 1–1200 m; *fl* & *fr* throughout the year.

Greater Antilles, Virgin Islands, Central America (Nicaragua to Panama). [L5: 283].

Erigeron jamaicensis L., Syst. Nat. (ed. 10) 2:1213. 1759 "jamaicense."

Low herb in Northern Limestone Hills, Toro Negro Forest Reserve; 160–1250 m; *fl* & *fr* Jan–Apr.

Greater Antilles, Mexico, southern Central America (Costa Rica). [L5: 283].

Flaveria bidentis (L.) Kuntze, Revis. Gen. Pl. 3(3):148. 1898.
Bas.: *Ethulia bidentis* L. (1767).

Suffrutescent herb in Southern Coastal Lowlands, Southern Coastal Limestone, Susúa Forest Reserve, Guánica Forest Reserve, Southwestern Coastal Lowlands; 1–100 m; *fl* & *fr* Feb, May–Nov.

Native from southern South America to Ecuador, Brazil. [L5: 296].

Flaveria trinervia (Spreng.) C. Mohr, Contr. U.S. Natl. Herb. 6:810. 1901.
Bas.: *Oedera trinervia* Spreng. (1800).

Suffrutescent herb in Southern Coastal Lowlands (Guayanilla), Southern Coastal Limestone (Peñuelas), Susúa Forest Reserve; 2–150 m; *fl* & *fr* Dec–Apr.

Bahamas, Greater Antilles, southern United States, Mexico, northern Central America (Belize). [L5: 296].

Fleischmannia microstemon (Cass.) R.M. King & H. Rob., Phytologia 19(4):204. 1970.
Bas.: *Eupatorium microstemon* Cass. (1822).

Suffrutescent herb in Northern Limestone Hills, Río Abajo Forest Reserve, Sierra de Luquillo, Caguas Valley (Caguas), Eastern and Central Cordillera, Susúa Forest Reserve, Maricao Forest Reserve; 175–1050 m; *fl* & *fr* Jan–May, Aug–Oct.

Antilles, southern Mexico (Yucatan), Central America, northern South America to Bolivia. [L5: 291].

Galinsoga parviflora Cav., Icon. 3(2):41, *t. 281*. 1795–1796.

Herb in Las Tetas de Cayey, Central Cordillera; 500–800 m; *fl* & *fr* Jan–Mar, Jun.

Greater Antilles, southwestern United States, Mexico, Central America, South America. [L5: 298].

Yerba pajarito.

Galinsoga quadriradiata Ruiz & Pav., Syst. Veg. Fl. Peruv. Chil. 1:198. 1798.

Syn.: *Galinsoga caracasana* (DC.) Sch. Bip. (1865) "caracassana."

Galinsoga ciliata (Raf.) S.F. Blake (1922).

Herb in Sierra de Luquillo, Sierra de Cayey, Eastern Cordillera (La Torrecilla), Central Cordillera, Western Cordillera (Maricao); 350–1300 m; *fl* & *fr* throughout the year.

Probably native to Mexico. [L5: 299].

Yerba pajarito.

Gamochaeta americana (Mill.) Wedd., Chlor. Andina 1(4–6):151. 1856.

Bas.: *Gnaphalium americanum* Mill. (1768).

Syn.: *Gnaphalium spicatum* Lam. (1788).

Herb in Northern Coastal Lowlands (Laguna Tortuguero), Eastern Cordillera (La Torrecilla), Central Cordillera (Orocovis); 2–900 m; *fl* & *fr* Feb–Apr.

Bahamas, Greater Antilles, Mexico, Central America, South America. [L5: 301].

Gamochaeta argyrinea G.L. Nesom, Sida 21(2):718, *ff. 1, 3–4*. 2004.

Herb in Las Tetas de Cayey, Eastern Cordillera (Barranquitas), Central Cordillera; 600–1200 m; *fl* & *fr* Nov–Jul.

Native to North America.

Notes: Specimens of this taxon were previously misidentified as *Gnaphalium americanum* [=*Gamochaeta americana*], which also occurs here.

Gamochaeta pensylvanica (Willd.) Cabrera, Bol. Soc. Argent. Bot. 9:375. 1961.

Bas.: *Gnaphalium pensylvanicum* Willd. (1809).

Herb in Northern Coastal Lowlands (Laguna Tortuguero, Barceloneta), Sierra de Cayey, Las Tetas de Cayey, Las Piedras Chiquitas, Southeastern Foothills (Coamo, Aibonito), Eastern Cordillera (Aibonito), Central Cordillera (Adjuntas); 2–1050 m; *fl* & *fr* Dec–Jun.

Probably native to southern United States. [L5: 301].

Notes: Britton & Wilson (1925), Liogier & Martorell (1982, 2000) and Liogier (1997) misapplied the name *Gnaphalium indicum* L. to this taxon.

Yerba lanuda.

Gundlachia corymbosa (Urb.) Britton ex Bold., Fl. Ned. West-Ind. 391. 1913.

Bas.: *Gundlachia domingensis* (Spreng.) A. Gray var. *corymbosa* Urb. (1903).

Shrub in Northern Coastal Lowlands; 1–50 m; *fl* & *fr* Aug–Feb, May.

Bahamas, Greater Antilles (Hispaniola), Lesser Antilles, Southern Caribbean Islands (Curaçao). [L5: 290].

Sereno.

Hebeclinium macrophyllum (L.) DC., Prodr. 5:136. 1836.

Bas.: *Eupatorium macrophyllum* L. (1763).

Suffrutescent herb in Northern Limestone Hills, Río Abajo Forest Reserve, Sierra de Cayey (San Lorenzo); 225–600 m; *fl* & *fr* May–Nov.

Antilles, Mexico, Central America, South America. [L5: 301].

Guerrero, Turma de toro.

Koanophyllon dolicholepis (Urb.) R.M. King & H. Rob., Phytologia 32(3):256. 1975.

Bas.: *Eupatorium villosum* Sw. var. *dolicholepis* Urb. (1899).

Syn.: *Eupatorium dolicholepis* (Urb.) Britton (1925).

Shrub in Northern Limestone Hills, Río Abajo Forest Reserve, Eastern Cordillera (Cerro La Peña), Maricao Forest Reserve; 50–350 m; *fl* & *fr* Nov–Apr.

Endemic. [L5: 287].

Oreganillo.

Koanophyllon droserolepis (B.L. Rob.) R.M. King & H. Rob., Phytologia 32(3):256. 1975.

Bas.: *Eupatorium droserolepis* B.L. Rob. (1918).

Suffrutescent herb in Eastern Cordillera (La Torrecilla), Central Cordillera (Ciales), Maricao Forest Reserve; 600–900 m; *fl* & *fr* Nov–Feb, Jul.

Endemic. [L5: 288].

Koanophyllon polyodon (Urb.) R.M. King & H. Rob., Phytologia 32(3):262. 1975.

Bas.: *Eupatorium polyodon* Urb. (1899).

Shrub in Northern Limestone Hills (Ciales), Río Abajo Forest Reserve, Sierra de Cayey, Las Tetas de Cayey, Eastern and Central Cordillera, Susúa Forest Reserve, Maricao Forest Reserve; 375–900 m; *fl* & *fr* Jun–Jul, Oct–Feb.

Endemic. [L5: 293; AR: 97].

Lactuca floridana (L.) Gaertn., Fruct. Sem. Pl. 2(3):362. 1791.
Bas.: *Sonchus floridanus* L. (1753).
Herb in Northern Limestone Hills (Florida), Sierra de Luquillo, Eastern Cordillera (La Torrecilla), Central Cordillera; 150–1250 m; *fl* & *fr* Jan–Mar, Jul–Sep.
Native to North America. [L5: 308].
Lechuga del monte.

Lagascea mollis Cav., Anales Ci. Nat. 6(18):332, *t. 44*. 1803 'Lagasca.'
Syn.: *Nocca mollis* (Cav.) Jacq. (1805) "Noccaea."
Suffrutescent herb in Northern Coastal Lowlands (Luquillo), Northern Limestone Hills (Quebradillas, Aguadilla), Las Piedras Chiquitas, Southeastern Foothills, Central Cordillera (Villalba, Ponce), Cerro de Las Cuevas, Southern Coastal Lowlands, Southwestern Coastal Lowlands, Sierra Bermeja, Vieques; 1–600 m; *fl* & *fr* throughout the year.
West Indes, Mexico, Central America, South America. [L5: 310].

Launaea intybacea (Jacq.) Beauverd, Bull. Soc. Bot. Genève, sér. 2 2:114. 1910.
Bas.: *Lactuca intybacea* Jacq. (1784).
Syn.: *Brachyramphus intybaceus* (Jacq.) DC. (1838).
Herb in Southeastern Foothills, Central Cordillera (Villalba), Southern Coastal Lowlands, Southern Coastal Limestone (Peñuelas), Guánica Forest Reserve, Southwestern Coastal Lowlands; 10–50 m; *fl* & *fr* throughout the year.
Pantropical and -subtropical. [L5: 312].
Achicoria azul.

Lepidaploa borinquensis (Urb.) H. Rob., Proc. Biol. Soc. Wash. 103(2):483. 1990.
Bas.: *Vernonia borinquensis* Urb. (1903).
Shrub in Northern Limestone Hills (Ciales), Sierra de Cayey, Las Tetas de Cayey, Southeastern Foothills (Cayey), Eastern Cordillera (Cerro La Peña, La Torrecilla), Central Cordillera, Western Cordillera, Cerro Las Mesas; 300–1200 m; *fl* & *fr* Sep–Mar.
Endemic. [L5: 371; AR: 97].

Lepidaploa glabra (Willd.) H. Rob., Proc. Biol. Soc. Wash. 103(2):487. 1990.
Bas.: *Conyza glabra* Willd. (1803).
Syn.: *Vernonia albicaulis* Pers. (1807).
Shrub in Northern Limestone Hills, Río Abajo Forest Reserve, Southern Coastal Limestone, Vieques, Desecheo; 1–350 m; *fl* & *fr* throughout the year.
Virgin Islands, Lesser Antilles. [L5: 371].
Santa María.

Lepidaploa proctorii (Urbatsch) H. Rob., Proc. Biol. Soc. Wash. 103(2):491. 1990.
Bas.: *Vernonia proctorii* Urbatsch (1989).
Shrub in Sierra Bermeja; 225–300 m; *fl* & *fr* Mar–Apr, Jul–Dec.
Endemic. [L5: 372].
Notes: This taxon is on the U.S. Fish & Wildlife Service list of threatened and endangered plants.

Lepidaploa sericea (Rich.) H. Rob., Proc. Biol. Soc. Wash. 103(2):492. 1990.
Bas.: *Vernonia sericea* Rich. (1792).
Shrub in Northern Coastal Lowlands (Laguna Tortuguero, Manatí), Northern Limestone Hills, Río Abajo Forest Reserve, Sierra de Luquillo, Eastern Coast, Sierra de Cayey, Las Tetas de Cayey, Las Piedras Chiquitas, Southeastern Foothills, Eastern and Central Cordillera, Susúa Forest Reserve, Maricao Forest Reserve, Southwestern Coastal Lowlands (Cabo Rojo), Cerro Las Mesas, Western Coast (Rincón); 1–900 m; *fl* & *fr* Sep–May.
Greater Antilles (Cuba, Hispaniola), Virgin Islands. [L5: 373].
Escobilla, Huye que te cojo, Tapacaminos, Yerba socialista.

Leucanthemum maximum (Ramond) DC., Prodr. 6:46. 1838.
Bas.: *Chrysanthemum maximum* Ramond (1800).
Herb in Toro Negro Forest Reserve; 1275–1300 m; *fl* & *fr* Apr, Aug. [*Liogier 28828, 33139* (UPR); *Worthington 33278* (UPRRP)].
Native to southwestern Europe (French and Spanish Pyrenees). [L5: 257].

Melampodium divaricatum (Rich.) DC., Prodr. 5:520. 1836.
Bas.: *Dysodium divaricatum* Rich. (1807).
Suffrutescent herb in Western Coast (Añasco); 30 m; *fl* & *fr* Mar (one specimen). [*Axelrod 790* (UPRRP)].
Native to Mexico, Central America, northern South America to Bolivia. [L5: 314].
Botón de oro.

Melanthera nivea (L.) Small, Fl. S.E. U.S. 1251, 1340. 1903.
Bas.: *Bidens nivea* L. (1753).
Syn.: *Melanthera aspera* (Jacq.) Steud. ex Small (1909).
Melanthera canescens (Kuntze) O.E. Schulz (1911).

Melanthera montana O.E. Schulz (1911).
Melanthera calcicola Britton (1925).
Melanthera confusa Britton (1925).
Melanthera aspera (Jacq.) Steud. ex Small var. *glabriuscula* (Kuntze) Parks (1973).

Suffrutescent herb in Northern Coastal Lowlands, Northern Limestone Hills, Río Abajo Forest Reserve, Eastern Coast, Caguas Valley (Caguas), Sierra de Cayey, Southeastern Foothills, Eastern and Central Cordillera, Southern Coastal Lowlands (Santa Isabel), Susúa Forest Reserve, Western Cordillera, Southwestern Coastal Lowlands (Lajas), Western Coast (Aguada), Vieques, Culebra; 1–1250 m; *fl* & *fr* throughout the year.

West Indies, southeastern United States (Florida), Mexico, Central America, northwestern and west-central South America. [L5: 315].

Cariaquillo blanco, Salaíllo, Yerba de cabra.

Mikania congesta DC., Prodr. 5:197. 1836.

Vine in Northern Coastal Lowlands (Loíza); 2–25 m; *fl* & *fr* Nov (one specimen). [*Axelrod 3118* (UPRRP)].

Greater Antilles (Jamaica, Hispaniola), Trinidad, northern Central America (Belize, El Salvador), western South America and Venezuela. [L5: 318; AR: 100].

Notes: Britton & Wilson (1925) misapplied the above name to *Mikania micrantha*. Liogier & Martorell (1982) used the name *Mikania micrantha* Kunth var. *congesta* (DC.) B.L. Rob., by which they probably meant *Mikania micrantha* Kunth f. *congesta* (DC.) B.L. Rob., a synonym of the above taxon.

Although the specimen cited above was determined as this taxon by W.C. Holmes (BAYLU), I am myself unable to distinguish this taxon from *Mikania micrantha* in Puerto Rican material.

Guaco falso.

Mikania cordifolia (L.f.) Willd., Sp. Pl. 3(3):1746. 1803.

Bas.: *Cacalia cordifolia* L.f. (1782).

Vine in Northern Coastal Lowlands, Northern Limestone Hills, Sierra de Luquillo, Eastern Coast, Sierra de Cayey, Las Tetas de Cayey, Southeastern Foothills (Cayey), Eastern Cordillera (La Torrecilla), Central Cordillera, Maricao Forest Reserve; 100–1050 m; *fl* & *fr* Nov–Mar.

Greater Antilles, Virgin Islands, Lesser Antilles (Guadeloupe, St. Lucia), southeastern United States (Gulf Coast), Mexico, Central America, South America. [L5: 319; AR: 100].

Guaco.

Mikania fragilis Urb., Symb. Antill. 1(3):464. 1899.

Vine in Sierra de Luquillo, Sierra de Cayey, Central Cordillera, Susúa Forest Reserve, Maricao Forest Reserve; 500–1335 m; *fl* & *fr* Sep–Jun.

Endemic. [L5: 319; AR: 100].

Guaco redondo.

Mikania micrantha Kunth in Humboldt, Bonpland & Kunth, Nov. Gen. Sp. 4:105 [folio]. 1818; 4:134 [quarto]. 1820.

Vine in Northern Coastal Lowlands, Río Abajo Forest Reserve, Caguas Valley (Caguas), Sierra de Cayey, Eastern Cordillera (Cerro La Peña, Aguas Buenas), Southern Coastal Lowlands (Ponce), Western Cordillera, Southwestern Coastal Lowlands (Lajas), Sierra Bermeja, Western Coast; 1–550 m; *fl* & *fr* Oct–May.

Greater Antilles (Cuba), Lesser Antilles, Trinidad, Mexico, Central America, South America. [L5: 320; AR: 103].

Guaco falso.

Mikania odoratissima Urb., Symb. Antill. 1(3):464. 1899.

Vine in Northern Limestone Hills (Florida), Río Abajo Forest Reserve, Las Tetas de Cayey, Central Cordillera, Maricao Forest Reserve; 250–1100 m; *fl* & *fr* Nov–Mar.

Endemic. [L5: 320; AR: 103].

Mikania pachyphylla Urb., Symb. Antill. 1(3):463. 1899.

Vine in Sierra de Luquillo, Toro Negro Forest Reserve; 750–1325 m; *fl* & *fr* Feb–Jun.

Endemic. [L5: 320; AR: 103].

Mikania porosa Urb., Symb. Antill. 1(3):465. 1899.

Vine in Northern Limestone Hills (Arecibo), Río Abajo Forest Reserve, Maricao Forest Reserve; 300–900 m; *fl* & *fr* Dec–Feb.

Endemic. [L5: 321; AR: 105].

Guaco de cabra.

Mikania stevensiana Britton, Bull. Torrey Bot. Club 43(9):458. 1916.

Vine in Western Cordillera; 475–900 m; *fl* & *fr* Dec–Feb.

Endemic. [L5: 321; AR: 105].

Neurolaena lobata (L.) Cass. in Cuvier, Dict. Sci. Nat. (ed. 2) 34:501. 1825.

Bas.: *Conyza lobata* L. (1753).

Shrub in Río Abajo Forest Reserve, Sierra de Luquillo, Eastern Coast (Maunabo), Caguas Valley (Caguas), Sierra de Cayey, Las Tetas de Cayey, Eastern and Central Cordillera, Maricao Forest Reserve; 200–1050 m; *fl* & *fr* Feb–Jul.

West Indies, Mexico, Central America, northern South America to Bolivia. [L5: 323].

Sepí.

Parthenium hysterophorus L., Sp. Pl. 2:988. 1753.

Herb in Northern Coastal Lowlands, Northern Limestone Hills, Las Piedras Chiquitas, Southeastern Foothills, Eastern Cordillera, Southern Coastal Lowlands, Southwestern Coastal Lowlands, Western Coast, Vieques; 1–600 m; *fl* & *fr* throughout the year.

West Indies, Mexico, Central America, South America. [L5: 324].

Ajenjo cimarrón, Artemisa cimarrona, Santa María, Yerba amarga.

Pectis ciliaris L., Syst. Nat. (ed. 10) 2:1221. 1759.

Herb in Southern Coastal Lowlands (Ponce); 100–200 m; *fl* & *fr* Dec (one specimen).

Greater Antilles, Lesser Antilles (Martinique). [L5: 327].

Romero cimarrón.

Pectis elongata Kunth var. **floribunda** (A. Rich.) D.J. Keil, Ann. Missouri Bot. Gard. 62(4):1229. 1976.

Bas.: *Pectis floribunda* A. Rich. (1853).

Herb in Northern Coastal Lowlands, Northern Limestone Hills (Hatillo), Río Abajo Forest Reserve, Eastern Coast, Southern Coastal Lowlands (Patillas), Southwestern Coastal Lowlands (Cabo Rojo), Western Coast, Vieques; 1–600 m; *fl* & *fr* Sep–Jun.

Greater Antilles, Lesser Antilles (Dominica, Martinique, St. Lucia), Mexico, southern Central America (Panama). [L5: 328].

Limoncillo del pasto, Romero cimarrón, San Rafael.

Pectis humifusa Sw., Prodr. 114. 1788.

Herb in Northern Coastal Lowlands, Eastern Coast (Cayo Santiago), Vieques; 1–60 m; *fl* & *fr* throughout the year.

Antilles. [L5: 328].

Yerba de San Juan.

Pectis linearis La Llave, Reg. Trimestre 1:451. 1832.

Syn.: *Pectis febrifuga* H.C. Hall (1861).

Herb in Southwestern Coastal Lowlands, Sierra Bermeja; 1–225 m; *fl* & *fr* Feb, Jul–Sep.

Greater Antilles (Jamaica), Lesser Antilles, Mexico, Central America, northern South America. [L5: 329].

Pectis linifolia L., Syst. Nat. (ed. 10) 2:1221. 1759.

Herb in Las Piedras Chiquitas, Southeastern Foothills, Southern Coastal Lowlands, Guánica Forest Reserve, Sierra Bermeja, Cerro Las Mesas, Mona, Desecheo; 1–450 m; *fl* & *fr* Sep–Mar, Jun.

West Indies, southwestern United States (Arizona), Central America, northern South America to Bolivia. [L5: 329].

Romero cimarrón, Romero macho.

Pectis prostrata Cav., Icon. 4(1):12, *t. 324*. 1797.

Syn.: *Pectis portoricensis* Urb. (1907).

Prostrate herb in Northern Coastal Lowlands, Southeastern Foothills, Southern Coastal Lowlands; 2–600 m; *fl* & *fr* May–Jan.

West Indies, southern United States, Mexico, Central America, northern South America. [L5: 329].

Piptocarpha tetrantha Urb., Symb. Antill. 1(3):457. 1899.

Woody vine/shrub in Sierra de Luquillo, Sierra de Cayey, Central Cordillera; 600–1325 m; *fl* & *fr* Jul–Mar.

Endemic. [L5: 331; AR: 107].

Piptocoma acevedoi Pruski, Novon 6(1):98, *f. 1*. 1996.

Woody vine in Guajataca Forest Reserve, Maricao Forest Reserve; 250–550 m; *fl* & *fr* Aug, Nov–Jan.

Endemic. [L5: 332; AR: 107].

Piptocoma antillana Urb., Ark. Bot. 23A(11):50. 1931.

Shrub in Northern Coastal Lowlands (Fajardo), Susúa Forest Reserve, Maricao Forest Reserve, Southwestern Coastal Lowlands (San Germán), Sierra Bermeja, Vieques; 50–900 m; *fl* & *fr* throughout the year.

Greater Antilles (Hispaniola), Virgin Islands. [L5: 332].

Notes: Britton & Wilson (1925) misapplied the name *Piptocoma rufescens* Cass. to this taxon.

Pluchea carolinensis (Jacq.) G. Don in Sweet, Hort. Brit. (ed. 3) 350. 1839.

Bas.: *Conyza carolinensis* Jacq. (1788).

Shrub in Northern Coastal Lowlands, Northern Limestone Hills, Río Abajo Forest Reserve, Eastern Coast, Caguas Valley (Caguas), Sierra de Cayey, Eastern Cordillera (Aibonito), Central Cordillera (southern slopes), Southern Coastal Lowlands, Susúa Forest Reserve, Guánica Forest Reserve,

Southwestern Coastal Lowlands (Cabo Rojo), Western Coast, Vieques, Culebra, Mona; 1–400 m; *fl* & *fr* throughout the year.

West Indies, southeastern United States (Florida), Mexico, Central America, northern South America to Ecuador. [L5: 334].

Notes: Britton & Wilson (1925) misapplied the name *Pluchea odorata* to this taxon; Liogier & Martorell (1982) misapplied *Pluchea symphytifolia* (Mill.) Gillis to it, but later (1997) used the correct name, while wrongly placing *Pluchea symphytifolia* in synonymy with it.

Salvia.

Pluchea odorata (L.) Cass. in Cuvier, Dict. Sci. Nat. (ed. 2) 42:3. 1826.

Bas.: *Conyza odorata* L. (1759).

Syn.: *Pluchea purpurascens* (Sw.) DC. (1836).

Herb in Northern Coastal Lowlands, Eastern Coast (Humacao), Southern Coastal Lowlands (Guayama), Guánica Forest Reserve, Southwestern Coastal Lowlands, Vieques, Mona; 1–35 m; *fl* & *fr* throughout the year.

West Indies, eastern and southern United States, Mexico, Central America, South America (Venezuela, Ecuador, Bolivia). [L5: 335].

Salvia.

Porophyllum leiocarpum (Urb.) Rydb., N. Amer. Fl. 34(3):184. 1916.

Bas.: *Porophyllum macrocephalum* DC. var. *leiocarpum* Urb. (1899).

Herb in Las Piedras Chiquitas, Southeastern Foothills, Cerro de Las Cuevas, Southern Coastal Lowlands (Coamo), Southern Coastal Limestone (Peñuelas), Guánica Forest Reserve, Southwestern Coastal Lowlands, Sierra Bermeja, Western Coast (Rincón); 1–550 m; *fl* & *fr* Mar, Jul–Dec.

Greater Antilles (Hispaniola), northern South America (Venezuela). [L5: 336].

Yerba de peo.

Porophyllum ruderale (Jacq.) Cass. in Cuvier, Dict. Sci. Nat. (ed. 2) 43:56. 1826.

Bas.: *Kleinia ruderalis* Jacq. (1760).

Syn.: *Cacalia porophyllum* L. (1753). *Porophyllum porophyllum* (L.) Kuntze (1898).

Herb in Northern Coastal Lowlands (Fajardo), Northern Limestone Hills (San Sebastián), Southeastern Foothills (Cayey), Cerro Las Mesas, Western Coast (Aguada); 1–300 m; *fl* & *fr* Apr–Aug.

West Indies, Trinidad and Tobago, southwestern United States, Mexico, Central America, South America. [L5: 337].

Yerba porosa.

Pseudelephantopus spicatus (B. Juss. ex Aubl.) C.F. Baker, Trans. Acad. Sci. St. Louis 12(5):55. 1902.

Bas.: *Elephantopus spicatus* B. Juss. ex Aubl. (1775).

Herb in Northern Coastal Lowlands, Northern Limestone Hills, Río Abajo Forest Reserve, Sierra de Luquillo, Eastern Coast (Cuchilla de Panduras), Central Cordillera (Ponce), Southern Coastal Lowlands (Coamo), Western Coast; 1–600 m; *fl* & *fr* Dec–Jun.

Antilles, Mexico, Central America, northern South America to Peru. [L5: 340].

Lengua de vaca, Yerba de burro.

Pseudelephantopus spiralis (Less.) Cronquist, Madroño 20:255. 1970.

Bas.: *Distreptus spiralis* Less. (1831).

Herb in Río Abajo Forest Reserve, Central Cordillera (Adjuntas, Yauco), Western Cordillera; 275–850 m; *fl* & *fr* Jul–Oct.

Native to Lesser Antilles (St. Lucia, St. Vincent, Grenada), southern Central America, northwestern and west-central South America. [L5: 342].

Pseudognaphalium elegans (Kunth) Kartesz in Kartesz and Meacham, Synthesis N. Amer. Fl., version 1, Nomencl. Innov.: No. 28. 1999.

Bas.: *Gnapalium elegans* Kunth (1818).

Syn.: *Gnaphalium portoricense* Urb. (1903).

Herb in Central Cordillera, Maricao Forest Reserve; 600–1300 m; *fl* & *fr* Nov–Feb.

Greater Antilles, Virgin Islands, Mexico, Central America, South America. [L5: 301].

Notes: Liogier & Martorell (1982, 2000) and Liogier (1997) misapplied the name *Gnaphalium domingense* Lam. to this taxon.

Botoncillo.

Pseudogynoxys chenopodioides (Kunth) Cabrera, Brittonia 7(2):56. 1950.

Bas.: *Senecio chenopodioides* Kunth (1818).

Syn.: *Senecio confusus* Britten (1898).

Vine in Northern Limestone Hills, Río Abajo Forest Reserve, Sierra de Luquillo, Sierra de Cayey, Southeastern Foothills (Cayey), Central Cordillera; 90–600 m; *fl* & *fr* Nov–May.

Native to Mexico, Central America, northern South America. [L5: 343; AR: 109].

Margarita mexicana, Orgullo de Mayagüez, Senecio.

Pterocaulon alopecuroides (Lam.) DC., Prodr. 5:454. 1836.

Bas.: *Conyza alopecuroides* Lam. (1786).

Herb in Northern Coastal Lowlands (Laguna Tortuguero), Eastern Cordillera (Barranquitas); 1–700 m; *fl* & *fr* Oct–Nov.

Antilles, South America (scattered). [L5: 345].

Escobilla.

Pterocaulon virgatum (L.) DC., Prodr. 5:454. 1836.
Bas.: *Gnaphalium virgatum* L. (1759).

Herb in Northern Coastal Lowlands (Laguna Tortuguero), Northern Limestone Hills (Ciales), Sierra de Cayey, Eastern Cordillera (Cerro La Peña, Barranquitas), Southwestern Coastal Lowlands (Lajas); 1–700 m; *fl* & *fr* Sep–Nov, Feb.

Greater Antilles, Virgin Islands, south-central United States (Louisiana, Texas), Mexico, Central America, South America (scattered). [L5: 345].

Rolandra fruticosa (L.) Kuntze, Revis. Gen. Pl. 1:360. 1891.
Bas.: *Echinops fruticosa* L. (1753).

Shrub in Northern Coastal Lowlands, Cambalache Forest Reserve, Sierra de Luquillo, Eastern Coast, Sierra de Cayey, Eastern Cordillera (Cerro La Peña, Cañón San Cristóbal), Central Cordillera, Susúa Forest Reserve, Maricao Forest Reserve; 1–700 m; *fl* & *fr* Jan–Jun.

Greater Antilles (Hispaniola), Lesser Antilles, Central America, northern South America to Brazil. [L5: 347].

Yerba de plata.

Salmea scandens (L.) DC., Cat. Pl. Horti Monsp. 141. 1813.
Bas.: *Bidens scandens* L. (1753).

Vine in Northern Limestone Hills, Río Abajo Forest Reserve, Sierra de Cayey, Toro Negro Forest Reserve; 200–1000 m; *fl* & *fr* Feb–May.

Greater Antilles, Mexico, Central America, South America. [L5: 348; AR: 109].

Bejuco de miel, Bejuco de muela, Tabacón.

Sclerocarpus africanus Jacq. ex Murray, Syst. Veg. (ed. 14) 783. 1784.

Herb in Vieques; 1–110 m; *fl* & *fr* Jun, Nov (two specimens). [*Proctor 48025, 48474* (SJ)].

Native to tropical Africa. [L5: 350].

Solidago sempervirens L. subsp. **mexicana** (L.) Semple, Sida 20(4):1615. 2003.
Bas.: *Solidago mexicana* L. (1753).
Syn.: *Solidago sempervirens* L. var. *mexicana* (L.) Fernald (1935).

Herb planted and persistent in Eastern Cordillera (Aibonito), Central Cordillera; 500–950 m; *fl* & *fr* Jul–Jan.

Native to Bahamas, along coast of eastern United States to Texas, Mexico. [L5: 350].

Notes: Britton & Wilson (1926) and Liogier & Martorell (1982) used *Solidago sempervirens* L., *sensu lato*, for the name of this taxon.

Solidago.

Sonchus asper (L.) Hill, Herb. Brit. 1:47. 1769.
Bas.: *Sonchus oleraceus* L. var. *asper* L. (1753).

Herb in Sierra de Cayey, Eastern and Central Cordillera, Maricao Forest Reserve, Southwestern Coastal Lowlands (Cabo Rojo), Desecheo; 50–1250 m; *fl* & *fr* Sep–Oct, Jan–Apr.

Native to Europe, Africa, temperate Asia. [L5: 353].

Achicoria silvestre, Chicoria.

Sonchus oleraceus L., Sp. Pl. 2:794. 1753.

Herb in Northern Coastal Lowlands, Northern Limestone Hills, Río Abajo Forest Reserve, Sierra de Luquillo, Sierra de Cayey, Las Piedras Chiquitas, Eastern Cordillera (La Torrecilla), Central Cordillera, Vieques; 5–1300 m; *fl* & *fr* throughout the year.

Native to Europe, northern Africa, temperate Asia, Indian Subcontinent. [L5: 353].

Achicoria, Achicoria silvestre, Chicoria.

Sphagneticola trilobata (L.) Pruski in Acevedo-Rodríguez, Mem. New York Bot. Gard. 78:114. 1996 17 Dec; et Novon 6(4):412. 1996 27 Dec.
Bas.: *Silphium trilobatum* L. (1759).
Syn.: *Wedelia trilobata* (L.) Hitchc. (1893).

Herb in Northern Coastal Lowlands, Northern Limestone Hills (San Sebastián), Río Abajo Forest Reserve, Sierra de Luquillo, Eastern Coast, Caguas Valley (Caguas), Sierra de Cayey, Central Cordillera, Susúa Forest Reserve, Maricao Forest Reserve, Southwestern Coastal Lowlands, Western Coast, Vieques; 1–900 m; *fl* & *fr* May–Feb.

West Indies, Central America, northern South America to Brazil, Ecuador. [L5: 378; AR: 111].

Manzanilla, Manzanilla de playa, Margarita amarilla, Margarita del pasto, Romerillo.

Spiracantha cornifolia Kunth in Humboldt, Bonpland & Kunth, Nov. Gen. Sp. 4:23, *t. 313* [folio]. 1818; 4:29, *t. 313* [quarto]. 1820.

Herb in Western Coast (Aguada, Rincón); 1–100 m; *fl* & *fr* Jan–Feb (one specimen). [*Liogier 35875* (UPR)].

Greater Antilles (Hispaniola), Mexico, Central America, northern South America (Venezuela, Colombia). [L5: 355].

Struchium sparganophorum (L.) Kuntze, Revis. Gen. Pl. 1:366. 1891.
Bas.: *Ethulia sparganophora* L. (1763).
Herb in the Northern Coastal Lowlands, Northern Limestone Hills (San Sebastián), Eastern Coast (Humacao), Caguas Valley (Caguas); 1–200 m; *fl* & *fr* Jan–May.
Antilles, Mexico, Central America, South America (scattered). [L5: 356].
Yerba de faja.

Symphyotrichum subulatum (Michx.) G.L. Nesom var. **parviflorum** (Nees) S.D. Sundb., Sida 21(2):907. 2004.
Bas.: *Tripolium subulatum* (Michx.) DC. var. *parviflorum* Nees (1833)
Syn.: *Aster subulatus* Michx. var. *cubensis* (DC.) Shinners (1953).
Herb in Northern Coastal Lowlands, Río Abajo Forest Reserve, Sierra de Luquillo, Eastern Coast (Ceiba), Sierra de Cayey, Eastern Cordillera (Aguas Buenas, Cidra), Toro Negro Forest Reserve, Southern Coastal Lowlands (Juana Díaz), Maricao Forest Reserve, Southwestern Coastal Lowlands, Western Coast (Aguada); 1–1300 m; *fl* & *fr* throughout the year.
West Indies, southern United States, Mexico, Central America, northern South America. [L5: 242].
Notes: Liogier (1997) and Liogier & Martorell (2000) used *Aster subulatus* Michx., *sensu lato*, for the name of this taxon.

Synedrella nodiflora (L.) Gaertn., Fruct. Sem. Pl. 2(3):456, *t. 171*, *f. 7*. 1791.
Bas.: *Verbesina nodiflora* L. (1755).
Suffrutescent herb in Northern Coastal Lowlands (San Juan), Northern Limestone Hills, Río Abajo Forest Reserve, Sierra de Luquillo, Eastern Coast, Caguas Valley (Juncos), Sierra de Cayey, Southeastern Foothills, Eastern and Central Cordillera, Western Cordillera, Southwestern Coastal Lowlands (Lajas), Western Coast (Aguada), Vieques, Mona; 1–800 m; *fl* & *fr* throughout the year.
West Indies, Mexico, Central America, South America. [L5: 358].
Cerbatana, Serbatana.

Tagetes erecta L., Sp. Pl. 2:887. 1753.
Syn.: *Tagetes patula* L. (1753).
Herb in Northern Limestone Hills (Toa Baja), Eastern Cordillera (Barranquitas); 50–600 m; *fl* & *fr* Jan, Aug (two specimens). [*Liogier 29164, 36159* (UPR)].
Native to Mexico, Central America, western South America to Bolivia. [L5: 360].
Clavel de muerto, Clavelitos de muerto, Ruda del pasto.

Taraxacum officinale Weber ex F. H. Wigg., Prim. Fl. Holsat. 56. 1780.
Syn.: *Leontodon taraxacum* L. (1753).
Herb in Eastern Cordillera (La Torrecilla), Central Cordillera; 825–950 m; *fl* & *fr* Mar, Sep.
Native to Eurasia. [L5: 362].
Diente de león.

Tithonia diversifolia (Hemsl.) A. Gray, Proc. Amer. Acad. Arts 19:5. 1883.
Bas.: *Mirasolia diversifolia* Hemsl. (1881).
Shrub in Piñones Forest Reserve, Northern Limestone Hills (Morovis, Lares), Las Tetas de Cayey, Central Cordillera, Cerro de Las Cuevas, Western Cordillera; 1–800 m; *fl* & *fr* Oct–Jun.
Native to Mexico, Central America. [L5: 364].
Girasol mexicano, Girasol pequeño, Mirasol mexicano, Mirasol pequeño.

Tithonia rotundifolia (Mill.) S.F. Blake, Contr. Gray. Herb. 52:41. 1917.
Bas.: *Tagetes rotundifolia* Mill. (1768).
Herb in Southern Coastal Lowlands (Coamo), Southwestern Coastal Lowlands (Cabo Rojo); 25–50 m; *fl* & *fr* Sep–Oct (two specimens). [*Liogier 29462, 29467* (UPR)].
Native to Mexico, Central America. [L5: 365].
Clavel de muerto, Escopeta.

Tridax procumbens L., Sp. Pl. 2:900. 1753.
Herb in Northern Coastal Lowlands, Guajataca Forest Reserve, Río Abajo Forest Reserve, Las Piedras Chiquitas, Southeastern Foothills, Southern Coastal Lowlands, Southern Coastal Limestone, Southwestern Coastal Lowlands, Sierra Bermeja, Vieques, Culebra, Mona, Desecheo; 1–550 m; *fl* & *fr* throughout the year.
Native to Mexico, Central America, South America (scattered). [L5: 366].
Pancha, Trídax, Yerba de San Juan del monte.

Verbesina alata L., Sp. Pl. 2:901. 1753.
Syn.: *Tepion alatum* (L.) Britton (1917).
Herb in Northern Coastal Lowlands (Piñones Forest Reserve, Trujillo Alto), Northern Limestone Hills, Río Abajo Forest Reserve, Eastern Cordillera (Corozal, Aibonito), Central Cordillera, Western Cordillera (Maricao); 10–500 m; *fl* & *fr* throughout the year.
Greater Antilles, Virgin Islands, Lesser Antilles (St. Kitts, Guadeloupe, Marie Galante). [L5: 368].
Capitaneja, Manzanilla cimarrona.

Verbesina encelioides (Cav.) Benth. & Hook. f. ex A. Gray in W.H. Brewer, S. Watson, & A. Gray, Bot. California 1:350. 1876.
Bas.: *Ximenesia encelioides* Cav. (1793–1794).

Suffrutescent herb in Northern Limestone Hills (Camuy), Southeastern Foothills (Salinas), Southern Coastal Lowlands (Santa Isabel), Southern Coastal Limestone, Guánica Forest Reserve, Southwestern Coastal Lowlands; 5–200 m; *fl* & *fr* Jun–Mar.

Native to southwestern and south-central United States, Mexico. [L5: 369].

Wedelia acapulcensis Kunth var. **parviceps** (S.F. Blake) Strother, Syst. Bot. Monogr. 33:58. 1991.
Bas.: *Wedelia parviceps* S.F. Blake (1922).

Herb in Central Cordillera (Adjuntas); 470 m; *fl* & *fr* Sep (one specimen). [*Breckon 6503* (MAPR)].

Native to Mexico, Central America.

Wedelia calycina Rich. in Persoon, Syn. Pl. 2(2):490. 1807.
Syn.: *Wedelia parviflora* Rich. (1807). *Wedelia calycina* Rich. var. *parviflora* (Rich.) Alain (1968).
Wedelia lanceolata DC. (1836).

Shrub in Northern Coastal Lowlands (Fajardo and adjacent islands), Northern Limestone Hills, Río Abajo Forest Reserve, Eastern Coast, Sierra de Cayey, Las Tetas de Cayey, Southeastern Foothills (Coamo), Southern Coastal Limestone, Quebrada de Los Cedros, Maricao Forest Reserve, Guánica Forest Reserve, Southwestern Coastal Lowlands, Sierra Bermeja, Vieques, Culebra, Mona, Desecheo; 1–825 m; *fl* & *fr* throughout the year.

Bahamas, Greater Antilles (Cuba, Hispaniola), Virgin Islands, Lesser Antilles, Trinidad, southern Central America (Panama), northern South America to Brazil, Peru. [L5: 376, 377].

Notes: *Wedelia calycina* is a highly variable taxon and neither *Wedelia parviflora* nor *Wedelia lanceolata* are sufficiently distinct from it as to merit separate taxonomic status; in fact, according to Gary Breckon (pers. comm.), the three form a continuum along a moisture gradient. V.A. Funk and J.F. Pruski in Acevedo-Rodríguez (1996) misapplied the name *Wedelia fruticosa* Jacq. to this taxon.

Margarita de las rocas.

Wedelia reticulata DC., Prodr. 5:540. 1836.

Shrub in Northern Coastal Lowlands (Manatí), Northern Limestone Hills, Sierra de Cayey, Southeastern Foothills, Eastern Cordillera (Aibonito), Central Cordillera, Southern Coastal Limestone, Susúa Forest Reserve; 20–875 m; *fl* & *fr* throughout the year.

Greater Antilles (Cuba, Hispaniola). [L5: 378].

Margarita de monte.

Xanthium strumarium L., Sp. Pl. 2:987. 1753.
Syn.: *Xanthium chinense* Mill. (1768).
Xanthium occidentale Bertol. (1822).

Herb in Eastern Coast (Punta Guayanés), Vieques; 1–50 m; *fl* & *fr* Oct (two specimens). [*Vivaldi a.n. 5539* (UPR); *Woodbury a.n. 151757* (UPR); *Proctor 48426* (SJ)].

Bahamas, Greater Antilles, Lesser Antilles (Antigua, Nevis, Martinique), northern temperate zone. [L5: 380].

Bardana.

Youngia japonica (L.) DC., Prodr. 7(1):194. 1838.
Bas.: *Prenanthes japonica* L. (1767).
Syn.: *Crepis japonica* (L.) Benth. (1861).

Herb in Northern Coastal Lowlands, Northern Limestone Hills, Río Abajo Forest Reserve, Sierra de Luquillo, Eastern Coast, Sierra de Cayey, Eastern and Central Cordillera, Southern Coastal Limestone, Western Cordillera, Western Coast (Mayagüez), Vieques; 10–1250 m; *fl* & *fr* throughout the year.

Native to Asia. [L5: 382].

Zinnia elegans Jacq., Icon. Pl. Rar. 3:15, *t. 589*. 1792, *nom. cons. prop.*
Syn.: *Crassina elegans* (Jacq.) Kuntze (1891).

Herb in Piñones Forest Reserve, Northern Limestone Hills (Hatillo), Sierra de Luquillo, Caguas Valley (Caguas), Sierra de Cayey, Western Coast (Aguada); 1–400 m; *fl* & *fr* Jan–Oct.

Native to northern and central Mexico. [L5: 384].

Clavel.

Zinnia peruviana (L.) L., Syst. Nat. (ed 10) 2:1221. 1759.
Bas.: *Chyrsogonum peruvianum* L. (1753).
Syn.: *Crassina multiflora* (L.) Kuntze (1891).

Herb in Central Cordillera (Ponce), Southern Coastal Lowlands (Coamo), Southwestern Coastal Lowlands; 50–400 m; *fl* & *fr* Jun, Sep, Jan.

Native to southwestern United States (Arizona), Mexico, Central America, western South America and Venezuela. [L5: 384].

Cabrón, Clavelón, Escopetón, Eterno, Zinia, Zinia del monte.

TAXA NOT RECENTLY COLLECTED

Baccharis dioica Vahl, Symb. Bot. 3:98, *t. 74*. 1794.

Shrub collected in Vieques on 17 Feb 1914 [*Shafer 2941* (NY)].

West Indies. [L5: 244].

Gnaphalium polycaulon Pers., Syn. Pl. 2(2):421. 1807 "policaulon."

Herb collected in Río Piedras on 10 Feb 1914 [*Johnson & Stevenson 1296* (UPR)] and on 12 Mar 1917 [*Stevenson 6281* (UPR)], above Villalba on 19 Feb 1922 [*Britton & Earle 6074* (UPR)] and in Cayey on 5 Jan 1938 [*Otero 424* (UPR)].

Native to India.

Pectis carthusianorum Less., Linnaea 6(4):712. 1831.

Herb collected in Cabo Rojo on 17 Jan 1885 [*Sintenis 676* (NY)], between Rincón and Aguada on 4 Dec 1886 [*Sintenis 5546* (NY)] and in Cabo Rojo on 28 Sep 1913 [*Stevens 3194* (NY)].

Greater Antilles (Cuba, Hispaniola), northern South America. [L5: 326].

Pectis tenuicaulis Urb., Symb. Antill. 1(3):468. 1899.

Prostrate herb collected in Cabo Rojo on 18 Feb 1885 [*Sintenis 561* (GOET), scan seen], in Ponce on 18 Nov 1902 [*Heller 6058* (NY)], in Lares on 22 Nov 1913 [*Stevens & Hess 4958* (NY)], at Punta Casabe (Cabo Rojo) on 26 Feb 1915 [*Britton et al. 4679* (NY)] and in Boquerón on 28 Feb 1929 [*Britton & Britton 9250* (NY)].

Greater Antilles (Jamaica), Lesser Antilles (Antigua, Guadeloupe, Martinique). [L5: 330].

Sphagneticola gracilis (Rich.) Pruski, Novon 6(4): 412. 1996.

Bas.: *Wedelia gracilis* Rich. (1807).

Herb collected in Río Piedras on 21 May 1912 [*Johnston 334* (NY)] and also, according to Britton & Wilson (1925), collected by Schwanecke in eastern Puerto Rico prior to 1849.

Greater Antilles. [L5: 377].

TAXA CITED BY PREVIOUS AUTHORS, BUT NOT INCLUDED

CITED DUE TO MISIDENTIFICATION

Emilia coccinea (Sims) G. Don.

Herb cited by Liogier (1997) and Liogier & Martorell (2000) as of sporadic occurrence. All specimens identified as this taxon that I have thus far seen are referable to *Emilia fosbergii*. [L5: 275].

Notes: Britton & Wilson (1926) misapplied the above name to *Emilia fosbergii*.

ONLY UNDER CULTIVATION

Ayapana triplinervis (Vahl) R.M. King & H. Rob.

Syn.: *Eupatorium triplinerve* Vahl.

Herb cited by Britton & Wilson (1925) as escaped from cultivation into woodlands at higher elevations; subsequently cited by Liogier & Martorell (1982, 2000) and Liogier (1997). [L5: 295].

Helianthus annuus L.

Herb cited by Liogier & Martorell (1982, 2000) and Liogier (1997) as cultivated and probably becoming naturalized. [L5: 305].

Lactuca sativa L.

Herb cited by Liogier & Martorell (1982, 2000) and Liogier (1997) as occasionally spontaneous after cultivation for its edible leaves. [L5: 309].

Xerochrysum bracteatum (Vent.) Tzvelev. Syn.: *Helichrysum bracteatum* (Vent.) Haw.

Herb cited by Liogier & Martorell (1982, 2000) and Liogier (1997) as cultivated and escaping. [L5: 307].

BALANOPHORACEAE

TAXA NOT RECENTLY COLLECTED

Scybalium jamaicense (Sw.) Schott & Endl., Melet. Bot. 12. 1832.

Bas.: *Cynomorium jamaicense* Sw. (1788).

Herb collected in Adjuntas on 28 Apr 1886 [*Sintenis 4238* (NY)].

Greater Antilles. [L1: 115].

BALSAMINACEAE

Impatiens balsamina L., Sp. Pl. 2:938. 1753.

Herb in Sierra de Luquillo, Eastern Cordillera (Aguas Buenas); 300–700 m; *fl* & *fr* Apr–May (two specimens). [*Liogier 28609* (UPR); *Ruiz 55* (UPRRP)].

Native to India, Myanmar. [L3: 73].

Espuela de galán.

Impatiens walleriana Hook. f. in Oliver, Fl. Trop. Afr. 1:302. 1868.

Herb in Northern Limestone Hills, Sierra de Luquillo, Central Cordillera, Maricao Forest Reserve; 225–1200 m; *fl* & *fr* Jul–Apr.

Native to eastern and southern tropical Africa. [L3: 74].

Alegría, Besos, Miramelinda.

BASELLACEAE

Anredera vesicaria (Lam.) C.F. Gaertn., Suppl. Carp. 2:176, *t. 213*, *no. 4*. 1807.

Bas.: *Basella vesicaria* Lam. (1785).

Syn.: *Boussingaultia leptostachys* Moq. (1849). *Anredera leptostachys* (Moq.) Steenis (1957).

Vine in Northern Coastal Lowlands, Northern Limestone Hills, Eastern Coast, Las Tetas de Cayey, Las Piedras Chiquitas, Southeastern Foothills, Eastern Cordillera (Cayey), Central Cordillera (Ponce, Yauco), Southwestern Coastal Lowlands (Lajas), Sierra Bermeja, Vieques, Culebra; 1–600 m; *fl* Apr–Jun, Sep–Nov, *fr* (?).

West Indies, southern United States (Florida, Texas), Mexico, Central America, northern South America and Peru. [L1: 211; AR: 114].

Consuelda, Suelda con suelda.

TAXA CITED BY PREVIOUS AUTHORS, BUT NOT INCLUDED

ONLY UNDER CULTIVATION

Anredera cordifolia (Ten.) Steenis.

Vine cited by Liogier & Martorell (1982, 2000), Liogier (1985) as grown as an ornamental and by Acevedo-Rodríguez (2005) as cultivated as an ornament and naturalized, though uncommon. [L1: 209; AR: 113].

Notes: Britton & Wilson (1924) misapplied the name *Boussingaultia baselloides* Kunth, and Liogier & Martorell (1982, 2000) and Liogier (1985) misapplied *Anredera baselloides* (Kunth) Baill., to this taxon.

Basella alba L. Syn.: *Basella rubra* L.

Vine cited by Liogier & Martorell (1982, 2000) and Liogier (1985) as sparingly cultivated at lower elevations and by Acevedo-Rodríguez (2005) as much cultivated as a vegetable. [L1: 212; AR: 104].

BATACEAE

Batis maritima L., Syst. Nat. (ed. 10) 2:1289. 1759.

Suffrutescent herb along coast in Northern Coastal Lowlands, Eastern Coast (Cayo Santiago), Southern Coastal Lowlands, Guánica Forest Reserve, Southwestern Coastal Lowlands, Vieques, Culebra; 0–1 m; *fl* & *fr* throughout the year.

West Indies, southern United States, Mexico, Central America, northern South America to Brazil, Peru. [L1: 45].

Barilla, Planta de sal.

BEGONIACEAE

Begonia decandra Pav. ex A. DC., Ann. Sci. Nat., Bot., sér. 4 11:122. 1859.

Herb in Río Abajo Forest Reserve, Sierra de Luquillo, Caguas Valley (Caguas), Sierra de Cayey, Eastern Cordillera (La Torrecilla), Central Cordillera; 200–1300 m; *fl* & *fr* throughout the year.

Endemic. [L3: 303].

Begonia, Begonia de sierra, Begonia de poeta, Begonia nativa, Corazón de poeta.

Begonia falcifolia Hook. f., Bot. Mag. 94: *t. 5707*. 1868.

Herb in Northern Limestone Hills (Ciales), Maricao Forest Reserve; 150–250 m; *fl* Aug–Oct (two specimens), *fr* (?). [*Proctor 47212* (SJ); *Acevedo-Rodríguez 11884* (UPRRP)].

Native to Peru.

Notes: Both of the above specimens were identified by S.F. Smith (US) as 'cf. *falcifolia*.'

Begonia heracleifolia Schltdl. & Cham., Linnaea 5(4):603. 1830.

Herb in Northern Limestone Hills, Río Abajo Forest Reserve, Southeastern Foothills (Cayey), Central Cordillera (Ponce); 200–500 m; *fl* & *fr* Jan–May.

Native to Mexico, northern Central America and Panama. [L3: 305].

Begonia.

Begonia hirtella Link, Enum. Hort. Berol. Alt. 2:396. 1822.

Herb in Northern Coastal Lowlands (San Juan), Sierra de Luquillo, Central Cordillera (Adjuntas); 3–1000 m; *fl* & *fr* Nov–Mar.

Native to western tropical South America. [L5: 409].

Begonia nelumbiifolia Schltdl. & Cham., Linnaea 5(4):604. 1830.

Herb in Northern Limestone Hills (Florida), Río Abajo Forest Reserve, Sierra de Luquillo, Central Cordillera (Ponce); 150–500 m; *fl* & *fr* Feb–Jun.

Native to Mexico, Central America, northern South America (Colombia). [L3: 305].

Begonia.

TAXA CITED BY PREVIOUS AUTHORS, BUT NOT INCLUDED

ONLY UNDER CULTIVATION

Begonia coccinea Hook.

Herb cited by Liogier & Martorell (1982, 2000) and Liogier (1994) as cultivated and escaped in the eastern mountains. [L3: 302].

Begonia convolvulacea (Klotzsch) A. DC.

Herb cited by Liogier & Martorell (1982, 2000) and Liogier (1994) as cultivated and in the Aibonito-Naranjito region. [L3: 303].

Begonia cucullata Willd. Syn.: *Begonia cucullata*

Willd. var. *hookeri* (Sweet) L.B. Sm. & B.G. Schub.

Herb cited by Liogier & Martorell (1982, 2000) and Liogier (1994) as cultivated and escaped. [L3: 303].

BIGNONIACEAE (including HYDROPHYLLACEAE)

Amphilophium paniculatum (L.) Kunth in Humboldt, Bonpland & Kunth, Nov. Gen. Sp. 3:116 [folio]; 3:149 [quarto]. 1819.
Bas.: *Bignonia paniculata* L. (1753).

Vine in Río Abajo Forest Reserve; 300–350 m; *fl* (?) & *fr* (?). [*Axelrod 10489* (UPRRP); *Acevedo-Rodríguez 10457, 10620* (US), neither seen].

Lesser Antilles, Trinidad, Mexico, Central America, South America. [L4: 515; AR: 117].

Liana de cuello.

Amphitecna latifolia (Mill.) A.H. Gentry, Taxon 25(1):108. 1976.
Bas.: *Crescentia latifolia* Mill. (1768).
Syn.: *Enallagma latifolia* (Mill.) Small (1913).

Tree in Northern Coastal Lowlands, Northern Limestone Hills (Ciales), Susúa Forest Reserve, Vieques; 1–285 m; *fl* Feb, May, Sep–Oct, *fr* Feb–Jun, Sep–Nov.

Greater Antilles, Virgin Islands, Lesser Antilles (Dominica, Martinique, St. Vincent), Trinidad and Tobago, southeastern United States (Florida), Mexico, Central America, northern South America. [L4: 516].

Cocoa, Higüerillo, Higüerita.

Arrabidaea chica (Humb. & Bonpl.) Verl., Rev. Hort. (Paris) 40:154. 1868.
Bas.: *Bignonia chica* Humb. & Bonpl. (1807).

Vine in Northern Limestone Hills (Bayamón); 75 m; *fl* (?), *fr* Apr (one specimen). [*Caraballo 1456, 1480, 2855* (UPR)].

Greater Antilles (Hispaniola), Lesser Antilles (Dominica), Trinidad and Tobago, Mexico, Central America, South America. [L4: 518; AR: 118].

Crescentia cujete L., Sp. Pl. 2:626. 1753.

Tree in Northern Coastal Lowlands, Northern Limestone Hills (Quebradillas), Eastern Coast, Southeastern Foothills, Southern Coastal Lowlands, Southwestern Coastal Lowlands, Vieques, Mona; 1–100 m; *fl* & *fr* throughout the year.

West Indies, Central America, South America (scattered). [L4: 522].

Notes: This tree is also often planted as an ornamental.

Higüera, Higüero.

Crescentia linearifolia Miers, Trans. Linn. Soc. London 26(1):172. 1868.

Tree in Guánica Forest Reserve, Southwestern Coastal Lowlands, Sierra Bermeja; 60–200 m; *fl* May–Feb, *fr* Aug–Feb, May.

Greater Antilles (Hispaniola), Virgin Islands, Lesser Antilles (St. Martin, St. Barts). [L4: 523].

Higüerita, Higüerito.

Crescentia portoricensis Britton, Bull. Torrey Bot. Club 43(9):457. 1916.

Shrub in Susúa Forest Reserve, Maricao Forest Reserve; 200–850 m; *fl* Jan–Apr, Jul, *fr* Feb–Oct.

Endemic. [L4: 524].

Notes: This taxon is on the U.S. Fish & Wildlife Service list of threatened and endangered plants.

Higüero de sierra.

Cydista aequinoctialis (L.) Miers, Proc. Roy. Hort. Soc. London 3:191. 1863.
Bas.: *Bignonia aequinoctialis* L. (1753).

Woody vine in the Northern Coastal Lowlands, Eastern Coast (Punta Guayanés), Caguas Valley (Gurabo), Southern Coastal Lowlands (Patillas); 1–50 m; *fl* Feb–Jul, *fr* (?).

Greater Antilles (Cuba, Hispaniola), Lesser Antilles, Mexico, Central America, northern South America to Brazil, Bolivia. [L4: 525; AR: 118].

Bejuco blanco, Liana de sierra.

Distictis lactiflora (Vahl) A. DC., Prodr. 9:191. 1845.
Bas.: *Bignonia lactiflora* Vahl (1794).

Woody vine in Northern Coastal Lowlands, Northern Limestone Hills, Río Abajo Forest Reserve, Las Piedras Chiquitas, Southeastern Foothills, Eastern Cordillera (Corozal, Aibonito), Southern Coastal Lowlands, Quebrada de Los Cedros, Susúa Forest Reserve, Guánica Forest Reserve, Southwestern Coastal Lowlands, Sierra Bermeja, Vieques, Culebra, Mona; 1–650 m; *fl* & *fr* throughout the year.

Greater Antilles (Cuba, Hispaniola), Virgin Islands. [L4: 526; AR: 120].

Liana fragrante, Viuda.

Macfadyena unguis-cati (L.) A.H. Gentry, Brittonia 25(3):236. 1973.
Bas.: *Bignonia unguis-cati* L. (1753).

Woody vine in Northern Coastal Lowlands, Northern Limestone Hills, Río Abajo Forest Reserve, Las Tetas de Cayey, Las Piedras Chiquitas, Southeastern Foothills, Eastern Cordillera (Aibonito), Southern Coastal Lowlands (Coamo, Caja de Muertos), Monte Guilarte Forest Reserve, Quebrada de Los Cedros, Susúa Forest Reserve, Maricao Forest

Reserve, Guánica Forest Reserve, Southwestern Coastal Lowlands, Sierra Bermeja, Western Coast (Mayagüez), Vieques, Culebra, Mona, Desecheo; 1–1050 m; *fl* & *fr* throughout the year.

West Indies, Mexico, Central America, South America. [L4: 532; AR: 122].

Notes: Britton & Wilson (1925) used *Batocydia unguis* (L. ex DC.) Mart. ex DC., a *nom. inval.*, for the name of this taxon.

Bejuco de gato, Liana uñada, Paz y justicia, Pegapalo, Uña de gato.

Mansoa hymenaea (DC.) A.H. Gentry, Ann. Missouri Bot. Gard. 66(4):782. 1980.
Bas.: *Bignonia hymenaea* DC. (1845).

Woody vine planted and persistent in Northern Limestone Hills (Barceloneta); 25–50 m; *fl* Jan–Feb (two specimens), *fr* Jan–Feb (one specimen).

Native to Mexico, Central America, northern South America to Brazil, Peru. [L4: 534; AR: 123].

Bejuco de ajo, Mata de ajo.

Nama jamaicensis L., Syst. Nat (ed. 10) 2:950. 1759.
Syn.: *Marilaunidium jamaicense* (L.) Kuntze (1891).

Herb in Northern Limestone Hills (Toa Baja), Southeastern Foothills (Salinas), Guánica Forest Reserve; 1–115 m; *fl* Nov, Mar–Apr, *fr* (?).

West Indies, Mexico, Central America, Venezuela. [L4: 297].

Podranea ricasoliana (Tanfani) Sprague in W.T. Thiselton-Dyer, Fl. Cap. 4(2):450. 1904.
Bas.: *Tecoma ricasoliana* Tanfani (1887).

Scandent shrub/vine planted and persistent in Northern Coastal Lowlands (Laguna Tortuguero), Eastern Cordillera (Cayey, Aibonito); 2–500 m; *fl* Oct–Feb, *fr* (?). [*Breckon 6288* (MAPR); *Acevedo-Rodríguez 7920* (US)].

Native to South Africa. [L4: 538; AR: 126].

Millonaria, Ricasolana.

Spathodea campanulata P. Beauv., Fl. Oware 1:47, *tt.* 27, 28. 1805.

Tree in Northern Coastal Lowlands, Sierra de Luquillo, Las Tetas de Cayey, Eastern Cordillera (Cayey, Comerío), Susúa Forest Reserve, Maricao Forest Reserve, Southwestern Coastal Lowlands, Western Coast; 1–825 m; *fl* throughout the year, *fr* Oct–Jun.

Native to western tropical Africa. [L4: 544].

Meaito, Tulipán africano.

Tabebuia haemantha (Bertero ex Spreng.) A. DC., Prodr. 9:214. 1845.
Bas.: *Bignonia haemantha* Bertero ex Spreng. (1825).

Tree in Las Tetas de Cayey, Las Piedras Chiquitas, Southeastern Foothills, Eastern Cordillera (Aibonito), Southern Coastal Lowlands (Juana Díaz), Susúa Forest Reserve, Maricao Forest Reserve, Southwestern Coastal Lowlands (Cabo Rojo), Cerro Las Mesas; 75–850 m; *fl* Dec–Sep, *fr* Apr–Sep.

Endemic. [L4: 549].

Roble bobo, Roble cimarrón, Roble colorado.

Tabebuia heterophylla (DC.) Britton, Ann. Missouri Bot. Gard. 2:48. 1915.
Bas.: *Raputia heterophylla* DC. (1822).
Syn.: *Tabebuia lucida* Britton (1915).

Tree in Northern Coastal Lowlands, Northern Limestone Hills (Bayamón, Toa Baja), Río Abajo Forest Reserve, Sierra de Luquillo, Eastern Coast, Sierra de Cayey, Southeastern Foothills, Central Cordillera (Los Tres Picachos), Southern Coastal Lowlands, Southern Coastal Limestone, Maricao Forest Reserve, Guánica Forest Reserve, Southwestern Coastal Lowlands, Sierra Bermeja, Vieques, Culebra, Mona; 1–800 m; *fl* & *fr* throughout the year.

West Indies, Tobago. [L4: 549].

Notes: Britton & Wilson (1925) misapplied the name *Tabebuia pallida* (Lindl.) Miers to this taxon. This tree is also often planted as an ornamental.

Prieto, Roble, Roble blanco, Roble prieto, Roble de costa, Roble de yugo, Roble de Mona.

Tabebuia karsoana Trejo, Kew Bull. 64(2):296, *f. 1.* 2009

Tree in Northern Limestone Hills, Río Abajo Forest Reserve; 25–350 m; *fl* Jun–Aug, Dec–Jan, *fr* Jul (one specimen).

Endemic.

Tabebuia rigida Urb., Symb. Antill. 1(3):404. 1899.

Tree in Sierra de Luquillo; 475–1000 m; *fl* throughout the year, *fr* Jul–Aug (two specimens).

Endemic. [L4: 551].

Roble de guayo, Roble de sierra.

Tabebuia schumanniana Urb., Symb. Antill. 1(3):404. 1899.

Tree in Sierra de Cayey, Central Cordillera, Maricao Forest Reserve; 500–1300 m; *fl* Jan–Oct, *fr* Mar–Apr, Aug–Oct.

Endemic. [L4: 551].

Roble cimarrón, Roble colorado, Roble de sierra.

Tecoma stans (L.) Juss. ex Kunth in Humboldt, Bonpland & Kunth, Nov. Gen. Sp. 3:112 [folio]; 3:144 [quarto]. 1819.
Bas.: *Bignonia stans* L. (1763).

Shrub/small tree in Northern Coastal Lowlands, Eastern Coast, Southeastern Foothills, Central Cordillera (lower southern slopes), Southern Coastal Lowlands, Guánica Forest Reserve, Southwestern Coastal Lowlands, Vieques; 1–300 m; *fl* throughout the year, *fr* Jun–Mar.

Native to southwestern United States, Mexico, Central America, South America. [L4: 553].

Roble amarillo, Ruibarba, Saúco amarillo.

TAXA CITED BY PREVIOUS AUTHORS, BUT NOT INCLUDED

ONLY UNDER CULTIVATION

Catalpa longissima (Jacq.) Dum. Cours.
Tree cited by Liogier & Martorell (1982, 2000) and Liogier (1995) as rare as an introduced ornamental and shade tree, western Puerto Rico and in Cambalache Forest Reserve. [L4: 520].

Crescentia alata Kunth.
Tree cited by Liogier & Martorell (1982, 2000) and Liogier (1995) as cultivated and persistent. [L4: 522].

Jacaranda mimosifolia D. Don.
Tree cited by Liogier & Martorell (1982, 2000) and Liogier (1995) as planted for ornament from cuttings and from seeds. [L4: 528].

Kigelia africana (Lam.) Benth.
Tree cited by Liogier & Martorell (1982, 2000) and Liogier (1995) as occasionally planted and persistent. [L4: 530].

Mansoa alliacea (Lam.) A.H. Gentry.
Vine cited by Liogier & Martorell (1982, 2000) and Liogier (1995) as cultivated and persistent, but cited by Acevedo-Rodríguez (2005) as only cultivated. [L4: 534; AR: 123].

Parmentiera aculeata (Kunth) Seem.
Tree cited by Liogier & Martorell (1982, 2000) and Liogier (1995) as occasionally planted in gardens, rarely naturalized. [L4: 536].

Parmentiera cereifera Seem.
Tree cited by Liogier & Martorell (1982, 2000) and Liogier (1995) as planted as an ornamental in gardens, rarely escaped. [L4: 536].

Pyrostegia venusta (Ker Gawl.) Miers.
Vine cited by Liogier & Martorell (1982, 2000) and Liogier (1995) as planted as an ornamental and persistent, but cited by Acevedo-Rodríguez (2005) as only cultivated. [L4: 540; AR: 126].

Saritaea magnifica (Sprague ex Steenis) Dugand.
Vine cited by Liogier & Martorell (1982, 2000) and Liogier (1995) as cultivated as an ornamental and persistent, but cited by Acevedo-Rodríguez (2005) as only cultivated. [L4: 540; AR: 128].

Tabebuia aurea (Silva Manso) Benth. & Hook. f. ex S. Moore. Syn.: *Tabebuia argentea* (Bureau & K. Schum.) Britton.
Tree cited by Liogier & Martorell (1982, 2000) and Liogier (1995) as planted as an ornamental and persistent. [L4: 547].

Tabebuia chrysantha (Jacq.) G. Nicholson. Syn.: *Tabebuia glomerata* Urb.
Tree cited by Liogier & Martorell (1982, 2000) and Liogier (1995) as planted as an ornamental and persistent. [L4: 548].

Tabebuia donnell-smithii Rose.
Tree cited by Liogier (1995) and Liogier & Martorell (2000) as introduced and rare, persistent and becoming naturalized. [L4: 548].

Tabebuia rosea (Bertol.) A. DC.
Tree cited by Liogier & Martorell (1982, 2000) and Liogier (1995) as planted as an ornamental and naturalized. [L4: 551].

Tabebuia serratifolia (Vahl) G. Nicholson.
Tree cited by Liogier & Martorell (1982, 2000) and Liogier (1995) as occasionally planted and naturalized. [L4: 552].

Tecoma capensis (Thunb.) Lindl. Syn.: *Tecomaria capensis* (Thunb.) Spach.
Scandent shrub cited by Liogier & Martorell (1982, 2000) and Liogier (1995) as planted for ornament and persistent and by Acevedo-Rodríguez (2005) as only cultivated. [L4: 553; AR: 130].

Tynanthus polyanthus (Bureau) Sandwith. Syn.: *Tynanthus caryophylleus* (Bello) Alain. *Bignonia caryophyllea* Bello.
Vine cited by Britton & Wilson (1925) from an old sterile collection in Añasco; subsequently cited by Liogier & Martorell (1982, 2000), and Liogier (1995) as rare in western districts and by Acevedo-Rodríguez (2005) as only cultivated. [L4: 556; AR: 130].

Notes: Liogier & Martorell (1982, 2000), and Liogier (1995) used the orthographic variant '*Tynnanthus*' for the name of the genus.

BIXACEAE (including COCHLOSPERMACEAE)

Bixa orellana L., Sp. Pl. 1:512. 1753.

Tree planted and persistent in Northern Coastal Lowlands, Northern Limestone Hills, Río Abajo Forest Reserve, Sierra de Luquillo, Eastern Coast (Naguabo), Caguas Valley (Caguas), Sierra de Cayey, Las Tetas de Cayey, Southeastern Foothills (Arroyo), Eastern Cordillera (Aguas Buenas), Central Cordillera (Ponce), Western Cordillera, Southwestern Coastal Lowlands (Lajas), Western Coast (Mayagüez); 1–825 m; *fl* May–Dec, *fr* throughout the year.

Native to Mexico, Central America, South America. [L3: 244].

Achiote, Bija.

Cochlospermum vitifolium (Willd.) Spreng., Syst. Veg. 2:596. 1825.

Bas.: *Bombax vitifolium* Willd. (1809).

Tree planted and persistent in Northern Coastal Lowlands, Central Cordillera (Adjuntas), Western Coast (Mayagüez); 1–500 m; *fl* Jan–Mar, *fr* Feb–Mar (two specimens).

Native to Mexico, Central America, northern South America to Brazil, Bolivia. [L3: 246].

Emperatriz de la selva, Rosa china, Rosa de Maximiliano, Rosa imperial.

BORAGINACEAE

Bourreria succulenta Jacq., Enum. Syst. Pl. 14. 1760.

Syn.: *Bourreria revoluta* Kunth (1818). *Bourreria succulenta* Jacq. var. *revoluta* (Kunth) O.E. Schulz (1911).

Tree in Northern Coastal Lowlands, Northern Limestone Hills (Isabela, Rincón), Eastern Coast, Las Piedras Chiquitas, Southeastern Foothills, Southern Coastal Lowlands, Southern Coastal Limestone, Susúa Forest Reserve, Guánica Forest Reserve, Southwestern Coastal Lowlands, Sierra Bermeja, Western Coast, Vieques, Culebra, Mona, Desecheo; 1–550 m; *fl* throughout the year, *fr* May–Feb.

West Indies, Trinidad and Tobago, southern Mexico. [L4: 302].

Capá, Cerecillo, Doncella, Guayabacón, Muñeco colorado, Palo de puerco, Palo de vaca, Roble de guayo.

Bourreria virgata (Sw.) G. Don, Gen. Hist. 4:389. 1838.

Bas.: *Ehretia virgata* Sw. (1788).

Syn.: *Bourreria domingensis* (DC.) Griseb. (1862).

Shrub/small tree in Northern Limestone Hills, Southern Coastal Limestone, Guánica Forest Reserve, Southwestern Coastal Lowlands (Cabo Rojo), Vieques; 1–250 m; *fl* throughout the year, *fr* Jul–Apr.

Greater Antilles (Cuba, Hispaniola), Lesser Antilles (Dominica), southeastern United States (Florida). [L4: 302, 303].

Notes: *Bourreria domingensis* has previously been treated as distinct from this taxon primarily because its leaves are smooth above rather than scabrous. The fact that the leaves of some specimens are only slightly scabrous, making them difficult to assign to one or the other taxon, calls into question the need for two distinct taxa. Britton & Wilson (1925) had already commented that the one was perhaps not distinct from the other; following suit, I treat them as conspecific.

Hoja menuda, Palo de vaca, Roble de guayo.

Cordia alliodora (Ruiz & Pav.) Oken, Allg. Naturgesch. 3(2):1098. 1841.

Bas.: *Cerdana alliodora* Ruiz & Pav. (1799).

Tree in Northern Coastal Lowlands, Northern Limestone Hills, Río Abajo Forest Reserve, Caguas Valley (Juncos), Sierra de Cayey, Las Tetas de Cayey, Southeastern Foothills, Central Cordillera (Adjuntas), Western Cordillera, Sierra Bermeja, Western Coast (Rincón), Vieques; 10–900 m; *fl* Jun–Nov, *fr* Aug–Nov.

Greater Antilles (Cuba, Hispaniola), Virgin Islands (Tortola), Lesser Antilles, Trinidad, Mexico, Central America, northern South America to Brazil, Bolivia. [L4: 307].

Capá, Capá prieto, Muñeco.

Cordia borinquensis Urb., Symb. Antill. 1(3):390. 1899.

Tree in Sierra de Luquillo, Sierra de Cayey, Toro Negro Forest Reserve, Maricao Forest Reserve; 300–1100 m; *fl* & *fr* throughout the year.

Endemic. [L4: 307].

Capá, Capá cimarrón, Moralcillo, Moral prieto, Muñeca, Muñeco, Palo de muñeco.

Cordia collococca L., Fl. Jamaic. 14. 1759 "callococca."

Syn.: *Cordia glabra* L. (1753).

Tree in Northern Coastal Lowlands, Eastern Coast (Cayo Santiago), Southeastern Foothills, Central Cordillera (Ponce), Southern Coastal Limestone (Yauco), Southwestern Coastal Lowlands, Sierra Bermeja, Western Coast, Vieques, Culebra, Mona; 5–475 m; *fl* Feb–Oct, *fr* Mar–Jun.

Antilles, Trinidad and Tobago, Mexico, Central America, northwestern South America and Venezuela. [L4: 308].

Cerezo, Palo de muñeca, Vaca.

Cordia dentata Poir., Encycl. 7:48. 1806.

Shrub/small tree in Northern Coastal Lowlands (Fajardo), Southeastern Foothills (Salinas), Vieques, Culebra; 1–10 m; *fl* Feb, Jun–Nov, *fr* Jun–Jul, Nov.

Antilles, Trinidad, northern South America. [L4: 308].

Notes: Britton & Wilson (1925) misapplied the name *Calyptracordia alba* (Jacq.) Britton, and Liogier & Martorell (1982) misapplied *Cordia alba* (Jacq.) Roem. & Schult., to this taxon.

Capá, Capá blanco, Cereza blanca.

Cordia laevigata Lam., Tabl. Encycl. 1:422. 1792.

Syn.: *Cordia nitida* Vahl (1793).

Tree in Northern Coastal Lowlands, Northern Limestone Hills, Las Piedras Chiquitas, Southeastern Foothills, Central Cordillera (Ponce), Southern Coastal Lowlands (Aguirre), Southwestern Coastal Lowlands, Sierra Bermeja, Vieques, Mona; 1–600 m; *fl* Aug–May, *fr* Oct–Mar.

Greater Antilles (Jamaica, Hispaniola), Virgin Islands, Lesser Antilles (Dominica, Martinique, St. Lucia). [L4: 311].

Capá, Capá colorado, Cerezo, Cerezo del país, Muñeco, Palo de goma.

Cordia obliqua Willd., Sp. Pl. 1(2):1072. 1798.

Tree in Northern Coastal Lowlands, Eastern Cordillera (Cayey), Central Cordillera (Villalba), Southern Coastal Lowlands (Juana Díaz), Guánica Forest Reserve, Southwestern Coastal Lowlands (Lajas), Vieques; 1–800 m; *fl* May–Jul, Nov, *fr* May–Jan.

Native to India. [L4: 312].

Cereza blanca.

Cordia rickseckeri Millsp., Publ. Field Columb. Mus., Bot. Ser. 1(7):522. 1902.

Syn.: *Sebesten rickseckeri* (Millsp.) Britton (1925).

Tree in Northern Coastal Lowlands, Southern Coastal Limestone, Guánica Forest Reserve, Southwestern Coastal Lowlands, Vieques, Culebra; 1–50 m; *fl* Dec–Sep, *fr* Apr–Sep, Dec–Jan.

Virgin Islands. [L4: 313].

Lija, San Bartolomé.

Cordia sulcata DC., Prodr. 9:488. 1845.

Tree in Northern Coastal Lowlands, Guajataca Forest Reserve, Río Abajo Forest Reserve, Sierra de Luquillo, Caguas Valley (Gurabo), Sierra de Cayey, Las Tetas de Cayey, Central Cordillera, Maricao Forest Reserve, Western Coast (Añasco), Vieques; 5–900 m; *fl* & *fr* Feb–Nov.

Greater Antilles (Cuba, Hispaniola), Virgin Islands, Lesser Antilles. [L4: 315].

Moral, Moral de paz.

Cynoglossum amabile Stapf & J.R. Drumm., Bull. Misc. Inform. Kew 1906(6):202. 1906.

Herb in Maricao Forest Reserve; 800 m; *fl* May–Jul (two specimens), *fr* (?). [*Liogier 9357* (MAPR); *Liogier 36780* (UPR)].

Native to Bhutan, western China. [L4: 318].

Heliotropium amplexicaule Vahl, Symb. Bot. 3:21. 1794.

Herb in Northern Coastal Lowlands (San Juan); 10 m; *fl* Jul (one specimen), *fr* (?). [*Proctor 40570* (SJ)].

Native to south-central South America. [L4: 321].

Heliotropium angiospermum Murray, Prodr. Stirp. Gött. 217. 1770.

Syn.: *Schobera angiosperma* (Murray) Britton (1925).

Suffrutescent herb in Northern Coastal Lowlands, Cambalache Forest Reserve, Eastern Coast (Cayo Santiago), Las Piedras Chiquitas, Southeastern Foothills, Eastern Cordillera (Aibonito), Southern Coastal Lowlands, Quebrada de Los Cedros, Guánica Forest Reserve, Southwestern Coastal Lowlands, Sierra Bermeja, Vieques, Culebra, Mona, Desecheo; 1–650 m; *fl* & *fr* throughout the year.

West Indies, Trinidad, south-central United States (Texas), Mexico, Central America, South America. [L4: 321].

Cotorrilla, Rabo de alacrán.

Heliotropium crispiflorum Urb., Symb. Antill. 1(3):394. 1899.

Syn: *Heliotropium microphyllum* Sw. ex Wikstr. (1828).

Heliotropium guanicense Urb. (1899).

Suffrutescent herb/shrub in Southern Coastal Limestone, Guánica Forest Reserve, Mona; 5–100 m; *fl* Oct–Jul, *fr* Oct–Mar.

Greater Antilles (Hispaniola), Virgin Islands (Anegada). [L4: 322, 324].

Notes: Liogier & Martorell (1982) used *Heliotropium microphyllum* Sw., a *nom. nud.*, for the name of this taxon.

Heliotropium curassavicum L., Sp. Pl. 1:130. 1753.

Herb in Northern Coastal Lowlands, Eastern Coast (Cayo Santiago), Southern Coastal Lowlands, Guánica Forest Reserve, Southwestern Coastal

Lowlands, Vieques, Culebra, Mona; 1–30 m; *fl* & *fr* throughout the year.

West Indies, southeastern and south-central United States, Mexico, Central America, South America. [L4: 322].

Cotorrera de playa.

Heliotropium fruticosum L., Syst. Nat. (ed. 10) 2:913. 1759.

Herb in Southwestern Coastal Lowlands (Cabo Rojo), Sierra Bermeja; 15–175 m; *fl* Aug–Feb, *fr* Sep–Feb.

Greater Antilles, south-central United States (Texas, Arizona), Mexico, Central America, northern South America (Venezuela, Colombia) and Brazil. [L4: 323].

Heliotropium indicum L., Sp. Pl. 1:130. 1753.

Syn.: *Tiaridium indicum* (L.) Lehm. (1818).

Herb in Northern Coastal Lowlands, Northern Limestone Hills (San Sebastián, Aguadilla), Eastern Coast, Eastern Cordillera (Corozal), Southern Coastal Lowlands, Southwestern Coastal Lowlands (Cabo Rojo), Western Coast, Vieques, Culebra; 1–200 m; *fl* & *fr* Jan–Oct.

Pantropical and -subtropical. [L4: 324].

Cotorrera, Moco de pavo, Pico de cotorra, Yerba cotorra.

Heliotropium lagoense (Warm.) Gürke in Engler & Prantl, Nat. Pflanzenfam. 4(3a):97. 1893.

Bas.: *Schleidenia lagoensis* Warm. (1868).

Syn.: *Heliotropium antillanum* Urb. (1910).

Herb in Northern Coastal Lowlands (Laguna Tortuguero), Northern Limestone Hills (San Sebastián); 1–200 m; *fl* & *fr* May, Aug. [*Woodbury a.n. 8453* (UPR); *Caudales a.n. 874326* (NY); *Proctor 49903* (SJ)].

Greater Antilles, Trinidad, Central America, South America. [L4: 325].

Heliotropium procumbens Mill., Gard. Dict. (ed. 8): *Heliotropium* no. 10. 1768.

Syn.: *Heliotropium inundatum* Sw. (1788).

Herb in Northern Coastal Lowlands (Laguna Tortuguero), Northern Limestone Hills (San Sebastián), Southern Coastal Lowlands, Southwestern Coastal Lowlands (Cabo Rojo), Western Coast (Mayagüez); 1–200 m; *fl* Feb–Jun, Sep–Nov, *fr* Apr–Jun, Sep–Nov.

Bahamas, Greater Antilles, Lesser Antilles (Antigua), Trinidad, southeasterrn and south-central United States, Mexico, Central America, South America. [L4: 325].

Cotorrera de agua.

Heliotropium ternatum Vahl, Symb. Bot. 3:21. 1794.

Suffrutescent herb in Southern Coastal Limestone (Peñuelas, Guayanilla), Southwestern Coastal Lowlands, Vieques, Culebra; 1–100 m; *fl* throughout the year, *fr* Mar, Jul–Sep, Dec.

West Indies, Trinidad, Mexico, Central America, northern South America. [L4: 326].

Rochefortia acanthophora (DC.) Griseb., Fl. Brit. W.I. 482. 1862.

Bas.: *Ehretia acanthophora* DC. (1845).

Tree in Southeastern Foothills (Salinas), Southern Coastal Lowlands (Salinas), Southern Coastal Limestone (Guayanilla), Guánica Forest Reserve, Southwestern Coastal Lowlands, Sierra Bermeja, Vieques, Mona, Desecheo; 1–200 m; *fl* Aug–Jan, *fr* Nov–Jun.

Antilles, Virgin Islands. [L4: 326].

Juso.

Rochefortia spinosa (Jacq.) Urb., Repert. Spec. Nov. Regni Veg. 13:472. 1915.

Bas.: *Ehretia spinosa* Jacq. (1760).

Syn.: *Rochefortia cuneata* Sw. (1788).

Tree in Northern Coastal Lowlands (Fajardo), Northern Limestone Hills, Guánica Forest Reserve; 5–75 m; *fl* (?), *fr* (?).

West Indies, Mexico, northern Central America, northern South America (Venezuela, Colombia) and Peru. [L4: 328].

Espino.

Tournefortia bicolor Sw., Prodr. 40. 1788.

Vine in Northern Coastal Lowlands (Vega Baja), Northern Limestone Hills, Río Abajo Forest Reserve, Caguas Valley (Caguas), Las Tetas de Cayey, Central Cordillera; 50–1000 m; *fl* Feb–Apr, *fr* Feb–Jun.

Antilles, Trinidad, Mexico, Central America, northern South America to Brazil, Bolivia. [L4: 329; AR: 135].

Nigüita.

Tournefortia filiflora Griseb., Fl. Brit. W.I. 483. 1862.

Tree in Northern Limestone Hills, Río Abajo Forest Reserve, Las Tetas de Cayey, Southeastern Foothills (Coamo), Eastern Cordillera (Cayey), Central Cordillera (Los Tres Picachos, Utuado, Yauco), Southern Coastal Limestone; 2–850 m; *fl* & *fr* throughout the year.

Virgin Islands (St. Croix, St. John), Lesser Antilles. [L4: 330].

Nigua.

Tournefortia gnaphalodes (L.) R. Br. ex Roem. & Schult., Syst. Veg. 4:538. 1819.
Bas.: *Heliotropium gnaphalodes* L. (1759).
Syn.: *Mallotonia gnaphalodes* (L.) Britton (1915). *Argusia gnaphalodes* (L.) Heine (1976).

Shrub in Northern Coastal Lowlands (islands adjacent to Fajardo, Isabela), Vieques, Culebra, Mona; 1 m; *fl* throughout the year, *fr* Feb–Sep.

West Indies, southeastern United States (Florida), Central America, northern South America (Venezuela). [L4: 300].

Nigua de playa, Té del mar, Temporana, Yerba cotorra.

Tournefortia hirsutissima L., Sp. Pl. 1:140. 1753.

Climbing shrub in Northern Coastal Lowlands, Northern Limestone Hills, Caguas Valley (Gurabo), Sierra de Cayey, Las Tetas de Cayey, Las Piedras Chiquitas, Southeastern Foothills (Cayey), Eastern Cordillera (Cayey), Central Cordillera, Southern Coastal Lowlands, Southern Coastal Limestone, Susúa Forest Reserve, Maricao Forest Reserve, Southwestern Coastal Lowlands (Cabo Rojo), Sierra Bermeja, Western Coast (Hormigueros), Vieques, Mona; 1–1000 m; *fl* Feb–Nov, *fr* Mar–Nov.

Antilles, southeastern United States (Florida), Mexico, Central America, northwestern South America and Venezuela. [L4: 330; AR: 135].

Nigua.

Tournefortia maculata Jacq., Enum. Syst. Pl. 14. 1760.
Syn.: *Tournefortia laurifolia* Vent. (1803). *Tournefortia peruviana* Poir. (1816).

Vine in Northern Limestone Hills, Río Abajo Forest Reserve, Sierra de Luquillo, Sierra de Cayey, Las Tetas de Cayey, Eastern Cordillera (Cerro La Peña), Central Cordillera, Susúa Forest Reserve, Maricao Forest Reserve; 50–1200 m; *fl* Jan–Jun, Oct, *fr* Jan–Oct.

Greater Antilles (Cuba, Jamaica), Lesser Antilles, Trinidad and Tobago, Mexico, Central America, northwestern, west-central South America and Venezuela. [L4: 330, 332; AR: 135].

Bejuco de masa.

Tournefortia scabra Lam., Tabl. Encycl. 1:417. 1792.

Vine in Northern Limestone Hills (Isabela), Southern Coastal Limestone; 2–15 m; *fl* Oct–Feb, *fr* Jan–Feb, May–Jun.

Greater Antilles (Cuba, Hispaniola). [L4: 332; AR: 139].

Tournefortia volubilis L., Sp. Pl. 1:140. 1753.
Syn.: *Tournefortia microphylla* Bertero ex Spreng. (1824).

Vine in Northern Coastal Lowlands, Northern Limestone Hills (Aguadilla), Eastern Coast, Las Piedras Chiquitas, Southeastern Foothills (Coamo), Eastern Cordillera (Cidra), Southern Coastal Lowlands (Patillas), Southern Coastal Limestone, Guánica Forest Reserve, Southwestern Coastal Lowlands, Vieques, Culebra, Mona, Desecheo; 1–650 m; *fl* & *fr* throughout the year.

West Indies, southeastern United States (Florida), Mexico, Central America, northwestern South America and Venezuela. [L4: 332, 333; AR: 138].

Notes: Because of the extreme variability of this taxon throught its range, James Miller (pers. comm.) sees no point in segregating a separate taxon primarily on the basis of leaf size and thus considers *Tournefortia microphylla* conspecific with it.

Nigua enradadera.

Varronia bellonis (Urb.) Britton in Britton & Wilson, Sci. Surv. Porto Rico & Virgin Islands 6(1):128. 1925.
Bas.: *Cordia bellonis* Urb. (1899).

Shrub in Northern Limestone Hills (Ciales), Río Abajo Forest Reserve, Susúa Forest Reserve, Maricao Forest Reserve; 250–900 m; *fl* Jun, Sep–Jan, *fr* Nov–Jan.

Endemic. [L4: 307; AR: 132].

Notes: This taxon is on the U. S. Fish & Wildlife Service list of threatened and endangered plants.

Varronia bullata L. subsp. **humilis** (Jacq.) Feuillet, J. Bot. Res. Inst. Texas 2(2):837. 2008.
Bas.: *Varronia humilis* Jacq. (1760).
Syn.: *Cordia globosa* (Jacq.) Kunth var. *humilis* (Jacq.) I.M. Johnst. (1949).

Shrub in Guánica Forest Reserve, Southwestern Coastal Lowlands, Sierra Bermeja, Vieques, Mona; 1–300 m; *fl* Feb–Nov, *fr* May–Feb.

West Indies, Trinidad, southeastern United States (Florida), Mexico, Central America, northern South America (Venezuela, Colombia) and Ecuador. [L4: 309].

Notes: Britton & Wilson (1925) used *Varronia globosa* Jacq., *sensu lato*, for the name of this taxon.

Cupeíllo, Saraguaso prieto.

Varronia curassavica Jacq., Enum. Syst. Pl. 14. 1760.
Syn.: *Cordia curassavica* (Jacq.) Roem. & Schult. (1819).
Varronia angustifolia H. West ex Willd. (1797).

Cordia stenophylla Alain (1956).

Shrub in Las Piedras Chiquitas, Southeastern Foothills, Susúa Forest Reserve, Guánica Forest Reserve, Southwestern Coastal Lowlands, Sierra Bermeja, Mona; 20–600 m; *fl* May–Feb, *fr* Jul–Nov.

Greater Antilles (Cuba), Lesser Antilles, Trinidad and Tobago, Mexico, Central America, South America. [L4: 315].

Notes: According to James Miller (pers. comm.), *Cordia stenophylla* is conspecific with this taxon.

Basora.

Varronia lima Desv., J. Bot. (Desvaux) 1:278. 1808.
Syn.: *Cordia lima* (Desv.) Roem. & Schult. (1819).

Shrub in Northern Limestone Hills (Camuy, Isabela), Southern Coastal Limestone (Guayanilla), Susúa Forest Reserve, Western Cordillera; 10–800 m; *fl* throughout the year, *fr* Sep–Jun.

Greater Antilles (Hispaniola). [L4: 311].

Lija.

Varronia polycephala Lam., Tabl. Encycl. 1:418. 1792.
Syn.: *Cordia polycephala* (Lam.) I.M. Johnst. (1935).
Varronia corymbosa (L.) Desv. (1808).

Shrub in Northern Coastal Lowlands, Northern Limestone Hills, Río Abajo Forest Reserve, Eastern Coast, Sierra de Cayey, Las Tetas de Cayey, Las Piedras Chiquitas, Southeastern Foothills, Eastern Cordillera (Cerro La Peña), Central Cordillera, Southern Coastal Lowlands, Susúa Forest Reserve, Maricao Forest Reserve, Sierra Bermeja, Vieques; 1–1150 m; *fl* throughout the year, *fr* May–Dec.

Greater Antilles (Hispaniola), Virgin Islands, Lesser Antilles, South America (scattered). [L4: 312; AR: 132].

Basora, Basora prieta, Jaraguaso, Saraguaso, Saragüero, Santa María.

Varronia rupicola (Urb.) Britton in Britton & Wilson, Sci. Surv. Porto Rico & Virgin Islands 6(1):128. 1925.
Bas.: *Cordia rupicola* Urb. (1899).

Shrub in Southern Coastal Limestone, Guánica Forest Reserve, Vieques; 5–50 m; *fl* Jun–Jan, *fr* May–Jan.

Virgin Islands (Anegada). [L4: 313].

Varronia wagnerorum (R.A. Howard) Borhidi, Acta Bot. Hung. 34:393. 1988.
Bas.: *Cordia wagnerorum* R. A. Howard (1966).

Shrub in Sierra de Luquillo; 800–850 m; *fl* Oct (one specimen), *fr* Dec–Jan (two specimens). [*Wagner 695* (NY); *Woodbury a.n. 9616, 9617* (SJ); *Axelrod 13333* (UPRRP)].

Endemic. [L4: 316].

TAXA CITED BY PREVIOUS AUTHORS, BUT NOT INCLUDED

NO SPECIMENS SEEN

Cordia gerascanthus L.
Tree cited by Liogier & Martorell (1982, 2000) and Liogier (1994) as having been observed in Yauco. [L4: 309].

ONLY UNDER CULTIVATION

Cordia sebestena L. Syn.: *Sebesten sebestena* (L.) Britton.
Tree cited by Britton & Wilson (1925) as occasional along roadsides and by Liogier & Martorell (1982, 2000) and Liogier (1994) as occasionally naturalized. [L4: 313].

Cynoglossum zeylanicum Brand.
Herb cited by Liogier & Martorell (1982, 2000) and Liogier (1994) as rarely naturalized. [L4: 319].

Heliotropium arborescens L.
Herb cited by Liogier & Martorell (1982, 2000) and Liogier (1994) as often grown in gardens. [L4: 322].

BRASSICACEAE (CRUCIFERAE)

Brassica juncea (L.) Czern., Consp. Pl. Charc. 8. 1859.
Bas.: *Sinapis juncea* L. (1753).
Syn.: *Brassica integrifolia* (H. West) O.E. Schulz (1903).

Herb in Northern Coastal Lowlands (Dorado), Northern Limestone Hills (Florida), Río Abajo Forest Reserve, Caguas Valley (Juncos), Eastern and Central Cordillera, Susúa Forest Reserve; 150–1060 m; *fl* & *fr* Jan–Aug.

Native to temperate Asia. [L1: 291].

Notes: Liogier & Martorell (1982) and Liogier (1985) misapplied the name *Brassica willdenovii* Boiss. to, and Liogier & Martorell (2000) wrongly placed it in synonymy with, this taxon.

Mostaza.

Cakile lanceolata (Willd.) O.E. Schulz in Urban, Symb. Antill. 3(3):504. 1903.
Bas.: *Raphanus lanceolatus* Willd. (1801).

Herb in Northern Coastal Lowlands, Southern Coastal Lowlands (Caja de Muertos), Guánica Forest Reserve, Western Coast, Vieques, Culebra, Mona; 0–50 m; *fl* & *fr* Feb–Nov.

West Indies, southeastern United States

(Florida, Mississippi), Mexico, Central America, northern South America. [L1: 293].

Mostacilla del mar.

Cardamine flexuosa With., Arr. Brit. Pl. (ed. 3) 3:578. 1796.

Herb in Northern Coastal Lowlands, Northern Limestone Hills (Arecibo), Sierra de Luquillo, Eastern Cordillera (Cayey, Aibonito), Central Cordillera, Susúa Forest Reserve, Vieques; 20–1050 m; *fl* & *fr* Nov–Jun.

Native to Eurasia. [L1: 295].

Notes: Liogier & Martorell (1982) and Liogier (1985) misapplied the name *Cardamine pensylvanica* Muhl. ex Willd. to this taxon.

Lepidium virginicum L., Sp. Pl. 2:645. 1753.

Herb in Northern Coastal Lowlands (Manatí), Northern Limestone Hills, Las Tetas de Cayey, Southeastern Foothills, Eastern and Central Cordillera, Southern Coastal Lowlands, Southwestern Coastal Lowlands, Mona; 1–1050 m; *fl* & *fr* Nov–Aug.

Native to Canada, United States, Mexico, Central America. [L1: 297].

Cresón, Lentejilla, Mastuerzo.

Nasturtium officinale W.T. Aiton, Hort. Kew. (ed. 2) 4:110. 1812.

Syn.: *Sisymbrium nasturtium-aquaticum* L. (1753) "Nasturtium [inverted triangle]um." *Rorippa nasturtium-aquaticum* (L.) Hayek (1905).

Herb in Northern Coastal Lowlands (Laguna Tortuguero), Central Cordillera (Adjuntas), Susúa Forest Reserve; 1–900 m; *fl* & *fr* Dec–Apr, Aug.

Native to Europe, northern Africa, Pakistan, temperate Asia. [L1: 298].

Notes: Liogier & Martorell (1982, 2000) and Liogier (1985) used *Nasturtium officinale* R. Br. for the name of this taxon. According to GRIN, although R. Brown is considered to be the author of the Brassicaceae in Aiton's Hortus Kew. ed. 2, no indication of his authorship appears in the work itself. Accordingly, W.T. Aiton is now treated as the author of any new names therein.

Berro.

Rorippa dubia (Pers.) H. Hara, J. Jap. Bot. 30(7):196. 1955.

Bas.: *Sisymbrium dubium* Pers. (1806).

Syn.: *Rorippa indica* (L.) Hiern var. *apetala* (DC.) Hochr. (1925).

Rorippa heterophylla (Blume) R. O. Williams (1929).

Herb in Northern Limestone Hills, Sierra de Cayey, Eastern Cordillera (Cañón San Cristóbal), Central Cordillera (Adjuntas); 10–1000 m; *fl* & *fr* Oct–Jul.

Native to Indian Subcontinent, China, Japan. [L1: 301].

Notes: Liogier & Martorell (1982, 2000) and Liogier (1985) used the homonym *Rorippa indica* (L.) Hochr. for the name of this taxon.

Rorippa palustris (L.) Besser, Enum. Pl. 27. 1822 "Roripa."

Bas.: *Sisymbrium amphibium* L. var. *palustre* L. (1753).

Syn.: *Radicula palustris* (L.) Moench (1794).

Herb in Northern Coastal Lowlands, Northern Limestone Hills (San Sebastián), Sierra de Cayey, Eastern Cordillera (Aibonito, Cañón San Cristóbal), Central Cordillera (Adjuntas), Western Cordillera (Maricao), Western Coast (Añasco); 1–800 m; *fl* & *fr* Oct, Jan–Jul.

Native to the north temperate zone. [L1: 301].

Notes: Liogier & Martorell (1982, 2000) and Liogier (1985) misapplied the name *Rorippa islandica* (Oeder ex Murray) Borbás var. *hispida* (Desv.) Butters & Abbe to this taxon.

TAXA NOT RECENTLY COLLECTED

Lepidium didymum L., Mant. Pl. 1:92. 1767.

Syn.: *Coronopus didymus* (L.) Sm. (1800). *Carara didyma* (L.) Britton (1913).

Herb collected above Villalba on 29 Jan 1930 [*Britton & Britton 9437* (GH)].

Possibly native to Europe or to South America. [L1: 295].

Rorippa portoricensis (Spreng.) Stehlé, Rev. Int. Bot. Appl. Agric. Trop. 26:103. 1946.

Bas.: *Nasturtium portoricense* Spreng. (1825).

Syn.: *Radicula portoricensis* (Spreng.) Britton (1924).

Nasturtium palustre (L.) DC. var. *brevipes* DC. (1821).

Herb collected in Cabo Rojo on 18 Jan 1885 [*Sintenis 699* (US)], in Adjuntas on 11 Apr 1886 [*Sintenis 4033* (US)] and on 21 Mar 1906 [*Britton & Cowell 1245* (NY)], in Añasco on 7 Dec 1886 [*Sintenis 5625* (US)], in Aibonito on 22 Mar 1899 [*Heller & Heller 855* (NY)] and in Barceloneta on 3 Mar 1923 [*Britton & Britton* 7798 (NY)].

Bahamas, Greater Antilles (Cuba, Hispaniola). [L1: 301].

Berro cimarrón, Mastuerzo amarillo.

Sinapis alba L., Sp. Pl. 2:668. 1753.

Syn.: *Brassica hirta* Moench. (1802).

Herb collected in Dorado [*Sintenis 7020* (US), not seen].

Native to Europe, northern Africa, western Asia, Pakistan. [L1: 291].

Sinapis arvensis L., Sp. Pl. 2:668. 1753.

Herb collected on the road from Caguas to Cayey on 25 Jun 1901 [*Underwood & Griggs 353* (US)].

Native to northern Africa, temperate Asia, Pakistan.

TAXA CITED BY PREVIOUS AUTHORS, BUT NOT INCLUDED

NO SPECIMENS SEEN

Lepidium pinnatisectum (O.E. Schulz) C.L. Hitchc.
Herb cited by Liogier & Martorell (1982, 2000) and Liogier (1985) as having been collected near Mayagüez. [L1: 297].

ONLY UNDER CULTIVATION

Brassica oleracea L.
Herb cited by Liogier & Martorell (1982, 2000) and Liogier (1985) as widely planted for food. [L1: 291].

Raphanus sativus L.
Herb cited by Liogier & Martorell (1982, 2000) and Liogier (1985) as grown for food. [L1: 300].

BROMELIACEAE

Aechmea fasciata (Lindl.) Baker, J. Bot. 17:231. 1879.
Bas.: *Billbergia fasciata* Lindl. (1828).

Epiphytic herb planted and persistent in Eastern Coast (Humacao); 1 m; *fl* Feb (one specimen), *fr* (?). [*Proctor 48637* (SJ)].

Native to Brazil. [AR&S: 201].

Aechmea lingulata (L.) Baker, J. Bot. 17:164. 1879.
Bas.: *Bromelia lingulata* L. (1753).
Syn.: *Wittmackia lingulata* (L.) Mez (1891).

Epiphytic herb in Vieques, Culebra; 65–290 m; *fl* (?), *fr* Jun–Oct.

West Indies, Trinidad, southern Central America, northern South America and Brazil. [AR&S: 202].

Aechmea nudicaulis (L.) Griseb., Fl. Brit. W.I. 593. 1864.
Bas.: *Bromelia nudicaulis* L. (1753).

Terrestrial herb in Las Tetas de Cayey, Eastern Cordillera (Cañón San Cristóbal); 600–725 m; *fl* Nov (one specimen), *fr* May, Oct (two specimens).

Greater Antilles (Cuba, Hispaniola), Trinidad, Mexico, southern Central America, northern South America to Peru. [AR&S: 202].

Ananas comosus (L.) Merr., Interpr. Herb. Amboin. 133. 1917.
Bas.: *Bromelia comosa* L. (1754).
Syn.: *Ananas ananas* (L.) Cockerell (1897).

Herb planted and persistent in Northern Coastal Lowlands; 1–10 m; *fl* Mar–Apr, *fr* (?).

Native to Brazil. [AR&S: 203].

Ananás, Piña.

Bromelia pinguin L., Sp. Pl. 1:285. 1753.

Erect robust herb in Northern Coastal Lowlands, Eastern Coast (Punta Guayanés), Las Piedras Chiquitas, Southeastern Foothills, Eastern Cordillera, Susúa Forest Reserve, Maricao Forest Reserve, Guánica Forest Reserve, Southwestern Coastal Lowlands (Cabo Rojo), Vieques; 1–600 m; *fl* Feb–Jul, Nov, *fr* Jul–Nov, Feb–Mar.

Greater Antilles, Lesser Antilles (Martinique), Mexico, Central America, northern South America to Ecuador. [AR&S: 204].

Maya.

Catopsis floribunda L.B. Sm., Contr. Gray Herb. 117:5. 1937.

Epiphytic herb in Northern Limestone Hills, Sierra de Luquillo, Sierra de Cayey, Las Tetas de Cayey, Eastern Cordillera (Barranquitas), Central Cordillera, Maricao Forest Reserve, Vieques; 100–1200 m; *fl* May–Aug, *fr* Jul–Jan.

West Indies, Trinidad and Tobago, southeastern United States (Florida), northern South America (Venezuela). [AR&S: 206].

Catopsis nitida (Hook.) Griseb., Fl. Brit. W.I. 599. 1864.
Bas.: *Tillandsia nitida* Hook. (1827).

Epiphytic herb in Sierra de Luquillo, Eastern Cordillera (La Torrecilla), Central Cordillera, Maricao Forest Reserve; 500–1250 m; *fl* Nov–Jul, *fr* Jul–Mar.

Antilles, Mexico, Central America. [AR&S: 207].

Catopsis sessiliflora (Ruiz & Pav.) Mez in C. de Candolle, Monogr. Phan. 9:625. 1896.
Bas.: *Tillandsia sessiliflora* Ruiz & Pav. (1802).

Epiphytic herb in Sierra de Luquillo, Eastern Coast (Maunabo), Sierra de Cayey; 1–700 m; *fl* Sep–Oct, *fr* Jan, Jun (two specimens).

Greater Antilles (Cuba, Hispaniola), Lesser Antilles (Guadeloupe, Martinique, Grenada), Tobago, Mexico, Central America, northwestern, west-central South America and Venezuela. [AR&S: 207].

Guzmania berteroniana (Schult. & Schult. f.) Mez in C. de Candolle, Monogr. Phan. 9:904. 1896 "berteroana."

Bas.: *Caraguata berteroniana* Schult. & Schult. f. (1830).

Epiphytic herb in Río Abajo Forest Reserve, Sierra de Luquillo, Eastern Coast (Cuchilla de Panduras), Sierra de Cayey, Las Tetas de Cayey, Eastern Cordillera (La Torrecilla), Central Cordillera, Susúa Forest Reserve, Maricao Forest Reserve; 200–1250 m; *fl* throughout the year, *fr* (?).

Greater Antilles (Cuba, Hispaniola), southern Central America (Panama). [AR&S: 208].

Parásita, Pirigallo.

Guzmania lingulata (L.) Mez var. **concolor** Proctor & Cedeño-Mald. in Acevedo-Rodríguez & Strong, Contr. U.S. Natl. Herb. 52:209, *f. 41A-C.* 2005.

Epiphytic herb in Northern Coastal Lowlands, Northern Limestone Hills (Arecibo), Río Abajo Forest Reserve, Sierra de Luquillo, Eastern Coast, Sierra de Cayey (San Lorenzo), Las Tetas de Cayey, Eastern Cordillera (Cerro La Peña), Monte Guilarte Forest Reserve, Maricao Forest Reserve; 300–1100 m; *fl* May–Oct, *fr* Sep–Jan.

Greater Antilles (Cuba, Hispaniola), Lesser Antilles, Trinidad and Tobago, Central America, northern South America to Bolivia. [AR&S: 209].

Notes: Britton & Wilson (1923) and Liogier & Martorell (1982, 2000) used *Guzmania lingulata* (L.) Mez, *sensu lato*, for the name of this taxon.

Guzmania monostachia (L.) Rusby ex Mez in C. de Candolle, Monogr. Phan. 9:905. 1896 "monostachya."

Bas.: *Renealmia monostachia* L. (1753).

Epiphytic herb in Northern Limestone Hills (Arecibo, Moca), Río Abajo Forest Reserve, Sierra de Cayey, Las Tetas de Cayey, Eastern and Central Cordillera, Western Cordillera; 25–850 m; *fl* May–Oct, *fr* Jul–Apr.

West Indies, southeastern United States (Florida), southern Central America, northwestern, west-central South America and Venezuela, Brazil. [AR&S: 209].

Hohenbergia antillana Mez in C. de Candolle, Monogr. Phan. 9:137. 1896.

Syn.: *Hohenbergia tetaensis* Proctor & Cedeño-Mald. (1999).

Terrestrial and epiphytic herb in Northern Coastal Lowlands, Northern Limestone Hills, Río Abajo Forest Reserve, Las Tetas de Cayey, Eastern Cordillera (Cidra), Southwestern Coastal Lowlands (Cabo Rojo); 1–825 m; *fl* (?), *fr* Oct–Apr.

Endemic. [AR&S: 212].

Hohenbergia portoricensis Mez in C. DC., Monogr. Phan. 9:136. 1896.

Syn.: *Hohenbergia attenuata* Britton (1923).

Epiphytic herb in Sierra de Luquillo, Eastern Coast (Cuchilla de Panduras), Sierra de Cayey, Las Tetas de Cayey, Eastern Cordillera (La Torrecilla), Central Cordillera (Los Tres Picachos), Maricao Forest Reserve; 300–1000 m; *fl* Jan–Jun, Oct, *fr* Mar–Sep.

Endemic. [AR&S: 212].

Pitcairnia angustifolia Aiton var. **angustifolia,** Hort. Kew. 1:401. 1789.

Syn.: *Pitcairnia latifolia* Aiton (1789).

Terrestrial, epiphytic and epilithic herb in Northern Limestone Hills, Río Abajo Forest Reserve, Eastern Coast (Naguabo), Sierra de Cayey, Las Tetas de Cayey, Las Piedras Chiquitas, Southeastern Foothills, Eastern Cordillera (Cerro La Peña), Central Cordillera (Los Tres Picachos), Susúa Forest Reserve, Maricao Forest Reserve, Sierra Bermeja, Vieques; 50–725 m; *fl* throughout the year, *fr* May–Dec.

Antilles. [AR&S: 214].

Notes: Britton & Wilson (1923) used the homonym *Pitcairnia angustifolia* (Sw.) Redouté for the name of this taxon.

Erizo, Piña cortadora, Piña de cuervo.

Pitcairnia angustifolia Aiton var. **simplicior** Proctor & Cedeño-Mald. in Acevedo-Rodríguez & Strong, Contr. U.S. Natl. Herb. 52:214. 2005.

Terrestrial, epiphytic and epilithic herb in Sierra de Luquillo, Sierra de Cayey; 265–1030 m; *fl* throughout the year, *fr* Aug–Oct.

Endemic. [AR&S: 214].

Notes: *Pitcairnia angustifolia* may very well be a single, highly variable taxon. In the Lesser Antilles, it has variants in both color and form of inflorescence, all with the same floral morphology, that have been designated as separate taxa. Thus, the designation of the above subspecies, which is distinguished primarily in its unbranched inflorescence, may very well be superfluous.

Erizo, Piña cortadora, Piña de cuervo.

Tillandsia ariza-juliae L.B. Sm. & J. Jiménez Alm., Phytologia 6:433, *t. 1, ff. 1, 2.* 1959.

Epiphytic herb in Maricao Forest Reserve; 600–800 m; *fl* Jan (one specimen), *fr* (?). [*Liogier 33783* (UPR)].

Greater Antilles (Hispaniola). [AR&S: 218].

Tillandsia borinquensis Cedeño-Mald. & Proctor, Harvard Pap. Bot. 4(1):113, *t.* 2. 1999.

Epiphytic herb in Northern Limestone Hills, Río Abajo Forest Reserve, Central Cordillera (Adjuntas), Susúa Forest Reserve, Maricao Forest Reserve; 125–650 m; *fl* May–Jul, Dec, *fr* Jul–Aug, Nov–Jan.

Endemic. [AR&S: 216].

Tillandsia bulbosa Hook., Exot. Fl. 3: *t. 173*. 1825.

Epiphytic herb in Northern Limestone Hills (Arecibo), Río Abajo Forest Reserve, Maricao Forest Reserve, Desecheo; 350–775 m; *fl* Jun–Aug, *fr* Aug–Feb.

Bahamas, Greater Antilles, Lesser Antilles (Guadeloupe, Martinique), Trinidad, Mexico, Central America, northwestern South America and Venezuela. [AR&S: 219].

Tillandsia fasciculata Sw., Prodr. 56. 1788.

Epiphytic herb in Northern Coastal Lowlands, Northern Limestone Hills, Río Abajo Forest Reserve, Las Tetas de Cayey, Las Piedras Chiquitas, Southeastern Foothills (Salinas), Eastern Cordillera (Aibonito), Central Cordillera (Ponce), Southern Coastal Limestone (Peñuelas), Susúa Forest Reserve, Maricao Forest Reserve, Guánica Forest Reserve, Vieques; 1–825 m; *fl* Sep–Dec, *fr* Mar–Aug.

West Indies, southeastern United States (Georgia, Florida), Trinidad, Mexico, Central America, northern South America (Venezuela, Colombia) and Brazil. [AR&S: 219].

Tillandsia festucoides Brongn. ex Mez in C. de Candolle, Monogr. Phan. 9:678. 1896.

Epiphytic herb in Susúa Forest Reserve; 150–175 m; *fl* Apr–May (two specimens), *fr* Feb–Apr.

Greater Antilles, Mexico, Central America. [AR&S: 220].

Tillandsia flexuosa Sw., Prodr. 56. 1788.

Syn.: *Tillandsia aloifolia* Hook. (1826).

Epiphytic herb in Northern Coastal Lowlands (Laguna Tortuguero), Susúa Forest Reserve, Guánica Forest Reserve; 1–325 m; *fl* Jan, Apr–Jun, Oct, *fr* Sep–Nov.

Bahamas, Greater Antilles, Lesser Antilles (St. Vincent, Grenadines, Grenada), Trinidad and Tobago, southeastern United States (Florida), southern Central America, northern South America (Venezuela, Colombia). [AR&S: 221].

Tillandsia lineatispica Mez in C. de Candolle, Monogr. Phan. 9:699. 1896.

Epiphytic herb in Vieques, Culebra; 50(?)-150 m; *fl* (?), *fr* Aug (one specimen). [*Woodbury V-121a, b* (UPR); *Hansen 9460* (UPR); *Proctor 39984* (SJ)]

Virgin Islands (St. John). [AR&S: 221].

Notes: Because of its presumed sterility, Cedeño-Maldonado in Acevedo-Rodríguez & Strong (2005), following Acevedo-Rodríguez (1996), treated this taxon as *Tillandsia ×lineatispica* Mez, possibly a hybrid between *Tillandsia utriculata* and *Tillandsia fasciculata*.

Tillandsia polystachia (L.) L., Sp. Pl. (ed. 2) 1:410. 1762.

Bas.: *Renealmia polystachia* L. (1753).

Epiphytic herb in Northern Coastal Lowlands, Northern Limestone Hills, Río Abajo Forest Reserve, Sierra de Cayey, Las Tetas de Cayey, Eastern and Central Cordillera, Susúa Forest Reserve, Maricao Forest Reserve, Mona; 1–950 m; *fl* Feb–Aug, *fr* Aug–Mar.

West Indies, Mexico, Central America, northwestern, west-central South America and Venezuela. [AR&S: 222].

Parásita, Piñón.

Tillandsia pruinosa Sw., Fl. Ind. Occid. 1:594. 1797.

Epiphytic herb in Susúa Forest Reserve; 130–250 m; *fl* Feb–Mar, *fr* Feb (two specimens).

Greater Antilles, southeastern United States (Florida), Mexico, Central America, northwestern South America and Venezuela. [AR&S: 222].

Tillandsia recurvata (L.) L., Sp. Pl. (ed. 2) 1:410. 1762.

Bas.: *Renealmia recurvata* L. (1753).

Epiphytic herb in Northern Coastal Lowlands (San Juan, Toa Baja), Northern Limestone Hills (Morovis), Río Abajo Forest Reserve, Southeastern Foothills, Eastern Cordillera (Cañón San Cristóbal), Central Cordillera (Ponce), Southern Coastal Lowlands, Susúa Forest Reserve, Maricao Forest Reserve, Guánica Forest Reserve, Southwestern Coastal Lowlands (Cabo Rojo), Sierra Bermeja, Western Coast (Mayagüez), Mona, Desecheo, Vieques; 1–700 m; *fl* Apr–Nov, *fr* Oct–Jun.

West Indies, southern United States, Mexico, Central America, South America. [AR&S: 223].

Nidos de gungulén.

Tillandsia setacea Sw., Fl. Ind. Occid. 1:593. 1797.

Epiphytic herb in Northern Limestone Hills, Río Abajo Forest Reserve, Eastern Cordillera, Central Cordillera (Adjuntas), Susúa Forest Reserve, Guánica Forest Reserve, Mona; 30–550 m; *fl* Sep–Oct, Jan–Feb, *fr* Oct–Feb, May.

Greater Antilles, southeastern United States (Georgia, Florida), Mexico, northern Central America. [AR&S: 223].

Tillandsia tenuifolia L., Sp. Pl. 1:286. 1753.

Epiphytic herb in Las Tetas de Cayey, Las Piedras

Chiquitas, Southeastern Foothills (Salinas), Eastern Cordillera (Cidra), Central Cordillera; 500–1150 m; *fl* Feb, May, *fr* (?).

Greater Antilles, Lesser Antilles (Martinique, St. Lucia), South America (scattered). [AR&S: 224].

Notes: Britton & Wilson (1923) misapplied the above name to *Tillandsia setacea*.

Tillandsia usneoides (L.) L., Sp. Pl. (ed. 2) 1:411. 1762.
Bas.: *Renealmia usneoides* L. (1753).
Syn.: *Dendropogon usneoides* (L.) Raf. (1825).

Epiphytic herb in Río Abajo Forest Reserve, Las Piedras Chiquitas, Southeastern Foothills, Southern Coastal Lowlands, Southwestern Coastal Lowlands (Lajas), Sierra Bermeja; 30–450 m; *fl* (?), *fr* (?).

West Indies, Trinidad, southeastern and south-central United States, Central America, South America. [AR&S: 225].

Barbas de úcar.

Tillandsia utriculata L., Sp. Pl. 1:286. 1753.

Epiphytic herb in Northern Coastal Lowlands, Guajataca Forest Reserve, Eastern Coast (Punta Guayanés), Sierra de Cayey, Las Tetas de Cayey, Eastern Cordillera (Cayey), Southern Coastal Lowlands, Susúa Forest Reserve, Maricao Forest Reserve, Guánica Forest Reserve, Southwestern Coastal Lowlands (Cabo Rojo), Sierra Bermeja, Vieques, Culebra, Mona, Desecheo; 1–800 m; *fl* Apr–Nov, *fr* Jul–Mar.

West Indies, Trinidad and Tobago, southeastern United States (Georgia, Florida), Mexico, Central America, northern South America (Venezuela). [AR&S: 225].

Tillandsia variabilis Schltdl., Linnaea 18(4):418. 1845.
Syn.: *Tillandsia valenzuelana* A. Rich. (1850). *Tillandsia sublaxa* Baker (1887).

Epiphytic herb in Northern Limestone Hills, Río Abajo Forest Reserve, Las Tetas de Cayey, Southeastern Foothills (Salinas), Eastern Cordillera (Cidra, Cañón San Cristóbal), Central Cordillera, Susúa Forest Reserve, Maricao Forest Reserve, Guánica Forest Reserve, Southwestern Coastal Lowlands (Cabo Rojo), Vieques; 20–875 m; *fl* Nov–Jun, *fr* Jan–Jul.

Bahamas, Greater Antilles, southeastern United States (Florida), southern Mexico, Central America, northwestern, west-central South America and Venezuela. [AR&S: 226].

Vriesea macrostachya (Bello) Mez in C. de Candolle, Monogr. Phan. 9:601. 1896 "Vriesia."
Bas.: *Caraguata macrostachya* Bello (1883).
Syn.: *Neovriesia macrostachya* (Bello) Britton (1923).

Epiphytic herb in Northern Limestone Hills (Arecibo), Sierra de Luquillo, Las Tetas de Cayey, Eastern Cordillera (La Torrecilla), Toro Negro Forest Reserve, Maricao Forest Reserve; 250–925 m; *fl* May, Aug–Nov, *fr* Jun–Sep, Dec–Feb.

Greater Antilles (Cuba, Hispaniola). [AR&S: 228].

Parásita, Pirigallo.

Werauhia proctorii Cedeño-Mald., Harvard Pap. Bot. 4(1):115, *t. 3.* 1999.

Epiphytic herb in Sierra de Luquillo, Sierra de Cayey, Eastern Cordillera (La Torrecilla), Toro Negro Forest Reserve; 850–1300 m; *fl* Feb–Jul, *fr* Jul–Oct.

Endemic. [AR&S: 229].

Werauhia ringens (Griseb.) J.R. Grant, Trop. Subtrop. Pflanzenwelt 91:35. 1995.
Bas.: *Tillandsia ringens* Griseb. (1866).
Syn.: *Vriesea ringens* (Griseb.) Harms (1929).

Epiphytic herb in Sierra de Luquillo; 400 m; *fl* (?), *fr* (?). [*Woodbury Jun 1963 & a.n. 5495* (UPR)].

Antilles, southern Central America, northwestern South America. [AR&S: 230].

Werauhia sintenisii (Baker) J.R. Grant, Trop. Subtrop. Pflanzenwelt 91:48. 1995.
Bas.: *Caraguata sintenisii* Baker (1889).
Syn.: *Thecophyllum sintenisii* (Baker) Mez (1903). *Vriesea sintenisii* (Baker) L.B. Sm. & Pittendr. (1954).

Epiphytic herb in Sierra de Luquillo, Sierra de Cayey, Eastern Cordillera (La Torrecilla), Central Cordillera, Maricao Forest Reserve; 775–1300 m; *fl* Nov, Feb–Jun, *fr* Nov–Aug.

Greater Antilles. [AR&S: 230].

TAXA CITED BY PREVIOUS AUTHORS, BUT NOT INCLUDED

NO SPECIMENS SEEN

Tillandsia juncea (Ruiz & Pav.) Poir.
Herb cited by Liogier & Martorell (1982, 2000) from one collection near Adjuntas.

DOUBTFUL OCCURRENCE

Bromelia plumieri (E. Morren) L.B. Sm. Syn.: *Bromelia karatas* L., *pro parte*.
Herb cited by Britton & Wilson (1923) as recorded as Puerto Rican by Bello, but with the caveat that it is not otherwise known within the geographical area of their book and may have been in cultivation;

subsequently cited by Liogier & Martorell (1982, 2000) without the caveat. [AR&S: 206].

CITED DUE TO MISIDENTIFICATION
Catopsis nutans (Sw.) Griseb.
Herb cited by Liogier & Martorell (1982, 2000) as epiphytic on trees in the central mountains, rare. According to Cedeño-Maldonado (pers. comm.), this citation is based on the misidentification of specimens of *Catopsis floribunda*. Oddly enough, Liogier & Martorell (1982) were aware of the fact that Britton & Wilson (1923) misapplied this name to *Catopsis floribunda*. [AR&S: 206].

Guzmania erythrolepis Brongn. ex Planch.
Herb cited by Britton & Wilson (1923) as occurring in Monte Allegrillo; subsequently cited by Liogier & Martorell (1982, 2000), who clarified the locality as being near Maricao. According to Cedeño-Maldonado in Acevedo-Rodríguez & Strong (2005), these citations were based on the misidentification of specimens of *Guzmania berteroniana*. [AR&S: 211].

ONLY UNDER CULTIVATION
Billbergia pyramidalis (Sims) Lindl.
Herb cited by Liogier & Martorell (1982, 2000) as grown as an ornamental and sometimes escaped.

BRUNELLIACEAE

Brunellia comocladiifolia Humb. & Bonpl. subsp. **domingensis** Cuatrec., Fl. Neotrop. Monogr. 2:74, *t. 30*. 1970 "comocladifolia."

Tree in Central Cordillera; 850–1300 m; *fl* Feb–May, Oct, *fr* Aug–Oct.

Greater Antilles, Lesser Antilles (Guadeloupe), northwestern South America and Venezuela. [L1: 312].

Cabra, Palo bobo.

BURMANNIACEAE

Apteria aphylla (Nutt.) Barnhart ex Small, Fl. S.E. U.S. 309. 1903.
Bas.: *Lobelia aphylla* Nutt. (1822).
Syn.: *Apteria hymenanthera* Miq. (1850).

Delicate herb in Sierra de Luquillo, Sierra de Cayey, Eastern Cordillera (Cerro La Peña); 400–600 m; *fl* Sep–Mar, *fr* Oct–Mar.

Antilles, Trinidad, southeastern and south-central United States, southern Central America, South America (scattered). [AR&S: 93].

Burmannia capitata (Walter ex J.F. Gmel.) Mart., Nov. Gen. Sp. Pl. 1:12. 1823–1824.
Bas.: *Vogelia capitata* Walter ex J.F. Gmel. (1791).

Delicate herb in Northern Coastal Lowlands (Laguna Tortuguero); 1–2 m; *fl* & *fr* Aug–Mar.

Greater Antilles, southeastern and south-central United States, Mexico, Central America, South America (scattered). [AR&S: 94].

Gymnosiphon niveus (Griseb.) Urb., Symb. Antill. 3(3):444. 1903.
Bas.: *Ptychomeria nivea* Griseb. (1866).
Syn.: *Ptychomeria portoricensis* (Urb.) Schltr. (1921).

Delicate herb in Maricao Forest Reserve; 775 m; *fl* & *fr* Nov (one specimen). [*Cedeño-Maldonado 207* (MAPR)].

Greater Antilles, Lesser Antilles. [AR&S: 95].

Gymnosiphon sphaerocarpus Urb., Symb. Antill. 3(3):442. 1903.

Delicate herb in Guajataca Forest Reserve, Sierra de Luquillo, Sierra de Cayey, Eastern Cordillera (La Torrecilla), Maricao Forest Reserve; 150–900 m; *fl* & *fr* Mar–Apr, Jul–Dec.

Greater Antilles, Lesser Antilles (Guadeloupe, Dominica). [AR&S: 96].

TAXA CITED BY PREVIOUS AUTHORS, BUT NOT INCLUDED

CITED DUE TO MISIDENTIFICATION
Cymbocarpa refracta Miers.
Herb cited by Liogier & Martorell (2000) as occurring in wet forests, Luquillo Mountains. All the specimens so identified that I have thus far seen are referable to other taxa.

BURSERACEAE

Bursera simaruba (L.) Sarg., Gard. & Forest 3(218):260. 1890.
Bas.: *Pistacia simaruba* L. (1753).
Syn.: *Elaphrium simaruba* (L.) Rose (1911).

Tree in Northern Coastal Lowlands, Northern Limestone Hills, Río Abajo Forest Reserve, Eastern Coast (Punta Guayanés), Las Piedras Chiquitas, Southeastern Foothills, Central Cordillera (Ponce), Southern Coastal Lowlands, Southern Coastal Limestone, Susúa Forest Reserve, Guánica Forest Reserve, Southwestern Coastal Lowlands, Sierra Bermeja, Vieques, Culebra, Mona, Desecheo; 1–500 m; *fl* Mar–Sep, *fr* Feb–Nov.

West Indies, Trinidad and Tobago, southeastern United States (Florida), Mexico, Central America, northern South America to Brazil. [L2: 289].

Almácigo.

Dacryodes excelsa Vahl, Skr. Naturhist.-Selsk. 6: 116. 1810.

Tree in Sierra de Luquillo, Sierra de Cayey, Eastern Cordillera (Cerro La Peña), Central Cordillera (Utuado), Maricao Forest Reserve; 100–650 m; *fl* May–Oct, *fr* Oct–May.

Lesser Antilles. [L2: 291].

Tabonuco.

Tetragastris balsamifera (Sw.) Oken, Allg. Naturgesch. 3(3):1764. 1841.

Bas.: *Hedwigia balsamifera* Sw. (1788).

Tree in Northern Coastal Lowlands (Carolina), Northern Limestone Hills, Central Cordillera (Utuado), Susúa Forest Reserve, Maricao Forest Reserve, Sierra Bermeja, Vieques; 100–600 m; *fl* May–Jul, *fr* Dec, Apr.

Greater Antilles (Hispaniola), Virgin Islands, Lesser Antilles (Guadeloupe). [L2: 292].

Notes: Britton & Wilson (1924), Liogier & Martorell (1982, 2000), and Liogier (1988) used the homonym *Tetragastris balsamifera* (Sw.) Kuntze for the name of this taxon.

Masa, Palo de aceite.

BUXACEAE

Buxus portoricensis Alain, Bull. Torrey Bot. Club 92(4):296. 1965.

Tree in Northern Limestone Hills, Río Abajo Forest Reserve, Susúa Forest Reserve, Maricao Forest Reserve; 135–900 m; *fl* Apr–Jan, *fr* May–Sep.

Endemic. [L2: 435].

Buxus vahlii Baill., Monogr. Buxac. 67. 1859.

Syn.: *Tricera vahlii* (Baill.) Britton (1915).

Small tree in Northern Limestone Hills; 20–325 m; *fl* Oct–Aug, *fr* Feb–Oct.

Virgin Islands (St. Croix). [L2: 437].

Notes: This taxon is on the U.S. Fish & Wildlife Service list of threatened and endangered plants.

TAXA CITED BY PREVIOUS AUTHORS, BUT NOT INCLUDED

CITED DUE TO MISIDENTIFICATION

Buxus citrifolia (Willd.) Spreng. Syn.: *Tricera citrifolia* Willd.

Shrub cited by Britton and Wilson (1924), known to them only as recorded by Mueller and by Urban, as occurring in eastern Puerto Rico, collected only by Wydler, or perhaps also by Bredemeyer. Whereas Liogier & Martorell (1982) and Liogier (1988) thought that Britton & Wilson had misapplied this name to specimens of *Buxus laevigata* (Sw.) Spreng., which is endemic to Jamaica, Liogier & Martorell (2000) corrected the error and treated the latter name as misapplied to specimens of *Buxus portoricensis*. [L2: 435].

CABOMBACEAE

TAXA NOT RECENTLY COLLECTED

Cabomba haynesii Wiersema, Ann. Missouri Bot. Gard. 76(4):1167. 1989.

Aquatic herb collected in Añasco on 7 Dec 1886 [*Sintenis 5601* (GH)], in Aguadilla on 14 Dec 1886 [*Sintenis 5766* (GH)] and in the vicinity of Vega Baja on 24 Mar 1922 [*Britton et al. 6761* (US)].

Greater Antilles, Central America, northern South America to Brazil, Ecuador. [L5: 396].

TAXA CITED BY PREVIOUS AUTHORS, BUT NOT INCLUDED

CITED DUE TO MISIDENTIFICATION

Cabomba furcata Schult. & Schult. f. Syn.: *Cabomba piauhyensis* Gardner.

Aquatic herb cited by Britton & Wilson (1924); subsequently cited by Liogier & Martorell (1982, 2000) and Liogier (1985). The citation was based on the misidentification of specimens of *Cabomba haynesii*. [L1: 221].

CACTACEAE

Harrisia portoricensis Britton, Bull. Torrey Bot. Club 35(12):563. 1909.

Succulent on Mona, Desecheo; 30–200 m; *fl* Feb–Mar, Aug, *fr* Aug, Nov, Feb–Mar.

Endemic. [L3: 312].

Notes: Although the type collection of this taxon is from the Southern Coastal Limestone region, it has not recently been found on the mainland of Puerto Rico.

This taxon is on the U.S. Fish & Wildlife Service list of threatened and endangered plants.

Higo chumbo.

Hylocereus trigonus (Haw.) Saff., Rep. (Annual) Board Regents Smithsonian Inst. 1908:553. 1909.

Bas.: *Cereus trigonus* Haw. (1812).

Succulent in Northern Coastal Lowlands, Northern Limestone Hills, Las Tetas de Cayey, Las Piedras Chiquitas, Guánica Forest Reserve, Southwestern Coastal Lowlands, Sierra Bermeja, Vieques, Culebra, Mona; 1–840 m; *fl* Jul–Oct, *fr* Jul–Dec.

Greater Antilles (Hispaniola), Virgin Islands, Lesser Antilles. [L3: 313; AR: 142].

Pitajaya.

Hylocereus undatus (Haw.) Britton & Rose in Britton, Fl. Bermuda 256, *t. 278*. 1918.

Bas.: *Cereus undatus* Haw. (1830).

Succulent planted and persistent in Northern Coastal Lowlands (Fajardo), Eastern Cordillera (Cayey), Southern Coastal Lowlands (Ponce); 1–300 m; *fl* Sep–Nov, *fr* Sep–Nov (two specimens).

Only cultivated; originating in the Neotropics. [L3: 313; AR: 142].

Leptocereus grantianus Britton, Cact. Succ. J. (Los Angeles) 5:469. 1933.

Succulent on Culebra; 2–10 m; *fl* Apr–May (one specimen), *fr* (?). [*Proctor 46823, 51211* (SJ)].

Endemic. [L3: 315].

Leptocereus quadricostatus (Bello) Britton & Rose, Contr. U.S. Natl. Herb. 16(7):242. 1913.

Bas.: *Cereus quadricostatus* Bello (1881).

Succulent in Southeastern Foothills (Salinas), Guánica Forest Reserve, Southwestern Coastal Lowlands; 1–100 m; *fl* Jan, Jun–Sep, *fr* Jun–Feb.

Virgin Islands (Anegada). [L3: 315].

Pitahaya.

Mammillaria nivosa Link ex Pfeiff., Enum. Diagn. Cact. 11. 1837.

Syn.: *Neomammillaria nivosa* (Link ex Pfeiff.) Britton & Rose (1923).

Succulent on Culebra, Mona 5–60 m; *fl* May, Nov (two specimens), *fr* May–Oct.

Bahamas, Turks and Caicos, Virgin Islands (Tortola), Lesser Antilles. [L3: 317].

Cacto lanoso, Erizo blanco.

Melocactus intortus (Mill.) Urb., Repert. Spec. Nov. Regni Veg. 16:35. 1919.

Bas.: *Cactus intortus* Mill. (1768).

Succulent in Northern Coastal Lowlands (Fajardo), Susúa Forest Reserve, Guánica Forest Reserve, Southwestern Coastal Lowlands, Sierra Bermeja, Vieques; 10–300 m; *fl* & *fr* Dec (one specimen).

West Indies. [L3: 319].

Erizo, Melón de costa.

Opuntia cochenillifera (L.) Mill., Gard. Dict. (ed. 8): *Opuntia* no. 6. 1768 "cochinelifera."

Bas.: *Cactus cochenillifer* L. (1753).

Syn.: *Nopalea cochenillifera* (L.) Salm-Dyck (1850) "coccinellifera."

Succulent planted and persistent in Northern Coastal Lowlands (Fajardo), Guánica Forest Reserve; 5–145 m; *fl* Mar–Apr (one specimen), *fr* Jul (one specimen).

Only cultivated; probably originating in Mexico. [L3: 322].

Nopal, Tuna mansa.

Opuntia ficus-indica (L.) Mill., Gard. Dict. (ed. 8): *Opuntia* no. 2. 1768.

Bas.: *Cactus ficus-indicus* L. (1753).

Succulent planted and persistent in Northern Coastal Lowlands (Manatí), Eastern Coast (Maunabo), Southern Coastal Lowlands (Santa Isabel), Southwestern Coastal Lowlands (Cabo Rojo), Vieques; 1–10 m; *fl* Jan, Apr–Jun, Oct, *fr* Jun (one specimen).

Possibly native to Mexico. [L3: 325].

Higo chumbo, Tuna, Tuna de España.

Opuntia moniliformis (L.) Haw. ex Steud., Nomencl. Bot. (ed. 2) 2(10):221. 1841.

Bas.: *Cactus moniliformis* L. (1753).

Syn.: *Consolea moniliformis* (L.) A. Berger (1926).

Succulent on Culebra, Mona, Desecheo; 2–150 m; *fl* Feb–Aug, *fr* (?).

Greater Antilles (Cuba, Hispaniola). [L3: 325].

Tuna.

Opuntia repens Bello, Anales Soc. Esp. Hist. Nat. 10:277. 1881.

Syn.: *Opuntia borinquensis* Britton & Rose (1919).

Small succulent in Southern Coastal Lowlands (Caja de Muertos), Guánica Forest Reserve, Sierra Bermeja, Vieques, Mona; 10–300 m; *fl* Apr–Jul, *fr* Jul (one specimen).

Virgin Islands. [L3: 327].

Gatos, Olaga.

Opuntia rubescens Salm-Dyck ex DC., Prodr. 3:474. 1828.

Syn.: *Consolea rubescens* (Salm-Dyck ex DC.) Lem. (1862).

Succulent in Northern Coastal Lowlands (Fajardo), Guánica Forest Reserve, Southwestern Coastal Lowlands, Sierra Bermeja, Vieques, Culebra; 1–100 m; *fl* throughout the year, *fr* Jan, Jul.

Virgin Islands, Lesser Antilles. [L3: 327].

Tuna, Tuna de petate, Tuna de yagua.

Opuntia stricta (Haw.) Haw., Syn. Pl. Succ. 191. 1812.

Bas.: *Cactus strictus* Haw. (1803).

Syn.: *Opuntia dillenii* (Ker Gawl.) Haw. (1819).

Succulent in Northern Coastal Lowlands,

Southern Coastal Lowlands (Caja de Muertos), Guánica Forest Reserve, Southwestern Coastal Lowlands, Sierra Bermeja, Vieques, Culebra, Mona, Desecheo; 1–50 m; *fl* throughout the year, *fr* Jul–Apr.

West Indies, southeastern and south-central United States, eastern Mexico. [L3: 322].

Higo de mar, Tuna, Tuna brava, Tuna de petate, Tuna de yagua.

Opuntia triacanthos (Willd.) Sweet, Hort. Brit. (ed. 1) 172. 1826 "triacantha."
Bas.: *Cactus triacanthos* Willd. (1814).

Succulent in Desecheo; 5–40 m; *fl* Mar, *fr* (?). [*Breckon 4745, 5020* (MAPR)].

Virgin Islands (St. Thomas, St. Croix), Lesser Antilles. [L3: 328].

Pereskia aculeata Mill., Gard. Dict. (ed. 8): *Pereskia* no. 1. 1768.
Syn.: *Pereskia pereskia* (L.) H. Karst. (1882).

Succulent in Northern Limestone Hills (Quebradillas), Las Piedras Chiquitas, Southeastern Foothills (Salinas), Vieques; 50–425 m; *fl* Aug–Nov, *fr* (?).

Greater Antilles (Cuba, Hispaniola), Lesser Antilles, South America; exact native range uncertain. [L3: 328; AR: 144].

Hortensia de bejuco, Uva espinosa.

Pilosocereus royenii (L.) Byles & G.D. Rowley, Cact. Succ. J. Gr. Brit. 19(3):67. 1957.
Bas.: *Cactus royenii* L. (1753).
Syn.: *Cephalocereus royenii* (L.) Britton & Rose (1909).

Succulent in Northern Coastal Lowlands (Fajardo), Eastern Coast (Ceiba), Las Piedras Chiquitas, Southern Coastal Lowlands (Caja de Muertos, Ponce), Susúa Forest Reserve, Guánica Forest Reserve, Southwestern Coastal Lowlands (Cabo Rojo), Sierra Bermeja, Vieques, Culebra, Mona, Desecheo; 1–450 m; *fl* Mar, Jun–Nov, *fr* Jun–Nov.

Greater Antilles (Jamaica, Hispaniola), Virgin Islands, Lesser Antilles, Tobago, Mexico. [L3: 330].

Dildo, Sebucán.

Rhipsalis baccifera (J.S. Muell.) Stearn, Cact. J. (Croydon) 7(4):107. 1939.
Bas.: *Cassytha baccifera* J.S. Muell. (1771) "Cassyta."
Syn.: *Rhipsalis cassytha* Gaertn. (1788) "cassutha."

Viny herb in Northern Limestone Hills (Arecibo, Utuado), Sierra de Luquillo, Caguas Valley (Caguas), Sierra de Cayey, Las Tetas de Cayey, Southeastern Foothills (Cayey), Central Cordillera, Western Cordillera; 125–825 m; *fl* & *fr* throughout the year.

Antilles, Trinidad and Tobago, southeastern United States (Florida), Mexico, Central America, South America, Africa. [L3: 332].

Barbas de palo, Pegapalo.

Selenicereus grandiflorus (L.) Britton & Rose, Contr. U.S. Natl. Herb. 12(10):430. 1909.
Bas.: *Cactus grandiflorus* L. (1753).

Succulent planted and persistent in Eastern Coast (Ceiba); 10–45 m; *fl* (?), *fr* (?).

Native to Greater Antilles (except Puerto Rico), Mexico, Central America (Belize, Nicaragua, Costa Rica). [L3: 334; AR: 144].

Cardón, Reina de la noche.

Stenocereus fimbriatus (Lam.) Lourteig, Bradea 5(44):408. 1991.
Bas.: *Cactus fimbriatus* Lam. (1785).
Syn.: *Lemairocereus hystrix* (Haw.) Britton & Rose (1909). *Stenocereus hystrix* (Haw.) Buxb. (1961).
Stenocereus peruvianus (Mill.) R. Kiesling (1982), *nom. illeg.*

Succulent in Southern Coastal Lowlands (Caja de Muertos), Guánica Forest Reserve, Mona, Desecheo; 10–135 m; *fl* Jun–Aug, *fr* May–Jul.

Greater Antilles. [L3: 336].

Dildo español.

TAXA CITED BY PREVIOUS AUTHORS, BUT NOT INCLUDED

DOUBTFUL OCCURRENCE

Opuntia cubensis Britton & Rose. Syn.: *Opuntia antillana* Britton & Rose.

Succulent cited by Britton & Wilson (1924) as occurring on plains near the southern coast from the vicinity of Salinas westward; subsequently cited by Liogier & Martorell (1982, 2000) and Liogier (1994). [L3: 321].

Notes: Howard (1989) noted that *Opuntia antillana* was based on a collection from St. Kitts that appeared to be a hybrid between *Opuntia dillenii* [=*Opuntia stricta*] and *Opuntia triacanthos.*

ONLY UNDER CULTIVATION

Cereus hexagonus (L.) Mill.

Succulent cited by Liogier & Martorell (1982, 2000) and Liogier (1994) as occasionally planted and persistent. [L3: 308].

Epiphyllum oxypetalum (DC.) Haw.

Succulent cited by Liogier & Martorell (1982, 2000) and Liogier (1994) as occasionally planted and persistent. [L3: 310].

Epiphyllum phyllanthus (L.) Haw. var. ***hookeri*** (Haw.) Kimnach. Syn.: *Epiphyllum hookeri* Haw.
Succulent cited by Liogier & Martorell (1982, 2000) and Liogier (1994) as occasionally planted and persistent. [L3: 310].

Opuntia brasiliensis (Willd.) Haw.
Succulent cited by Liogier (1994) and Liogier & Martorell (2000) as planted and naturalized. [L3: 322].

Pereskia grandifolia Haw.
Shrub/small tree cited by Britton & Wilson (1924) as spontaneous after cultivation; subsequently cited by Liogier & Martorell (1982, 2000) and Liogier (1994). [L3: 330].

CAMPANULACEAE (including LOBELIACEAE)

Hippobroma longiflora (L.) G. Don, Gen. Hist. 3:717. 1834.
Bas.: *Lobelia longiflora* L. (1753).
Syn.: *Isotoma longiflora* (L.) C. Presl (1836).

Herb in Northern Limestone Hills, Río Abajo Forest Reserve, Sierra de Luquillo, Sierra de Cayey, Central Cordillera, Southern Coastal Lowlands (Salinas), Western Coast, Vieques; 10–725 m; *fl* & *fr* Jan–Oct.

Antilles, Trinidad, southern Mexico, Central America, northern South America to Brazil, Peru. [L5: 210].

Cipril, Tibey, Tibey blanco.

Lobelia assurgens L. var. **portoricensis** (A. DC.) Urb., Symb. Antill. 1(3):454. 1899.
Bas.: *Tupa assurgens* A. DC. var. *portoricensis* A. DC. (1839).
Syn.: *Tupa robusta* (Graham) A. DC. (1839).

Herb in Northern Limestone Hills, Río Abajo Forest Reserve, Sierra de Luquillo, Sierra de Cayey, Las Tetas de Cayey, Eastern Cordillera (La Torrecilla), Central Cordillera, Western Cordillera; 50–1150 m; *fl* & *fr* throughout the year.

Greater Antilles (Hispaniola). [L5: 212].

Chicoria cimarrona.

Lobelia cliffortiana L., Sp. Pl. 2:931. 1753.

Herb in Northern Limestone Hills (Arecibo), Río Abajo Forest Reserve, Caguas Valley (Caguas), Sierra de Cayey, Las Tetas de Cayey, Southeastern Foothills, Eastern Cordillera (Naranjito, Aibonito), Central Cordillera, Western Cordillera (Maricao); 50–900 m; *fl* & *fr* Nov–Jul.

Greater Antilles, Lesser Antilles (St. Lucia), Trinidad. [L5: 213].

Lobelia portoricensis (Vatke) Urb., Symb. Antill. 1(3):453. 1899.
Bas.: *Tupa portoricensis* Vatke (1874).

Herb in Sierra de Luquillo, Eastern Coast (Cuchilla de Panduras), Sierra de Cayey, Monte Guilarte Forest Reserve; 475–1100 m; *fl* & *fr* throughout the year.

Greater Antilles (Hispaniola). [L5: 213].

Tibey-tupa, Tupa.

Lobelia rotundifolia Juss. in A. de Candolle, Prodr. 7(2):383. 1839.

Herb in Toro Negro Forest Reserve; 1300 m; *fl* May–Aug, *fr* Jul–Aug (two specimens).

Greater Antilles (Hispaniola). [L5: 213].

Lobelia vivaldii Lammers & Proctor, Brittonia 46(4):274. 1994.

Shrub in Mona; 60 m; *fl* & *fr* Jan, May–Sep.

Endemic. [L5: 215].

CANELLACEAE

Canella winterana (L.) Gaertn., Fruct. Sem. Pl. 1:373, *t. 77, f. 2.* 1788.
Bas.: *Laurus winterana* L. (1753).

Tree in Northern Coastal Lowlands (Fajardo and adjacent islands), Southern Coastal Lowlands (Caja de Muertos, Ponce), Guánica Forest Reserve, Southwestern Coastal Lowlands, Sierra Bermeja, Vieques, Culebra, Mona; 1–300 m; *fl* & *fr* throughout the year.

West Indies, southeastern United States (Florida), Mexico, northern Central America (Honduras), northern South America. [L3: 249].

Barbasco, Canela.

Pleodendron macranthum (Baill.) Tiegh., J. Bot. (Morot) 13:272. 1899.
Bas.: *Cinnamodendron macranthum* Baill. (1882).

Tree in Río Abajo Forest Reserve, Sierra de Luquillo; 250–500 m; *fl* Dec–Mar, *fr* Jun–Jan.

Endemic. [L3: 251].

Notes: This tree is on the U.S. Fish & Wildlife Service list of threatened and endangered plants.

Aceitillo, Chupacallos.

CANNABACEAE

TAXA CITED BY PREVIOUS AUTHORS, BUT NOT INCLUDED

NO SPECIMENS SEEN
Cannabis sativa L.

Herb cited by Liogier (1997) and Liogier & Martorell (2000) as cultivated and probably escaping. [L5: 391].

CANNACEAE

Canna glauca L., Sp. Pl. 1:1. 1753.

Herb in Northern Coastal Lowlands (Río Grande, vicinity of Laguna Tortuguero); 1–2 m; *fl* Apr–Jun, Dec, *fr* Apr–Jun (two specimens). [AR&S: 192].

Antilles, Trinidad, south-central United States (Louisiana, Texas), Mexico, Central America, South America.

Maraca amarilla, Maraca boba, Maraca de pantano.

Canna indica L., Sp. Pl. 1:1. 1753.
Syn.: *Canna coccinea* Mill. (1768).
Canna sylvestris Roscoe (1828).

Herb in Northern Coastal Lowlands, Northern Limestone Hills, Río Abajo Forest Reserve, Sierra de Luquillo, Caguas Valley, Sierra de Cayey (San Lorenzo), Las Tetas de Cayey, Southeastern Foothills, Eastern Cordillera (Aibonito, Barranquitas), Central Cordillera, Susúa Forest Reserve, Western Cordillera (Maricao), Western Coast; 5–900 m; *fl* & *fr* throughout the year.

Probably native to the Neotropics. [AR&S: 192].

Bandera española, Gruya, Maraca, Maraca morada, Maraca roja.

Canna jaegeriana Urb., Repert. Spec. Nov. Regni Veg. 15(418–421):102. 1917, *nom. cons.*
Syn.: *Canna pertusa* Urb. (1917).

Herb in Central Cordillera; 600–1100 m; *fl* & *fr* Jan–Mar.

Greater Antilles (Hispaniola), northwestern and west-central South America. [AR&S: 194].

Maraca montuna.

CAPPARACEAE

Capparidastrum frondosum (Jacq.) Cornejo & Iltis, Harvard Pap. Bot. 13(2):232. 2008.
Bas.: *Capparis frondosa* Jacq. (1760).

Tree in Northern Limestone Hills (Quebradillas), Las Piedras Chiquitas, Southeastern Foothills, Southern Coastal Lowlands (Salinas), Southwestern Coastal Lowlands (Lajas), Sierra Bermeja, Vieques, Culebra; 70–550 m; *fl* Nov, Mar–Jun, *fr* Feb–Jul, Oct.

Antilles, Mexico, southern Central America, northern South America to Brazil, Peru. [L1: 280].

Notes: Liogier & Martorell (2000) used the name *Capparis baducca* L. for this taxon. That name, though recently used as the accepted one, turns out to be controversial since it has problems of lectotypification.

Burro, Sapo.

Cynophalla amplissima (Lam.) Iltis & Cornejo, Rodriguésia 61(1):154. 2010.
Bas.: *Capparis amplissima* Lam. (1785).
Syn.: *Capparis portoricensis* Urb. (1899).

Tree in Northern Coastal Lowlands (Carolina), Northern Limestone Hills, Sierra de Cayey (Patillas), Las Tetas de Cayey, Las Piedras Chiquitas, Southeastern Foothills, Eastern Cordillera (Aibonito), Southern Coastal Limestone (Yauco), Guánica Forest Reserve, Vieques; 10–825 m; *fl* Jun–Mar, *fr* May–Aug.

Greater Antilles (Hispaniola), Virgin Islands, Lesser Antilles (Dominica), southern Central America, northern South America to Brazil, Bolivia. [L1: 277].

Burro, Burro blanco, Palinguán, Sapo.

Cynophalla flexuosa (L.) J. Presl in Berchtold & J. Presl, Přir. Rostlin 2:275. 1825.
Bas.: *Morisonia flexuosa* L. (1759).
Syn.: *Capparis flexuosa* (L.) L. (1762).

Shrub/small tree in Northern Coastal Lowlands, Northern Limestone Hills (Bayamón to Dorado), Eastern Coast, Las Piedras Chiquitas, Southeastern Foothills (Salinas), Southern Coastal Lowlands, Southern Coastal Limestone, Guánica Forest Reserve, Southwestern Coastal Lowlands, Sierra Bermeja, Vieques, Culebra, Mona, Desecheo; 1–550 m; *fl* & *fr* throughout the year.

West Indies, southeastern United States (Florida), Central America, South America. [L1: 279; AR: 147].

Burro, Palinguán, Palo de burro.

Cynophalla hastata (Jacq.) J. Presl in Berchtold & J. Presl, Přir. Rostlin 2:275. 1825.
Bas.: *Capparis hastata* Jacq. (1865).
Syn.: *Capparis coccolobifolia* Mart. ex Eichler (1865).

Tree in Northern Coastal Lowlands, Southeastern Foothills, Guánica Forest Reserve, Southwestern Coastal Lowlands, Sierra Bermeja, Vieques, Culebra; 1–250 m; *fl* Jan, Apr–Oct, *fr* May–Jan.

Greater Antilles (Hispaniola), Virgin Islands, Lesser Antilles, Southern Caribbean Islands (Curaçao). Trinidad, northern South America (Venezuela, Colombia). [L1: 280].

Burro, Sapo.

Forchhammeria brevipes Urb., Symb. Antill. 7(2):225. 1912.

Tree in Northern Limestone Hills, Río Abajo Forest Reserve; 40–350 m; *fl* Oct–Mar, *fr* Dec–May.

Greater Antilles (Hispaniola). [L1: 286].

Morisonia americana L., Sp. Pl. 1:503. 1753.

Tree in Eastern Coast (Cuchilla de Panduras) Vieques, Culebra, Mona; 4–450 m; *fl* Nov–Mar, *fr* Aug–Nov, Feb.

Antilles, Mexico, Central America, northern South America amd Ecuador. [L1: 288].

Palo de peo.

Quadrella cynophallophora (L.) Hutch., Gen. Fl. Pl. 2:309. 1967.

Syn.: *Capparis cynophallophora* L. (1753).

Tree in Northern Coastal Lowlands (islands adjacent to Fajardo), Northern Limestone Hills (Manatí, Ciales), Southeastern Foothills, Southern Coastal Lowlands (Ponce), Southern Coastal Limestone, Guánica Forest Reserve, Southwestern Coastal Lowlands, Sierra Bermeja, Vieques, Culebra, Mona, Desecheo; 1–100 m; *fl* & *fr* throughout the year.

West Indies, southeastern United States (Florida), Mexico, Central America, northern South America. [L1: 277].

Bejuco inglés, Burro prieto, Sapo.

Quadrella indica (L.) Iltis & Cornejo, J. Bot. Res. Inst. Texas 4(1):126. 2010.

Bas.: *Breynia indica* L. (1753).

Syn.: *Capparis indica* (L.) Druce (1917).

Tree in Northern Coastal Lowlands (Fajardo), Las Piedras Chiquitas, Southeastern Foothills, Southern Coastal Lowlands (Juana Díaz), Southern Coastal Limestone, Guánica Forest Reserve, Southwestern Coastal Lowlands, Sierra Bermeja, Vieques, Culebra, Desecheo; 15–500 m; *fl* Mar–Jul, *fr* throughout the year.

Greater Antilles (Hispaniola), Lesser Antilles, southern Mexico, Central America, northern South America. [L1: 281].

Notes: Britton & Wilson (1924), Liogier & Martorell (1982, 2000) and Liogier (1985) used the homonym *Capparis indica* (L.) Fawc. & Rendle for the name of this taxon.

Burro, Sapo prieto.

CAPRIFOLIACEAE

Lonicera japonica Thunb. in Murray, Syst. Veg. (ed. 14) 216. 1784 May–Jun; et Fl. Jap. 89. 1784 Aug.

Woody vine in Central Cordillera; 800–1220 m; *fl* throughout the year, *fr* Nov (one specimen).

Native to China, Japan, Korea. [L5: 177; AR: 147].

Madreselva.

CARICACEAE

Carica papaya L., Sp. Pl. 2:1036. 1753.

Small tree cultivated and sometimes persistent in Northern Coastal Lowlands, Eastern Cordillera (Barranquitas), Vieques, Mona; 1–800 m; *fl* Feb–Nov, *fr* May–Jul (two specimens).

West Indies, Trinidad and Tobago, Mexico, Central America, South America; exact native range uncertain. [L3: 298].

Lechosa, Papaya.

CARYOPHYLLACEAE

Cerastium fontanum Baumg., Enum. Stirp. Transsilv. 1:425. 1816.

Herb in Toro Negro Forest Reserve; 1200 m; *fl* Feb–Mar (one specimen), *fr* (?). [*Vives a.n. 45077* (UPRRP)].

Native to Europe, northwestern Africa, Turkey, western Russia.

Drymaria cordata (L.) Willd. ex Schult. in Roemer & Schultes, Syst. Veg. 5:406. 1819.

Bas.: *Holosteum cordatum* L. (1753).

Scandent herb in Northern Coastal Lowlands, Northern Limestone Hills, Sierra de Luquillo, Eastern Coast (Cuchilla de Panduras), Caguas Valley (Caguas), Las Tetas de Cayey, Eastern Cordillera (Cañón San Cristóbal), Central Cordillera, Susúa Forest Reserve, Western Cordillera (Maricao), Vieques; 5–1200 m; *fl* & *fr* Aug–May.

West Indies, southeastern United States (Florida), Mexico, Central America, South America. [L1: 213].

Drimaria, Yerba de estrella.

Stellaria antillana Urb., Symb. Antill 1(2):303. 1899.

Syn.: *Alsine antillana* (Urb.) Britton & P. Wilson (1924).

Scandent herb in Northern Limestone Hills (Ciales, Florida), Las Tetas de Cayey, Southeastern Foothills (Coamo), Eastern Cordillera (Cayey), Central Cordillera; 200–975 m; *fl* throughout the year, *fr* Oct–Jun.

Greater Antilles (Hispaniola). [L1: 215].

Pega pollo.

CASUARINACEAE

Casuarina equisetifolia L., Amoen Acad. 4:143. 1759 "equisefolia."

Tree planted and persistent in Northern Coastal Lowlands, Northern Limestone Hills (Morovis), Sierra de Luquillo, Eastern Coast, Eastern Cordillera (Corozal), Southern Coastal Lowlands (Caja de Muertos), Susúa Forest Reserve, Maricao Forest Reserve, Guánica Forest Reserve, Western Coast

(Aguada), Vieques, Culebra, Mona; 1–300 m; *fl* Aug–May, *fr* throughout the year.

Native to southeastern Asia, Australia, islands of the Indo-Pacific. [L1: 14].

Notes: Liogier & Martorell (1982, 2000) and Liogier (1985) used the homonym *Casuarina equisetifolia* J.R. Forst. & G. Forst. for the name of this taxon.

Casuarina, Pino, Pino australiano.

TAXA CITED BY PREVIOUS AUTHORS, BUT NOT INCLUDED

ONLY UNDER CULTIVATION

Casuarina cristata Miq.
Tree cited by Liogier & Martorell (1982, 2000) and Liogier (1985) as planted and persistent. [L1: 14].

Casuarina cunninghamiana Miq.
Tree cited by Liogier & Martorell (1982, 2000) and Liogier (1985) as planted and persistent. [L1: 14].

CELASTRACEAE (including HIPPOCRATEACEAE)

Crossopetalum rhacoma Cranz, Inst. Rei Herb. 2:321. 1766.
Syn.: *Rhacoma crossopetalum* L. (1759).

Shrub in Northern Coastal Lowlands, Northern Limestone Hills, Río Abajo Forest Reserve, Southern Coastal Lowlands (Caja de Muertos), Southern Coastal Limestone, Guánica Forest Reserve, Southwestern Coastal Lowlands, Sierra Bermeja, Vieques, Culebra, Mona; 1–280 m; *fl* & *fr* throughout the year.

West Indies, southeastern United States (Florida), Mexico, northern Central America, northern South America. [L3: 23].

Coral, Manto, Maravedí, Palo de paloma, Pico de paloma.

Elaeodendron xylocarpum (Vent.) DC., Prodr. 2:11. 1825.
Bas.: *Cassine xylocarpa* Vent. (1803).

Tree in Northern Coastal Lowlands, Northern Limestone Hills, Southern Coastal Lowlands (Caja de Muertos), Southern Coastal Limestone, Susúa Forest Reserve, Maricao Forest Reserve, Guánica Forest Reserve, Southwestern Coastal Lowlands (Cabo Rojo), Vieques, Culebra; 1–650 m; *fl* & *fr* throughout the year.

West Indies, Mexico, southern Central America (Panama), northern South America (Venezuela). [L3: 21].

Aceituno, Cipote, Coscorrón, Guayabote, Guayarote, Manzanillo.

Gyminda latifolia (Sw.) Urb., Symb. Antill. 5(1):80. 1904.
Bas.: *Myginda latifolia* Sw. (1788).

Shrub in Northern Coastal Lowlands (islands adjacent to Fajardo), Northern Limestone Hills, Río Abajo Forest Reserve, Central Cordillera (Utuado), Southern Coastal Lowlands (Caja de Muertos), Southern Coastal Limestone, Susúa Forest Reserve, Maricao Forest Reserve, Southwestern Coastal Lowlands (Cabo Rojo), Sierra Bermeja, Vieques, Mona; 1–900 m; *fl* & *fr* throughout the year.

West Indies, southeastern United States (Florida), Mexico. [L3: 25].

Araña, Coscorroncito, Mala mujer.

Hippocratea volubilis L., Sp. Pl. 2:1191. 1753.

Woody vine in Northern Coastal Lowlands, Northern Limestone Hills, Río Abajo Forest Reserve, Sierra de Luquillo, Eastern Coast (Punta Guayanés), Sierra de Cayey, Eastern Cordillera (Cidra, Comerío), Central Cordillera (Ciales), Southern Coastal Lowlands, Quebrada de Los Cedros, Susúa Forest Reserve, Maricao Forest Reserve, Southwestern Coastal Lowlands (Cabo Rojo), Sierra Bermeja, Cerro Las Mesas, Western Coast (Mayagüez), Vieques; 1–800 m; *fl* throughout the year, *fr* Jan–Oct.

Antilles, Trinidad, southeastern United States (Florida), Mexico, Central America, South America. [L3: 34; AR: 149].

Bejuco prieto.

Maytenus cymosa Krug & Urb. in Urban, Symb. Antill. 5(1):61. 1904.

Tree in Northern Coastal Lowlands (Fajardo and adjacent islands), Vieques; 1–150 m; *fl* Feb–Aug, *fr* Jul–Aug.

Virgin Islands. [L3: 27].

Maytenus domingensis Krug & Urb. in Urban, Symb. Antill. 5(1):56. 1904.
Syn.: *Maytenus ponceana* Britton (1924).

Tree in Northern Limestone Hills, Río Abajo Forest Reserve, Las Tetas de Cayey, Central Cordillera (Yauco), Susúa Forest Reserve, Maricao Forest Reserve; 50–885 m; *fl* Feb, May–Nov, *fr* throughout the year.

Greater Antilles (Hispaniola). [L3: 29].

Maytenus elongata (Urb.) Britton in Britton & Wilson, Sci. Surv. Porto Rico & Virgin Islands 5(4):516. 1924.
Bas.: *Maytenus elliptica* (Lam.) Krug & Urb. var. *elongata* Urb. (1904).

Tree in Northern Limestone Hills, Sierra de Luquillo, Caguas Valley, Central Cordillera, Maricao Forest Reserve; 150–1100 m; *fl* May–Jul, Nov, *fr* Feb–Aug.

Greater Antilles (Hispaniola). [L3: 27].

Cuero de sapo.

Maytenus laevigata (Vahl) Griseb. ex Eggers, Vidensk. Meddel. Dansk Naturhist. Foren. Kjøbenhavn 1876:109. 1876.
Bas.: *Rhamnus laevigata* Vahl (1794).
Syn.: *Maytenus elliptica* (Lam.) Krug & Urb. (1897).

Tree in Northern Coastal Lowlands (Fajardo and adjacent islands), Northern Limestone Hills, Sierra de Luquillo, Eastern Coast (Ceiba), Southeastern Foothills, Vieques, Culebra; 1–350 m; *fl* May–Jul, *fr* May–Aug.

Antilles. [L3: 27].

Alfilerillo, Cuero de sapo.

Pristimera caribaea (Urb.) A.C. Sm., Brittonia 3(3):378. 1940.
Bas.: *Hippocratea caribaea* Urb. (1909).

Woody vine in Northern Coastal Lowlands (Fajardo), Northern Limestone Hills, Río Abajo Forest Reserve, Sierra de Cayey, Las Tetas de Cayey, Eastern Cordillera (Cañón San Cristóbal), Maricao Forest Reserve, Sierra Bermeja, Western Coast (Mayagüez), Vieques; 35–825 m; *fl* Jul–Aug, Nov–Jan, *fr* (?).

Greater Antilles (Hispaniola), Lesser Antilles, northern South America (Guyana). [L3: 35; AR: 151].

Schaefferia frutescens Jacq., Enum. Syst. Pl. 33. 1760.

Shrub/small tree in Northern Coastal Lowlands, Northern Limestone Hills, Las Tetas de Cayey, Las Piedras Chiquitas, Southeastern Foothills (Coamo), Southern Coastal Lowlands (Caja de Muertos), Guánica Forest Reserve, Southwestern Coastal Lowlands (Cabo Rojo), Sierra Bermeja, Vieques, Culebra, Mona; 1–825 m; *fl* Apr–Jun, Oct, *fr* throughout the year.

West Indies, southeastern United States (Florida), southern Mexico, southern Central America (Costa Rica), northern South America (Venezuela, Colombia). [L3: 29].

Cafeíllo, Jibá.

Torralbasia cuneifolia (C. Wright ex A. Gray) Krug & Urb. in Urban, Symb. Antill. 5(1):50. 1904.
Bas.: *Euonymus cuneifolius* C. Wright ex A. Gray (1860).

Tree in Sierra de Luquillo, Central Cordillera, Maricao Forest Reserve; 900–1300 m; *fl* Oct–Jan, Apr–Jun, *fr* Jan–Mar.

Greater Antilles (Cuba, Hispaniola). [L3: 31].

Boje.

CERATOPHYLLACEAE

Ceratophyllum demersum L., Sp. Pl. 2:992. 1753.

Aquatic herb in Northern Coastal Lowlands, Northern Limestone Hills (Arecibo), Guánica Forest Reserve, Southwestern Coastal Lowlands (Guánica); 1–50 m; *fl* Apr (one specimen), *fr* Apr (one specimen).

Nearly cosmopolitan. [L1: 223].

Candelabro de agua, Celestina de agua.

CHLORANTHACEAE

Hedyosmum arborescens Sw., Prodr. 84. 1788.

Tree in Sierra de Luquillo, Sierra de Cayey; 500–1000 m; *fl* throughout the year, *fr* Jan–Sep.

Antilles. [L1: 35].

Azafrán.

CHRYSOBALANACEAE

Chrysobalanus icaco L., Sp. Pl. 1:513. 1753.
Syn.: *Chrysobalanus pellocarpus* G. Mey. (1818). *Chrysobalanus icaco* L. var. *pellocarpus* (G. Mey.) Hook. f. (1867).

Shrub/small tree in Northern Coastal Lowlands, Sierra de Luquillo, Eastern Coast, Susúa Forest Reserve, Southwestern Coastal Lowlands (Cabo Rojo), Sierra Bermeja, Cerro Las Mesas, Western Coast; 1–300 m; *fl* & *fr* throughout the year.

West Indies, southeastern United States (Florida), Mexico, Central America, northern South America to Brazil, Ecuador. [L1: 324, 326].

Hicaco, Icaco, Jicaco.

Hirtella rugosa Pers., Syn. Pl. 1:250. 1805.

Tree in Sierra de Luquillo, Sierra de Cayey, Eastern Cordillera (Cerro La Peña), Susúa Forest Reserve, Maricao Forest Reserve; 200–1000 m; *fl* & *fr* throughout the year.

Greater Antilles (Hispaniola). [L1: 326].

Hicaquillo, Icaquillo, Jicaquillo, Juanilla, Teta de burra cimarrona.

Hirtella triandra Sw., Prodr. 51. 1788.

Tree in Northern Limestone Hills (Vega Baja, San Sebastián), Sierra de Luquillo, Sierra de Cayey; 100–600 m; *fl* Sep–Apr, *fr* Nov–Feb.

Antilles, Mexico, Central America, northwestern, west-central South America and Venezuela. [L1: 327].

CLEOMACEAE

Arivela viscosa (L.) Raf., Sylva Tellur. 110. 1838.
Bas.: *Cleome viscosa* L. (1753).
Syn.: *Cleome icosandra* L. (1753).

Herb in Northern Coastal Lowlands (Manatí), Southeastern Foothills, Southern Coastal Lowlands, Southern Coastal Limestone (Ponce), Guánica Forest Reserve, Southwestern Coastal Lowlands (Cabo Rojo), Western Coast (Añasco), Vieques; 2–250 m; *fl* & *fr* throughout the year.

Probably native to tropical Asia. [L1: 286].

Cleome stenophylla Klotzsch ex Urb., Symb. Antill. 4(2):251. 1905.

Herb in Southern Coastal Lowlands (Guayama), Guánica Forest Reserve, Southwestern Coastal Lowlands, Sierra Bermeja, Vieques; 1–200 m; *fl* & *fr* Aug–May.

Lesser Antilles, Southern Caribbean Islands, Central America (Belize, Nicaragua), northern South America. [L1: 285].

Notes: This taxon will eventually be placed in the genus *Physostemon* by Hugh Iltis and Theodore Cochrane.

Cleoserrata serrata (Jacq.) Iltis, Novon 17(4):448. 2007.
Bas.: *Cleome serrata* Jacq. (1760).

Herb in Northern Coastal Lowlands (San Juan), Northern Limestone Hills (San Sebastián), Western Coast (Mayagüez); 1–400 m; *fl* & *fr* Apr–Jun, Oct.

Antilles, Trinidad, southern Mexico, Central America, South America. [L1: 283].

Cleoserrata speciosa (Raf.) Iltis, Novon 17(4):448. 2007.
Bas.: *Cleome speciosa* Raf. (1817).
Syn.: *Gynandropsis speciosa* (Raf.) DC. (1824).

Herb in Northern Coastal Lowlands (Fajardo, Río Grande), Eastern Coast (Ceiba), Susúa Forest Reserve, Western Cordillera, Vieques; 10–400 m; *fl* Jan–Mar, Jun–Oct, *fr* Jan–Mar, Jun, Sep–Oct.

Native to southern Mexico, Central America, northern South America and Ecuador; exact native range uncertain. [L1: 284].

Notes: Britton & Wilson (1924), Liogier & Martorell (1982, 2000) and Liogier (1985) used the homonym *Cleome speciosa* Kunth for the name of this taxon.

Volantines preciosos.

Gynandropsis gynandra (L.) Briq., Annuaire Conserv. Jard. Bot. Genève 17:382. 1914.
Bas.: *Cleome gynandra* L. (1753).

Herb in Northern Coastal Lowlands (San Juan), Southern Coastal Lowlands, Southwestern Coastal Lowlands (Cabo Rojo), Mona; 1–75 m; *fl* & *fr* Jan–Sep.

Native to Africa, southern Arabian peninsula, tropical Asia, China; exact native range uncertain. [L1: 283].

Jazmín del río.

Hemiscola aculeata (L.) Raf., Sylva Tellur. 111. 1838.
Bas.: *Cleome aculeata* L. (1768).

Herb in Northern Coastal Lowlands, Eastern Coast, Caguas Valley, Southeastern Foothills (Salinas), Western Coast (Mayagüez), Vieques; 2–200 m; *fl* & *fr* throughout the year.

Greater Antilles (Jamaica, Hispaniola), Lesser Antilles, Trinidad, south-central United States (Texas), Mexico, Central America, South America. [L1: 282].

Tarenaya spinosa (Jacq.) Raf., Sylva Tellur. 111. 1838.
Bas.: *Cleome spinosa* Jacq. (1838).

Herb in Northern Coastal Lowlands, Northern Limestone Hills (Florida), Río Abajo Forest Reserve, Caguas Valley (Caguas), Southeastern Foothills (Coamo), Eastern Cordillera (Comerío), Central Cordillera (Utuado), Southern Coastal Lowlands, Southwestern Coastal Lowlands (Lajas), Western Coast, Culebra; 5–200 m; *fl* & *fr* throughout the year.

Antilles, southern Mexico, Central America, northern South America. [L1: 285].

Aromo, Flor de perro, Gusanera, Jazmín apestoso, Sambo, Volantines punzantes.

TAXA CITED BY PREVIOUS AUTHORS, BUT NOT INCLUDED

ONLY UNDER CULTIVATION

Tarenaya hassleriana (Chodat) Iltis. Syn.: *Cleome hassleriana* Chodat.

Herb cited by Liogier (1985) as cultivated and escaped in the vicinity of Aibonito and by Liogier & Martorell (2000) as cultivated and naturalized in the Central Cordillera. [L1: 283].

CLUSIACEAE (GUTTIFERAE)

Calophyllum antillanum Britton in Britton & Wilson, Sci. Surv. Porto Rico & Virgin Islands 5(4):584. 1924.
Syn.: *Calophyllum brasiliense* Cambess. var. *antillanum* (Britton) Standl. (1932).

Tree in Northern Coastal Lowlands, Northern Limestone Hills, Río Abajo Forest Reserve, Sierra de Luquillo, Eastern Coast, Caguas Valley (San Lorenzo),

Susúa Forest Reserve, Western Cordillera, Vieques; 1–900 m; *fl* Apr–Oct, *fr* Sep–Jun.

West Indies, Trinidad and Tobago. [L3: 228].

Notes: According to Nicolson (1991), there is a problem in the typification of *Calophyllum calaba* L. which prevents the use of this name, commomly employed in previous floras of the region; indeed, the Flora of Ceylon (1980) claims the taxon of this name as an endemic. Consequently, that name was misapplied by Liogier (1994) and Liogier & Martorell (2000) to the above taxon; Liogier & Martorell (1982) used *Calophyllum brasiliense* Cambess., *sensu lato*, for it. This tree is also often planted as an ornamental.

Aceite de María, Calaba, Galba, María, Palo de María, Santa María.

Calophyllum inophyllum L., Sp. Pl. 1:513. 1753.
Syn.: *Balsamaria inophyllum* (L.) Lour. (1790).

Tree planted and persistent in Northern Coastal Lowlands (Río Grande, Carolina), Sierra de Luquillo, Eastern Coast (Naguabo), Western Coast (Mayagüez); 10–350 m; *fl* Feb–Jun, Sep, *fr* Oct–Dec.

Native to eastern Africa, tropical Asia, Japan, Australia, Fiji. [L3: 230].

Kamani, María grande, Santa María.

Clusia clusioides (Griseb.) D'Arcy, Ann. Missouri Bot. Gard. 67(4):990. 1981.
Bas.: *Tovomita clusioides* Griseb. (1860).
Syn.: *Clusia krugiana* Urb. (1899).
Clusia grisebachiana (Triana & Planch.) Alain (1953).

Tree in Río Abajo Forest Reserve, Sierra de Luquillo, Central Cordillera, Maricao Forest Reserve; 300–1100 m; *fl* May–Feb, *fr* throughout the year.

Greater Antilles (Cuba, Hispaniola), northern South America (Colombia). [L3: 231].

Cupeíllo.

Clusia gundlachii A. Stahl, Estud. Fl. Puerto-Rico 2:122. 1884.

Woody vine/shrub in Río Abajo Forest Reserve, Sierra de Luquillo, Sierra de Cayey, Las Tetas de Cayey, Eastern Cordillera (Cerro La Peña), Central Cordillera, Maricao Forest Reserve, Guánica Forest Reserve; 150–950 m; *fl* & *fr* throughout the year.

Endemic. [L3: 232; AR: 152].

Cupeíllo de altura, Cupey, Cupey de altura.

Clusia minor L., Sp. Pl. 1:510. 1753.

Woody vine/shrub in Northern Limestone Hills (Ciales, Lares), Río Abajo Forest Reserve, Sierra de Luquillo, Southeastern Foothills (Salinas), Central Cordillera, Susúa Forest Reserve, Southwestern Coastal Lowlands (Cabo Rojo), Sierra Bermeja; 5–1335 m; *fl* Jul–Jan, *fr* May, Aug–Sep.

Greater Antilles (Cuba, Hispaniola), Virgin Islands, Lesser Antilles (St. Vincent, Grenadines), Trinidad and Tobago, Central America, northwestern, west-central South America and Venezuela. [L3: 232].

Cupeíllo, Cupey, Cupey de monte, Cupey trepador.

Clusia rosea Jacq., Enum. Syst. Pl. 34. 1760.

Tree in Northern Coastal Lowlands, Northern Limestone Hills, Río Abajo Forest Reserve, Sierra de Luquillo, Eastern Coast (Punta Guayanés), Las Tetas de Cayey, Central Cordillera, Susúa Forest Reserve, Western Cordillera, Guánica Forest Reserve, Southwestern Coastal Lowlands (Cabo Rojo), Sierra Bermeja, Vieques, Culebra, Mona, Desecheo; 5–850 m; *fl* throughout the year, *fr* throughout the year.

Bahamas, Greater Antilles, Virgin Islands, Lesser Antilles (Anguilla, St. Martin), southeastern United States (Florida), northwestern, west-central South America and Venezuela. [L3: 232].

Notes: In Puerto Rico this tree is known only with pistillate, apomictic flowers.

Cupey.

Garcinia hessii (Britton) Alain, Phytologia 61(6):353. 1986.
Bas.: *Rheedia hessii* Britton (1915).

Shrub/small tree in Susúa Forest Reserve, Western Cordillera (Maricao); 180–400 m; *fl* Dec–May, *fr* Jan–Sep.

Endemic. [L3: 234].

Hinca barriga.

Garcinia portoricensis (Urb.) Alain, Phytologia 61(6):353. 1986.
Bas.: *Rheedia portoricensis* Urb. (1899).

Tree in Northern Limestone Hills, Río Abajo Forest Reserve, Sierra de Luquillo, Eastern Coast (Punta Guayanés), Sierra de Cayey, Sierra Bermeja, Vieques; 10–930 m; *fl* throughout the year, *fr* Mar–Dec.

Endemic. [L3: 236].

Notes: Britton & Wilson (1924) used *Rheedia acuminata* (Spreng.) Triana & Planch. for the name of this taxon.

Guayabacoa, Palo de cruz.

Mammea americana L., Sp. Pl. 1:512. 1753.

Tree in Northern Limestone Hills, Río Abajo Forest Reserve, Sierra de Luquillo, Sierra de Cayey (Patillas), Southeastern Foothills (Salinas), Central Cordillera (Ponce), Cerro Las Mesas; 200–650 m; *fl* throughout the year, *fr* Oct–Feb, May.

West Indies, Trinidad and Tobago, Mexico, Central America; exact native range uncertain. [L3: 238].

Mamey.

TAXA CITED BY PREVIOUS AUTHORS, BUT NOT INCLUDED

ONLY UNDER CULTIVATION

Garcinia dulcis (Roxb.) Kurz.
Tree cited by Liogier & Martorell (1982, 2000) and Liogier (1994) as planted and persistent. [L3: 234].

Garcinia mangostana L.
Tree cited by Liogier & Martorell (1982, 2000) and Liogier (1994) as planted and persistent. [L3: 236].

COMBRETACEAE

Buchenavia tetraphylla (Aubl.) R.A. Howard, J. Arnold Arbor. 64(2):266. 1983.
Bas.: *Cordia tetraphylla* Aubl. (1775).
Syn.: *Buchenavia capitata* (Vahl) Eichler (1866).

Tree in Sierra de Luquillo, Sierra de Cayey, Central Cordillera, Susúa Forest Reserve, Western Cordillera; 350–900 m; *fl* Feb–Mar, *fr* Oct–Mar.

Antilles, southern Central America, northern South America to Brazil, Bolivia. [L3: 367].

Granadillo.

Bucida buceras L., Syst. Nat. (ed. 10) 2:1025. 1759.

Tree in Northern Coastal Lowlands, Northern Limestone Hills, Sierra de Luquillo, Caguas Valley (Caguas), Las Piedras Chiquitas, Southeastern Foothills, Southern Coastal Lowlands, Guánica Forest Reserve, Southwestern Coastal Lowlands, Sierra Bermeja, Vieques, Culebra, Mona; 1–500 m; *fl* Mar–Nov, *fr* Apr–Dec.

West Indies, Mexico, northern Central America. [L3: 368].

Notes: This tree is also often planted as an ornamental.

Bucaro, Úcar.

Combretum grandiflorum G. Don, Edinburgh Philos. J. 11:346. 1824.

Woody vine planted and persistent in Northern Coastal Lowlands (Toa Baja); 15 m; *fl* Nov–Mar, Aug, *fr* Nov, Feb (two specimens).

Native to western tropical Africa. [AR: 154].

Conocarpus erectus L., Sp. Pl. 1:176. 1753 "erecta."

Tree in Northern Coastal Lowlands, Eastern Coast, Southern Coastal Lowlands, Guánica Forest Reserve, Southwestern Coastal Lowlands, Western Coast (Rincón), Vieques, Culebra, Mona, Desecheo; 1–10 m; *fl* throughout the year, *fr* Aug–Feb.

West Indies, Trinidad and Tobago, southeastern United States (Florida), Central America, northern South America to Brazil, Peru, western tropical Africa. [L3: 370].

Botoncillo, Mangle botón.

Laguncularia racemosa (L.) C.F. Gaertn., Suppl. Carp. 2:209, *t. 217, f.* 3. 1807.
Bas.: *Conocarpus racemosus* L. (1759) "racemosa."

Tree in coastal swamps of Northern Coastal Lowlands, Eastern Coast (Cayo Santiago), Southern Coastal Lowlands (Caja de Muertos and offshore islands), Guánica Forest Reserve, Southwestern Coastal Lowlands, Vieques, Culebra, Mona; 0 m; *fl* Nov–Aug, *fr* Jun–Feb.

Coastal swamps of West Indies, southeastern United States (Florida), Mexico, Central America, northwestern, west-central South America and Venezuela. [L3: 371].

Mangle blanco, Mangle bobo.

Quisqualis indica L., Sp. Pl. (ed. 2) 1:556. 1762.
Syn.: *Combretum indicum* (L.) DeFilipps (1998).

Shrub/vine planted and persistent in Northern Coastal Lowlands (Manatí), Northern Limestone Hills, Southern Coastal Lowlands (Caja de Muertos), Susúa Forest Reserve, Western Coast (Mayagüez), Vieques; 5–250 m; *fl* Jul–Apr, *fr* (?).

Native to Africa, eastern tropical Asia, China. [L3: 373; AR: 155].

Notes: Acevedo-Rodríguez (2005) used the combination *Combretum indicum* (L.) Jongkind for the name of the synonym.

Cocuisa, Corazón de hombre, Cuiscualis.

Terminalia catappa L., Syst. Nat. (ed. 12) 2:674; et Mant. Pl. 1:128. 1767.

Tree planted and persistent in Northern Coastal Lowlands, Northern Limestone Hills, Sierra de Luquillo, Eastern Cordillera (Comerío), Southern Coastal Lowlands (Caja de Muertos), Susúa Forest Reserve, Guánica Forest Reserve, Western Coast, Vieques, Mona, Desecheo; 1–500 m; *fl* Feb–Nov, *fr* Aug–May.

Native to Madagascar, eastern tropical Asia, China, Australia, New Caledonia. [L3: 375].

Almendra, Almendro.

TAXA CITED BY PREVIOUS AUTHORS, BUT NOT INCLUDED

ONLY UNDER CULTIVATION

Terminalia ivorensis A. Chev.

Tree cited by Liogier (1984) and Liogier & Martorell (2000) as planted and becoming naturalized. [L3: 375].

Terminalia myriocarpa Huerck & Müll. Arg.

Tree cited by Liogier (1984) and Liogier & Martorell (2000) as planted and spontaneous in western Puerto Rico. [L3: 376].

Terminalia oblonga (Ruiz & Pav.) Steud.

Tree cited by Liogier (1984) and Liogier & Martorell (2000) as planted and becoming naturalized. [L3: 377].

COMMELINACEAE

Callisia fragrans (Lindl.) Woodson, Ann. Missouri Bot. Gard. 29(3):154. 1942.

Bas.: *Spironema fragrans* Lindl. (1840).

Herb in Northern Coastal Lowlands (Manatí), Northern Limestone Hills, Eastern Cordillera, Central Cordillera (Ponce), Maricao Forest Reserve, Guánica Forest Reserve; 25–800 m; *fl* Jan–Jun, *fr* (?).

Native to Mexico. [AR&S: 156].

Callisia monandra (Sw.) Schult. & Schult. f., Syst. Veg. 7(2):1179. 1830.

Bas.: *Tradescantia monandra* Sw. (1788).

Syn.: *Aploleia monandra* (Sw.) H.E. Moore (1961).

Herb in Northern Limestone Hills (Ciales), Río Abajo Forest Reserve, Sierra de Luquillo, Eastern Coast (Yabucoa), Caguas Valley (Caguas), Sierra de Cayey, Las Tetas de Cayey, Eastern and Central Cordillera, Western Cordillera; 100–1050 m; *fl* & *fr* Jan–Jun.

Greater Antilles (Cuba, Hispaniola), Lesser Antilles (Guadeloupe, Martinique), Mexico, Central America, South America (scattered). [AR&S: 155].

Cohítre morado.

Callisia repens (Jacq.) L., Sp. Pl. (ed. 2) 1:62. 1762.

Bas.: *Hapalanthus repens* Jacq. (1760).

Herb in Northern Coastal Lowlands, Las Tetas de Cayey, Las Piedras Chiquitas, Eastern Cordillera (Comerío), Sierra Bermeja, Vieques, Mona; 1–825 m; *fl* Feb–Mar (two specimens), *fr* Feb–Mar.

Antilles, south-central United States (Louisiana, Texas), Mexico, Central America, South America. [AR&S: 156].

Belleza, Cohítre enano, Lengua de gallina.

Commelina benghalensis L., Sp. Pl. 1:41. 1753, *nom. cons.*

Herb in Eastern Cordillera (Cayey); 525 m; *fl* & *fr* Feb (one specimen). [*Axelrod 730* (UPRRP)].

Native to tropical Africa, Arabia, tropical Asia, China, Japan. [AR&S: 158].

Commelina diffusa Burm. f., Fl. Indica 18: *t. 7, f.* 2. 1768.

Syn.: *Commelina longicaulis* Jacq. (1791).

Herb in Northern Coastal Lowlands, Río Abajo Forest Reserve, Sierra de Luquillo, Eastern Coast, Caguas Valley (Caguas), Sierra de Cayey, Southeastern Foothills (Cayey), Eastern Cordillera (Corozal, Aibonito), Central Cordillera, Susúa Forest Reserve, Western Cordillera (Maricao), Southwestern Coastal Lowlands (Lajas), Vieques; 1–1300 m; *fl* & *fr* throughout the year.

Pantropical and -subtropical. [AR&S: 158].

Cohítre, Cohítre azul, Cojitre.

Commelina erecta L., Sp. Pl. 1:41. 1753.

Syn.: *Commelina elegans* Kunth (1816).

Herb in Northern Coastal Lowlands, Northern Limestone Hills (Utuado), Río Abajo Forest Reserve, Eastern Coast (Ceiba), Las Piedras Chiquitas, Southeastern Foothills, Central Cordillera, Southern Coastal Limestone, Susúa Forest Reserve, Guánica Forest Reserve, Southwestern Coastal Lowlands, Sierra Bermeja, Vieques, Culebra, Mona, Desecheo; 1–1000 m; *fl* & *fr* throughout the year.

West Indies, Trinidad, eastern and central United States, Mexico, Central America, South America. [AR&S: 159].

Cohítre azul.

Commelina rufipes Seub. var. **glabrata** (D.R. Hunt) Faden & D.R. Hunt, Ann. Missouri Bot. Gard. 74(1):122. 1987.

Bas.: *Commelinopsis glabrata* D.R. Hunt (1981).

Herb in Northern Coastal Lowlands (Loíza, Dorado), Northern Limestone Hills (Quebradillas), Río Abajo Forest Reserve, Sierra de Luquillo, Las Tetas de Cayey, Sierra de Cayey, Eastern Cordillera (Aibonito), Central Cordillera (Utuado), Maricao Forest Reserve; 1–825 m; *fl* & *fr* Jan–Jun.

Greater Antilles (Cuba, Hispaniola), Trinidad, northwestern, west-central South America and Venezuela. [AR&S: 159].

Notes: Britton & Wilson (1923) misapplied the name *Athyrocarpus persicariifolius* (DC.) Hemsl., and Liogier & Martorell (1982) misapplied *Commelinopsis persicariifolia* (Delile) Pichon, to this taxon.

Cohítre blanco.

Gibasis geniculata (Jacq.) Rohweder, Abh. Auslandsk., Reihe C, Naturwiss. 18:143. 1956.
Bas.: *Tradescantia geniculata* Jacq. (1763).

Herb in Northern Coastal Lowlands (Vega Alta), Northern Limestone Hills, Río Abajo Forest Reserve, Sierra de Luquillo, Eastern Coast (Cuchilla de Panduras), Caguas Valley (Caguas), Sierra de Cayey, Las Tetas de Cayey, Southeastern Foothills, Eastern Cordillera (Cayey), Central Cordillera, Western Cordillera, Vieques; 10–800 m; *fl* & *fr* throughout the year.

Greater Antilles (Hispaniola), Lesser Antilles, Trinidad, Mexico, Central America, South America. [AR&S: 161].

Cohítre azul.

Murdannia nudiflora (L.) Brenan, Kew Bull. 7(2):189. 1952.
Bas.: *Commelina nudiflora* L. (1753).

Herb in Northern Coastal Lowlands, Sierra de Luquillo, Eastern Coast (Naguabo), Sierra de Cayey, Southwestern Coastal Lowlands (Cabo Rojo); 1–800 m; *fl* & *fr* Sep–Jun.

Pantropical. [AR&S: 162].

Tradescantia pallida (Rose) D.R. Hunt, Kew Bull. 30(3):452. 1975.
Bas.: *Setcreasea pallida* Rose (1911).

Herb planted and persistent in Northern Coastal Lowlands; 5–50 m; *fl* & *fr* Feb–May.

Native to Mexico. [AR&S: 164].

Cohítre morado.

Tradescantia spathacea Sw., Prodr. 57. 1788.
Syn.: *Rhoeo spathacea* (Sw.) Stearn (1957). *Tradescantia discolor* L'Hér. (1789). *Rhoeo discolor* (L'Hér.) Hance (1852).

Herb in Northern Coastal Lowlands, Northern Limestone Hills, Río Abajo Forest Reserve; 2–300 m; *fl* Nov–Jun, *fr* Apr–Jun, Nov.

Native to Lesser Antilles, Mexico, northern Central America. [AR&S: 165].

Notes: When Liogier & Martorell (2000) used *Tradescantia bicolor* L'Hér. for the name of this taxon, I presume they made an orthographic error in the name of the above synonym.

Sangría, Sanguinaria.

Tradescantia zanonia (L.) Sw., Prodr. 57. 1788.
Bas.: *Commelina zanonia* L. (1753).
Syn.: *Campelia zanonia* (L.) Kunth (1816).

Herb in Northern Limestone Hills (Ciales, Arecibo), Río Abajo Forest Reserve, Las Tetas de Cayey, Central Cordillera; 225–800 m; *fl* Dec–Apr, *fr* Nov–Jul.

Greater Antilles, Trinidad, Mexico, Central America, northwestern, west-central South America and Venezuela, Brazil. [AR&S: 165].

Cohítre blanco.

Tradescantia zebrina hort. ex Bosse, Vollst. Handb. Bl.-gärtn. 4:655. 1849.
Syn.: *Zebrina pendula* Schnizl. (1849).

Herb in Northern Coastal Lowlands (San Juan), Northern Limestone Hills (Florida, Arecibo), Río Abajo Forest Reserve, Sierra de Luquillo, Sierra de Cayey, Las Tetas de Cayey, Southeastern Foothills, Eastern Cordillera (Cerro La Peña), Central Cordillera, Western Cordillera (Maricao); 200–975 m; *fl* throughout the year, *fr* (?).

Native to Mexico. [AR&S: 165].

Cohítre morado, Judío errante.

Tripogandra serrulata (Vahl) Handlos, Baileya 17:33. 1970.
Bas.: *Commelina serrulata* Vahl (1798).
Syn.: *Tradescantia elongata* G. Mey. (1818).

Herb in Northern Coastal Lowlands, Río Abajo Forest Reserve, Sierra de Luquillo, Eastern Coast (Maunabo), Caguas Valley (Caguas), Sierra de Cayey (San Lorenzo), Southeastern Foothills (Arroyo), Eastern Cordillera (Corozal), Central Cordillera (Orocovis), Maricao Forest Reserve; 10–750 m; *fl* & *fr* Sep, Dec–May.

Greater Antilles (Hispaniola), Lesser Antilles, Trinidad, Mexico, Central America, northwestern, west-central South America and Venezuela. [AR&S: 166].

CONNARACEAE

Rourea surinamensis Miq., Linnaea 26(2):221. 1854.

Woody vine in Northern Coastal Lowlands, Guajataca Forest Reserve, Sierra de Luquillo, Eastern Coast (Yabucoa), Western Cordillera (Maricao); 1–600 m; *fl* throughout the year, *fr* Sep–May.

Greater Antilles (Hispaniola), Lesser Antilles (St. Lucia, Grenada), Tobago, northern South America. [L1: 329; AR: 157].

Bejuco de garrote, Juan caliente.

CONVOLVULACEAE

Aniseia martinicensis (Jacq.) Choisy, Mém. Soc. Phys. Genève 8:144. 1837 (Convolv. Diss. Sec.).
Bas.: *Convolvulus martinicensis* Jacq. (1763).

Vine in Northern Coastal Lowlands, Eastern Coast (Humacao); 1–5 m; *fl* throughout the year, *fr* Jan–Jul.

Greater Antilles, Lesser Antilles (Guadeloupe, Martinique), Trinidad, Mexico, Central America, South America. [L4: 250; AR: 159].

Argyreia nervosa (Burm. f.) Bojer, Hortus Maurit. 224. 1837.
Bas.: *Convolvulus nervosus* Burm. f. (1768).
Vine in Southeastern Foothills (Salinas), Eastern Cordillera (Aguas Buenas), Culebra; 5–300 m; *fl* Jun–Sep, *fr* Jun, Sep (two specimens).
Native to India. [L4: 252; AR: 160].

Convolvulus nodiflorus Desr. in Lamarck, Encycl. 3(2):557. 1792.
Syn.: *Jacquemontia nodiflora* (Desr.) G. Don (1838).
Vine in Northern Coastal Lowlands (Fajardo and adjacent islands), Northern Limestone Hills, Eastern Coast, Las Piedras Chiquitas, Southeastern Foothills, Eastern Cordillera (Cidra), Cerro de Las Cuevas, Southern Coastal Limestone, Susúa Forest Reserve, Guánica Forest Reserve, Southwestern Coastal Lowlands, Sierra Bermeja, Western Coast (Rincón), Vieques, Culebra, Desecheo; 1–650 m; *fl* Nov–Jun, *fr* Oct–Jun.
Antilles, Mexico, Central America, northwestern, west-central South America and Venezuela. [L4: 254; AR: 161].
Aguinaldo blanco.

Cuscuta americana L., Sp. Pl. 1:124. 1753.
Parasitic vine in Northern Coastal Lowlands, Northern Limestone Hills (Aguadilla), Eastern Coast, Southeastern Foothills (Salinas), Sierra Bermeja, Western Coast (Mayagüez), Desecheo; 5–300 m; *fl* & *fr* throughout the year.
West Indies, southeastern United States (Florida), South America (scattered). [L4: 256; AR: 222].
Bejuco de mona, Fideíllo, Fideos, Tente en el aire.

Cuscuta boldinghii Urb., Repert. Spec. Nov. Regni. Veg. 16:38. 1919.
Parasitic vine in Northern Coastal Lowlands (islands adjacent to Fajardo: Ramos); 1–5 m; *fl* & *fr* Nov (one specimen). [*Liogier 34653, 34662* (UPR)].
Greater Antilles (Hispaniola), Lesser Antilles (Grenada), Southern Caribbean Islands (Bonaire), Mexico, Central America.

Cuscuta campestris Yunck., Mem. Torrey Bot. Club 18(2):138. 1932.
Parasitic vine in Northern Coastal Lowlands (San Juan, Arecibo), Río Abajo Forest Reserve, Toro Negro Forest Reserve; 1–1000 m; *fl* & *fr* May–Jun, Sep.
Bahamas, Greater Antilles (Cuba, Jamaica), Lesser Antilles (Guadeloupe, Martinique), Canada, United States, Mexico. [L4: 256].

Cuscuta globulosa Benth., Bot. Voy. Sulphur 138. 1845.
Parasitic vine in Northern Coastal Lowlands (Fajardo to San Juan), Northern Limestone Hills (Isabela), Southeastern Foothills, Central Cordillera (Ponce), Guánica Forest Reserve, Sierra Bermeja, Cerro Las Mesas, Western Coast (Rincón), Vieques, Culebra; 1–300 m; *fl* & *fr* Nov–Aug.
West Indies, Mexico. [L4: 256].

Cuscuta indecora Choisy, Mém. Soc. Phys. Genève 9(2):278, *t.* 3, *f.* 3. 1841 (Convolv. Diss. Tert.).
Parasitic vine in Southwestern Coastal Lowlands (Lajas); 10 m; *fl* & *fr* Mar–Jul. [*Liogier 30562* (UPR); *Proctor 40573* (SJ); *Oviedo a.n. 25647* (MAPR)].
Greater Antilles (Cuba, Jamaica), Lesser Antilles (Barbados), United States, Mexico, Central America. [L4: 256].

Cuscuta umbellata Kunth in Humboldt, Bonpland & Kunth, Nov. Gen. Sp. 3:95 [folio]; 3:121 [quarto]. 1819.
Parasitic vine in Southern Coastal Lowlands (Caja de Muertos), Guánica Forest Reserve, Southwestern Coastal Lowlands (Lajas), Mona; 5–50 m; *fl* & *fr* throughout the year.
West Indies, southern United States, Mexico, Central America, northern South America. [L4: 258].

Evolvulus convolvuloides (Willd.) Stearn, Taxon 21(5–6):649. 1972.
Bas.: *Nama convolvuloides* Willd. (1820).
Syn.: *Evolvulus glaber* Spreng. (1824).
Herb in Southern Coastal Lowlands (Caja de Muertos), Guánica Forest Reserve, Southwestern Coastal Lowlands, Vieques, Culebra, Mona; 1–100 m; *fl* & *fr* Nov–Aug.
Bahamas, Virgin Islands, Lesser Antilles, southern Mexico (Yucatan), Central America (Nicaragua), South America (scattered). [L4: 261].

Evolvulus nummularius (L.) L., Sp. Pl. (ed. 2) 1:391. 1762.
Bas.: *Convolvulus nummularius* L. (1753).
Herb in Northern Coastal Lowlands (islands adjacent to Fajardo), Northern Limestone Hills, Sierra Bermeja, Western Coast (Mayagüez); 5–300 m; *fl* Nov–Jun, *fr* Jun (one specimen).
Widespread in the tropics; native range uncertain. [L4: 261].

Agracejo rastrero.

Evolvulus sericeus Sw., Prodr. 55. 1788.

Herb in Southern Coastal Limestone, Guánica Forest Reserve, Cerro Las Mesas, Vieques; 2–300 m; *fl* & *fr* Sep–Jun.

West Indies, southeastern United States, Mexico, Central America, South America. [L4: 261].

Ipomoea alba L., Sp. Pl. 1:161. 1753.

Syn.: *Calonyction aculeatum* (L.) House (1904).

Vine in Northern Coastal Lowlands, Río Abajo Forest Reserve, Caguas Valley (Caguas), Eastern Cordillera, Southern Coastal Lowlands (Arroyo), Guánica Forest Reserve, Sierra Bermeja, Cerro Las Mesas, Western Coast (Hormigueros); 1–650 m; *fl* Oct–Apr, *fr* Mar (one specimen).

West Indies, Mexico, Central America, South America. [L4: 266; AR: 164].

Bejuco de puerco, Bejuco de vaca, Claro de luna, Flor de luna, Gloria de la mañana blanca.

Ipomoea aquatica Forssk., Fl. Aegypt.-Arab. 44. 1775.

Aquatic herb in Northern Coastal Lowlands (islands adjacent to Fajardo: Piñeros); 1–10 m; *fl* (?), *fr* (?). [*Woodbury a.n. 2914* (UPR)].

Native to tropical Africa, tropical Asia, Papua New Guinea, Australia. [L4: 266].

Notes: This citation is based on a single undated specimen with fruit.

Ipomoea batatas (L.) Lam., Tabl. Encycl. 1:465. 1793.

Bas.: *Convolvulus batatas* L. (1753).

Vine escaped from cultivation in Sierra de Cayey, Central Cordillera (Peñuelas), Western Coast (Mayagüez); 25–1000 m; *fl* Jan–Apr, Oct, *fr* (?).

Originating in the Neotropics; exact native range uncertain. [L4: 268; AR: 164].

Batata, Camote.

Ipomoea carnea Jacq. subsp. **fistulosa** (Mart. ex Choisy) D.F. Austin, Taxon 26(2–3):237, *f.* 2. 1977.

Bas.: *Ipomoea fistulosa* Mart. ex Choisy (1845).

Shrub planted and persistent in Northern Coastal Lowlands, Southern Coastal Lowlands (Ponce), Southern Coastal Limestone (Peñuelas), Southwestern Coastal Lowlands (Lajas); 2–200 m; *fl* Mar–Jul, Nov, *fr* Jun–Jul, Nov.

Native to Mexico, Central America, northern South America to Bolivia, Paraguay. [L4: 268].

Aguinaldo morado, Batatilla carnosa.

Ipomoea eggersii (House) D.F. Austin, Ann. Missouri Bot. Gard. 64(2):335. 1978.

Bas.: *Exogonium eggersii* House (1908).

Vine in Vieques, Culebra; 1–20 m; *fl* Dec–Feb (two specimens), *fr* (?). [*Woodbury V-68* & *a.n. 22022* (UPR); *Proctor 48577* (SJ)].

Virgin Islands, Lesser Antilles. [AR: 166].

Ipomoea hederifolia L., Syst. Nat. (ed. 10) 2:925. 1759.

Vine in Las Piedras Chiquitas, Southeastern Foothills, Central Cordillera (Villalba), Southern Coastal Lowlands (Guayama), Southern Coastal Limestone (Peñuelas), Western Coast; 3–500 m; *fl* Oct–Feb, *fr* Nov–Feb.

West Indies (except Cuba), southeastern United States and Texas, Mexico, Central America, northern South America to Bolivia, Paraguay. [L4: 269; AR: 166].

Notes: Britton & Wilson (1925) misapplied the name *Quamoclit coccinea* (L.) Moench to this taxon.

Cambustera.

Ipomoea horsfalliae Hook., Bot. Mag. 61:*t.3315*. 1834.

Vine in Northern Coastal Lowlands (Arecibo), Toro Negro Forest Reserve; 200–1100 m; *fl* Jan (one specimen), *fr* (?).

Native to northern South America and Brazil. [L4: 269; AR: 167].

Campanitas de coral.

Ipomoea imperati (Vahl) Griseb., Cat. Pl. Cub. 203. 1866.

Bas.: *Convolvulus imperati* Vahl (1790).

Syn.: *Ipomoea stolonifera* J.F. Gmel. (1791), *nom. illeg. superfl.*

Vine in Northern Coastal Lowlands, Eastern Coast (Humacao), Southern Coastal Lowlands, Vieques; 1–5 m; *fl* Aug–May, *fr* (?).

Pantropical; native range uncertain. [L4: 269; AR: 167].

Batatilla, Bejuco de costa, Bejuco de puerco de costa.

Ipomoea indica (Burm.) Merr. var. **indica**, Interpr. Herb. Amboin. 445. 1917.

Bas.: *Convolvulus indicus* Burm. (1755).

Vine in Las Tetas de Cayey, Las Piedras Chiquitas; 500–825 m; *fl* Dec–Jan, May–Jul, *fr* (?).

Neotropical. [AR: 169].

Ipomoea indica (Burm.) Merr. var. **acuminata** (Vahl) Fosberg, Bot. Not. 129:38. 1976.

Bas.: *Convolvulus acuminatus* Vahl (1794).

Syn.: *Ipomoea cathartica* Poir. (1816).

Vine in Northern Coastal Lowlands, Northern

Limestone Hills, Río Abajo Forest Reserve, Eastern Coast, Las Tetas de Cayey, Eastern Cordillera (Cayey, Aibonito), Monte Guilarte Forest Reserve, Vieques, Culebra, Desecheo; 1–1200 m; *fl* throughout the year, *fr* (?).

Neotropical. [L4: 270; AR 169].

Bejuco de gloria.

Ipomoea microdactyla Griseb., Cat. Pl. Cub. 204. 1866.

Vine in Mona; 1–60 m; *fl* Sep–May, *fr* (?).

Bahamas, Greater Antilles (Cuba), southeastern United States (Florida). [L4: 272; AR: 171].

Ipomoea nil (L.) Roth, Catal. Bot. 1:36. 1797.

Bas.: *Convolvulus nil* L. (1762).

Vine in Las Piedras Chiquitas, Southeastern Foothills, Central Cordillera (Ponce), Southern Coastal Lowlands, Southern Coastal Limestone, Guánica Forest Reserve, Southwestern Coastal Lowlands, Cerro Las Mesas, Vieques, Culebra; 10–550 m; *fl* Oct–Mar, *fr* Dec–Mar, Jun.

Neotropical. [L4: 272; AR: 171].

Bejuco de vaca.

Ipomoea ochracea (Lindl.) G. Don var. **curtissii** (House) Stearn, Proc. Linn. Soc. Lond. 170:145. 1959.

Bas.: *Ipomoea curtissii* House (1908).

Vine in Susúa Forest Reserve, Maricao Forest Reserve, Guánica Forest Reserve, Southwestern Coastal Lowlands, Western Coast (Mayagüez); 1–800 m; *fl* & *fr* Nov–Apr.

Native to tropical Africa. [L4: 272; AR: 173].

Notes: Liogier & Martorell (1982) misapplied the name *Ipomoea ochroleuca* Span. to this taxon.

Ipomoea pes-caprae (L.) R. Br. subsp. **brasiliensis** (L.) Ooststr., Blumea 3(3):533. 1940.

Bas.: *Convolvulus brasiliensis* L. (1753).

Vine in Northern Coastal Lowlands, Eastern Coast, Southern Coastal Lowlands, Guánica Forest Reserve, Southwestern Coastal Lowlands (Cabo Rojo), Western Coast, Vieques, Culebra, Mona, Desecheo; 0–5 m; *fl* throughout the year, *fr* Jul–Feb.

Pantropical. [L4: 273; AR: 173].

Notes: Britton & Wilson (1925) used the homonym *Ipomoea pes-caprae* (L.) Roth for the name of this taxon.

Bejuco de playa, Bejuco de puerco playero.

Ipomoea purpurea (L.) Roth, Bot. Abh. Beobacht. 27. 1787.

Bas.: *Convolvulus purpureus* L. (1762).

Vine in Northern Coastal Lowlands (Manatí), Eastern Cordillera (Aibonito); 20–500 m; *fl* Dec, Mar, *fr* (?). [*Axelrod 11575* (UPRRP)].

Neotropical; probably native to Mexico. [L4: 273; AR: 173].

Ipomoea quamoclit L., Sp. Pl. 1:159. 1753.

Syn.: *Quamoclit quamoclit* (L.) Britton (1898).

Vine in Northern Coastal Lowlands, Central Cordillera (Utuado), Western Coast (Mayagüez); 1–400 m; *fl* Oct–Apr, *fr* Oct, Mar (two specimens).

Neotropical, usually cultivated. [L4: 274; AR: 175].

Bejuco de ciprés, Bejuco de coral, Cambustera.

Ipomoea repanda Jacq., Enum. Syst. Pl. 13. 1760.

Syn.: *Exogonium repandum* (Jacq.) Choisy (1838).

Vine in Northern Limestone Hills, Río Abajo Forest Reserve, Sierra de Luquillo, Sierra de Cayey, Maricao Forest Reserve; 50–1000 m; *fl* throughout the year, *fr* Mar, Dec.

Greater Antilles (Hispaniola), Virgin Islands, Lesser Antilles. [L4: 274; AR: 175].

Bejuco colorado, Bejuco rosado, Mari-de-Lugo.

Ipomoea setifera Poir., Encycl. 6(1):17. 1804.

Syn.: *Ipomoea rubra* (Vahl) Millsp. (1900).
Ipomoea palustris Urb. (1925).

Vine in Northern Coastal Lowlands, Northern Limestone Hills, Sierra de Luquillo, Eastern Coast, Caguas Valley, Sierra de Cayey, Eastern Cordillera, Maricao Forest Reserve, Western Coast (Mayagüez); 1–1000 m; *fl* Sep–Apr, *fr* Sep, Feb–Apr.

West Indies (except Cuba), Trinidad, Central America, South America (scattered). [L4: 274; AR: 177].

Bejuco de puerco, Gloria de la mañana silvestre.

Ipomoea steudelii Millsp., Publ. Field Columb. Mus., Bot. Ser. 2(1):86. 1900.

Syn.: *Exogonium arenarium* Choisy (1838).

Vine in Northern Limestone Hills, Eastern Coast (Ceiba), Las Piedras Chiquitas, Southeastern Foothills, Central Cordillera (Ponce), Southern Coastal Lowlands, Southern Coastal Limestone, Susúa Forest Reserve, Western Cordillera, Guánica Forest Reserve, Southwestern Coastal Lowlands, Sierra Bermeja; 3–700 m; *fl* Oct–May, *fr* Nov–May.

Endemic. [L4: 275; AR: 177].

Ipomoea tiliacea (Willd.) Choisy in A. de Candolle, Prodr. 9:375. 1845.

Bas.: *Convolvulus tiliaceus* Willd. (1809).

Vine in Northern Coastal Lowlands, Northern

Limestone Hills, Río Abajo Forest Reserve, Sierra de Luquillo, Eastern Coast, Sierra de Cayey, Las Tetas de Cayey, Eastern and Central Cordillera, Southern Coastal Lowlands (Guayama), Maricao Forest Reserve, Southwestern Coastal Lowlands, Cerro Las Mesas, Western Coast, Vieques, Culebra, Mona; 1–1200 m; *fl* throughout the year, *fr* Nov–Mar.

West Indies, Trinidad and Tobago, Mexico, Central America, northern South America to Brazil. [L4: 275; AR: 179].

Bejuco blanco, Bejuco de puerco.

Ipomoea triloba L., Sp. Pl. 1:161. 1753.

Vine in Northern Coastal Lowlands, Eastern Coast (Humacao including Cayo Santiago), Caguas Valley, Eastern Cordillera (Cayey), Southern Coastal Lowlands, Guánica Forest Reserve, Southwestern Coastal Lowlands, Western Coast (Añasco), Vieques, Culebra, Mona, Desecheo; 1–400 m; *fl* Sep–Jun, *fr* Jun–Mar.

Probably originating in the Neotropics. [L4: 277; AR: 179].

Bejuquillo de puerco.

Ipomoea violacea L., Sp. Pl. 1:161. 1753.

Syn.: *Ipomoea macrantha* Roem. & Schult. (1819).

Calonyction tuba (Schltdl.) Colla (1840).

Vine in Northern Coastal Lowlands, Southern Coastal Lowlands (offshore islands: Cayo de Ratones), Guánica Forest Reserve, Vieques, Culebra, Mona, Desecheo; 1–15 m; *fl* & *fr* throughout the year.

Pantropical. [L4: 277; AR: 181].

Bejuco de luna, Bejuco de vaca, Flor de luna.

Ipomoea wrightii A. Gray, Syn. Fl. N. Amer. 2(1): 213. 1878.

Syn.: *Ipomoea heptaphylla* (Rottler & Willd.) Voight (1845).

Vine in Northern Coastal Lowlands (Dorado), Western Coast (Mayagüez); 1–100 m; *fl* & *fr* May, Sep (two specimens). [*Woodbury a.n. 9309* (MAPR), *Martin 184* (MAPR), *Breckon 7759* (MAPR)].

Antilles, south-central United States, Mexico, Central America, South America. [L4: 279; AR: 181].

Jacquemontia cayensis Britton in Britton & Millspaugh, Fl. Bahamas 349. 1920.

Vine in Northern Coastal Lowlands (islands adjacent to Fajardo: Cayo Icacos), Southern Coastal Lowlands (Ponce), Southern Coastal Limestone, Guánica Forest Reserve, Mona; 5–100 m; *fl* Nov–Dec, Mar, *fr* Dec–Mar.

West Indies. [L4: 280; AR: 184].

Jacquemontia cumanensis (Kunth) Kuntze, Revis. Gen. Pl. 2:441. 1891.

Bas.: *Convolvulus cumanensis* Kunth (1819).

Vine in Las Piedras Chiquitas, Southeastern Foothills, Southern Coastal Lowlands (Caja de Muertos), Southern Coastal Limestone (Peñuelas), Guánica Forest Reserve, Southwestern Coastal Lowlands (Cabo Rojo), Sierra Bermeja, Vieques, Culebra; 1–550 m; *fl* Sep–Jun, *fr* Nov–Mar.

Virgin Islands, Lesser Antilles, northern South America (Venezuela, Colombia). [L4: 280; AR: 184].

Jacquemontia havanensis (Jacq.) Urb., Symb. Antill. 3(2):342. 1902.

Bas.: *Convolvulus havanensis* Jacq. (1767).

Syn.: *Jacquemontia jamaicensis* (Jacq.) Hallier f. ex Soler. (1899).

Vine in Northern Coastal Lowlands, Northern Limestone Hills, Southern Coastal Lowlands, Southern Coastal Limestone, Guánica Forest Reserve, Western Coast (Mayagüez), Vieques, Culebra, Mona; 1–100 m; *fl* Oct–Apr, *fr* Nov–May.

Bahamas, Greater Antilles, Lesser Antilles (Antigua, Barbuda), southeastern United States (Florida), Mexico, northern Central America (Belize). [L4: 281; AR: 185].

Aguinaldo de costa.

Jacquemontia ovalifolia (Vahl ex H. West) Hallier f. subsp. **obcordata** (Millsp.) K.R. Robertson, Ann. Missouri Bot. Gard. 61(2):508. 1974.

Bas.: *Convolvulus obcordatus* Millsp. (1900).

Syn.: *Jacquemontia subsalina* Britton (1925).

Prostrate herb in Southern Coastal Lowlands (Caja de Muertos); 1–25 m; *fl* Jan (one specimen), *fr* (?). [*Woodbury M-184, M-216* (UPR)].

Antilles, Mexico, Central America. [L4: 281].

Jacquemontia pentanthos (Jacq.) G. Don, Gen. Hist. 4:283. 1837–1838 "pentantha."

Bas.: *Convolvulus pentanthos* Jacq. (1791).

Vine in Northern Coastal Lowlands, Northern Limestone Hills, Eastern Coast, Sierra de Cayey (Patillas), Eastern Cordillera (Comerío), Central Cordillera (Ponce, Peñuelas), Southern Coastal Lowlands, Susúa Forest Reserve, Maricao Forest Reserve, Southwestern Coastal Lowlands (Cabo Rojo), Cerro Las Mesas, Western Coast (Rincón), Vieques, Culebra, Mona, Desecheo; 1–900 m; *fl* throughout the year, *fr* Nov–Apr.

West Indies, southeastern United States (Florida), Mexico, Central America, northwestern, west-central South America and Venezuela. [L4: 281; AR: 185].

Aguinaldo azul, Coquetitas.

Jacquemontia solanifolia (L.) Hallier f., Bot. Jahrb. Syst. 16(4–5):542. 1893.
Bas.: *Ipomoea solanifolia* L. (1753).
Syn.: *Exogonium solanifolium* (L.) Britton (1918).
Vine in Northern Coastal Lowlands, Cambalache Forest Reserve, Río Abajo Forest Reserve, Eastern Coast (Punta Guayanés), Guánica Forest Reserve, Vieques, Culebra, Mona; 1–225 m; *fl* Aug–May, *fr* Nov–Aug.
Virgin Islands, Lesser Antilles. [L4: 283; AR: 185].
Cambustera de costa.

Jacquemontia tamnifolia (L.) Griseb., Fl. Brit. W.I. 474. 1862.
Bas.: *Ipomoea tamnifolia* L. (1753).
Syn.: *Thyella tamnifolia* (L.) Raf. (1838).
Vine in Northern Coastal Lowlands, Northern Limestone Hills (Quebradillas), Eastern Coast, Susúa Forest Reserve, Southwestern Coastal Lowlands (Cabo Rojo); 1–200 m; *fl* Sep–Feb, *fr* Nov–Feb.
Bahamas, Greater Antilles (Cuba, Hispaniola), southeastern United States, Mexico, Central America, South America. [L4: 283; AR: 187].
Aguinaldo peludo.

Jacquemontia verticillata (L.) Urb., Symb. Antill. 3(2):339. 1902.
Bas.: *Ipomoea verticillata* L. (1759).
Vine in Northern Limestone Hills, Río Abajo Forest Reserve; 50–300 m; *fl* Nov–Dec, *fr* Nov–Jan, May.
West Indies, Mexico, northern Central America (Guatemala, Belize). [L4: 283; AR: 187].
Aguinaldo rosado.

Merremia aegyptia (L.) Urb., Symb. Antill. 4(3):505. 1910.
Bas.: *Ipomoea aegyptia* L. (1753).
Vine in Northern Coastal Lowlands, Eastern Coast (Yabucoa), Southern Coastal Lowlands, Guánica Forest Reserve, Southwestern Coastal Lowlands, Western Coast, Culebra, Mona; 1–50 m; *fl* Nov–May, *fr* Dec–Apr.
Pantropical. [L4: 284; AR: 189].

Merremia cissoides (Lam.) Hallier f., Bot. Jahrb. Syst. 16(4–5):552. 1893.
Bas.: *Convolvulus cissoides* Lam. (1793).
Vine in Eastern Coast; 1–50 m; *fl* Nov–Mar, *fr* Feb (two specimens).
Greater Antilles (Cuba), Lesser Antilles (Antigua, Guadeloupe), Mexico, Central America, South America. [L4: 285; AR: 190].

Merremia dissecta (Jacq.) Hallier f., Bot. Jahrb. Syst. 16(4–5):552. 1893.
Bas.: *Convolvulus dissectus* Jacq. (1767).
Syn.: *Ipomoea dissecta* (Jacq.) Pursh (1813).
Vine in Northern Coastal Lowlands, Northern Limestone Hills, Southern Coastal Lowlands (Ponce), Southern Coastal Limestone (Peñuelas), Guánica Forest Reserve, Southwestern Coastal Lowlands, Western Coast (Rincón), Vieques; 1–250 m; *fl* throughout the year, *fr* Sep–May.
West Indies, Trinidad and Tobago, south-central United States (Texas), Mexico, Central America, northern South America. [L4: 285; AR: 190].
Noyó.

Merremia quinquefolia (L.) Hallier f., Bot. Jahrb. Syst. 16(4–5):552. 1893.
Bas.: *Ipomoea quinquefolia* L. (1753).
Vine in Northern Coastal Lowlands, Eastern Coast, Caguas Valley (Juncos), Las Piedras Chiquitas, Southeastern Foothills, Eastern and Central Cordillera, Southern Coastal Lowlands, Susúa Forest Reserve, Western Cordillera (Maricao), Southwestern Coastal Lowlands, Sierra Bermeja, Western Coast, Vieques, Culebra, Mona; 1–1150 m; *fl* Jul–Apr, *fr* throughout the year.
West Indies, Trinidad and Tobago, Mexico, Central America, northern and northwestern South America to Peru, Paraguay. [L4: 287; AR: 190].
Batatilla blanca.

Merremia tuberosa (L.) Rendle in W.T. Thiselton-Dyer, Fl. Trop. Afr. 4(2):104. 1905.
Bas.: *Ipomoea tuberosa* L. (1753).
Vine in Sierra de Cayey (San Lorenzo), Southwestern Coastal Lowlands (Lajas); 50–400 m; *fl* Oct–Jan, *fr* (?).
Pantropical, probably originating in Mexico, Central America. [L4: 287; AR: 192].
Batatilla ventruda, Ferrocarril, Flor de palo.

Merremia umbellata (L.) Hallier f., Bot. Jahrb. Syst. 16(4–5):552. 1893.
Bas.: *Convolvulus umbellatus* L. (1753).
Syn.: *Ipomoea polyanthes* Roem. & Schult. (1819).
Vine in Northern Coastal Lowlands, Northern Limestone Hills, Eastern Coast, Caguas Valley (Caguas), Southeastern Foothills, Eastern Cordillera (Aibonito), Central Cordillera, Southern Coastal Lowlands (Guayama), Susúa Forest Reserve, Southwestern Coastal Lowlands, Western Coast, Vieques; 1–750 m; *fl* Oct–Apr, *fr* Nov–Apr.
Pantropical. [L4: 288; AR: 192].
Aguinaldo amarillo.

Operculina turpethum (L.) Silva Manso var. **ventricosa** (Bertero) Staples & D.F. Austin, Brittonia 33(4):595. 1981.

Bas.: *Convolvulus ventricosus* Bertero (1824).

Vine planted and persistent in Central Cordillera (Ponce); 300–400 m; *fl* Jan–Feb, *fr* (?).

Native to the Old World. [L4: 288; AR: 194].

Poranopsis paniculata (Roxb.) Roberty, Candollea 14:26. 1952.

Bas.: *Porana paniculata* Roxb. (1819).

Vine planted and persistent in Northern Coastal Lowlands (San Juan), Central Cordillera (Ponce); 5–400 m; *fl* Jan–Feb (two specimens), *fr* (?). [*Liogier 34879* (UPR); *Acevedo-Rodríguez 9293* (UPR)].

Native to Indian Subcontinent, Myanmar, China. [L4: 289; AR: 195].

Velo de novia.

Stictocardia tiliifolia (Desr.) Hallier f., Bot. Jahrb. Syst. 18(1–2):159. 1894 "tiliaefolia."

Bas.: *Convolvulus tiliifolius* Desr. (1792) "tiliaefolius."

Syn.: *Rivea campanulata* (L.) House (1909).

Vine in Northern Coastal Lowlands, Eastern Coast (Cayo Santiago); 1–25 m; *fl* Aug, Dec–Jan, *fr* Jan–Feb.

Native to southern tropical Asia, Taiwan, Japan, Australia. [L4: 291; AR: 197].

Turbina corymbosa (L.) Raf., Fl. Tellur. 4:81. 1838.

Bas.: *Convolvulus corymbosus* L. (1759).

Vine in Northern Coastal Lowlands (Toa Alta, Dorado), Northern Limestone Hills (Arecibo, Camuy), Río Abajo Forest Reserve, Eastern Cordillera; 1–500 m; *fl* Oct–May, *fr* Jan–Feb.

West Indies, south-central United States (Texas), Mexico, Central America, northwestern, west-central South America and Venezuela. [L4: 292; AR: 198].

Aguinaldo, Aguinaldo blanco, Corona de novia.

Xenostegia tridentata (L.) D.F. Austin & Staples, Brittonia 32(4):533. 1981.

Bas.: *Convolvulus tridentatus* L. (1753).

Syn.: *Ipomoea angustifolia* Jacq. (1788).

Vine in Northern Coastal Lowlands; 1–2 m; *fl* & *fr* Sep–Jun.

Native to tropical Africa, tropical Asia, Taiwan, Australia, Micronesia. [L4: 294; AR: 198].

TAXA NOT RECENTLY COLLECTED

Cuscuta obtusiflora Kunth var. **glandulosa** Engelm., Trans. Acad. Sci. St. Louis 1(3):492. 1859.

Syn.: *Cuscuta glandulosa* (Englem.) Small (1903).

Parasitic vine collected in the Sierra de Naguabo on 7 Mar 1914 [*Britton et al. 2109* (NY)].

Greater Antilles (Cuba, Jamaica), United States, Mexico, Central America. [L4: 258].

Bejuco amarillo.

Ipomoea calantha Griseb., Cat. Pl. Cub. 202. 1866.

Vine collected in Coamo on 4 Jan 1886 [*Sintenis 3128* (US)] and in Guánica on 2 Feb 1886 [*Sintenis 3684* (NY)].

Greater Antilles (Cuba Hispaniola), Southern Caribbean Islands (Curaçao), Central America (Nicaragua, Costa Rica), northern South America (Venezuela, Colombia) and Ecuador. [L4: 268; AR: 166].

Ipomoea meyeri (Spreng.) G. Don, Gen. Hist. 4:275. 1838.

Bas.: *Convolvulus meyeri* Spreng. (1824).

Vine collected in Rincón on 1 Dec 1886 [*Sintenis 5553* (US)], in Coamo on 23 Nov 1899 [*Goll 745* (US)] and in Ponce on 6 Dec 1902 [*Heller 6225* (US)].

Greater Antilles, Trinidad, Mexico, Central America, northern South America (Venezuela, Colombia) and Ecuador. [L4: 271; AR: 171].

Ipomoea tricolor Cav., Icon. 3(1):5, *t. 208*. 1795–1796.

Vine collected by Krug in Mayagüez (according to Britton & Wilson, but no specimen seen), in Ponce on 26 Nov 1899 [*Goll 797* (US)] and in Juana Díaz on 2 Jun 1929 [*Britton & Britton 9117* (NY)].

Neotropical. [L4: 277; AR: 179].

TAXA CITED BY PREVIOUS AUTHORS, BUT NOT INCLUDED

NO SPECIMENS SEEN

Ipomoea tenuissima Choisy.

Vine cited by Britton & Wilson (1925) as occurring in a sandy coconut grove, Joyuda, in Apr 1913, but no collector is specified. It was subsequently cited by Liogier & Martorell (1982, 2000), Liogier (1995) and Acevedo-Rodríguez (2005). [L4: 275; AR: 177].

DOUBTFUL OCCURRENCE

Ipomoea krugii Urb.

Vine cited by Britton & Wilson (1925) as occurring near Mayagüez, collected only by Krug and known to them by description only; subsequently cited by Liogier & Martorell (1982, 2000) and Liogier (1995). According to Acevedo-Rodríguez (2005), it is now known only by description and from a photograph of the type specimen (which has itself been lost) that do not definitively distinguish it as a new taxon rather than some previously known one. [L4: 270].

CITED DUE TO MISIDENTIFICATION
Evolvulus alsinoides (L.) L. Syn.: *Evolvulus alsinoides* (L.) L. var. *grisebachianus* Meisn. *Evolvulus alsinoides* (L.) L. var. *debilis* (Kunth) Ooststr. *Evolvulus linifolius* L.

Herb cited by Liogier & Martorell (1982, 2000) and Liogier (1995) as occurring in Morillos de Cabo Rojo. All the specimens from Morillos de Cabo Rojo that I have seen identified as this taxon are referable to *Evolvulus convolvuloides*. [L4: 260].

ONLY UNDER CULTIVATION
Dichondra repens J.R.Forst & G. Forst.

Herb cited by Liogier & Martorell (1982, 2000) and Liogier (1995) as cultivated and persistent as a ground cover. [L4: 258].

COSTACEAE

Costus guanaiensis Rusby var. **macrostrobilus** (K. Schum.) Maas, Fl. Neotrop. Monogr. 8:52, *tt.* *25–26*. 1972.
Bas.: *Costus macrostrobilus* K. Schum. (1903).

Robust herb in Northern Limestone Hills (Ciales, Florida), Río Abajo Forest Reserve, Sierra de Luquillo, Central Cordillera (Utuado), Western Cordillera (Maricao); 80–900 m; *fl* May–Jan, *fr* (?).

Native to Trinidad, Mexico, Central America, northwestern, west-central South America and Venezuela. [AR&S: 187].

Caña India.

Costus scaber Ruiz & Pav., Fl. Peruv. 1:2, *t. 3*. 1798.

Herb in Northern Limestone Hills (Arecibo), Sierra de Luquillo, Eastern Coast (Ceiba); 150–650 m; *fl* Jan, Jun–Jul, *fr* (?).

Native to southern Central America, northwestern, west-central South America and Venezuela. [AR&S: 188].

Costus speciosus (J. Koenig) Sm., Trans. Linn. Soc. London 1:249. 1791.
Bas.: *Banksea speciosa* J. Koenig (1783).

Robust herb in Northern Coastal Lowlands (Río Grande), Río Abajo Forest Reserve, Sierra de Luquillo, Eastern Coast (Ceiba, Naguabo), Caguas Valley (Caguas); 50–225 m; *fl* Jul–Dec, *fr* (?).

Native to tropical Asia, China. [AR&S: 190].

Costus spicatus (Jacq.) Sw., Prodr. 11. 1788.
Bas.: *Alpinia spicata* Jacq. (1763).
Syn.: *Costus cylindricus* Jacq. (1809).

Robust herb planted and persistent in Eastern Coast (Humacao); 1–5 m; *fl* Jul (one specimen), *fr* (?). [*Woodbury a.n. 4871* (UPR)].

Greater Antilles (Hispaniola), Lesser Antilles (Guadeloupe, Dominica, Martinique). [AR&S: 190].

Caña amarga.

TAXA CITED BY PREVIOUS AUTHORS, BUT NOT INCLUDED

ONLY UNDER CULTIVATION
Costus malortieanus H. Wendl.

Herb cited in Acevedo-Rodríguez & Strong (2005) as found by Proctor, who, on the basis of a single farm site in Río Grande, considered it to be naturalized. [AR&S: 188].

CRASSULACEAE

Kalanchoe blossfeldiana Poell., Repert. Spec. Nov. Regni Veg. 35:159. 1934.

Herb in Northern Coastal Lowlands (Guaynabo), Central Cordillera (Adjuntas); 100–500 m; *fl* Jan–Mar, *fr* (?).

Native to Madagascar. [L1: 309].

Kalanchoe daigremontiana Raym.-Hamet & H. Perrier, Ann. Mus. Colon. Marseille, sér. 3 2:128. 1914.

Herb in Northern Coastal Lowlands (Laguna Tortuguero, Manatí), Northern Limestone Hills (Quebradillas), Sierra de Cayey, Maricao Forest Reserve, Guánica Forest Reserve, Southwestern Coastal Lowlands (Guánica), Vieques; 1–825 m; *fl* Oct–Jul, *fr* (?).

Native to Madagascar. [L1: 309].

Kalanchoe delagoensis Eckl. & Zeyh., Enum. Pl. Afric. Austral. 305. 1837.
Syn.: *Bryophyllum delagoense* (Eckl. & Zeyh.) Schinz (1900).
Bryophyllum tubiflorum Harv. (1862). *Kalanchoe tubiflora* (Harv.) Raym.-Hamet & H. Perrier (1914).

Herb in Northern Coastal Lowlands (Loíza, Quebradillas), Southwestern Coastal Lowlands, Vieques; 1–50 m; *fl* Feb (two specimens), *fr* (?).

Native to Madagascar. [L1: 311].

Brujilla de jardín, Malá falsa.

Kalanchoe pinnata (Lam.) Pers., Syn. Pl. 1:446. 1805.
Bas.: *Cotyledon pinnata* Lam. (1786).
Syn.: *Bryophyllum pinnatum* (Lam.) Oken (1841).

Herb in Northern Coastal Lowlands (Luquillo), Northern Limestone Hills, Río Abajo Forest Reserve, Sierra de Cayey, Las Tetas de Cayey, Las Piedras

Chiquitas, Southeastern Foothills, Eastern and Central Cordillera, Susúa Forest Reserve, Maricao Forest Reserve, Southwestern Coastal Lowlands, Sierra Bermeja, Vieques; 50–1150 m; *fl* Jan–Apr, *fr* (?).

Native to Madagascar. [L1: 311].

Bruja, Yerba de bruja.

TAXA CITED BY PREVIOUS AUTHORS, BUT NOT INCLUDED

ONLY UNDER CULTIVATION

Kalanchoe fedtschenkoi Raym.-Hamet & H. Perrier.
Syn.: *Bryophyllum fedtschenkoi* (Raym.-Hamet & H. Perrier) Lauz.-March.

Herb cited by Liogier & Martorell (1982, 2000) and Liogier (1985) as planted as an ornamental and escaped. [L1: 309].

Kalanchoe integra (Medik.) Kuntze.

Herb cited by Liogier & Martorell (1982, 2000) and Liogier (1985) as cultivated as an ornamental and spontaneous. [L1: 311].

Notes: *Kalanchoe brasiliensis* Cambess., which Liogier & Martorell (1982, 2000) and Liogier (1985) wrongly gave as a synonym of the above, was cited by Britton & Wilson (1924) solely for flower gardens of St. Thomas. Not having seen specimens, I cannot vouch for the correct name of the ornamental that grows in Puerto Rico.

CUCURBITACEAE

Cayaponia americana (Lam.) Cogn. in A. & C. de Candolle, Monogr. Phan. 3:785. 1881.
Bas.: *Bryonia americana* Lam. (1785).

Vine in Northern Coastal Lowlands, Northern Limestone Hills (Ciales, Utuado), Río Abajo Forest Reserve, Sierra de Luquillo, Eastern Coast (Ceiba, Punta Guayanés), Sierra de Cayey, Southeastern Foothills (Coamo), Central Cordillera, Cerro de Las Cuevas, Maricao Forest Reserve, Vieques, Culebra; 1–1300 m; *fl* & *fr* throughout the year.

West Indies. [L5: 185; AR: 201].

Bejuco de torero.

Cayaponia racemosa (Mill.) Cogn. in A. & C. de Candolle, Monogr. Phan. 3:768. 1881.
Bas.: *Bryonia racemosa* Mill. (1768).

Vine in Northern Coastal Lowlands, Northern Limestone Hills, Eastern Coast, Caguas Valley, Sierra de Cayey (San Lorenzo), Eastern Cordillera (Corozal), Central Cordillera (Ponce), Susúa Forest Reserve; 5–325 m; *fl* Apr, Dec, *fr* Dec–Jul.

Bahamas, Greater Antilles, Trinidad, Mexico, Central America, northwestern, west-central South America and Venezuela. [L5: 185; AR: 202].

Coloquintilla.

Citrullus lanatus (Thunb.) Matsum. & Nakai, Cat. Sem. Spor. Hort. Bot. Univ. Imp. Tokyo 30, no. 854. 1916.
Bas.: *Momordica lanata* Thunb. (1794).
Syn.: *Citrullus citrullus* (L.) H. Karst. (1882).

Vine planted and persistent in Eastern Coast, Southwestern Coastal Lowlands (Lajas); 1–5 m; *fl* Jan–Feb, *fr* (?). [*Liogier 31610* (UPR); *Acevedo-Rodríguez 9347* (UPR)].

Native to southern Africa. [L5: 187; AR: 202].

Melón, Patilla, Sandía.

Cucumis anguria L., Sp. Pl. 2:1011. 1753.

Vine in Northern Coastal Lowlands (Fajardo), Las Piedras Chiquitas, Southeastern Foothills, Southern Coastal Lowlands, Guánica Forest Reserve, Southwestern Coastal Lowlands, Vieques, Mona; 1–550 m; *fl* Jul–Apr, *fr* Oct–Apr, Jul.

Native to southern Africa. [L5: 189; AR: 206].

Cohombro, Pepinillo, Pepinillo silvestre, Pepineto.

Cucumis dipsaceus Ehrenb. ex Spach, Hist. Nat. Vég. 6:211. 1838.

Vine in Southern Coastal Lowlands (Peñuelas, Guayanilla); 1–20 m; *fl* & *fr* Jun–Jul.

Native to eastern Africa. [AR: 206].

Cucumis melo L., Sp. Pl. 2:1011. 1753.

Vine in Northern Coastal Lowlands, Caguas Valley (Gurabo), Southern Coastal Lowlands, Southwestern Coastal Lowlands (Lajas); 2–50 m; *fl* Sep–Feb, *fr* Oct–Mar.

Native to the Paleotropics; native range uncertain. [L5: 189; AR: 206].

Melón.

Cucurbita moschata Duchesne, Essai Hist. Nat. Courges 7. 1786.
Syn.: *Pepo moschata* (Duchesne) Britton (1918).

Vine in Northern Coastal Lowlands, Northern Limestone Hills (Utuado), Río Abajo Forest Reserve, Caguas Valley (Caguas), Southeastern Foothills (Arroyo), Eastern Cordillera, Southern Coastal Lowlands, Quebrada de Los Cedros; 1–950 m; *fl* Nov–Jul, *fr* (?).

Originating in the Neotropics, but now only cultivated. [L5: 191; AR: 207].

Calabaza.

Cucurbita pepo L., Sp. Pl. 2:1010. 1753.

Vine planted and persistent in Sierra de Cayey; 450 m; *fl* Feb–Mar, Oct, *fr* (?).

Native to central United States. [L5: 192].

Calabacín.

Doyerea emetocathartica Grosourdy, Méd. Bot. Criollo 1(2):338; 2(1):289. 1864.

Syn.: *Corallocarpus emetocatharticus* (Grosourdy) Cogn. (1891).

Vine in Las Piedras Chiquitas, Southeastern Foothills (Salinas), Southern Coastal Lowlands (Guayanilla), Guánica Forest Reserve, Southwestern Coastal Lowlands (Cabo Rojo), Culebra; 15–500 m; *fl* Dec–Feb, Jun, *fr* (?).

Greater Antilles (Jamaica), Lesser Antilles (Guadeloupe, Martinique, Grenadines), Trinidad, Mexico, Central America (Nicaragua), northern South America (Venezuela, Colombia). [L5: 193; AR: 209].

Fevillea cordifolia L., Sp. Pl. 2:1013. 1753.

Syn.: *Nhandiroba cordifolia* (L.) Kuntze (1891).

Vine in Northern Coastal Lowlands (Bayamón), Northern Limestone Hills (Ciales), Eastern Cordillera, Central Cordillera (Orocovis), Maricao Forest Reserve; 10–550 m; *fl* Sep–Mar, *fr* Mar, Jun (two specimens).

Greater Antilles, southern Central America, northwestern, west-central South America and Venezuela. [L5: 195; AR: 209].

Calabaza de Castilla, Pepita amarga, Secua, Uyama.

Lagenaria siceraria (Molina) Standl., Publ. Field Mus. Nat. Hist., Bot. Ser. 3(3):435. 1930.

Bas.: *Cucurbita siceraria* Molina (1782).

Syn.: *Cucurbita lagenaria* L. (1753).

Vine planted and persistent in Southwestern Coastal Lowlands (Guánica); 150 m; *fl* & *fr* Sep (one specimen). [*Acevedo-Rodríguez 11421* (MAPR)].

Native to Africa (Zimbabwe). [L5: 196; AR: 211].

Calabozo, Candungo amargo, Carracho, Güícharo, Güiro, Marimbo.

Luffa aegyptiaca Mill., Gard. Dict. (ed. 8): *Luffa*. 1768.

Syn.: *Luffa cylindrica* (L.) M. Roem. (1846).

Vine in Northern Coastal Lowlands, Eastern Coast, Central Cordillera (Orocovis), Southern Coastal Lowlands; 1–450 m; *fl* Jul–Apr, *fr* Nov–Apr.

Native to tropical Asia or possibly Africa. [L5: 198; AR: 213].

Esponja, Estropajo, Pepinillo de esponja, Servilleta de pobre.

Melothria pendula L., Sp. Pl. 1:35. 1753.

Syn.: *Melothria guadalupensis* (Spreng.) Cogn. (1881).

Vine in Northern Coastal Lowlands, Northern Limestone Hills, Sierra de Luquillo, Eastern Coast, Caguas Valley (Caguas), Sierra de Cayey, Susúa Forest Reserve, Western Cordillera (Maricao), Western Coast (Añasco); 5–600 m; *fl* & *fr* throughout the year.

West Indies, eastern and central United States, Central America, South America. [L5: 200; AR: 215].

Pepinillo cimarrón, Pepinito.

Momordica charantia L., Sp. Pl. 2:1009. 1753.

Vine in Northern Coastal Lowlands, Northern Limestone Hills, Río Abajo Forest Reserve, Sierra de Luquillo, Southern Coastal Lowlands (Ponce), Western Cordillera (Maricao, San Germán), Southwestern Coastal Lowlands (Lajas), Western Coast, Vieques, Mona; 1–700 m; *fl* & *fr* throughout the year.

Native to tropical Africa, tropical Asia, China, Australia. [L5: 202; AR: 215].

Cundeamor.

Psiguria pedata (L.) R.A. Howard, J. Arnold Arbor. 54:441. 1973.

Bas.: *Cucumis pedatus* L. (1759).

Syn.: *Anguria pedata* (L.) Jacq. (1760).

Anguria trilobata (L.) Jacq. (1760). *Psiguria trilobata* (L.) R.A. Howard (1973).

Anguria trifoliata L. (1763). *Psiguria trifoliata* (L.) Alain (1980).

Anguria ottoniana Schltdl. (1851). *Psiguria ottoniana* (Schltdl.) C. Jeffrey (1978).

Anguria cookiana Britton (1925).

Vine in Northern Limestone Hills (Toa Baja, Arecibo), Río Abajo Forest Reserve, Sierra de Cayey, Las Tetas de Cayey, Southeastern Foothills (Cayey), Central Cordillera (Peñuelas); 25–750 m; *fl* Jan–Oct, *fr* May–Jul, Dec.

Bahamas, Greater Antilles (Cuba, Hispaniola), Trinidad. [L5: 203, 205; AR: 217, 218].

Notes: Roxi Steele (unpublished ms.) has found that there is only one species of *Psiguria* in the Greater Antilles, a conclusion that is supported by both molecular data and morphology. Confusion has resulted from the fact that the leaf morphology of an individual can change drastically over its lifetime.

Sechium edule (Jacq.) Sw., Fl. Ind. Occid. 2:1150. 1800.

Bas.: *Sicyos edulis* Jacq. (1760).

Vine planted and persistent in Northern Limestone Hills (Vega Baja, Utuado), Río Abajo Forest Reserve, Eastern Cordillera (Aibonito, Barranquitas), Central Cordillera (Adjuntas, Guayanilla), Maricao Forest Reserve; 50–1000 m; *fl* Nov–Jun, Sep, *fr* Mar, Sep.

Native to central Mexico. [L5: 206; AR: 220].

Chayote, Tallón, Tallote.

Sicana odorifera (Vell.) Naudin, Ann. Sci. Nat., Bot., sér. 4 18:181, *t. 8.* 1862.

Bas.: *Cucurbita odorifera* Vell. (1831).

Vine in Western Coast (Aguada); 5–20 m; *fl* & *fr* Nov (one specimen). [*Román 19* (MAPR)].

Native to South America. [L5: 207; AR: 220].

Casabana, Curuba, Pepino, Pepino angolo.

TAXA CITED BY PREVIOUS AUTHORS, BUT NOT INCLUDED

NO SPECIMENS SEEN

Coccinia grandis (L.) Voigt.

Vine cited by Acevedo-Rodríguez (2005) as occurring in the southwestern region and in Quebradillas. [AR: 204].

ONLY UNDER CULTIVATION

Cucumis *sativus* L.

Vine cited by Liogier & Martorell (1982, 2000) and Liogier (1997) as planted for its fruits and sometimes spontaneous. [L5: 190].

Cucurbita maxima Duchesne.

Vine cited by Liogier (1997) and Liogier & Martorell (2000) as cultivated and escaped. [L5: 191].

Luffa acutangula (L.) Roxb.

Vine cited by Britton & Wilson (1925) as occasionally spontaneous after cultivation; subsequently cited by Liogier & Martorell (1982, 2000) and Liogier (1997); cited by Acevedo-Rodríguez (2005) as only cultivated and uncommon. [L5: 198; AR: 213].

CUNONIACEAE

Weinmannia pinnata L., Syst. Nat. (ed. 10) 2:1005. 1759.

Tree in Sierra de Cayey, Central Cordillera; 600–1300 m; *fl* Jun–Aug (two specimens), *fr* (?).

Antilles, Mexico, northern South America. [L1: 314].

Oreganillo.

CYCLANTHACEAE

TAXA CITED BY PREVIOUS AUTHORS, BUT NOT INCLUDED

ONLY UNDER CULTIVATION

Carludovica palmata Ruiz & Pav.

Shrub cited by Liogier & Martorell (1982, 2000) as occasionally planted for ornament and persistent after cultivation.

CYMODOCEACEAE

Halodule wrightii Asch., Sitzungsber. Ges. Naturf. Freunde Berlin 1868:19. 1868.

Aquatic herb in coastal waters off Northern Coastal Lowlands, Guánica Forest Reserve, Southwestern Coastal Lowlands, Vieques; <0 m; *fl* (?), *fr* Jun (one specimen).

Coastal waters of Bahamas, Greater Antilles (Cuba, Jamaica), Virgin Islands, southern coastal United States, Mexico, Central America, northern South America (Venezuela). [AR&S: 72].

Syringodium filiforme Kütz. in Hohenacker, Alg. Marin. Sicc. 9: no. 426. 1860.

Syn.: *Cymodocea manatorum* Asch. (1868).

Aquatic herb in coastal waters off Northern Coastal Lowlands (Fajardo to Luquillo), Eastern Coast (Ceiba, Cayo Santiago), Southwestern Coastal Lowlands, Culebra; <0 m; *fl* Mar, *fr* (?).

Coastal waters of West Indies, south-central United States and Florida. [AR&S: 73].

Yerba manatí.

CYPERACEAE

Abildgaardia ovata (Burm. f.) Kral, Sida 4(2):72. 1971.

Bas.: *Carex ovata* Burm. f. (1768).

Syn.: *Fimbristylis ovata* (Burm. f.) J. Kern (1967).

Abildgaardia monostachya (L.) Vahl (1805).

Herb in Northern Coastal Lowlands, Northern Limestone Hills (Vega Alta, San Sebastián), Río Abajo Forest Reserve, Eastern Coast, Eastern Cordillera (Aguas Buenas), Susúa Forest Reserve, Guánica Forest Reserve, Southwestern Coastal Lowlands (Cabo Rojo), Western Coast, Vieques; 1–550 m; *fl* & *fr* throughout the year.

Pantropical. [AR&S: 241].

Bulbostylis antillana (Britton) Fernald, Rhodora 40:392. 1938.

Bas.: *Stenophyllus antillanus* Britton (1916).

Syn.: *Bulbostylis capillaris* (L.) Kunth ex C.B.

Clarke subsp. *antillana* (Britton) T. Koyama (1979).

Herb in Northern Coastal Lowlands (Laguna Tortuguero), Eastern Coast (Punta Guayanés), Susúa Forest Reserve, Sierra Bermeja; 1–300 m; *fl* & *fr* May, Aug–Oct, Jan.

Lesser Antilles. [AR&S: 243].

Bulbostylis capillaris (L.) C.B. Clarke subsp. **insulana** M.T. Strong in Acevedo-Rodríguez & Strong, Contr. U.S. Natl. Herb. 52:243. 2005.

Herb in Northern Coastal Lowlands (Laguna Tortuguero); 1–2 m; *fl* & *fr* Jan–Jun, Sep–Oct.

Greater Antilles (Cuba, Hispaniola). [AR&S: 243].

Notes: Liogier & Martorell (1982, 2000) misapplied the name *Bulbostylis juncoides* (Vahl) Kük. ex Osten to this taxon.

Bulbostylis curassavica (Britton) Kük. ex Ekman, Ark. Bot. 22A(16):8; et 22A(17):6. 1929.
Bas.: *Stenophyllus curassavica* Britton (1916).

Herb in Southern Coastal Lowlands (Caja de Muertos), Maricao Forest Reserve, Guánica Forest Reserve, Mona; 1–600 m; *fl* & *fr* May–Jan.

Greater Antilles (Hispaniola), Virgin Islands (Anegada), Southern Caribbean Islands (Curaçao). [AR&S: 244].

Bulbostylis pauciflora (Liebm.) C.B. Clarke in Urban, Symb. Antill. 5(2):290. 1907, *nom. cons.*
Bas.: *Oncostylis pauciflora* Liebm. (1850).
Syn.: *Stenophyllus portoricensis* Britton (1913).

Herb in Southern Coastal Limestone, Susúa Forest Reserve, Guánica Forest Reserve, Vieques; 5–65 m; *fl* & *fr* May–Dec.

Greater Antilles (Cuba, Hispaniola), Virgin Islands (St. John, St. Croix), Lesser Antilles (Anguilla). [AR&S: 245].

Bulbostylis stenocarpa Kük., Bot. Jahrb. Syst. 56(Beibl. 125):15. 1921.

Herb in Northern Limestone Hills (Moca); 100–125 m; *fl* & *fr* Oct (one specimen). [*Delgado-Montado 108* (MAPR)].

Native to Mexico, South America (scattered). [AR&S: 245].

Bulbostylis subaphylla C.B. Clarke in Urban, Symb. Antill. 2(1):86. 1900.

Herb in Southern Coastal Limestone, Guánica Forest Reserve; 5–65 m; *fl* & *fr* Jun, Sep–Feb.

Greater Antilles. [AR&S: 247].

Bulbostylis vestita (Kunth) C.B. Clarke in Urban, Symb. Antill. 2(1):87. 1900.
Bas.: *Isolepis vestita* Kunth (1837).
Syn.: *Stenophyllus vestitus* (Kunth) Britton (1916).

Herb in Northern Coastal Lowlands (Laguna Tortuguero), Western Cordillera (Maricao), Cerro Las Mesas, Western Coast (Moca); 1–400 m; *fl* & *fr* May–Feb.

Greater Antilles, Mexico, Central America, northern South America to Bolivia, Paraguay. [AR&S: 247].

Carex longii Mack., Bull. Torrey Bot. Club 49(12):373. 1923.

Herb in Sierra de Cayey; 1000 m; *fl* & *fr* Jul (one specimen). [*Worthington 31243* (UTEP), not seen].

Native to eastern Canada, eastern and south-central United States, Mexico, Central America (Guatemala, Costa Rica), northwestern South America and Venezuela. [AR&S: 249].

Carex polystachya Sw. ex Wahlenb., Kongl. Vetensk. Acad. Nya Handl. 24:149. 1803.

Herb in Río Abajo Forest Reserve, Sierra de Luquillo, Sierra de Cayey, Las Tetas de Cayey, Toro Negro Forest Reserve, Maricao Forest Reserve; 475–1250 m; *fl* & *fr* throughout the year.

Greater Antilles, Lesser Antilles (St. Kitts, Martinique), Mexico, Central America, northwestern, west-central South America and Venezuela. [AR&S: 250].

Cladium jamaicense Crantz, Inst. Rei Herb. 1:362. 1776.
Syn.: *Mariscus jamaicensis* (Crantz) Britton (1913).

Herb in Northern Coastal Lowlands, Susúa Forest Reserve; 1–150 m; *fl* & *fr* Feb–Sep.

West Indies, southern and south-central United States, Mexico, Central America, northwestern and west-central South America. [AR&S: 251].

Cortadora de ciénaga, Serrucho.

Cyperus aggregatus (Willd.) Endl., Cat. Horti Vindob. 1:93. 1842.
Bas.: *Mariscus aggregatus* Willd. (1809).
Syn.: *Cyperus flavus* (Vahl) Nees (1847), *nom. illeg.*
Mariscus cayennensis (Lam.) Urb. (1900).
Cyperus cayennensis (Lam.) Britton (1907).

Herb in Northern Coastal Lowlands, Northern Limestone Hills (San Sebastián), Eastern Coast (Humacao), Las Piedras Chiquitas, Central Cordillera, Susúa Forest Reserve, Western Cordillera (Maricao, San Germán), Southwestern Coastal Lowlands (Cabo Rojo), Sierra Bermeja, Western Coast (Mayagüez); 1–1200 m; *fl* & *fr* throughout the year.

Greater Antilles, Lesser Antilles (Guadeloupe, Martinique, St. Vincent), south-central United States, Mexico, Central America, South America. [AR&S: 258].

Cyperus articulatus L., Sp. Pl. 1:44. 1753.

Herb in Northern Coastal Lowlands, Sierra de Luquillo, Eastern Coast (Naguabo), Southern Coastal Lowlands (Santa Isabel, Juana Díaz), Southwestern Coastal Lowlands (Lajas); 1–700 m; *fl* & *fr* throughout the year.

Pantropical and subtropical. [AR&S: 259].

Junco cimarrón.

Cyperus brunneus Sw., Fl. Ind. Occid. 1:116. 1797.

Syn.: *Mariscus brunneus* (Sw.) C.B. Clarke (1900).

Herb in Northern Coastal Lowlands, Eastern Coast, Southern Coastal Lowlands (Caja de Muertos, Ponce), Guánica Forest Reserve, Southwestern Coastal Lowlands (Cabo Rojo), Vieques, Mona, Desecheo; 1–20 m; *fl* & *fr* throughout the year.

West Indies, southeastern United States (Florida), eastern Mexico, northern Central America. [AR&S: 260].

Cyperus compressus L., Sp. Pl. 1:46. 1753.

Herb in Northern Coastal Lowlands, Eastern Coast, Caguas Valley, Southern Coastal Lowlands (Juana Díaz), Southern Coastal Limestone (Guayanilla), Southwestern Coastal Lowlands (Cabo Rojo), Western Coast (Mayagüez), Vieques; 1–100 m; *fl* & *fr* Aug–Feb, May.

Throughout tropic, subtropic and warm temperate regions. [AR&S: 261].

Cyperus confertus Sw., Prodr. 20. 1788.

Syn.: *Mariscus confertus* (Sw.) C.B. Clarke (1900).

Herb in Northern Coastal Lowlands, Las Piedras Chiquitas, Southeastern Foothills (Salinas), Eastern Cordillera (Barranquitas), Southern Coastal Lowlands (Caja de Muertos); 5–800 m; *fl* & *fr* Oct–Feb, May.

Bahamas, Greater Antilles, Lesser Antilles (St. Barts, Martinique), northern South America (Venezuela, Colombia) and Galapagos Islands. [AR&S: 262].

Cyperus croceus Vahl, Enum. Pl. 2:357. 1805.

Herb in Northern Coastal Lowlands, Sierra Bermeja; 1–300 m; *fl* & *fr* throughout the year.

Native to southeastern and south-central United States, Mexico, Central America, northern South America (Venezuela). [AR&S: 262].

Notes: Britton & Wilson (1923) wrongly placed this taxon in synonymy with *Cyperus globulosus* Aubl.

Cyperus difformis L., Cent. Pl. II 6. 1756.

Herb in Northern Coastal Lowlands, Río Abajo Forest Reserve, Eastern Coast, Caguas Valley (Caguas), Western Coast; 1–250 m; *fl* & *fr* throughout the year.

Native to the Paleotropics. [AR&S: 263].

Cyperus digitatus Roxb., Fl. Ind. 1:209. 1820.

Herb in Northern Coastal Lowlands (Toa Baja), Northern Limestone Hills (Florida), Southwestern Coastal Lowlands; 1–10 m; *fl* & *fr* May–Aug, Nov.

Pantropical and -subtropical. [AR&S: 264].

Cyperus distans L.f., Suppl. Pl. 103. 1782.

Herb in Northern Coastal Lowlands, Sierra de Luquillo, Eastern Coast, Caguas Valley, Sierra de Cayey, Southeastern Foothills, Eastern and Central Cordillera, Southern Coastal Lowlands, Western Cordillera, Vieques; 5–1000 m; *fl* & *fr* throughout the year.

Pantropical. [AR&S: 264].

Cyperus eggersii Boeck., Beitr. Cyper. 1:53. 1888.

Herb in Northern Coastal Lowlands, Northern Limestone Hills (San Sebastián); 10–200 m; *fl* & *fr* May–Jul, Oct.

West Indies, southwestern United States, Mexico, Central America, and northwestern, west-central South America and Venezuela. [AR&S: 265].

Notes: Although G.C. Tucker in The Flora of North America, Vol. 23 (2002), included this taxon within the circumscription of *Cyperus odoratus*, he left the door open for the future recogniton of segregates such as this taxon.

Cyperus elegans L., Sp. Pl. 1:45. 1753.

Herb in Northern Coastal Lowlands, Southern Coastal Lowlands, Southwestern Coastal Lowlands, Mona; 1–50 m; *fl* & *fr* throughout the year.

West Indies, southeastern (Florida) and south-central United States, Mexico, Central America, west-central South America and Argentina. [AR&S: 266].

Cyperus esculentus L., Sp. Pl. 1:45. 1753.

Herb in Northern Coastal Lowlands, Western Coast (Mayagüez), Vieques; 1–100 m; *fl* & *fr* Apr–Dec.

Tropical, subtropical and warm-temperate regions. [AR&S: 267].

Chufas.

Cyperus flavescens L., Sp. Pl. 1:46. 1753.
Syn.: *Pycreus flavescens* (L.) P. Beauv. ex Rchb. (1830).

Herb in Sierra de Luquillo, Central Cordillera (Orocovis); 800–1050 m; *fl* & *fr* Aug–Oct, Feb, May.

Tropical, subtropical and warm-temperate regions. [AR&S: 268].

Cyperus floridanus Britton in Small, Fl. S.E. U.S. 170, 1327. 1903.

Herb in Guánica Forest Reserve, Mona; 1–70 m; *fl* & *fr* Mar, Jul, Oct–Nov.

Bahamas, Greater Antilles, Virgin Islands (St. Thomas, St. Croix), southeastern United States (Florida). [AR&S: 269].

Cyperus fuligineus Chapm., Fl. South. U.S. 511. 1860.
Syn.: *Mariscus fuligineus* (Chapm.) C.B. Clarke (1907).

Herb in Northern Coastal Lowlands (islands adjacent to Fajardo: Cayo Icacos); 1–10 m; *fl* & *fr* Aug–Sep (two specimens). [*Woodbury I-9, I-30* (UPR); *Proctor 42114* (SJ)].

Bahamas, Greater Antilles (Cuba, Hispaniola), Virgin Islands (Anegada), Lesser Antilles (Anguilla), southeastern United States (Florida Keys). [AR&S: 270].

Cyperus giganteus Vahl, Enum. Pl. 2:364. 1805.

Herb in Northern Coastal Lowlands, Western Coast; 1–2 m; *fl* & *fr* Sep–Dec.

Greater Antilles, Trinidad, Mexico, Central America, South America. [AR&S: 271].

Junco de ciénaga.

Cyperus haspan L., Sp. Pl. 1:45. 1753.

Herb in Northern Coastal Lowlands, Sierra de Luquillo, Sierra de Cayey; 1–1050 m; *fl* & *fr* throughout the year.

Pantropical. [AR&S: 271].

Cyperus imbricatus Retz., Observ. Bot. 5:12. 1788.
Syn.: *Cyperus radiatus* Vahl (1805).

Herb in Northern Coastal Lowlands, Northern Limestone Hills, Eastern Coast (Naguabo), Eastern Cordillera (Aibonito), Southern Coastal Lowlands (Guayama), Western Coast; 1–500 m; *fl* & *fr* Sep–Jun.

Pantropical. [AR&S: 272].

Cyperus involucratus Rottb., Descr. Pl. Rar. 22. 1772.

Herb in Northern Coastal Lowlands, Río Abajo Forest Reserve, Sierra de Luquillo, Eastern Coast (Naguabo), Southeastern Foothills, Eastern Cordillera (Aibonito), Central Cordillera, Southern Coastal Lowlands, Southern Coastal Limestone (Guayanilla), Susúa Forest Reserve, Maricao Forest Reserve, Southwestern Coastal Lowlands (Guánica), Western Coast (Mayagüez), Vieques; 10–1300 m; *fl* & *fr* throughout the year.

Native to tropical Africa, Yemen. [AR&S: 273].

Notes: Britton & Wilson (1923) and Liogier & Martorell (1982, 2000) misapplied the name *Cyperus alternifolius* L., which refers to a rare endemic of Madagascar, to this taxon.

Paragüita.

Cyperus iria L., Sp. Pl. 1:45. 1753.

Herb in Northern Coastal Lowlands, Eastern Coast, Caguas Valley, Central Cordillera (Utuado), Southwestern Coastal Lowlands (Lajas), Western Coast, Vieques; 1–100 m; *fl* & *fr* throughout the year.

Native to the Paleotropics and -subtropics. [AR&S: 274].

Cyperus laevigatus L., Mant. Pl. 2:179. 1771.

Herb in Southern Coastal Lowlands, Guánica Forest Reserve, Southwestern Coastal Lowlands (Guánica); 1–5 m; *fl* & *fr* Aug–Feb, May.

Pantropical and -subtropical. [AR&S: 274].

Cyperus ligularis L., Syst. Nat. (ed. 10) 2:867; et Fl. Jamaic. Pug. 3. 1759.
Syn.: *Mariscus ligularis* (L.) Urb. (1900).

Herb in Northern Coastal Lowlands, Sierra de Luquillo, Eastern Coast, Sierra de Cayey, Central Cordillera (Utuado), Southern Coastal Lowlands, Guánica Forest Reserve, Southwestern Coastal Lowlands, Western Coast, Vieques, Culebra, Mona, Desecheo; 1–1050 m; *fl* & *fr* throughout the year.

West Indies, southeastern United States (Florida, Alabama) and California, Mexico, Central America, northern South America to Bolivia, western tropical Africa. [AR&S: 275].

Junco de agua.

Cyperus mutisii (Kunth) Andersson, Galapagos Veg. 53. 1854.
Bas.: *Mariscus mutisii* Kunth (1816).

Herb in Central Cordillera; 250–925 m; *fl* & *fr* Jun, Sep, Dec.

Greater Antilles, Lesser Antilles (Martinique, Grenada), Mexico, Central America, northwestern, west-central South America and Venezuela, Brazil. [AR&S: 276].

Notes: Britton and Wilson (1923) misapplied the name *Cyperus incompletus* (Jacq.) Link to this taxon; Liogier & Martorell (2000) used the homonym *Cyperus mutisii* (Kunth) Griesb. for it.

Cyperus nanus Willd., Sp. Pl. 1(1):272. 1797.
Syn.: *Mariscus capillaris* (Sw.) Vahl (1805).
Cyperus granularis (Desf. ex Boeck.) Britton (1907).
Herb in Northern Coastal Lowlands (islands adjacent to Fajardo: Palomino), Sierra Bermeja, Western Coast (Mayagüez); 1–300 m; *fl* & *fr* Oct–Nov.
Antilles, Mexico. [AR&S: 277].

Cyperus ochraceus Vahl, Enum. Pl. 2:325. 1805.
Herb in Northern Coastal Lowlands, Northern Limestone Hills (Florida), Río Abajo Forest Reserve, Eastern Cordillera (Cayey), Southern Coastal Lowlands, Southwestern Coastal Lowlands (Cabo Rojo), Western Coast (Añasco); 1–350 m; *fl* & *fr* Sep–Jun.
Bahamas, Greater Antilles, Virgin Islands, Lesser Antilles (Antigua, Guadeloupe), southeastern and south-central United States, Mexico, Central America, South America. [AR&S: 279].

Cyperus odoratus L., Sp. Pl. 1:46. 1753.
Syn.: *Torulinium odoratum* (L.) S.S. Hooper (1972).
Cyperus ferax Rich. (1792).
Herb in Northern Coastal Lowlands, Northern Limestone Hills, Río Abajo Forest Reserve, Sierra de Luquillo, Eastern Coast, Caguas Valley, Sierra de Cayey, Eastern Cordillera, Central Cordillera (Adjuntas), Southern Coastal Lowlands, Southern Coastal Limestone (Sabana Grande), Maricao Forest Reserve, Guánica Forest Reserve, Southwestern Coastal Lowlands, Western Coast, Vieques; 1–600 m; *fl* & *fr* throughout the year.
Throughout tropical, subtropical and warm-temperate regions. [AR&S: 280].
Notes: Britton & Wilson (1923) misapplied the above name to *Cyperus polystachyos*.

Cyperus papyrus L., Sp. Pl. 1:47. 1753.
Herb planted and persistent in Northern Limestone Hills (Ciales, Lares), Sierra de Luquillo; 300–450 m; *fl* & *fr* May–Aug.
Native to tropical Africa. [AR&S: 281].

Cyperus planifolius Rich., Actes Soc. Hist. Nat. Paris 1:106. 1792.
Syn.: *Mariscus planifolius* (Rich.) Urb. (1900).
Herb in Northern Coastal Lowlands, Eastern Coast (Punta Guayanés), Southwestern Coastal Lowlands (Cabo Rojo), Vieques, Culebra, Mona, Desecheo; 1–40 m; *fl* & *fr* May–Nov.
West Indies, southeastern United States (Georgia, Florida, Alabama), eastern Mexico, northern Central America, South America (Brazil). [AR&S: 282].

Cyperus polystachyos Rottb., Descr. Pl. Rar. 21. 1772.
Syn.: *Pycreus polystachyos* (Rottb.) P. Beauv. (1816).
Herb in Northern Coastal Lowlands, Río Abajo Forest Reserve, Sierra de Luquillo, Eastern Coast, Sierra de Cayey, Eastern Cordillera (Cerro La Peña), Central Cordillera (Ciales), Southern Coastal Lowlands, Southwestern Coastal Lowlands, Western Coast, Vieques, Mona; 1–1050 m; *fl* & *fr* throughout the year.
Pantropical and -subtropical. [AR&S: 283].

Cyperus pulguerensis M.T. Strong in Acevedo-Rodríguez & Strong, Contr. U.S. Natl. Herb. 52:284, *f. 49*. 2005.
Herb in Northern Coastal Lowlands (Manatí); 5 m; *fl* & *fr* Jun (one specimen). [*Proctor 43804* (SJ)].
Endemic. [AR&S: 284].
Notes: This taxon is known only from the type specimen.

Cyperus rotundus L., Sp. Pl. 1:45. 1753.
Herb in Northern Coastal Lowlands, Eastern Coast, Caguas Valley (Las Piedras), Southeastern Foothills (Coamo), Southern Coastal Lowlands, Southwestern Coastal Lowlands, Western Coast, Vieques, Culebra, Mona, Desecheo; 1–150 m; *fl* & *fr* throughout the year.
Pantropical. [AR&S: 284].
Coquí, Coquillo, Yerba coquí.

Cyperus sphacelatus Rottb., Descr. Pl. Rar. 21. 1772.
Herb in Northern Coastal Lowlands, Northern Limestone Hills, Río Abajo Forest Reserve, Sierra de Luquillo, Eastern Coast, Caguas Valley, Sierra de Cayey, Las Tetas de Cayey, Eastern and Central Cordillera, Southern Coastal Lowlands (Patillas), Maricao Forest Reserve, Cerro Las Mesas, Western Coast, Vieques; 1–825 m; *fl* & *fr* throughout the year.
Antilles, Trinidad and Tobago, southern Central America, northern South America to Bolivia, tropical Africa. [AR&S: 286].

Cyperus subtenuis (Kük.) M.T. Strong in Acevedo-Rodríguez & Strong, Contr. U.S. Natl. Herb. 52:288. 2005.
Bas.: *Cyperus nanus* Willd. var. *subtenuis* Kük. (1936).

Herb in Southwestern Coastal Lowlands (Lajas); 30–40 m; *fl* & *fr* Oct–Nov (two specimens). [*McKenzie a.n. 12919* (SJ); *Proctor 44138* (SJ)].

Greater Antilles, Virgin Islands. [AR&S: 288].

Cyperus surinamensis Rottb., Descr. Pl. Rar. 20. 1772; 35, *t. 6.* 1773.

Herb in Northern Coastal Lowlands, Sierra de Luquillo, Eastern Coast, Caguas Valley, Sierra de Cayey, Eastern Cordillera, Southern Coastal Lowlands (Arroyo, Juana Díaz), Southwestern Coastal Lowlands, Western Coast, Vieques; 1–1000 m; *fl* & *fr* throughout the year.

West Indies, southeastern and south-central United States, Mexico, Central America, South America. [AR&S: 289].

Cyperus tenuis Sw., Prodr. 20. 1788.

Northern Coastal Lowlands (Dorado), Maricao Forest Reserve, Western Coast (Mayagüez); 1–900 m; *fl* & *fr* Mar–Nov.

Greater Antilles, Trinidad, Mexico, Central America, northwestern, west-central South America and Venezuela. [AR&S: 290].

Notes: Britton & Wilson (1923) misapplied the above name to *Cyperus subtenuis*.

Cyperus unifolius Boeck., Linnaea 36(3):374. 1870.

Syn.: *Cyperus calcicola* Britton (1915).

Herb in Southern Coastal Lowlands (Caja de Muertos), Southern Coastal Limestone (Guayanilla), Guánica Forest Reserve, Southwestern Coastal Lowlands, Mona; 1–65 m; *fl* & *fr* throughout the year.

Virgin Islands. [AR&S: 291].

Notes: Britton & Wilson (1923) and Liogier & Martorell (2000) misapplied the name *Cyperus filiformis* Sw., and Liogier & Martorell (1982) misapplied *Torulinium filiforme* (Sw.) C.B. Clarke, to this taxon.

Pajón de costa.

Cyperus unioloides R. Br., Prodr. 216. 1810.

Syn.: *Pycreus unioloides* (R. Br.) Urb. (1900).

Herb in Río Abajo Forest Reserve; 300–325 m; *fl* & *fr* Jul–Sep (two specimens). [*Acevedo-Rodríguez 332* (SJ); *Axelrod 10915* (UPRRP)].

Greater Antilles, Central America (scattered), South America (scattered), Paleotropics. [AR&S: 292].

Cyperus urbanii Boeck., Beitr. Cyper. 1:10. 1888 "Urbani."

Syn.: *Mariscus urbanii* (Boeck.) C.B. Clarke (1900) "Urbani."

Herb in Las Tetas de Cayey, Las Piedras Chiquitas, Southeastern Foothills (Coamo), Eastern Cordillera (Cañón San Cristóbal), Central Cordillera (Villalba), Sierra Bermeja, Vieques; 150–825 m; *fl* & *fr* Aug–Jan.

Endemic. [AR&S: 292].

Cyperus virens Michx., Fl. Bor.-Amer. 1:28. 1803.

Herb in Northern Coastal Lowlands (Fajardo, Toa Baja); 1 m; *fl* & *fr* Apr–May, Nov (two specimens). [*González-Más 1142* (MAPR); *Axelrod 10220* (UPRRP)].

Greater Antilles, southeastern, south-central United States and California, Mexico, Central America, South America (scattered). [AR&S: 293].

Eleocharis cellulosa Torr., Ann. Lyceum Nat. Hist. New York 3:298. 1836.

Herb in Northern Coastal Lowlands (Laguna Tortuguero); 1–2 m; *fl* & *fr* Nov, Feb–Jun.

Bahamas, Greater Antilles, southeastern and south-central United States, Mexico, Central America, northern South America and Ecuador. [AR&S: 296].

Junco fino.

Eleocharis debilis Kunth, Enum. Pl. 2:143. 1837.

Syn.: *Eleocharis yunquensis* Britton (1923).

Herb in Sierra de Luquillo; 425–1025 m; *fl* & *fr* throughout the year.

Greater Antilles (Cuba), northern Central America (Belize, Honduras), northern South America to Paraguay. [AR&S: 297].

Eleocharis elegans (Kunth) Roem. & Schult., Syst. Veg. 2:150. 1817.

Bas.: *Scirpus elegans* Kunth (1816).

Herb in Northern Coastal Lowlands, Río Abajo Forest Reserve, Sierra de Luquillo, Caguas Valley (Caguas), Sierra de Cayey, Eastern Cordillera (Aguas Buenas, Corozal), Central Cordillera, Southwestern Coastal Lowlands (Cabo Rojo), Western Coast; 10–1050 m; *fl* & *fr* throughout the year.

Greater Antilles, Lesser Antilles (Dominica, Martinique), Central America, South America (scattered). [AR&S: 297].

Junquillo.

Eleocharis fallax Weath., Rhodora 24(278):23. 1922.

Syn.: *Eleocharis ambigens* Fernald (1935).

Herb in Southwestern Coastal Lowlands (Lajas); 11–12 m; *fl* & *fr* May–Jul (two specimens). [*Proctor 46385* (SJ)].

Greater Antilles (Cuba), Virgin Islands (St. Croix), eastern and south-central United States. [AR&S: 298].

Notes: S. Galen Smith in the Flora of North America, Vol. 23 (2002), treats *Eleocharis fallax* and

Eleocharis ambigens (based on the style branches:3–fid in the former vs. 2–fid in the latter) as separate taxa: *Eleocharis fallax* as a very local northeastern United States species and *Eleocharis ambigens* as a more widespread one. Mark Strong (pers. comm.) does not consider this character difference important within the genus and it is his treatment in Acevedo-Rodríguez & Strong (2005) that I follow.

Eleocharis flavescens (Poir.) Urb., Symb. Antill. 4(1):116. 1903.
Bas.: *Scirpus flavescens* Poir. (1805).
Syn.: *Eleocharis flaccida* (Rchb. ex A. Spreng.) Urb. (1900).

Herb in Northern Coastal Lowlands, Northern Limestone Hills (Florida, San Sebastián), Río Abajo Forest Reserve, Sierra de Luquillo, Eastern Coast, Caguas Valley (San Lorenzo), Sierra de Cayey, Eastern Cordillera (Aibonito, Cañón San Cristóbal), Central Cordillera; 1–1050 m; *fl* & *fr* throughout the year.

West Indies, southern and east-central United States, Mexico, Central America (scattered), South America (scattered). [AR&S: 299].

Eleocharis geniculata (L.) Roem. & Schult., Syst. Veg. 2:150. 1817.
Bas.: *Scirpus geniculatus* L. (1753).
Syn.: *Eleocharis sintenisii* Boeck. (1888).
Eleocharis caribaea (Rottb.) S.F. Blake (1918).

Herb in Northern Coastal Lowlands, Northern Limestone Hills, Sierra de Luquillo, Eastern Coast, Caguas Valley (Caguas), Eastern Cordillera, Toro Negro Forest Reserve, Southern Coastal Lowlands, Maricao Forest Reserve, Guánica Forest Reserve, Southwestern Coastal Lowlands, Western Coast (Mayagüez), Vieques; 1–600 m; *fl* & *fr* throughout the year.

Pantropical. [AR&S: 300].

Notes: Britton & Wilson (1923) misapplied the above name to *Eleocharis elegans* and González-Más (1964) wrongly placed it in synonymy with the latter. Britton & Wilson (1923) misapplied the name *Eleocharis atropurpurea* (Retz.) Kunth, a homonym of *Eleocharis atropurpurea* (Retz.) J. Presl & C. Presl, to this taxon; however, they later (1926) corrected themselves and used instead *Eleocharis sintenisii*. González-Más (1964) and Liogier & Martorell (1982) wrongly placed *Eleocharis yunquensis* in synonymy with *Eleocharis sintenisii*.

Eleocharis interstincta (Vahl) Roem. & Schult., Syst. Veg. 2:149. 1817.
Bas.: *Scirpus interstinctus* Vahl (1805).

Herb in Northern Coastal Lowlands, Northern Limestone Hills, Río Abajo Forest Reserve, Sierra de Luquillo, Eastern Coast, Caguas Valley (Caguas), Southern Coastal Lowlands (Guayama), Southern Coastal Limestone (Peñuelas), Guánica Forest Reserve, Southwestern Coastal Lowlands (Guánica), Western Coast; 1–350 m; *fl* & *fr* throughout the year.

West Indies, Trinidad, southern United States (Florida, Alabama, Texas), Mexico, Central America, South America. [AR&S: 301].

Junco, Junco de aparejos, Junco de ciénaga, Junco de espiga.

Eleocharis montana (Kunth) Roem. & Schult., Syst. Veg. 2:153. 1817.
Bas.: *Scirpus montanus* Kunth (1816).
Syn.: *Eleocharis nodulosa* (Roth) Schult. (1824).

Herb in Sierra de Luquillo; 700–1000 m; *fl* & *fr* Sep, Jan.

Greater Antilles (Cuba), Virgin Islands (St. Thomas), Lesser Antilles (Nevis, Antigua, Guadeloupe), southern United States, Mexico, Central America, South America (scattered). [AR&S: 302].

Eleocharis mutata (L.) Roem. & Schult., Syst. Veg. 2:155. 1817.
Bas.: *Scirpus mutatus* L. (1759).

Herb in Northern Coastal Lowlands, Eastern Coast, Southern Coastal Lowlands (Guayama, Salinas), Southwestern Coastal Lowlands (Cabo Rojo), Western Coast (Añasco); 1–15 m; *fl* & *fr* throughout the year.

Antilles, southern Mexico, Central America, northwestern, west-central South America and Venezuela, tropical Africa. [AR&S: 302].

Eleocharis oligantha C.B. Clarke in Urban, Symb. Antill. 2(1):69. 1900.

Herb in Northern Coastal Lowlands (Vega Alta); 5 m; *fl* & *fr* Mar, Jul, Nov. [*Proctor 44145, 44572, 45796* (SJ)].

Greater Antilles (Cuba, Jamaica). [AR&S: 303].

Notes: Liogier & Martorell (2000) misapplied the name *Websteria submersa* (Miq.) Britton, by which they probably meant *Websteria submersa* (C. Wright) Britton, to this taxon.

Eleocharis pachystyla (C. Wright) C.B. Clarke in Urban, Symb. Antill. 2(1):72. 1900.
Bas.: *Scirpus pachystylus* C. Wright (1871).

Herb in Northern Coastal Lowlands (Vega Alta to Laguna Tortuguero); 1–10 m; *fl* & *fr* Dec–Apr.

Greater Antilles (Cuba, Hispaniola), Central America, northern South America (Venezuela). [AR&S: 304].

Eleocharis radicans (A. Dietr.) Kunth, Enum. Pl. 2:142. 1837.

Bas.: *Eleogiton radicans* A. Dietr. (1833).

Herb in Northern Coastal Lowlands (San Juan); 25 m; *fl* & *fr* May, Oct (two specimens). [*Liogier 29030, 32839* (UPR)].

Native to southern United States (scattered) and Michigan, Mexico, Central America, South America (scattered). [AR&S: 304].

Notes: Specimens of this taxon were collected only in the Univeristy of Puerto Rico Botanical Garden and thus it is most likely an inadvertent and temporary introduction.

Eleocharis retroflexa (Poir.) Urb., Symb. Antill. 2(2):165. 1900.

Bas.: *Scirpus retroflexus* Poir. (1805).

Herb in Río Abajo Forest Reserve, Sierra de Luquillo, Caguas Valley (Caguas), Sierra de Cayey, Eastern Cordillera (Cidra), Central Cordillera (Adjuntas); 325–1000 m; *fl* & *fr* Sep–May.

Antilles, southeastern United States (Alabama), Mexico, Central America, northwestern South America and Venezuela. [AR&S: 305].

Eleocharis rostellata (Torr.) Torr., Fl. New York 2:347. 1843.

Bas.: *Scirpus rostellatus* Torr. (1836).

Herb in Northern Coastal Lowlands (Vega Baja to Arecibo); 1–2 m; *fl* & *fr* Feb–Jun, Sep.

Native to southern Canada, United States, Mexico, northern Central America (Guatemala), northwestern and west-central South America. [AR&S: 305].

Fimbristylis complanata (Retz.) Link, Hort. Berol. 1:292. 1827.

Bas.: *Scirpus complanatus* Retz. (1788).

Herb in Northern Coastal Lowlands, Northern Limestone Hills (San Sebastián, Moca), Sierra de Luquillo, Caguas Valley (Caguas), Sierra de Cayey, Sierra Bermeja, Western Coast (Añasco); 2–900 m; *fl* & *fr* throughout the year.

Pantropical. [AR&S: 308].

Fimbristylis cymosa R. Br., Prodr. 228. 1810.

Syn.: *Fimbristylis spathacea* Roth (1821). *Fimbristylis cymosa* R. Br. subsp. *spathacea* (Roth) T. Koyama (1964).

Fimbristylis glomerata (Retz.) Urb. (1900).

Herb in Northern Coastal Lowlands, Northern Limestone Hills (Moca), Eastern Coast, Southern Coastal Lowlands, Western Cordillera (Maricao), Guánica Forest Reserve, Southwestern Coastal Lowlands, Western Coast, Vieques, Mona; 1–425 m; *fl* & *fr* throughout the year.

West Indies, Trinidad and Tobago, southeastern United States (Florida), Mexico, Central America, northern South America to Brazil, Ecuador, Australia. [AR&S: 308].

Fimbristylis dichotoma (L.) Vahl, Enum. Pl. 2:287. 1805.

Bas.: *Scirpus dichotomus* L. (1753).

Syn.: *Fimbristylis diphylla* (Retz.) Vahl (1805).

Herb in Northern Coastal Lowlands, Northern Limestone Hills, Río Abajo Forest Reserve, Sierra de Luquillo, Eastern Coast, Caguas Valley, Sierra de Cayey, Las Tetas de Cayey, Southeastern Foothills (Coamo), Eastern and Central Cordillera, Southern Coastal Lowlands (Guayama), Western Cordillera, Southwestern Coastal Lowlands (Cabo Rojo), Cerro Las Mesas, Western Coast, Vieques, Mona; 1–1200 m; *fl* & *fr* throughout the year.

Warm-temperate and tropical regions. [AR&S: 309].

Junquito.

Fimbristylis ferruginea (L.) Vahl, Enum. Pl. 2:291. 1805.

Bas.: *Scirpus ferrugineus* L. (1753).

Herb in Northern Coastal Lowlands, Southern Coastal Lowlands, Southwestern Coastal Lowlands (Cabo Rojo), Vieques; 1–3 m; *fl* & *fr* throughout the year.

Pantropical. [AR&S: 311].

Fimbristylis littoralis Gaudich., Voy. Uranie 413. 1829.

Herb in Northern Coastal Lowlands (Carolina, San Juan), Eastern Coast, Caguas Valley (Caguas), Sierra de Cayey; 10–600 m; *fl* & *fr* Jun, Sep–Jan.

Native to the Paleotropics. [AR&S: 312].

Notes: Britton & Wilson (1923) and Liogier & Martorell (1982, 2000) used *Fimbristylis miliacea* (L.) Vahl, a *nom. rej.*, for the name of this taxon.

Fimbristylis quinquangularis (Vahl) Kunth, Enum. Pl. 2:229. 1837.

Bas.: *Scirpus quinquangularis* Vahl (1805).

Herb in Northern Coastal Lowlands, Sierra de Luquillo, Eastern Coast, Caguas Valley (Gurabo), Southwestern Coastal Lowlands (Cabo Rojo), Western Coast (Añasco, Mayagüez); 1–400 m; *fl* & *fr* May–Jan.

Native to the Paleotropics. [AR&S: 313].

Fimbristylis spadicea (L.) Vahl, Enum. Pl. 2:294. 1805.

Bas.: *Scirpus spadiceus* L. (1753).

Syn.: *Fimbristylis castanea* (Michx.) Vahl (1805).

Herb in Southern Coastal Lowlands, Guánica Forest Reserve, Southwestern Coastal Lowlands (Cabo Rojo), Vieques; 1–5 m; *fl* & *fr* throughout the year.

Bahamas, Greater Antilles, Lesser Antilles (Barbuda, Martinique, Grenada), Mexico, Central America, northern South America to Paraguay. [AR&S: 314].

Fuirena robusta Kunth, Enum. Pl. 2:185. 1837.

Herb in Northern Limestone Hills (Florida); 50–200 m; *fl* & *fr* May (one specimen). [*Woodbury a.n. 28920* (UPR)].

Native to Mexico, Central America (Honduras, Nicaragua, Panama), South America (scattered). [AR&S: 315].

Fuirena umbellata Rottb., Descr. Icon. Rar. Pl. 70, *t. 19, f. 3*. 1773.

Herb in Northern Coastal Lowlands, Northern Limestone Hills (San Sebastián), Río Abajo Forest Reserve, Sierra de Luquillo, Eastern Coast (Punta Guayanés), Sierra de Cayey, Western Coast; 1–925 m; *fl* & *fr* throughout the year.

Pantropical. [AR&S: 316].

Kyllinga brevifolia Rottb., Descr. Icon. Rar. Pl. 13, *t. 4, f. 3*. 1773.

Syn.: *Cyperus brevifolius* (Rottb.) Endl. ex Hassk. (1844).

Herb in Northern Coastal Lowlands, Northern Limestone Hills (San Sebastián), Sierra de Luquillo, Eastern Coast, Caguas Valley (Caguas), Eastern Cordillera (Aguas Buenas), Central Cordillera, Southern Coastal Lowlands (Guayama), Western Cordillera, Southwestern Coastal Lowlands (Cabo Rojo), Western Coast; 1–1330 m; *fl* & *fr* throughout the year.

Tropical, subtropical and warm-temperate regions. [AR&S: 318].

Kyllinga nemoralis (J.R. Forst. & G. Forst.) Dandy ex Hutch. & Dalziel, Fl. W. Trop. Afr. 2:486. 1936.

Bas.: *Thryocephalon nemorale* J.R. Forst. & G. Forst. (1775).

Herb in Northern Coastal Lowlands, Northern Limestone Hills (Utuado), Río Abajo Forest Reserve, Eastern Coast (Naguabo), Caguas Valley (Gurabo), Southeastern Foothills (Arroyo), Eastern Cordillera (Comerío), Western Coast (Mayagüez); 5–600 m; *fl* & *fr* Jul–Apr.

Pantropical. [AR&S: 318].

Kyllinga odorata Vahl, Enum. Pl. 2:382. 1805.

Syn.: *Cyperus sesquiflorus* (Torr.) Mattf. & Kük. (1936).

Herb in Northern Coastal Lowlands (Loíza to Bayamón), Las Tetas de Cayey, Eastern Cordillera (Corozal), Western Coast (Mayagüez); 5–825 m; *fl* & *fr* Nov–May.

Pantropical. [AR&S: 319].

Kyllinga pumila Michx., Fl. Bor.-Amer. 1:28. 1803.

Syn.: *Cyperus densicaespitosus* Mattf. & Kük. (1936).

Cyperus tenuifolius (Steud.) Dandy (1944).

Herb in Northern Coastal Lowlands, Sierra de Luquillo, Eastern Coast (Humacao), Caguas Valley, Eastern Cordillera (Cañón San Cristóbal), Central Cordillera, Southern Coastal Lowlands (Coamo), Western Cordillera; 25–1200 m; *fl* & *fr* throughout the year.

Antilles, eastern and south-central United States, Mexico, Central America, South America, tropical Africa. [AR&S: 321].

Kyllinga tibialis Poit. ex Ledeb. in Ledebour & Alderstam, Diss. Bot. Pl. Doming. 6 (no. I). 1805.

Herb in Northern Coastal Lowlands, Sierra de Luquillo, Eastern Coast (Humacao); 1–1050 m; *fl* & *fr* throughout the year.

Antilles, Central America, northern South America to Brazil, western tropical Africa. [AR&S: 321].

Kyllinga vaginata Lam., Tabl. Encycl. 1:148. 1791.

Syn.: *Kyllinga peruviana* Lam. (1792). *Cyperus peruvianus* (Lam.) F. N. Williams (1907).

Kyllinga pungens Link (1827).

Cyperus obtusatus (Presl) Mattf. & Kük. (1936).

Herb in Eastern Coast (Humacao); 1–5 m; *fl* & *fr* Jul (one specimen). [*González-Más 109* (MAPR)].

Greater Antilles, Central America, South America, tropical Africa. [AR&S: 322].

Notes: Britton & Wilson (1923) and Liogier & Martorell (1982) misapplied the name *Kyllinga peruviana*, and Liogier & Martorell (2000) misapplied *Cyperus peruvianus*, to *Kyllinga tibialis*.

Lagenocarpus guianensis Lindl., Linnaea 9(3):304. 1834.

Syn.: *Lagenocarpus portoricensis* Britton (1923).

Herb in Northern Coastal Lowlands (Vega Alta to Manatí); 1–2 m; *fl* & *fr* Jan, May–Sep.

Greater Antilles (Cuba), Trinidad, Central America (Belize, Honduras, Nicaragua), northwestern, west-central South America and Venezuela. [AR&S: 323].

Lipocarpha micrantha (Vahl) G.C. Tucker, J. Arnold Arbor. 68(4):410. 1987.
Bas.: *Scirpus micranthus* Vahl (1805).
Syn.: *Hemicarpha micrantha* (Vahl) Pax (1887).

Herb in Northern Coastal Lowlands (Laguna Tortuguero); 1–2 m; *fl* & *fr* Feb–Mar. [*Woodbury 12 Mar 1960 & Feb 1967* (UPR); *Vélez-Gavilán 614* (MAPR)].

Greater Antilles, Lesser Antilles (Guadeloupe, Martinique), Canada, United States, Mexico, Central America, South America (scattered), tropical Africa. [AR&S: 324].

Machaerina restioides (Sw.) Vahl, Enum. Pl. 2:238. 1805.
Bas.: *Schoenus restioides* Sw. (1788).
Syn.: *Cladium restioides* (Sw.) Benth. ex C.B. Clarke (1900).

Herb in Sierra de Luquillo; 900–1025 m; *fl* & *fr* Apr–Jul, Oct.

Antilles. [AR&S: 327].

Cortadera, Cortadora, Cortadora de altura, Lambedora.

Oxycaryum cubense (Poepp. & Kunth) Palla, Denkschr. Kaiserl. Akad. Wiss., Math.-Naturwiss. Kl. 79(1):169. 1908.
Bas.: *Scirpus cubensis* Poepp. & Kunth (1837).

Herb in Northern Coastal Lowlands (Toa Baja, Dorado); 3–5 m; *fl* & *fr* Dec, Mar. [*Axelrod 13273* (UPRRP)].

Native to Greater Antilles (Cuba), southern United States (Florida to Texas), Mexico, Central America, South America, Africa.

Remirea maritima Aubl., Hist. Pl. Guiane 1:45, *t.* 16. 1775.
Syn.: *Remirea pedunculata* R. Br. (1810). *Cyperus pedunculatus* (R. Br.) J. Kern (1958). *Mariscus pedunculatus* (R. Br.) T. Koyama (1977).

Herb in Northern Coastal Lowlands (Arecibo to Camuy); 1–3 m; *fl* & *fr* Oct–Jan.

Pantropical. [AR&S: 327].

Junco de playa.

Rhynchospora alba (L.) Vahl, Enum. Pl. 2:236. 1805 "Rynchospora."
Bas.: *Schoenus albus* L. (1753).
Syn.: *Rhynchospora luquillensis* Britton (1923).

Herb in Sierra de Luquillo, Toro Negro Forest Reserve; 800–1300 m; *fl* & *fr* Feb–Sep.

Native to temperate regions of North America, Europe and Asia. [AR&S: 333].

Rhynchospora berteroi (Spreng.) C.B. Clarke in Urban, Symb. Antill. 2(1):119. 1900 "Berterii."
Bas.: *Hypolytrum berteroi* Spreng. (1820) "Berterii."
Syn.: *Rhynchospora pusilla* (Sw.) Griseb. (1857).

Herb in Northern Coastal Lowlands, Northern Limestone Hills, Río Abajo Forest Reserve, Central Cordillera (Utuado), Susúa Forest Reserve, Western Cordillera, Western Coast (Rincón); 25–825 m; *fl* & *fr* Feb–Nov.

Greater Antilles, Lesser Antilles (Guadeloupe), northern Central America (Belize). [AR&S: 335].

Rhynchospora breviseta (Gale) Channell, Rhodora 58(696):336. 1956.
Bas.: *Rhynchospora oligantha* A. Gray var. *breviseta* Gale (1944).

Herb in Northern Coastal Lowlands (Dorado to Laguna Tortuguero), Sierra de Luquillo; 1–900 m; *fl* & *fr* Jan, May.

Greater Antilles, southeastern United States. [AR&S: 336].

Notes: Liogier & Martorell (1982, 2000) misapplied the name *Rhynchospora oligantha* A. Gray var. *oligantha* to this taxon; also, Liogier & Martorell (2000) wrongly had Griseb. as author of the basionym.

Rhynchospora ciliata (G. Mey.) Kük., Bot. Jahrb. Syst. 56(Beibl. 125):16. 1921.
Bas.: *Schoenus ciliatus* G. Mey. (1818).
Syn.: *Dichromena ciliata* Vahl (1805), *nom. illeg. Rhynchospora nervosa* (Vahl) Boeck. subsp. *ciliata* T. Koyama (1970).

Herb in Northern Coastal Lowlands, Northern Limestone Hills, Río Abajo Forest Reserve, Sierra de Luquillo, Eastern Coast, Caguas Valley (Caguas), Sierra de Cayey, Las Tetas de Cayey, Eastern and Central Cordillera, Susúa Forest Reserve, Western Cordillera, Southwestern Coastal Lowlands (Cabo Rojo), Cerro Las Mesas, Western Coast (Añasco); 1–850 m; *fl* & *fr* throughout the year.

Antilles, Mexico, Central America, South America. [AR&S: 337].

Botoncillo, Coquí blanco, Yerba de estrella.

Rhynchospora colorata (L.) H. Pfeiff., Repert. Spec. Nov. Regni Veg. 38(6–12):89. 1935.
Bas.: *Schoenus coloratus* L. (1753).
Syn.: *Dichromena colorata* (L.) Hitchc. (1893). *Rhynchospora stellata* (Lam.) Griseb. (1857).

Herb in Northern Coastal Lowlands, Eastern Coast (Humacao), Río Abajo Forest Reserve, Western

Coast (Aguada); 1–300 m; *fl* & *fr* throughout the year.

West Indies, southeastern and south-central United States, Mexico, northern Central America. [AR&S: 340].

Rhynchospora contracta (Nees) J. Raynal, Adansonia, n.s. 17(3):277. 1978.

Bas.: *Haloschoenus contractus* Nees (1842).

Herb in Northern Limestone Hills (Moca), Eastern Coast, Maricao Forest Reserve, Western Coast; 10–825 m; *fl* & *fr* Jun–Jul, Oct–Jan.

Antilles, Mexico, Central America, northern South America (Venezuela, Ecuador). [AR&S: 340].

Notes: Britton & Wilson (1923) and Liogier & Martorell (1982, 2000) used *Rhynchospora micrantha* Vahl, *sensu lato*, for the name of this taxon.

Rhynchospora corymbosa (L.) Britton, Trans. New York. Acad. Sci. 11:84. 1892.

Bas.: *Scirpus corymbosus* L. (1756).

Herb in Northern Coastal Lowlands, Río Abajo Forest Reserve, Eastern Coast (Naguabo), Western Coast (Añasco); 1–325 m; *fl* & *fr* Apr–Dec.

Native to South Africa, Sri Lanka, Malaysia, Philippines. [AR&S: 341].

Junco pico.

Rhynchospora depressirostris M.T. Strong, Brittonia 52(3):241, *ff. 1A-E, 2A, B*. 2000.

Herb in Northern Coastal Lowlands (Cataño to Laguna Tortuguero), Eastern Coast (Punta Guayanés); 1–25 m; *fl* & *fr* Sep–May.

Endemic. [AR&S: 342].

Notes: Britton & Wilson (1923) and Liogier & Martorell (1982, 2000) referred specimens of this taxon to *Rhynchospora tenuis* Link.

Rhynchospora domingensis Urb., Symb. Antill. 7(2):170. 1912.

Herb in Central Cordillera; 1000–1200 m; *fl* & *fr* Mar (two specimens). [*Woodbury a.n. 4652* (UPR); *Liogier 30576* (UPR)].

Greater Antilles (Cuba, Hispaniola). [AR&S: 344].

Notes: Liogier & Martorell (1982, 2000) misapplied the name *Rhynchospora longiflora* C. Presl to this taxon.

Rhynchospora fascicularis (Michx.) Vahl, Enum. Pl. 2:234. 1805 "Rynchospora."

Bas.: *Schoenus fascicularis* Michx. (1803).

Syn.: *Rhynchospora distans* (Michx.) Vahl (1805) "Rynchospora." *Rhynchospora fascicularis* (Michx.) Vahl subsp. *fascicularis* var. *distans* (Michx.) Chapm. (1860).

Herb in Northern Coastal Lowlands; 1–10 m; *fl* & *fr* throughout the year.

Greater Antilles, southeastern and south-central United States, Mexico, northern Central America. [AR&S: 346].

Rhynchospora gigantea Link, Jahrb. Gewächsk. 1(3):76. 1820.

Herb in Northern Coastal Lowlands (Dorado to Laguna Tortuguero); 1–2 m; *fl* & *fr* Mar–Jun, Oct–Nov.

Greater Antilles (Cuba, Jamaica), Trinidad, Mexico, Central America (Nicaragua), northern South America to Ecuador, Brazil. [AR&S: 347].

Rhynchospora holoschoenoides (Rich.) Herter, Revista Sudamer. Bot. 9(5):157. 1953.

Bas.: *Schoenus holoschoenoides* Rich. (1792).

Syn.: *Rhynchospora cyperoides* Mart. (1817).

Herb in Northern Coastal Lowlands, Northern Limestone Hills (San Sebastián), Río Abajo Forest Reserve, Sierra de Luquillo, Eastern Coast, Sierra de Cayey, Maricao Forest Reserve, Western Coast; 1–1050 m; *fl* & *fr* throughout the year.

West Indies, Mexico, Central America, northern South America to Brazil, tropical Africa. [AR&S: 348].

Rhynchospora jamaicensis Britton, Bull. Torrey Bot. Club 41(1):1. 1914.

Herb in Sierra de Luquillo, Sierra de Cayey, Central Cordillera; 600–1100 m; *fl* & *fr* Jan, Apr–Oct.

Greater Antilles (Jamaica). [AR&S: 349].

Rhynchospora lindeniana Griseb., Cat. Pl. Cub. 244. 1866.

Syn.: *Rhynchospora lindeniana* Griseb. var. *bahamensis* (Britton) Gale (1944).

Herb in Susúa Forest Reserve, Maricao Forest Reserve; 500–850 m; *fl* & *fr* throughout the year.

Bahamas, Greater Antilles, southern Mexico, northern Central America (Guatemala). [AR&S: 350].

Rhynchospora marisculus Lindl. ex Nees in Martius, Fl. Bras. 2(1):142. 1842.

Syn.: *Rhynchospora borinquensis* Britton (1915).

Herb in Sierra de Luquillo; 700–1050 m; *fl* & *fr* Nov–Aug.

Greater Antilles, Lesser Antilles (Dominica, St. Lucia), Central America, South America. [AR&S: 351].

Rhynchospora microcarpa Baldwin ex A. Gray, Ann. Lyceum Nat. Hist. New York 3:202. 1835.

Herb in Northern Coastal Lowlands (Laguna Tortuguero); 1–2 m; *fl* & *fr* Oct–May.

West Indies, southeastern and south-central United States, northern Central America (Belize). [AR&S: 352].

Rhynchospora nitens (Vahl) A. Gray, Manual (ed. 5) 568. 1867.
Bas.: *Scirpus nitens* Vahl (1805).
Syn.: *Psilocarya nitens* (Vahl) A.W. Wood (1870).
Psilocarya portoricensis Britton (1915).

Herb in Northern Coastal Lowlands (Vega Baja to Laguna Tortuguero); 1–2 m; *fl* & *fr* Aug–Mar.

Greater Antilles, eastern and south-central United States, Central America (Belize, Nicaragua). [AR&S: 352].

Rhynchospora odorata C. Wright ex Griseb., Cat. Pl. Cub. 242. 1866.

Herb in Northern Coastal Lowlands (Laguna Tortuguero); 1–2 m; *fl* & *fr* May (one specimen). [*Woodbury a.n. 4666, 5565* (UPR)].

Bahamas, Greater Antilles, southeastern United States. [AR&S: 353].

Rhynchospora plumosa Elliott, Sketch Bot. S. Carolina 1(1):58. 1816.

Herb in Northern Coastal Lowlands (Dorado to Laguna Tortuguero); 1–2 m; *fl* & *fr* Jan–Jun.

Greater Antilles (Cuba), southeastern United States, Central America (Belize, Honduras, Nicaragua). [AR&S: 354].

Rhynchospora pusilla Chapm. ex M.A. Curtis, Amer. J. Sci. Arts, ser. 2 7(21)[=57]:409. 1849.
Syn.: *Rhynchospora intermixta* C. Wright (1871).
Rhynchospora bruneri Britton (1923).

Herb in Northern Coastal Lowlands (Vega Alta to Laguna Tortuguero), Sierra de Luquillo; 1–1025 m; *fl* & *fr* throughout the year.

Greater Antilles, Lesser Antilles (Guadeloupe, Marie Galante), Mexico, Central America (Belize, Nicaragua). [AR&S: 355].

Rhynchospora racemosa C. Wright in Sauvalle, Anales Acad. Ci. Méd. Habana 8:86. 1871.

Herb in Susúa Forest Reserve, Maricao Forest Reserve; 180–550 m; *fl* & *fr* May–Aug.

Greater Antilles. [AR&S: 355].

Rhynchospora radicans (Schltdl. & Cham.) H. Pfeiff. subsp. **microcephala** (Bertero ex Spreng.) W.W. Thomas, Mem. New York Bot. Gard. 37:60. 1984.
Bas.: *Dichromena microcephala* Bertero ex Spreng. (1824).

Herb in Northern Limestone Hills, Río Abajo Forest Reserve, Sierra de Luquillo, Eastern Coast, Sierra de Cayey, Central Cordillera, Western Cordillera, Western Coast (Añasco); 50–1000 m; *fl* & *fr* throughout the year.

Antilles, southern Mexico, Central America, northern and western South America to northern Chile. [AR&S: 356].

Notes: Britton & Wilson (1923) used *Dichromena radicans* Schltdl. & Cham., *sensu lato*, and Liogier & Martorell (1982, 2000) used *Rhynchospora radicans* (Schltdl. & Cham.) H. Pfeiff., *sensu lato*, for the name of this taxon.

Rhynchospora rariflora (Michx.) Elliott, Sketch Bot. S. Carolina 1(1):58. 1816.
Bas.: *Schoenus rariflorus* Michx. (1803).

Herb in Northern Coastal Lowlands (Dorado to Laguna Tortuguero); 1–2 m; *fl* & *fr* Mar–May (two specimens). [*Woodbury a.n. 4684, 4885, 4686, 4887* (UPR)].

Greater Antilles, southeastern and south-central United States, Central America (Belize, Nicaragua). [AR&S: 357].

Rhynchospora recognita (Gale) Kral, Novon 9(2): 205. 1999.
Bas.: *Rhynchospora globularis* (Chapm.) Small var. *recognita* Gale (1944).

Herb in Northern Coastal Lowlands (Vega Baja to Laguna Tortuguero), Sierra de Luquillo, Eastern Coast (Punta Guayanés), Maricao Forest Reserve, Cerro Las Mesas; 1–900 m; *fl* & *fr* Aug–Oct, Feb–May.

Antilles, United States (south from New York to Kansas and California), Central America. [AR&S: 358].

Notes: Britton & Wilson (1923) used *Rhynchospora cymosa* Elliott, *sensu lato*, for the name of this taxon; Liogier & Martorell (1982, 2000) misapplied the name *Rhynchospora globularis* (Chapm.) Small var. *pinetorum* (Britton & Small) Gale to it.

Rhynchospora rugosa (Vahl) Gale, Rhodora 46(546):275, *t. 835, f. 1A, B.* 1944.
Bas.: *Schoenus rugosus* Vahl (1798).
Syn.: *Rhynchospora glauca* Vahl (1805).

Herb in Northern Coastal Lowlands (Laguna Tortuguero), Sierra de Luquillo, Eastern Cordillera (Cerro La Peña), Central Cordillera (Adjuntas), Cerro Las Mesas, Western Coast; 1–1000 m; *fl* & *fr* through-out the year.

Greater Antilles, Lesser Antilles (Nevis, Dominica), Trinidad, Central America, South America, tropical Africa. [AR&S: 359].

Rhynchospora tenerrima Nees ex Spreng., Syst. Veg. 4(2, Cur. Post.):26. 1827.
Syn.: *Rhynchospora setacea* (P. J. Bergius) Boeck. (1869).

Herb in Northern Coastal Lowlands, Northern Limestone Hills (San Sebastián), Sierra de Luquillo, Eastern Coast, Sierra de Cayey, Southwestern Coastal Lowlands (Cabo Rojo); 1–800 m; *fl* & *fr* throughout the year.

Antilles, Central America, northwestern, west-central South America and Venezuela. [AR&S: 360].

Rhynchospora uniflora Boeck., Flora 63(28):439. 1880.
Syn.: *Rhynchospora elongata* Boeck. (1888).

Herb in Sierra de Luquillo; 850–1000 m; *fl* & *fr* Mar–Apr, Jul–Aug.

Greater Antilles. [AR&S: 361].

Rhynchospora wrightiana Boeck., Flora 64(5):78. 1881.

Herb in Northern Coastal Lowlands (Laguna Tortuguero), Sierra de Luquillo; 1–1050 m; *fl* & *fr* Feb–Jun, Sep.

Greater Antilles (Cuba), southeastern United States. [AR&S: 362].

Schoenoplectus americanus (Pers.) Volkart ex Schinz & R. Keller, Fl. Schweiz (ed. 2) 1:75. 1905.
Bas.: *Scirpus americanus* Pers. (1805).
Syn.: *Scirpus chilensis* Nees & Meyen (1843).
Scirpus olneyi A. Gray (1845).

Herb in Northern Coastal Lowlands, Southern Coastal Lowlands (Guayama, Santa Isabel); 1–10 m; *fl* & *fr* Jan–Aug.

Greater Antilles, Lesser Antilles (Nevis), Canada, United States, Mexico, Central America, South America. [AR&S: 365].

Schoenoplectus validus (Vahl) Á. Löve & D. Löve, Bull. Torrey Bot. Club 81(1):33. 1954.
Bas.: *Scirpus validus* Vahl (1805).

Herb in Northern Coastal Lowlands (Toa Baja, Vega Alta), Río Abajo Forest Reserve, Southern Coastal Lowlands (Salinas), Southwestern Coastal Lowlands (Guánica, San Germán); 1–325 m; *fl* & *fr* throughout the year.

Bahamas, Greater Antilles, North America, Mexico, Central America, South America. [AR&S: 365].

Notes: S. Galen Smith in the Flora of North America, Vol. 23 (2002), places this taxon in synonymy with *Schoenoplectus tabernaemontani* (C.C. Gmel.) Palla, thus making of the latter a large, complex cosmopolitan taxon.

Junco.

Scleria brittonii Core in Small, Man. S.E. Fl. 190. 1933; et Brittonia 1(5–6):243. 1934.
Syn.: *Scleria ciliata* Michx. var. *glabra* (Chapm.) Fairey (1967).

Herb in Northern Coastal Lowlands (Laguna Tortuguero); 1–2 m; *fl* & *fr* May, Oct (two specimens). [*Woodbury a.n. 4698* (UPR); *Liogier 33563* (UPR)].

Greater Antilles (Cuba), southeastern and south-central United States. [AR&S: 371].

Scleria canescens Boeck., Beitr. Cyper. 1:37. 1888.

Viny herb in Sierra de Luquillo, Sierra de Cayey, Central Cordillera, Maricao Forest Reserve; 550–1250 m; *fl* & *fr* throughout the year.

Endemic. [AR&S: 370; AR: 419].

Cortadera, Cortadora, Cortadora de altura, Lambedora.

Scleria ciliata Michx. var. **ciliata**, Fl. Bor.-Amer. 2:167. 1803.

Herb in Maricao Forest Reserve, Cerro Las Mesas; 400–750 m; *fl* & *fr* Mar, Jun–Jul. [*Liogier 30532, 30692* (UPR)].

Greater Antilles (Cuba), southeastern and south-central United States, Central America. [AR&S: 371].

Scleria distans Poir., Encycl. 7:4. 1806.

Herb in Northern Coastal Lowlands (Vega Alta to Laguna Tortuguero), Northern Limestone Hills (Moca), Sierra de Luquillo, Sierra de Cayey, Toro Negro Forest Reserve, Cerro Las Mesas, Western Coast (Moca); 2–1050 m; *fl* & *fr* throughout the year.

Greater Antilles, southeastern and south-central United States, Mexico, Central America, northern South America to Brazil, tropical Africa, Madagascar. [AR&S: 371].

Scleria eggersiana Boeck., Beitr. Cyper. 2:41. 1890.
Syn.: *Scleria mitis* P.J. Bergius subsp. *eggersiana* (Boeck.) T. Koyama (1986).
Scleria grisebachii C.B. Clarke (1900).

Herb in Northern Coastal Lowlands, Eastern Coast; 1–5 m; *fl* & *fr* Jan–Feb, May–Jun, Sep–Oct.

Antilles, Trinidad and Tobago, Mexico, Central America, northern South America to Brazil, Ecuador. [AR&S: 372].

Scleria georgiana Core, Brittonia 1(5–6):243. 1934.

Herb in Northern Coastal Lowlands (Vega Baja to Laguna Tortuguero); 1–2 m; *fl* & *fr* Feb–May, Sep.

Greater Antilles, southeastern and south-central United States, Central America (Belize, Nicaragua). [AR&S: 373].

Scleria havanensis Britton, Bull. Torrey Bot. Club 42(9):492. 1915.

Herb in Susúa Forest Reserve, Western Cordillera, Cerro Las Mesas; 120–800 m; *fl* & *fr* May–Nov.

Greater Antilles (Cuba, Hispaniola). [AR&S: 373].

Notes: Liogier & Martorell (2000) misapplied the name *Scleria triglomerata* Michx. to this taxon.

Scleria hirtella Sw., Prodr. 19. 1788.

Syn.: *Scleria doradoensis* Britton (1923).

Herb in Northern Coastal Lowlands (Dorado to Laguna Tortuguero), Río Abajo Forest Reserve, Sierra de Luquillo, Eastern Coast (Cuchilla de Panduras); 1–900 m; *fl* & *fr* Sep–Jun.

Antilles, Trinidad, southeastern and south-central United States, northern Central America (Honduras), South America (scattered), tropical Africa. [AR&S: 374].

Scleria lithosperma (L.) Sw., Prodr. 18. 1788.

Bas.: *Scirpus lithospermus* L. (1753).

Herb in Northern Coastal Lowlands, Northern Limestone Hills, Río Abajo Forest Reserve, Eastern Coast, Southeastern Foothills (Guayama), Southern Coastal Limestone, Susúa Forest Reserve, Guánica Forest Reserve, Sierra Bermeja, Western Coast (San Germán, Rincón), Vieques, Mona; 5–325 m; *fl* & *fr* throughout the year.

Pantropical. [AR&S: 375].

Scleria melaleuca Rchb. ex Schltdl. & Cham., Linnaea 6(1):29. 1831.

Herb in Northern Coastal Lowlands, Northern Limestone Hills, Río Abajo Forest Reserve, Sierra de Luquillo, Eastern Coast, Caguas Valley (Caguas), Sierra de Cayey, Eastern Cordillera (Aguas Buenas), Central Cordillera (Adjuntas), Maricao Forest Reserve, Western Coast; 1–700 m; *fl* & *fr* throughout the year.

Antilles, Mexico, Central America, South America. [AR&S: 376].

Notes: Britton & Wilson (1923) and Liogier & Martorell (1982, 2000) used *Scleria pterota* C. Presl, a *nom. nud.*, for the name of this taxon.

Cortadora blanca.

Scleria microcarpa Nees ex Kunth, Enum. Pl. 2:341. 1837.

Herb in Northern Coastal Lowlands, Northern Limestone Hills (Florida), Río Abajo Forest Reserve, Sierra de Luquillo, Sierra de Cayey, Western Coast (Mayagüez); 1–800 m; *fl* & *fr* throughout the year.

Antilles, Trinidad, Mexico, Central America, northern South America to Paraguay. [AR&S: 377].

Scleria mucronata Poir., Encycl. 7:3. 1806.

Syn.: *Scleria cubensis* Boeck. (1890).

Herb in Sierra de Luquillo, Sierra de Cayey, Eastern Cordillera (La Torrecilla), Maricao Forest Reserve; 350–850 m; *fl* & *fr* Oct–Jul.

Greater Antilles. [AR&S: 378].

Scleria muehlenbergii Steud., Nomencl. Bot. (ed. 2) 2:543. 1841.

Herb in Northern Coastal Lowlands (Laguna Tortuguero), Northern Limestone Hills (Añasco, Moca), Western Coast (Rincón); 1–300 m; *fl* & *fr* May–Nov, Feb.

Antilles, Trinidad, eastern and south-central United States, Central America, South America. [AR&S: 378].

Scleria secans (L.) Urb., Symb. Antill. 2(2):169. 1900.

Bas.: *Schoenus secans* L. (1759).

Viny herb in Northern Coastal Lowlands (Laguna Tortuguero), Sierra de Luquillo, Sierra de Cayey, Eastern Cordillera (Cerro La Peña); 1–700 m; *fl* & *fr* Oct–Dec, Mar–May.

Antilles, Trinidad and Tobago, Mexico, Central America, South America (scattered). [AR&S: 380; AR: 419].

Cortadora de altura.

Scleria verticillata Muhl. ex Willd., Sp. Pl. 4(1):317. 1805.

Herb in Northern Coastal Lowlands (Laguna Tortuguero); 1–2 m; *fl* & *fr* Dec–May.

Bahamas, Greater Antilles (Cuba), Canada, eastern and south-central United States, Mexico, northern Central America (Honduras), northern South America (Venezuela). [AR&S: 382].

TAXA NOT RECENTLY COLLECTED

Cyperus swartzii Boeck., Beitr. Cyper. 1:10. 1888.

Syn.: *Mariscus gracilis* Vahl (1805).

Herb collected in Guánica on 17 May 1886 [*Sintenis 3808* (B), probably destroyed] and in Peñuelas on 6 Jul 1886 [*Sintenis 4742* (GH)].

Greater Antilles, Lesser Antilles (Guadeloupe), southern Mexico. [AR&S: 289].

Rhynchospora brachychaeta C. Wright in Sauvalle, Anales Acad. Ci. Méd. Habana 8:85. 1871 "brachychata."

Syn.: *Rhynchospora blauneri* Britton (1923).

Herb collected in the Sierra de Luquillo sometime during 1852–1853 [*Blauner 247* (NY), scan seen].

Native to southeastern United States (Florida, Alabama, Mississippi), Central America (Belize, Nicaragua), northern South America (Guyana). [AR&S: 336].

Rhynchospora filiformis Vahl, Enum. Pl. 2:232. 1805.

Syn.: *Rhynchospora podosperma* C. Wright (1871).

Herb collected in Puerto Rico without additional information [*Ledru s.n.* (C), scan seen].

Greater Antilles (Cuba, Hispaniola), Trinidad, Mexico, Central America, northern South America and Brazil. [AR&S: 346].

Notes: This taxon is known in Puerto Rico from the type specimen only.

Rhynchospora velutina (Kunth) Boeck., Vidensk. Meddel. Dansk Naturhist. Foren. Kjøbenhavn 1869:149. 1869.

Bas.: *Dichromena velutina* Kunth (1837).

Syn.: *Rhynchospora rufa* (Nees) Boeck. (1869).

Herb collected in Río Piedras in Martin Peña on 22 Jun 1913 [*Johnston 842* (NY), scan seen] and on 25 Apr 1914 [*Johnston & Stevenson 1883* (NY)].

Mexico, Central America, Greater Antilles, South America (scattered). [AR&S: 362].

TAXA CITED BY PREVIOUS AUTHORS, BUT NOT INCLUDED

NO SPECIMENS SEEN

Bulbostylis junciformis (Kunth) C.B. Clarke. Syn.: *Bulbostylis papillosa* Kük. *Fimbristylis papillosa* (Kük.) Alain.

Herb cited by Liogier & Martorell (1982, 2000), based on a citation by Roy Woodbury from Laguna Tortuguero on siliceous sand. [AR&S: 248].

Notes: Liogier & Martorell (2000) used the homonym *Bulbostylis junciformis* (Kunth) Lindman for the name of this taxon.

Rhynchospora divergens Chapm. ex M.A. Curtis

Herb cited by Liogier & Martorell (1982, 2000) as occurring in El Yunque, Luquillo Mountains and at Laguna Tortuguero, collected by Roy Woodbury. [AR&S: 364].

Scleria oligantha Michx.

Herb cited by Liogier & Martorell (1982, 2000) as occurring at Laguna Tortuguero, Vega Baja, collected by Roy Woodbury. [AR&S: 383].

DOUBTFUL OCCURRENCE

Cyperus luzulae (L.) Retz.

Herb cited by Britton & Wilson (1923) as occurring in moist grounds, collected by Bertero and near Bayamón by Brother Hioram; subsequently cited by Liogier & Martorell (1982, 2000) as doubtful. [AR&S: 294].

Fuirena squarrosa Michx.

Herb cited by Britton & Wilson (1923) as occurring in the western part of island, determined by Clarke from a drawing by Consul King, otherwise unknown from the island. It was not cited by Liogier & Martorell. [AR&S: 316].

Rhynchospora divaricata (Ham.) M.T. Strong

Syn.: *Fimbristylis divaricata* Ham. *Rhynchospora hispidula* (Vahl) Boeck. *Rhynchospora trichodes* C.B. Clarke.

Herb cited by Britton & Wilson (1923) as represented in the Delessert Herbarium by a specimen indicated to be from Puerto Rico; subsequently cited by Liogier & Martorell (1982, 2000) as a doubtful record. Strong in Acevedo-Rodríguez & Strong (2005) remarks that the locality of the type specimen in Puerto Rico is questionable. [AR&S: 343].

Rhynchospora miliacea (Lam.) A. Gray.

Herb mentioned by Britton & Wilson (1923) as recorded as Puerto Rican by Schultes; subsequently cited by Liogier & Martorell (1982, 2000) as a doubtful record, not collected recently. [AR&S: 364].

CITED DUE TO MISIDENTIFICATION

Cyperus flexuosus Vahl.

Herb cited by Liogier & Martorell (2000) as occurring in wet areas. This citation was based upon my misidentification of an immature specimen of *Cyperus odoratus*. [AR&S: 269].

Eleocharis microcarpa Torr.

Herb mentioned by Adams (1972) as occurring in Puerto Rico; subsequently cited by Liogier & Martorell (1982, 2000). This taxon was cited by Britton & Wilson (1923) as occurring in St. Thomas; but, according to M.T. Strong (2005), the name was misapplied to specimens of *Eleocharis minutiflora*. [AR&S: 301].

Scleria mitis P. J. Bergius.

Herb cited by Britton & Wilson (1923) as occurring on wet ground near Bayamón and in the eastern mountains; subsequently cited by Liogier & Martorell (1982, 2000) as a doubtful record. González-Más (1964) remarked that all the identifiable specimens that he had seen annotated as this taxon were definitely *Scleria eggersiana*. [AR&S: 382].

Scleria pauciflora Muhl. ex Willd.
Herb cited by Liogier & Martorell (1982, 2000) as occurring on wet sand, Laguna Tortuguero, collected by Roy Woodbury. According to M.T. Strong (2005), the citation is most likely based on a misidentification of specimens of *Scleria brittonii*. [AR&S: 383].

CYRILLACEAE

Cyrilla racemiflora L., Syst. Nat. (ed. 12) 2:182; et Mant. Pl. 1:50. 1767.
Syn.: *Cyrilla antillana* Michx. (1803).

Tree in Sierra de Luquillo, Sierra de Cayey, Central Cordillera, Maricao Forest Reserve, Cerro Las Mesas; 300–1300 m; *fl* Mar–Oct, *fr* Jul–Jan.

Antilles, Trinidad, Central America (Belize, Honduras, Panama). [L3: 11].

Colorado, Palo colorado.

DILLENIACEAE

Pinzona coriacea Mart. & Zucc., Abh. Math.-Phys. Cl. Königl. Bayer. Akad. Wiss. 1:371. 1832.
Syn.: *Doliocarpus coriaceus* (Mart. & Zucc.) Gilg (1893).
Doliocarpus calineoides (Eichler) Gilg (1893).

Woody vine in Northern Coastal Lowlands (Guaynabo), Sierra de Luquillo, Sierra de Cayey, Central Cordillera (Utuado); 300–750 m; *fl* (?), *fr* Dec (one specimen).

Greater Antilles (Hispaniola), Lesser Antilles (Guadeloupe, Dominica), Trinidad, Central America, northern South America to Brazil, Peru. [L3: 210; AR: 225].

Bejuco de agua.

TAXA NOT RECENTLY COLLECTED

Doliocarpus brevipedicellatus Garcke, Linnaea 22(1):47. 1849.

Woody vine collected in Bayamón near Toa Baja on 24 Mar 1885 [*Sintenis 1013* (US)].

Greater Antilles (Hispaniola), Central America, Trinidad, west-central South America. [L3: 207; AR: 225].

Notes: According to Acevedo-Rodríguez (2005), this taxon was wrongly treated by Liogier (1994) and Liogier & Martorell (2000) as a synonym of *Doliocarpus major* J.F. Gmel., a distinct South American taxon.

TAXA CITED BY PREVIOUS AUTHORS, BUT NOT INCLUDED

CITED DUE TO MISIDENTIFICATION

Doliocarpus olivaceus Sprague & R.O. Williams ex Standl.
Woody vine cited by Liogier (1994) and Liogier & Martorell (2000) as collected near Patillas, just south of Carite, a sterile specimen. According to Acevedo-Rodríguez (2005), this report was based upon the misidentification of a specimen of *Pinzona coriacea*. [L3: 207].

ONLY UNDER CULTIVATION

Dillenia indica L.
Tree cited by Liogier & Martorell (1982, 2000) and Liogier (1994) as occasionally planted and persistent. [L3: 205].

DIOSCOREACEAE

Dioscorea alata L., Sp. Pl. 2:1033. 1753.

Vine in Northern Coastal Lowlands, Northern Limestone Hills, Río Abajo Forest Reserve, Sierra de Luquillo, Southeastern Foothills, Central Cordillera (Yauco), Western Cordillera (Maricao), Western Coast (Mayagüez), Vieques; 15–450 m; *fl* Nov–Jan, *fr* Nov.

Native to tropical Asia. [AR&S: 83; AR: 421].

Bejuco de guaraguao, Ñame, Ñame blanco, Ñame de agua, Ñame de mina, Ñame gulembo.

Dioscorea bulbifera L., Sp. Pl. 2:1033. 1753.
Syn.: *Dioscorea latifolia* Benth. (1849).

Vine in Northern Coastal Lowlands (Vega Alta), Northern Limestone Hills (Arecibo), Río Abajo Forest Reserve, Caguas Valley (Caguas); 60–350 m; *fl* Mar–Aug, *fr* Oct (one specimen).

Native to the Paleotropics. [AR&S: 85; AR: 422].

Gunda.

Dioscorea chondrocarpa Griseb. in Martius, Fl. Bras. 3(1):34. 1842.

Vine in Northern Limestone Hills (Manatí), Sierra de Luquillo; 100–350 m; *fl* (?), *fr* (?). [*Acevedo-Rodríguez 12296* (US); *Raz 358* (UPRRP); *Axelrod 14118* (UPRRP)].

Greater Antilles (Hispaniola), Lesser Antilles, Tobago, northern South America to Bolivia. [AR&S: 85; AR: 422].

Notes: According to Pedro Acevedo-Rodríguez

(pers. comm.), the type of *Dioscorea altissima* Lam., the name previously used for this taxon, belongs to *Dioscorea cayenensis*; the next available name for this taxon is the above.

Dunguey, Ñame de gunda, Ñame de mina, Ñame dunguey.

Dioscorea esculenta (Lour.) Burkhill, Gard. Bull. Straits Settlem., ser. 3 1:396, *t. 17*. 1917.
Bas.: *Oncus esculentus* Lour. (1790).

Vine planted and persistent in Sierra de Luquillo; 325 m; *fl* (?), *fr* (?). [*Axelrod 10051* (UPRRP)].

Native to tropical Asia.

Dioscorea pilosiuscula Bertero ex Spreng., Syst. Veg. (ed. 16) 2:152. 1825.

Vine in Northern Limestone Hills (Bayamón), Sierra de Luquillo, Western Coast (Mayagüez); 50–350 m; *fl* (?), *fr* (?).

Antilles, Mexico, Central America, northwestern, west-central South America and Venezuela. [AR&S: 86; AR: 424].

Dunguey, Gunda, Ñame de gunda.

Dioscorea polygonoides Humb. & Bonpl. ex Willd., Sp. Pl. 4(2):795. 1806.

Vine in Northern Coastal Lowlands (Fajardo), Northern Limestone Hills, Río Abajo Forest Reserve, Sierra de Luquillo, Sierra de Cayey, Eastern Cordillera (Guaynabo, Bayamón); 1–1000 m; *fl* Sep–Oct, Jan–May, *fr* Feb–Apr, Sep.

Greater Antilles (Jamaica, Hispaniola), Lesser Antilles, Trinidad and Tobago, Mexico, Central America, northwestern, west-central South America and Venezuela. [AR&S: 86; AR: 424].

Gunda, Hícamo, Matagallina.

Rajania cordata L. var. **cordata**, Sp. Pl. 2:1032. 1753.

Vine in Northern Coastal Lowlands (Fajardo, Laguna Tortuguero), Northern Limestone Hills, Río Abajo Forest Reserve, Sierra de Luquillo, Eastern Coast (Ceiba, Punta Guayanés), Sierra de Cayey, Las Tetas de Cayey, Eastern Cordillera (Cerro La Peña), Central Cordillera, Southern Coastal Limestone (Guayanilla), Susúa Forest Reserve, Western Cordillera, Guánica Forest Reserve; 10–1300 m; *fl* throughout the year, *fr* Apr, Jul–Jan.

Greater Antilles (Jamaica, Hispaniola), Lesser Antilles, Trinidad. [AR&S: 90; AR: 427].

Bejuco de guaraguao, Guáyaro, Ñame gulembo, Ñame de la India.

Rajania cordata L. var. **cymulifera** Uline ex R. Knuth, Notizbl. Bot. Gart. Berlin-Dahlem 7:220. 1917.

Vine in Northern Limestone Hills, Central Cordillera (Villalba), Vieques; 130–600 m; *fl* Aug–Nov, *fr* (?).

Endemic.

Notes: Britton & Wilson (1924) treated this taxon as a synonym of *Rajania cordata*, *sensu lato*.

TAXA CITED BY PREVIOUS AUTHORS, BUT NOT INCLUDED

ONLY UNDER CULTIVATION

Dioscorea cayenensis Lam.

Herb cited by Britton & Wilson (1924) as spontaneous after cultivation; subsequently cited by Liogier & Martorell (1982, 2000) and by Acevedo-Rodríguez (2005) as naturalized. [AR&S: 86; AR: 422].

Dioscorea rotundata Poir.

Herb cited by Britton & Wilson (1930) as collected many years ago by Riedlé, according to Knuth; subsequently cited by Liogier & Martorell (1982, 2000), who misspelled the attributor as 'Kunth,' and by Acevedo-Rodríguez (2005) who mentioned it as cultivated, but not naturalized.

Dioscorea trifida L.f.

Herb cited by Britton & Wilson (1924) as spontaneous after cultivation and subsequently by Liogier & Martorell (1982, 2000); cited by Acevedo-Rodríguez (2005), and also by him in Acevedo-Rodríguez & Strong (2005), as apparently naturalized. [AR&S: 88; AR: 427].

DROSERACEAE

Drosera capillaris Poir., Encycl. 6(1):299. 1804.

Low herb in Northern Coastal Lowlands (Vega Baja to Laguna Tortuguero); 1–2 m; *fl & fr* Feb–Jun, Sep.

Greater Antilles (Cuba, Jamaica), Trinidad, southeastern and south-central United States, Mexico, Central America, northern South America and Brazil to Paraguay, Uruguay. [L1: 307].

EBENACEAE

Diospyros revoluta Poir., Encycl. 5:435. 1804.

Tree in Northern Limestone Hills (Arecibo), Río Abajo Forest Reserve; 250–350 m; *fl* (?), *fr* Jan–Feb (two specimens).

Greater Antilles (Hispaniola), Lesser Antilles. [L4: 158].

Notes: Britton & Wilson (1924) misapplied the name *Diospyros ebenaster* Retz. to this taxon.

Guayabota, Zapote negro.

Diospyros sintenisii (Krug & Urb.) Standl., Publ. Carnegie Inst. Wash. 461(4):80. 1935.
Bas.: *Maba sintenisii* Krug & Urb. (1892).

Tree in Northern Limestone Hills, Río Abajo Forest Reserve, Susúa Forest Reserve, Maricao Forest Reserve; 50–800 m; *fl* Apr–Jun, *fr* Oct–Jul.

Endemic. [L4: 158].

Guayabota, Guayabota níspero, Múcaro, Tabeiba.

ELAEOCARPACEAE

Sloanea amygdalina Griseb., Pl. Wright. 1:165. 1860 (preprint of Mem. Amer. Acad. Arts, n.s. 8(1):165. 1861).

Tree in Northern Limestone Hills, Central Cordillera (Utuado), Susúa Forest Reserve, Maricao Forest Reserve, Cerro Las Mesas; 300–850 m; *fl* (?), *fr* Jan, Jul (two specimens).

Greater Antilles (Cuba, Hispaniola). [L3: 105].

Sloanea berteroana Choisy ex DC., Prodr. 1:516. 1824 "Berteriana."

Tree in Sierra de Luquillo, Sierra de Cayey, Eastern Cordillera (Guaynabo), Central Cordillera, Maricao Forest Reserve; 300–850 m; *fl* Feb, May–Oct, *fr* throughout the year.

Greater Antilles (Hispaniola), Lesser Antilles. [L3: 106].

Cacaíllo, Cacao motillo, Cacao roseta, Cacaotillo, Motillo, Roseta.

Sloanea ilicifolia Urb., Repert. Spec. Nov. Regni Veg. 18:365. 1922.

Tree in Toro Negro Forest Reserve; 1000–1200 m; *fl* (?); *fr* Feb (one specimen). [*Padron s.n.* (US), not seen].

Greater Antilles (Hispaniola).

ERICACEAE

Gonocalyx concolor Nevling, J. Arnold Arbor. 51:222. 1970.

Vine in Sierra de Cayey; 600–950 m; *fl* Nov–Sep, *fr* Sep–Nov (two specimens).

Endemic. [L4: 106; AR: 227].

Gonocalyx portoricensis (Urb.) A.C. Sm., Contr. U.S. Natl. Herb. 28(2):354, *t. 6.* 1932.
Bas.: *Thibaudia portoricensis* Urb. (1899).
Syn.: *Ceratostema portoricensis* (Urb.) Hoerold (1909).

Vine in Sierra de Luquillo; 550–970 m; *fl* Feb–Aug, *fr* Jul–Sep.

Endemic. [L4: 106; AR: 228].

Lyonia stahlii Urb. var. **stahlii,** Symb. Antill. 5(3):453. 1908.
Syn.: *Lyonia rubiginosa* (Pers.) G. Don var. *stahlii* (Urb.) Judd (1981). *Xolisma stahlii* (Urb.) Small (1914).

Shrub in Northern Limestone Hills (Lares), Sierra de Cayey, Eastern Cordillera (Cerro La Peña), Toro Negro Forest Reserve; 300–1340 m; *fl* Oct–Apr, *fr* throughout the year.

Endemic. [L4: 108].

Notes: Britton & Wilson (1925) used *Xolisma rubiginosa* (Pers.) Small, *sensu lato*, for the name of this taxon.

Lyonia truncata Urb. var. **proctorii** Judd, J. Arnold Arbor. 71(1):129. 1990.

Shrub in Sierra Bermeja; 225–300 m; *fl* Aug–Apr, *fr* Aug–Mar.

Endemic. [L4: 109].

Symphysia racemosa (Vahl) Stearn, Taxon 21(1):111. 1972.
Bas.: *Hornemannia racemosa* Vahl (1810).
Syn.: *Vaccinium racemosum* (Vahl) Wilbur & Luteyn (1977).
Thibaudia krugii Urb. & Hoerold (1908).

Shrub in Sierra de Luquillo, Sierra de Cayey, Toro Negro Forest Reserve; 500–1300 m; *fl* & *fr* throughout the year.

Greater Antilles, Lesser Antilles (Guadeloupe, Dominica, Martinique). [L4: 110; AR: 228].

ERYTHROXYLACEAE

Erythroxylum areolatum L., Syst. Nat. (ed. 10) 2:1035. 1759.

Tree in Southern Coastal Lowlands, Southern Coastal Limestone, Guánica Forest Reserve, Southwestern Coastal Lowlands (Cabo Rojo), Mona; 1–150 m; *fl* Apr–Nov, *fr* throughout the year.

Bahamas, Greater Antilles, Central America. [L2: 243].

Notes: Little & Wadsworth (1964) used the orthographic variant *Erythroxylon areolatum* for the name of this taxon.

Cocaína falsa, Indio.

Erythroxylum brevipes DC., Prodr. 1:573. 1824.

Tree in Northern Coastal Lowlands, Northern Limestone Hills, Las Tetas de Cayey, Las Piedras Chiquitas, Southeastern Foothills, Eastern Cordillera (Cañón San Cristóbal), Central Cordillera (Ponce), Southern Coastal Lowlands, Susúa Forest Reserve, Guánica Forest Reserve, Southwestern Coastal Lowlands (Cabo Rojo), Sierra Bermeja, Vieques, Culebra, Desecheo; 2–825 m; *fl* throughout the year, *fr* Mar–Nov.

Greater Antilles (Hispaniola), Virgin Islands, Lesser Antilles (St. Barts). [L2: 245].

Jibá, Ratón, Rocío.

Erythroxylum rotundifolium Lunan, Hort. Jamaic. 2:116. 1814 "Erythroxylon."

Tree in Southern Coastal Limestone (Ponce), Guánica Forest Reserve; 1–150 m; *fl* Sep, Dec–Jan, *fr* Jul, Oct–Feb.

Bahamas, Greater Antilles. [L2: 245].

Jibá, Ratón, Rocío.

Erythroxylum rufum Cav., Diss. 404, *t. 232.* 1789.

Tree in Northern Coastal Lowlands (Fajardo), Northern Limestone Hills, Eastern Cordillera (Cañón San Cristóbal), Susúa Forest Reserve, Maricao Forest Reserve; 10–600 m; *fl* Mar–Apr, Sep, *fr* Jan–Aug.

Greater Antilles (Cuba, Hispaniola), northern South America. [L2: 245].

Erythroxylum urbanii O.E. Schulz in Urban, Symb. Antill. 4(2):314. 1905.

Tree in Northern Coastal Lowlands (Laguna Tortuguero), Northern Limestone Hills, Southern Coastal Limestone, Susúa Forest Reserve, Maricao Forest Reserve; 25–900 m; *fl* Feb–Aug, *fr* Jun–Sep, Dec–Jan.

Greater Antilles (Hispaniola). [L2: 245].

EUPHORBIACEAE

Acalypha alopecuroidea Jacq., Collectanea 3:196. 1791; et Icon. Pl. Rar. 3(9): *t.* 620. 1792 "alopecuroides."

Suffrutescent herb in Northern Coastal Lowlands; 3–50 m; *fl* & *fr* May–Jan.

Bahamas, Greater Antilles, Lesser Antilles (Guadeloupe, Grenada), Trinidad and Tobago, Mexico, Central America, northwestern, west-central South America and Venezuela. [L2: 345].

Acalypha arvensis Poepp., Nov. Gen. Sp. Pl. 3:21. 1841.

Herb in Northern Coastal Lowlands (Carolina); 1 m; *fl* & *fr* Apr (one specimen). [*Proctor 48728* (SJ)].

Native to Mexico, Central America, northern South America to Brazil, Bolivia.

Acalypha bisetosa Bertero ex Spreng., Syst. Veg. 3:879. 1826.

Shrub in Northern Limestone Hills, Río Abajo Forest Reserve, Sierra de Luquillo, Eastern Coast (Maunabo), Sierra de Cayey (San Lorenzo), Las Tetas de Cayey, Central Cordillera, Vieques; 225–1000 m; *fl* Jan–Oct, *fr* Apr–May.

Endemic. [L2: 346].

Acalypha chamaedrifolia (Lam.) Müll. Arg. in de Candolle, Prodr. 15(2):879. 1866.

Bas.: *Croton chamaedrifolius* Lam. (1786).

Suffrutescent herb in Northern Coastal Lowlands, Cerro Las Mesas, Western Coast (Mayagüez); 1–300 m; *fl* & *fr* Mar–Dec.

West Indies, southeastern United States (Florida). [L2: 346].

Acalypha indica L., Sp. Pl. 2:1003. 1753.

Suffrutescent herb in Vieques, Culebra; 1–15 m; *fl* & *fr* May–Jun, Dec.

Native to tropical Africa, tropical Asia. [L2: 348].

Notes: Liogier & Martorell (2000) wrongly placed this taxon in synonymy with *Acalypha poiretii* Spreng., which does not occur in Puerto Rico.

Acalypha ostryifolia Riddell, Syn. Fl. West. States 33. 1835 "ostryaefolia."

Herb in Northern Coastal Lowlands (Dorado), Northern Limestone Hills, Southern Coastal Lowlands (Santa Isabel); 1–350 m; *fl* & *fr* Feb–Sep.

Bahamas, Greater Antilles, southern and central United States, Mexico. [L2: 348].

Acalypha portoricensis Müll. Arg., Linnaea 34(1): 22. 1865.

Suffrutescent herb in Northern Coastal Lowlands (Carolina), Northern Limestone Hills, Río Abajo Forest Reserve, Las Tetas de Cayey, Las Piedras Chiquitas, Southeastern Foothills (Coamo), Eastern Cordillera (Cañón San Cristóbal, Aibonito), Quebrada de Los Cedros, Susúa Forest Reserve, Maricao Forest Reserve, Guánica Forest Reserve, Southwestern Coastal Lowlands (Lajas), Sierra Bermeja, Vieques; 3–900 m; *fl* throughout the year, *fr* Feb, May–Nov.

Virgin Islands (St. Croix). [L2: 348].

Rabito.

Acalypha setosa A. Rich. in Sagra, Hist. Fis. Cuba, Bot. 11:204. 1850.

Herb in Northern Limestone Hills (Florida, Arecibo), Central Cordillera, Southwestern Coastal Lowlands (Cabo Rojo), Western Coast (Aguada); 1–500 m; *fl* & *fr* Apr–Aug, Nov–Jan.

Bahamas, Greater Antilles (Cuba, Hispaniola), Virgin Islands (Anegada), Mexico, Central America, South America (Venezuela, Ecuador). [L2: 349].

Adelia ricinella L., Syst. Nat. (ed. 10) 2:1298. 1759.

Syn.: *Ricinella ricinella* (L.) Britton (1917).

Shrub/small tree in Northern Coastal Lowlands,

Northern Limestone Hills, Eastern Coast (Naguabo), Las Piedras Chiquitas, Southeastern Foothills, Southern Coastal Lowlands, Susúa Forest Reserve, Guánica Forest Reserve, Southwestern Coastal Lowlands, Sierra Bermeja, Vieques; 5–450 m; *fl* & *fr* Jan–Jun.

Greater Antilles, Virgin Islands, Lesser Antilles (St. Martin, Grenada), northern South America (Venezuela). [L2: 350].

Cotorro, Escambrón, Espinillo.

Alchornea latifolia Sw., Prodr. 98. 1788.

Tree in Northern Coastal Lowlands (Dorado), Northern Limestone Hills (Lares), Río Abajo Forest Reserve, Sierra de Luquillo, Sierra de Cayey, Las Tetas de Cayey, Eastern Cordillera (Cerro La Peña, Corozal), Central Cordillera, Maricao Forest Reserve; 5–1000 m; *fl* & *fr* Feb–Jul.

Greater Antilles, Virgin Islands (Tortola), Mexico, Central America, northwestern, west-central South America and Venezuela, Brazil. [L2: 351].

Achiotillo, Jobillo, Palo de cotorra.

Alchorneopsis floribunda (Benth.) Müll. Arg., Linnaea 34(2):156. 1865.

Bas.: *Alchornea glandulosa* Poepp. var. *floribunda* Benth. (1854).

Tree in Sierra de Luquillo, Eastern Cordillera (Cerro La Peña), Central Cordillera (Utuado); 300–750 m; *fl* Feb–Jul, *fr* Mar, Aug–Oct.

Greater Antilles (Hispaniola), Central America (Honduras, Costa Rica, Panama), northwestern, west-central South America and Venezuela, Brazil. [L2: 352].

Palo de gallina.

Aleurites moluccanus (L.) Willd., Sp. Pl. 4(1):590. 1805 "moluccana."

Bas.: *Jatropha moluccana* L. (1753).

Tree planted and persistent in Northern Coastal Lowlands (Loíza), Northern Limestone Hills (Florida), Sierra de Luquillo, Maricao Forest Reserve; 5–500 m; *fl* Mar–Apr, Oct, *fr* Mar–Jun (two specimens).

Native to tropical Asia, China, Australia. [L2: 355].

Nogal, Nuez, Nuez de la India, Palo de nuez.

Argythamnia candicans Sw., Prodr. 39. 1788.

Suffrutescent herb/shrub in Northern Coastal Lowlands, Northern Limestone Hills, Southeastern Foothills, Southern Coastal Limestone, Guánica Forest Reserve, Southwestern Coastal Lowlands, Sierra Bermeja, Vieques, Mona; 1–300 m; *fl* & *fr* throughout the year.

West Indies. [L2: 357].

Argythamnia fasciculata (Vahl ex A. Juss.) Müll. Arg., Linnaea 34(2):146. 1865.

Bas.: *Ditaxis fasciculata* Vahl ex A. Juss. (1824).

Shrub in Southern Coastal Lowlands, Southern Coastal Limestone, Guánica Forest Reserve, Southwestern Coastal Lowlands (Cabo Rojo), Sierra Bermeja; 25–300 m; *fl* Aug–May, *fr* Sep–May.

Greater Antilles (Cuba), Virgin Islands, Lesser Antilles (St. Martin). [L2: 357].

Argythamnia stahlii Urb., Symb. Antill. 1(2):336. 1899.

Shrub in Northern Coastal Lowlands (islands adjacent to Fajardo), Southern Coastal Limestone, Southwestern Coastal Lowlands (Cabo Rojo), Vieques; 1–150 m; *fl* & *fr* throughout the year.

Virgin Islands (St. John, Anegada). [L2: 359].

Astraea lobata (L.) Klotzsch, Arch. Naturgesch. 7(1):194. 1841.

Bas.: *Croton lobatus* L. (1753) "lobatum."

Shrub in Northern Coastal Lowlands, Eastern Coast, Sierra de Cayey (San Lorenzo, Patillas), Southeastern Foothills, Southern Coastal Lowlands, Southern Coastal Limestone, Southwestern Coastal Lowlands, Sierra Bermeja, Western Coast (Añasco); Vieques; 1–250 m; *fl* & *fr* throughout the year.

West Indies (except Jamaica), Mexico, Central America, southern South America to Brazil. [L2: 380].

Croton lobulado.

Bernardia dichotoma (Willd.) Müll. Arg., Linnaea 34(2):172. 1865.

Bas.: *Croton dichotomus* Willd. (1805).

Syn.: *Adelia bernardia* L. (1759).

Tree in Southern Coastal Limestone, Guánica Forest Reserve; 1–150 m; *fl* Feb, May–Jun, Sep–Nov, *fr* Sep–Jan.

West Indies. [L2: 359].

Mierda de gallina.

Caperonia palustris (L.) A. St.-Hil., Hist. Pl. Remarq. Brésil 245. 1826.

Bas.: *Croton palustris* L. (1753).

Suffrutescent herb in Northern Coastal Lowlands, Northern Limestone Hills (San Sebastián), Río Abajo Forest Reserve, Eastern Coast (Ceiba), Caguas Valley, Southwestern Coastal Lowlands, Western Coast; 5–200 m; *fl* & *fr* throughout the year.

Greater Antilles (Cuba, Hispaniola), Lesser Antilles, Trinidad, Mexico, Central America, northwestern South America (scattered). [L2: 361].

Sacatrapo.

Cnidoscolus aconitifolius (Mill.) I.M. Johnst. subsp. **aconitifolius**, Contr. Gray Herb. 68:86. 1923. Bas.: *Jatropha aconitifolia* Mill. (1768).

Tree planted and persistent in Northern Coastal Lowlands (Manatí), Southern Coastal Lowlands (Ponce), Southern Coastal Limestone, Guánica Forest Reserve; 5–25 m; *fl* Feb–Jun, Sep–Oct, *fr* (?).

Native to southern Mexico, Central America. [L2: 373].

Notes: According to Gary Breckon (pers. comm.), the Puerto Rican plants of this taxon are sterile and reproduce only by cuttings.

Lechoso, Papayo macho, Papayuelo.

Croton astroites Aiton, Hort. Kew. 3:375. 1789.

Shrub in Northern Coastal Lowlands (Fajardo), Southwestern Coastal Lowlands, Sierra Bermeja, Vieques, Culebra; 1–300 m; *fl* & *fr* Jun–Feb.

Virgin Islands, Lesser Antilles. [L2: 377].

Maná, Marán.

Croton betulinus Vahl, Symb. Bot. 2:98. 1791 "betulinum."

Shrub in Northern Coastal Lowlands (Fajardo and adjacent islands), Southeastern Foothills (Cayey), Central Cordillera (Ponce), Southern Coastal Lowlands (Caja de Muertos), Southern Coastal Limestone, Susúa Forest Reserve, Guánica Forest Reserve, Southwestern Coastal Lowlands (Cabo Rojo), Sierra Bermeja, Vieques, Culebra, Mona, Desecheo; 1–450 m; *fl* & *fr* throughout the year.

Greater Antilles (Cuba, Hispaniola), Virgin Islands, Lesser Antilles. [L2: 377].

Croton discolor Willd., Sp. Pl. 4(1):532. 1805.

Shrub in Southern Coastal Limestone, Guánica Forest Reserve, Southwestern Coastal Lowlands, Vieques, Mona, Desecheo; 1–150 m; *fl* throughout the year, *fr* Apr–Dec.

Bahamas, Greater Antilles, Virgin Islands. [L2: 377].

Lechecillo.

Croton flavens L., Syst. Nat. (ed. 10) 2:1276. 1759.

Syn.: *Croton flavens* L. var. *rigidus* Müll. Arg. (1866).

Croton rigidus (Müll. Arg.) Britton (1924).

Shrub in Northern Coastal Lowlands, Northern Limestone Hills, Central Cordillera (Ponce), Southern Coastal Limestone, Susúa Forest Reserve, Guánica Forest Reserve, Southwestern Coastal Lowlands (Cabo Rojo), Sierra Bermeja, Western Coast, Vieques, Culebra, Mona; 1–400 m; *fl* & *fr* throughout the year.

Virgin Islands. [L2: 381].

Adormidera, Guayacanillo.

Croton glabellus L., Syst. Nat. (ed. 10) 2:1275. 1759 "glabellum."

Shrub in Guajataca Forest Reserve, Southern Coastal Lowlands (Caja de Muertos), Southern Coastal Limestone, Susúa Forest Reserve, Guánica Forest Reserve, Southwestern Coastal Lowlands, Sierra Bermeja, Mona; 1–300 m; *fl* throughout the year, *fr* May–Dec.

Bahamas, Greater Antilles, Mexico. [L2: 380].

Notes: Britton & Wilson (1924), Liogier & Martorell (1982, 2000) and Liogier (1988) misapplied the name *Croton lucidus* L. to this taxon; according to Berry et al. (2009) the type of the former belongs to a taxon endemic to Jamaica.

Croton glandulosus L., Syst. Nat. (ed. 10) 2:1275. 1759 "glandulosum."

Shrub in Northern Coastal Lowlands, Eastern Coast, Southeastern Foothills (Coamo), Cerro Las Mesas, Western Coast (Mayagüez); 1–300 m; *fl* & *fr* throughout the year.

Greater Antilles, Southern Caribbean Islands (Curaçao), Trinidad, eastern and central United States, Mexico, Central America, southern South America to Brazil. [L2: 379].

Vente conmigo.

Croton humilis L., Syst. Nat. (ed. 10) 2:1276. 1759 "humile."

Shrub in Southern Coastal Limestone, Southwestern Coastal Lowlands, Mona; 1–90 m; *fl* & *fr* May–Oct.

Bahamas, Greater Antilles (Jamaica, Hispaniola), Virgin Islands, Lesser Antilles (Grenadines), southeastern United States (Florida), Mexico. [L2: 379].

Yerba bellaca.

Croton impressus Urb., Symb. Antill. 1(2):335. 1899.

Shrub in Guajataca Forest Reserve, Susúa Forest Reserve, Western Cordillera; 150–800 m; *fl* & *fr* throughout the year.

Greater Antilles (Hispaniola). [L2: 380].

Croton microcarpus Ham., Prodr. Pl. Ind. Occid. 55. 1825.

Syn.: *Croton nummulariifolius* A. Rich. (1850).

Low herb in Southern Coastal Limestone (Ponce), Guánica Forest Reserve; 40–50 m; *fl* Sep–Mar, *fr* Nov (one specimen).

Greater Antilles (Cuba, Hispaniola), Lesser Antilles (Anguilla, St. Barts). [L2: 381].

Croton poecilanthus Urb., Symb. Antill. 1(2):334. 1899.

Tree in Sierra de Luquillo; 350–900 m; *fl* throughout the year, *fr* Jan–Mar, Jun–Jul.

Endemic. [L2: 381].

Sabinón.

Croton stenophyllus Griseb., Pl. Wright. 1:158. 1860 (preprint of Mem. Amer. Acad. Arts, n.s. 8(1):158. 1861).

Shrub in Northern Coastal Lowlands (islands adjacent to Fajardo: Piñeros Island); 1–25 m; *fl* (?), *fr* (?). [*Woodbury a.n. 1640* (UPR)].

Greater Antilles (Cuba). [L2: 382].

Dalechampia scandens L., Sp. Pl. 2:1054. 1753.

Vine in Northern Coastal Lowlands (Fajardo), Northern Limestone Hills (Quebradillas), Las Piedras Chiquitas, Southeastern Foothills (Salinas), Southern Coastal Lowlands, Guánica Forest Reserve, Southwestern Coastal Lowlands, Sierra Bermeja, Vieques; 5–550 m; *fl* Sep–May, *fr* throughout the year.

Bahamas, Greater Antilles (Cuba, Hispaniola), Lesser Antilles, Trinidad, Mexico, Central America, South America. [L2: 383; AR: 231].

Ditta myricoides Griseb., Pl. Wright. 1:160. 1860 (preprint of Mem. Amer. Acad. Arts, n.s. 8(1):160. 1861).

Tree in Sierra de Luquillo, Sierra de Cayey; 550–850 m; *fl* Feb–May, *fr* Aug–Dec.

Greater Antilles (Cuba, Hispaniola). [L2: 383].

Jaboncillo.

Euphorbia adenoptera Bertol. subsp. **pergamena** (Small) Oudejans, Phytologia 67(1):44. 1989.

Bas.: *Euphorbia pergamena* Small (1898).

Syn.: *Chamaesyce pergamena* (Small) Small (1903). *Chamaesyce adenoptera* (Bertol.) Small subsp. *pergamena* (Small) D.G. Burch (1966). *Chamaesyce monensis* Millsp. (1914).

Herb in Guánica Forest Reserve, Mona; 10–70 m; *fl* & *fr* Feb–Mar, Jul–Oct.

Greater Antilles (Cuba, Hispaniola), southeastern United States (southern Florida). [L2: 365].

Euphorbia articulata Aubl., Hist. Pl. Guiane 1:480. 1775.

Syn.: *Chamaesyce articulata* (Aubl.) Britton (1916).

Chamaesyce vahlii (Willd. ex Klotzsch & Garcke) P. Wilson (1917).

Shrub in Northern Coastal Lowlands (Fajardo and adjacent islands), Northern Limestone Hills, Eastern Coast (Ceiba), Southern Coastal Lowlands, Susúa Forest Reserve, Guánica Forest Reserve, Southwestern Coastal Lowlands (Cabo Rojo), Sierra Bermeja, Vieques, Culebra, Mona, Desecheo; 1–300 m; *fl* & *fr* throughout the year.

Bahamas, Virgin Islands, Lesser Antilles. [L2: 365].

Euphorbia berteroana Balb. ex Spreng., Syst. Veg. 3:794. 1826 "berteriana."

Syn.: *Chamaesyce berteroana* (Balb. ex Spreng.) Millsp. (1909).

Herb in Southern Coastal Limestone, Guánica Forest Reserve, Southwestern Coastal Lowlands; 5–150 m; *fl* & *fr* throughout the year.

Bahamas, Greater Antilles (Cuba, Hispaniola), Mexico, northern South America (Venezuela). [L2: 367].

Euphorbia cowellii (Millsp. ex Britton) Oudejans, Phytologia 67(1):45. 1989.

Bas.: *Chamaesyce cowellii* Millsp. ex Britton (1916).

Herb in Northern Coastal Lowlands (islands adjacent to Fajardo), Southern Coastal Lowlands (Caja de Muertos), Guánica Forest Reserve, Southwestern Coastal Lowlands (Guánica), Mona; 1–70 m; *fl* & *fr* throughout the year.

Greater Antilles (Hispaniola). [L2: 368].

Euphorbia cyathophora Murray, Commentat. Soc. Regiae Sci. Gott. 7:81, *t. 1.* 1786.

Syn.: *Poinsettia cyathophora* (Murray) Klotzsch & Garcke (1859).

Herb in Northern Coastal Lowlands, Northern Limestone Hills, Río Abajo Forest Reserve, Southeastern Foothills (Salinas), Eastern Cordillera (Corozal), Southern Coastal Limestone, Guánica Forest Reserve, Southwestern Coastal Lowlands, Cerro Las Mesas, Western Coast (Aguada), Vieques, Culebra, Mona; 1–300 m; *fl* & *fr* throughout the year.

West Indies, southern and central United States, Mexico, Central America, western and southern South America; exact native range uncertain. [L2: 390].

Maravilla.

Euphorbia heterophylla L., Sp. Pl. 1:453. 1753.

Syn.: *Poinsettia heterophylla* (L.) Klotzsch & Garcke (1859).

Poinsettia geniculata (Ortega) Klotzsch & Garcke (1859).

Herb in Northern Coastal Lowlands, Northern Limestone Hills, Río Abajo Forest Reserve, Sierra de Luquillo, Eastern Coast, Las Piedras Chiquitas,

Southeastern Foothills (Coamo), Eastern Cordillera (Aibonito), Central Cordillera, Southern Coastal Lowlands, Southern Coastal Limestone (Ponce), Susúa Forest Reserve, Southwestern Coastal Lowlands (Cabo Rojo), Sierra Bermeja, Cerro Las Mesas, Western Coast, Vieques, Culebra; 1–650 m; *fl* & *fr* throughout the year.

Native to the Neotropics and -subtropics. [L2: 391].

Acerca amor, Leche vana, Lechecillo.

Euphorbia hirta L., Sp. Pl. 1:454. 1753.
Syn.: *Chamaesyce hirta* (L.) Millsp. (1909).

Herb in Northern Coastal Lowlands, Río Abajo Forest Reserve, Eastern Coast (Cayo Santiago), Caguas Valley, Las Piedras Chiquitas, Southeastern Foothills, Eastern Cordillera, Southern Coastal Lowlands (Caja de Muertos, Ponce), Susúa Forest Reserve, Southwestern Coastal Lowlands (Lajas), Western Coast, Vieques, Culebra, Mona, Desecheo; 1–740 m; *fl* & *fr* throughout the year.

Pantropical. [L2: 369].

Lechecillo.

Euphorbia hypericifolia L., Sp. Pl. 1:454. 1753.
Syn.: *Chamaesyce hypericifolia* (L.) Millsp. (1909).
Chamaesyce glomerifera Millsp. (1913).
Euphorbia glomerifera (Millsp.) L.C. Wheeler (1939).

Herb in Northern Coastal Lowlands, Northern Limestone Hills, Río Abajo Forest Reserve, Eastern Coast (Ceiba), Caguas Valley (Juncos), Eastern Cordillera, Southern Coastal Lowlands, Susúa Forest Reserve, Southwestern Coastal Lowlands, Western Coast (Aguada), Mona; 1–600 m; *fl* & *fr* throughout the year.

West Indies, southern United States (Florida, Texas), Mexico, Central America, northern South America to Brazil, Bolivia. [L2: 368].

Yerba niña.

Euphorbia hyssopifolia L., Syst. Nat. (ed. 10) 2:1048. 1759.
Syn.: *Chamaesyce hyssopifolia* (L.) Small (1905).

Herb in Northern Coastal Lowlands, Northern Limestone Hills, Eastern Coast, Caguas Valley, Sierra de Cayey (San Lorenzo), Las Piedras Chiquitas, Southeastern Foothills, Eastern Cordillera (Cayey, Comerío), Southern Coastal Lowlands, Susúa Forest Reserve, Maricao Forest Reserve, Southwestern Coastal Lowlands, Cerro Las Mesas, Western Coast, Vieques, Culebra, Mona; 1–500 m; *fl* & *fr* throughout the year.

West Indies, southeastern United States (Florida), Central America, South America. [L2: 369].

Notes: Britton & Wilson (1924) misapplied the name *Chamaesyce nutans* (Lag.) Small to this taxon; Liogier (1988) wrongly placed the former in synonymy with it.

Lechera.

Euphorbia lactea Haw., Syn. Pl. Succ. 127. 1812.

Shrub planted and persistent in Southern Coastal Lowlands (Caja de Muertos), Guánica Forest Reserve; 1–20 m; *fl* Aug (one specimen), *fr* (?). [*Breckon 6968* (MAPR); *Oquendo 107* (UPR)].

Probably native to tropical Asia. [L2: 391].

Candelero, Escambrón, Moteado.

Euphorbia lasiocarpa Klotzsch, Nov. Actorum Acad. Caes. Leop.-Carol. Nat. Cur. 19(Suppl. 1):414. 1843.
Syn.: *Chamaesyce lasiocarpa* (Klotzsch) Arthur (1912).

Herb in Northern Coastal Lowlands, Central Cordillera (Villalba, Peñuelas), Southern Coastal Lowlands (Juana Díaz), Southern Coastal Limestone (Guayanilla); 3–500 m; *fl* & *fr* Feb–Jul, Oct.

Bahamas, Greater Antilles (Jamaica, Hispaniola), Lesser Antilles (Antigua, Barbados), Central America, northwestern, west-central South America and Venezuela. [L2: 370].

Euphorbia mesembrianthemifolia Jacq., Enum. Syst. Pl. 22. 1760.
Syn.: *Chamaesyce mesembrianthemifolia* (Jacq.) Dugand (1966).
Chamaesyce buxifolia (Lam.) Small (1903).

Herb in Northern Coastal Lowlands, Eastern Coast (Punta Guayanés), Southern Coastal Lowlands (Caja de Muertos), Guánica Forest Reserve, Southwestern Coastal Lowlands (Cabo Rojo), Vieques, Culebra, Mona; 1–50 m; *fl* & *fr* throughout the year.

West Indies, southeastern United States (Florida), Central America, northern South America (Venezuela, Colombia). [L2: 370].

Euphorbia oerstediana (Klotzsch & Garcke) Boiss. in de Candolle, Prodr. 15(2):59. 1862.
Bas.: *Poinsettia oerstediana* Klotzsch & Garcke (1859).
Syn.: *Dichylium oerstedianum* (Klotzsch & Garcke) Britton (1924).

Suffrutescent herb in Northern Limestone Hills (Dorado, Vega Alta); 1–15 m; *fl* & *fr* Jun–Mar.

Greater Antilles (Jamaica), Lesser Antilles, Trinidad, Central America, northern South America. [L2: 392].

Euphorbia ophthalmica Pers., Syn. Pl. 2(1):13. 1806.
Syn.: *Chamaesyce ophthalmica* (Pers.) D.G. Burch (1966).

Herb in Northern Coastal Lowlands (San Juan), Guánica Forest Reserve, Southwestern Coastal Lowlands (Cabo Rojo), Vieques, Mona; 1–10 m; *fl* & *fr* Oct–Feb, May.

West Indies, southeastern United States, southern Central America (Costa Rica), South America. [L2: 371].

Euphorbia orbifolia (Alain) Oudejans, Phytologia 67(1):47. 1989.
Bas.: *Chamaesyce orbifolia* Alain (1980).

Herb in Mona; 55–60 m; *fl* Mar, Jul–Aug, *fr* Feb–Mar.

Endemic. [L2: 371].

Euphorbia petiolaris Sims, Bot. Mag. 23: *t.* 883. 1805.
Syn.: *Aklema petiolare* (Sims) Millsp. (1915).

Shrub/small tree in Las Piedras Chiquitas, Guánica Forest Reserve, Vieques, Culebra, Mona; 10–550 m; *fl* Jan–Jun, Oct–Nov, *fr* Jan, Apr–Jul, Oct.

Bahamas, Greater Antilles (Jamaica, Hispaniola), Virgin Islands, Lesser Antilles (Guadeloupe, Martinique). [L2: 393].

Indio desnudo, Rascaso.

Euphorbia prostrata Aiton, Hort. Kew. 2:139. 1789.
Syn.: *Chamaesyce prostrata* (Aiton) Small (1903).

Herb in Northern Coastal Lowlands, Northern Limestone Hills (Florida, Arecibo), Southern Coastal Lowlands, Southern Coastal Limestone, Guánica Forest Reserve, Southwestern Coastal Lowlands (Cabo Rojo), Culebra, Mona; 1–70 m; *fl* & *fr* throughout the year.

Probably originating in the Neotropics or -subtropics. [L2: 371].

Lechecillo.

Euphorbia serpens Kunth in Humboldt, Bonpland & Kunth, Nov. Gen. Sp. 2:41 [folio]; 2:52 [quarto]. 1818.
Syn.: *Chamaesyce serpens* (Kunth) Small (1903).

Herb in Northern Coastal Lowlands, Eastern Coast, Southern Coastal Lowlands, Southwestern Coastal Lowlands, Vieques, Culebra, Mona; 1–300 m; *fl* & *fr* throughout the year.

Antilles, United States, Mexico, northern Central America (Belize), South America. [L2: 372].

Sanguinaria.

Euphorbia thymifolia L., Sp. Pl. 1:454. 1753.
Syn.: *Chamaesyce thymifolia* (L.) Millsp. (1916).

Herb in Northern Coastal Lowlands, Eastern Coast, Caguas Valley, Central Cordillera (Adjuntas), Guánica Forest Reserve, Western Coast (Mayagüez), Vieques; 1–740 m; *fl* & *fr* throughout the year.

Pantropical and -subtropical. [L2: 372].

Euphorbia tithymaloides L. subsp. **tithymaloides**, Sp. Pl. 1:453. 1753.
Syn.: *Pedilanthus tithymaloides* (L.) Poit. subsp. *tithymaloides* (1812).

Shrub planted and persistent in Northern Coastal Lowlands (Manatí), Guánica Forest Reserve, Vieques; 20–145 m; *fl* Feb (two specimens), *fr* (?). [*Axelrod 11499* (UPRRP); *Monsegur 668* (MAPR)].

Native to Mexico, Central Ameica, northern South America. [L2: 412].

Ipecacuana, Itamo real.

Euphorbia tithymaloides L. subsp. **angustifolia** (Poit.) V.W. Steinm., Acta Bot. Mex. 65:49. 2003.
Bas.: *Pedilanthus angustifolius* Poit. (1812).
Syn.: *Pedilanthus tithymaloides* (L.) Poit. subsp. *angustifolius* (Poit.) Dressler (1957).

Shrub in Northern Coastal Lowlands (islands adjacent to Fajardo: Cayo Icacos; Manatí), Guánica Forest Reserve, Southwestern Coastal Lowlands, Sierra Bermeja; 25–150 m; *fl* Mar–Apr, Sep, *fr* Mar–May (two specimens).

Greater Antilles (Cuba, Hispaniola), Virgin Islands, Lesser Antilles. [L2: 411].

Bejuco de estrella.

Euphorbia tithymaloides L. subsp. **padifolia** (L.) V.W. Steinm., Acta Bot. Mex. 65:49. 2003.
Bas.: *Euphorbia tithymaloides* L. var. *padifolia* L. (1753).
Syn.: *Pedilanthus padifolius* (L.) Poit. (1812). *Pedilanthus tithymaloides* (L.) Poit. subsp. *padifolius* (L.) Dressler (1957).

Shrub planted and persistent in Northern Coastal Lowlands (Carolina, Manatí), Vieques; 5–10 m; *fl* Feb–Mar, *fr* (?). [*Axelrod 11540* (UPRRP); *Breckon 8229* (MAPR)].

Native to Virgin Islands (St. Croix), Lesser Antilles. [L2: 411].

Euphorbia tithymaloides L. subsp. **parasitica** (Klotzsch & Garcke) V.W. Steinm, Acta Bot. Mex. 65:50. 2003.
Bas.: *Pedilanthus parasiticus* Klotzsch & Garcke (1860).
Syn.: *Pedilanthus tithymaloides* (L.) Poit. subsp.

parasiticus (Klotzsch & Garcke) Dressler (1957).
Pedilanthus latifolius Millsp. & Britton (1915).

Shrub planted and persistent in Northern Coastal Lowlands (Vega Alta), Southwestern Coastal Lowlands (Lajas); 25 m; *fl* Mar, *fr* (?). [*Axelrod 13983* (UPRRP)].

Native to Mexico, northern Central America (Belize). [L2: 411].

Euphorbia turpinii Boiss., Cent. Euphorb. 12. 1860.
Syn.: *Chamaesyce turpinii* (Boiss.) Millsp. (1916).
Chamaesyce portoricensis (Urb.) Millsp. (1914).

Herb in Southern Coastal Limestone, Susúa Forest Reserve, Guánica Forest Reserve, Southwestern Coastal Lowlands, Vieques, Mona; 5–150 m; *fl* & *fr* throughout the year.

Greater Antilles (Cuba, Hispaniola), Virgin Islands (Anegada). [L2: 372].

Gymnanthes lucida Sw., Prodr. 96. 1788.
Syn.: *Ateramnus lucidus* (Sw.) Rothm. (1944).

Tree in Northern Coastal Lowlands (islands adjacent to Fajardo), Northern Limestone Hills, Las Piedras Chiquitas, Southern Coastal Limestone, Susúa Forest Reserve, Guánica Forest Reserve, Southwestern Coastal Lowlands (Cabo Rojo), Sierra Bermeja, Vieques, Culebra, Mona, Desecheo; 10–500 m; *fl* Dec–Sep, *fr* Mar–Dec.

West Indies, southeastern United States (Florida), Central America. [L2: 396].

Ramón, Tabaco, Yaití.

Hippomane mancinella L., Sp. Pl. 2:1191. 1753.

Tree in Northern Coastal Lowlands (Fajardo and adjacent islands), Eastern Coast (Cayo Santiago), Southern Coastal Lowlands (Caja de Muertos), Guánica Forest Reserve, Southwestern Coastal Lowlands (Lajas), Vieques, Culebra, Mona, Desecheo; 1–20 m; *fl* May–Oct, *fr* May–Jan.

West Indies, southeastern United States (Florida), Mexico, Central America, northern South America (Venezuela, Colombia). [L2: 398].

Manzanillo.

Hura crepitans L., Sp. Pl. 2:1008. 1753.

Tree in Northern Coastal Lowlands, Northern Limestone Hills, Southern Coastal Lowlands (Coamo), Vieques; 10–120 m; *fl* Apr–Dec, *fr* Jul–Mar.

West Indies, Mexico, Central America, northern South America to Brazil, Bolivia. [L2: 399].

Havilla, Javilla, Molinillo.

Jatropha curcas L., Sp. Pl. 2:1006 "Gurcas;" 2: errata. 1753.
Syn.: *Curcas curcas* (L.) Britton & Millsp. (1920).

Shrub in Northern Coastal Lowlands (Carolina, Barceloneta), Western Cordillera, Guánica Forest Reserve, Southwestern Coastal Lowlands (Cabo Rojo), Sierra Bermeja, Western Coast (Rincón), Vieques; 5–300 m; *fl* Feb–Oct, *fr* Dec–Jul.

Native to Mexico, Central America, central and southern South America. [L2: 403].

Piñón, Piñón purgante, Tártago.

Jatropha gossypiifolia L., Sp. Pl. 2:1006. 1753 "gossypifolia."
Syn.: *Adenoropium gossypiifolium* (L.) Pohl (1826).

Shrub in Northern Coastal Lowlands, Eastern Coast, Southeastern Foothills, Eastern Cordillera (Comerío), Southern Coastal Lowlands, Southern Coastal Limestone (Yauco), Guánica Forest Reserve, Southwestern Coastal Lowlands, Sierra Bermeja, Western Coast (Mayagüez), Vieques, Culebra, Mona, Desecheo; 1–600 m; *fl* & *fr* throughout the year.

West Indies, Mexico, Central America, South America (scattered). [L2: 403].

Higuereta cimarrona, Tautuba, Tuatúa.

Jatropha hernandiifolia Vent., Jard. Malmaison 1: *sub t.* 52. 1803 "hernandiaefolia."
Syn.: *Curcas hernandiifolius* (Vent.) Britton (1924).

Shrub in Northern Limestone Hills, Eastern Coast (Cuchilla de Panduras), Cerro de Las Cuevas, Susúa Forest Reserve, Guánica Forest Reserve, Southwestern Coastal Lowlands (Cabo Rojo); 1–450 m; *fl* & *fr* throughout the year.

Greater Antilles (Jamaica, Hispaniola). [L2: 404].

Papayo, Tabaiba.

Jatropha multifida L., Sp. Pl. 2:1006. 1753.
Syn.: *Adenoropium multifidum* (L.) Pohl (1826).

Shrub/small tree planted and persistent in Eastern Cordillera (Aibonito), Guánica Forest Reserve, Mona; 25–400 m; *fl* Jan–Mar, Jun–Sep, *fr* Jul, Oct, Jan–Feb.

Neotropical. [L2: 405].

Don Tomás, Maná, Tártago, Tártago emético.

Manihot esculenta Crantz, Inst. Rei Herb. 1:167. 1766.
Syn.: *Manihot manihot* (L.) Cockerell (1892).

Herb planted and persistent in Eastern Coast

(Naguabo), Central Cordillera, Western Coast (Mayagüez); 1–950 m; *fl* May–Jul, *fr* Feb, Jul (two specimens).

Native from northern South America to Brazil, Bolivia. [L2: 407].

Yuca.

Microstachys corniculata (Vahl) Griseb., Fl. Brit. W.I. 49. 1859.
Bas.: *Tragia corniculata* Vahl (1798).
Syn.: *Sebastiania corniculata* (Vahl) Müll. Arg. (1866).

Herb in Northern Coastal Lowlands (Dorado to Manatí), Southwestern Coastal Lowlands (Cabo Rojo); 1–10 m; *fl* & *fr* throughout the year.

Greater Antilles (Cuba, Hispaniola), Lesser Antilles (Guadeloupe, Grenada), Central America, South America. [L2: 427].

Pera bumeliifolia Griseb., Nachr. Königl. Ges. Wiss. Georg-Augusts-Univ. 1:180. 1865.

Tree planted and persistent in Cambalache Forest Reserve; 30 m; *fl* (?), *fr* Aug (one specimen). [*Liogier 29177* (UPR)].

Native to Bahamas, Greater Antilles (Cuba, Hispaniola). [L2: 412].

Ricinus communis L., Sp. Pl. 2:1007. 1753.

Shrub in Northern Coastal Lowlands, Northern Limestone Hills, Southeastern Foothills (Guayama), Toro Negro Forest Reserve, Southern Coastal Lowlands (Ponce), Southern Coastal Limestone, Susúa Forest Reserve, Southwestern Coastal Lowlands, Western Coast, Vieques; 1–1100 m; *fl* & *fr* throughout the year.

Probably native to Africa. [L2: 422].

Higuereta, Higüerito, Ricino.

Sapium jamaicense Sw., Adnot. Bot. 62. 1829.

Tree in Northern Limestone Hills (Lares), Sierra de Luquillo, Maricao Forest Reserve; 200–500 m; *fl* Nov–Dec (two specimens), *fr* (?).

Greater Antilles, Mexico, Central America. [L2: 425].

Tabaiba.

Sapium laurocerasus Desf., Tabl. École Bot. (ed. 3) 411. 1829.

Tree in Northern Limestone Hills (Florida, Arecibo), Río Abajo Forest Reserve, Sierra de Luquillo, Eastern Coast (Cuchilla de Panduras), Sierra de Cayey, Las Tetas de Cayey, Eastern Cordillera (La Torrecilla), Central Cordillera, Maricao Forest Reserve; 200–1300 m; *fl* & *fr* throughout the year.

Endemic. [L2: 425].

Hinchahuevos, Jinchahuevos, Lechecillo, Manzanillo, Tabaiba.

Tragia volubilis L., Sp. Pl. 2:980. 1753.

Vine in Northern Coastal Lowlands, Northern Limestone Hills, Río Abajo Forest Reserve, Las Tetas de Cayey, Las Piedras Chiquitas, Southeastern Foothills, Eastern Cordillera (Cayey), Cerro de Las Cuevas, Southern Coastal Lowlands (Ponce), Susúa Forest Reserve, Maricao Forest Reserve, Guánica Forest Reserve, Sierra Bermeja, Vieques; 1–825 m; *fl* throughout the year, *fr* Feb–Nov.

Antilles, Mexico, Central America, South America. [L2: 432; AR: 231].

Pringamosa.

TAXA CITED BY PREVIOUS AUTHORS, BUT NOT INCLUDED

DOUBTFUL OCCURRENCE

Croton ovalifolius Vahl.

Shrub cited by Liogier & Martorell (2000) for Puerto Rico, according to Pedro Acevedo-Rodríguez (1996); but they had, in fact, cited it for the Greater Antilles in general, not specifically for Puerto Rico. Thus far it has not been collected here.

CITED DUE TO MISIDENTIFICATION

Acalypha berteroana Müll. Arg.

Suffrutescent herb cited by Britton & Wilson (1924) as occurring in maritime thickets, collected by Bertero on a bank above Guayama at 250 m; subsequently cited by Liogier & Martorell (1982) and Liogier (1988). [L2: 346].

Notes: Liogier & Martorell (1982) and Liogier (1988) speculated that this is perhaps a short-leaved form of *Acalypha portoricensis* Müll. Arg. Having seen specimens on which the citation was based [*Britton et al. 6428* (US, GH)], I agree with them.

Euphorbia blodgettii Engelm. ex Hitchc.
Syn.: *Chamaesyce blodgettii* (Engelm. ex Hitchc.) Small.

Herb cited by Liogier (1988) as occurring in southwestern dry regions, on coastal limestone, and similarly by Liogier & Martorell (2000), except for the replacement of 'southwestern' by 'southeastern.' This citation was based upon the misidentification of specimens of *Euphorbia cowellii*. [L2: 367].

ONLY UNDER CULTIVATION

Acalypha hispida Burm. f.

Shrub cited by Liogier & Martorell (1982, 2000) and Liogier (1988) as planted for ornament. [L2: 347].

Acalypha wilkesiana Müll. Arg. Syn.: *Acalypha amentacea* Roxb. subsp. *wilkesiana* (Müll. Arg.) Fosberg.

Shrub cited by Liogier & Martorell (1982, 2000) and Liogier (1988) as planted for ornament and persistent. [L2: 345].

Codiaeum variegatum (L.) A. Juss.

Shrub cited by Liogier & Martorell (1982, 2000) and Liogier (1988) as grown for ornament with a tendency to become naturalized. [L2: 375].

Euphorbia cotinifolia L.

Shrub/small tree cited by Liogier & Martorell (1982, 2000) and Liogier (1988) as planted for its ornamental foliage and permanent after cultivation. [L2: 390].

Euphorbia leucocephala Lotsy.

Shrub cited by Liogier & Martorell (1982, 2000) and Liogier (1988) as planted and persistent as an ornamental plant. [L2: 391].

Euphorbia milii Des Moul.

Shrub cited by Liogier & Martorell (1982, 2000) and Liogier (1988) as grown for ornament and persistent. [L2: 392].

Euphorbia neriifolia L.

Shrub cited by Liogier & Martorell (1982, 2000) and Liogier (1988) as occasionally planted for ornament and persistent. [L2: 392].

Euphorbia pulcherrima Willd. ex Klotzsch. Syn.: *Poinsettia pulcherrima* (Willd. ex Klotzsch.) Graham.

Shrub cited by Liogier & Martorell (1982, 2000) and Liogier (1988) as widely planted and persistent. [L2: 393].

Euphorbia tirucalli L.

Shrub cited by Liogier & Martorell (1982, 2000) and Liogier (1988) as planted for ornament. [L2: 394].

Jatropha integerrima Jacq. Syn.: *Adenoropium hastatum* (Jacq.) Britton & P. Wilson.

Shrub cited by Britton & Wilson (1924) as grown for ornament in gardens; subsequently cited by Liogier & Martorell (1982, 2000) and Liogier (1988). [L2: 405].

Jatropha podagrica Hook.

Shrub cited by Liogier & Martorell (1982, 2000) and Liogier (1988) as occasionally planted as an ornament and persistent. [L2: 406].

Vernicia fordii (Hemsl.) Airy Shaw. Syn.: *Aleurites fordii* Hemsl.

Tree cited by Liogier & Martorell (1982, 2000) and Liogier (1988) as planted and probably escaped. [L2: 355].

FABACEAE-CAESALPINIOIDEAE (LEGUMINOSAE-CAESALPINIOIDEAE)

Bauhinia monandra Kurz, J. Asiat. Soc. Bengal, Pt. 2, Nat. Hist. 42(2):73. 1873.
Syn.: *Caspareopsis monandra* (Kurz) Britton & Rose (1930).

Tree planted and persistent in Northern Coastal Lowlands, Northern Limestone Hills, Caguas Valley (Caguas), Southeastern Foothills, Southern Coastal Limestone, Southwestern Coastal Lowlands (San Germán), Sierra Bermeja, Vieques; 1–250 m; *fl* Apr–Oct, *fr* Apr–Nov.

Native to Madagascar. [L2: 55].

Alas de ángel, Baujinia, Flamboyán blanco, Mariposa, Seplina, Varital variable.

Bauhinia pauletia Pers., Syn. Pl. 1:455. 1805.

Shrub/small tree in Southwestern Coastal Lowlands (Lajas); 5 m; *fl* (?), *fr* Feb (one specimen). [*Liogier 32679* (NY)].

Mexico, Central America, northern South America (Venezuela, Colombia). [L2: 57].

Araña gato.

Bauhinia purpurea L., Sp. Pl. 1:375. 1753.
Syn.: *Phanera purpurea* (L.) Benth. (1852).

Tree planted and persistent in Northern Coastal Lowlands (San Juan, Cataño), Eastern Cordillera, Toro Negro Forest Reserve; 1–1000 m; *fl* Oct–Jan, *fr* Dec–Feb.

Native to Indian Subcontinent, southeastern Asia. [L2: 57].

Palo de orquídeas.

Bauhinia variegata L. var. **variegata**, Sp. Pl. 1:375. 1753.
Syn.: *Phanera variegata* (L.) Benth. (1852).

Tree planted and persistent in Northern Limestone Hills (Manatí, Isabela), Eastern Coast (Naguabo), Caguas Valley, Western Coast (Mayagüez); 1–900 m; *fl* Oct–Jun, *fr* Jan–Apr.

Native to China. [L2: 58].

Palo de orquídeas.

Bauhinia variegata L. var. **candida** Voigt, Hort. Suburb. Calcutt. 253. 1845.

Tree planted and persistent in Northern Coastal Lowlands (Laguna Tortuguero, Manatí), Northern Limestone Hills (Florida, Isabela), Eastern Cordillera (Barranquitas), Toro Negro Forest Reserve, Vieques; 10–1000 m; *fl* Jan–Apr, *fr* Feb–Apr (two specimens).

Native to Asia.

Caesalpinia decapetala (Roth) Alston in Trimen, Handb. Fl. Ceylon 6(Suppl.):89. 1931.

Bas.: *Reichardia decapetala* Roth (1821).
Syn.: *Biancaea sepiaria* (Roxb.) Todaro (1876).

Shrub in Northern Limestone Hills (Florida, San Sebastián), Eastern Cordillera (Cañón San Cristóbal), Central Cordillera, Western Cordillera (Maricao); 200–1300 m; *fl* Dec–Mar, *fr* Mar–Apr.

Native to tropical Asia, China, Japan, Korea. [L2: 62; AR: 235].

Zarza de cercas.

Caesalpinia major (Medik.) Dandy & Exell, J. Bot. 76:180. 1938.
Bas.: *Bonduc majus* Medik. (1786).
Syn.: *Caesalpinia globulorum* Bakh. f. & P. Royen (1963).

Shrub in Northern Limestone Hills, Río Abajo Forest Reserve; 10–350 m; *fl* (?), *fr* (?).

Pantropical. [L2: 63; AR: 235].

Mato amarillo.

Caesalpinia monensis Britton in Britton & Wilson, Sci. Surv. Porto Rico & Virgin Islands 5(3):377. 1924.

Shrub in Mona; 40–75 m; *fl* May–Nov, *fr* Apr–Jul, Nov.

Endemic. [L2: 64].

Caesalpinia pulcherrima (L.) Sw., Observ. Bot. 166. 1791.
Bas.: *Poinciana pulcherrima* L. (1753).

Shrub planted and persistent in Northern Coastal Lowlands, Southeastern Foothills (Guayama), Southern Coastal Lowlands (Ponce), Guánica Forest Reserve, Southwestern Coastal Lowlands (Cabo Rojo), Sierra Bermeja; 1–400 m; *fl* Mar–Nov, *fr* Mar–Nov.

Cultivated and naturalized throughout the tropics, of unknown origin. [L2: 64].

Clavellina.

Cassia fistula L., Sp. Pl. 1:377. 1753.

Tree planted and persistent in Northern Coastal Lowlands (San Juan), Southern Coastal Lowlands (Juana Díaz); 5–25 m; *fl* Mar–Apr, Sep, *fr* May, Sep–Nov.

Possibly native to tropical Asia. [L2: 65].

Cañafistula.

Cassia grandis L.f., Suppl. Pl. 230. 1782.

Tree in Guánica Forest Reserve, Western Coast (Mayagüez); 5–150 m; *fl* Apr (one specimen), *fr* (?). [*Quigley 482* (SJ); *Liogier 35568* (UPR)].

Greater Antilles, Virgin Islands (St. Croix), Lesser Antilles (Montserrat), Trinidad, Mexico, northern South America to Brazil, Peru. [L2: 67].

Cañafistula cimarrona, Gigantón.

Cassia javanica L., Sp. Pl. 1:379. 1753.

Tree planted and persistent in Northern Coastal Lowlands, Western Coast (Mayagüez); 5–80 m; *fl* May–Nov, *fr* Oct–Nov (two specimens).

Native to tropical Asia, China. [L2: 67].

Acacia rosada, Casia rosada.

Chamaecrista diphylla (L.) Greene, Pittonia 4:28. 1899.
Bas.: *Cassia diphylla* L. (1753).

Suffrutescent herb in Northern Coastal Lowlands, Eastern Coast, Sierra de Cayey, Eastern Cordillera (Cerro La Peña); 1–800 m; *fl* & *fr* throughout the year.

Greater Antilles (Cuba, Hispaniola), Lesser Antilles, Trinidad and Tobago, Mexico, Central America, northern South America to Brazil. [L2: 68].

Hediondilla.

Chamaecrista glandulosa (L.) Greene var. **mirabilis** (Pollard) H.S. Irwin & Barneby, Mem. New York Bot. Gard. 35(2):787. 1982.
Bas.: *Chamaecrista mirabilis* Pollard (1902).
Syn.: *Cassia mirabilis* (Pollard) Urb. (1905).

Suffrutescent herb in white sand in Northern Coastal Lowlands (Vega Baja to Manatí); 1–5 m; *fl* Sep–Jun, *fr* throughout the year.

Virgin Islands. [L2: 69].

Notes: C. Raunkiaer collected a specimen of this taxon, '*ex Ind. occid.* 1905–06,' probably somewhere in what is now the U.S. Virgin Islands [*Raunkiaer a.n. 32718* (UPRRP)]; it is the only specimen that I have seen from outside the mainland of Puerto Rico.

This taxon is on the U.S. Fish & Wildlife Service list of threatened and endangered plants.

Chamaecrista glandulosa (L.) Greene var. **swartzii** (Wikstr.) H.S. Irwin & Barneby, Mem. New York Bot. Gard. 35(2):784. 1982.
Bas.: *Cassia swartzii* Wikstr. (1826).
Syn.: *Chamaecrista swartzii* (Wikstr.) Britton (1917). *Cassia glandulosa* Wikstr. var. *swartzii* (Wikstr.) J.F. Macbr. (1919).

Suffrutescent herb in Northern Coastal Lowlands (Fajardo and adjacent islands), Northern Limestone Hills, Río Abajo Forest Reserve, Sierra de Luquillo, Eastern Coast, Sierra de Cayey, Las Tetas de Cayey, Southeastern Foothills (Coamo), Eastern and Central Cordillera, Southern Coastal Lowlands (Caja de Muertos), Southwestern Coastal Lowlands (Cabo Rojo), Sierra Bermeja, Vieques, Culebra; 1–1050 m; *fl* & *fr* throughout the year.

West Indies. [L2: 70].

Tamarindillo.

Chamaecrista lineata (Sw.) Greene var. **brachyloba** (Griseb.) H.S. Irwin & Barneby, Mem. New York Bot. Gard. 35(2):757. 1982.
Bas.: *Cassia lineata* Sw. var. *brachyloba* Griseb. (1860).
Syn.: *Chamaecrista grammica* (Spreng.) Pollard (1900).
Chamaecrista granulata (Urb.) Britton (1915).

Shrub in Southern Coastal Lowlands (Caja de Muertos), Guánica Forest Reserve, Southwestern Coastal Lowlands, Sierra Bermeja, Mona; 1–80 m; *fl* throughout the year, *fr* Sep–May.

Greater Antilles (Cuba, Hispaniola). [L2: 70].

Escobilla.

Chamaecrista nictitans (L.) Moench subsp. **nictitans** var. **diffusa** (DC.) H.S. Irwin & Barneby, Mem. New York Bot. Gard. 35(2):833. 1982.
Bas.: *Cassia diffusa* DC. (1824).
Syn.: *Cassia exunguis* Urb. (1918). *Chamaecrista exunguis* (Urb.) Britton (1930).

Suffrutescent herb in Northern Coastal Lowlands, Eastern Coast (Humacao), Sierra de Cayey, Las Piedras Chiquitas, Southeastern Foothills (Cayey), Southern Coastal Lowlands (Caja de Muertos), Southern Coastal Limestone, Susúa Forest Reserve, Maricao Forest Reserve, Southwestern Coastal Lowlands (Cabo Rojo), Vieques; 1–750 m; *fl* & *fr* Sep–Feb.

West Indies, northern South America (Venezuela, Colombia). [L2: 71].

Notes: Britton & Wilson (1924) misapplied the name *Chamaecrista chamaecrista* (L.) Britton, a synonym of *Chamaecrista fasciculata* (Michx.) Greene, to this taxon; Liogier & Martorell (1982) misapplied *Cassia chamaecrista* L. to it and Liogier (1988) wrongly placed *Chamaecrista chamaecrista* in synonymy with it.

Chamaecrista nictitans (L.) Moench subsp. **patellaria** (Collad.) H.S. Irwin & Barneby var. **glabrata** (Vogel) H.S. Irwin & Barneby, Mem. New York Bot. Gard. 35(2):822. 1982.
Bas.: *Cassia patellaria* DC. ex Collad. var. *glabrata* Vogel (1837).
Syn.: *Cassia aeschinomene* DC. ex Collad. (1816). *Chamaecrista aeschinomene* (DC. ex Collad.) Greene (1899).

Suffrutescent herb in Northern Coastal Lowlands, Northern Limestone Hills, Sierra de Luquillo, Eastern Coast, Caguas Valley, Sierra de Cayey, Las Piedras Chiquitas, Eastern and Central Cordillera, Susúa Forest Reserve, Maricao Forest Reserve, Sierra Bermeja, Western Coast, Vieques, Desecheo; 1–1200 m; *fl* & *fr* throughout the year.

Greater Antilles, Lesser Antilles (Guadeloupe, Martinique), Mexico, South America (lower Orinoco and Amazon valleys). [L2: 71].

Notes: Britton & Wilson (1930) misapplied the name *Chamaecrista patellaria* (DC. ex Collad.) Greene, and Liogier & Martorell (1982) misapplied *Cassia ramosa* Vogel, to this taxon.

Moriviví bobo, Tamarindillo.

Chamaecrista portoricensis (Urb.) O.F. Cook & G.N. Collins var. **portoricensis**, Contr. U.S. Natl. Herb. 8:113. 1903.
Bas.: *Cassia portoricensis* Urb. (1899).

Suffrutescent herb/shrub in Southern Coastal Limestone, Susúa Forest Reserve, Western Cordillera (San Germán), Southwestern Coastal Lowlands, Cerro Las Mesas, Vieques; 10–300 m; *fl* throughout the year, *fr* Dec–Sep.

Greater Antilles (Hispaniola), Virgin Islands. [L2: 72].

Retama.

Chamaecrista rotundifolia (Pers.) Greene var. **rotundifolia,** Pittonia 4:31. 1899.
Bas.: *Cassia rotundifolia* Pers. (1805).

Suffrutescent herb in Northern Coastal Lowlands, Vieques; 1–25 m; *fl* & *fr* Sep–Jun.

Greater Antilles (Cuba, Jamaica), Mexico, Central America, South America. [L2: 72].

Cynometra portoricensis Krug & Urb. in Urban, Symb. Antill. 1(2):312. 1899.

Tree in Northern Limestone Hills, Cerro de Las Cuevas, Southwestern Coastal Lowlands; 15–625 m; *fl* Feb, Jun, Sep–Oct, *fr* Mar, Jun, Oct–Dec.

Greater Antilles (Hispaniola). [L2: 74].

Oreganillo.

Delonix regia (Bojer) Raf., Fl. Tellur. 2:92. 1837.
Bas.: *Poinciana regia* Bojer (1829).

Tree planted and persistent in Northern Coastal Lowlands, Eastern Coast (Naguabo), Southeastern Foothills, Southern Coastal Lowlands, Susúa Forest Reserve, Guánica Forest Reserve, Southwestern Coastal Lowlands, Vieques, Mona; 1–300 m; *fl* Apr–Dec, *fr* Jul–Feb.

Native to Madagascar. [L2: 75].

Flamboyán, Flamboyán colorado, Flamboyán rojo.

Guilandina bonduc L., Sp. Pl. 1:381. 1753.
Syn: *Caesalpinia bonduc* (L.) Roxb. (1832).

Shrub in Northern Coastal Lowlands, Southern

Coastal Lowlands, Guánica Forest Reserve, Southwestern Coastal Lowlands (Cabo Rojo), Vieques, Culebra, Mona; 1–25 m; *fl* Aug–Mar, *fr* Aug, Nov–Mar.

Pantropical. [L2: 59; AR: 234].

Notes: Britton & Wilson (1924) misapplied the name *Guilandina crista* (L.) Small to this taxon; they also misapplied the name of this taxon to *Caesalpinia major*.

Haba de San Antonio, Mato azul, Mato de playa.

Guilandina ciliata Bergius ex Wikstr., Kongl. Vetensk. Acad. Handl. 1825(2):431. 1825.
Syn.: *Caesalpinia ciliata* (Bergius ex Wikstr.) Urb. (1900).
Guilandina melanosperma Eggers (1879). *Caesalpinia melanosperma* (Eggers) Urb. (1900).
Caesalpinia divergens Urb. (1900). *Guilandina divergens* (Urb.) Britton (1915).
Guilandina portoricensis Britton & P. Wilson (1924). *Caesalpinia portoricensis* (Britton & P. Wilson) Alain (1963).

Shrub in Northern Coastal Lowlands (islands adjacent to Fajardo: Cayo Ratones), Vieques, Culebra, Mona; 1–5 m; *fl* May–Jun, Sep–Oct, *fr* May–Jul, Oct–Nov.

Bahamas, Greater Antilles (Cuba, Hispaniola), Virgin Islands, Lesser Antilles. [L2: 62, 63; AR: 234, 237].

Notes: I have placed *Guilandina portoricensis* in synonymy with *Guilandina ciliata*, since these taxa differ only in seed color and that character, according to Howard (1988), is highly variable.

Matitos amarillos, Mato de playa, Mato negro.

Guilandina culebrae Britton & P. Wilson in Britton & Rose, N. Amer. Fl. 23(5):339. 1930.
Syn.: *Caesalpinia culebrae* (Britton & P. Wilson) Alain (1963).

Shrub in Northern Coastal Lowlands (islands adjacent to Fajardo: Cayo Diablo); 1–5 m; *fl* (?), *fr* May (one specimen). [*Woodbury I-28, I-39 a & b* (UPR)].

Endemic. [L2: 60; AR: 235].

Haematoxylum campechianum L., Sp. Pl. 1:384. 1753.

Tree in Southern Coastal Limestone (Peñuelas), Guánica Forest Reserve, Southwestern Coastal Lowlands (Cabo Rojo), Western Coast (Añasco), Mona; 5–150 m; *fl* Oct–Mar, *fr* Oct–Jan, May.

Native to southern Mexico, northern Central America (Guatemala, Belize). [L2: 77].

Campeche, Palo de campeche.

Hymenaea courbaril L., Sp. Pl. 2:1192. 1753.

Tree in Northern Coastal Lowlands, Eastern Coast (Yabucoa), Las Piedras Chiquitas, Southeastern Foothills, Susúa Forest Reserve, Sierra Bermeja, Vieques, Culebra; 5–450 m; *fl* May–Oct, *fr* Aug–Feb.

Antilles, Mexico, Central America, northern South America to Brazil, Bolivia. [L2: 79].

Algarroba, Algarrobo.

Libidibia coriaria (Jacq.) Schltdl., Linnaea 5:193. 1830.
Bas.: *Poinciana coriaria* Jacq. (1830).
Syn.: *Caesalpinia coriaria* (Jacq.) Willd. (1799).

Tree in Southwestern Coastal Lowlands (Cabo Rojo); 1 m; *fl* (?), *fr* Feb (one specimen). [*Liogier 34935* (UPR)].

West Indies, Mexico, Central America, northern South America (Venezuela, Colombia). [L2: 60].

Dividivi, Guatapaná.

Parkinsonia aculeata L., Sp. Pl. 1:375. 1753.

Tree planted and persistent in Southeastern Foothills, Southern Coastal Lowlands, Southern Coastal Limestone, Western Cordillera (Maricao), Guánica Forest Reserve, Southwestern Coastal Lowlands, Vieques; 1–600 m; *fl* Mar–Jun, Sep–Dec, *fr* Mar–Nov.

Native to south-central United States (Texas, Arizona), Mexico, southern and west-central South America. [L2: 80].

Flor de mayo, Flor de rayo, Palo de rayo.

Peltophorum pterocarpum (DC.) Backer ex K. Heyne, Nutt. Pl. Ned.-Ind. (ed. 2) 2:755. 1927.
Bas.: *Inga pterocarpa* DC. (1825).
Syn.: *Peltophorum inerme* (Roxb.) Náves ex Fern.-Vill. (1880).

Tree planted and persistent in Northern Coastal Lowlands, Eastern Coast (Ceiba, Humacao), Vieques; 1–25 m; *fl* Mar–May, Aug–Sep, *fr* Mar–Nov.

Native to southeastern Asia, Australia. [L2: 82].

Flamboyán amarillo.

Senna alata (L.) Roxb., Fl. Ind. (ed. 1832) 2:349. 1832.
Bas.: *Cassia alata* L. (1753).
Syn.: *Herpetica alata* (L.) Raf. (1838).

Shrub in Northern Coastal Lowlands, Northern Limestone Hills (Florida), Eastern Coast, Caguas Valley, Southwestern Coastal Lowlands (Lajas), Western Coast (Rincón), Vieques; 1–300 m; *fl* Nov–Mar, *fr* Dec–Mar.

Native from northern South America to Brazil. [L2: 86].

Talantala, Talantalán, Talantro.

Senna atomaria (L.) H.S. Irwin & Barneby, Mem. New York Bot. Gard. 35(2):588. 1982.

Bas.: *Cassia atomaria* L. (1767).

Shrub in Southeastern Foothills; 50–300 m; *fl* Apr–Oct, *fr* Jun–Jan.

West Indies, Mexico, Central America, northern South America (Venezuela, Colombia). [L2: 87].

Notes: Britton & Wilson (1924) misapplied the name *Isandrina emarginata* (L.) Britton & Rose to this taxon.

Flor de San José, Velamuerto.

Senna bicapsularis (L.) Roxb. var. **bicapsularis**, Fl. Ind. (ed. 1832) 2:342. 1832.
Bas.: *Cassia bicapsularis* L. (1753).
Syn.: *Adipera bicapsularis* (L.) Britton & Rose (1924).
Cassia emarginata L. (1753).

Shrub in Northern Coastal Lowlands, Eastern Coast, Sierra de Cayey (San Lorenzo), Southeastern Foothills, Eastern Cordillera (Corozal), Vieques; 1–400 m; *fl* & *fr* Oct–Jun.

Greater Antilles (Jamaica), Lesser Antilles, Trinidad and Tobago, southern Central America (Panama), northern South America (Venezuela, Colombia). [L2: 88; AR: 237].

Notes: Little, Woodbury, & Wadsworth (1974) misapplied the name *Cassia emarginata* to *Senna atomaria*.

Hoja de sen, Sen del país.

Senna hirsuta (L.) H.S. Irwin & Barneby var. **hirsuta**, Phytologia 44(7):499. 1979.
Bas.: *Cassia hirsuta* L. (1753).
Syn.: *Ditremexa hirsuta* (L.) Britton & Rose (1924).

Shrub in Northern Coastal Lowlands (Guaynabo), Northern Limestone Hills, Río Abajo Forest Reserve, Caguas Valley (Caguas), Sierra de Cayey, Southeastern Foothills, Eastern and Central Cordillera, Western Cordillera (Maricao), Southwestern Coastal Lowlands (Cabo Rojo), Western Coast; 1–925 m; *fl* Oct–Apr, *fr* Nov–May.

Lesser Antilles (Guadeloupe, Martinique, St. Vincent), Trinidad, southern Central America (Panama), northern South America. [L2: 88].

Hedionda cimarrona.

Senna nitida (Rich.) H.S. Irwin & Barneby, Mem. New York Bot. Gard. 35(1):159. 1982.
Bas.: *Cassia nitida* Rich. (1792).
Syn.: *Chamaefistula antillana* Britton & Rose (1924).

Shrub in Northern Coastal Lowlands, Northern Limestone Hills (Hatillo, Isabela), Sierra de Luquillo, Caguas Valley (Caguas), Sierra de Cayey, Las Tetas de Cayey, Eastern Cordillera (Cañón San Cristóbal), Central Cordillera (Adjuntas), Western Cordillera, Cerro Las Mesas; 10–850 m; *fl* Jul–Feb, *fr* Nov–Mar.

Virgin Islands (St. Thomas, Tortola), Lesser Antilles (St. Kitts, Nevis). [L2: 89; AR: 239].

Hediondilla.

Senna obtusifolia (L.) H.S. Irwin & Barneby, Mem. New York Bot. Gard. 35(1):252. 1982.
Bas.: *Cassia obtusifolia* L. (1753).

Suffrutescent herb in Northern Coastal Lowlands, Northern Limestone Hills (Camuy, San Sebastián), Río Abajo Forest Reserve, Eastern Coast, Caguas Valley, Las Piedras Chiquitas, Southeastern Foothills (Salinas), Eastern Cordillera (Barranquitas), Southern Coastal Lowlands (Guayama including Caja de Muertos), Southwestern Coastal Lowlands (Cabo Rojo), Sierra Bermeja, Western Coast (Rincón), Vieques, Culebra; 1–550 m; *fl* Mar–Dec, *fr* Apr–Dec.

Pantropical, though probably native to the New World. [L2: 90].

Notes: Britton & Wilson (1924) misapplied the name *Emelista tora* (L.) Britton & Rose to this taxon.

Dormidera, Dormilona, Hediondilla.

Senna occidentalis (L.) Link, Handbuch 2:140. 1829.
Bas.: *Cassia occidentalis* L. (1753).
Syn.: *Ditremexa occidentalis* (L.) Britton & Rose (1924).

Suffrutescent herb in Northern Coastal Lowlands, Northern Limestone Hills, Sierra de Luquillo, Eastern Coast, Caguas Valley (Caguas), Las Piedras Chiquitas, Eastern Cordillera (Aibonito, Cañón San Cristóbal), Central Cordillera (Utuado), Southern Coastal Lowlands (Caja de Muertos, Ponce), Southwestern Coastal Lowlands, Sierra Bermeja, Western Coast, Vieques, Culebra, Desecheo; 1–800 m; *fl* & *fr* throughout the year.

Tropical and warm temperate regions. [L2: 90].

Hedionda.

Senna pendula (Willd.) H.S. Irwin & Barneby var. **stahlii** (Urb.) H.S. Irwin & Barneby, Mem. New York Bot. Gard. 35(1):392. 1982.
Bas.: *Cassia stahlii* Urb. (1899).
Syn.: *Adipera stahlii* (Urb.) Britton & Rose (1924).

Shrub/small tree in Sierra de Cayey, Las Tetas de Cayey, Toro Negro Forest Reserve; 200–1000 m; *fl* Sep–Dec, *fr* Feb (one specimen).

Endemic. [L2: 91].

Notes: Liogier & Martorell (1982) used *Cassia*

pendula Humb. & Bonpl. ex Willd., *sensu lato*, for the name of this taxon.

Velamuerto.

Senna polyphylla (Jacq.) H.S. Irwin & Barneby var. **polyphylla**, Mem. New York Bot. Gard. 35(2):517, 519. 1982.
Bas.: *Cassia polyphylla* Jacq. (1791).
Syn.: *Peiranisia polyphylla* (Jacq.) Britton & Rose (1924).

Shrub in Las Piedras Chiquitas, Southeastern Foothills, Southern Coastal Lowlands (Coamo), Southern Coastal Limestone, Guánica Forest Reserve, Southwestern Coastal Lowlands, Sierra Bermeja; 20–550 m; *fl* throughout the year, *fr* Sep–Jun.

Virgin Islands (St. Thomas, St. Croix). [L2: 91].

Hediondilla, Retama, Retama prieta.

Senna septemtrionalis (Viv.) H.S. Irwin & Barneby, Mem. New York Bot. Gard. 35(1):365. 1982.
Bas.: *Cassia septemptrionalis* Viv. (1802).
Syn.: *Adipera laevigata* (Willd.) Britton & Rose (1924).

Shrub/small tree in Sierra de Cayey, Central Cordillera, Western Cordillera (Maricao); 500–825 m; *fl* & *fr* Mar, Jun–Sep.

Native to Mexico, Central America. [L2: 93].

Notes: Liogier & Martorell (1982) misapplied the name *Cassia floribunda* Cav. to this taxon.

Hedionda macho, Yerba hedionda.

Senna siamea (Lam.) H.S. Irwin & Barneby, Mem. New York Bot. Gard. 35(1):98. 1982.
Bas.: *Cassia siamea* Lam. (1785).
Syn.: *Sciacassia siamea* (Lam.) Britton (1930).

Tree in Northern Coastal Lowlands, Northern Limestone Hills (San Sebastián), Eastern Coast (Naguabo), Caguas Valley (Caguas), Eastern Cordillera (Aibonito), Southern Coastal Lowlands, Guánica Forest Reserve, Southwestern Coastal Lowlands (Cabo Rojo), Western Coast, Vieques; 1–650 m; *fl* Jun–Nov, *fr* Jul–Nov, Feb.

Native to southeastern Asia. [L2: 93].

Casia, Casia amarilla, Casia de Siam, Casia siamea.

Senna sophera (L.) Roxb., Fl. Ind. (ed. 1832) 2:347. 1832.
Bas.: *Cassia sophera* L. (1753).
Syn.: *Ditremexa sophera* (L.) Britton & Rose (1924).

Suffrutescent herb/shrub in Southeastern Foothills (Coamo); 250 m; *fl* & *fr* Mar (one specimen). [*Liogier 32836* (UPR)].

Pantropical. [L2: 94].

Algarrobilla.

Senna spectabilis (DC.) H.S. Irwin & Barneby var. **spectabilis**, Mem. New York Bot. Gard. 35(2):600, 603. 1982.
Bas.: *Cassia spectabilis* DC. (1813).
Syn.: *Pseudocassia spectabilis* (DC.) Britton & Rose (1930).

Tree planted and persistent in Northern Coastal Lowlands (Fajardo), Caguas Valley (Gurabo), Sierra de Cayey (San Lorenzo), Eastern Cordillera, Central Cordillera (Ponce), Southern Coastal Lowlands (Coamo), Western Coast; 50–1000 m; *fl* Aug–Nov, *fr* Sep–Feb.

Greater Antilles, Lesser Antilles (Guadeloupe, Martinique, Barbados), Mexico, Central America (probably introduced), South America (scattered). [L2: 94].

Casia amarilla.

Senna uniflora (Mill.) H.S. Irwin & Barneby, Mem. New York Bot. Gard. 35(1):258. 1982.
Bas.: *Cassia uniflora* Mill. (1768).

Herb in Southeastern Foothills, Central Cordillera (Ponce), Southern Coastal Lowlands, Southern Coastal Limestone, Guánica Forest Reserve, Southwestern Coastal Lowlands (Guánica); 1–250 m; *fl* & *fr* Jul–Feb.

Bahamas, Greater Antilles, Lesser Antilles (Nevis, Barbados), Mexico, Central America, South America (Venezuela, Brazil, Ecuador). [L2: 95].

Stahlia monosperma (Tul.) Urb., Symb. Antill. 2(2):285. 1900.
Bas.: *Caesalpinia monosperma* Tul. (1844).

Tree in Northern Coastal Lowlands (Río Grande), Southern Coastal Limestone (Sabana Grande), Southwestern Coastal Lowlands, Sierra Bermeja, Vieques; 0–200 m; *fl* Mar–Apr, *fr* Feb–Jul.

Greater Antilles (Hispaniola), Virgin Islands. [L2: 96].

Notes: This taxon is on the U.S. Fish & Wildlife Service list of threatened and endangered plants.

Cóbana, Cóbana negra, Polisandro.

Tamarindus indica L., Sp. Pl. 1:34. 1753.

Tree in Northern Coastal Lowlands (Fajardo and adjacent islands), Caguas Valley (Caguas), Southeastern Foothills, Eastern Cordillera (Comerío), Southern Coastal Lowlands, Southern Coastal Limestone, Guánica Forest Reserve, Southwestern Coastal Lowlands, Sierra Bermeja, Vieques, Culebra, Mona; 1–350 m; *fl* Jun–Nov, *fr* throughout the year.

Native to tropical Africa, Yemen. [L2: 97].

Tamarindo.

TAXA NOT RECENTLY COLLECTED

Bauhinia tomentosa L., Sp. Pl. 1:375. 1753.
Syn.: *Alvesia tomentosa* (L.) Britton & Rose (1930).

Shrub/small tree collected roadside in Fajardo on 19 Apr 1885 [*Sintenis 1155* (US)] and also in Aug 1885 [*Sintenis 1889* (US)].

Native to southern and eastern Africa, Yemen, India. [L2: 58].

Mariposa, Petán.

Senna bacillaris (L.f.) H.S. Irwin & Barneby var. **bacillaris**, Mem. New York Bot. Gard. 35(1):111, 113. 1982.
Bas.: *Cassia bacillaris* L.f. (1782).
Syn.: *Chamaefistula bacillaris* (L.f.) G. Don (1832).

Shrub collected in the Sierra de Luquillo in Sep 1883 [*Eggers 1395* (NY)].

Native to northern South America. [L2: 87].

TAXA CITED BY PREVIOUS AUTHORS, BUT NOT INCLUDED

ONLY UNDER CULTIVATION

Bauhinia galpinii N.E. Br.
Shrub cited by Liogier & Martorell (1982, 2000) and Liogier (1988) as planted and persistent. [L2: 57].

Notes: Liogier (1988) and Liogier & Martorell (2000) misapplied the name *Bauhinia punctata* Bolle to this taxon and placed *Bauhinia galpinii* in synonymy with it.

Bauhinia multinervia (Kunth) DC. Syn.: *Bauhinia megalandra* Griseb.
Shrub cited by Liogier & Martorell (1982, 2000) and Liogier (1988) as planted and spontaneous. [L2: 55].

Caesalpinia gilliesii (Wall. ex Hook.) D. Dietr.
Shrub cited by Liogier & Martorell (1982, 2000) and Liogier (1988) as planted in gardens and persistent. [L2: 62].

Copaifera officinalis (Jacq.) L. Syn.: *Copaiva officinalis* Jacq.
Tree cited by Liogier (1988) as planted and naturalized in several areas. [L2: 479].

Schizolobium parahyba (Vell.) S.F. Blake "parahybum."
Tree cited by Liogier (1997) and Liogier & Martorell (2000) as planted as an experimental timber and becoming naturalized. [L5: 400].

Senna ligustrina (L.) H.S. Irwin & Barneby. Syn.: *Ditremexa ligustrina* (L.) Britton & Rose.
Shrub cited by Liogier & Martorell (1982, 2000) and Liogier (1988) as introduced and probably escaped. [L2: 89].

Senna multijuga (Rich.) H.S. Irwin & Barneby subsp. ***multijuga***. Syn.: *Cassia multijuga* Rich.
Tree cited by Liogier & Martorell (1982, 2000) and Liogier (1988) as planted as an ornamental with a tendency to become naturalized. [L2: 89].

Senna surattensis (Burm. f.) H.S. Irwin & Barneby.
Syn.: *Cassia surattensis* Burm. f.
Tree cited by Liogier & Martorell (1982, 2000) and Liogier (1988) as sparingly planted for ornament at lower elevations, tending to become naturalized. [L2: 95].

FABACEAE-FABOIDEAE (LEGUMINOSAE-PAPILIONOIDEAE)

Abrus precatorius L., Syst. Nat. (ed. 12) 2:472. 1767.
Syn.: *Abrus abrus* (L.) W. Wight (1905).

Vine in Northern Coastal Lowlands, Northern Limestone Hills (Bayamón), Las Piedras Chiquitas, Southeastern Foothills (Coamo), Southern Coastal Lowlands (Caja de Muertos), Susúa Forest Reserve, Southwestern Coastal Lowlands (Lajas), Sierra Bermeja, Vieques, Culebra; 1–500 m; *fl* Oct, Jun, *fr* Feb–Oct.

Native to the Paleotropics. [L2: 106; AR: 242].

Ojos de cangrejo, Peronía.

Aeschynomene americana L., Sp. Pl. 2:713. 1753.
Syn.: *Aeschynomene glandulosa* Poir. (1816). *Aeschynomene americana* var. *glandulosa* (Poir.) Rudd (1955).

Suffrutescent herb in Northern Coastal Lowlands, Northern Limestone Hills, Sierra de Luquillo, Eastern Coast, Sierra de Cayey, Las Piedras Chiquitas, Southeastern Foothills, Eastern Cordillera, Central Cordillera (Peñuelas), Western Cordillera (Maricao, San Germán), Cerro Las Mesas, Western Coast (Añasco), Vieques; 1–700 m; *fl* & *fr* Jul–Mar.

West Indies, Trinidad and Tobago, southeastern United States (Florida), Mexico, Central America, South America. [L2: 109].

Moriviví bobo, Yerba rosario.

Aeschynomene portoricensis Urb., Symb. Antill. 1(2):325. 1899.

Herb in Northern Coastal Lowlands (Dorado, Vega Alta), Las Tetas de Cayey, Las Piedras Chiquitas, Central Cordillera (Peñuelas); 5–825 m; *fl* Oct–Mar, *fr* Oct–Mar, Jun.

Endemic. [L2: 109].

Aeschynomene rudis Benth., Pl. Hartw. 116. 1843.
Shrub in Southwestern Coastal Lowlands (Lajas), Western Coast; 1–20 m; *fl* & *fr* Apr–Jul, Oct–Nov.
Native to Greater Antilles (Cuba), Mexico, Central America, South America (scattered). [L2: 110].

Aeschynomene sensitiva Sw., Prodr. 107. 1788.
Herb in Northern Coastal Lowlands, Río Abajo Forest Reserve, Sierra de Luquillo, Eastern Coast, Sierra de Cayey, Central Cordillera (Utuado), Southern Coastal Lowlands (Guayama), Southwestern Coastal Lowlands (Lajas), Western Coast; 1–900 m; *fl* & *fr* throughout the year.
Antilles, Trinidad, Mexico, Central America, South America. [L2: 110].
Moriviví bobo, Yerba de ciénaga.

Aeschynomene villosa Poir., Encycl. Suppl. 4(1):76. 1816.
Herb in Caguas Valley (Juncos), Sierra de Cayey, Las Piedras Chiquitas, Eastern Cordillera (Barranquitas), Toro Negro Forest Reserve; 250–1200 m; *fl* Nov, Feb–May, *fr* Nov, Feb–Apr.
Antilles, southwestern United States (Arizona), Mexico, Central America, northwestern, west-central South America and Venezuela. [L2: 110].

Alysicarpus vaginalis (L.) DC., Prodr. 2:353. 1825.
Bas.: *Hedysarum vaginale* L. (1753).
Prostrate herb in Northern Coastal Lowlands, Río Abajo Forest Reserve, Eastern Coast, Southeastern Foothills (Salinas), Southern Coastal Lowlands (Salinas), Southwestern Coastal Lowlands, Sierra Bermeja, Western Coast (Rincón), Vieques, Mona; 1–475 m; *fl* & *fr* throughout the year.
Native to tropical Africa, Yemen, tropical Asia, Japan, Australia. [L2: 112].
Yerba de contrabando.

Andira inermis (W. Wright) Kunth ex DC., Prodr. 2:475. 1825.
Bas.: *Geoffroea inermis* W. Wright (1787).
Tree in Northern Coastal Lowlands, Northern Limestone Hills, Río Abajo Forest Reserve, Sierra de Luquillo, Eastern Coast, Sierra de Cayey, Las Piedras Chiquitas, Southeastern Foothills, Eastern Cordillera (Comerío, Aibonito), Central Cordillera (Villalba), Southern Coastal Limestone (Guayanilla), Susúa Forest Reserve, Maricao Forest Reserve, Southwestern Coastal Lowlands, Sierra Bermeja, Western Coast (Añasco), Vieques; 1–600 m; *fl* Mar–Nov, *fr* Jul–Mar.
Antilles, Trinidad, Mexico, Central America, South America. [L2: 112].
Moca, Moca blanca.

Arachis repens Handro, Arq. Bot. Estado São Paulo 3(4):180, *t*. 47. 1958.
Herb planted and persistent in Northern Coastal Lowlands (Loíza), Western Coast (Mayagüez); 2–100 m; *fl* Feb–Jul, *fr* (?).
Native to Brazil. [L2: 115].
Notes: Liogier & Martorell (1982) misapplied the name *Arachis glabrata* Benth., and Liogier (1988) and Liogier & Martorell (2000) misapplied *Arachis prostrata* Benth., to this taxon.

Barbieria pinnata (Pers.) Baill., Hist. Pl. 2:263. 1870.
Bas.: *Galactia pinnata* Pers. (1807).
Vine in Sierra de Luquillo, Eastern Cordillera (Cerro La Peña, Aguas Buenas), Central Cordillera (Utuado), Western Cordillera (Maricao); 150–575 m; *fl* Nov–Jan, *fr* Dec–Apr.
Greater Antilles (Cuba, Hispaniola), Mexico, Central America (Guatemala, Belize, Costa Rica), northwestern, west-central South America and Brazil. [L2: 117; AR: 243].
Enredadera.

Brya ebenus (L.) DC., Prodr. 2:421. 1825.
Bas.: *Aspalathus ebenus* L. (1759).
Shrub/small tree between Caguas and Cayey; 200–400 m; *fl* Mar (one specimen), *fr* (?). [*Liogier 30457* (UPR)].
Native to Greater Antilles (Cuba, Jamaica). [L2: 117].

Cajanus cajan (L.) Millsp., Publ. Field Columb. Mus., Bot. Ser. 2(1):53. 1900.
Bas.: *Cytisus cajan* L. (1753).
Shrub planted and persistent in Northern Coastal Lowlands, Guajataca Forest Reserve, Southeastern Foothills (Coamo), Eastern Cordillera (Cayey), Central Cordillera, Southern Coastal Lowlands, Southwestern Coastal Lowlands (San Germán, Cabo Rojo), Western Coast (Mayagüez), Vieques; 1–1100 m; *fl* & *fr* throughout the year.
Probably native to India. [L2: 119].
Gandúl, Gandures.

Calopogonium caeruleum (Benth.) C. Wright in Sauvalle, Anales Acad. Ci. Méd. Habana 5:337. 1869.
Bas.: *Stenolobium caeruleum* Benth. (1837).
Vine in Northern Coastal Lowlands, Northern Limestone Hills (San Sebastián), Western Coast (Mayagüez); 1–250 m; *fl* & *fr* Nov–Mar.

Greater Antilles (Cuba, Hispaniola), Lesser Antilles (St. Vincent), Trinidad and Tobago, Mexico, Central America, South America. [L2: 121; AR: 245].

Notes: Britton & Wilson (1924), Liogier & Martorell (1982, 2000), Liogier (1988) and Acevedo-Rodríguez (2005) used the orthographic variant *Calopogonium coeruleum* for the name of this taxon.

Calopogonium mucunoides Desv., Ann. Sci. Nat. (Paris) 9:423. 1826.

Syn.: *Calopogonium orthocarpum* Urb. (1899).

Vine in Northern Coastal Lowlands, Caguas Valley (Caguas), Río Abajo Forest Reserve, Western Coast (Mayagüez); 1–350 m; *fl* Nov–Apr, *fr* Dec–Apr.

Antilles, Mexico, Central America, northern South America to Bolivia. [L2: 121; AR: 245].

Jícama.

Canavalia nitida (Cav.) Piper, Contr. U.S. Natl. Herb. 20(14):562. 1925.

Bas.: *Clementea nitida* Cav. (1804).

Vine in Eastern Cordillera (Cidra), Central Cordillera (Peñuelas); 300–500 m; *fl* (?), *fr* Nov (two specimens). [*Liogier 33718* (UPR); *Axelrod 7320* (UPRRP)].

Bahamas, Greater Antilles (Cuba, Hispaniola), Virgin Islands (St. Thomas). [L2: 124; AR: 248].

Haba de burro, Haba de burro de jardín, Mato colorado.

Canavalia rosea (Sw.) DC., Prodr. 2:404. 1825.

Bas.: *Dolichos roseus* Sw. (1788).

Syn.: *Canavalia maritima* Thouars (1813) "Canavali."

Canavalia maritima (Aubl.) Urb. (1919).

Vine in the Northern Coastal Lowlands, Eastern Coast (Punta Guayanés), Southern Coastal Lowlands, Guánica Forest Reserve, Southwestern Coastal Lowlands (Cabo Rojo), Western Coast, Vieques, Culebra, Mona; 1–20 m; *fl* Jun–Apr, *fr* Jul–Apr.

Widespread on tropical and subtropical shores. [L2: 124; AR: 248].

Notes: Britton & Wilson (1924), Liogier & Martorell (1982) and Liogier (1988) used *Canavalia maritima* (Aubl.) Thouars for the name of this taxon; but, according to TROPICOS, Thouars did not indicate that he based his name on *Dolichos maritimus* Aubl.

Canavalia, Haba de playa, Habichuela playera, Mato de playa.

Centrosema plumieri (Turpin ex Pers.) Benth., Comm. Legum. Gen. 54. 1837.

Bas.: *Clitoria plumieri* Turpin ex Pers. (1807).

Syn.: *Bradburya plumieri* (Turpin ex Pers.) Kuntze (1891).

Vine in Río Abajo Forest Reserve, Central Cordillera (Ponce), Western Coast (Mayagüez); 10–325 m; *fl* Nov–Jan, *fr* Dec–Jan (two specimens).

Antilles, Mexico, Central America, northern South America to Brazil, Bolivia. [L2: 126; AR: 250].

Flor de conchitas.

Centrosema pubescens Benth., Comm. Legum. Gen. 55. 1837, *nom. cons. prop.*

Syn.: *Bradburya pubescens* (Benth.) Kuntze (1891).

Vine in Northern Coastal Lowlands, Río Abajo Forest Reserve, Sierra de Luquillo, Eastern Coast, Caguas Valley (Caguas), Sierra de Cayey, Southeastern Foothills (Guayama), Eastern Cordillera (Aibonito), Susúa Forest Reserve, Western Cordillera (Maricao), Western Coast, Vieques, Mona; 1–725 m; *fl* Sep–Apr, *fr* Dec–Apr.

Antilles, Mexico, Central America, northern South America to Brazil, Bolivia. [L2: 127; AR: 250].

Flor de conchitas, Flor de pito, Flor de pito blanca.

Centrosema virginianum (L.) Benth., Comm. Legum. Gen. 56. 1837.

Bas.: *Clitoria virginiana* L. (1753).

Syn.: *Bradburya virginiana* (L.) Kuntze (1891).

Vine in Northern Coastal Lowlands, Northern Limestone Hills, Río Abajo Forest Reserve, Eastern Coast (Yabucoa), Las Piedras Chiquitas, Southeastern Foothills, Eastern Cordillera (Cayey, Cañón San Cristóbal), Southern Coastal Lowlands (Caja de Muertos), Susúa Forest Reserve, Western Cordillera, Guánica Forest Reserve, Southwestern Coastal Lowlands, Sierra Bermeja, Cerro Las Mesas, Vieques, Culebra, Mona; 1–600 m; *fl* & *fr* throughout the year.

West Indies, east-central to south-central United States, Mexico, Central America, South America. [L2: 127; AR: 251].

Conchita de Virginia, Flor de conchitas.

Clitoria falcata Lam., Encycl. 2(1):51. 1786.

Syn.: *Martiusia rubiginosa* (Juss. ex Pers.) Britton (1924).

Vine in Northern Coastal Lowlands, Northern Limestone Hills (Ciales), Río Abajo Forest Reserve, Sierra de Luquillo, Eastern Coast, Sierra de Cayey, Eastern Cordillera (Cerro La Peña, Aguas Buenas), Central Cordillera (Juana Díaz); 1–750 m; *fl* Aug–May, *fr* Sep–May.

Greater Antilles, Lesser Antilles (Guadeloupe,

Martinique), Mexico, Central America, South America. [L2: 131; AR: 253].

Ahogagallina, Conchita, Flor de pito.

Clitoria laurifolia Poir., Encycl. Suppl. 2(1):301. 1811.

Syn.: *Martiusia laurifolia* (Poir.) Britton (1924).

Shrub in Northern Coastal Lowlands (Dorado to Manatí); 1–8 m; *fl* Apr–Oct, *fr* Mar–Nov.

Greater Antilles (Cuba, Hispaniola), northern South America and Brazil. [L2: 131].

Clitoria ternatea L., Sp. Pl. 2:753. 1753.

Vine in Northern Coastal Lowlands, Northern Limestone Hills (Utuado), Southeastern Foothills, Southern Coastal Lowlands, Guánica Forest Reserve, Southwestern Coastal Lowlands, Sierra Bermeja, Western Coast, Vieques; 1–250 m; *fl* throughout the year, *fr* Oct–Jun.

Native to the Paleotropics; native range uncertain. [L2: 131; AR: 253].

Conchitas, Bejuco de conchitas, Deleite, Papito.

Coursetia caribaea (Jacq.) Lavin in Stirton, Adv. Legume Syst. 3:63. 1987.

Bas.: *Galega caribaea* Jacq. (1763).

Syn.: *Cracca caribaea* (Jacq.) Benth. (1854). *Benthamantha caribaea* (Jacq.) Kuntze (1898).

Suffrutescent herb in Northern Coastal Lowlands (Fajardo), Eastern Coast (Ceiba), Las Piedras Chiquitas, Southeastern Foothills, Southern Coastal Lowlands, Southern Coastal Limestone, Guánica Forest Reserve, Southwestern Coastal Lowlands, Sierra Bermeja, Vieques, Culebra, Desecheo; 1–500 m; *fl* & *fr* throughout the year.

Greater Antilles (Jamaica, Hispaniola), Virgin Islands, Lesser Antilles, northwestern, west-central South America and Venezuela. [L2: 134].

Añil falso.

Crotalaria brevidens Benth., London J. Bot. 2:585. 1843.

Herb in Sierra de Cayey, Eastern and Central Cordillera, Western Cordillera; 350–1100 m; *fl* & *fr* throughout the year.

Native to tropical Africa. [L2: 137].

Crotalaria incana L., Sp. Pl. 2:716. 1753.

Herb in Northern Coastal Lowlands, Las Piedras Chiquitas, Southeastern Foothills, Southern Coastal Lowlands, Western Coast (Mayagüez), Vieques, Culebra; 1–500 m; *fl* Sep–Jun, *fr* Sep–Apr.

Pantropical. [L2: 138].

Cascabelillo vacío.

Crotalaria lotifolia L., Sp. Pl. 2:715. 1753, *nom. cons.*

Suffrutescent herb in Northern Coastal Lowlands (Fajardo), Southeastern Foothills (Salinas), Southern Coastal Limestone, Susúa Forest Reserve, Guánica Forest Reserve, Sierra Bermeja, Culebra; 1–300 m; *fl* Oct–Jan, *fr* Oct–Feb, Jun.

West Indies, Mexico, Central America. [L2: 138].

Cascabelillo axilar.

Crotalaria pallida Aiton var. **obovata** (G. Don) Polhill, Kew Bull. 22(2):265. 1968.

Bas.: *Crotalaria obovata* G. Don (1832).

Syn.: *Crotalaria falcata* Vahl ex DC. (1825).

Herb in Northern Coastal Lowlands, Northern Limestone Hills (Florida), Sierra de Luquillo, Eastern Coast, Caguas Valley, Las Tetas de Cayey, Las Piedras Chiquitas, Eastern Cordillera (Aguas Buenas, Comerío), Central Cordillera, Southern Coastal Lowlands (Ponce), Western Cordillera (Maricao), Southwestern Coastal Lowlands (Lajas), Sierra Bermeja, Cerro Las Mesas, Western Coast, Vieques; 1–1000 m; *fl* & *fr* throughout the year.

Cultivated and naturalized throughout the tropics; native range obscure. [L2: 137].

Matraca, Sonajuelas.

Crotalaria retusa L., Sp. Pl. 2:715. 1753.

Herb in Northern Coastal Lowlands, Northern Limestone Hills (Toa Baja, Moca), Sierra de Luquillo, Eastern Coast, Caguas Valley (Caguas), Southeastern Foothills, Eastern Cordillera (Comerío), Southern Coastal Lowlands, Southern Coastal Limestone (Guayanilla), Susúa Forest Reserve, Maricao Forest Reserve, Southwestern Coastal Lowlands (Lajas, San Germán), Sierra Bermeja, Cerro L as Mesas, Western Coast, Vieques; 1–300 m; *fl* & *fr* throughout the year.

Widely naturalized in tropics; native range obscure. [L2: 140].

Cascabelillo, Matraca, Sonajuelas.

Crotalaria rotundifolia J.F. Gmel., Syst. Nat. 2(2):1095. 1792.

Syn.: *Crotalaria maritima* Chapm. (1878).

Prostrate herb in Northern Coastal Lowlands (Dorado); 3 m; *fl* & *fr* May, Dec (two specimens). [*Liogier 34864, 36296* (UPR)].

Greater Antilles (Hispaniola), southeastern and south-central United States, Mexico, Central America. [L2: 140].

Crotalaria stipularia Desv., J. Bot. Agric. 3:76. 1814.

Herb in Northern Coastal Lowlands (Río

Grande), Río Abajo Forest Reserve, Sierra de Luquillo, Eastern Coast, Sierra de Cayey, Eastern Cordillera (Aibonito), Central Cordillera, Maricao Forest Reserve, Cerro Las Mesas; 1–900 m; *fl* Apr–Jan, *fr* Apr–Nov.

Greater Antilles (Hispaniola), Lesser Antilles, Trinidad, South America. [L2: 141].

Cascabelillo, Cascabelillo alado.

Crotalaria trichotoma Bojer, Ann. Sci. Nat., Bot., sér. 2 4:265. 1835.
Syn.: *Crotalaria zanzibarica* Benth. (1843).

Herb in Northern Coastal Lowlands, Northern Limestone Hills (Arecibo), Eastern Cordillera (Aibonito), Central Cordillera (Ciales), Western Cordillera (Maricao), Cerro Las Mesas, Western Coast; 1–600 m; *fl* & *fr* Feb–Jun, Sep–Oct.

Native to Africa (Mozambique, Tanzania). [L2: 141].

Crotalaria verrucosa L., Sp. Pl. 2:715. 1753.

Herb in Northern Coastal Lowlands, Eastern Coast (Humacao); 1–5 m; *fl* Jan–Oct, *fr* Jan–Feb, May–Oct.

Native to tropical Asia, China. [L2: 141].

Crotalaria lengua de vaca.

Dalbergia ecastaphyllum (L.) Taub. in Engler & Prantl, Nat. Pflanzenfam. 3(3):335. 1894.
Bas.: *Hedysarum ecastaphyllum* L. (1759).
Syn.: *Ecastaphyllum ecastaphyllum* (L.) Britton (1918).

Shrub in Northern Coastal Lowlands, Eastern Coast, Western Coast, Vieques; 1–50 m; *fl* throughout the year, *fr* Jun–Mar.

West Indies, Trinidad, southeastern United States (Florida), Central America, northern South America to Brazil, Peru. [L2: 142; AR: 255].

Maraimaray, Palo de pollo.

Dalbergia monetaria L.f., Suppl. Pl. 317. 1782.
Syn.: *Securidaca volubilis* L., *pro parte* (1753).

Woody vine in Northern Coastal Lowlands (Bayamón), Northern Limestone Hills, Sierra de Luquillo, Southeastern Foothills (Cayey), Eastern Cordillera (Cerro La Peña), Central Cordillera, Susúa Forest Reserve, Western Cordillera; 175–550 m; *fl* Jun–Feb, *fr* Oct–May.

Greater Antilles, Lesser Antilles (Guadeloupe, Martinique, St. Vincent), northern South America to Brazil, Bolivia. [L2: 144; AR: 256].

Membrillo, Palo de brasilete.

Dalbergia sissoo Roxb. ex DC., Prodr. 2:416. 1825.

Tree planted and persistent in Sierra de Luquillo, Southeastern Foothills (Salinas), Southwestern Coastal Lowlands; 1–350 m; *fl* Mar–May, Aug–Sep, *fr* Apr–May, Aug–Nov.

Native to Indian Subcontinent, Myanmar. [L2: 144].

Siso, Sisu.

Dalea carthagenensis (Jacq.) J.F. Macbr. var. **portoricana** Barneby, Mem. New York Bot. Gard. 27:518, *t. 125*. 1977.

Shrub in Las Piedras Chiquitas, Southeastern Foothills (Salinas), Southern Coastal Limestone (Peñuelas), Guánica Forest Reserve; 25–475 m; *fl* Nov–Jun, *fr* Nov–Mar.

Endemic. [L2: 145].

Notes: Britton & Wilson (1924) misapplied the name *Parosela domingensis* (DC.) Millsp. to this taxon.

Desmodium adscendens (Sw.) DC., Prodr. 2:332. 1825.
Bas.: *Hedysarum adscendens* Sw. (1788).
Syn.: *Meibomia adscendens* (Sw.) Kuntze (1891).

Herb in Northern Coastal Lowlands, Guajataca Forest Reserve, Río Abajo Forest Reserve, Sierra de Luquillo, Eastern Coast, Caguas Valley (Caguas), Sierra de Cayey, Central Cordillera, Western Cordillera, Culebra; 1–1050 m; *fl* & *fr* throughout the year.

Pantropical. [L2: 148].

Zarzabacoa galana.

Desmodium affine Schltdl., Linnaea 12(3):312. 1838.
Syn.: *Meibomia affinis* (Schltdl.) Kuntze (1891).

Herb in Las Piedras Chiquitas, Southeastern Foothills (Salinas); 75–500 m; *fl* Oct–Jan, *fr* Dec–Jan (two specimens).

Greater Antilles, Trinidad, Mexico, Central America, South America. [L2: 149].

Amor seco.

Desmodium axillare (Sw.) DC. var. **axillare**, Prodr. 2:333. 1825.
Bas.: *Hedysarum axillare* Sw. (1788).
Syn.: *Meibomia axillaris* (Sw.) Kuntze (1891). *Nephromeria axillaris* (Sw.) Schindl. (1924).

Herb in Northern Limestone Hills, Río Abajo Forest Reserve, Sierra de Cayey, Las Piedras Chiquitas, Southeastern Foothills (Salinas), Eastern Cordillera (Cañón San Cristóbal), Central Cordillera, Western Cordillera (Maricao), Western Coast (Aguada); 1–900 m; *fl* & *fr* May–Jan.

Antilles, Mexico, Central America, northern South America. [L2: 150; AR: 256, 258].

Zarzabacoa del monte.

Desmodium axillare (Sw.) DC. var. **acutifolium** (Kuntze) Urb., Symb. Antill. 4(2):292. 1905.

Bas.: *Meibomia axillaris* (Sw.) Kuntze var. *acutifolia* Kuntze (1891).

Syn.: *Meibomia umbrosa* Britton (1910). *Nephromeria umbrosa* (Britton) Britton & P. Wilson (1930).

Herb in Northern Limestone Hills (Camuy, Isabela), Sierra de Luquillo, Central Cordillera (Ciales); 250–450 m; *fl* & *fr* Jul–Nov.

Greater Antilles, Lesser Antilles (Guadeloupe, Dominica, Martinique), Trinidad, Central America, South America (Colombia, Brazil, Bolivia). [L2: 150; AR: 256, 258].

Desmodium axillare (Sw.) DC. var. **stoloniferum** (Rich. ex Poir.) B.G. Schub., J. Arnold Arbor. 44(2):289. 1963.

Bas.: *Hedysarum stoloniferum* Rich. ex Poir. (1805).

Syn.: *Meibomia sintenisii* (Urb.) Britton (1924). *Nephromeria sintenisii* (Urb.) Britton & P. Wilson (1930).

Herb in Sierra de Luquillo, Sierra de Cayey (San Lorenzo), Maricao Forest Reserve; 350–600 m; *fl* Jul–Aug, Nov, *fr* Jul–Aug, Nov–Jan.

Greater Antilles, Central America, west-central South America. [L2: 150; AR: 256, 258].

Desmodium barbatum (L.) Benth. in Miquel, Pl. Jungh. 2:224. 1852.

Bas.: *Hedysarum barbatum* L. (1759).

Syn.: *Meibomia barbata* (L.) Kuntze (1891).

Herb in Northern Coastal Lowlands, Sierra de Luquillo, Eastern Cordillera (Cerro La Peña), Central Cordillera, Cerro Las Mesas; 1–1200 m; *fl* & *fr* Feb–Nov.

Widely distributed throughout the tropics and subtropics. [L2: 151].

Zarzabacoa peluda.

Desmodium glabrum (Mill.) DC., Prodr. 2:338. 1825.

Bas.: *Hedysarum glabrum* Mill. (1768).

Syn.: *Meibomia mollis* (Vahl) Kuntze (1891).

Herb in Southeastern Foothills, Southern Coastal Lowlands, Guánica Forest Reserve, Southwestern Coastal Lowlands, Sierra Bermeja, Vieques, Mona; 1–400 m; *fl* & *fr* Oct–Mar, Jun.

West Indies, Mexico, Central America, South America. [L2: 152].

Zarzabacoa dulce.

Desmodium incanum DC., Prodr. 2:332. 1825.

Syn.: *Desmodium incanum* DC. var. *angustifolium* Griseb. (1859). *Desmodium canum* Schinz & Thell. var. *angustifolium* (Griseb.) León & Alain (1950).

Meibomia supina (Sw.) Britton (1892).

Desmodium canum Schinz & Thell. (1913).

Herb in Northern Coastal Lowlands, Northern Limestone Hills, Río Abajo Forest Reserve, Sierra de Luquillo, Eastern Coast, Sierra de Cayey, Las Tetas de Cayey, Las Piedras Chiquitas, Southeastern Foothills, Eastern Cordillera, Susúa Forest Reserve, Maricao Forest Reserve, Southwestern Coastal Lowlands (Cabo Rojo), Sierra Bermeja, Western Coast, Vieques, Culebra, Mona; 1–900 m; *fl* & *fr* throughout the year.

West Indies, Mexico, Central America, South America. [L2: 153; AR: 258].

Notes: Liogier (1988) and Liogier & Martorell (2000) recognized two varieties of this taxon: var. *incanum*, which is trifoliate and broad, and var. *angustifolium*, which is usually unifoliate and narrow. Howard (1988) said that in the Lesser Antilles var. *angustifolium* may have trifoliate leaves as well as single leaflets and shows a great variation in leaflet size and shape on one individual. Consequently, he does not make a varietal distinction, nor do I.

Zarzabacoa común.

Desmodium intortum (Mill.) Urb., Symb. Antill. 8(1):292. 1920.

Bas.: *Hedysarum intortum* Mill. (1768).

Herb in Central Cordillera; 760–1300 m; *fl* & *fr* Dec–Jul.

Greater Antilles (Jamaica, Hispaniola), Mexico, Central America, northwestern, west-central South America and Venezuela, Brazil. [L2: 154; AR: 258].

Desmodium procumbens (Mill.) Hitchc., Rep. (Annual) Missouri Bot. Gard. 4:76. 1893.

Bas.: *Hedysarum procumbens* Mill. (1768).

Syn.: *Meibomia procumbens* (Mill.) Britton (1924).

Herb in Northern Coastal Lowlands (Fajardo), Eastern Coast, Las Piedras Chiquitas, Southeastern Foothills (Salinas), Cerro de Las Cuevas, Southern Coastal Lowlands (Caja de Muertos), Guánica Forest Reserve, Southwestern Coastal Lowlands (Cabo Rojo), Sierra Bermeja, Western Coast (Rincón), Vieques, Mona; 1–550 m; *fl* & *fr* Oct–Mar.

Antilles, southwestern United States (Arizona), Mexico, Central America, northern South America to Brazil, Ecuador. [L2: 154].

Zarzabacoa espiral.

Desmodium scorpiurus (Sw.) Desv., J. Bot. Agric. 1:122, *t. 5, f. 13*. 1813.
Bas.: *Hedysarum scorpiurus* Sw. (1788).
Syn.: *Meibomia scorpiurus* (Sw.) Kuntze (1891).
Herb in Northern Coastal Lowlands, Eastern Coast, Caguas Valley (Caguas), Eastern Cordillera (Aibonito), Southern Coastal Lowlands (Juana Díaz), Western Coast (San Germán); 1–300 m; *fl* Oct–Aug, *fr* Nov–Aug.
Antilles, Trinidad and Tobago, Mexico, Central America, northern South America to Bolivia. [L2: 155].
Zarzabacoa cola de escorpión.

Desmodium tortuosum (Sw.) DC., Prodr. 2:332. 1825.
Bas.: *Hedysarum tortuosum* Sw. (1788).
Syn.: *Meibomia purpurea* (Mill.) Vail (1903).
Herb in Northern Coastal Lowlands, Eastern Coast, Caguas Valley (Juncos), Sierra de Cayey, Southeastern Foothills, Eastern and Central Cordillera, Southern Coastal Lowlands, Southern Coastal Limestone (Yauco), vicinity of Susúa Forest Reserve (Sabana Grande), Western Coast, Vieques, Mona; 5–900 m; *fl* & *fr* throughout the year.
West Indies, southern United States, Mexico, Central America, South America. [L2: 155].
Cadillo, Cadillo pegajoso, Junquillo, Zarzabacoa junquillo, Zarzabacoa tortuosa.

Desmodium triflorum (L.) DC., Prodr. 2:334. 1825.
Bas.: *Hedysarum triflorum* L. (1753).
Syn.: *Sagotia triflora* (L.) Duchass. & Walp. (1850).
Herb in Northern Coastal Lowlands, Eastern Coast (Cayo Santiago), Western Coast (Hormigueros), Vieques; 1–175 m; *fl* Nov–Apr, *fr* Dec–May.
Pantropical. [L2: 156].
Zarzabacoa de tres flores.

Desmodium wydlerianum Urb., Symb. Antill. 2(2):302. 1900.
Syn.: *Meibomia wydleriana* (Urb.) Britton (1924).
Herb in Northern Limestone Hills, Sierra de Luquillo, Sierra de Cayey, Central Cordillera (Ponce), Vieques; 50–600 m; *fl* Oct–Jan, Apr–Jun, *fr* Sep–Jan, Apr–Jun.
Greater Antilles (Cuba, Hispaniola), southern Central America, northern South America (Venezuela, Colombia) and Brazil. [L2: 156].

Dioclea reflexa Hook. f. in Hooker, Niger Fl. 306. 1849.
Vine in Northern Coastal Lowlands (Fajardo, Loíza), Eastern Coast; 1–200 m; *fl* Nov–Dec, *fr* Jun, Sep (two specimens).
Greater Antilles, Central America, northern South America to Peru, western tropical Africa. [L2: 157; AR: 259].
Notes: Liogier & Martorell (1982, 2000) and Liogier (1988) misapplied the name *Dioclea hexandra* (Ralph) Mabb., which is now considered a distinct Asian taxon, to this taxon.
Bejuco de mato, Maya prieta.

Eriosema crinitum (Kunth) G. Don, Gen. Hist. 2:348. 1832.
Bas.: *Glycine crinita* Kunth (1824).
Herb in Northern Coastal Lowlands (Laguna Tortuguero); 1–2 m; *fl* & *fr* Aug (one specimen). [*Woodbury a.n. 5152* (UPR)].
Greater Antilles (Cuba, Hispaniola), Central America, South America. [L2: 159].

Erythrina berteroana Urb., Symb. Antill. 5(3):370. 1908.
Tree in Northern Limestone Hills (Vega Baja), Central Cordillera, Maricao Forest Reserve; 100–1025 m; *fl* Jan–Apr, *fr* Mar, Aug–Sep.
Greater Antilles (Cuba), Mexico, Central America, northwestern, west-central South America and Venezuela. [L2: 160].
Bucare enano, Bucayo enano, Bucayo sin espinas, Machete.

Erythrina corallodendrum L., Sp. Pl. 2:706. 1753.
Tree in Guajataca Forest Reserve, Sierra de Luquillo, Central Cordillera (Ciales); 200–550 m; *fl* Feb–Mar, *fr* (?).
Greater Antilles (Jamaica, Hispaniola), Lesser Antilles. [L2: 160].
Notes: The specific epithet is often changed to '*corallodendron*,' but Linnaeus used the above spelling in the protologue.
Bucare, Piñón espinoso.

Erythrina eggersii Krukoff & Moldenke, Phytologia 1(8):289. 1938.
Syn.: *Erythrina horrida* Eggers (1879).
Tree in Northern Limestone Hills, Vieques; 10–100 m; *fl* Mar–Apr, *fr* (?).
Virgin Islands. [L2: 161].
Búcar, Bucare, Bucayo, Coral, Coral vegetal, Espuela de gallo, Piñón espinoso.

Erythrina fusca Lour., Fl. Cochinch. 2:427. 1790.
Syn.: *Erythrina glauca* Willd. (1801).
Tree planted and persistent in Northern Coastal Lowlands; 1–15 m; *fl* Jan–Feb, *fr* Apr (two specimens).
Pantropical. [L2: 161].
Brucayo, Búcar, Bucare, Bucayo.

Erythrina poeppigiana (Walp.) O.F. Cook, U.S.D.A. Div. Bot. Bull. 25:57. 1901.
Bas.: *Micropteryx poeppigiana* Walp. (1851).

Tree planted and persistent in Río Abajo Forest Reserve, Sierra de Luquillo, Central Cordillera, Maricao Forest Reserve; 250–1050 m; *fl* Jan–Mar, *fr* (?).

Native to southern Central America (Panama), northwestern, west-central South America and Venezuela, Brazil. [L2: 163].

Brucayo, Búcar, Bucare, Bucare gigante, Bucayo, Bucayo gigante, Palo de boya.

Erythrina variegata L., Herb. Amb. 10. 1754.
Syn.: *Erythrina variegata* L. var. *orientalis* (L.) Merr. (1917).
Erythrina indica Lam. (1786).

Tree planted and persistent in Northern Coastal Lowlands (Fajardo, Canovanas), Eastern Coast (Naguabo), Caguas Valley (Juncos), Western Coast (Añasco); 5–250 m; *fl* Jan–Mar, *fr* (?).

Native to the Paleotropics. [L2: 163].

Bucare, Bucayo, Bucayo haitiano, Pompón haitiano.

Flemingia macrophylla (Willd.) Merr., Philipp. J. Sci., C 5:130. 1910.
Bas.: *Crotalaria macrophylla* Willd. (1802).

Shrub planted and persistent in Central Cordillera (Jajuya); 700 m; *fl* & *fr* Jan, Aug (two specimens). [*Worthington 33299* (UTEP), not seen; *Axelrod 13963* (UPRRP)].

Native to Indian Subcontinenet, Southeast Asia, Malesia, China.

Flemingia strobilifera (L.) W.T. Aiton, Hort. Kew. (ed. 2) 4:350. 1812.
Bas.: *Hedysarum strobiliferum* L. (1753).

Shrub planted and persistent in Northern Coastal Lowlands (Río Grande, Dorado), Northern Limestone Hills (Arecibo), Caguas Valley (Caguas); 75–175 m; *fl* Apr (one specimen), *fr* Mar–Apr.

Native to tropical Asia, China. [L2: 164].

Notes: Britton & Wilson (1924) used *Moghania strobilifera* (R. Br.) J. St.-Hil. for the name of this taxon.

Galactia dubia DC., Prodr. 2:238. 1825.

Vine in Northern Coastal Lowlands, Northern Limestone Hills (Arecibo, Camuy), Río Abajo Forest Reserve, Las Piedras Chiquitas, Southeastern Foothills, Southern Coastal Lowlands, Southern Coastal Limestone, Susúa Forest Reserve, Guánica Forest Reserve, Southwestern Coastal Lowlands (Cabo Rojo), Sierra Bermeja, Vieques, Mona, Desecheo; 1–500 m; *fl* throughout the year, *fr* Oct–Jul.

West Indies, Mexico, South America (scattered). [L2: 166; AR: 262].

Galactia longifolia (Jacq.) Benth., Comm. Legum. Gen. 63. 1837.
Bas.: *Galega longifolia* Jacq. (1789).

Vine in Southern Coastal Lowlands (Guayama); 1–100 m; *fl* & *fr* Mar–May. [*Liogier 37048* (UPR); *Axelrod 9753* (UPRRP)].

Native to southern South America (Argentina, Paraguay). [L2: 166; AR: 263].

Galactia striata (Jacq.) Urb., Symb. Antill. 2(2):320. 1900.
Bas.: *Glycine striata* Jacq. (1771).

Vine in Northern Coastal Lowlands, Northern Limestone Hills, Río Abajo Forest Reserve, Caguas Valley (Caguas), Las Piedras Chiquitas, Southeastern Foothills, Eastern Cordillera, Southern Coastal Lowlands, Southern Coastal Limestone, Guánica Forest Reserve, Southwestern Coastal Lowlands, Sierra Bermeja, Vieques, Culebra, Mona, Desecheo; 1–550 m; *fl* & *fr* throughout the year.

West Indies, Tobago, southeastern United States (Florida), Mexico, Central America, South America. [L2: 166; AR: 263].

Gliricidia sepium (Jacq.) Kunth in Walpers, Repert. Bot. Syst. 1(4):679. 1842.
Bas.: *Robinia sepium* Jacq. (1760).

Tree in Northern Coastal Lowlands, Northern Limestone Hills, Sierra de Luquillo, Eastern Cordillera, Central Cordillera (Orocovis), Southern Coastal Lowlands, Quebrada de Los Cedros, Susúa Forest Reserve, Southwestern Coastal Lowlands (Lajas, San Germán), Western Coast (Aguada), Vieques; 1–725 m; *fl* throughout the year, *fr* Jan–Jun.

Native to Mexico, Central America. [L2: 168].

Madre de cacao, Mata ratón.

Hebestigma cubense (Kunth) Urb., Symb. Antill. 2(2):289. 1900.
Bas.: *Robinia cubensis* Kunth (1824).

Tree planted and persistent in Northern Coastal Lowlands (Río Grande); 50–200 m; *fl* (?), *fr* Sep–Nov (two specimens). [*Liogier 35673* (UPR); *Mercado 138* (UPRRP)].

Native to Greater Antilles (Cuba). [L2: 170].

Frijolillo.

Indigofera hirsuta L., Sp. Pl. 2:751. 1753.

Herb in Northern Coastal Lowlands (Manatí, Isabela), Southern Coastal Lowlands (Guayama), Southwestern Coastal Lowlands (Lajas), Western Coast (Mayagüez); 10–100 m; *fl* & *fr* Jan–May.

Native to the Paleotropics. [L2: 172].

Indigofera spicata Forssk., Fl. Aegypt.-Arab. 138. 1775.

Herb in Northern Coastal Lowlands (Isabela), Sierra de Cayey, Southeastern Foothills (Salinas), Central Cordillera (Ponce), Southern Coastal Lowlands (Guayanilla), Southern Coastal Limestone (Sabana Grande), Susúa Forest Reserve, Guánica Forest Reserve, Western Coast, Vieques, Mona; 1–350 m; *fl* & *fr* Jan–Oct.

Native to tropical Africa, Yemen. [L2: 172].

Indigofera suffruticosa Mill., Gard. Dict. (ed. 8): *Indigofera* no. 2. 1768.

Suffrutescent herb in Northern Coastal Lowlands, Northern Limestone Hills (San Sebastián), Río Abajo Forest Reserve, Eastern Coast, Las Piedras Chiquitas, Southeastern Foothills, Eastern Cordillera (Aibonito), Central Cordillera, Southern Coastal Lowlands (Caja de Muertos), Southern Coastal Limestone (Yauco), Southwestern Coastal Lowlands, Sierra Bermeja, Western Coast, Mona; 1–1100 m; *fl* & *fr* throughout the year.

West Indies, southern United States, Mexico, Central America, South America. [L2: 173].

Añil, Añil de pasto, Indigo.

Indigofera tinctoria L., Sp. Pl. 2:751. 1753.

Herb in Northern Coastal Lowlands, Sierra de Luquillo, Western Coast (Mayagüez), Vieques, Culebra; 1–300 m; *fl* & *fr* throughout the year.

Native to the Paleotropics; native range uncertain. [L2: 173].

Añil verdadero.

Lablab purpureus (L.) Sweet, Hort. Brit. (ed. 1) 481. 1826.
Bas.: *Dolichos purpureus* L. (1763).
Syn.: *Dolichos lablab* L. (1753).

Vine in Northern Coastal Lowlands, Northern Limestone Hills, Río Abajo Forest Reserve, Sierra de Luquillo, Eastern Coast (Ceiba), Sierra de Cayey, Eastern Cordillera (Cidra, Cayey), Central Cordillera, Southern Coastal Lowlands, Western Coast (Mayagüez), Vieques; 1–725 m; *fl* & *fr* throughout the year.

Native to tropical Africa, Madagascar. [L2: 175; AR: 265].

Chícharos, Chícharos de jardín, Frijol caballero, Guisantes de jardín, Habas de caballo, Haba de tocón, Látiros.

Lonchocarpus glaucifolius Urb., Symb. Antill. 1(2):326. 1899.

Tree in Northern Limestone Hills (Quebradillas, Isabela); 50–200 m; *fl* Apr–Nov, *fr* Jun–Nov.

Greater Antilles (Cuba). [L2: 177].

Geno.

Lonchocarpus heptaphyllus (Poir.) Kunth ex DC., Prodr. 2:259. 1825.
Bas.: *Dalbergia heptaphylla* Poir. (1812).
Syn.: *Lonchocarpus latifolius* (Willd.) Kunth ex DC. (1825).
Lonchocarpus pentaphyllus (Poir.) Kunth ex DC. (1825).

Tree in Northern Coastal Lowlands, Northern Limestone Hills, Río Abajo Forest Reserve, Sierra de Luquillo, Eastern Coast (Naguabo), Susúa Forest Reserve, Maricao Forest Reserve, Mona; 1–480 m; *fl* Mar–Apr, *fr* Apr–Jul.

Antilles, Mexico, Central America, northern South America and Brazil. [L2: 177].

Forteventura, Genogeno, Hediondo, Palo hediondo, Palo seco, Retama.

Lonchocarpus sericeus (Poir.) Kunth ex DC., Prodr. 2:260. 1825.
Bas.: *Robinia sericea* Poir. (1804).
Syn.: *Lonchocarpus domingensis* (Turpin ex Pers.) DC. (1825).

Tree in Northern Coastal Lowlands (Fajardo), Susúa Forest Reserve, Southwestern Coastal Lowlands, Western Coast (Mayagüez), Vieques, Mona; 1–130 m; *fl* Sep–Jan, Apr–Jul, *fr* Nov–Mar, Jun.

Bahamas, Greater Antilles, Lesser Antilles (Martinique). [L2: 177].

Genogeno.

Machaerium lunatum (L.f.) Ducke, Arch. Jard. Bot. Rio de Janeiro 4:310. 1925.
Bas.: *Pterocarpus lunatus* L.f. (1782).
Syn.: *Drepanocarpus lunatus* (L.f.) G. Mey. (1818).

Shrub in wet areas of Northern Coastal Lowlands, Eastern Coast (Humacao), Western Coast (Añasco); 1–2 m; *fl* Dec–Sep, *fr* throughout the year.

Greater Antilles, Lesser Antilles (Martinique, St. Lucia, St. Vincent), Trinidad, southern Central America, northern South America to Brazil. [L2: 179; AR: 266].

Escambrón, Palo de hoz.

Macroptilium atropurpureum (DC.) Urb., Symb. Antill. 9(4):457. 1928.
Bas.: *Phaseolus atropurpureus* DC. (1825).

Vine in Northern Coastal Lowlands (Carolina), Southern Coastal Lowlands (Guayanilla), Southwestern Coastal Lowlands (Cabo Rojo),

Western Coast (Mayagüez); 1–60 m; *fl* Jan–Feb, May–Sep, *fr* Jan–Feb, May–Jul.

Bahamas, Greater Antilles (Cuba, Hispaniola), Lesser Antilles (Guadeloupe, Martinique, St. Lucia), south-central United States (Texas), Mexico, Central America, northern South America to Brazil, Peru. [L2: 181; AR: 266].

Macroptilium lathyroides (L.) Urb., Symb. Antill. 9(4):457. 1928.
Bas.: *Phaseolus lathyroides* L. (1763).

Herb in Northern Coastal Lowlands, Río Abajo Forest Reserve, Sierra de Luquillo, Eastern Coast, Caguas Valley (Juncos), Eastern Cordillera (Aibonito, Cayey), Central Cordillera (Ponce, Adjuntas), Southern Coastal Lowlands, Guánica Forest Reserve, Western Coast, Vieques, Culebra; 1–500 m; *fl* & *fr* throughout the year.

West Indies, Mexico, Central America, South America. [L2: 181; AR: 268].

Habichuela parada, Látiros del pasto.

Millettia pinnata (L.) Panigrahi in Panigrahi & Murti, Fl. Bilaspur Distr. 1:210. 1989.
Bas.: *Cytisus pinnatus* L. (1753).
Syn.: *Pongamia pinnata* (L.) Pierre (1898).

Tree planted and persistent in Río Abajo Forest Reserve, Southern Coastal Lowlands (Aguirre); 1–300 m; *fl* Apr–Jun, Oct, *fr* Apr–Jun.

Native to tropical Asia, China, Japan. [L5: 402].

Mucuna pruriens (L.) DC., Prodr. 2:405. 1825.
Bas.: *Dolichos pruriens* L. (1754).
Syn.: *Stizolobium pruriens* (L.) Medik. (1787).
Stizolobium pruritum (Wight) Piper & Tracy (1910).

Vine in Northern Coastal Lowlands, Northern Limestone Hills (Florida, Utuado), Río Abajo Forest Reserve, Southern Coastal Lowlands (Ponce), Susúa Forest Reserve, Maricao Forest Reserve, Western Coast, Vieques; 25–700 m; *fl* Oct–Mar, Jun, *fr* Oct–Apr.

Native to the Paleotropics. [L2: 213; AR: 270].

Pica-pica.

Mucuna sloanei Fawc. & Rendle, J. Bot. 55:36. 1917.

Vine in Susúa Forest Reserve, Maricao Forest Reserve; 150–700 m; *fl* Apr (one specimen), *fr* (?).

Antilles, southeastern United States (Florida), Central America, South America (scattered). [L2: 185; AR: 270].

Mato, Mato de monte, Ojo de buey.

Mucuna urens (L.) Medik., Vorles. Churpfälz. Phys.-Öcon. Ges. 2:399. 1787.
Bas.: *Dolichos urens* L. (1759).

Vine in Northern Coastal Lowlands (Río Grande), Northern Limestone Hills, Río Abajo Forest Reserve, Sierra de Cayey, Eastern Cordillera (Cerro La Peña), Central Cordillera, Maricao Forest Reserve; 5–725 m; *fl* Oct, Jan–Jun, *fr* May–Nov, Feb.

Antilles, Trinidad, Central America, northern South America to Brazil, Bolivia. [L2: 185; AR: 270].

Mato, Pica-pica, Tortera.

Neorudolphia volubilis (Willd.) Britton in Britton & Wilson, Sci. Surv. Porto Rico & Virgin Islands 5(3):426. 1924.
Bas.: *Rudolphia volubilis* Willd. (1801).

Vine in Northern Limestone Hills, Río Abajo Forest Reserve, Sierra de Luquillo, Sierra de Cayey, Las Tetas de Cayey, Central Cordillera, Susúa Forest Reserve, Maricao Forest Reserve; 100–1325 m; *fl* May–Jan, *fr* Sep–May.

Endemic. [L2: 187; AR: 272].

Bejuco colorado, Bejuco de alambac, Bejuco de alambre, Bejuco prieto.

Ormosia krugii Urb., Symb. Antill. 1(2):320. 1899.

Tree in Northern Coastal Lowlands (Guaynabo), Sierra de Luquillo, Caguas Valley (Caguas), Sierra de Cayey, Central Cordillera, Maricao Forest Reserve; 200–650 m; *fl* Sep–Jan, May, *fr* Nov–Mar, Jun.

Greater Antilles (Hispaniola), Virgin Islands (Guadeloupe, Dominica). [L2: 187].

Mato, Matillo, Palo de matos, Palo de peronías, Peronía.

Pachyrhizus erosus (L.) Urb., Symb. Antill. 4(2):311. 1905.
Bas.: *Dolichos erosus* L. (1753).
Syn.: *Cacara erosa* (L.) Kuntze (1891).

Vine in Northern Coastal Lowlands (San Juan), Northern Limestone Hills, Eastern Cordillera (Corozal), Western Coast (Aguada); 1–325 m; *fl* May–Oct, *fr* May–Sep, Jan.

Native to Mexico, Central America. [L2: 189; AR: 274].

Habilla, Jícama.

Phaseolus lunatus L., Sp. Pl. 2:724. 1753.

Vine in Northern Coastal Lowlands (San Juan), Sierra de Cayey, Southeastern Foothills, Maricao Forest Reserve, Western Coast (Rincón); 2–975 m; *fl* Dec–Apr, Jul–Aug, *fr* Dec–Jan, Apr, Jul–Aug.

Native to Mexico, Central America, northwestern, west-central South America and Venezuela, Brazil. [L2: 192; AR: 275].

Haba, Haba lima.

Phaseolus vulgaris L., Sp. Pl. 2:723. 1753.

Vine in Northern Coastal Lowlands (San Juan), Las Piedras Chiquitas, Southeastern Foothills (Cayey), Southern Coastal Lowlands (Caja de Muertos), Western Coast (Mayagüez); 25–550 m; *fl* Sep–Nov, *fr* Jul, Nov.

Native to Mexico, Central America, South America (scattered). [L2: 194; AR: 275].

Frijole, Habichuela.

Pictetia aculeata (Vahl) Urb., Symb. Antill. 2(2):294. 1900.

Bas.: *Robinia aculeata* Vahl (1793).

Shrub/small tree in Northern Coastal Lowlands (Fajardo and adjacent islands), Las Piedras Chiquitas, Southeastern Foothills, Southern Coastal Lowlands, Southern Coastal Limestone, Guánica Forest Reserve, Southwestern Coastal Lowlands (Cabo Rojo), Sierra Bermeja, Vieques, Culebra; 1–500 m; *fl* May–Oct, Jan, *fr* Jul, Oct–Nov.

Virgin Islands. [L2: 195].

Tachuelo.

Piscidia carthagenensis Jacq., Enum. Syst. Pl. 27. 1760.

Tree in Northern Coastal Lowlands (Fajardo), Cambalache Forest Reserve, Eastern Coast (Cayo Santiago), Southern Coastal Limestone (Peñuelas), Southwestern Coastal Lowlands, Sierra Bermeja, Vieques, Culebra; 1–200 m; *fl* Mar–May, *fr* Dec, Mar–Jul.

Virgin Islands, Lesser Antilles, Mexico, Central America, northwestern, west-central South America and Venezuela. [L2: 195].

Notes: Britton & Wilson (1924) misapplied the name *Ichthyomethia piscipula* (L.) Hitchc. to this taxon.

Ventura.

Poitea florida (Vahl) Lavin, Syst. Bot. Monogr. 37:61. 1993.

Bas.: *Robinia florida* Vahl (1793).

Syn.: *Sabinea florida* (Vahl) DC. (1825).

Shrub/small tree in Northern Coastal Lowlands, Northern Limestone Hills (Arecibo, Isabela), Sierra de Luquillo, Eastern Coast (Naguabo), Central Cordillera (Ponce, Villalba), Susúa Forest Reserve, Guánica Forest Reserve, Southwestern Coastal Lowlands (Cabo Rojo), Western Coast (Rincón), Vieques, Culebra; 25–550 m; *fl* throughout the year, *fr* Feb–Oct.

Virgin Islands. [L2: 205].

Retama, San José.

Poitea paucifolia (DC.) Lavin, Syst. Bot. Monogr. 37:69. 1993.

Bas.: *Corynella paucifolia* DC. (1825).

Shrub in Southern Coastal Limestone, Susúa Forest Reserve, Guánica Forest Reserve, Southwestern Coastal Lowlands (Cabo Rojo), Sierra Bermeja; 25–250 m; *fl* Sep–May, *fr* Feb–May.

Greater Antilles (Hispaniola). [L2: 133].

Notes: Britton & Wilson (1924, 1926) used the variant *Corynella pauciflora* DC. for the name of this taxon, but subsequently (1930) corrected it.

Retama.

Poitea punicea (Urb.) Lavin, Syst. Bot. Monogr. 37:66. 1993.

Bas.: *Sabinea punicea* Urb. (1899).

Shrub in Northern Limestone Hills (Arecibo, San Sebastián), Sierra de Luquillo (La Condesa), Eastern Cordillera (Aibonito), Susúa Forest Reserve, Maricao Forest Reserve, Cerro Las Mesas; 150–900 m; *fl* throughout the year, *fr* Dec–May.

Endemic. [L2: 207].

Caracolillo.

Pterocarpus indicus Willd., Sp. Pl. 3(2):904. 1802.

Tree planted and persistent in Northern Coastal Lowlands, Northern Limestone Hills (Florida), Southern Coastal Lowlands; 1–200 m; *fl* Apr–May, Sep–Oct, *fr* Apr–Dec.

Native to tropical Asia, Japan. [L2: 198].

Terocarpo.

Pterocarpus officinalis Jacq., Select. Stirp. Amer. Hist. 283, *t. 183*, *f. 92*. 1763.

Tree in Northern Coastal Lowlands, Sierra de Luquillo, Eastern Coast (Naguabo), Maricao Forest Reserve, Western Coast (Aguada); 0–550 m; *fl* Feb–Aug, *fr* Jul–Jan.

Antilles, Trinidad, Central America, northwestern South America and Venezuela, Brazil. [L2: 199].

Palo de pollo, Sangre de drago.

Pueraria phaseoloides (Roxb.) Benth., J. Linn. Soc., Bot. 9:125. 1867.

Bas.: *Dolichos phaseoloides* Roxb. (1832).

Vine in Northern Coastal Lowlands, Northern Limestone Hills (San Sebastián), Río Abajo Forest Reserve, Sierra de Luquillo, Eastern Coast (Cuchilla de Panduras), Caguas Valley (Caguas), Sierra de Cayey, Eastern Cordillera (Comerío), Central Cordillera, Western Coast (Mayagüez); 1–800 m; *fl* Nov–Apr, *fr* Dec–Apr.

Native to tropical Asia, China. [L2: 200; AR: 277].

Corsú, Kudzu tropical, Yerba kudzu.

Rhynchosia minima (L.) DC., Prodr. 2:385. 1825.
Bas.: *Dolichos minimus* L. (1753).
Syn.: *Dolicholus minimus* (L.) Medik. (1787).

Vine in Northern Coastal Lowlands, Northern Limestone Hills, Eastern Coast (Naguabo), Sierra de Cayey, Southeastern Foothills (Coamo), Central Cordillera (Orocovis), Southern Coastal Lowlands, Southwestern Coastal Lowlands, Western Coast, Vieques, Culebra, Mona, Desecheo; 1–600 m; *fl* & *fr* throughout the year.

Pantropical. [L2: 200; AR: 278].

Frijolillo, Habilla.

Rhynchosia phaseoloides (Sw.) DC., Prodr. 2:385. 1825.
Bas.: *Glycine phaseoloides* Sw. (1788).

Vine in Río Abajo Forest Reserve, Southeastern Foothills (Salinas), Eastern Cordillera (Cerro La Peña), Central Cordillera (Adjuntas), Maricao Forest Reserve; 200–900 m; *fl* Feb–Mar, *fr* Feb–May.

West Indies, Trinidad, southern Central America (Panama), northern South America to Bolivia, Paraguay. [L2: 203; AR: 279].

Bejuco de paloma, Peonías.

Rhynchosia pyramidalis (Lam.) Urb., Repert. Spec. Nov. Regni Veg. 15:318. 1918.
Bas.: *Dolichos pyramidalis* Lam. (1786).

Vine in Northern Coastal Lowlands (Río Grande), Northern Limestone Hills (Florida), Río Abajo Forest Reserve; 1–300 m; *fl* (?), *fr* Feb, May–Jul.

Greater Antilles, eastern Mexico, northern Central America.

Notes: Britton & Wilson (1924) used the name *Dolicholus pyramidalis* (Lam.) Britton & P. Wilson, in part, for this taxon and misapplied it, in part, to *Rhynchosia phaseoloides*.

Rhynchosia reticulata (Sw.) DC., Prodr. 2:385. 1825.
Bas.: *Glycine reticulata* Sw. (1788).
Syn.: *Dolicholus reticulatus* (Sw.) Millsp. (1900).

Vine in Northern Coastal Lowlands, Northern Limestone Hills, Río Abajo Forest Reserve, Eastern Coast, Las Tetas de Cayey, Las Piedras Chiquitas, Southeastern Foothills, Eastern Cordillera, Central Cordillera (Yauco), Susúa Forest Reserve, Maricao Forest Reserve, Southwestern Coastal Lowlands (Cabo Rojo), Sierra Bermeja, Cerro Las Mesas, Vieques, Culebra, Mona, Desecheo; 1–825 m; *fl* throughout the year, *fr* Aug–May.

West Indies, Mexico, Central America, northwestern, west-central South America and Venezuela, Paraguay. [L2: 204; AR: 279].

Frijolillo, Habilla.

Sesbania emerus (Aubl.) Urb., Repert. Spec. Nov. Regni Veg. 16:149. 1919.
Bas.: *Aeschynomene emerus* Aubl. (1775).

Suffrutescent herb/shrub in Northern Coastal Lowlands (Toa Baja, Arecibo), Sierra de Cayey, Southwestern Coastal Lowlands (Lajas); 1–800 m; *fl* & *fr* Nov–Jan, Apr.

Bahamas, Greater Antilles (Cuba, Hispaniola), Lesser Antilles, southeastern United States (Florida), Central America, northeastern South America (Brazil). [L2: 208].

Sesbania grandiflora (L.) Pers., Syn. Pl. 2(2):316. 1807.
Bas.: *Robinia grandiflora* L. (1753).
Syn.: *Sesban grandiflorus* (L.) Poir. (1806). *Agati grandiflora* (L.) Desv. (1813).

Small tree planted and persistent in Northern Coastal Lowlands (Canóvanas, Arecibo), Sierra de Cayey, Southwestern Coastal Lowlands (San Germán), Western Coast (Añasco); 1–600 m; *fl* Oct–Dec, Apr–May, *fr* Dec, May, Aug.

Possibly native to Indonesia. [L2: 209].

Agati, Báculo, Cresta de gallo, Gallito.

Sesbania sericea (Willd.) Link, Enum. Hort. Berol. Alt. 2:244. 1822.
Bas.: *Coronilla sericea* Willd. (1809).
Syn.: *Sesban sericea* (Willd.) DC. (1825).

Suffrutescent herb in Northern Coastal Lowlands, Eastern Coast, Caguas Valley, Eastern Cordillera (Cayey), Southern Coastal Lowlands, Guánica Forest Reserve, Southwestern Coastal Lowlands, Western Coast, Vieques, Culebra; 1–400 m; *fl* Oct–Apr, *fr* Oct–Jun.

Native to tropical Africa, Arabia, Sri Lanka, southeastern Asia. [L2: 209].

Papagayo.

Sesbania sesban (L.) Merr., Philipp. J. Sci., C 7(4): 235. 1912.
Bas.: *Aeschynomene sesban* L. (1753).
Syn.: *Sesban sesban* (L.) Britton (1918).

Suffrutescent herb/shrub in Northern Coastal Lowlands, Northern Limestone Hills (San Sebastián), Eastern Coast (Ceiba), Sierra de Cayey; 1–800 m; *fl* & *fr* Oct–Dec, Apr.

Native to tropical Africa, temperate and tropical Asia, Australia. [L2: 210].

Sesbania.

Sophora tomentosa L., Sp. Pl. 1:373. 1753.
Suffrutescent herb/shrub in Northern Coastal Lowlands (Fajardo), Eastern Coast (Humacao), Southwestern Coastal Lowlands (Guánica), Vieques; 0–1 m; *fl* Feb, May–Jun, *fr* Feb, May (two specimens).
Pantropical. [L2: 212].

Stylosanthes guianensis (Aubl.) Sw., Kongl. Vetensk. Acad. Nya Handl. 10:301. 1789.
Bas.: *Trifolium guianense* Aubl. (1775).
Suffrutescent herb in Sierra de Luquillo; 600–650 m; *fl* Jan–Mar (two specimens), *fr* Mar (one specimen). [*Proctor 43208* (SJ); *Axelrod 12415* (UPRRP)].
Native to Mexico, Central America, northern South America to Brazil, Bolivia. [L2: 215].

Stylosanthes hamata (L.) Taub., Verh. Bot. Vereins Prov. Brandenburg 32(1):22. 1890.
Bas.: *Hedysarum hamatum* L. (1759).
Prostrate herb in Northern Coastal Lowlands, Northern Limestone Hills (Camuy), Río Abajo Forest Reserve, Eastern Coast (Ceiba), Las Piedras Chiquitas, Southeastern Foothills (Salinas), Eastern Cordillera (Corozal), Southern Coastal Lowlands, Southern Coastal Limestone (Guayanilla), Susúa Forest Reserve, Guánica Forest Reserve, Southwestern Coastal Lowlands, Sierra Bermeja, Western Coast, Vieques, Mona; 1–550 m; *fl* & *fr* throughout the year.
West Indies, Central America (Guatemala, Nicaragua), northern South America (Venezuela, Colombia) and Brazil. [L2: 215].
Zarzabacoa enana.

Stylosanthes viscosa (L.) Sw., Prodr. 108. 1788.
Bas.: *Hedysarum hamatum* L. var. *viscosum* L. (1759) "Viscosa."
Suffrutescent herb in Southwestern Coastal Lowlands (Cabo Rojo), Sierra Bermeja; 25–300 m; *fl* Feb–Mar, Jun, Sep–Nov, *fr* Feb–Mar, Jun.
Greater Antilles, Mexico, Central America, South America. [L5: 403].

Tephrosia candida DC., Prodr. 2:249. 1825.
Shrub planted and persistent in Central Cordillera (Los Tres Picachos); 800 m; *fl* (?), *fr* Mar (one specimen). [*Liogier 34079* (UPR)].
Native to Indian Subcontinent. [L2: 217].
Tefrosia.

Tephrosia cinerea (L.) Pers., Syn. Pl. 2(2):328. 1807.
Bas.: *Galega cinerea* L. (1759).
Syn.: *Cracca cinerea* (L.) Morong (1892).
Herb in Northern Coastal Lowlands, Eastern Coast (Ceiba), Southeastern Foothills, Southern Coastal Lowlands, Guánica Forest Reserve, Southwestern Coastal Lowlands, Sierra Bermeja, Cerro Las Mesas, Vieques, Culebra, Mona, Desecheo; 1–300 m; *fl* & *fr* throughout the year.
West Indies, Mexico, Central America, South America. [L2: 218].
Añil cenizo.

Tephrosia noctiflora Bojer ex Baker in Oliver, Fl. Trop. Afr. 2:112. 1871.
Suffrutescent herb in Northern Limestone Hills (Vega Baja); 50 m; *fl* & *fr* Oct (one specimen). [*Proctor 51452* (SJ)].
Native to tropical Africa, India.

Tephrosia senna Kunth in Humboldt, Bonpland & Kunth, Nov. Gen. Sp. 6:359 [folio]; 6:458 [quarto]. 1824.
Syn.: *Cracca cathartica* (Sessé & Moc.) Britton & Millsp. (1920).
Suffrutescent herb in Northern Coastal Lowlands, Southern Coastal Lowlands (Ponce), Southwestern Coastal Lowlands (Cabo Rojo), Sierra Bermeja; 1–175 m; *fl* & *fr* Nov–Jun.
Bahamas, Greater Antilles (Hispaniola), Lesser Antilles, Mexico, northern Central America, northern South America (Colombia). [L2: 219].
Añil racemillo.

Teramnus labialis (L.f.) Spreng., Syst. Veg. 3:235. 1826.
Bas.: *Glycine labialis* L.f. (1782).
Vine in Northern Coastal Lowlands, Eastern Coast, Las Piedras Chiquitas, Southeastern Foothills, Central Cordillera (Coamo), Southern Coastal Lowlands (Caja de Muertos, Ponce), Guánica Forest Reserve, Vieques, Culebra, Mona; 1–550 m; *fl* May, Aug–Feb, *fr* May–Feb.
West Indies, Central America. [L2: 220; AR: 282].
Frijolillo.

Teramnus uncinatus (L.) Sw., Prodr. 105. 1788.
Bas.: *Dolichos uncinatus* L. (1763).
Vine in Río Abajo Forest Reserve, Caguas Valley (Caguas), Sierra de Cayey, Las Piedras Chiquitas, Southeastern Foothills, Eastern Cordillera (Barranquitas), Central Cordillera (Orocovis), Maricao Forest Reserve; 50–925 m; *fl* Aug, Nov–Jan, *fr* Nov–May.
Bahamas, Greater Antilles, Mexico, Central America, tropical South America. [L2: 221; AR: 283].
Cresta de gallo blanco, Frijolillo peludo.

Vigna adenantha (G. Mey.) Maréchal, Mascherpa & Stainier, Taxon 27(2–3):202. 1978.
Bas.: *Phaseolus adenanthus* G. Mey. (1818).

Vine in Northern Coastal Lowlands, Northern Limestone Hills (Arecibo, Utuado), Sierra de Luquillo, Eastern Coast, Southwestern Coastal Lowlands (Lajas), Western Coast; 1–250 m; *fl* Nov–Apr, *fr* Dec–Apr.

Antilles, Mexico, Central America, South America. [L2: 224; AR: 285].

Habichuela cimarrona.

Vigna antillana (Urb.) Fawc. & Rendle, Fl. Jamaica 4(2):69. 1920.
Bas.: *Phaseolus antillanus* Urb. (1905).

Vine in Central Cordillera (Adjuntas), Southern Coastal Lowlands (Caja de Muertos), Vieques, Culebra; 1–600 m; *fl* Nov–Feb, *fr* Jan (two specimens).

Greater Antilles, Virgin Islands, Lesser Antilles (Martinique, St. Vincent). [L2: 192; AR: 285].

Vigna hosei (Craib) Backer in Backer & van Slooten, Geill. Handb. Jav. Theeonkr. 153. 1924.
Bas.: *Dolichos hosei* Craib (1914).

Vine in Northern Coastal Lowlands, Río Abajo Forest Reserve, Sierra de Luquillo, Eastern Coast (Humacao), Sierra de Cayey, Eastern Cordillera (Cañón San Cristóbal); 5–950 m; *fl* Jan–Oct, *fr* May–Jul.

Probably native to Africa. [L2: 225; AR: 286].

Frijol de Abisinia.

Vigna longifolia (Benth.) Verdc., Kew Bull. 24(3):541. 1970.
Bas.: *Phaseolus longifolius* Benth. (1837).
Syn.: *Phaseolus trichocarpus* C. Wright (1869).

Vine in Northern Coastal Lowlands (Laguna Tortuguero); 1–2 m; *fl* Nov, Feb–Mar, *fr* Mar–May (two specimens).

Greater Antilles (Cuba), Central America (Nicaragua, Panama), southern South America to Brazil, Ecuador. [L2: 225; AR: 286].

Habichuela cimarrona.

Vigna luteola (Jacq.) Benth. in Martius, Fl. Bras. 15(1B):194, *t. 50, f.* 2. 1859.
Bas.: *Dolichos luteolus* Jacq. (1771).
Syn.: *Vigna repens* (L.) Kuntze (1891).

Vine in Northern Coastal Lowlands, Northern Limestone Hills, Río Abajo Forest Reserve, Sierra de Luquillo, Eastern Coast, Caguas Valley (Gurabo), Sierra de Cayey, Southeastern Foothills (Salinas), Eastern Cordillera (Cerro La Peña, Aguas Buenas), Toro Negro Forest Reserve, Southern Coastal Lowlands (Ponce), Maricao Forest Reserve, Southwestern Coastal Lowlands (Lajas), Western Coast, Vieques, Culebra; 1–1300 m; *fl* & *fr* throughout the year.

Naturalized throughout the tropics and subtropics; native range uncertain. [L2: 226; AR: 288].

Frijol silvestre.

Vigna marina (Burm.) Merr., Interpr. Herb. Amboin. 285. 1917.
Bas.: *Phaseolus marinus* Burm. (1755).
Syn.: *Vigna retusa* (E. Mey.) Walp. (1840).

Vine in Northern Coastal Lowlands (San Juan); 10 m; *fl* Oct (one specimen), *fr* (?). [*Molina 56* (UPRRP)].

Probably native to the Paleotropics. [L2: 226; AR: 288].

Vigna peduncularis (Kunth) Fawc. & Rendle, Fl. Jamaica. 4(2):68. 1920.
Bas.: *Phaseolus peduncularis* Kunth (1824).

Vine in Río Abajo Forest Reserve, Central Cordillera (Utuado); 250–350 m; *fl* Jan, Jun–Aug, *fr* Jan, Jun–Jul.

Greater Antilles, Virgin Islands, Trinidad, Central America, South America. [L2: 192; AR: 290].

Vigna unguiculata (L.) Walp., Repert. Bot. Syst. 1(5):779. 1843.
Bas.: *Dolichos unguiculatus* L. (1753).

Vine in Northern Coastal Lowlands (Fajardo, San Juan), Northern Limestone Hills (Dorado to Vega Baja), Río Abajo Forest Reserve, Sierra de Luquillo, Caguas Valley (Gurabo), Southern Coastal Lowlands (Coamo); 10–700 m; *fl* Feb–Nov, *fr* Feb–Mar, Aug–Nov.

Native to Africa. [L2: 228; AR: 290].

Frijol, Lentejas.

Vigna vexillata (L.) A. Rich. in Sagra, Hist. Fis. Cuba, Bot. 10:191. 1845.
Bas.: *Phaseolus vexillatus* L. (1753).

Vine in Northern Coastal Lowlands, Northern Limestone Hills, Río Abajo Forest Reserve, Eastern Coast, Caguas Valley, Southeastern Foothills (Arroyo), Eastern Cordillera (Aguas Buenas), Central Cordillera (Utuado), Southern Coastal Lowlands, Western Coast (Mayagüez), Vieques; 1–900 m; *fl* & *fr* throughout the year.

Throughout the tropics and subtropics. [L2: 228; AR: 291].

Frijol cimarrón.

Zornia gemella Vogel, Linnaea 12(1):61. 1838.

Herb in Northern Limestone Hills (Quebradillas), Susúa Forest Reserve; 50–185 m; *fl* & *fr* May, Nov (two specimens). [*Breckon 4612, 6098* (MAPR)].

Greater Antilles (Cuba), Lesser Antilles

(Martinique), south-central United States (Texas), Mexico, Central America (Nicaragua, Costa Rica), South America. [L2: 229].

Zornia reticulata Sm. in Rees, Cycl. 39: *Zornia* no. 2. 1818.

Herb in Northern Coastal Lowlands, Eastern Coast (Punta Guayanés), Maricao Forest Reserve, Southwestern Coastal Lowlands (Cabo Rojo), Sierra Bermeja, Cerro Las Mesas; 1–300 m; *fl* & *fr* throughout the year.

Greater Antilles, Lesser Antilles (St. Kitts, Montserrat), south-central United States (Texas, Arizona), Mexico, Central America, South America. [L2: 230].

Notes: Britton & Wilson (1924) misapplied the name *Zornia diphylla* (L.) Pers. to this taxon.

Zarzabacoa de dos hojas.

TAXA NOT RECENTLY COLLECTED

Crotalaria sagittalis L., Sp. Pl. 2:714. 1753.

Syn.: *Crotalaria sagittalis* L. var. *fruticosa* (Mill.) Fawc. & Rendle (1920).

Herb collected at the shore in Bayamón on 27 Mar 1885 [*Sintenis 1093* (NY)].

Greater Antilles, eastern United States, Mexico, Central America, South America (Colombia, Brazil, Bolivia, Peru). [L2: 140].

Desmodium cubense Griseb., Cat. Pl. Cub. 73. 1866.

Syn.: *Meibomia cubensis* (Griseb.) Schindl. (1924).

Herb cited by Britton & Wilson (1926) as, according to Urban, having been collected by Sintenis at Monte Santana, near Coamo (no specimen seen).

Greater Antilles (Cuba, Hispaniola), northern Central America (Guatemala). [L2: 152].

Indigofera micheliana Rose, Contr. U.S. Natl. Herb. 8(4):310. 1905.

Shrub collected in Guánica on 10 Feb 1886 [*Sintenis 3682* (US)] and in Rincón on 12 Dec 1886 [*Sintenis 5604* (US)].

Virgin Islands (St. Thomas), Mexico, Central America, western-central South America (Ecuador, Peru). [L2: 172].

Notes: Britton & Wilson (1924) used *Indigofera guatimalensis* Moc., Sessé & Cerv. ex Prain & Backer f. ['*guatemalensis*'], a *nom. inval.*, for the name of this taxon, as did Liogier & Martorell (1982, 2000) ['ex Prain & Backer' omitted in both] and Liogier (1988).

Sesbania bispinosa (Jacq.) W. Wight, U.S.D.A. Bur. Pl. Industr. Bull. 137:15. 1909.

Bas.: *Aeschynomene bispinosa* Jacq. (1795).

Syn.: *Sesban bispinosus* (Jacq.) Rydb. (1924).

Suffrutescent herb/shrub collected in Coamo on 9 Dec 1885 [*Sintenis 3012* (US)] and in Guayanilla on 13 Nov 1913 [*Stevens 5926* (NY)].

Native to tropical Africa, Iran, Afghanistan, Indian Subcontinent, southeastern Asia. [L2: 208].

Notes: Liogier & Martorell (1982, 2000) and Liogier (1988) used the homonym *Sesbania bispinosa* (Jacq.) Spreng. ex Steud. for the name of this taxon.

Sesbania exasperata Kunth in Humboldt, Bonpland & Kunth., Nov. Gen. Sp. 6:417 [folio]; 6:534 [quarto] 1824.

Suffrutescent herb/shrub collected in Guánica in the water of a lagoon on 18 Feb 1886 [*Sintenis 3802* (NY), scan seen] and at Cartagena Lagoon in May 1924 [*Danforth a.n. 855807* (NY), scan seen].

Native to Trinidad and Tobago, southern Central America (Panama), South America. [B&W5: 395].

Notes: Britton and Wilson (1924) probably misapplied the name *Sesban emerus* (Aubl.) Urb. to specimens of this taxon.

Trifolium pratense L., Sp. Pl. 2:768. 1753.

Herb collected between Ponce and Adjuntas on 11 Jul 1901 [*Underwood & Griggs 760* (US)] and in the Insular Fair Grounds, San Juan, on 11 Jul 1919 [*Shafer & Watson 3818* (NY), scan seen].

Native to Europe, northern Africa, temperate western and central Asia, northern Indian Subcontinent. [L2: 222].

TAXA CITED BY PREVIOUS AUTHORS, BUT NOT INCLUDED

DOUBTFUL OCCURRENCE

Aeschynomene evenia C. Wright.

Herb cited by Liogier & Martorell (1982, 2000) and Liogier (1988) as doubtfully recorded from the Sierra de Luquillo. [L2: 109].

CITED DUE TO MISIDENTIFICATION

Crotalaria pallida Aiton var. ***pallida.*** Syn.: *Crotalaria striata* DC.

Herb cited by Liogier & Martorell (1982, 2000) and Liogier (1988) as occurring in Cerro Las Mesas, Mayagüez. The specimens that I have seen identified as this taxon are referable to *Crotalaria pallida* var. *obovata*. [L2: 138].

Crotalaria spectabilis Roth.

Herb cited by Liogier & Martorell (1982, 2000) and Liogier (1988) as occurring in Cerro Las Mesas, Mayagüez and Luquillo, in mangrove. The specimens that I have seen identified as this taxon are referable to *Crotalaria retusa*. [L2: 141].

Phaseolus polystachios (L.) Britton, Sterns & Poggenb.
Vine cited by Liogier & Martorell (1982, 2000) and Liogier (1988) as occurring in thickets at lower elevations, Rincón; but, according to Acevedo-Rodríguez (2005), this citation is based on the misidentification of a specimen of *Phaseolus lunatus*. [L2: 194; AR: 277].

Tephrosia sessiliflora (Poir.) Hassl. Syn.: *Cracca brevipes* (Benth.) Kuntze.
Suffrutescent herb citied by Britton & Wilson (1924), as having been attributed to Puerto Rico by Poiret, in error; even so, it was still cited by Liogier (1988) and Liogier & Martorell (2000) as recorded for Puerto Rico by Poiret. [L2: 219].

Vigna juruana (Harms) Verdc.
Herb cited in Liogier & Martorell (1982, 2000) and Liogier (1988) as occurring in the eastern region; this citation was based on the misidentification of a specimen of *Vigna hosei*. [L2: 225].

ONLY UNDER CULTIVATION

Arachis hypogaea L.
Herb cited by Liogier & Martorell (1982, 2000) and Liogier (1988) as subspontaneous after cultivation. [L2: 115].

Canavalia ensiformis (L.) DC.
Herb cited by Liogier & Martorell (1982, 2000) and Liogier (1988) as subspontaneous after cultivation. [L2: 123; AR: 247].

Cicer arietinum L.
Herb cited by Britton & Wilson (1924) as occurring on waste grounds, San Juan, in 1914; subsequently cited by Liogier & Martorell (1982, 2000) and Liogier (1988) as perhaps not persistent. [L2: 129].

Clitoria fairchildiana R.A. Howard.
Herb cited by Liogier & Martorell (2000) as planted and naturalized.

Codariocalyx gyroides (Roxb. ex Link) Hassk. Syn.: *Desmodium gyroides* (Roxb. ex Link) DC.
Shrub cited by Liogier (1988) and Liogier & Martorell (2000) as introduced as a forage plant and escaped in Corozal. [L2: 153].

Crotalaria juncea L.
Herb cited by Britton & Wilson (1924) as cultivated as a nitrogen-fixing plant; subsequently cited by Liogier & Martorell (1982, 2000) and Liogier (1988) as cultivated and probably escaped. [L2: 138].

Crotalaria micans Link.
Shrub cited by Liogier (1988) and Liogier & Martorell (2000) as planted as experimental fodder at Mayagüez. [L2: 480].

Erythrina crista-galli L.
Tree cited by Liogier & Martorell (1982, 2000) and Liogier (1988) as occasionally planted and escaped. [L2: 161].

Glycine max (L.) Merr. Syn.: *Soja max* (L.) Piper.
Herb cited by Liogier & Martorell (1982, 2000) and Liogier (1988) as cultivated and escaped in western Puerto Rico. [L2: 168].

Medicago sativa L.
Herb cited Britton & Wilson (1924) as occurring in waste grounds, Mayagüez, and occasionally cultivated for fodder and persistent for a few years; subsequently cited by Liogier & Martorell (1982, 2000) and Liogier (1988). [L2: 182].

Melilotus albus Medik.
Herb cited by Liogier & Martorell (1982, 2000) and Liogier (1988) as introduced as a forage crop and perhaps not persistent. [L2: 183].

Myrospermum frutescens Jacq.
Shrub/small tree cited by Britton & Wilson (1924) as occurring in roadsides and woodlands, spontaneous after cultivation and naturalized; subsequently cited by Liogier & Martorell (1982, 2000) and Liogier (1988). [L2: 186].

Pachyrhizus tuberosus (Lam.) Spreng.
Vine cited by Liogier & Martorell (1982, 2000) and Liogier (1988) as occasionally cultivated. [L2: 190].

Pterocarpus macrocarpus Kurz.
Tree cited by Britton & Wilson (1926) as planted in Mayagüez and by Liogier & Martorell (1982, 2000) and Liogier (1988) as sparingly planted for ornament and shade. [L2: 198].

Rhynchosia caribaea (Jacq.) DC.
Vine cited by Liogier & Martorell (1982) as collected by Sintenis in Adjuntas, and by Liogier (1988) and Liogier & Martorell (2000), who add that it is introduced and perhaps not permanent; also cited by Acevedo-Rodríguez (2005), who did not believe that it had become naturalized. [L2: 202; AR: 282].

Sesbania tomentosa Hook. & Arn.
Shrub cited by Liogier (1988) and Liogier & Martorell (2000) as introduced and escaped in Gurabo and Lajas. [L2: 211].

FABACEAE-MIMOSOIDEAE (LEGUMINOSAE-MIMOSOIDEAE)

Acacia laeta R. Br. ex Benth., London J. Bot. 1:508. 1842.

Tree in Vieques; 2–3 m; *fl* & *fr* Sep (one specmen). [*Breckon 8343* (MAPR)].

Native to tropical Africa, Arabian peninsula.

Adenanthera pavonina L., Sp. Pl. 1:384. 1753.

Tree in Northern Coastal Lowlands (Carolina), Northern Limestone Hills, Cerro Las Mesas; 100–300 m; *fl* Jul (two specimens), *fr* Apr–Jun, Oct–Nov.

Native to tropical Asia, China, Australia. [L2: 20].

Coral, Coralitos, Mato colorado, Palo de mato, Peronías, Peronías chatas.

Albizia carbonaria Britton in Britton & Wilson, Sci. Surv. Porto Rico & Virgin Islands 6(3):348. 1926.
Syn.: *Pithecellobium carbonarium* (Britton) Niezgoda & Nevling (1979) "carbonaria."

Tree planted and persistent in Central Cordillera (Adjuntas); 550–600 m; *fl* & *fr* Jan, May (two specimens). [*Axelrod 11366* (UPRRP); *González 273* (UPR)].

Native to southern Central America (Panama), northern South America (Venezuela, Colombia). [L2: 45].

Carbonero.

Albizia lebbeck (L.) Benth., London J. Bot. 3:87. 1844.
Bas.: *Mimosa lebbeck* L. (1753).

Tree in Northern Coastal Lowlands, Southern Coastal Lowlands (Juana Díaz, Villalba), vicinity of Susúa Forest Reserve (Sabana Grande), Southwestern Coastal Lowlands, Vieques; 1–225 m; *fl* Apr–Jul, Oct, *fr* Oct–Jul.

Probably native to tropical Asia. [L2: 22].

Acacia, Acacia amarilla, Amor platónico, Aroma, Casia amarilla, Lengua viperina, Lengua de mujer.

Albizia procera (Roxb.) Benth., London J. Bot. 3:89. 1844.
Bas.: *Mimosa procera* Roxb. (1799).

Tree in Northern Coastal Lowlands, Northern Limestone Hills (San Sebastián), Sierra de Luquillo, Eastern Coast (Punta Guayanés), Western Coast (Añasco); 1–800 m; *fl* Jun–Sep, *fr* Sep–Mar.

Native to tropical Asia, China, Australia. [L2: 24].

Albicia.

Anadenanthera peregrina (L.) Speg., Physis (Buenos Aires) 6:314. 1923.
Bas.: *Mimosa peregrina* L. (1753).
Syn.: *Piptadenia peregrina* (L.) Benth. (1841). *Niopa peregrina* (L.) Britton & Rose (1927).

Tree in Southeastern Foothills (Cayey), Central Cordillera (Villalba, Ponce), Western Cordillera (Maricao), Southwestern Coastal Lowlands (Cabo Rojo); 50–350 m; *fl* Mar–May, *fr* Oct–Jan.

Greater Antilles (Hispaniola), Lesser Antilles, Trinidad, northern and central South America. [L2: 43].

Cohoba, Cojoba, Cojóbana, Cojobilla, Cojobo.

Calliandra haematomma (Bertero ex DC.) Benth. var. **haematomma**, London J. Bot. 3:103. 1844.
Bas.: *Acacia haematomma* Bertero ex DC. (1825) "haematoma."
Syn.: *Anneslia haematomma* (Bertero ex DC.) Britton (1918) "haematostoma."

Shrub in Southern Coastal Limestone; 10–25 m; *fl* (?), *fr* Aug–Oct.

Greater Antilles (Hispaniola), Virgin Islands (St. Thomas). [L2: 25].

Calliandra haematomma (Bertero ex DC.) Benth. var. **locoensis** (R.G. García & Kolt.) Barneby, Mem. New York Bot. Gard. 74(3):144. 1998.
Bas.: *Calliandra locoensis* R.G. García & Kolt. (1992).

Shrub in Susúa Forest Reserve; 175 m; *fl* Oct–Nov, Mar–May, *fr* Nov, Apr–May.

Endemic. [L5: 398].

Cojobilla colorada.

Calliandra surinamensis Benth., London J. Bot. 3:105. 1844.

Shrub in Northern Coastal Lowlands (Laguna Tortuguero), Northern Limestone Hills (Morovis), Eastern Coast (Naguabo), Eastern Cordillera (Aibonito); 25–500 m; *fl* & *fr* Dec–Mar.

Native from northern South America to Brazil, Peru. [L2: 26].

Caliandra de Surinam.

Cojoba arborea (L.) Britton & Rose, N. Amer. Fl. 23(1):29. 1928.
Bas.: *Mimosa arborea* L. (1753).
Syn.: *Pithecellobium arboreum* (L.) Urb. (1900) "Pithecolobium."

Tree in Northern Limestone Hills, Río Abajo Forest Reserve, Susúa Forest Reserve, Maricao Forest Reserve; 25–550 m; *fl* Feb–Apr, Sep, *fr* Jan, Jun–Jul.

Greater Antilles, Mexico, Central America, northwestern and west-central South America. [L2: 45].

Acacia silvestre, Cohoba, Cojoba, Cojóbana, Cojóbana negra, Cojobanilla, Tamarindillo.

Desmanthus leptophyllus Kunth in Humboldt, Bonpland & Kunth, Nov. Gen. Sp. 6:208 [folio]; 6:264 [quarto]. 1824.

Suffrutescent herb in Northern Coastal

Lowlands, Northern Limestone Hills (Quebradillas, Isabela), Eastern Coast (Ceiba), Southeastern Foothills, Southern Coastal Lowlands, Southwestern Coastal Lowlands (Lajas), Western Coast (Mayagüez), Vieques, Mona, Desecheo; 1–600 m; *fl* & *fr* Jun–Feb.

Antilles, Mexico, Central America, northwestern, west-central South America and Venezuela. [L5: 399].

Desmanto.

Desmanthus pernambucanus (L.) Thell., Mém Soc. Sci. Nat. Cherbourg 38(sér. 4, vol. 8):296. 1912 (Fl. Adv. Montpellier).
Bas.: *Mimosa pernambucana* L. (1753).

Suffrutescent herb in Northern Coastal Lowlands, Northern Limestone Hills (Quebradillas), Caguas Valley (Caguas), Southeastern Foothills (Cayey), Southern Coastal Lowlands, Guánica Forest Reserve, Southwestern Coastal Lowlands (Cabo Rojo), Sierra Bermeja, Vieques, Culebra, Mona, Desecheo; 1–600 m; *fl* & *fr* throughout the year.

Bahamas, Greater Antilles (Cuba, Hispaniola), Virgin Islands, Lesser Antilles, Tobago, South America (scattered). [L5: 399].

Desmanto.

Desmanthus virgatus (L.) Willd., Sp. Pl. 4(2):1047. 1806.
Bas.: *Mimosa virgata* L. (1753).
Syn.: *Acuan virgatum* (L.) Medik. (1786). *Desmanthus depressus* Humb. & Bonpl. ex Willd. (1806). *Acuan depressum* (Humb. & Bonpl. ex Willd.) Kuntze (1891) "depressa." *Acuan insulare* Britton & Rose (1928).

Suffrutescent herb in Northern Coastal Lowlands, Río Abajo Forest Reserve, Caguas Valley (Juncos), Southeastern Foothills (Salinas), Eastern Cordillera (Aibonito), Central Cordillera (Ponce), Southern Coastal Lowlands, Southern Coastal Limestone, Southwestern Coastal Lowlands, Western Coast, Vieques, Culebra, Desecheo; 1–350 m; *fl* & *fr* throughout the year.

Antilles, south-central United States (Texas), Mexico, Central America, South America. [L2: 27, 28; L5: 399].

Desmanto.

Enterolobium cyclocarpum (Jacq.) Griseb., Fl. Brit. W.I. 226. 1860.
Bas.: *Mimosa cyclocarpa* Jacq. (1800).

Tree planted and persistent in Northern Limestone Hills (Arecibo), Eastern Coast (Ceiba); 5–150 m; *fl* May (one specimen), *fr* Jan, Apr–May (two specimens).

Native to Mexico, Central America, northwestern South America and Venezuela, Brazil. [L2: 30].

Dormilón, Guanacaste, Oreja de mono.

Inga laurina (Sw.) Willd., Sp. Pl. 4(2):1018. 1806.
Bas.: *Mimosa laurina* Sw. (1788).
Syn.: *Inga fagifolia* (L.) Willd. ex Benth. (1875).

Tree in Northern Coastal Lowlands, Northern Limestone Hills, Sierra de Luquillo, Caguas Valley (Caguas), Sierra de Cayey, Las Tetas de Cayey, Southeastern Foothills (Salinas), Eastern Cordillera, Susúa Forest Reserve, Maricao Forest Reserve, Sierra Bermeja, Vieques; 1–925 m; *fl* Aug–May, *fr* Feb–Jun, Sep–Nov.

Greater Antilles (Cuba, Hispaniola), Virgin Islands, Lesser Antilles, Trinidad, Mexico, Central America, South America (scattered). [L2: 33].

Guamá.

Inga nobilis Willd. subsp. **quaternata** (Poepp. & Endl.) T.D. Penn., Gen. *Inga*, Bot. 383. 1997.
Bas.: *Inga quaternata* Poepp. & Endl. (1845).

Tree in Northern Limestone Hills (Ciales, Hatillo), Río Abajo Forest Reserve, Sierra de Luquillo, Sierra de Cayey, Las Tetas de Cayey, Eastern Cordillera (Aibonito), Central Cordillera, Western Cordillera; 100–1000 m; *fl* & *fr* throughout the year.

Native to Mexico, Central America, northern South America to Brazil, Bolivia. [L2: 33].

Guamá venezolano.

Inga vera Willd., Sp. Pl. 4(2):1010. 1806.
Syn.: *Inga inga* (L.) Britton (1918).

Tree in Northern Coastal Lowlands (Río Grande), Northern Limestone Hills, Río Abajo Forest Reserve, Sierra de Luquillo, Sierra de Cayey, Las Tetas de Cayey, Southeastern Foothills (Arroyo), Eastern Cordillera (Barranquitas), Central Cordillera, Western Cordillera, Western Coast (Añasco); 100–800 m; *fl* throughout the year, *fr* Jan–Jul.

Greater Antilles, Mexico, Central America, South America. [L2: 34].

Guaba, Guaba del país, Guaba nativa.

Leucaena leucocephala (Lam.) de Wit, Taxon 10(2): 54. 1961.
Bas.: *Mimosa leucocephala* Lam. (1783).

Shrub/small tree in Northern Coastal Lowlands, Northern Limestone Hills, Eastern Coast, Las Piedras Chiquitas, Southeastern Foothills (Guayama), Southern Coastal Lowlands, Southern Coastal Limestone, Susúa Forest Reserve, Maricao Forest Reserve, Guánica Forest Reserve, Southwestern

Coastal Lowlands, Western Coast (Aguada), Vieques, Culebra, Mona; 1–475 m; *fl* & *fr* throughout the year.

Native to southern Mexico, northern Central America (Guatemala, Belize). [L2: 35].

Notes: Britton & Wilson (1924) misapplied the name *Leucaena glauca* (L.) Benth. to this taxon.

Acacia, Acacia pálida, Campeche, Hediondilla, Tamarindillo, Zarcilla.

Mimosa arenosa (Willd.) Poir., Encycl. Suppl. 1(1):66. 1810.

Bas.: *Acacia arenosa* Willd. (1806).

Shrub in Northern Coastal Lowlands (Fajardo), Eastern Coast (Ceiba); 1–70 m; *fl* Dec–Mar, *fr* Jan–Mar.

Native to Lesser Antilles (Dominica, Martinique), South America (Venezuela, Brazil, Bolivia). [L2: 37].

Mimosa casta L., Sp. Pl. 1:518. 1753.

Shrub in Northern Coastal Lowlands, Eastern Coast, Caguas Valley, Sierra de Cayey, Eastern Cordillera (Aguas Buenas); 1–800 m; *fl* Oct–Feb, *fr* Dec–May.

Native to Lesser Antilles, Trinidad and Tobago, southern Central America (Panama), northern South America to Brazil. [L2: 37; AR: 296].

Zarza.

Mimosa ceratonia L., Sp. Pl. 1:523. 1753.

Syn.: *Lomoplis ceratonia* (L.) Raf. (1838).

Scandent shrub in Northern Coastal Lowlands, Río Abajo Forest Reserve, Sierra de Luquillo, Eastern Coast (Cuchilla de Panduras), Eastern Cordillera (Aibonito, Cayey), Central Cordillera, Southern Coastal Lowlands (Patillas), Susúa Forest Reserve, Western Cordillera, Sierra Bermeja, Western Coast (Mayagüez), Vieques; 1–800 m; *fl* Aug–Jan, *fr* Oct–Mar.

Greater Antilles (Hispaniola), Virgin Islands, Lesser Antilles, South America (Brazil). [L2: 37; AR: 298].

Lamedora, Zarza.

Mimosa diplotricha C. Wright in Sauvalle, Anales Acad. Ci. Méd. Habana 5:405. 1869.

Scandent shrub in Northern Coastal Lowlands (Guaynabo), Guajataca Forest Reserve, Río Abajo Forest Reserve, Eastern Coast, Caguas Valley, Sierra de Cayey, Southeastern Foothills, Eastern Cordillera (Aguas Buenas), Vieques, Culebra; 1–300 m; *fl* Oct–Apr, *fr* Dec, Mar–Apr.

Greater Antilles (Cuba, Hispaniola), Mexico, Central America, northern South America to Bolivia, Paraguay. [L2: 39; AR: 298].

Mimosa pigra L., Cent. Pl. I 13. 1755, *nom. cons.*

Syn.: *Mimosa pellita* Humb. & Bonpl. ex Willd. (1805).

Shrub in Northern Coastal Lowlands, Río Abajo Forest Reserve, Eastern Coast (Humacao), Eastern Cordillera (Cayey), Southwestern Coastal Lowlands (Lajas), Western Coast; 1–400 m; *fl* & *fr* Mar–Dec.

Native to Greater Antilles (Cuba), Lesser Antilles, Trinidad and Tobago, Mexico, Central America, South America. [L2: 39].

Mimosa pudica L. var. **unijuga** (Walp. & Duchass.) Griseb., Abh. Königl. Ges. Wiss. Göttingen 7:211. 1857.

Bas.: *Mimosa unijuga* Walp. & Duchass. (1850).

Herb in Northern Coastal Lowlands, Northern Limestone Hills, Río Abajo Forest Reserve, Sierra de Luquillo, Eastern Coast, Caguas Valley (Juncos), Sierra de Cayey, Las Tetas de Cayey, Las Piedras Chiquitas, Southeastern Foothills (Coamo), Toro Negro Forest Reserve, vicinity of Susúa Forest Reserve (Sabana Grande), Western Cordillera, Southwestern Coastal Lowlands (Cabo Rojo), Western Coast, Vieques, Culebra; 1–875 m; *fl* & *fr* throughout the year.

Probably native to the Neotropics. [L2: 40].

Mimosa, Moriviví, Sensitiva.

Mimosa quadrivalvis L. var. **urbaniana** Barneby, Mem. New York Bot. Gard. 65:297. 1991.

Syn.: *Morongia portoricensis* (Urb.) Britton (1924). *Leptoglottis portoricensis* (Urb.) Britton & Rose (1928).

Shrub in Las Piedras Chiquitas; 400–500 m; *fl* Jun, Nov (two specimens), *fr* Nov–Dec, Mar.

Endemic. [L2: 51; AR: 298].

Notes: Liogier & Martorell (1982) and Liogier (1988) misapplied the name *Schrankia leptocarpa* DC., and Liogier & Martorell (2000) misapplied *Mimosa quadrivalvis* L. var. *leptocarpa* (DC.) Barneby, to this taxon.

Cuernecillo, Zarcilla, Zarzaparilla.

Neptunia plena (L.) Benth. J. Bot. (Hooker) 4(31): 355. 1841.

Bas.: *Mimosa plena* L. (1753).

Herb in Northern Coastal Lowlands, Eastern Coast, Southern Coastal Lowlands, Southwestern Coastal Lowlands, Western Coast (Rincón), Vieques, Mona; 1–100 m; *fl* & *fr* throughout the year.

West Indies, south-central United States (Texas), Mexico, Central America, northern, northwestern and western-central South America. [L2: 41].

Desmanto amarillo.

Pithecellobium dulce (Roxb.) Benth., London J. Bot. 3:199. 1844 "Pithecolobium."

Bas.: *Mimosa dulcis* Roxb. (1798).

Tree planted and persistent in Northern Coastal Lowlands, Eastern Coast (Humacao), Southern Coastal Lowlands, Southwestern Coastal Lowlands, Sierra Bermeja; 1–250 m; *fl* Nov–May, *fr* Dec–May.

Native to Mexico, Central America, northwestern South America and Venezuela. [L2: 47].

Guamá americano.

Pithecellobium unguis-cati (L.) Benth., London J. Bot. 3:200. 1844 "Pithecolobium."

Bas.: *Mimosa unguis-cati* L. (1753).

Tree in Northern Coastal Lowlands (Fajardo and adjacent islands), Southeastern Foothills, Southern Coastal Lowlands, Southern Coastal Limestone, Guánica Forest Reserve, Southwestern Coastal Lowlands, Sierra Bermeja, Vieques, Culebra, Mona, Desecheo; 1–400 m; *fl* throughout the year, *fr* Nov–Jul.

West Indies, Central America, northern South America (Venezuela, Colombia). [L2: 47].

Escambrón colorado, Rolón, Rolón escambrón colorado, Uña de gato.

Prosopis juliflora (Sw.) DC., Prodr. 2:447. 1825.

Bas.: *Mimosa juliflora* Sw. (1788).

Syn.: *Neltuma juliflora* (Sw.) Raf. (1838).

Tree in Northern Coastal Lowlands (islands adjacent to Fajardo), Eastern Coast (Cayo Santiago), Southeastern Foothills (Guayama), Southern Coastal Lowlands, Southern Coastal Limestone (Ponce), Guánica Forest Reserve, Southwestern Coastal Lowlands, Sierra Bermeja, Vieques, Culebra, Mona; 1–150 m; *fl* throughout the year, *fr* Sep–Jun.

Antilles (possibly), Mexico, Central America, northwestern, west-central South America and Venezuela; exact native range uncertain. [L2: 47].

Notes: Liogier & Martorell (1982) and Liogier (1988) misapplied the name *Prosopis pallida* (Humb. & Bonpl. ex Willd.) Kunth to this taxon, but subsequently (2000) corrected it. Little and Wadsworth initially (1964) used the above name, but subsequently (1989) changed it to *Prosopis pallida*.

Algarrobo americano, Algarrobo de Hawaii, Aroma americana, Bayahonda, Mesquite.

Samanea saman (Jacq.) Merr., J. Washington Acad. Sci. 6(2):47. 1916.

Bas.: *Mimosa saman* Jacq. (1800).

Syn.: *Pithecellobium saman* (Jacq.) Benth. (1844). *Albizia saman* (Jacq.) F. Muell. (1876).

Tree planted and persistent in Northern Coastal Lowlands (Fajardo), Eastern Coast (Ceiba), Southeastern Foothills, Southern Coastal Lowlands, Southwestern Coastal Lowlands, Vieques; 1–400 m; *fl* May–Nov, *fr* Jan–Feb.

Native to southern Central America, northern South America (Venezuela, Colombia). [L2: 49].

Dormilón, Guango, Samán.

Senegalia muricata (L.) Britton & Rose, N. Amer. Fl. 23(2):113. 1928.

Bas.: *Mimosa muricata* L. (1759).

Syn.: *Acacia muricata* (L.) Willd. (1806).

Tree in Northern Coastal Lowlands (Fajardo), Eastern Coast, Southeastern Foothills, Vieques; 10–350 m; *fl* Mar–Jul, *fr* throughout the year.

Antilles. [L2: 17].

Acacia nudosa, Cohoba, Cojoba, Tamarindo cimarrón.

Senegalia riparia (Kunth) Britton & Rose ex Britton & Killip, Ann. New York Acad. Sci. 35(3):144. 1936.

Bas.: *Acacia riparia* Kunth (1824).

Syn.: *Acacia westiana* DC. (1825). *Senegalia westiana* (DC.) Britton & Rose (1928).

Shrub/small tree in Northern Coastal Lowlands (Río Grande), Northern Limestone Hills, Eastern Coast (Ceiba), Las Tetas de Cayey, Las Piedras Chiquitas, Southeastern Foothills, Southern Coastal Lowlands (Santa Isabel), Southern Coastal Limestone (Peñuelas), Susúa Forest Reserve, Guánica Forest Reserve, Southwestern Coastal Lowlands, Sierra Bermeja, Vieques, Culebra; 10–825 m; *fl* Jul–Oct, *fr* Jul–Feb.

Greater Antilles (Hispaniola), Virgin Islands, Lesser Antilles, Trinidad, northern South America. [L2: 19; AR: 293].

Notes: According to David Seigler (pers. comm.), because of the inadequacy of type material, the name *Acacia retusa* (Jacq.) R.A. Howard cannot, as yet, be shown to be conspecific with the above taxon.

Acacia zarza, Zarza, Zarza brava.

Senegalia vogeliana (Steud.) Britton & Rose, N. Amer. Fl. 23(2):116. 1928.

Bas.: *Acacia vogeliana* Steud. (1840).

Syn.: *Lysiloma vogelianum* (Britton & Rose) Stehlé (1946) "vogeliana."

Acacia ambigua Vogel (1836). *Lysiloma ambiguum* Urb. (1929) "ambigua," *nom. illeg.*

Shrub/small tree in Southwestern Coastal Lowlands (Lajas); 25 m; *fl* Jun–Aug (two specimens), *fr* Jun (one specimen). [*Acevedo-Rodríguez a.n. 34752* (UPR)].

Greater Antilles (Hispaniola), Virgin Islands (St. John). [L2: 19; AR: 294].

Vachellia farnesiana (L.) Wight & Arn., Prodr. Fl. Ind. Orient. 1:272. 1834.
Bas.: *Mimosa farnesiana* L. (1753).
Syn.: *Acacia farnesiana* (L.) Willd. (1806).

Tree in Northern Coastal Lowlands, Eastern Coast (Humacao), Las Piedras Chiquitas, Southeastern Foothills, Southern Coastal Lowlands, Susúa Forest Reserve, Guánica Forest Reserve, Southwestern Coastal Lowlands, Sierra Bermeja, Western Coast (Rincón), Vieques; 1–500 m; *fl* & *fr* throughout the year.

Neotropics and -subtropics; native range uncertain. [L2: 16].

Aroma.

Vachellia macracantha (Humb. & Bonpl. ex Willd.) Seigler & Ebinger, Phytologia 87(3):160. 2006.
Bas.: *Acacia macracantha* Humb. & Bonpl. ex Willd. (1806).
Syn.: *Poponax macracanthoides* (Bertero ex DC.) Britton & Rose (1928).

Tree in Northern Coastal Lowlands (Fajardo), Cambalache Forest Reserve, Eastern Coast (Ceiba), Vieques, Culebra; 1–70 m; *fl* Jun–Jul, Oct–Nov, *fr* Jan–Feb (two specimens).

West Indies, southeastern United States (Florida), Mexico, western, southern South America and Venezuela. [L2: 16].

Tamarindo silvestre.

Vachellia tortuosa (L.) Seigler & Ebinger, Phytologia 87(3):168. 2006.
Bas.: *Mimosa tortuosa* L. (1759).
Syn.: *Acacia tortuosa* (L.) Willd. (1806).

Tree in Northern Coastal Lowlands (Fajardo and adjacent islands), Southern Coastal Lowlands (Caja de Muertos), Vieques, Culebra; 1–100 m; *fl* Feb–Sep, *fr* May–Jan.

Bahamas, Greater Antilles (Jamaica, Hispaniola), Virgin Islands, Lesser Antilles, southeastern United States (Florida), southern Central America (Panama), northwestern South America and Venezuela. [L2: 19].

Notes: Britton & Wilson (1930) used the homonym *Poponax tortuosa* (L.) Britton & Rose for the name of this taxon.

Casia.

Zapoteca portoricensis (Jacq.) H.M. Hern., Ann. Missouri Bot. Gard. 73(4):758. 1987.
Bas.: *Mimosa portoricensis* Jacq. (1791).
Syn.: *Calliandra portoricensis* (Jacq.) Benth. (1844). *Anneslia portoricensis* (Jacq.) Britton (1918).

Shrub in Northern Limestone Hills, Las Tetas de Cayey, Las Piedras Chiquitas, Southeastern Foothills (Coamo), Eastern Cordillera (Aibonito, Cañón San Cristóbal), Cerro de Las Cuevas, Southern Coastal Limestone (Guayanilla), Guánica Forest Reserve, Southwestern Coastal Lowlands (Cabo Rojo), Sierra Bermeja, Vieques, Culebra; 5–825 m; *fl* throughout the year, *fr* Oct–Jun.

Greater Antilles (Jamaica, Hispaniola), Virgin Islands, Mexico, Central America, South America (scattered). [L2: 25].

Cojobillo, Moriviví cimarrón, Zarza boba.

TAXA NOT RECENTLY COLLECTED

Entada polystachya (L.) DC. var. **polyphylla** (Benth.) Barneby, Brittonia 48(2):175. 1996.
Bas.: *Entada polyphylla* Benth. (1840).
Syn.: *Entadopsis polyphylla* (Benth.) Britton (1928).

Vine collected near Fajardo on 2 May 1885 [*Sintenis 1240* (US)].

Native to South America. [L2: 29; AR 294].

Tamarindillo.

TAXA CITED BY PREVIOUS AUTHORS, BUT NOT INCLUDED

CITED DUE TO MISIDENTIFICATION

Inga ingoides (Rich.) Willd.
Shrub/small tree cited by Liogier & Martorell (1982) and Liogier (1988) as collected near Naguabo, in forest, probably plante d and escaped. According to Gary Breckon (pers. comm.), the specimen upon which this citation was based was misidentified. [L2: 32].

Neptunia oleracea Lour.
Herb cited by Liogier & Martorell (1982) as occurring in shallow water, Arecibo, reported by Woodbury; subsequently cited by Liogier (1988) as also occurring in Laguna Cartagena, Lajas, and by Liogier & Martorell (2000) as occurring in wet places, in western Puerto Rico. All the specimens that I have seen identified as this taxon are referable to *Neptunia plena*. [L2: 40].

ONLY UNDER CULTIVATION

Acacia nilotica (L.) Delile.
Shrub/small tree cited by Liogier & Martorell (1982, 2000) and Liogier (1988) as planted and naturalized. [L2: 18].

Acacia polyacantha Willd.
Shrub/small tree cited by Liogier (1988) and Liogier & Martorell (2000) as naturalized. [L2: 18].

Albizia adinocephala (Donn. Sm.) Britton & Rose ex Record.
Tree cited by Liogier (1988) as introduced and escaping in Mayagüuez and by Liogier & Martorell (2000) as introduced and naturalized in the Mayagüez area. [L2: 22].

Inga fastuosa (Jacq.) Willd.
Shrub/small tree cited by Liogier (1988) and Liogier & Martorell (2000) as planted as coffee shade in the mountains. [L2: 32].

GENTIANACEAE

Lisianthius laxiflorus Urb., Symb. Antill. 3(2):332. 1902 "Lisianthus."

Herb in Sierra de Luquillo, Sierra de Cayey, Central Cordillera, Maricao Forest Reserve; 475–1300 m; *fl* Sep–May, *fr* Jan–Mar (two specimens).

Endemic. [L4: 192].

Campanilla, Campanilla amarilla.

Voyria aphylla (Jacq.) Pers., Syn. Pl. 1:284. 1805 "Vohiria."
Bas.: *Gentiana aphylla* Jacq. (1760).
Syn.: *Leiphaimos aphylla* (Jacq.) Gilg (1895).

Herb in Sierra de Luquillo, Central Cordillera, Maricao Forest Reserve; 900–1000 m; *fl* & *fr* Jun–Mar.

Antilles, Mexico, Central America, northwestern, west-central South America and Venezuela. [L4: 195].

Voyria parasitica (Schltdl. & Cham.) Ruyters & Maas, Acta Bot. Neerl. 30(1–2):143. 1981.
Bas.: *Leiphaimos parasitica* Schltdl. & Cham. (1831).

Herb in Northern Coastal Lowlands (Laguna Tortuguero), Northern Limestone Hills (Ciales, Arecibo), Río Abajo Forest Reserve, Maricao Forest Reserve; 300–750 m; *fl* & *fr* Aug–May.

Bahamas, Greater Antilles, southeastern United States (Florida), Mexico, Central America. [L4: 195].

Voyria tenella Hook., Bot. Misc. 1:47, *t. 25*, *f. B.* 1829.
Syn.: *Leiphaimos portoricensis* Britton (1925).

Herb in Eastern Cordillera (Cerro La Peña), Maricao Forest Reserve; 280–850 m; *fl* & *fr* Sep–Mar.

Greater Antilles (Cuba, Hispaniola), Lesser Antilles, Mexico, Central America, northwestern, west-central South America and Venezuela. [L4: 197].

TAXA NOT RECENTLY COLLECTED

Schultesia brachyptera Cham., Linnaea 8(1):8. 1833.
Syn.: *Schultesia heterophylla* Miq. (1846).

Herb collected on wet sand near Bayamón [*Stahl 116*, not seen].

Greater Antilles, Mexico, Central America, northern South America to Paraguay. [L4: 193].

TAXA CITED BY PREVIOUS AUTHORS, BUT NOT INCLUDED

DOUBTFUL OCCURRENCE

Enicostema verticillatum (L.) Engl. ex Gilg.
Herb cited by Britton & Wilson (1925) as recorded in 1845 by Grisebach for Puerto Rico; subsequently cited by Liogier & Martorell (1982, 2000) and Liogier (1995). [L4: 190].

GERANIACEAE

TAXA CITED BY PREVIOUS AUTHORS, BUT NOT INCLUDED

ONLY UNDER CULTIVATION

Pelargonium graveolens L'Hér.
Herb cited by Liogier & Martorell (1982, 2000) and Liogier (1988) as cultivated and sometimes escaped in the mountains. [L2: 239].

Pelargonium odoratissimum (L.) L'Hér.
Herb cited by Liogier & Martorell (1982, 2000) and Liogier (1988) as cultivated and sometimes escaped. [L2: 240].

Pelargonium peltatum (L.) L'Hér.
Herb cited by Liogier & Martorell (1982, 2000) and Liogier (1988) as cultivated and sometimes escaped. [L2: 240].

Pelargonium zonale (L.) L'Hér.
Herb cited by Liogier & Martorell (1982, 2000) and Liogier (1988) as cultivated and sometimes escaped, usually in mountainous areas. [L2: 240].

GESNERIACEAE

Columnea ambigua (Urb.) B.D. Morley, Proc. Roy. Irish Acad. 74B(24):412. 1974.
Bas.: *Alloplectus ambiguus* Urb. (1899).
Syn.: *Crantzia ambigua* (Urb.) Britton (1925).
Trichantha ambigua (Urb.) Wiehler (1975).

Epiphytic herb in Sierra de Luquillo, Sierra de Cayey, Las Tetas de Cayey, Eastern Cordillera (La Torrecilla), Central Cordillera, Maricao Forest Reserve; 200–1300 m; *fl* throughout the year, *fr* Oct–Jul.

Endemic. [L4: 565].

Tibey de cresta.

Columnea scandens L., Sp. Pl. 2:638. 1753.
Syn.: *Columnea tulae* Urb. (1899).
Epiphytic herb in Río Abajo Forest Reserve, Sierra de Luquillo, Eastern Coast, Sierra de Cayey, Las Tetas de Cayey, Central Cordillera, Maricao Forest Reserve; 200–925 m; *fl* & *fr* throughout the year.
Greater Antilles (Hispaniola), Lesser Antilles, Trinidad and Tobago, southern Central America (Costa Rica), northern South America (Venezuela). [L4: 569].
Tibey parásito.

Gesneria citrina Urb., Symb. Antill. 1(4):477. 1900.
Suffrutescent herb in Northern Limestone Hills, Río Abajo Forest Reserve; 200–350 m; *fl* Apr–Jan, *fr* Apr–May, Oct.
Endemic. [L4: 572].

Gesneria cuneifolia (DC.) Fritsch in Engler & Prantl, Nat. Pflanzenfam. 4(3b):184. 1894.
Bas.: *Conradia cuneifolia* DC. (1839).
Herb in Northern Limestone Hills, Río Abajo Forest Reserve; 50–350 m; *fl* Feb–Nov, *fr* Feb, May–Nov.
Endemic. [L4: 573].
Yerba parrera, Yerba de cueva.

Gesneria pauciflora Urb., Symb. Antill. 1(4):478. 1900.
Suffrutescent herb in Maricao Forest Reserve; 500–600 m; *fl* Oct–Jul, *fr* Apr (one specimen). [*Ackerman 1900* (UPRRP)].
Endemic. [L4: 573].
Notes: This taxon is on the U.S. Fish & Wildlife Service list of threatened and endangered plants.

Gesneria pedunculosa (DC.) Fritsch in Engler & Prantl, Nat. Pflanzenfam. 4(3b):184. 1894.
Bas.: *Conradia pedunculosa* DC. (1839).
Syn.: *Pentarhaphia albiflora* Decne. (1846).
Shrub in Northern Limestone Hills, Río Abajo Forest Reserve, Sierra de Luquillo, Caguas Valley (Caguas), Sierra de Cayey, Las Tetas de Cayey, Las Piedras Chiquitas, Southeastern Foothills (Coamo), Eastern Cordillera (La Torrecilla), Central Cordillera, Southern Coastal Lowlands (Ponce), Susúa Forest Reserve, Western Cordillera, Guánica Forest Reserve; 10–950 m; *fl* & *fr* throughout the year.
Endemic. [L4: 574].
Árbol de Navidad.

Gesneria reticulata (Griseb.) Urb., Symb. Antill. 1(4):478. 1900.
Bas.: *Conradia reticulata* Griseb. (1866).
Suffrutescent herb in Sierra de Luquillo; 450–940 m; *fl* Dec, Mar (two specimens), *fr* (?). [*Kolterman a.n. 19025* (MAPR)].
Greater Antilles (Cuba, Hispaniola). [L4: 574].
Yerba de cueva.

Gesneria viridiflora (Decne.) Kuntze subsp. **sintenisii** (Urb.) L.E. Skog, Smithsonian Contr. Bot. 29:142. 1976.
Bas.: *Gesneria sintenisii* Urb. (1901).
Syn.: *Duchartrea sintenisii* (Urb.) Britton (1925).
Shrub in Sierra de Luquillo, Sierra de Cayey; 600–1000 m; *fl* Oct–Mar, *fr* Sep, Dec–Apr.
Greater Antilles (Hispaniola). [L4: 576].

Rhytidophyllum auriculatum Hook., Bot. Mag. 64: *t. 3562*. 1837.
Shrub in Northern Limestone Hills (Arecibo), Río Abajo Forest Reserve, Central Cordillera, Western Cordillera (Maricao); 50–1300 m; *fl* & *fr* throughout the year.
Greater Antilles (Hispaniola). [L4: 579].
Tabacón, Tibey amarillo.

TAXA CITED BY PREVIOUS AUTHORS, BUT NOT INCLUDED

ONLY UNDER CULTIVATION

Chrysothemis pulchella (Donn ex Sims) Decne.
Herb cited by Liogier (1995) and Liogier & Martorell (2000) as planted and naturalized in humid situations. [L4: 567].

Episcia cupreata (Hook.) Hanst.
Herb cited by Liogier & Martorell (1982, 2000) and Liogier (1988) as cultivated and persistent in wet places. [L4: 569].

Gloxinia perennis (L.) Fritsch.
Herb cited by Liogier (1995) and Liogier & Martorell (2000) as occasionally naturalized from cultivation. [L4: 577].

GOODENIACEAE

Scaevola plumieri (L.) Vahl, Symb. Bot. 2:36. 1791.
Bas.: *Lobelia plumieri* L. (1753).
Shrub in Northern Coastal Lowlands, Southern Coastal Lowlands (Caja de Muertos), Guánica Forest Reserve, Southwestern Coastal Lowlands (Cabo Rojo), Vieques, Culebra; 0–1 m; *fl* May–Dec, *fr* Jun–Jan.
Widespread on tropical shores. [L5: 218].
Bosborín.

Scaevola taccada (Gaertn.) Roxb., Hort. Bengal. 15. 1814.
Bas.: *Lobelia taccada* Gaertn. (1788).
Herb in Northern Coastal Lowlands (Fajardo, Loíza), Vieques; 0–5 m; *fl* & *fr* Jun, Sep–Oct.
Native to the Paleotropics and -subtropics.

HAEMODORACEAE

Xiphidium caeruleum Aubl., Hist. Pl. Guiane 1:33, *t. 11*. 1775.
Herb in Sierra de Luquillo, Caguas Valley (Caguas); 100–500 m; *fl* Sep–Feb, *fr* Dec–Mar.
Antilles, Trinidad, Mexico, Central America, northern South America to Bolivia. [AR&S: 168].
Cola de paloma, Mano poderosa.

HELICONIACEAE

Heliconia bihai (L.) L., Mant. Pl. 2:211. 1771.
Bas.: *Musa bihai* L. (1753).
Syn.: *Bihai bihai* (L.) Griggs (1904).
Herb planted and persistent in Río Abajo Forest Reserve, Sierra de Luquillo, Southern Coastal Lowlands (Santa Isabel); 40–350 m; *fl* Apr–May, *fr* (?).
Native to Greater Antilles (Jamaica), Lesser Antilles, northern South America and Brazil. [AR&S: 174].

Heliconia caribaea Lam., Encycl. 1(2):426. 1785.
Herb in Northern Limestone Hills (San Sebastián), Sierra de Luquillo, Eastern Coast (Cuchilla de Panduras), Caguas Valley (Caguas), Sierra de Cayey, Southeastern Foothills (Guayama), Central Cordillera; 100–860 m; *fl* Feb–Aug, *fr* May–Jun, Sep–Dec.
Greater Antilles (Jamaica, Hispaniola), Virgin Islands (Tortola), Lesser Antilles. [AR&S: 175].
Botecitos, Guineo cimarrón, Guineo silvestre, Pámpano, Plátano de indio.

Heliconia psittacorum L.f., Suppl. Pl. 158. 1782.
Herb planted and persistent in Northern Coastal Lowlands (San Juan), Río Abajo Forest Reserve, Sierra de Luquillo, Sierra de Cayey, Central Cordillera (Orocovis); 20–860 m; *fl* Oct–Jun, *fr* Mar, Jun, Sep–Oct.
Native to Trinidad, northern South America and Brazil. [AR&S: 175].

Heliconia rostrata Ruiz & Pav., Fl. Peruv. 3:71, *t. 305*. 1802.
Herb planted and persistent in Sierra de Luquillo; 700 m; *fl* Apr (one specimen), *fr* (?). [*Rodríguez 46* (UPRRP)].
Native to northwestern, west-central South America and Venezuela. [AR&S: 177].

TAXA CITED BY PREVIOUS AUTHORS, BUT NOT INCLUDED

ONLY UNDER CULTIVATION

Heliconia stricta Huber.
Robust herb cited in Acevedo-Rodríguez & Strong (2005) as occurring in swamps and wet forests, but the only specimen cited occurs in cultivation. [AR&S: 177].

Heliconia subulata Ruiz & Pav.
Robust herb cited by Liogier & Martorell (2000) as cultivated and naturalized.

HERNANDIACEAE

Hernandia sonora L., Sp. Pl. 2:981. 1753.
Tree in Northern Coastal Lowlands (Luquillo), Northern Limestone Hills, Western Coast (Mayagüez); 1–200 m; *fl* Jun–Sep, *fr* Feb–Mar, Jun–Jul.
Antilles. [L1: 270].
Mago, Toporite.

HYDROCHARITACEAE

Egeria densa Planch., Ann. Sci. Nat., Bot., sér. 3 11:80. 1849.
Syn.: *Elodea densa* (Planch.) Casp. (1857). *Anacharis densa* (Planch.) Vict. (1931).
Aquatic herb in Northern Coastal Lowlands (Dorado); 10–15 m; *fl* (?), *fr* (?). [*Liogier 33402, 33769* (UPR)].
Native to Brazil. [AR&S: 64].
Tomillo de agua.

Elodea canadensis Michx., Fl. Bor.-Amer. 1:20. 1803.
Syn.: *Anacharis canadensis* (Michx.) Planch. (1848).
Aquatic herb in Northern Coastal Lowlands (Río Grande), Northern Limestone Hills (Florida); 50–150 m; *fl* Jan (one specimen), *fr* (?). [*Liogier 32599* (UPR); *Aragon a.n. 38512* (UPRRP)].
Native to Canada, United States. [AR&S: 64].

Halophila baillonis Asch., J. Linn. Soc., Bot. 14:317. 1874.
Syn.: *Halophila aschersonii* Ostenf. (1902).
Aquatic herb in coastal waters off Northern Coast, Western Coast (Mayagüez), Southwestern Coast; -3 to -17 m; *fl* (?), *fr* (?).
Coastal waters of Greater Antilles (Jamaica), Virgin Islands (St. Thomas, St. Croix), Lesser Antilles (Guadeloupe), Southern Caribbean Islands (Curaçao),

southern Central America (Panama: Pacific coast), South America (Brazil). [AR&S: 65].

Halophila decipiens Ostenf., Bot. Tidsskr. 24(3):260, *t. s.n.* 1902.
Syn.: *Halophila decipiens* Ostenf. var. *pubescens* Hartog (1957).
Aquatic herb in coastal waters off Southwestern Coast; -5 to -17 m; *fl* (?), *fr* (?).
Coastal waters of Greater Antilles (Jamaica), southeastern United States (Florida), Central America (Nicaragua, Costa Rica), northern South America (Venezuela, Colombia), Indian and Pacific Oceans. [AR&S: 66].
Notes: Britton & Wilson (1923) misapplied the name *Halophila baillonis* to this taxon.

Halophila engelmannii Asch. in Neumayer, Anl. Wiss. Beobacht. Reisen 368. 1875.
Aquatic herb in coastal waters off Southern Coast (Salinas, Peñuelas); -2 m; *fl* (?), *fr* (?).
Coastal waters of Bahamas, Greater Antilles (Cuba, Cayman Islands), southeastern and south-central United States. [AR&S: 66].

Limnobium laevigatum (Humb. & Bonpl. ex Willd.) Heine, Adansonia, n.s. 8(3):315. 1968.
Bas.: *Salvinia laevigata* Humb. & Bonpl. ex Willd. (1810).
Syn.: *Hydromystria laevigata* (Humb. & Bonpl. ex Willd.) Hunz. (1981). *Limnobium spongia* (Bosc) Steud. subsp. *laevigatum* (Humb. & Bonpl. ex Willd.) Lowden (1992).
Hydromystria stolonifera G. Mey. (1818).
Aquatic herb in Northern Coastal Lowlands, Northern Limestone Hills (Vega Baja, Florida); 1–150 m; *fl* Oct–May, *fr* (?).
Greater Antilles (Cuba, Hispaniola), Lesser Antilles, Trinidad, Mexico, Central America, South America. [AR&S: 67].
Cuchara.

Najas guadalupensis (Spreng.) Magnus, Beitr. Kenntn. *Najas* viii. 1870.
Bas.: *Caulinia guadalupensis* Spreng. (1824).
Aquatic herb in Northern Coastal Lowlands, Northern Limestone Hills (Ciales), Southwestern Coastal Lowlands (Cabo Rojo); 1–150 m; *fl* (?), *fr* (?).
Bahamas, Greater Antilles, Lesser Antilles (Antigua, Guadeloupe), Southern Caribbean Islands (Aruba, Curaçao), Canada, United States, Mexico, Central America, South America (scattered). [AR&S: 68].
Notes: Britton & Wilson (1923) used the homonym *Najas guadalupensis* (Spreng.) Morong for the name of this taxon.

Najas marina L., Sp. Pl. 2:1015. 1753.
Aquatic herb in Northern Coastal Lowlands (Laguna Tortuguero, Arecibo), Eastern Coast (Humacao); 1 m; *fl* (?), *fr* (?).
Bahamas, Greater Antilles, Canada, United States, Mexico, Central America, South America, Europe, Asia, Australia. [AR&S: 68].

Thalassia testudinum K.D. Koenig, Ann. Bot. (König & Sims) 2:96. 1805.
Aquatic herb in coastal waters off Northern Coast, Eastern Coast (Cayo Santiago), Guánica Forest Reserve, Southwestern Coast, Culebra; <0 m; *fl* Apr, *fr* (?).
Coastal waters of West Indies, southeastern and south-central United States, Central America, northern South America. [AR&S: 70].
Hierba de manatí, Palma de mar.

TAXA CITED BY PREVIOUS AUTHORS, BUT NOT INCLUDED

CITED DUE TO MISIDENTIFICATION

Limnobium spongia (Bosc) Rich. ex Steud.
Herb cited by Liogier & Martorell (1982, 2000) as collected in Añasco by Sintenis and in Loíza Aldea and Florida by Roy Woodbury. This citation was based on the misidentification of specimens of *Limnobium laevigatum*.

HYPERICACEAE

Hypericum diosmoides Griseb., Cat. Pl. Cub. 40. 1866.
Herb in Northern Limestone Hills (Ciales), Sierra de Cayey, Las Tetas de Cayey, Las Piedras Chiquitas, Eastern Cordillera, Central Cordillera (Yauco), Western Cordillera (Maricao); 280–925 m; *fl* & *fr* throughout the year.
Greater Antilles (Cuba, Hispaniola). [L3: 240].

Hypericum hypericoides (L.) Crantz, Inst. Rei Herb. 2:520. 1766.
Bas.: *Ascyrum hypericoides* L. (1753).
Herb in Northern Coastal Lowlands (Laguna Tortuguero), Eastern Cordillera (La Torrecilla), Central Cordillera; 1–1250 m; *fl* & *fr* Aug–May.
Bahamas, Greater Antilles, eastern and central United States, Mexico, northern Central America (Guatemala, Honduras). [L3: 240].
Arrayanillo, Cruz de San Andrés.

HYPOXIDACEAE

Curculigo scorzonerifolia (Lam.) Baker, J. Linn. Soc., Bot. 17(1):124. 1878 "scorzoneraefolia."
Bas.: *Hypoxis scorzonerifolia* Lam. (1789) "scorzoneraefolia."

Herb in Northern Coastal Lowlands (Dorado, Laguna Tortuguero), Cerro Las Mesas; 1–400 m; *fl* & *fr* Mar, Jun.

Antilles, Trinidad, Mexico, Central America, South America (Venezuela, Brazil). [AR&S: 97].

Gorgojo.

Hypoxis decumbens L., Syst. Nat. (ed. 10) 2:986. 1759.

Herb in Northern Coastal Lowlands, Northern Limestone Hills, Río Abajo Forest Reserve, Sierra de Luquillo, Sierra de Cayey, Las Tetas de Cayey, Eastern and Central Cordillera, Western Cordillera, Cerro Las Mesas; 1–1150 m; *fl* throughout the year, *fr* Jan–Oct.

Antilles, Mexico, Central America, South America. [AR&S: 100].

Coquí.

Hypoxis wrightii (Baker) Brackett, Rhodora 25(296): 140, *t. 11.* 1923.
Bas.: *Hypoxis juncea* Sm. var. *wrightii* Baker (1878).

Herb in Northern Coastal Lowlands (Laguna Tortuguero); 1 m; *fl* Oct–Nov, Feb, *fr* Oct–Nov, Feb.

Bahamas, Greater Antilles, southeastern and south-central United States. [AR&S: 100].

TAXA CITED BY PREVIOUS AUTHORS, BUT NOT INCLUDED

ONLY UNDER CULTIVATION

Molineria capitulata (Lour.) Herb. Syn.: *Curculigo capitulata* (Lour.) Kuntze. *Molineria hortensis* Britton.

Herb cited by Liogier & Martorell (1982, 2000) as cultivated and persistent. [AR&S: 97].

ICACINACEAE

Mappia racemosa Jacq., Pl. Hort. Schoenbr. 1:22, *t. 47.* 1797.

Shrub/small tree in Northern Limestone Hills (Quebradillas, Isabela), Río Abajo Forest Reserve, Eastern Cordillera (Cayey); 10–545 m; *fl* Jan–Apr, Jul, Oct, *fr* (?).

Greater Antilles, Mexico, Central America (Costa Rica). [L3: 39].

Ottoschulzia rhodoxylon (Urb.) Urb., Symb. Antill. 7(2):274. 1912.
Bas.: *Poraqueiba rhodoxylon* Urb. (1908).

Tree in Northern Coastal Lowlands (Fajardo), Northern Limestone Hills, Río Abajo Forest Reserve, Southern Coastal Limestone (Guayanilla), Susúa Forest Reserve, Maricao Forest Reserve, Guánica Forest Reserve, Southwestern Coastal Lowlands (Cabo Rojo); 1–450 m; *fl* Jun–Jul, Dec, *fr* Mar, Jul–Oct.

Greater Antilles (Hispaniola). [L3: 42].

Notes: This taxon is on the U.S. Fish & Wildlife Service list of threatened and endangered plants.

Palo de rosa.

IRIDACEAE

Crocosmia ×crocosmiiflora (Lemoine) N.E. Br., Trans. Roy. Soc. South Africa 20(3):264. 1932 "crocosmiflora."
Bas.: *Montbretia ×crocosomiiflora* Lemoine (1880).

Herb in Sierra de Cayey, Eastern Cordillera (Aibonito), Central Cordillera; 500–1300 m; *fl* throughout the year, *fr* Aug (one specimen).

Only cultivated. [AR&S: 102].

Notes: This plant is a horticultural hybrid between *Crocosmia aurea* (Pappe) Planch., a native of Zimbabwe, and *Crocosmia pottsii* (Baker) N.E. Br., a native of Nepal.

Iris domestica Goldblatt & Mabb., Novon 15:129. 2005.
Syn.: *Belamcanda chinensis* (L.) DC. (1805).

Herb planted and persistent in Eastern Coast (Cuchilla de Panduras), Southwestern Coastal Lowlands (San Germán); 25–200 m; *fl* & *fr* Sep, Nov (two specimens). [*Liogier 30981* (UPR)].

Native to eastern temperate Asia and tropical Asia. [AR&S: 101].

Maravilla.

Sisyrinchium rosulatum E.P. Bicknell, Bull. Torrey Bot. Club 26(5):228. 1899.
Syn.: *Sisyrinchium exile* E.P. Bicknell (1901).

Herb in Central Cordillera; 800–1300 m; *fl* & *fr* Sep–Jun.

Greater Antilles (Hispaniola), southeastern and south-central United States. [AR&S: 105].

Trimezia steyermarkii R.C. Foster, Rhodora 64:310. 1962.

Herb planted and persistent in Northern Coastal Lowlands, Río Abajo Forest Reserve, Sierra de Luquillo, Sierra de Cayey, Eastern Cordillera (Corozal), Central Cordillera (Adjuntas), Maricao Forest Reserve; 5–1000 m; *fl* & *fr* Dec–May, Aug–Sep.

Native to southern Mexico, Central America, northern South America (Colombia). [AR&S: 106].

Notes: Britton & Wilson (1930) and Liogier & Martorell (1982, 2000) misapplied the name *Trimezia martinicensis* (Jacq.) Herb. to this taxon.

TAXA NOT RECENTLY COLLECTED

Eleutherine bulbosa (Mill.) Urb., Repert. Spec. Nov. Regni Veg. 15:305. 1918.

Bas.: *Sisyrinchium bulbosum* Mill. (1768).

Syn.: *Galatea bulbosa* (Mill.) Britton (1918).

Herb collected in Sabana Llana (Juana Díaz) on 5 Nov 1914 [*Stevenson 2307* (US)].

West Indies, Central America, South America. [AR&S: 104].

Lágrimas de la Virgen, Mariposa, Ninfa.

TAXA CITED BY PREVIOUS AUTHORS, BUT NOT INCLUDED

ONLY UNDER CULTIVATION

Neomarica northiana (Schneev.) Sprague.

Herb cited by Liogier & Martorell (1982, 2000) as planted as an ornamental and escaped in wet forest in Jajome; subsequently cited by Acevedo-Rodríguez & Strong (2005). [AR&S: 104].

JUGLANDACEAE

Juglans jamaicensis C. DC. in A. de Candolle, Prodr. 16(2):138. 1864.

Tree in Central Cordillera (Utuado, Monte Guilarte Forest Reserve); 200–1000 m; *fl* (?), *fr* Apr–Jul, Nov. [*Acevedo 9, 40* (SJ); *Liogier 36248* (UPR); *Stone 4262* (MAPR)].

Greater Antilles (Cuba, Hispaniola). [L1: 43].

Notes: This taxon is on the U.S. Fish & Wildlife Service list of threatened and endangered plants.

Nogal, Nuez, Palo de nuez.

JUNCACEAE

Juncus dichotomus Elliott, Sketch Bot. S. Carolina 1:406. 1817.

Syn.: *Juncus tenuis* Willd. var. *dichotomus* (Elliott) A.W. Wood (1861).

Herb in Sierra de Luquillo; 800–1050 m; *fl* & *fr* Jun–Oct.

Native to eastern and south-central United States, Mexico, Central America, South America. [AR&S: 236].

Juncus tenuis Willd., Sp. Pl. 2(1):214. 1799.

Herb in Toro Negro Forest Reserve; 900–1325 m; *fl* & *fr* Sep–Jun.

Native to Canada, United States, Mexico, Central America (Guatemala, Honduras, Costa Rica), northwestern, west-central South America and Brazil. [AR&S: 236].

TAXA CITED BY PREVIOUS AUTHORS, BUT NOT INCLUDED

DOUBTFUL OCCURRENCE

Juncus effusus L.

Herb cited by Liogier & Martorell (1982) as collected in the Toro Negro area, a doubtful record, and by Liogier & Martorell (2000) as occurring in open grounds at high elevations, Toro Negro area.

KRAMERIACEAE

Krameria ixine L., Syst. Nat. (ed. 10) 2:899. 1759.

Shrub in Southern Coastal Limestone, Susúa Forest Reserve, Guánica Forest Reserve, Southwestern Coastal Lowlands, Sierra Bermeja, Mona; 1–300 m; *fl* & *fr* throughout the year.

Greater Antilles (Hispaniola), Lesser Antilles, Mexico, Central America (Guatemala, Honduras, Costa Rica), northern South America. [L2: 339].

Notes: Britton & Wilson (1925) and Liogier & Martorell (1982) used the orthographic variant *Krameria ixina* L., published in 1762, for the name of this taxon.

Abrojo colorado, Ratama falsa.

LAMIACEAE (LABIATAE)

Aegiphila martinicensis Jacq., Observ. Bot. 2:3, *t.* 27. 1767.

Syn.: *Aegiphila oligoneura* Urb. (1919).

Tree in Northern Coastal Lowlands, Northern Limestone Hills (Bayamón to Manatí), Sierra de Luquillo; 1–450 m; *fl* Sep–Mar, *fr* Jan–Mar.

Greater Antilles (Jamaica), Lesser Antilles, Mexico, Central America (scattered), northern South America (Colombia). [L4: 337].

Capaíllo.

Callicarpa ampla Schauer in A. de Candolle, Prodr. 11:642. 1847.

Tree in Sierra de Luquillo; 150–500 m; *fl* Sep–Nov, *fr* Oct–Nov (two specimens).

Virgin Islands (St. Thomas, but only historically). [L4: 342].

Notes: This taxon is on the U.S. Fish & Wildlife Service list of threatened and endangered plants.

Capá rosa, Caparrosa, Péndula cimarrona.

Clerodendrum chinense (Osbeck) Mabb., Pl.-Book (corr. repr. 1989) 707. 1989.

Bas.: *Cryptanthus chinensis* Osbeck (1757).
Syn.: *Clerodendrum fragrans* Vent. (1804).
Clerodendrum philippinum Schauer (1847) "Clerodendron."

Shrub in Northern Coastal Lowlands, Northern Limestone Hills, Caguas Valley, Sierra de Cayey, Las Tetas de Cayey, Southeastern Foothills, Eastern Cordillera (Cañón San Cristóbal), Central Cordillera (Ponce, Utuado), Western Cordillera (Maricao), Western Coast (Mayagüez); 50–825 m; *fl* throughout the year, *fr* (?).

Native to China. [L4: 349].

Flor de muerto, Jazmín hediondo, Yapaná.

Clerodendrum paniculatum L., Mant. Pl. 1:90. 1767 "paniculata."

Shrub planted and persistent in Northern Coastal Lowlands (Río Grande), Sierra de Luquillo, Caguas Valley (Caguas); 1–450 m; *fl* Aug–Sep, *fr* (?).

Native to tropical Asia, China.

Clerodendrum ×speciosum Dombrain (pro sp.), Fl. Mag. (London) 8: *t. 432*. 1869.

Woody vine planted and persistent in Northern Coastal Lowlands, Río Abajo Forest Reserve, Las Tetas de Cayey, Southwestern Coastal Lowlands (San Germán); 1–640 m; *fl* Sep–Mar, *fr* Dec (one specimen).

Only cultivated. [L4: 350; AR: 391].

Notes: This taxon, when considered as a cross between between *Clerodendrum splendens* G. Don and *Clerodendrum thomsoniae* Balf., has been named *Clerodendrum ×speciosum* both by Dombrain (1869), as above, and by Teijs. & Binn. ex Lemaire (Ill. Hort. 16: *t. 593*. 1869 May). Moldenke & Moldenke, in the Flora of Ceylon (1983), think that, barring future genetic research to the contrary, those are just names applied to a cultivated form of *Clerodendrum umbellatum* Poir.

Clerodendrum wallichii Merr., J. Arnold Arbor. 33:220. 1952 "Clerodendron."

Shrub planted and persistent in Maricao Forest Reserve; 470 m; *fl* Aug (one specimen), *fr* (?). [*Vélez Gavilán 464* (MAPR)]

Native to India, Bangladesh, Myanmar, Vietnam, China. [L4: 351].

Clinopodium vimineum (L.) Kuntze, Revis. Gen. Pl. 2:516. 1891.

Bas.: *Satureja viminea* L. (1753).

Shrub in Susúa Forest Reserve; 150–375 m; *fl* Feb–Sep, *fr* Feb (one specimen).

Greater Antilles. [L4: 418].

Congea tomentosa Roxb., Pl. Coromandel 3:90. 1820.

Woody vine planted and persistent in Central Cordillera (Villalba, Utuado), Cerro Las Mesas; 275–500 m; *fl* Dec–Mar, *fr* (?).

Native to tropical Asia, China. [L4: 351; AR: 392].

Lluvia de orquídeas, Terciopelo.

Cornutia obovata Urb., Symb. Antill. 1(3):395. 1899.

Tree in Northern Limestone Hills, Río Abajo Forest Reserve, Eastern Cordillera (La Torrecilla); 250–925 m; *fl* Jun–Jul, *fr* Sep (one specimen).

Endemic. [L4: 353].

Notes: This taxon is on the U.S. Fish & Wildlife Service list of threatened and endangered plants.

Capá jigüerilla, Nigua, Palo de nigua.

Holmskioldia sanguinea Retz., Observ. Bot. 6:31. 1791.

Shrub planted and persistent in Eastern Cordillera (Aibonito), Central Cordillera (Peñuelas); 400–500 m; *fl* Oct–Apr, *fr* (?).

Native to northern India. [L4: 359; AR: 394].

Paragüita chino, Platillo.

Hyptis americana (Poir.) Briq. in Engler & Prantl, Nat. Pflanzenfam. 4(3a):338. 1897.

Bas.: *Satureja americana* Poir. (1805).
Syn.: *Hyptis scoparia* Poit. (1806).

Shrub in Monte Guilarte Forest Reserve; 1000 m; *fl* (?), *fr* (?). [*Liogier 10016* (MAPR)].

Greater Antilles (Cuba, Hispaniola), Mexico, southern Central America (Panama). [L4: 395].

Hyptis atrorubens Poit., Ann. Mus. Nat. Hist. 7:466, *t. 27, f. 3*. 1806.

Herb in Northern Coastal Lowlands (Dorado to Manatí), Sierra de Luquillo, Eastern Coast, Sierra de Cayey, Eastern Cordillera (Cerro La Peña), Central Cordillera; 1–1335 m; *fl* & *fr* throughout the year.

Lesser Antilles, Mexico, Central America, northern South America to Bolivia, Paraguay. [L4: 395].

Marubio oscuro.

Hyptis capitata Jacq., Collectanea 1:102; et Icon. Pl. Rar. 1: *t. 114*. 1787.

Herb in Northern Coastal Lowlands, Northern Limestone Hills, Caguas Valley (Caguas), Sierra de Cayey, Las Tetas de Cayey, Central Cordillera, Maricao Forest Reserve, Southwestern Coastal Lowlands (Lajas), Western Coast (Añasco); 1–950 m; *fl* & *fr* Dec–May.

West Indies, Trinidad and Tobago, Mexico,

Central America, northwestern, west-central South America and Venezuela. [L4: 395].

Botoncillo negro, Marubio botón.

Hyptis lantanifolia Poit., Ann. Mus. Nat. Hist. 7:468, *t. 29, f. 1.* 1806 "lantanaefolia."

Herb in Northern Coastal Lowlands (Vega Baja to Laguna Tortuguero), Central Cordillera, Maricao Forest Reserve, Cerro Las Mesas; 1–825 m; *fl* Aug–Sep, Dec–Mar, *fr* Dec–Mar, Jun–Aug.

Greater Antilles (Cuba, Hispaniola), Trinidad, northwestern, west-central South America and Venezuela. [L4: 396].

Marubio blanco.

Hyptis pectinata (L.) Poit., Ann. Mus. Nat. Hist. 7:474, *t. 30.* 1806.

Bas.: *Nepeta pectinata* L. (1759).

Herb in Northern Coastal Lowlands, Río Abajo Forest Reserve, Caguas Valley, Sierra de Cayey, Las Tetas de Cayey, Las Piedras Chiquitas, Southeastern Foothills (Coamo), Eastern and Central Cordillera, Susúa Forest Reserve, Sierra Bermeja, Vieques; 1–1250 m; *fl* Nov–May, Aug, *fr* Jan–Aug.

West Indies, Trinidad and Tobago, Central America, northern South America to Brazil, Peru. [L4: 397].

Marubio falso.

Hyptis suaveolens (L.) Poit., Ann. Mus. Nat. Hist. 7:472, *t. 29, f. 2.* 1806.

Bas.: *Ballota suaveolens* L. (1759).

Suffrutescent herb in Northern Coastal Lowlands (Loíza), Eastern Coast, Caguas Valley (Juncos), Southeastern Foothills, Southern Coastal Lowlands (Salinas), Susúa Forest Reserve, Southwestern Coastal Lowlands, Sierra Bermeja, Western Coast (Aguada). Vieques; 1–425 m; *fl* & *fr* Oct–Apr.

Neotropical. [L4: 398].

Marubio.

Hyptis verticillata Jacq., Collectanea 1:101; et Icon. Pl. Rar. 1: *t. 113.* 1787.

Shrub in Northern Coastal Lowlands, Eastern Coast (Naguabo), Caguas Valley (Caguas), Eastern Cordillera (Aguas Buenas); 1–100 m; *fl* Jul–Oct, Feb–Mar, *fr* Jul–Oct, Feb.

Antilles, Trinidad, Mexico, Central America, northwestern South America. [L4: 399].

Leonotis nepetifolia (L.) R. Br. in W.T. Aiton, Hort. Kew. (ed. 2) 3:409. 1811.

Bas.: *Phlomis nepetifolia* L. (1753) "nepetaefolia."

Herb in Northern Coastal Lowlands, Caguas Valley, Las Piedras Chiquitas, Southeastern Foothills, Eastern Cordillera (Aibonito), Cerro de Las Cuevas, Southern Coastal Lowlands (Ponce), Susúa Forest Reserve, Southwestern Coastal Lowlands, Western Coast (Rincón), Vieques; 25–500 m; *fl* throughout the year, *fr* Jul–Nov, Mar.

Native to tropical Africa. [L4: 400].

Botón de cadete, Molinillo, Quinino del pasto.

Leonurus japonicus Houtt., Nat. Hist. 2(9):336, *t. 57, f. 1.* 1778.

Herb in Northern Limestone Hills (Vega Alta), Las Tetas de Cayey, Southeastern Foothills, Eastern and Central Cordillera, Southern Coastal Lowlands (Ponce); 5–1175 m; *fl* Nov–Aug, *fr* Feb–Jun.

Native to tropical Asia, eastern Russia, China, Korea. [L4: 401].

Notes: Britton & Wilson (1925), Liogier & Martorell (1982, 2000) and Liogier (1995) misapplied the name *Leonurus sibiricus* L., which refers to a taxon endemic to a narrow region of Siberia, to this taxon.

Agripalma, Cebadilla, Rabo de león.

Marsypianthes chamaedrys (Vahl) Kuntze, Revis. Gen. Pl. 2:524. 1891.

Bas.: *Clinopodium chamaedrys* Vahl (1794).

Prostrate herb in Northern Coastal Lowlands, Eastern Coast (Humacao), Vieques; 1–10 m; *fl* Oct–Nov, Feb–Jun, *fr* Jan–Jun, Oct.

Greater Antilles (Cuba), Lesser Antilles, Trinidad, Mexico, Central America, South America. [L4: 402].

Ortelá.

Mentha ×piperita L. (pro sp.), Sp. Pl. 2:576. 1753.

Syn.: *Mentha ×citrata* Ehrh. (1792). *Mentha ×piperita* L. var. *citrata* (Ehrh.) Briq. (1899).

Herb in Río Abajo Forest Reserve, Sierra de Luquillo; 325–1065 m; *fl* Apr, Aug–Sep (two specimens), *fr* (?). [*Woodbury a.n. 10779* (SJ); *Liogier 33200* (UPR); *Axelrod 13979* (UPRRP)].

Native to the Old World. [L4: 405].

Agua florida, Menta piperita.

Mentha suaveolens Ehrh., Beitr. Naturk. 7:149. 1792.

Herb in Sierra de Luquillo; 700 m; *fl* (?), *fr* (?). [*Liogier 33333* (UPR)].

Native to Europe, northwestern Africa, Turkey.

Ocimum basilicum L., Sp. Pl. 2:597. 1753.

Herb in Southern Coastal Lowlands (Santa Isabel), Southern Coastal Limestone (Sabana Grande), Southwestern Coastal Lowlands (Lajas);

1–100 m; *fl* Apr–May, Sep–Nov, *fr* Sep–Nov (two specimens).

Possibly originating in Africa. [L4: 407].

Albahaca, Albahaca blanca, Orégano falso.

Ocimum campechianum Mill., Gard. Dict. (ed. 8): *Ocimum* no. 5. 1768.

Syn.: *Ocimum micranthum* Willd. (1809).

Herb in Northern Coastal Lowlands, Northern Limestone Hills, Southeastern Foothills, Eastern Cordillera (Aibonito), Guánica Forest Reserve, Southwestern Coastal Lowlands, Sierra Bermeja, Western Coast (Aguada), Vieques, Culebra, Mona; 1–600 m; *fl* & *fr* throughout the year.

West Indies, southeastern United States (Florida), Mexico, Central America, northern South America. [L4: 407].

Albahaca cimarrona, Albahaca de puerco.

Ocimum tenuiflorum L., Sp. Pl. 2:597. 1753.

Syn.: *Ocimum sanctum* L. (1767).

Herb in Southwestern Coastal Lowlands (Lajas), Mona; 1–5 m; *fl* & *fr* Feb–Jul.

Native to India, Sri Lanka, Malesia. [L4: 409].

Albahaca morada.

Petitia domingensis Jacq., Enum. Syst. Pl. 12. 1760.

Tree in Northern Coastal Lowlands, Northern Limestone Hills (Isabela to Moca), Río Abajo Forest Reserve, Eastern Cordillera (Comerío), Susúa Forest Reserve, Western Cordillera, Southwestern Coastal Lowlands (Cabo Rojo), Sierra Bermeja, Cerro Las Mesas, Western Coast (Rincón); 2–475 m; *fl* throughout the year, *fr* Feb–Nov.

Bahamas, Greater Antilles. [L4: 371].

Capá amarillo, Capá blanco, Capá rosado, Capá de sabana.

Plectranthus amboinicus (Lour.) Spreng., Syst. Veg. 2:690. 1825.

Bas.: *Coleus amboinicus* Lour. (1790).

Herb planted and persistent in Southern Coastal Lowlands (Salinas), Southwestern Coastal Lowlands (Lajas), Vieques; 1–75 m; *fl* Feb, May (two specimens), *fr* (?).

Native to eastern and southern tropical Africa. [L4: 410].

Orégano brujo, Orégano de España.

Plectranthus verticillatus (L.f.) Druce, Rep. Bot. Soc. Exch. Club Brit. Isles 4(Suppl. 2):640. 1917.

Bas.: *Ocimum verticillatum* L.f. (1782).

Syn.: *Plectranthus nummularius* Briq. (1903).

Herb in Maricao Forest Reserve; 800 m; *fl* Oct (one specimen), *fr* (?). [*Acevedo-Rodríguez* 7713 (MAPR)].

Native to South Africa.

Salvia coccinea Buc'hoz ex Etl., Comm. *Salvia* 23. 1777.

Herb in Northern Limestone Hills (San Sebastián), Río Abajo Forest Reserve, Las Piedras Chiquitas, Southeastern Foothills, Eastern Cordillera (Aibonito), Central Cordillera (Villalba, Ponce), Cerro de Las Cuevas, Maricao Forest Reserve; 150–500 m; *fl* & *fr* Apr–Jan.

West Indies, southern United States, Mexico, Central America, South America (scattered). [L4: 414].

Cizaña, Moradilla colorada.

Salvia misella Kunth in Humboldt, Bonpland & Kunth, Nov. Gen. Sp. 2:234 [folio]; 2:290 [quarto]. 1818.

Herb in Northern Coastal Lowlands (Laguna Tortuguero); 1 m; *fl* Jan–Feb (one specimen), *fr* (?). [*Liogier 32603* (UPR)].

Greater Antilles, Mexico, Central America, northerwestern South America. [L4: 415].

Salvia occidentalis Sw., Prodr. 14. 1788.

Herb in Northern Coastal Lowlands, Northern Limestone Hills (Quebradillas), Río Abajo Forest Reserve, Eastern Coast, Sierra de Cayey (San Lorenzo), Las Piedras Chiquitas, Southeastern Foothills, Eastern Cordillera (Aibonito, Cayey), Central Cordillera, Western Coast (Rincón); 1–800 m; *fl* & *fr* Dec–Jun.

West Indies, Mexico, Central America, northwestern, west-central South America and Venezuela. [L4: 416].

Moradilla.

Salvia serotina L., Mant. Pl. 1:25. 1767.

Suffrutescent herb in Southern Coastal Lowlands (Caja de Muertos), Maricao Forest Reserve, Mona, Desecheo; 1–800 m; *fl* Feb–Nov, *fr* throughout the year.

West Indies, southeastern United States (Florida). [L4: 416].

Moradilla azul.

Salvia splendens Sellow ex Schult., Mantissa 1:185. 1822.

Suffrutescent herb in Central Cordillera; 600–1100 m; *fl* Feb–Jul, *fr* (?).

Native to Brazil. [L4: 416].

Salvia.

Scutellaria havanensis Jacq., Enum. Syst. Pl. 25. 1760.

Herb in Northern Limestone Hills, Southern Coastal Limestone (Guayanilla), Susúa Forest Reserve, Maricao Forest Reserve, Mona; 30–900 m; *fl* & *fr* throughout the year.

Bahamas, Greater Antilles, southeastern United States (Florida). [L4: 420].

Solenostemon scutellarioides (L.) Codd, Bothalia 11(4):439. 1975.

Bas.: *Ocimum scutellarioides* L. (1763).

Syn.: *Plectranthus scutellarioides* (L.) R. Br. (1810). *Coleus scutellarioides* (L.) Benth. (1830).

Coleus blumei Benth. (1832). *Plectranthus blumei* (Benth.) Launert (1968).

Herb in Río Abajo Forest Reserve, Sierra de Luquillo, Sierra de Cayey, Southeastern Foothills (Arroyo), Central Cordillera, Western Cordillera, Western Coast (Mayagüez); 175–1300 m; *fl* throughout the year, *fr* Jul–Apr.

Native to tropical Asia, China, Australia. [L4: 410].

Cóleo, Nazareno, Tocador, Vergüenza.

Tectona grandis L.f., Suppl. Pl. 151. 1782.

Tree planted and persistent in Río Abajo Forest Reserve, Sierra de Luquillo, Sierra de Cayey, Susúa Forest Reserve; 300–500 m; *fl* Aug, Nov, *fr* Jul–Mar.

Native to tropical Asia. [L4: 383].

Teca.

Vitex divaricata Sw., Prodr. 93. 1788.

Tree in Northern Coastal Lowlands (Río Grande, San Juan), Northern Limestone Hills (Toa Baja, Quebradillas), Sierra de Luquillo, Caguas Valley (Gurabo), Sierra de Cayey, Central Cordillera, Southern Coastal Limestone (Guayanilla), Susúa Forest Reserve, Western Cordillera, Southwestern Coastal Lowlands, Sierra Bermeja, Western Coast (Mayagüez); 5–975 m; *fl* Apr–Oct, *fr* Jun–Sep.

Greater Antilles (Cuba, Hispaniola), Virgin Islands, Lesser Antilles, Trinidad, northern South America (Venezuela, Colombia). [L4: 388].

Higüerillo, Péndula, Péndula blanca.

Vitex parviflora Juss., Ann. Mus. Nat. Hist. 7:76. 1806.

Tree planted and persistent in Sierra de Luquillo, Western Cordillera (Maricao); 450–700 m; *fl* Aug–Oct (two specimens), *fr* Oct (one specimen).

Native to Thailand, Malesia. [L4: 388].

Higüerillo, Molave.

Volkameria aculeata L., Sp. Pl. 2:637. 1753.

Syn.: *Clerodendrum aculeatum* (L.) Schltdl. (1831) "Clerodendron."

Shrub in Northern Coastal Lowlands, Eastern Coast (Cayo Santiago), Southern Coastal Lowlands, Guánica Forest Reserve, Southwestern Coastal Lowlands, Vieques, Culebra, Mona; 1–15 m; *fl* throughout the year, *fr* Jun–Apr.

West Indies, Trinidad and Tobago, northern South America. [L4: 347; AR: 390].

Botón de oro, Descamisador, Escambrón blanco.

TAXA NOT RECENTLY COLLECTED

Cornutia pyramidata L., Sp. Pl. 2:628. 1753.

Shrub/small tree collected at the Río Blanco near Naguabo in Jun 1881 [*Eggers 426* (US)] and in Jul 1881 [*Eggers 626* (US)], possibly from the same collection, numbers and dates notwithstanding.

Antilles, Central America. [L4: 355].

Hyptis spicigera Lam., Encycl. 3(1):185. 1789.

Herb collected between Cabo Rojo and Mayagüez on 23 Jan 1885 [*Sintenis 684* (US)].

Greater Antilles, Mexico, Central America, northern South America to Bolivia. [L4: 397].

Notes: Britton & Wilson (1925) used the homonym *Hyptis americana* (Aubl.) Urb. for the name of this taxon.

Mentha spicata L., Sp. Pl. 2:576. 1753.

Herb collected near Adjuntas on 20 Jun 1886 [*Sintenis 4602* (US)].

Native to southeastern Europe, western Asia. [L4: 405].

Notes: Urban had determined the above specimen as *Mentha nemorosa* Willd.

Yerba buena.

Stachys arvensis (L.) L., Sp. Pl. (ed. 2) 2:814. 1763.

Bas.: *Glechoma arvensis* L. (1753).

Herb collected in Cayey on 5 Jan 1938 [*Otero 403* (GH)].

Native to Europe, northwestern Africa, Middle East. [L4: 421].

TAXA CITED BY PREVIOUS AUTHORS, BUT NOT INCLUDED

NO SPECIMENS SEEN

Aegiphila elata Sw.

Woody vine/shrub cited by Liogier & Martorell (1982, 2000) and Liogier (1995) as, according to Moldenke, collected by Blake in Puerto Rico. [L4: 337].

CITED DUE TO MISIDENTIFICATION

Hyptis escobilla Urb.

Shrub cited by Liogier & Martorell (1982, 2000) and Liogier (1995) as occurring near Bayamón; the citation is based upon the misidentification of a specimen of *Hyptis verticillata*. [L4: 396].

Salvia micrantha Vahl.

Shrub cited by Liogier & Martorell (1982, 2000) and Liogier (1995) as collected on Mona Island and probably present on the mainland; the citation is based upon the misidentification of specimens of *Salvia serotina*. [L4: 415].

Notes: Britton & Wilson (1925) placed this taxon in synonymy with *Salvia serotina*.

ONLY UNDER CULTIVATION

Clerodendrum bungei Steud.

Herb cited by Liogier & Martorell (1982, 2000) and Liogier (1995) as planted and escaped. [L4: 349].

Clerodendrum speciosissimum C. Morren "Clerodendron."

Herb cited by Liogier & Martorell (1982, 2000) and Liogier (1995) as occurring on roadsides, escaped from cultivation. [L4: 350].

Clerodendrum thomsoniae Balf. "thomsonae."

Vine cited by Liogier & Martorell (1982, 2000) and Liogier (1995) as planted and persistent and by Acevedo-Rodríguez (2005) as only cultivated. [L4: 351; AR: 392].

Gmelina arborea Roxb.

Tree cited by Liogier & Martorell (1982, 2000) and Liogier (1995) as scarcely planted and sometimes escaped. [L4: 357].

Gmelina asiatica L.

Shrub cited by Liogier & Martorell (1982, 2000) and Liogier (1995) as planted as an ornamental and escaped from cultivation, rare. [L4: 357].

Plectranthus parviflorus Willd. Syn: *Plectranthus australis* R. Br.

Herb cited by Liogier & Martorell (1982) as cultivated and escaped.

Rosmarinus officinalis L.

Suffrutescent herb/shrub cited by Liogier & Martorell (1982, 2000) and Liogier (1995) as occasionally grown in gardens and escaped or at least persistent. [L4: 413].

Vitex agnus-castus L.

Tree cited by Liogier & Martorell (1982, 2000) and Liogier (1995) as cultivated and persistent. [L4: 388].

LAURACEAE

Aniba bracteata (Nees) Mez, Jahrb. Königl. Bot. Gart. Berlin 5:66, *t. 3*, *f. 32*. 1889.

Bas.: *Aydendron bracteatum* Nees (1836).

Tree in Sierra de Luquillo, Sierra de Cayey; 300–800 m; *fl* Feb, Aug (two specimens), *fr* Jan–Apr, Jul.

Lesser Antilles, southern Central America (Panama), South America (Venezuela, Ecuador, Bolivia). [L1: 249].

Canelillo.

Beilschmiedia pendula (Sw.) Hemsl., Biol. Cent.-Amer., Bot. 3(14):70. 1882.

Bas.: *Laurus pendula* Sw. (1788).

Syn.: *Hufelandia pendula* (Sw.) Nees (1833).

Tree in Northern Coastal Lowlands (Luquillo), Northern Limestone Hills (Vega Baja), Río Abajo Forest Reserve, Central Cordillera (Adjuntas), Maricao Forest Reserve; 15–750 m; *fl* Aug (one specimen), *fr* May, Aug–Sep.

Antilles, Central America, northwestern, west-central South America and Venezuela. [L1: 252].

Aguacate cimarrón, Aguacatillo, Cedro macho, Guajón, Palo colorado.

Cassytha filiformis L., Sp. Pl. 1:35. 1753.

Parasitic vine in Northern Coastal Lowlands, Northern Limestone Hills (Bayamón), Eastern Coast, Guánica Forest Reserve, Sierra Bermeja, Western Coast (Rincón), Vieques, Culebra, Mona; 1–100 m; *fl* & *fr* throughout the year.

Pantropical. [L1: 252; AR: 300].

Bejuco dorado, Cabellos de ángel, Fideíllo, Fideos, Tente en el aire.

Cinnamomum elongatum (Nees) Kosterm., Reinwardtia 6:21. 1961.

Bas.: *Phoebe elongata* Nees (1836).

Tree in Northern Coastal Lowlands, Northern Limestone Hills (Bayamón to Dorado), Sierra de Luquillo, Eastern Coast, Caguas Valley, Southeastern Foothills, Eastern Cordillera, Central Cordillera (Ciales, Utuado), Maricao Forest Reserve; 70–900 m; *fl* throughout the year, *fr* Feb–Jul.

Greater Antilles (Cuba, Hispaniola), Lesser Antilles, Trinidad and Tobago. [L1: 255].

Avispillo, Laurel avispillo, Laurel bobo.

Cinnamomum montanum (Sw.) J. Presl in Berchtold & J. Presl, Přir. Rostlin 2:36. 1825.

Bas.: *Laurus montana* Sw. (1788).

Syn.: *Phoebe montana* (Sw.) Griseb. (1860).

Tree in Northern Limestone Hills, Río Abajo

Forest Reserve, Maricao Forest Reserve; 200–700 m; *fl* Feb (one specimen), *fr* (?).

Greater Antilles. [L1: 256].

Avispillo, Laurel.

Licaria brittoniana C.K. Allen & L.E. Greg., Brittonia 7(4):267, *ff. 1–2*. 1951.

Tree in Northern Limestone Hills, Río Abajo Forest Reserve, Susúa Forest Reserve, Maricao Forest Reserve; 300–875 m; *fl* May–Jun, Sep, *fr* Oct, Jan, Apr–Jun.

Endemic. [L1: 257].

Cacao macho, Canela, Canela amarilla, Canelón, Casa María laurel.

Licaria parvifolia (Lam.) Kosterm., J. Sci. Res. (Jakarta) 1:149. 1952.

Bas.: *Laurus parvifolia* Lam. (1792).

Syn.: *Acrodiclidium salicifolium* (Sw.) Griseb. (1861). *Licaria salicifolia* (Sw.) Kosterm. (1937).

Tree in Northern Coastal Lowlands, Northern Limestone Hills, Las Piedras Chiquitas, Southeastern Foothills, Central Cordillera (Ponce), Southern Coastal Lowlands (Patillas), Susúa Forest Reserve, Guánica Forest Reserve, Southwestern Coastal Lowlands (Cabo Rojo), Sierra Bermeja, Cerro Las Mesas, Vieques; 1–550 m; *fl* Aug–May, *fr* Sep–May.

Lesser Antilles. [L1: 257].

Cacaíllo, Canela, Canela del país, Canelilla, Canelillo.

Licaria triandra (Sw.) Kosterm., Meded. Bot. Mus. Herb. Rijks Univ. Utrecht 42:588; et Recueil Trav. Bot. Néerl. 34:588. 1937.

Bas.: *Laurus triandra* Sw. (1788).

Syn.: *Misanteca triandra* (Sw.) Mez (1889).

Tree in Northern Coastal Lowlands (Río Grande, Laguna Tortuguero), Northern Limestone Hills, Southeastern Foothills (Salinas), Eastern Cordillera (Cayey); 1–535 m; *fl* May, *fr* Apr–May, Sep.

Antilles, southeastern United States (Florida), Mexico, Central America, northwestern, west-central South America and Venezuela. [L1: 258].

Palo de misanteco.

Nectandra coriacea (Sw.) Griseb., Fl. Brit. W.I. 281. 1860.

Bas.: *Laurus coriacea* Sw. (1788).

Syn.: *Ocotea coriacea* (Sw.) Britton (1920).

Tree in Northern Coastal Lowlands (Fajardo), Northern Limestone Hills, Río Abajo Forest Reserve, Maricao Forest Reserve, Guánica Forest Reserve, Vieques, Mona; 10–350 m; *fl* Mar–Nov, *fr* Jun–Jan.

West Indies, Trinidad, southeastern United States (Florida), Mexico, Central America. [L1: 261].

Avispillo, Cigua, Laurel, Laurel cigua.

Nectandra hihua (Ruiz & Pav.) Rohwer, Fl. Neotrop. Monogr. 60:196. 1993.

Bas.: *Laurus hihua* Ruiz & Pav. (1804).

Syn.: *Nectandra antillana* Meisn. (1864).

Tree in Northern Coastal Lowlands; 1–100 m; *fl* Dec–Jun, *fr* Feb–Jul.

Greater Antilles, Mexico, Central America, South America. [L1: 263].

Notes: Liogier & Martorell (1982) and Liogier (1985) misapplied the name *Ocotea globosa* (Aubl.) Schltdl. & Cham. to this taxon.

Aguacatillo, Geo rojo, Laurel bobo.

Nectandra krugii Mez, Jahrb. Königl. Bot. Gart. Berlin 5:422. 1889.

Syn.: *Ocotea krugii* (Mez) R.A. Howard (1981).

Tree in Sierra de Cayey, Eastern Cordillera (Cañón San Cristóbal), Toro Negro Forest Reserve, Western Cordillera; 500–1000 m; *fl* May–Nov, *fr* Apr (one specimen).

Greater Antilles (Hispaniola), Lesser Antilles. [L1: 263].

Laurel canelón.

Nectandra membranacea (Sw.) Griseb., Fl. Brit. W.I. 282. 1860.

Bas.: *Laurus membranacea* Sw. (1788).

Syn.: *Ocotea membranacea* (Sw.) R.A. Howard (1981).

Tree in Sierra de Luquillo, Eastern Coast (Yabucoa), Sierra de Cayey, Eastern Cordillera (La Torrecilla), Central Cordillera; 250–950 m; *fl* Mar–Nov, *fr* Nov, Mar–Jun.

Antilles, Trinidad and Tobago, southern Central America, South America. [L1: 264].

Laurelillo, Laurel prieto.

Nectandra patens (Sw.) Griseb., Fl. Br. W.I. 281. 1860.

Bas.: *Laurus patens* Sw. (1788).

Syn.: *Ocotea patens* (Sw.) Nees (1833).

Tree in Northern Limestone Hills, Río Abajo Forest Reserve, Sierra de Luquillo, Las Tetas de Cayey, Eastern Cordillera (Cañón San Cristóbal), Central Cordillera (Ciales, Adjuntas), Maricao Forest Reserve, Guánica Forest Reserve, Cerro Las Mesas; 200–1000 m; *fl* & *fr* throughout the year.

Antilles. [L1: 265].

Laurel roseta, Laurel geo colorado.

Nectandra turbacensis (Kunth) Nees, Syst. Laur. 316. 1836.

Bas.: *Ocotea turbacensis* Kunth (1817).
Syn.: *Nectandra sintenisii* Mez (1889). *Ocotea sintenisii* (Mez) Alain (1982).

Tree in Northern Coastal Lowlands (Río Grande), Northern Limestone Hills, Sierra de Luquillo, Las Tetas de Cayey, Eastern and Central Cordillera, Southern Coastal Lowlands (Ponce); 50–1000 m; *fl* Aug–Nov, Mar, *fr* Nov–Jul.

Greater Antilles, Virgin Islands, Trinidad and Tobago, Mexico, Central America, northwestern, west-central South America and Venezuela, Brazil. [L1: 266].

Laurel amarillo, Laurel blanco, Laurel geo, Laurel macho.

Ocotea acarina C.K. Allen, Mem. New York Bot. Gard. 21(2):111, *f*. 2. 1971.

Tree in Río Abajo Forest Reserve; 300–875 m; *fl* Mar (one specimen), *fr* (?).

Greater Antilles (Hispaniola).

Ocotea cuneata (Griseb.) M. Gómez, Dicc. Bot. Nombres Vulg. Cub. Puerto-Riq. 30. 1889.
Bas.: *Nectandra cuneata* Griseb. (1860).
Syn.: *Ocotea nemodaphne* Mez (1889), *nom. illeg.*

Tree in Northern Limestone Hills (Arecibo, San Sebastián), Río Abajo Forest Reserve, Las Tetas de Cayey, Maricao Forest Reserve; 200–850 m; *fl* May–Jun, *fr* throughout the year.

Greater Antilles (Cuba, Hispaniola). [L1: 265].

Canela, Laurel sasafrás, Sasafrás.

Ocotea floribunda (Sw.) Mez, Jahrb. Königl. Bot. Gart. Berlin 5:325. 1889.
Bas.: *Laurus floribunda* Sw. (1788).

Tree in Northern Limestone Hills, Río Abajo Forest Reserve, Sierra de Luquillo, Sierra de Cayey, Eastern and Central Cordillera, Maricao Forest Reserve; 250–1000 m; *fl* Aug–Jan, *fr* Feb–May.

Greater Antilles, Lesser Antilles (Antigua, Montserrat, Guadeloupe), Central America, northern South America to Brazil, Bolivia. [L1: 261].

Laurel, Laurel espada.

Ocotea foeniculacea Mez, Jahrb. Königl. Bot. Gart. Berlin 5:266. 1889.

Tree in Central Cordillera, Western Cordillera (Maricao); 600–1200 m; *fl* (?), *fr* Mar (one specimen). [*Woodbury Mar 1971* (SJ); *Vivaldi a.n. 14021* (UPR); *Liogier 35090* (UPR)].

Greater Antilles (Cuba, Hispaniola). [L1: 262].

Laurel, Palo santo.

Ocotea leucoxylon (Sw.) Laness., Pl. Util. Col. Franç. 158. 1886.
Bas.: *Laurus leucoxylon* Sw. (1788).

Tree in Northern Coastal Lowlands (Río Grande, Dorado), Guajataca Forest Reserve, Sierra de Luquillo, Sierra de Cayey, Las Tetas de Cayey, Eastern Cordillera (Cerro La Peña, Aibonito), Central Cordillera, Western Cordillera, Cerro Las Mesas; 20–1050 m; *fl* & *fr* throughout the year.

Antilles, Trinidad and Tobago, Mexico, Central America, northwestern, west-central South America and Venezuela, Brazil. [L1: 263].

Cacaíllo, Geo, Laurel geo, Laurel geogeo.

Ocotea moschata (Meisn.) Mez, Jahrb. Königl. Bot. Gart. Berlin 5:269. 1889.
Bas.: *Mespilodaphne moschata* Meisn. (1864).

Tree in Sierra de Luquillo, Caguas Valley (Juncos), Sierra de Cayey; 400–750 m; *fl* Feb, May (two specimens), *fr* Mar, Nov.

Endemic. [L1: 264].

Nemocá, Nuez moscada del país, Nuez moscada macho.

Ocotea portoricensis Mez, Jahrb. Königl. Bot. Gart. Berlin 5:364. 1889.

Tree in Sierra de Luquillo, Central Cordillera (Yauco), Western Cordillera (Maricao); 400–875 m; *fl* May–Oct, *fr* May (one specimen).

Endemic. [L1: 265].

Avispillo, Geo, Laurel, Laurel avispillo, Laurel geo, Laurel de paloma.

Ocotea spathulata Mez, Jahrb. Königl. Bot. Gart. Berlin 5:268. 1889.

Tree in Sierra de Luquillo, Sierra de Cayey; 775–1025 m; *fl* Jan, Apr–Sep, *fr* throughout the year.

Greater Antilles (Cuba), Virgin Islands (Tortola). [L1: 266].

Cabalonga cimarrona, Granadillo, Nemocá cimarrona, Nemocá macho, Nuez moscada cimarrona.

Ocotea wrightii (Meisn.) Mez, Jahrb. Königl. Bot. Gart. Berlin 5:267. 1889.
Bas.: *Acrodiclidium wrightii* Meisn. (1864).

Tree in Monte Guilarte Forest Reserve, Maricao Forest Reserve; 700–1150 m; *fl* Jun (two specimens), *fr* Oct–Apr.

Greater Antilles (Cuba, Hispaniola). [L1: 266].

Canela, Canelón, Laurel canelón.

Persea americana Mill., Gard. Dict. (ed. 8): *Persea* no. 1. 1768.
Syn.: *Persea persea* (L.) Cockerell (1892).

Tree cultivated and persistent in Northern

Coastal Lowlands, Northern Limestone Hills, Eastern Cordillera (Corozal), Central Cordillera (Adjuntas), Maricao Forest Reserve; 5–850 m; *fl* Jan–Apr, *fr* May–Sep.

Native to Mexico, Central America. [L1: 267].

Aguacate.

Persea krugii Mez, Jahrb. Königl. Bot. Gart. Berlin 5:177. 1889.

Tree in Sierra de Luquillo, Toro Negro Forest Reserve, Maricao Forest Reserve; 800–900 m; *fl* May–Aug, *fr* Mar, Jun (two specimens).

Greater Antilles (Hispaniola). [L1: 269].

Canela, Canela fina.

Persea urbaniana Mez, Jahrb. Königl. Bot. Gart. Berlin 5:143. 1889.

Syn.: *Persea portoricensis* Britton & P. Wilson (1926).

Tree in Sierra de Cayey, Central Cordillera, Maricao Forest Reserve; 600–1325 m; *fl* Jun–Oct, *fr* Jun–Nov.

Greater Antilles (Jamaica), Lesser Antilles. [L1: 269].

Aguacatillo.

TAXA CITED BY PREVIOUS AUTHORS, BUT NOT INCLUDED

ONLY UNDER CULTIVATION

Cinnamomum camphora (L.) J. Presl.

Tree cited by Liogier & Martorell (1982, 2000) and Liogier (1985) as occasionally planted. [L1: 254].

Cinnamomum verum J. Presl.

Tree cited by Liogier & Martorell (1982, 2000) and Liogier (1985) as occasionally planted and persistent. [L1: 256].

LAXMANNIACEAE

TAXA CITED BY PREVIOUS AUTHORS, BUT NOT INCLUDED

ONLY UNDER CULTIVATION

Cordyline fruticosa (L.) A. Chev. Syn.: *Dracaena terminalis* (L.) L.

Shrub cited by Liogier & Martorell (1982, 2000) as cultivated and persistent and by Acevedo-Rodríguez & Strong (2005) as cultivated. [AR&S: 129].

LECYTHIDACEAE

TAXA CITED BY PREVIOUS AUTHORS, BUT NOT INCLUDED

ONLY UNDER CULTIVATION

Barringtonia asiatica (L.) Kurz.

Tree cited by Liogier & Martorell (1982, 2000) and Liogier (1994) as sparingly planted and persistent. [L3: 358].

Couroupita guianensis Aubl.

Tree cited by Liogier & Martorell (1982, 2000) and Liogier (1994) as a rare ornamental, planted and persistent in parks and gardens. [L3: 360].

LENTIBULARIACEAE

Utricularia gibba L., Sp. Pl. 1:18. 1753.

Syn.: *Utricularia obtusa* Sw. (1788).

Herb in Northern Coastal Lowlands (Vega Alta, Vega Baja); 1–5 m; *fl* Sep–Nov (two specimens), *fr* Apr, Sep (two specimens).

Pantropical and -subtropical. [L4: 582].

Utricularia guyanensis A. DC., Prodr. 8:11. 1844.

Herb in Sierra de Luquillo; 800–900 m; *fl* & *fr* Mar (one specimen). [*Proctor 40304* (SJ)].

Trinidad, Central America (Belize, Honduras, Nicaragua), northern South America. [L4: 581].

Utricularia juncea Vahl, Enum. Pl. 1:202. 1804.

Syn.: *Stomoisia juncea* (Vahl) Barnhart (1913).

Herb in Northern Coastal Lowlands (Vega Baja, Laguna Tortuguero), Sierra de Luquillo; 1–1000 m; *fl* & *fr* Mar–Sep.

Greater Antilles (Cuba, Hispaniola), Trinidad, eastern and south-central United States, Mexico, Central America, northern South America to Brazil. [L4: 582].

Utricularia pusilla Vahl, Enum. Pl. 1:202. 1804.

Syn.: *Setiscapella pusilla* (Vahl) Barnhart ex Britton & P. Wilson (1925).

Herb in Northern Coastal Lowlands (Laguna Tortuguero), Sierra de Luquillo; 1–925 m; *fl* Apr, Aug–Sep, *fr* Feb–Apr, Aug–Sep.

Antilles, Trinidad, Mexico, Central America, South America. [L4: 583].

Utricularia subulata L., Sp. Pl. 1:18. 1753.

Syn.: *Setiscapella subulata* (L.) Barnhart (1913). *Utricularia subulata* L. f. *cleistogama* (A. Gray) Fernald (1922).

Herb in Northern Coastal Lowlands (Vega Alta to Manatí), Sierra de Luquillo; 1–1000 m; *fl* Aug, Nov–Jan, Apr, *fr* Mar–Apr, Aug.

Pantropical. [L4: 583].

Grasilla.

LIMNOCHARITACEAE

TAXA NOT RECENTLY COLLECTED

Hydrocleys nymphoides (Humb. & Bonpl. ex Willd.) Buchenau, Index Crit. Butom. Alism. Juncag. 2, 9. 1868 (Abh. Naturwiss. Vereins Bremen 2:2, 9. 1869).
Bas.: *Stratiotes nymphoides* Humb. & Bonpl. ex Willd. (1806).

Aquatic herb collected in Añasco [*Sintenis 5779* (U), not seen] and at Villa Espinosa, Vega Alta, on 7 Mar 1932 [*Britton & Britton 9941* (NY)].

Native to Southern Caribbean Islands (Curaçao), Trinidad, northern Central America (Guatemala), South America (scattered). [AR&S: 58].

Lirio de agua amarillo.

LINDERNIACEAE

Lindernia crustacea (L.) F. Muell., Syst. Census Austral. Pl. 1:97. 1882–1883.
Bas.: *Capraria crustacea* L. (1767).

Herb in Northern Coastal Lowlands, Sierra de Luquillo, Eastern Coast, Caguas Valley, Sierra de Cayey, Western Coast (Mayagüez); 1–925 m; *fl* & *fr* throughout the year.

Native to tropical Asia, China, Japan, Korea, Australia. [L4: 494].

Lindernia diffusa (L.) Wettst. in Engler & Prantl, Nat. Pflanzenfam. 4(3b):79. 1891.
Bas.: *Vandellia diffusa* L. (1767).

Herb in Northern Coastal Lowlands, Sierra de Luquillo, Eastern Coast (Maunabo), Sierra de Cayey, Central Cordillera (Ciales); 1–1000 m; *fl* & *fr* Feb–Nov.

Greater Antilles (Jamaica, Hispaniola), Lesser Antilles, Trinidad, Mexico, Central America, northwestern, west-central South America and Venezuela. [L4: 495].

Lindernia dubia (L.) Pennell, Acad. Nat. Sci. Philadelphia Monogr. 1:141. 1935 (Scroph. E. N. Amer.).
Bas.: *Gratiola dubia* L. (1753).
Syn.: *Ilysanthes dubia* (L.) Barnhart (1899).

Herb in Northern Coastal Lowlands, Eastern Coast, Caguas Valley (San Lorenzo), Southwestern Coastal Lowlands (Cabo Rojo), Western Coast (Añasco); 1–85 m; *fl* & *fr* Jan–Oct.

Greater Antilles, Canada, United States, Mexico, Central America, South America. [L4: 496].

Yerba graciosa.

LOASACEAE

Mentzelia aspera L., Sp. Pl. 1:516. 1753.

Straggling herb in Las Piedras Chiquitas, Southeastern Foothills, Southern Coastal Lowlands, Southern Coastal Limestone (Peñuelas); 50–550 m; *fl* Sep–Feb, *fr* Nov–Feb.

Greater Antilles, Lesser Antilles (St. Lucia), Southern Caribbean Islands (Aruba, Curaçao), southwestern United States, Mexico, Central America, South America. [L3: 300].

Pegadora.

LOGANIACEAE

Mitreola petiolata (J.F. Gmel.) Torr. & A. Gray, Fl. N. Amer. 2(1):45. 1841.
Bas.: *Cynoctonum petiolatum* J.F. Gmel. (1791).
Syn.: *Cynoctonum mitreola* (L.) Britton (1894).

Herb in Northern Coastal Lowlands (Vega Alta to Laguna Tortuguero), Northern Limestone Hills (San Sebastián), Río Abajo Forest Reserve; 1–300 m; *fl* & *fr* throughout the year.

Bahamas, Greater Antilles, Trinidad, southeastern and south-central United States, Mexico, Central America, northwestern, west-central South America and Venezuela. [L4: 184].

Spigelia anthelmia L., Sp. Pl. 1:149. 1753.

Herb in Northern Coastal Lowlands, Northern Limestone Hills, Río Abajo Forest Reserve, Eastern Coast, Caguas Valley (Las Piedras), Southeastern Foothills (Coamo), Guánica Forest Reserve, Western Coast, Vieques; 1–250 m; *fl* & *fr* throughout the year.

West Indies, southeastern United States (Florida), Mexico, Central America, northern South America to Brazil, Peru. [L4: 188].

Lombricera, Yerba lombricera.

LORANTHACEAE

Dendropemon bicolor Krug & Urb., Bot. Jahrb. Syst. 24(1):24. 1897.
Syn.: *Phthirusa bicolor* (Krug & Urb.) Engl. (1897).

Parasitic shrub in Northern Limestone Hills, Río Abajo Forest Reserve, Eastern Cordillera (Aibonito), Central Cordillera, Maricao Forest Reserve, Cerro Las Mesas; 5–1025 m; *fl* & *fr* throughout the year.

Endemic. [L1: 99].

Hicaquillo, Muérdago.

Dendropemon caribaeus Krug & Urb., Bot. Jahrb. Syst. 24(1):27. 1897.
Syn.: *Phthirusa caribaea* (Krug & Urb.) Engl. (1897).

Parasitic shrub in Northern Coastal Lowlands,

Northern Limestone Hills, Eastern Coast (Humacao, Punta Guayanés), Las Tetas de Cayey, Las Piedras Chiquitas, Southeastern Foothills (Salinas), Central Cordillera (Yauco), Southern Coastal Lowlands, Southern Coastal Limestone, Susúa Forest Reserve, Maricao Forest Reserve, Guánica Forest Reserve, Southwestern Coastal Lowlands, Sierra Bermeja, Vieques; 1–875 m; *fl* & *fr* throughout the year.

Greater Antilles (Hispaniola), Virgin Islands, Lesser Antilles. [L1: 100].

Capitana, Parásita.

Dendropemon purpureus (L.) Krug & Urb., Bot. Jahrb. Syst. 24(1):26. 1897.
Bas.: *Viscum purpureum* L. (1753).
Syn.: *Phthirusa purpurea* (Krug & Urb.) Engl. (1897).

Parasitic shrub in Northern Coastal Hills (Arecibo), Guánica Forest Reserve, Southwestern Coastal Lowlands (Lajas); 10–200 m; *fl* & *fr* Sep–Nov, Mar–Apr.

Bahamas, Greater Antilles (Cuba, Hispaniola). [L1: 101].

Caballero, Capitana, Hicaquillo, Pegapalo.

TAXA NOT RECENTLY COLLECTED

Dendropemon sintenisii Krug & Urb., Bot. Jahrb. Syst. 24(1):25. 1897.
Syn.: *Phthirusa sintenisii* (Krug & Urb.) Engl. (1897).

Parasitic shrub collected in Barrio del Pasto near Aibonito on 25 Nov 1885 [*Sintenis 2856* (US), not seen].

Endemic. [L1: 101].

Notes: This taxon is known only from the type specimen, which, according to Job Kuijt (pers. comm.), is probably referable to *Dendropemon caribaeus*.

LYTHRACEAE (including PUNICACEAE)

Ammannia baccifera L., Sp. Pl. 1:120. 1753.

Herb in Northern Coastal Lowlands, Southern Coastal Lowlands (Salinas), Southern Coastal Limestone (Sabana Grande); 1–100 m; *fl* & *fr* Nov–Jun.

Native to India. [L3: 343].

Ammannia coccinea Rottb., Pl. Horti Univ. Rar. Progr. 7. 1773.

Herb in Northern Coastal Lowlands, Western Coast (Mayagüez); 1–5 m; *fl* & *fr* Dec–Sep.

West Indies, southern and central United States, Mexico, Central America, northwestern and west-central South America. [L3: 343].

Ammannia latifolia L., Sp. Pl. 1:119. 1753.

Herb in Northern Coastal Lowlands, Sierra de Luquillo, Eastern Coast, Southern Coastal Lowlands (Santa Isabel), Southern Coastal Limestone (Sabana Grande), Southwestern Coastal Lowlands, Vieques; 1–400 m; *fl* & *fr* Dec–Sep.

West Indies, southeastern United States, Mexico, Central America, South America (scattered). [L3: 345].

Yerba de cangrejo.

Cuphea hyssopifolia Kunth in Humboldt, Bonpland & Kunth, Nov. Gen. Sp. 6:157 [folio]. 1823; 6:199 [quarto]. 1824.

Suffrutescent herb in Northern Limestone Hills, Eastern Cordillera (Barranquitas), Central Cordillera, Maricao Forest Reserve; 100–800 m; *fl* Jan–May, *fr* Jan (one specimen).

Native to Mexico, northern Central America. [L3: 346].

Cuphea parsonsia (L.) R. Br. ex Steud., Nomencl. Bot. (ed. 1) 1:245. 1821.
Bas.: *Lythrum parsonsia* L. (1759).
Syn.: *Parsonsia parsonsia* (L.) Britton (1902).

Herb in Northern Coastal Lowlands, Northern Limestone Hills (Arecibo, Moca), Eastern Coast (Ceiba), Central Cordillera, Maricao Forest Reserve; 1–1250 m; *fl* & *fr* throughout the year.

Bahamas, Greater Antilles, Lesser Antilles (Dominica, Martinique), Mexico. [L3: 346].

Chiagari.

Cuphea strigulosa Kunth in Humboldt, Bonpland & Kunth, Nov. Gen. Sp. 6:161 [folio]. 1823; 6:204 [quarto]. 1824.

Herb in Northern Coastal Lowlands, Sierra de Luquillo, Eastern Coast, Caguas Valley (Juncos), Sierra de Cayey, Eastern Cordillera, Central Cordillera (Utuado), Southern Coastal Limestone (Guayanilla), Western Coast (Añasco), Vieques, Culebra; 1–925 m; *fl* & *fr* throughout the year.

Greater Antilles, Lesser Antilles (Dominica), west-central South America and Brazil. [L3: 348].

Ginoria rohrii (Vahl) Koehne, Bot. Jahrb. Syst. 3:351. 1882.
Bas.: *Antherylium rohrii* Vahl (1792).

Tree in Northern Coastal Lowlands (Fajardo), Eastern Coast (Ceiba), Eastern Cordillera (Cañón San Cristóbal), Southwestern Coastal Lowlands (Cabo Rojo), Vieques; 1–450 m; *fl* Mar–Sep, *fr* Apr–Jul.

Greater Antilles (Hispaniola), Virgin Islands. [L3: 349].

Cereza, Rosa de ciénaga, Serrazuela, Ucarillo.

Lagerstroemia indica L., Syst. Nat. (ed. 10) 2:1076. 1759.

Tree planted and persistent in Northern Coastal Lowlands, Eastern Coast (Naguabo), Caguas Valley (Caguas); 1–200 m; *fl* Aug–Oct, *fr* Jul–Oct.

Native to southeastern Asia, China. [L3: 350].

Astromelia, Astromero, Stromelia.

Lagerstroemia speciosa (L.) Pers., Syn. Pl. 2(1):72. 1806.

Bas.: *Munchausia speciosa* L. (1770).

Tree planted and persistent in Northern Coastal Lowlands, Western Coast (Mayagüez); 1–110 m; *fl* Apr–Sep, *fr* Aug–Dec, Mar.

Native to Indo-China, China. [L3: 351].

Astromelia, Flor de la reina, Reina de las flores.

Punica granatum L., Sp. Pl. 1:472. 1753.

Shrub/small tree planted and persistent in Southwestern Coastal Lowlands (Guánica), Culebra; 10 m; *fl* Apr–Jul, Oct, *fr* Jun–Jul, Oct (two specimens).

Native to temperate Asia. [L3: 356].

Granada, Granado, Granadero.

Rotala ramosior (L.) Koehne in Martius, Fl. Bras. 13(2):194, *t. 39*, *f. 1.* 1877.

Bas.: *Ammannia ramosior* L. (1753).

Herb in Northern Coastal Lowlands, Eastern Coast (Maunabo); 1–10 m; *fl & fr* Dec–Jun.

Antilles, southern and eastern United States, Mexico, Central America, northern South America to Brazil, Bolivia. [L3: 354].

Yerba de cáncer.

TAXA NOT RECENTLY COLLECTED

Ammannia auriculata Willd., Hort. Berol. 1:7, *t. 7.* 1803.

Herb collected in a mangrove swamp in the vicinity of Aguirre on 11 Mar 1922 [*Britton & Earle 6490* (NY), scan seen].

Antilles, United States, Mexico, Central America, South America, Africa, Asia, Australia. [L3: 343].

Cuphea micrantha Kunth in Humboldt, Bonpland & Kunth, Nov. Gen. Sp. 6:155 [folio]. 1823; 6:196 [quarto]. 1824.

Syn.: *Parsonsia micrantha* (Kunth) Jennings (1917).

Herb collected, probably in Dorado, in Sep 1886 [*Stahl 311* (S), not seen] and in Guanajibo, Mayagüez, on 26 Dec 1929 [*Britton & Britton* 9396 (NY)].

Greater Antilles (Cuba, Hispaniola), Lesser Antilles, southern Mexico, Central America, South America (Venezuela, Guyana, Brazil, Paraguay). [L3: 346].

TAXA CITED BY PREVIOUS AUTHORS, BUT NOT INCLUDED

CITED DUE TO MISIDENTIFICATION

Cuphea carthagenensis (Jacq.) J.F. Macbr.

Herb cited by Liogier & Martorell (1982) as common on the northern and northeastern slopes in wet districts, but subsequently Liogier (1994) stated that what had been reported as this taxon was actually *Cuphea strigulosa*.

ONLY UNDER CULTIVATION

Lawsonia inermis L.

Shrub/small tree cited by Britton & Wilson (1925) as widely planted for ornament and spontaneous after cultivation; subsequently cited by Liogier & Martorell (1982, 2000) and Liogier (1994). [L3: 352].

MAGNOLIACEAE

Magnolia portoricensis Bello, Anales. Soc. Esp. Hist. Nat. 10:233. 1880.

Tree in Central Cordillera, Maricao Forest Reserve; 600–1100 m; *fl* Feb, May–Aug, *fr* Jul–Nov.

Endemic. [L1: 234].

Alceiba, Alciba, Anonillo, Burro, Burro mauricio, Jagüilla, Mauricio, Ortegón.

Magnolia splendens Urb., Symb. Antill. 1(2):306. 1899.

Tree in Sierra de Luquillo; 400–850 m; *fl* Mar, Jun–Aug, *fr* Jul–Oct.

Endemic. [L1: 235].

Bella, Laurel sabino, Sabino.

TAXA CITED BY PREVIOUS AUTHORS, BUT NOT INCLUDED

ONLY UNDER CULTIVATION

Magnolia grandiflora L.

Tree cited by Liogier & Martorell (1982, 2000) and Liogier (1985) as planted for ornament and persistent. [L1: 233].

MALPIGHIACEAE

Bunchosia glandulosa (Cav.) DC., Prodr. 1:581. 1824.

Bas.: *Malpighia glandulosa* Cav. (1789).

Tree in Northern Coastal Lowlands (Fajardo and adjacent islands), Northern Limestone Hills, Southeastern Foothills (Guayama), Eastern Cordillera, Southern Coastal Lowlands (Salinas), Quebrada de Los Cedros, Susúa Forest Reserve, Guánica Forest Reserve, Southwestern Coastal Lowlands (Cabo Rojo), Vieques, Culebra; 5–650 m; *fl* Feb–Mar, Jun–Jul, Oct, *fr* Jun–Jul, Oct–Nov.

Bahamas, Greater Antilles (Hispaniola), Virgin Islands, Lesser Antilles. [L2: 312].

Café forastero, Cafeíllo, Yeso.

Byrsonima crassifolia (L.) Kunth in Humboldt, Bonpland & Kunth, Nov. Gen. Sp. 5:115 [folio]; 5:149 [quarto]. 1822.
Bas.: *Malpighia crassifolia* L. (1753).

Tree in Eastern Cordillera (Corozal), Central Cordillera (Utuado), Western Cordillera (Maricao, Las Marías), Southwestern Coastal Lowlands (Cabo Rojo), Sierra Bermeja, Cerro Las Mesas, Western Coast (Añasco); 25–425 m; *fl* & *fr* throughout the year.

Antilles, Mexico, Central America, South America (scattered). [L2: 315, 316].

Peralejo, Peralejo blanco, Maricao cimarrón, Maricao verde.

Byrsonima lucida (Mill.) DC., Prodr. 1:580. 1824.
Bas.: *Malpighia lucida* Mill. (1768).
Syn.: *Byrsonima cuneata* (Turcz.) P. Wilson (1917).

Tree in Northern Coastal Lowlands (islands adjacent to Fajardo: Cayo Icacos; Piñones Forest Reserve), Northern Limestone Hills (Lares, Rincón), Río Abajo Forest Reserve, Eastern Coast (Punta Guayanés), Southern Coastal Limestone (Ponce), Susúa Forest Reserve, Western Cordillera, Guánica Forest Reserve, Southwestern Coastal Lowlands (Cabo Rojo), Vieques, Mona; 1–700 m; *fl & fr* throughout the year.

Bahamas, Greater Antilles (Cuba, Hispaniola), Lesser Antilles, southeastern United States (Florida). [L2: 315].

Notes: Liogier & Martorell (1982) and Liogier (1988) thought *Byrsonima ophiticola* Britton & Small to be a synonym of this taxon, which it may very well be; Pedro González Gutiérrez (pers. comm.) thinks that *Byrsonima cuneata* is not conspecific with this taxon, but is likely to be conspecific with a Cuban endemic, *Byrsonima biflora* Griseb.

Aceituna, Maricao, Palo de doncella, Sangre de doncella.

Byrsonima lucida ×spicata.

Tree in Eastern Coast (Punta Guayanés), Southwestern Coastal Lowlands (San Germán); 25 m; *fl* May–Jun, Sep–Oct, Jan, *fr* Oct, Jan.

Virgin Islands (St. John), Lesser Antilles (Dominica). [L2: 315].

Notes: According to Nicolson (1991), *Byrsonima lucida* and *Byrsonima spicata* occasionally produce a hybrid intermediate in leaf characters and with yellow flowers in Dominica; Little, Woodbury, and Wadsworth (1974) claimed that both the names *Byrsonima ophiticola* Britton & Small and *Byrsonima horneana* Small referred to either similar hybrids or to variations of *Byrsonima spicata* in western Puerto Rico; on the other hand, W.R. Anderson (pers. comm.) thinks that the type of *Byrsonima horneana* is actually a small-leaved, glabrate form of *Byrsonima crassifolia*. Although I have kept the above name for this taxon, I believe that its identity, whether hybrid or not, requires further study.

Byrsonima spicata (Cav.) DC., Prodr. 1:580. 1824.
Bas.: *Malpighia spicata* Cav. (1789).

Tree in Northern Coastal Lowlands, Northern Limestone Hills, Sierra de Luquillo, Eastern Coast (Yabucoa including Punta Guayanés), Caguas Valley (Gurabo), Southeastern Foothills (Cayey), Sierra de Cayey, Eastern Cordillera (Cerro La Peña), Central Cordillera (Orocovis, Utuado), Western Cordillera, Southwestern Coastal Lowlands (Cabo Rojo); 1–650 m; *fl* May–Nov, *fr* Aug–Apr.

Greater Antilles (Cuba, Hispaniola), Virgin Islands, Lesser Antilles, southern Central America, northern South America to Bolivia. [L2: 316].

Notes: Little & Wadsworth (1964) misapplied the name *Byrsonima coriacea* Sw. to this taxon; Liogier & Martorell (1982) and Liogier (1988) wrongly thought *Byrsonima horneana* to be a synonym of it.

Doncella, Maricao.

Byrsonima wadsworthii Little, Phytologia 4:417. 1953.

Tree in Sierra de Luquillo, Central Cordillera, Maricao Forest Reserve; 650–1300 m; *fl* Dec–Mar, Jun–Aug, *fr* Dec–Sep.

Endemic. [L2: 316].

Almendrillo, Maricao.

Galphimia gracilis Bartl., Linnaea 13(5):552. 1840.

Shrub/small tree in Northern Limestone Hills (Arecibo), Las Piedras Chiquitas; 200–450 m; *fl* May, Aug–Nov, *fr* Aug (one specimen).

Native to Mexico. [L2: 318].

Notes: Britton & Wilson (1924) misapplied the name *Thryallis glauca* (Cav.) Kuntze to this taxon.

Consuelita, Lluvia de oro, Palo de doncella, Resedá, Sangre de doncella.

Heteropterys laurifolia (L.) A. Juss., Ann. Sci. Nat., Bot., sér. 2 13:276. 1840 "Heteropteris."
Bas.: *Banisteria laurifolia* L. (1762).

Woody vine in Northern Coastal Lowlands, Northern Limestone Hills, Eastern Coast, Central

Cordillera (Jayuya), Susúa Forest Reserve, Maricao Forest Reserve, Southwestern Coastal Lowlands (Lajas), Sierra Bermeja, Western Coast; 1–425 m; *fl* Feb–Aug, Nov, *fr* Jun–Feb.

Greater Antilles, Mexico, Central America, northwestern, west-central South America and Brazil. [L2: 320; AR: 302].

Notes: Liogier & Martorell (1982) and Liogier (1988) used the orthographic variant *Heteropteris laurifolia* (L.) A. Juss. for the name of this taxon. The genus name was originally published as 'Heteropteris' but is now conserved with the above spelling.

Bejuco de buey, Peralejo, Peralejo velludo.

Heteropterys purpurea (L.) Kunth in Humboldt, Bonpland & Kunth, Nov. Gen. Sp. 5:126 [folio]; 5:164 [quarto]. 1822 "Heteropteris." Bas.: *Banisteria purpurea* L. (1753).

Vine in Northern Coastal Lowlands, Northern Limestone Hills, Las Piedras Chiquitas, Southeastern Foothills, Eastern Cordillera (Aibonito, Cañón San Cristóbal), Southern Coastal Lowlands, Southern Coastal Limestone (Guayanilla), Susúa Forest Reserve, Guánica Forest Reserve, Southwestern Coastal Lowlands (Lajas), Sierra Bermeja, Vieques, Culebra; 1–550 m; *fl* & *fr* throughout the year.

Antilles (except Cuba), Trinidad, northern South America (Venezuela). [L2: 320; AR: 302].

Notes: Liogier & Martorell (1982) and Liogier (1988) used the orthographic variant *Heteropteris purpurea* (L.) Kunth for the name of this taxon.

Bejuco de peralejo, Bejuco de toro.

Heteropterys wydleriana A. Juss., Ann. Sci. Nat., Bot., sér. 2 13:275. 1840 "Heteropteris wydlerana."
Syn.: *Banisteria wydleriana* (A. Juss.) C.B. Rob. (1910).

Woody vine in Northern Limestone Hills, Río Abajo Forest Reserve, Susúa Forest Reserve, Western Cordillera; 325–800 m; *fl* Aug–Jan, *fr* Jan–Apr.

Endemic. [L2: 321; AR: 302].

Notes: Liogier & Martorell (1982) and Liogier (1988) used the orthographic variant *Heteropteris wydleriana* A. Juss. for the name of this taxon.

Ahorcacaballo.

Malpighia coccigera L., Sp. Pl. 1:426. 1753.

Shrub in Northern Coastal Lowlands, Northern Limestone Hills, Río Abajo Forest Reserve, Eastern Coast (Naguabo), Central Cordillera, Southern Coastal Limestone (Peñuelas), Susúa Forest Reserve, Maricao Forest Reserve, Mona; 1–1025 m; *fl* Mar–Dec, *fr* Apr–Oct.

Greater Antilles (Cuba, Hispaniola), Lesser Antilles. [L2: 322].

Árbol de la dicha, Azotacaballo.

Malpighia emarginata DC., Prodr. 1:578. 1824.

Tree planted and persistent in Northern Coastal Lowlands, Southwestern Coastal Lowlands (Cabo Rojo); 10–200 m; *fl* Apr–Aug, Dec, *fr* Apr–Jun.

Antilles, Mexico, Central America, northern South America to Bolivia; exact native range uncertain. [L2: 323].

Notes: The name *Malpighia punicifolia* L. has frequently been misapplied to this taxon. José Vivaldi looked at the type of that name and found it to be conspecific with *Malpighia glabra* L.

Acerola, Cereza, Cereza colorada.

Malpighia fucata Ker Gawl., Bot. Reg. 3: *t. 189*. 1817.
Syn.: *Malpighia shaferi* Britton & P. Wilson (1916).

Shrub/small tree in Northern Coastal Lowlands (Cataño), Sierra de Luquillo, Eastern Coast (Maunabo), Southern Coastal Lowlands (Patillas), Vieques; 25–300 m; *fl* Mar–Nov, *fr* Jun–Dec.

Endemic. [L2: 323].

Olaga, Palo bronco.

Malpighia linearis Jacq., Enum. Syst. Pl. 21. 1760.

Shrub in Vieques, Culebra; 1–10 m; *fl* & *fr* Sep (one specimen). [*Woodbury a.n. 26789* (UPR); *Griffs a.n. 881814* (NY)].

Virgin Islands, Lesser Antilles. [L2: 325].

Notes: Since recently collected specimens are all sterile, the flowering and fruiting time was taken from a specimen collected historically.

Malpighia setosa Spreng., Neue Entdeck. Pflanzenk. 3:50. 1822.

Shrub in Mona; 1–60 m; *fl* (?), *fr* Jun (one specimen). [*Woodbury a.n. 580* (UPR); *Sustache 1067* (SJ)].

Greater Antilles (Hispaniola). [L2: 325].

Malpighia woodburyana Vivaldi in Acevedo-Rodríguez, Brittonia 45(2):130, *t. 1*. 1993.

Shrub/small tree in Northern Coastal Lowlands (Fajardo), Northern Limestone Hills (Quebradillas), Southwestern Coastal Lowlands (Lajas), Vieques, Culebra; 10–150 m; *fl* May–Aug, *fr* Jul, Oct–Dec.

Virgin Islands. [L2: 323, L5: 404].

Notes: Little, Woodbury & Wadsworth (1974) and Liogier & Martorell (1982) misapplied the name *Malpighia infestissima* (A. Juss.) Rich. ex Nied. to this taxon; the former taxon occurs only in the Virgin Islands.

Olaga.

Stigmaphyllon bannisterioides (L.) C.E. Anderson, Taxon 41(2):328. 1992.
Bas.: *Malpighia bannisterioides* L. (1775).
Syn.: *Stigmaphyllon ovatum* (Cav.) Nied. (1900). *Brachypterys ovata* (Cav.) Small (1910).

Viny shrub in Eastern Coast, Vieques; 1 m; *fl* throughout the year, *fr* Dec–Feb.

Antilles, Mexico, Central America, northern South America to Brazil. [L2: 327; AR: 304].

Stigmaphyllon emarginatum (Cav.) A. Juss., Ann. Sci. Nat., Bot., sér 2 13:290. 1840.
Bas.: *Banisteria emarginata* Cav. (1790).
Syn.: *Stigmaphyllon periplocifolium* (Desf. ex DC.) A. Juss. (1840).
Stigmaphyllon lingulatum (Poir.) Small (1910).

Woody vine in Northern Coastal Lowlands, Eastern Coast (Cayo Santiago), Las Piedras Chiquitas, Southeastern Foothills, Cerro de Las Cuevas, Southern Coastal Lowlands, Southern Coastal Limestone, Guánica Forest Reserve, Southwestern Coastal Lowlands (San Germán, Cabo Rojo), Sierra Bermeja, Western Coast, Vieques, Culebra, Mona, Desecheo; 1–550 m; *fl* & *fr* throughout the year.

Antilles. [L2: 328; AR: 305].

Bejuco de San Juan, Bejuco de San Pedro.

Stigmaphyllon floribundum (DC.) C.E. Anderson, Syst. Bot. 11(1):128, *f.* 5. 1986.
Bas.: *Banisteria floribunda* DC. (1824).
Syn.: *Stigmaphyllon tomentosum* (Desf. ex DC.) Nied. (1899).

Woody vine in Northern Coastal Lowlands, Northern Limestone Hills, Río Abajo Forest Reserve, Las Tetas de Cayey, Southeastern Foothills, Eastern Cordillera (La Torrecilla), Central Cordillera, Southern Coastal Limestone, Susúa Forest Reserve, Maricao Forest Reserve, Guánica Forest Reserve, Southwestern Coastal Lowlands; 1–1050 m; *fl* throughout the year, *fr* Oct–Jun.

Virgin Islands. [L2: 327; AR: 305].

Notes: Britton & Wilson (1924) misapplied the name *Stigmaphyllon ledifolium* (Kunth) Small to this taxon.

Bejuco de menta, Bejuco de toro, Bejuco de toro de hoja grande, Bejuco sabanero.

Stigmaphyllon puberum (Rich.) A. Juss., Ann. Sci. Nat., Bot., sér 2 13:289. 1840.
Bas.: *Banisteria pubera* Rich. (1792).

Vine in Northern Coastal Lowlands (Fajardo); 5–50 m; *fl* Sep–Dec, *fr* Nov (one specimen).

Antilles, Trinidad, Central America, northern South America to Brazil, Peru. [L2: 328; AR: 306].

Tetrapterys inaequalis Cav., Diss. 9:433, *t. 260.* 1790 "Tetrapteris."

Vine in Northern Coastal Lowlands (Dorado), Northern Limestone Hills, Sierra de Cayey, Southeastern Foothills, Eastern Cordillera (Cañón San Cristóbal), Vieques; 1–700 m; *fl* Oct (one specimen), *fr* Dec–Mar.

Virgin Islands (St. Croix), Lesser Antilles. [L2: 330; AR: 308].

Notes: Liogier & Martorell (1982) and Liogier (1988) used *Tetrapterys citrifolia* (Sw.) Pers. for the name of this taxon since they thought *Tetrapterys inaequalis* to be a synonym of it. W.R. Anderson, in his treatment of the latter in Howard (1988), reestablished *Tetrapterys citrifolia* (Sw.) Pers. as a distinct taxon, endemic to Jamaica.

Bejuco de peralejo, Bejuco de sopla.

TAXA NOT RECENTLY COLLECTED

Bunchosia glandulifera (Jacq.) Kunth in Humboldt, Bonpland & Kunth, Nov. Gen. Sp. 5:119 [folio]; 5:154 [quarto]. 1822.
Bas.: *Malpighia glandulifera* Jacq. (1791).

Shrub/small tree collected in the Sierra de Yabucoa on 4 Sep 1885 [*Sintenis 1880* (NY)] and near Añasco on 7 Dec 1886 [*Sintenis 5690* (GH)].

Native to northern South America (Venezuela, Colombia).

Café, Café falso, Café forastero.

Bunchosia polystachia (Andrews) DC., Prodr. 1:581. 1824 "polystachya."
Bas.: *Malpighia polystachia* Andrews (1810).

Shrub/small tree collected near Guánica on 2 Mar 1886 [*Sintenis 3927* (GH)].

Greater Antilles (Cuba, Hispaniola), Virgin Islands, Lesser Antilles, southern Mexico, Central America (Nicaragua, Costa Rica). [L2: 314].

Notes: Britton & Wilson (1924), Liogier & Martorell (1982) and Liogier (1988) misapplied the name *Bunchosia nitida* (Jacq.) DC. to this taxon.

TAXA CITED BY PREVIOUS AUTHORS, BUT NOT INCLUDED

DOUBTFUL OCCURRENCE

Banisteriopsis lucida (Rich.) Small. Syn.: *Banisteria lucida* Rich.

Climbing shrub cited by Britton & Wilson (1924) as collected by Kuntze (according to Niedenzu); subsequently cited by Liogier & Martorell (1982,

2000) and Liogier (1988), who thought that the collection probably came from a cultivated plant. Acevedo-Rodríguez (2005) regards Kuntze's report as an error. [L2: 311; AR 308].

Heteropterys platyptera DC. Syn.: *Banisteria longifolia* Sw.
Woody vine cited by Britton & Wilson (1930) as, recorded by Niedenzu in 1928, collected many years before in Puerto Rico by Ventanat. [B&W6: 545].

MALVACEAE (including BOMBACACEAE, STERCULIACEAE AND TILIACEAE)

Abelmoschus esculentus (L.) Moench, Methodus 1:617. 1794.
Bas.: *Hibiscus esculentus* L. (1753).

Herb cultivated and escaped in Northern Coastal Lowlands (Cataño), Southwestern Coastal Lowlands (Cabo Rojo); 1–50 m; *fl* Jun, Dec (two specimens), *fr* Jun (one specimen).

Cultivated throughout the tropics and subtropics, unknown in the wild. [L3: 120].

Guimgombó, Guingombó, Molondrón, Quimbombó, Quingambó.

Abutilon abutiloides (Jacq.) Garcke ex Britton & P. Wilson, Sci. Surv. Porto Rico & Virgin Islands 5(4):547. 1924.
Bas.: *Sida abutiloides* Jacq. (1764).
Syn.: *Abutilon americanum* (L.) Sweet (1826).

Suffrutescent herb in Southwestern Coastal Lowlands (Guánica), Mona; 1–100 m; *fl* Jun, Nov, *fr* Nov–Jun.

Bahamas, Greater Antilles, Virgin Islands, south-central United States (Texas, Arizona), Mexico. [L3: 122].

Abutilon hirtum (Lam.) Sweet, Hort. Brit. (ed. 1) 53. 1826.
Bas.: *Sida hirta* Lam. (1783).

Shrub in Northern Coastal Lowlands (Fajardo), Southern Coastal Limestone (Ponce), Southwestern Coastal Lowlands; 1–15 m; *fl* & *fr* Oct–Apr.

Native to tropical Africa, tropical Asia. [L3: 123].

Buenos días.

Anoda cristata (L.) Schltdl., Linnaea 11(2):210. 1837.
Bas.: *Sida cristata* L. (1753).

Suffrutescent herb in Northern Limestone Hills (Isabela, San Sebastián), Las Piedras Chiquitas, Southeastern Foothills (Coamo), Eastern Cordillera (Aibonito), Cerro de Las Cuevas, Southern Coastal Limestone, Southwestern Coastal Lowlands, Western Coast; 1–600 m; *fl* & *fr* Oct–Jun.

Greater Antilles, southwestern United States, Mexico, Central America, western South America. [L3: 127].

Notes: Britton & Wilson (1924), Liogier & Martorell (1982, 2000) and Liogier (1994) misapplied the name *Anoda acerifolia* DC. to this taxon.

Violeta.

Ayenia insulicola Cristóbal, Opera Lilloana 4:164, *t. 43G-M.* 1960 "insulaecola."

Low suffrutescent herb in Southern Coastal Lowlands, Southern Coastal Limestone, Guánica Forest Reserve, Southwestern Coastal Lowlands, Vieques, Culebra, Mona, Desecheo; 1–100 m; *fl* & *fr* throughout the year.

Bahamas, Greater Antilles (Hispaniola), Virgin Islands, Lesser Antilles. [L3: 180].

Notes: Britton & Wilson (1924) misapplied the name *Ayenia pusilla* L. to this taxon.

Bastardia viscosa (L.) Kunth var. **viscosa** in Humboldt, Bonpland & Kunth, Nov. Gen. Sp. 5:199 [folio]; 5:256 [quarto]. 1822.
Bas.: *Sida viscosa* L. (1759).

Suffrutescent herb in Northern Coastal Lowlands (Fajardo), Northern Limestone Hills (Lares), Eastern Coast (Ceiba, Humacao), Southeastern Foothills (Salinas), Southern Coastal Lowlands, Southern Coastal Limestone (Peñuelas), Guánica Forest Reserve, Southwestern Coastal Lowlands, Sierra Bermeja, Vieques, Mona, Desecheo; 1–60 m; *fl* & *fr* throughout the year.

West Indies, south-central United States (Texas), Mexico, Central America, northwestern, west-central South America and Venezuela. [L3: 129].

Escoba babosa, Malva viscosa.

Ceiba pentandra (L.) Gaertn., Fruct. Sem. Pl. 2(2): 244, *t. 133.* 1791.
Bas.: *Bombax pentandrum* L. (1753).

Tree in Las Tetas de Cayey, Southern Coastal Lowlands (Ponce), Southwestern Coastal Lowlands, Sierra Bermeja, Western Coast (Hormigueros), Vieques; 10–825 m; *fl* Jan–Feb, *fr* Mar (one specimen).

West Indies, Mexico, Central America, northern South America to Brazil, Bolivia, tropical Africa; exact native range uncertain. [L3: 178].

Ceiba.

Corchorus aestuans L., Syst. Nat. (ed. 10) 2:1079. 1759.

Suffrutescent herb in Northern Coastal Lowlands, Eastern Coast (Cayo Santiago), Southeastern Foothills, Southern Coastal Lowlands, Guánica

Forest Reserve, Southwestern Coastal Lowlands, Sierra Bermeja, Western Coast (Añasco), Vieques; 1–300 m; *fl* May–Feb, *fr* throughout the year.

Pantropical, possibly originating in the Neotropics. [L3: 108].

Corchorus hirsutus L., Sp. Pl. 1:530. 1753.

Suffrutescent herb in Northern Coastal Lowlands, Eastern Coast (Punta Guayanés), Las Piedras Chiquitas, Southeastern Foothills, Eastern Cordillera (Aibonito), Southern Coastal Lowlands, Southern Coastal Limestone, Susúa Forest Reserve, Guánica Forest Reserve, Southwestern Coastal Lowlands (Cabo Rojo), Sierra Bermeja, Western Coast (Mayagüez), Vieques, Culebra, Mona, Desecheo; 1–650 m; *fl* & *fr* throughout the year.

West Indies, Mexico, Central America, South America, northeastern tropical Africa. [L3: 109].

Malvavisco.

Corchorus hirtus L., Syst. Nat. (ed. 2) 1:747. 1762.

Syn.: *Corchorus orinocensis* Kunth (1823).

Suffrutescent herb in Northern Coastal Lowlands, Río Abajo Forest Reserve, Eastern Coast (Maunabo), Caguas Valley (Gurabo), Sierra de Cayey, Las Piedras Chiquitas, Southeastern Foothills, Eastern Cordillera, Central Cordillera (Orocovis), Southern Coastal Lowlands, Southwestern Coastal Lowlands, Western Coast (Aguada), Vieques; 1–700 m; *fl* & *fr* throughout the year.

Antilles, southern United States, Mexico, Central America, northwestern, west-central South America and Venezuela, Argentina. [L3: 109].

Malvaté.

Corchorus siliquosus L., Sp. Pl. 1:529. 1753.

Suffrutescent herb in Northern Coastal Lowlands, Northern Limestone Hills, Río Abajo Forest Reserve, Eastern Coast (Humacao including Cayo Santiago), Las Piedras Chiquitas, Southeastern Foothills, Central Cordillera (Utuado), Southwestern Coastal Lowlands (Cabo Rojo), Western Coast, Vieques; 1–850 m; *fl* throughout the year, *fr* Sep–Jun.

West Indies, southeastern United States (Florida, Alabama, Mississippi), Mexico, Central America, western South America (Ecuador, Peru). [L3: 111].

Escoba blanca.

Fioria vitifolia (L.) Mattei, Boll. Reale Orto Bot. Palermo, n.s. 2:72. 1917.

Bas.: *Hibiscus vitifolius* L. (1753).

Shrub in Vieques; 1–10 m; *fl* & *fr* Jun (one specimen). [*Woodbury V-67* (UPR)].

Widespread in the Paleotropics. [L3: 132].

Gossypium hirsutum L., Sp. Pl. (ed. 2) 2:975. 1763.

Syn.: *Gossypium purpurascens* Poir. (1811).

Gossypium janiphifolium Bello (1881).

Gossypium hirsutum L. var. *punctatum* (Schumach.) Roberty (1945).

Gossypium hirsutum L. var. *marie-galante* (G. Watt) J.B. Hutch. (1947).

Shrub in Northern Coastal Lowlands, Northern Limestone Hills (Utuado), Eastern Coast, Eastern Cordillera (Cayey), Southern Coastal Lowlands, Guánica Forest Reserve, Southwestern Coastal Lowlands, Sierra Bermeja, Western Coast, Vieques, Mona; 1–500 m; *fl* Oct–Jun, *fr* throughout the year.

Antilles. [L3: 133–135].

Algodón, Algodón arbóreo, Algodón del país, Algodón silvestre.

Guazuma ulmifolia Lam., Encycl. 3(1):52. 1789.

Syn.: *Guazuma tomentosa* Kunth (1823).

Guazuma guazuma (L.) Cockerell (1892).

Tree in Guajataca Forest Reserve, Eastern Coast (Ceiba, Humacao), Las Piedras Chiquitas, Southeastern Foothills, Central Cordillera (lower southern slopes), Southern Coastal Lowlands, Southern Coastal Limestone, Susúa Forest Reserve, Maricao Forest Reserve, Guánica Forest Reserve, Southwestern Coastal Lowlands, Sierra Bermeja, Vieques; 1–500 m; *fl* Feb–Nov, *fr* Sep–May.

Antilles, Mexico, Central America, South America. [L3: 190].

Guácima.

Helicteres jamaicensis Jacq., Enum. Syst. Pl. 30. 1760.

Shrub in Northern Coastal Lowlands, Northern Limestone Hills, Southern Coastal Lowlands, Susúa Forest Reserve, Guánica Forest Reserve, Southwestern Coastal Lowlands (Cabo Rojo), Vieques, Mona; 1–210 m; *fl* & *fr* throughout the year.

Bahamas, Greater Antilles, Virgin Islands, Southern Caribbean Islands (Curaçao), Trinidad, northern South America (Colombia). [L3: 191].

Cuernecillo, Gato, Gato soga, Huevo de gato, Soga.

Herissantia crispa (L.) Brizicky, J. Arnold Arbor. 49(2):279. 1968.

Bas.: *Sida crispa* L. (1753).

Syn.: *Gayoides crispum* (L.) Small (1903).

Shrub in Southern Coastal Limestone (Peñuelas), Guánica Forest Reserve, Southwestern Coastal Lowlands (Cabo Rojo), Mona; 2–50 m; *fl* & *fr* throughout the year.

West Indies, southern United States, Mexico, Central America, South America. [L3: 136].

Hibiscus acetosella Welw. ex Hiern, Cat. Afr. Pl. 1:73. 1896.

Shrub in Central Cordillera (Adjuntas); 720 m; *fl* Sep (one specimen), *fr* (?). [*Ackerman 3961* (UPRRP)].

Native to tropical Africa. [L3: 138].

Hibiscus bifurcatus Cav., Diss. 3:146, *t. 51, f. 1.* 1787.

Shrub in Northern Coastal Lowlands, Northern Limestone Hills (Isabela), Sierra de Luquillo, Eastern Coast, Sierra de Cayey, Eastern Cordillera (Cerro La Peña), Western Cordillera (Maricao); 1–575 m; *fl* & *fr* Sep–May.

Greater Antilles, Lesser Antilles (Guadeloupe, Martinique, St. Vincent), northern Central America, northern South America. [L3: 139].

Buenas tardes.

Hibiscus clypeatus L., Syst. Nat. (ed. 10) 2:1149. 1759.

Shrub in Southern Coastal Limestone (Guayanilla), Guánica Forest Reserve; 25–150 m; *fl* Dec–Jan (one specimen), *fr* Oct, Feb–Mar (two specimens). [*Woodbury a.n. 1007345* (NY); *Monsegur 146, 619* (MAPR)].

Greater Antilles, Virgin Islands (St. Croix), Mexico, northern Central America (Guatemala). [L3: 140].

Huevo de gato.

Hibiscus furcellatus Desr. in Lamarck, Encycl. 3(1):358. 1789.

Shrub in Northern Coastal Lowlands, Río Abajo Forest Reserve, Eastern Cordillera (Cerro La Peña), Central Cordillera (Utuado), Southern Coastal Limestone (Guayanilla); 1–800 m; *fl* & *fr* throughout the year.

Greater Antilles, Lesser Antilles (Dominica), southeastern United States (Florida), southern Mexico, Central America, South America (scattered). [L3: 140].

Hibiscus phoeniceus Jacq., Hort. Bot. Vindob. 3:11(-12, *t. 14*). 1776.

Suffrutescent herb in Las Piedras Chiquitas, Southeastern Foothills, Southern Coastal Lowlands, Southern Coastal Limestone, Guánica Forest Reserve, Southwestern Coastal Lowlands, Sierra Bermeja, Vieques; 1–550 m; *fl* Jul–Apr, *fr* Sep–May.

Antilles, Mexico, Central America, northwestern, west-central South America and Venezuela. [L3: 142].

Notes: Britton & Wilson (1924) misapplied the name *Hibiscus brasiliensis* L. to this taxon.

Hibiscus rosa-sinensis L., Sp. Pl. 2:694. 1753.

Shrub planted and persistent in Northern Coastal Lowlands, Sierra de Luquillo, Eastern Coast, Sierra de Cayey, Maricao Forest Reserve, Vieques; 1–825 m; *fl* Mar–Dec, *fr* (?).

Widely cultivated, probably originating in tropical Asia. [L3: 142].

Amapola, Candelá, Candelada, Carta abierta, Hibisco, Marimoña, Pavona.

Hibiscus trilobus Aubl., Hist. Pl. Guiane 2:708. 1775.

Shrub in Piñones Forest Reserve, Southwestern Coastal Lowlands (Lajas), Western Coast (Añasco); 1–25 m; *fl* Sep–Nov, Feb, *fr* (?).

Greater Antilles (Jamaica, Hispaniola). [L3: 143].

Malachra alceifolia Jacq., Collectanea 2:350. 1789.

Herb in Northern Coastal Lowlands, Northern Limestone Hills, Río Abajo Forest Reserve, Sierra de Luquillo, Eastern Coast, Caguas Valley (Caguas), Southeastern Foothills (Coamo), Eastern Cordillera (Aibonito, Cayey), Southern Coastal Lowlands, Susúa Forest Reserve, Guánica Forest Reserve, Southwestern Coastal Lowlands, Western Coast, Vieques, Culebra; 1–400 m; *fl* & *fr* throughout the year.

Antilles, southern Mexico, Central America, northern South America to Ecuador. [L3: 145].

Malva de caballo.

Malachra capitata (L.) L., Syst. Nat. (ed. 12) 2:458. 1767 "capitat."

Bas.: *Sida capitata* L. (1753).

Herb in Northern Coastal Lowlands, Río Abajo Forest Reserve, Eastern Cordillera (Cayey), Southwestern Coastal Lowlands (Lajas); 1–400 m; *fl* & *fr* Jan–Oct.

West Indies, south-central United States (Texas), Mexico, northern Central America to Nicaragua, northwestern and west-central South America. [L3: 145].

Malva.

Malachra fasciata Jacq., Collectanea 2:352. 1789.

Herb in Northern Coastal Lowlands (San Juan, Arecibo), Caguas Valley (Caguas), Eastern Cordillera (Cidra, Cayey), Southwestern Coastal Lowlands (Cabo Rojo), Western Coast (Mayagüez); 10–600 m; *fl* & *fr* Nov–Apr.

Antilles, Trinidad and Tobago, Mexico, Central America, northern South America to Bolivia, Paraguay. [L3: 145].

Malva blanca.

Malachra urens Poit. ex Ledeb. in Ledebour &

Alderstam, Diss. Bot. Pl. Doming. 22 (no. IX). 1805.

Herb in Caguas Valley (Juncos), Southwestern Coastal Lowlands (Cabo Rojo); 1–55 m; *fl* & *fr* Dec–Feb (two specimens). [*Liogier 31604* (UPR)].

Bahamas, Greater Antilles, southeastern United States (Florida). [L3: 147].

Malvavisco.

Malvastrum americanum (L.) Torr. in Emory, Rep. U.S. Mex. Bound. 2(1):38. 1859.
Bas.: *Malva americana* L. (1753).

Suffrutescent herb in Northern Coastal Lowlands (Fajardo), Eastern Coast (Cayo Santiago), Southeastern Foothills, Southern Coastal Lowlands, Southwestern Coastal Lowlands, Sierra Bermeja, Western Coast (Mayagüez), Vieques, Culebra, Mona; 1–250 m; *fl* & *fr* Dec–Sep.

West Indies, southern United States, Mexico, Central America, South America. [L3: 148].

Notes: Britton & Wilson (1924) misapplied the name *Malvastrum spicatum* (L.) A. Gray to this taxon.

Malva silvestre.

Malvastrum corchorifolium (Desr.) Britton ex Small, Fl. Miami 119. 1913.
Bas.: *Malva corchorifolia* Desr. (1792).

Suffrutescent herb in Northern Coastal Lowlands, Caguas Valley (Juncos), Las Piedras Chiquitas, Southeastern Foothills, Southern Coastal Lowlands, Southern Coastal Limestone (Peñuelas), Guánica Forest Reserve, Southwestern Coastal Lowlands, Western Coast, Vieques, Culebra, Mona, Desecheo; 1–550 m; *fl* & *fr* throughout the year.

West Indies, southeastern United States (Florida), southern Mexico, northern Central America to Nicaragua. [L3: 148].

Malvastrum coromandelianum (L.) Garcke, Bonplandia 5:295. 1857.
Bas.: *Malva coromandeliana* L. (1753).

Suffrutescent herb in Northern Coastal Lowlands, Eastern Cordillera (Cayey, Aibonito), Southern Coastal Lowlands (Coamo, Juana Díaz), Western Coast, Vieques, Mona; 1–500 m; *fl* & *fr* throughout the year.

Pantropical. [L3: 149].

Escoba blanca, Malvavisco.

Malvaviscus penduliflorus DC., Prodr. 1:445. 1824.

Shrub in Northern Coastal Lowlands, Northern Limestone Hills, Sierra de Luquillo, Eastern Coast, Caguas Valley (Caguas), Sierra de Cayey, Eastern and Central Cordillera; 20–900 m; *fl* throughout the year, *fr* (?).

Widely cultivated; origin unknown. [L3: 152].

Capucha de monje.

Melochia nodiflora Sw., Prodr. 97. 1788.

Shrub in Northern Coastal Lowlands, Northern Limestone Hills, Eastern Coast, Caguas Valley (Juncos), Las Piedras Chiquitas, Southeastern Foothills, Eastern Cordillera (Comerío, Aibonito), Central Cordillera (Ponce), Southern Coastal Lowlands, Susúa Forest Reserve, Southwestern Coastal Lowlands (Lajas), Western Coast, Vieques; 1–600 m; *fl* & *fr* Nov–May.

West Indies, Mexico, Central America, northern South America. [L3: 195].

Bretónica prieta.

Melochia pyramidata L., Sp. Pl. 2:674. 1753.
Syn.: *Moluchia pyramidata* (L.) Britton (1918).

Shrub in Northern Coastal Lowlands, Northern Limestone Hills, Eastern Coast, Las Piedras Chiquitas, Southeastern Foothills, Eastern Cordillera (Aibonito), Southern Coastal Lowlands, Southern Coastal Limestone (Guayanilla), Southwestern Coastal Lowlands, Sierra Bermeja, Western Coast, Vieques, Culebra, Mona, Desecheo; 1–600 m; *fl* & *fr* throughout the year.

West Indies, southern United States (Florida, Louisiana, Texas), Mexico, Central America (except Panama), South America (scattered). [L3: 195].

Bretónica piramidal.

Melochia spicata (L.) Fryxell, Syst. Bot. Monogr. 25:457. 1988.
Bas.: *Malva spicata* L. (1759).
Syn.: *Riedlea hirsuta* (Cav.) DC. (1824). *Melochia villosa* (Mill.) Fawc. & Rendle (1926). *Riedlea villosa* (Mill.) Britton & P. Wilson (1930).

Shrub in Northern Coastal Lowlands, Eastern Coast (Humacao), Eastern Cordillera (Barranquitas), Maricao Forest Reserve, Cerro Las Mesas, Western Coast; 1–600 m; *fl* & *fr* throughout the year.

Greater Antilles, Trinidad, southeastern United States (Florida), Mexico, Central America, west-central South America. [L3: 197].

Bretónica aserrada, Bretónica peluda.

Melochia tomentosa L., Syst. Nat. (ed. 10) 2:1140. 1759.
Syn.: *Moluchia tomentosa* (L.) Britton (1918).

Shrub in Northern Coastal Lowlands (Fajardo), Las Piedras Chiquitas, Southeastern Foothills, Southern Coastal Lowlands, Southern Coastal Limestone (Guayanilla), Southwestern Coastal Lowlands, Guánica Forest Reserve, Southwestern

Coastal Lowlands, Sierra Bermeja, Vieques, Culebra, Mona, Desecheo; 1–500 m; *fl* & *fr* throughout the year.

West Indies, southern United States (Florida, Texas), Central America, northwestern, west-central, South America and Venezuela. [L3: 195].

Bretónica afelpada.

Ochroma pyramidale (Cav. ex Lam.) Urb., Repert. Spec. Nov. Regni Veg. Beih. 5:123. 1920.
Bas.: *Bombax pyramidale* Cav. ex Lam. (1788).
Syn.: *Ochroma lagopus* Sw. (1788).

Tree in Northern Coastal Lowlands (Fajardo), Northern Limestone Hills, Sierra de Luquillo, Eastern Coast (Naguabo), Caguas Valley (Caguas), Central Cordillera (Ponce), Southwestern Coastal Lowlands (San Germán), Western Coast (Mayagüez); 50–500 m; *fl* Nov–Mar, Jun, *fr* Feb–Jun.

Antilles, Mexico, Central America, northern South America to Brazil, Bolivia. [L3: 180].

Balsa, Corcho, Guano.

Pavonia fruticosa (Mill.) Fawc. & Rendle, Fl. Jamaica 5:130. 1926.
Bas.: *Sida fruticosa* Mill. (1768).
Syn.: *Typhalaea fruticosa* (Mill.) Britton (1924).

Suffrutescent herb in Northern Coastal Lowlands, Northern Limestone Hills, Río Abajo Forest Reserve, Sierra de Luquillo, Eastern Coast (Maunabo), Sierra de Cayey, Las Tetas de Cayey, Southeastern Foothills (Arroyo), Eastern Cordillera (Corozal, Cañón San Cristóbal), Central Cordillera, Quebrada de Los Cedros, Western Cordillera (Maricao); 10–825 m; *fl* & *fr* throughout the year.

Greater Antilles, Lesser Antilles (Dominica, St. Vincent), southern Central America, northwestern, west-central South America and Venezuela. [L3: 153].

Cadillo pequeque.

Pavonia paludicola Nicolson ex Fryxell in Howard, Fl. Lesser Antilles 5:241, *f. 94*. 1989.
Syn.: *Malache scabra* B. Vogel (1772).
Pavonia spicata Cav. (1787).

Shrub in Northern Coastal Lowlands, Western Coast (Añasco); 0–40 m; *fl* May–Jan, *fr* Aug–Feb.

West Indies, southeastern United States (Florida), Central America, northern South America. [L3: 153].

Cadillo de ciénaga.

Pavonia spinifex (L.) Cav., Diss. 3:133, *t. 45*, *f. 2*. 1787.
Bas.: *Hibiscus spinifex* L. (1759).

Suffrutescent herb in Northern Coastal Lowlands, Northern Limestone Hills, Río Abajo Forest Reserve, Sierra de Luquillo, Las Tetas de Cayey, Las Piedras Chiquitas, Southeastern Foothills, Central Cordillera, Southern Coastal Lowlands (Salinas), Southern Coastal Limestone (Peñuelas), vicinity of Susúa Forest Reserve (Sabana Grande), Guánica Forest Reserve, Southwestern Coastal Lowlands (San Germán, Cabo Rojo), Sierra Bermeja, Vieques, Culebra, Mona; 10–1200 m; *fl* & *fr* throughout the year.

West Indies, Trinidad and Tobago, southeastern United States (Georgia, Florida). [L3: 155].

Abutilón espinoso, Cadillo espinoso.

Pseudabutilon umbellatum (L.) Fryxell, Contr. Univ. Michigan Herb. 21:190. 1997.
Bas.: *Sida umbellata* L. (1759).
Syn.: *Abutilon umbellatum* (L.) (1826).

Suffrutescent herb in Northern Coastal Lowlands, Eastern Cordillera (Aibonito), Southern Coastal Lowlands, Guánica Forest Reserve, Southwestern Coastal Lowlands, Vieques, Culebra, Mona, Desecheo; 1–650 m; *fl* & *fr* throughout the year.

West Indies, south-central United States (Texas), Mexico, Central America, northwestern, west-central South America and Venezuela. [L3: 125].

Malvavisco cimarrón, Pichona mocha.

Quararibea turbinata (Sw.) Poir., Encycl. Suppl. 4(2):636. 1816.
Bas.: *Myrodia turbinata* Sw. (1788).

Small tree in Northern Limestone Hills, Río Abajo Forest Reserve, Sierra de Luquillo, Central Cordillera (Ponce); 10–350 m; *fl* throughout the year, *fr* Feb–Sep.

Greater Antilles (Hispaniola), Virgin Islands, Lesser Antilles. [L3: 184].

Asubillo, Garrocha, Garrocho, Palo de garrocha, Palo de percha.

Sida abutifolia Mill., Gard. Dict. (ed. 8): *Sida* no. 12. 1768.
Syn.: *Sida procumbens* Sw. (1788).

Decumbent suffrutescent herb in Southern Coastal Lowlands, Southern Coastal Limestone, Guánica Forest Reserve, Southwestern Coastal Lowlands, Sierra Bermeja, Vieques, Mona, Desecheo; 1–100 m; *fl* & *fr* throughout the year.

West Indies, southern United States, Mexico, Central America, west-central South America and Venezuela. [L3: 157].

Notes: Liogier (1982) used the variant *Sida abutilifolia* Mill. for the name of this taxon; it is, according to TROPICOS, the name given on the type specimen.

Escoba tendida.

Sida acuta Burm. f., Fl. Indica 147. 1768.

Syn.: *Sida carpinifolia* L.f. (1782).
Sida stipulata Cav. (1785).

Herb in Northern Coastal Lowlands, Northern Limestone Hills, Río Abajo Forest Reserve, Sierra de Luquillo, Eastern Coast, Sierra de Cayey, Las Piedras Chiquitas, Eastern Cordillera (Cañón San Cristóbal), Southern Coastal Lowlands, Southern Coastal Limestone (Yauco), Susúa Forest Reserve, Western Cordillera, Southwestern Coastal Lowlands, Sierra Bermeja, Western Coast, Vieques, Culebra, Mona, Desecheo; 1–900 m; *fl* & *fr* throughout the year.

Pantropical. [L3: 162].

Notes: Liogier (1994) used *Sida stipularis* Cav., probably an orthographic error for the synonym *Sida stipulata*, for the name for this taxon.

Escoba blanca, Escobita dulce.

Sida ciliaris L., Syst. Nat. (ed. 10) 2:1145. 1759.

Prostrate herb in Northern Coastal Lowlands (Fajardo), Eastern Coast (Ceiba), Guánica Forest Reserve, Southwestern Coastal Lowlands (Cabo Rojo), Sierra Bermeja, Vieques, Culebra, Mona; 1–100 m; *fl* & *fr* throughout the year.

West Indies, southern United States (Florida, Texas), Mexico, Central America, South America. [L3: 158].

Escoba peluda, Escoba pestañosa.

Sida cordifolia L., Sp. Pl. 2:684. 1753.

Suffrutescent herb in Northern Coastal Lowlands, Eastern Coast, Southern Coastal Lowlands (Caja de Muertos, Ponce), Southwestern Coastal Lowlands (Cabo Rojo), Sierra Bermeja, Vieques, Desecheo; 1–275 m; *fl* & *fr* Jul–Apr.

Pantropical. [L3: 158].

Escoba acorazonada.

Sida glabra Mill., Gard. Dict. (ed. 8): *Sida* no. 14. 1768.

Herb in Northern Coastal Lowlands (Fajardo), Las Piedras Chiquitas, Southeastern Foothills, Eastern Cordillera (Cayey), Southern Coastal Lowlands, Guánica Forest Reserve, Southwestern Coastal Lowlands, Western Coast (Rincón), Vieques, Mona, Desecheo; 20–500 m; *fl* & *fr* Oct–Jun.

West Indies, Antilles, Mexico, Central America, northwestern, west-central South America and Venezuela. [L3: 159].

Escoba dulce.

Sida glomerata Cav., Diss. 1:18, *t.* 2, *f.* 6. 1785.

Suffrutescent herb in Northern Coastal Lowlands, Eastern Coast, Las Piedras Chiquitas, Southeastern Foothills (Coamo), Eastern Cordillera (Aibonito), Southern Coastal Lowlands, Southwestern Coastal Lowlands, Sierra Bermeja, Vieques; 1–550 m; *fl* & *fr* Sep–Jun.

Antilles, Central America, northwestern, west-central South America and Venezuela. [L3: 159].

Escoba de palma.

Sida glutinosa Comm. ex Cav., Diss. 1:16, *t.* 2, *f.* 8. 1785.

Herb in Northern Coastal Lowlands, Northern Limestone Hills (Florida), Southeastern Foothills, Eastern Cordillera (Aibonito), Central Cordillera (Ponce), Western Coast (Rincón); 1–550 m; *fl* & *fr* Dec–Mar.

West Indies, Mexico, Central America, South America. [L3: 160].

Sida jamaicensis L., Syst. Nat. (ed. 10) 2:1145. 1759.

Suffrutescent herb in Southwestern Coastal Lowlands (Cabo Rojo), Sierra Bermeja, Culebra; 1–300 m; *fl* Aug–Dec, *fr* Nov–Dec.

Antilles, Mexico, Central America, northern South America. [L3: 160].

Sida repens Dombey ex Cav., Diss. 1:7, *t.* 1, *f.* 2. 1785.

Syn.: *Sida javensis* Cav. subsp. *expilosa* Borss. Waalk. (1966).

Creeping herb in Northern Coastal Lowlands, Sierra de Cayey, Las Tetas de Cayey, Las Piedras Chiquitas, Southeastern Foothills (Coamo), Eastern Cordillera (Cayey), Central Cordillera (Ponce), Southwestern Coastal Lowlands (Cabo Rojo), Culebra; 450–825 m; *fl* Nov–Apr, *fr* Nov–May.

Virgin Islands, Lesser Antilles, Central America, west-central South America. [L3: 161].

Notes: Britton & Wilson (1924) misapplied the name *Sida humilis* Cav. to this taxon.

Sida rhombifolia L., Sp. Pl. 2:684. 1753.

Suffrutescent herb in Northern Coastal Lowlands, Northern Limestone Hills, Río Abajo Forest Reserve, Sierra de Luquillo, Eastern Coast (Ceiba), Caguas Valley (Juncos), Sierra de Cayey, Southeastern Foothills, Eastern Cordillera (Cerro La Peña, Aibonito), Central Cordillera, Western Cordillera, Southwestern Coastal Lowlands (Cabo Rojo), Western Coast (Mayagüez), Vieques; 1–1000 m; *fl* & *fr* throughout the year.

Pantropical. [L3: 161].

Escoba colorada, Escoba dulce.

Sida salviifolia C. Presl, Reliq. Haenk. 2(2):110. 1835 "salviaefolia."
Syn.: *Sida erecta* Macfad. (1837).

Suffrutescent herb in Southern Coastal Lowlands, Southwestern Coastal Lowlands (Cabo Rojo); 1–200 m; *fl* & *fr* May–Feb.

Greater Antilles, Lesser Antilles (St. Barts, Montserrat), Mexico, northwestern South America and Venezuela. [L3: 161].

Escoba parada.

Sida spinosa L., Sp. Pl. 2:683. 1753.

Suffrutescent herb in Northern Coastal Lowlands, Eastern Coast (Ceiba), Las Piedras Chiquitas, Southeastern Foothills, Southern Coastal Lowlands, Southern Coastal Limestone, Susúa Forest Reserve, Maricao Forest Reserve, Southwestern Coastal Lowlands, Sierra Bermeja, Vieques, Mona; 2–600 m; *fl* & *fr* throughout the year.

West Indies, central United States, Mexico, Central America, South America. [L3: 162].

Escoba espinosa.

Sida urens L., Syst. Nat. (ed. 10) 2:1145. 1759.

Suffrutescent herb in Northern Coastal Lowlands, Northern Limestone Hills (Ciales), Río Abajo Forest Reserve, Sierra de Cayey, Las Tetas de Cayey, Southeastern Foothills, Eastern Cordillera (Cerro La Peña, Cayey), Central Cordillera, Southwestern Coastal Lowlands, Western Coast, Vieques; 1–925 m; *fl* & *fr* Dec–Apr.

West Indies, Mexico, Central America, South America. [L3: 163].

Sidastrum multiflorum (Jacq.) Fryxell, Brittonia 31(2):298. 1979.
Bas.: *Sida multiflora* Jacq. (1767).
Syn.: *Sida acuminata* DC. (1824).

Suffrutescent herb in Northern Coastal Lowlands (Fajardo and adjacent islands: Cayo Icacos), Las Piedras Chiquitas, Southeastern Foothills, Susúa Forest Reserve, Guánica Forest Reserve, Southwestern Coastal Lowlands, Sierra Bermeja, Vieques, Culebra, Mona, Desecheo; 1–550 m; *fl* Oct–Aug, *fr* Oct–Jul.

West Indies, South America. [L3: 164].

Sterculia apetala (Jacq.) H. Karst., Fl. Columb. 2(1):35, *t. 118*. 1862.
Bas.: *Helicteres apetala* Jacq. (1760).

Tree planted and persistent in Northern Coastal Lowlands (San Juan, Guaynabo), Central Cordillera (Villalba), Southern Coastal Lowlands (Ponce), Western Coast (Rincón); 15–75 m; *fl* Sep, Dec–Mar, *fr* Feb, Aug (two specimens).

Native to southern Mexico, Central America, northwestern, west-central South America and Venezuela, Brazil. [L3: 199].

Anacagüita.

Talipariti elatum (Sw.) Fryxell, Contr. Univ. Michigan Herb. 23:241. 2001.
Bas.: *Hibiscus elatus* Sw. (1788).

Tree in Northern Limestone Hills (Toa Alta), Guajataca Forest Reserve, Río Abajo Forest Reserve, Sierra de Luquillo, Eastern Cordillera (Cayey), Central Cordillera; 50–800 m; *fl* throughout the year, *fr* Jan–Apr, Aug.

Native to Greater Antilles (Cuba, Jamaica). [L3: 140].

Emajagua excelsa, Majagua excelsa, Majó.

Talipariti tiliaceum (L.) Fryxell var. **pernambucense** (Arruda) Fryxell, Contr. Univ. Michigan Herb. 23:262. 2001.
Bas.: *Hibiscus pernambucensis* Arruda (1810).

Tree in Northern Coastal Lowlands, Sierra de Luquillo, Eastern Coast (Cuchilla de Panduras), Caguas Valley (Caguas), Sierra de Cayey, Southeastern Foothills (Coamo), Eastern Cordillera (Barranquitas), Central Cordillera, Southwestern Coastal Lowlands (San Germán), Culebra, Mona; 1–1000 m; *fl* & *fr* throughout the year.

Native to Brazil. [L3: 140].

Notes: Britton & Wilson (1924) used *Pariti tiliaceum* (L.) A. Juss. for the name of this taxon; Liogier & Martorell (1982) used *Hibiscus tiliaceus* L., *sensu lato*, for it.

Emajagua, Majagua.

Theobroma cacao L., Sp. Pl. 2:782. 1753.

Tree planted and persistent in Eastern Coast (Cuchilla de Panduras), Central Cordillera (Ponce); 175–370 m; *fl* Jun–Sep, *fr* Apr, Sep (two specimens).

Native to southern Mexico, northern Central America, northern South America to Brazil, Peru. [L3: 202].

Cacao.

Thespesia grandiflora DC., Prodr. 1:456. 1824.
Syn.: *Montezuma speciosissima* DC. (1824).

Tree in Northern Coastal Lowlands, Northern Limestone Hills, Eastern Coast (Humacao), Caguas Valley (Gurabo), Sierra de Cayey (San Lorenzo), Central Cordillera; 1–1000 m; *fl* throughout the year, *fr* Nov–Jul.

Endemic. [L3: 166].

Notes: This tree is also often planted as an ornamental.

Maga, Maga colorada.

Thespesia populnea (L.) Sol. ex Corrêa, Ann. Mus. Nat. Hist. 9:290, *t. 8*, *f.* 2. 1807.
Bas.: *Hibiscus populneus* L. (1753).

Tree along coast in Northern Coastal Lowlands, Eastern Coast, Southern Coastal Lowlands (Caja de Muertos, Guayanilla), Guánica Forest Reserve, Southwestern Coastal Lowlands, Vieques, Culebra, Mona; 1–50 m; *fl* & *fr* throughout the year.

Pantropical. [L3: 166].

Caraña, Clamor, Emajagüilla, Frescura, Jaqueca, Majagüilla, Palo de jaqueca, Santa María.

Triumfetta lappula L., Sp. Pl. 1:444. 1753.

Suffrutescent herb/shrub in Northern Coastal Lowlands, Eastern Coast (Humacao including Cayo Santiago), Las Piedras Chiquitas, Southeastern Foothills, Cerro de Las Cuevas, Sierra Bermeja, Western Coast, Vieques; 1–550 m; *fl* Nov–Feb, *fr* Dec–Mar.

Antilles, Mexico, Central America, west-central South America. [L3: 113].

Cadillo, Mabo.

Triumfetta rhomboidea Jacq., Enum. Syst. Pl. 22. 1760.
Syn.: *Triumfetta bartramia* L. (1759).
Triumfetta excisa Urb. (1908).

Suffrutescent herb/shrub in Northern Coastal Lowlands, Eastern Coast (Humacao), Río Abajo Forest Reserve, Southeastern Foothills (Coamo), Western Coast (Añasco); 1–600 m; *fl* & *fr* Oct–May.

Pantropical, possibly originating in the Paleotropics. [L3: 114].

Cadillo.

Triumfetta semitriloba Jacq., Enum. Syst. Pl. 22. 1760.

Suffrutescent herb/shrub in Northern Coastal Lowlands, Northern Limestone Hills, Sierra de Luquillo, Eastern Coast, Sierra de Cayey, Las Tetas de Cayey, Las Piedras Chiquitas, Southeastern Foothills, Eastern and Central Cordillera, Susúa Forest Reserve, Western Cordillera, Western Coast, Vieques; 1–850 m; *fl* & *fr* Oct–Apr.

West Indies, southeastern United States (Georgia, Florida), Mexico, Central America, northwestern, west-central South America and Venezuela, Argentina. [L3: 114].

Cadillo de perro.

Urena lobata L., Sp. Pl. 2:692. 1753.
Syn.: *Urena sinuata* L. (1753).
Urena trilobata Vell. (1825).

Herb in Northern Coastal Lowlands, Northern Limestone Hills (Barceloneta, Aguadilla), Río Abajo Forest Reserve, Sierra de Luquillo, Eastern Coast, Caguas Valley, Sierra de Cayey, Southeastern Foothills (Arroyo), Eastern Cordillera (Comerío), Central Cordillera, Susúa Forest Reserve, Western Cordillera, Southwestern Coastal Lowlands (Lajas, San Germán), Cerro Las Mesas, Western Coast, Vieques; 1–1200 m; *fl* & *fr* Sep–May.

Pantropical. [L3: 168, 170].

Notes: Since *Urena sinuata* differs from the above taxon in but a single character, the lobed form of the leaves, I do not consider it to be distinct.

Cadillo, Cadillo pata de perro (*Urena sinuata*).

Waltheria calcicola Urb., Symb. Antill. 1(4):475. 1900.

Shrub in Southern Coastal Limestone, Southwestern Coastal Lowlands (Cabo Rojo); 25–85 m; *fl* Dec, Apr, *fr* (?). [*Woodbury a.n. 2039* (UPR); *Breckon 4512* (MAPR)].

Greater Antilles (Jamaica, Hispaniola). [L3: 202].

Waltheria indica L., Sp. Pl. 2:673. 1753.
Syn.: *Waltheria americana* L. (1753).

Shrub in Northern Coastal Lowlands, Eastern Coast, Caguas Valley (Gurabo), Las Piedras Chiquitas, Southeastern Foothills, Southern Coastal Lowlands, Southern Coastal Limestone (Peñuelas), Susúa Forest Reserve, Western Cordillera, Guánica Forest Reserve, Southwestern Coastal Lowlands, Sierra Bermeja, Cerro Las Mesas, Western Coast (Aguada), Vieques, Mona, Desecheo; 1–700 m; *fl* & *fr* throughout the year.

West Indies, southern United States, Mexico, Central America, South America. [L3: 204].

Basora prieta, Malvavisco.

Wissadula hernandioides (L'Hér.) Garcke, Z. Naturwiss. 63(2, 3):122. 1890.
Bas.: *Sida hernandioides* L'Hér. (1789).

Suffrutescent herb in Southeastern Foothills, Cerro de Las Cuevas, Southern Coastal Lowlands, Guánica Forest Reserve, Southwestern Coastal Lowlands, Sierra Bermeja, Western Coast (Rincón), Vieques; 25–300 m; *fl* & *fr* Sep–Feb.

Antilles, south-central United States (Texas), Mexico, Central America, west-central South America. [L3: 170].

Notes: According to Paul Fryxell (pers. comm.), the name *Wissadula amplissima* (L.) R.E. Fr. was misapplied to this taxon by Fries; it must be applied to a taxon long known as *Wissadula hirsuta* C. Presl from Central and South America.

Pichona.

Wissadula periplocifolia (L.) C. Presl ex Thwaites in Thwaites & Hooker, Enum. Pl. Zeyl. 27. 1858.

Bas.: *Sida periplocifolia* L. (1753).

Suffrutescent herb in Northern Coastal Lowlands (Fajardo), Eastern Coast (Ceiba), Southeastern Foothills, Southern Coastal Lowlands, Southwestern Coastal Lowlands, Sierra Bermeja; 1–300 m; *fl* & *fr* throughout the year.

Antilles, Mexico, Central America, South America. [L3: 172].

TAXA NOT RECENTLY COLLECTED

Abutilon indicum (L.) Sweet, Hort. Brit. (ed. 1) 54. 1826.

Bas.: *Sida indica* L. (1756).

Shrub collected in Cabo Rojo on 26 Jun 1885 [*Sintenis 799* (US)], in Coamo on 8 Dec 1885 [*Sintenis 3184* (US)], in Ponce near the coast on 3 Sep 1886 [*Sintenis 4930* (US)] and in the vicinity of Coamo Springs on 15 Feb 1922 [*Britton et al. 6037* (US)].

Native to the Paleotropics. [L3: 125].

Allosidastrum pyramidatum (Cav.) Krapov., Fryxell & D.M. Bates in Fryxell, Syst. Bot. Monogr. 25:74. 1988.

Bas.: *Sida pyramidata* Cav. (1785).

Shrub collected in Yauco in 1880 [*A.P. Garber 26* (NY)].

Greater Antilles, Mexico, Central America, northern South America. [L3: 160].

Bastardia bivalvis (Cav.) Kunth ex Griseb., Fl. Brit. W.I. 80. 1859.

Bas.: *Sida bivalvis* Cav. (1785).

Herb collected, according to Britton & Wilson (1924), by Sintenis between Serillos and Salinas.

Greater Antilles, Mexico, Central America, America (scattered). [L3: 129].

Escoba babosa.

Bastardiopsis eggersii (Baker f.) Fuertes & Fryxell in Acevedo-Rodríguez, Mem. New York Bot. Gard. 78:310. 1996.

Bas.: *Sida eggersii* Baker f. (1892).

Tree collected in Culebra between 3–12 Mar 1906 [*Britton & Wheeler 178* (US)].

Virgin Islands (Tortola). [L3: 150].

Malachra radiata (L.) L. Syst. Nat. (ed. 12) 2:459. 1767.

Bas.: *Sida radiata* L. (1763).

Herb collected in Canóvanas on 1 Jun 1916 [*Stevenson 5379* (US)] and, according to Liogier & Martorell (1982), in San Sebastián by Vélez (specimen not seen).

Greater Antilles, Central America, South America. [L3: 146].

Pavonia paniculata Cav., Diss. 3:135, *t. 46*, *f. 2*. 1787.

Syn.: *Lebretonia paniculata* (Cav.) Britton (1924).

Herb collected in Bayamón on 21 Mar 1885 [*Sintenis 1081* (US)] and also, according to Britton & Wilson (1924), in Lares.

Antilles, Mexico, Central America, northern South America to Bolivia. [L3: 155].

Cadillo altea, Cadillo anaranjado.

Pentapetes phoenicea L., Sp. Pl. 2:698. 1753.

Herb collected in Ceiba on 25 Apr 1885 [*Sintenis 1005* (US)].

Native to Myanmar, southeastern Asia. [L3: 198].

Flor impía.

TAXA CITED BY PREVIOUS AUTHORS, BUT NOT INCLUDED

NO SPECIMENS SEEN

Abutilon hulseanum (Torr. & A. Gray) Torr.

Shrub cited by Britton & Wilson (1924) as occurring in thickets and hillsides, vicinity of Guánica; subsequently cited by Liogier & Martorell (1982, 2000) and Liogier (1988). [L3: 123].

Notes: Britton & Wilson (1924) and Liogier & Martorell (1982) misapplied the name *Abutilon commutatum* K. Schum. to this taxon.

CITED DUE TO MISIDENTIFICATION

Abutilon permolle (Willd.) Sweet.

Suffrutescent herb/shrub cited by Liogier (1997) and Liogier & Martorell (2000) as cited from Puerto Rico by Paul Fryxell, who now (pers. comm.) regards this citation to be in error. [L5: 408].

ONLY UNDER CULTIVATION

Abelmoschus moschatus Medik. Syn.: *Abelmoschus abelmoschus* (L.) Karst.

Herb cited by Liogier & Martorell (1982, 2000) and Liogier (1988) as cultivated and sometimes escaped. [L3: 120].

Adansonia digitata L.

Tree cited by Liogier & Martorell (1982, 2000) and Liogier (1994) as planted for shade and ornament and persistent. [L3: 175].

Alcea rosea L. Syn.: *Althaea rosea* (L.) Cav.

Herb cited by Liogier & Martorell (1982, 2000) and Liogier (1988) as grown as a garden flower, sometimes escaped. [L3: 126].

Ceiba aesculifolia (Kunth) Britton & Baker f.

Tree cited by Liogier & Martorell (1982, 2000) and Liogier (1994) as occasionally planted as an ornament. [L3: 178].

Chorisia speciosa A. St.-Hil.
Tree cited by Liogier & Martorell (1982, 2000) and Liogier (1994) as planted as an ornamental and shade tree, rare. [L3: 178].

Cola acuminata (P. Beauv.) Schott & Endl.
Tree cited by Liogier & Martorell (1982, 2000) and Liogier (1994) as sparingly planted and persistent. [L3: 188].

Notes: Britton & Wilson (1926) misapplied the name *Cola vera* K. Schum. to this taxon.

Gossypium barbadense L. Syn.: *Gossypium brasiliense* Macfad. *Gossypium peruvianum* Cav. *Gossypium barbadense* L. var. *acuminatum* (Roxb. ex G. Don) Triana & Planch.
Shrub cited by Britton & Wilson (1924) as occurring in waste grounds and thickets, spontaneous after cultivation; subsequently cited by Liogier & Martorell (1982, 2000) and Liogier (1994). [L3: 133].

Hibiscus cannabinus L.
Herb cited by Liogier & Martorell (1982, 2000) and Liogier (1988) as grown in gardens, sometimes escaped. [L3: 139].

Hibiscus sabdariffa L., *nom. cons.*
Herb cited by Britton & Wilson (1924) as occasionally spontaneous after cultivation; subsequently cited by Liogier & Martorell (1982, 2000) and Liogier (1988). [L3: 143].

Hibiscus schizopetalus Dyer. Syn.: *Hibiscus rosa-sinensis* L. var. *schizopetalus* (Dyer) Hook. f.
Shrub cited by Liogier & Martorell (1982, 2000) and Liogier (1988) as commonly grown as an ornamental and persistent after cultivation. [L3: 143].

Notes: Liogier & Martorell (1982) and Liogier (1988) had *Hibiscus schizopetalus* Mast. as the name of the taxon and Liogier & Martorell (2000) had *Hibiscus rosa-sinensis* L. var. *schizopetalus* Mast. as the name of the synonym (basionym).

Kleinhovia hospita L.
Tree cited by Liogier & Martorell (1982, 2000) and Liogier (1994) as uncommon as a planted ornamental and persistent. [L3: 193].

Luehea speciosa Willd.
Tree cited by Liogier (1994) and Liogier & Martorell (2000) as introduced and naturalized in Cambalache Forest. [L3: 112].

Malvaviscus arboreus Cav.
Shrub cited by Liogier & Martorell (1982, 2000) and Liogier (1988) as cultivated and escaped in wet regions. [L3: 150].

Pachira aquatica Aubl.
Tree cited by Liogier & Martorell (1982, 2000) and Liogier (1994) as sparingly planted as an ornamental. [L3: 182].

Pachira insignis (Sw.) Savigny.
Tree cited by Liogier & Martorell (1982, 2000) and Liogier (1994) as rare, planted as an ornamental and shade tree. [L3: 182].

Sterculia foetida L.
Tree cited by Liogier & Martorell (1982, 2000) and Liogier (1994) as occasionally grown for ornament and often naturalized. [L3: 200].

MARANTACEAE

Calathea allouia (Aubl.) Lindl., Bot. Reg. 14: *sub t. 1210*. 1829.
Bas.: *Maranta allouia* Aubl. (1775).

Herb in Northern Limestone Hills, Eastern Coast (Cuchilla de Panduras), Sierra de Cayey; 175–400 m; *fl* Sep–Jan, *fr* (?).

Greater Antilles (Jamaica, Hispaniola), Lesser Antilles, Trinidad, Mexico, Central America, northern South America (Venezuela, Colombia). [AR&S: 195].

Lerén, Lerenes, Llerenes.

Calathea lutea (Aubl.) Schult., Mant. 1:8. 1822.
Bas.: *Maranta lutea* Aubl. (1775).

Herb planted and persistent in Sierra de Luquillo, Eastern Coast (Cuchilla de Panduras), Sierra de Cayey; 125–600 m; *fl* Jan–May, *fr* (?).

Greater Antilles (Cuba, Jamaica), Lesser Antilles (Guadeloupe, Dominica, Martinique), Trinidad, Central America, northern South America to Brazil, Peru. [AR&S: 196].

Notes: Britton & Wilson (1924) and Liogier & Martorell (1982, 2000) used *Calathea lutea* (Aubl.) G. Mey., a *nom. inval.*, for the name of this taxon.

Hoja de sal, Pámpano.

Maranta arundinacea L., Sp. Pl. 1:2. 1753.

Herb planted and persistent in Sierra de Luquillo (Fajardo), Eastern Coast (Naguabo); 10–300 m; *fl* May, Sep–Oct, *fr* (?).

Greater Antilles (Cuba, Hispaniola), Virgin Islands (St. Croix), Lesser Antilles, Mexico, Central America, northern South America to Ecuador. [AR&S: 197].

Amaranta, Maranta, Pitisilén, Yuquilla.

Thalia geniculata L., Sp. Pl. 2.1193. 1753.
Syn.: *Thalia trichocalyx* Gagnep. (1904).

Herb in Northern Coastal Lowlands (vicinity of Laguna Tortuguero), Western Coast; 1–2 m; *fl* Jun–Sep, *fr* Jun–Dec.

Greater Antilles, Trinidad, southern United States (Florida, Alabama, Louisiana), Mexico, Central America, South America. [AR&S: 199].

Pámpano.

TAXA CITED BY PREVIOUS AUTHORS, BUT NOT INCLUDED

ONLY UNDER CULTIVATION

Calathea crotalifera S. Watson.

Herb cited in Acevedo-Rodríguez & Strong (2005) as found by Proctor, who, on the basis of a single farm site in Río Grande, considered it to be naturalized. [AR&S: 196].

Calathea zebrina (Sims) Lindl.

Herb cited by Liogier & Martorell (1982, 2000) as cultivated and escaped in wet regions.

Myrosma cannifolia L.f.

Herb cited by Liogier & Martorell (1982, 2000) as cultivated and escaped.

MARCGRAVIACEAE

Marcgravia rectiflora Triana & Planch., Ann. Sci. Nat., Bot., sér 4 17:364. 1862.

Syn.: *Marcgravia brittoniana* Alain (1994).

Woody vine in Northern Limestone Hills (Florida, Arecibo), Río Abajo Forest Reserve, Sierra de Luquillo, Sierra de Cayey, Eastern Cordillera (La Torrecilla), Central Cordillera, Maricao Forest Reserve; 200–1250 m; *fl* Oct–Jun, *fr* throughout the year.

Greater Antilles (Cuba, Hispaniola), Virgin Islands (Tortola). [L3: 219; AR: 310].

Notes: Both Bedell in Howard (1989) and Liogier & Martorell (2000) wrongly considered the above to be a synonym of *Marcgravia trinitatis* C. Presl.

Bejuco de lira, Bejuco de paloma, Bejuco de rana, Bejuco de sapo, Pegapalma.

Marcgravia sintenisii Urb., Jahrb. Königl. Bot. Gart. Berlin 4:245. 1886.

Woody vine in Sierra de Luquillo, Sierra de Cayey, Central Cordillera; 500–1325 m; *fl* Jun–Mar, *fr* Aug–Apr.

Endemic. [L3: 219; AR: 311].

Bejuco de lira, Bejuco de rana, Lira del Yunque, Pegapalma.

MARTYNIACEAE

Craniolaria annua L., Sp. Pl. 2:618. 1753.

Herb in Northern Coastal Lowlands; 1–100 m; *fl* & *fr* throughout the year.

Greater Antilles (Cuba, Hispaniola), Trinidad, South America. [L4: 561].

Escorzonera.

Martynia annua L., Sp. Pl. 2:618. 1753.

Herb in Southeastern Foothills, Cerro de Las Cuevas, Southern Coastal Lowlands, Southern Coastal Limestone (Peñuelas), Guánica Forest Reserve, Southwestern Coastal Lowlands, Sierra Bermeja; 10–550 m; *fl* & *fr* Jul–Nov.

West Indies, Mexico, Central America. [L4: 562].

Escorzonera, Uña de gato.

MELASTOMATACEAE

Acisanthera quadrata Pers., Syn. Pl. 1:477. 1805.

Syn.: *Acisanthera acisanthera* (L.) Britton (1925).

Herb in Northern Coastal Lowlands (Vega Baja to Manatí), Eastern Coast (Yabucoa), Caguas Valley (San Lorenzo), Southwestern Coastal Lowlands (Cabo Rojo); 1–200 m; *fl* & *fr* Mar–Nov.

Antilles, Mexico, Central America, northwestern, west-central South America and Venezuela. [L4: 19].

Camasey de charco.

Arthrostemma ciliatum Pav. ex D. Don, Mem. Wern. Nat. Hist. Soc. 4:299. 1823.

Syn.: *Arthrostemma fragile* Lindl. (1847).

Scandent herb in Sierra de Luquillo, Sierra de Cayey, Western Cordillera (Maricao); 550–925 m; *fl* Aug–Feb, May, *fr* Aug–Feb.

Native to Greater Antilles (Cuba, Jamaica), Mexico, Central America, northwestern and west-central South America. [L4: 21].

Clidemia domingensis (DC.) Cogn., Jahrb. Königl. Bot. Gart. Berlin 4:283. 1886.

Bas.: *Sagraea domingensis* DC. (1828).

Shrub in Sierra de Luquillo, Sierra de Cayey; 500–700 m; *fl* & *fr* May–Jul.

Greater Antilles (Hispaniola). [L4: 57].

Clidemia eggersii (Cogn.) F.S. Axelrod, *comb. nov.*

Bas.: *Heterotrichum eggersii* Cogn., Jahrb. Königl. Bot. Gart. Berlin 4:282. 1886.

Syn.: *Heterotrichum cymosum* (Schrad. & J.C. Wendl. ex Spreng.) Urb. (1910).

Shrub in Río Abajo Forest Reserve, Sierra de Luquillo, Sierra de Cayey, Las Tetas de Cayey, Central

Cordillera, Maricao Forest Reserve; 50–1250 m; *fl* & *fr* throughout the year.

Greater Antilles. [L4: 23].

Notes: Liogier (1995) and Liogier & Martorell (2000) used the combination *Clidemia cymosa* (Schrad. & J.C. Wendl. ex Spreng.) Alain, published by Alain in 1994, for the name of this taxon; however, it turns out to be a homonym of *Clidemia cymosa* Gleason, which was validly published in 1925 and refers to a different taxon.

Camasey colorado, Camasey de paloma, Camasey peludo, Camasey terciopelo, Pelúa, Terciopelo.

Clidemia hirta (L.) D. Don, Mem. Wern. Nat. Hist. Soc. 4:309. 1823.
Bas.: *Melastoma hirtum* L. (1753).

Shrub in Northern Coastal Lowlands, Northern Limestone Hills, Río Abajo Forest Reserve, Sierra de Luquillo, Eastern Coast, Caguas Valley, Sierra de Cayey, Las Tetas de Cayey, Southeastern Foothills (Arroyo), Eastern Cordillera (Cerro La Peña, Cañón San Cristóbal), Central Cordillera, Susúa Forest Reserve, Western Cordillera (Maricao), Southwestern Coastal Lowlands (Cabo Rojo), Cerro Las Mesas, Western Coast (Mayagüez); 1–950 m; *fl* & *fr* throughout the year.

Antilles, Mexico, Central America, northern and central South America. [L4: 25].

Camasey, Camasey peludo, Nigua.

Clidemia pusilliflora Cogn. in de Candolle, Monogr. Phan. 7:1010. 1891.
Syn.: *Sagraea pusilliflora* (Cogn.) Alain (1999).
Clidemia portoricensis Alain (1982). *Sagraea portoricensis* (Alain) Alain (1994).

Shrub in Río Abajo Forest Reserve; 225–350 m; *fl* & *fr* May–Jul, Nov.

Greater Antilles (Hispaniola). [L4: 58].

Clidemia scabrosa (L.) Griseb., Fl. Brit. W.I. 248. 1860.
Bas.: *Melastoma scabrosum* L. (1759).
Syn.: *Ossaea scabrosa* (L.) DC. (1828). *Sagraea scabrosa* (L.) Alain. (1994).

Shrub in Río Abajo Forest Reserve, Central Cordillera (Yauco), Maricao Forest Reserve; 300–850 m; *fl* Feb, Jun (two specimens), *fr* Feb, Jun–Jul.

Greater Antilles. [L4: 58].

Clidemia scalpta (Vent.) F.S. Axelrod, *comb. nov.*
Bas.: *Maieta scalpta* Vent., Mém. Cl. Sci. Math. Inst. Natl. France. 8:12; et Choix Pl. *t. 33*. 1807.
Syn.: *Ossaea scalpta* (Vent.) DC. (1828). *Sagraea scalpta* (Vent.) Naudin (1852).
Ossaea domingensis Cogn. (1913).

Shrub in Central Cordillera (Yauco); 850 m; *fl* (?), *fr* Jul (one specimen). [*Caraballo 2654* (UPR)].

Greater Antilles (Cuba, Hispaniola). [L4: 59].

Clidemia strigillosa (Sw.) DC., Prodr. 3:159. 1828.
Bas.: *Melastoma strigillosum* Sw. (1788).

Shrub in Northern Coastal Lowlands (Carolina, San Juan), Eastern Cordillera (Cerro La Peña), Central Cordillera, Western Cordillera, Southwestern Coastal Lowlands (Cabo Rojo), Cerro Las Mesas, Western Coast (Mayagüez); 50–700 m; *fl* & *fr* throughout the year.

Greater Antilles, Central America, northern South America and Brazil. [L4: 25].

Terciopelo.

Clidemia umbrosa (Sw.) Cogn. in de Candolle, Monogr. Phan. 7:1019. 1891.
Bas.: *Melastoma umbrosum* Sw. (1788).
Syn.: *Sagraea umbrosa* (Sw.) DC. (1828).

Shrub in Sierra de Luquillo; 125–900 m; *fl* Apr–Aug, *fr* Apr–Oct.

Lesser Antilles, Trinidad. [L4: 59].

Conostegia rufescens Naudin, Ann. Sci. Nat., Bot., sér. 3 16(2):107. 1851.
Syn.: *Conostegia hotteana* Urb. & Ekman (1929).

Tree in Sierra de Luquillo, Sierra de Cayey, Central Cordillera (Utuado); 450–1000 m; *fl* & *fr* Sep–Nov, Mar.

Greater Antilles, southern Central America, northwestern South America (Colombia, Ecuador). [L4: 26].

Camasey, Camasey bobo, Camasey de oro, Camasey de paloma, Camasey peludo, Camasey simple.

Henriettea fascicularis (Sw.) M. Gómez, Anales Soc. Esp. Hist. Nat. 23:68. 1894.
Bas.: *Melastoma fasciculare* Sw. (1788).
Syn.: *Henriettella fascicularis* (Sw.) C. Wright (1869).

Tree in Northern Limestone Hills (Lares, San Sebastian), Sierra de Luquillo, Sierra de Cayey, Eastern Cordillera (Bayamón including Cerro La Peña), Central Cordillera, Western Cordillera (Maricao), Western Coast (Mayagüez); 10–950 m; *fl* Mar–May, Aug, *fr* throughout the year.

Greater Antilles, Central America. [L4: 31].

Henriettea macfadyenii (Triana) Alain, Bull. Torrey Bot. Club 92(4):300. 1965.
Bas.: *Henriettella macfadyenii* Triana (1871).

Tree in Sierra de Luquillo, Sierra de Cayey, Toro

Negro Forest Reserve; 500–1000 m; *fl* Jul (one specimen), *fr* Jun–Jul.

Greater Antilles (Jamaica). [L4: 32].

Camasey.

Henriettea squamulosa (Cogn.) Judd, Brittonia 38(3):240. 1986.

Bas.: *Calycogonium squamulosum* Cogn. (1886).

Tree in Sierra de Luquillo, Sierra de Cayey, Central Cordillera, Maricao Forest Reserve; 500–1300 m; *fl* May–Sep, *fr* Sep–Mar.

Endemic. [L4: 32].

Camasey jusillo, Camasey negro, Granadilla cimarrona, Husillo, Jusillo.

Heterotis rotundifolia (Sm.) Jacques-Fél., Adansonia, sér. 2 20(4):417. 1981.

Bas.: *Osbeckia rotundifolia* Sm. (1813).

Syn.: *Dissotis rotundifolia* (Sm.) Triana (1872).

Herb in Northern Coastal Lowlands (Fajardo), Northern Limestone Hills, Sierra de Luquillo, Eastern Coast, Caguas Valley, Sierra de Cayey, Eastern Cordillera (Cerro La Peña), Central Cordillera (Yauco), Maricao Forest Reserve; 40–770 m; *fl* & *fr* throughout the year.

Native to tropical Africa. [L4: 28].

Mecranium latifolium (Cogn.) Skean, Syst. Bot. Monogr. 39:66. 1993.

Bas.: *Mecranium amygdalinum* (Desr.) C. Wright f. *latifolium* Cogn. (1886).

Tree in Sierra de Luquillo, Sierra de Cayey, Central Cordillera, Maricao Forest Reserve; 250–1300 m; *fl* & *fr* throughout the year.

Virgin Islands (St. Thomas). [L4: 37].

Notes: Britton & Wilson (1924) and Liogier & Martorell (1982) used *Mecranium amygdalinum* (Desr.) C. Wright, *sensu lato*, for the name of this taxon.

Camasey, Camasey almendro.

Miconia affinis DC., Prodr. 3:187. 1828.

Syn.: *Miconia microcarpa* DC. (1828).

Shrub/small tree in Northern Limestone Hills (Lares), Sierra de Luquillo, Eastern Cordillera (Cerro La Peña), Central Cordillera (Ciales, Utuado), Western Cordillera (Las Marías); 150–400 m; *fl* Mar–May (two specimens), *fr* Mar–Sep.

Greater Antilles (Cuba, Hispaniola), Trinidad, Mexico, Central America, northwestern, west-central South America and Venezuela. [L4: 41].

Camasey.

Miconia calycina Cogn. in Urban, Symb. Antill. 7(3):312. 1912.

Syn.: *Miconia subcorymbosa* Britton (1925).

Tree in Northern Limestone Hills (Lares), Sierra de Cayey, Eastern and Central Cordillera; 250–1300 m; *fl* Aug–Mar, *fr* throughout the year.

Greater Antilles (Cuba, Hispaniola). [L4: 49].

Camasey.

Miconia foveolata Cogn., Jahrb. Königl. Bot. Gart. Berlin 4:281. 1886.

Shrub/small tree in Sierra de Luquillo; 775–1065 m; *fl* & *fr* throughout the year.

Endemic. [L4: 41].

Camasey.

Miconia impetiolaris (Sw.) D. Don ex DC., Prodr. 3:183. 1828.

Bas.: *Melastoma impetiolare* Sw. (1788).

Tree in Northern Coastal Lowlands (Río Grande, Guaynabo), Northern Limestone Hills, Río Abajo Forest Reserve, Sierra de Luquillo, Caguas Valley (Las Piedras), Central Cordillera, Maricao Forest Reserve; 10–850 m; *fl* Feb–Nov, *fr* Mar–Nov.

Antilles, Mexico, Central America, northwestern, west-central South America and Venezuela. [L4: 41].

Camasey, Camasey colorado, Camasey de costilla.

Miconia laevigata (L.) D. Don in Sweet, Hort. Brit. (ed. 1) 159. 1826.

Bas.: *Melastoma laevigatum* L. (1759).

Tree in Northern Coastal Lowlands, Northern Limestone Hills, Río Abajo Forest Reserve, Sierra de Luquillo, Eastern Coast (Cuchilla de Panduras), Sierra de Cayey, Las Tetas de Cayey, Las Piedras Chiquitas, Southeastern Foothills, Eastern and Central Cordillera, Susúa Forest Reserve, Western Cordillera, Vieques; 10–950 m; *fl* & *fr* throughout the year.

Antilles, Mexico, Central America, northwestern, west-central South America and Venezuela. [L4: 42].

Camasey, Camasey de paloma, Palo de costilla.

Miconia mirabilis (Aubl.) L.O. Williams, Fieldiana, Bot. 29(10):574. 1963.

Bas.: *Fothergilla mirabilis* Aubl. (1775).

Syn.: *Tamonea guianensis* Aubl. (1775).

Tree in Northern Coastal Lowlands (Guaynabo), Guajataca Forest Reserve, Sierra de Luquillo, Sierra de Cayey, Eastern Cordillera (Cerro La Peña), Central Cordillera, Maricao Forest Reserve; 250–1000 m; *fl* & *fr* throughout the year.

Antilles, Trinidad, Mexico, Central America, northern South America (Venezuela, Colombia). [L4: 43].

Camasey, Camasey blanco, Camasey cuatrocanales, Camasey de costilla, Camasey prieto.

Miconia pachyphylla Cogn., Jahrb. Königl. Bot. Gart. Berlin 4:279. 1886.

Tree in Sierra de Luquillo, Maricao Forest Reserve; 750–1050 m; *fl* Jan–May, *fr* Mar–Sep.

Endemic. [L4: 43].

Camasey racimoso.

Miconia prasina (Sw.) DC., Prodr. 3:188. 1828.

Bas.: *Melastoma prasinum* Sw. (1788).

Tree in Northern Coastal Lowlands, Northern Limestone Hills, Río Abajo Forest Reserve, Sierra de Luquillo, Eastern Coast, Sierra de Cayey, Eastern Cordillera (Cerro La Peña, Cañón San Cristóbal), Toro Negro Forest Reserve, Western Cordillera, Cerro Las Mesas, Western Coast (Mayagüez); 1–900 m; *fl* & *fr* throughout the year.

Antilles, Trinidad, Mexico, Central America, northwestern, west-central South America and Venezuela. [L4: 44].

Camasey, Camasey blanco.

Miconia pycnoneura Urb., Repert. Spec. Nov. Regni Veg. 17:162. 1921.

Tree in Sierra de Luquillo, Central Cordillera; 800–1300 m; *fl* Aug–Mar, *fr* Feb–May.

Endemic. [L4: 46].

Camasey.

Miconia pyramidalis (Desr.) DC., Prodr. 3:188. 1828.

Bas.: *Melastoma pyramidale* Desr. (1797).

Tree in Northern Limestone Hills, Río Abajo Forest Reserve, Eastern Cordillera (La Torrecilla), Central Cordillera, Maricao Forest Reserve, 60–950 m; *fl* Nov–Jul, *fr* Nov–Apr, Jul–Aug.

Greater Antilles (Cuba, Hispaniola).

Notes: Britton & Wilson (1925) and Liogier (1995) treated this taxon as conspecific with *Miconia laevigata*. Judd & Kabat (2005) distinguish it from the latter by the larger number of internodes (usually 5 or more vs. 1 or 2) of its basal branches of its inflorescence; by distal branchess with successively fewer internodes and with the flowers usually in 3–flowered dichasia (vs. scorpioid-secund clusters); and, by the adaxial leaf surface drying much darker (vs. only slightly darker) than the abaxial one.

Camasey, Camasey de paloma, Palo de costilla.

Miconia racemosa (Aubl.) DC., Prodr. 3:179. 1828.

Bas.: *Melastoma racemosum* Aubl. (1775).

Tree in Northern Coastal Lowlands, Northern Limestone Hills (Utuado), Río Abajo Forest Reserve, Sierra de Luquillo, Eastern Coast, Sierra de Cayey, Eastern Cordillera (Cerro La Peña, La Torrecilla), Central Cordillera, Western Cordillera, Western Coast (Mayagüez); 1–925 m; *fl* & *fr* throughout the year.

Greater Antilles (Hispaniola), Virgin Islands, Lesser Antilles, Trinidad and Tobago, southern Central America (Panama), northern South America. [L4: 46].

Camasey de felpa, Terciopelo.

Miconia rubiginosa (Bonpl.) DC., Prodr. 3:183. 1828.

Bas.: *Melastoma rubiginosum* Bonpl. (1816).

Shrub/small tree in Northern Limestone Hills (Lares), Western Cordillera, Cerro Las Mesas, Western Coast (Mayagüez); 300–375 m; *fl* Jan, May–Aug, *fr* Jun–Sep.

Greater Antilles (Hispaniola), Trinidad, southern Central America, northwestern, west-central South America and Venezuela. [L4: 47].

Camasey.

Miconia serrulata (DC.) Naudin, Ann. Sci. Nat., Bot., sér. 3 16(2):118. 1851.

Bas.: *Diplochita serrulata* DC. (1828).

Syn.: *Tamonea macrophylla* (D. Don) Krasser (1893).

Shrub/small tree in Northern Limestone Hills (Florida, Lares), Sierra de Luquillo, Sierra de Cayey, Eastern Cordillera (Cerro La Peña), Central Cordillera, Western Cordillera (Maricao), Cerro Las Mesas; 100–800 m; *fl* throughout the year, *fr* Jan–Oct.

Antilles, Trinidad, Mexico, Central America, South America. [L4: 47].

Camasey.

Miconia sintenisii Cogn., Jahrb. Königl. Bot. Gart. Berlin 4:281. 1886.

Tree in Sierra de Luquillo, Northern Limestone Hills (Lares), Sierra de Cayey, Central Cordillera, Maricao Forest Reserve; 475–1300 m; *fl* & *fr* throughout the year.

Endemic. [L4: 48].

Camasey.

Miconia tetrandra (Sw.) D. Don in Loudon, Hort. Brit. (ed. 2) 174. 1830.

Bas.: *Melastoma tetrandrum* Sw. (1788).

Tree in Sierra de Luquillo, Sierra de Cayey, Eastern Cordillera (Cerro La Peña), Central Cordillera, Maricao Forest Reserve; 150–850 m; *fl* Nov–May, *fr* Feb–Aug.

Greater Antilles, Lesser Antilles (Guadeloupe, Dominica, Grenada). [L4: 49].

Camasey.

Miconia tetrastoma Naudin, Ann. Sci. Nat., Bot., sér. 3 16(2):236. 1851.

Tree in Central Cordillera; 1250–1300 m; *fl* May (one specimen), *fr* May, Aug (two specimens). [*Woodbury a.n.* 2594 (UPR); *Liogier 28976* (UPR)].

Greater Antilles (Hispaniola). [L4: 49].

Notes: Liogier & Martorell (2000) misapplied the name *Miconia virescens* (Vahl) Triana, and Liogier (1995) and Liogier & Martorell (2000) misapplied *Miconia theizans* (Bonpl.) Cogn., for which they used the orthographic variant *Miconia theaezans* (Bonpl.) Cogn., to this taxon.

Miconia thomasiana DC., Prodr. 3:189. 1828.

Tree in Northern Coastal Lowlands, Eastern Coast, Eastern Cordillera (Cañón San Cristóbal), Susúa Forest Reserve, Maricao Forest Reserve; 1–625 m; *fl* & *fr* throughout the year.

Virgin Islands (Tortola). [L4: 50].

Camasey tomaso.

Mouriri domingensis (Tussac) Spach, Hist. Nat. Vég. 4:276. 1835.

Bas.: *Petaloma domingensis* Tussac (1824).

Tree in Northern Coastal Lowlands (Carolina, San Juan, Dorado), Eastern Coast (Maunabo), Vieques; 15–100 m; *fl* & *fr* May–Jul.

Greater Antilles (Hispaniola), Virgin Islands (St. Croix, Antigua), Lesser Antilles (Guadeloupe). [L4: 51].

Caimitillo, Guasávara, Murta.

Mouriri helleri Britton, Torreya 2:10. 1902.

Shrub/small tree in Northern Coastal Lowlands (Dorado), Northern Limestone Hills, Río Abajo Forest Reserve, Susúa Forest Reserve, Maricao Forest Reserve; 150–800 m; *fl* Apr–Jan, *fr* throughout the year.

Endemic. [L4: 51].

Mameyuelo, Palo prieto.

Nepsera aquatica (Aubl.) Naudin, Ann. Sci. Nat., Bot., sér. 3 13(1): *t. 14, f. 1*. 1850.

Bas.: *Melastoma aquaticum* Aubl. (1775).

Suffrutescent herb in Northern Coastal Lowlands, Northern Limestone Hills (Lares), Río Abajo Forest Reserve, Sierra de Luquillo, Eastern Coast, Sierra de Cayey, Eastern Cordillera (Cerro La Peña), Central Cordillera, Western Cordillera, Western Coast (Mayagüez); 1–1150 m; *fl* & *fr* throughout the year.

Antilles, Trinidad, Central America, northern South America to Brazil, Ecuador. [L4: 53].

Altea, Altea amarilla, Camasey enano, Camasey de ciénaga.

Ossaea krugiana Cogn. in de Candolle, Monogr. Phan. 7:1048. 1891.

Syn.: *Leandra krugiana* (Cogn.) Judd & Skean (1991).

Shrub/small tree in Central Cordillera; 1000–1200 m; *fl* May-Aug, *fr* Apr–May (one specimen).

Endemic. [L4: 34].

Notes: According to Fabian Michelangeli (pers. comm.), the genus *Leandra* is restricted to eastern Brazil.

Ossaea krugii (Cogn.) F.S. Axelrod, *comb. nov.*

Bas.: *Calycogonium krugii* Cogn., Jahrb. Königl. Bot. Gart. Berlin 4:278. 1886.

Syn.: *Leandra krugii* (Cogn.) Judd & Skean (1991).

Shrub/small tree in Central Cordillera, Susúa Forest Reserve, Maricao Forest Reserve; 150–1150 m; *fl* throughout the year, *fr* Jan–Aug.

Endemic. [L4: 36].

Pterolepis glomerata (Rottb.) Miq., Comm. Phytogr. 2:78. 1840.

Bas.: *Rhexia glomerata* Rottb. (1778).

Herb in Central Cordillera; 650–1250 m; *fl* & *fr* Sep–Mar.

Greater Antilles (Hispaniola), Lesser Antilles, Trinidad, northern South America to Paraguay. [L4: 54].

Rhexia cubensis Griseb., Cat. Pl. Cub. 104. 1866.

Suffrutescent herb in Northern Coastal Lowlands (Vega Alta to Laguna Tortuguero); 1–5 m; *fl* Mar–Nov, *fr* Apr–Nov.

Greater Antilles (Cuba, Hispaniola), southeastern United States (Florida). [L4: 56].

Camasey mariana.

Tetrazygia angustifolia (Sw.) DC., Prodr. 3:172. 1828.

Bas.: *Melastoma angustifolium* Sw. (1788).

Tree in Northern Limestone Hills, Eastern Coast (Punta Guayanés); 10–200 m; *fl* May–Dec, *fr* May–Feb.

Greater Antilles (Hispaniola), Virgin Islands, Lesser Antilles. [L4: 61].

Tetrazygia biflora (Cogn.) Urb., Repert. Spec. Nov. Regni Veg. 17:405. 1921.

Bas.: *Calycogonium biflorum* Cogn. (1886).

Syn.: *Menendezia biflora* (Cogn.) Britton (1925).

Tree in Northern Limestone Hills (Lares), Sierra de Luquillo, Eastern Coast (Cuchilla de Panduras), Sierra de Cayey, Las Tetas de Cayey, Eastern Cordillera (La Torrecilla), Central Cordillera; 300–950 m; *fl* & *fr* Jun–Sep.

Endemic. [L4: 62].

Camasey.

Tetrazygia crotonifolia (Desr.) DC., Prodr. 3:172. 1828.
Bas.: *Melastoma crotonifolium* Desr. (1797).
Shrub in Central Cordillera; 500–1300 m; *fl* & *fr* throughout the year.
Greater Antilles (Hispaniola). [L4: 62].

Tetrazygia elaeagnoides (Sw.) DC., Prodr. 3:172. 1828 "eloeagnoides."
Bas.: *Melastoma elaeagnoides* Sw. (1788) "eloeagnoides."
Tree in Northern Coastal Lowlands, Northern Limestone Hills, Río Abajo Forest Reserve, Las Piedras Chiquitas, Southeastern Foothills, Southern Coastal Lowlands (Patillas), Guánica Forest Reserve, Sierra Bermeja, Western Coast (Añasco, Rincón), Vieques; 20–600 m; *fl* & *fr* throughout the year.
Greater Antilles (Hispaniola), Virgin Islands. [L4: 63].
Camasey cenizo, Cenizo, Verdiseco.

Tetrazygia urbanii Cogn., Jahrb. Königl. Bot. Gart. Berlin 4:278. 1886.
Syn.: *Tetrazygia stahlii* Cogn. (1886). *Menendezia stahlii* (Cogn.) Britton (1925).
Tree in Sierra de Luquillo, Sierra de Cayey, Eastern Cordillera (Cerro La Peña, La Torrecilla), Toro Negro Forest Reserve 250–1300 m; *fl* Apr–Oct, *fr* Jun–Aug.
Endemic. [L4: 63].
Camasey, Cenizo.

Tibouchina granulosa (Desr.) Cogn. in Martius, Fl. Bras. 14(3):340. 1885.
Bas.: *Melastoma granulosum* Desr. (1797).
Tree planted and persistent in Caguas Valley (Caguas), Central Cordillera (Adjuntas); 350–850 m; *fl* Apr, Oct (two specimens), *fr* May, Oct (two specimens).
Native to South America (Brazil, Bolivia). [L4: 65].
Tibuchina.

Tibouchina pilosa ×fothergilliae.
Tree planted and persistent in Sierra de Cayey, Central Cordillera, Western Cordillera (Maricao); 300–1325 m; *fl* throughout the year, *fr* (?).
Native to Brazil. [L4: 65].
Notes: According to Frank Almeda (pers. comm.), Liogier & Martorell (1982, 2000) and Liogier (1995) misapplied the name *Tibouchina urvilleana* (DC.) Cogn. to this hybrid.
Tibuchina.

TAXA NOT RECENTLY COLLECTED

Henriettea membranifolia (Cogn.) Alain, Bull. Torrey Bot. Club 92(4):300. 1965.
Bas.: *Henriettella membranifolia* Cogn. (1891).
Shrub/small tree collected in Aguada on 18 Dec 1886 [*Sintenis 5730* (GOET), not seen].
Endemic. [L4: 32].
Camasey.

Miconia punctata (Desr.) D. Don ex DC., Prodr. 3:184. 1828.
Bas.: *Melastoma punctatum* Desr. (1797).
Shrub/small tree collected in Utuado on 8 Feb 1887 [*Sintenis 6201* (US)] and later in Feb 1887 [*Sintenis 6426* (US)].
Greater Antilles (Cuba, Hispaniola), Trinidad, Mexico, Central America, northern South America to Brazil and Peru. [L4: 44].
Camasey.

TAXA CITED BY PREVIOUS AUTHORS, BUT NOT INCLUDED

DOUBTFUL OCCURRENCE

Clidemia angustilamina Judd & Skean. Syn.: *Heterotrichum angustifolium* DC.
Shrub cited by Britton & Wilson (1925) as recorded by Cogniaux as having been collected in Puerto Rico by Riedlé; subsequently cited by Liogier & Martorell (1982, 2000) and Liogier (1995). [L4: 23].

Clidemia polystachya (Naudin) Cogn. Syn.: *Sagraea polystachya* (Naudin) Triana.
Shrub cited by Britton & Wilson (1925) as collected only by Plée, and known to them from descriptions only; subsequently cited by Liogier & Martorell (1982, 2000) and Liogier (1995). [L4: 58].

Graffenrieda ottoschulzii (Urb. & Ekman) Urb. & Ekman. Syn.: *Miconia ottoschulzii* Urb. & Ekman.
Tree cited by Britton & Wilson (1930) as occurring in primeval forest in Isabón, determined by Urban from a barren specimen collected by Sintenis many years ago; subsequently cited by Liogier & Martorell (1982, 2000) and Liogier (1995). [L4: 29].

Miconia lanata (DC.) Triana.
Shrub cited by Britton & Wilson (1925) as recorded by Cogniaux as having been collected in Puerto Rico by Bertero; subsequently cited by Liogier & Martorell (1982, 2000) and Liogier (1995). [L4: 42].

Miconia splendens (Sw.) Griseb.
Shrub/small tree cited by Britton & Wilson (1930) as determined by Urban from a barren specimen from

the Sierra de Yabucoa; subsequently cited by Liogier & Martorell (1982, 2000) and Liogier (1995). [L4: 48].

CITED DUE TO MISIDENTIFICATION
Henriettea triflora (Vahl) Alain.
Tree cited by Liogier & Martorell (1982) as rare in mountain forests in Carite. Liogier (1995) subsequently stated that the citation was based upon the misidentification of a specimen of *Henriettea fascicularis*.

MELIACEAE

Cedrela odorata L., Syst. Nat. (ed. 10) 2:940. 1759.

Tree in Northern Limestone Hills, Río Abajo Forest Reserve, Las Tetas de Cayey, Eastern Cordillera (Cayey), Central Cordillera, Maricao Forest Reserve; 130–1000 m; *fl* (?), *fr* Mar, Jul–Aug, Nov.

West Indies, Trinidad and Tobago, Mexico, Central America, South America. [L2: 297].

Cedro, Cedro del país, Cedro español, Cedro hembra, Cedro hembra del país, Cedro mexicano, Cedro oloroso.

Guarea glabra Vahl, Eclog. Amer. 3:8. 1807.
Syn.: *Guarea ramiflora* Vent. (1807).

Tree in Northern Limestone Hills (Arecibo), Río Abajo Forest Reserve, Sierra de Luquillo, Sierra de Cayey, Las Tetas de Cayey, Eastern and Central Cordillera, Maricao Forest Reserve; 225–1050 m; *fl* Jan–Oct, *fr* throughout the year.

Greater Antilles (Jamaica), Lesser Antilles, Tobago, Mexico, Central America, northwestern, west-central South America and Venezuela. [L2: 298].

Guaraguadillo, Guaraguaíllo, Guaraguao macho.

Guarea guidonia (L.) Sleumer, Taxon 5(8):194. 1956.
Bas.: *Samyda guidonia* L. (1753).
Syn.: *Guarea trichilioides* L. (1771), *nom. illeg. superfl.*
Guarea guara (Jacq.) P. Wilson (1924).

Tree in Northern Coastal Lowlands, Northern Limestone Hills, Río Abajo Forest Reserve, Sierra de Luquillo, Sierra de Cayey, Eastern Cordillera (La Torrecilla), Central Cordillera, Susúa Forest Reserve, Maricao Forest Reserve, Western Coast (Mayagüez); 25–950 m; *fl* & *fr* throughout the year.

Greater Antilles, Trinidad and Tobago, southern Central America, northwestern, west-central South America and Venezuela. [L2: 299].

Guaraguao.

Melia azedarach L., Sp. Pl. 1:384. 1753.

Tree planted and persistent in Northern Coastal Lowlands, Southeastern Foothills (Salinas), Southern Coastal Lowlands, Southwestern Coastal Lowlands, Sierra Bermeja, Vieques, Culebra; 1–350 m; *fl* throughout the year, *fr* Jul–Mar.

Native to Indian Subcontinent, Malesia, China, Japan, Australia. [L2: 301].

Alelaila, Lilaila, Pasilla.

Swietenia macrophylla King, Hooker's Icon Pl. 16: *t. 1550*. 1886.

Tree planted and persistent; *fl* Apr (one specimen), *fr* Nov (one specimen).

Native to Mexico, Central America, northwestern and west-central South America. [L2: 302].

Caoba de Honduras, Caoba de Venezuela, Caoba hondureña.

Notes: Although no specimens of this taxon have yet been collected outside of cultivation, hybrids of it with *Swietenia mahagoni* do so occur. The phenology is taken from cultivated specimens.

Swietenia mahagoni (L.) Jacq., Enum. Syst. Pl. 20. 1760.
Bas.: *Cedrela mahagoni* L. (1759) "Cedrela Mahag."

Tree planted and persistent in Northern Coastal Lowlands (San Juan), Guajataca Forest Reserve, Sierra de Luquillo, Eastern Coast (Cayo Santiago), Southern Coastal Limestone, Susúa Forest Reserve, Guánica Forest Reserve, Southwestern Coastal Lowlands (Cabo Rojo), Vieques, Mona; 5–250 m; *fl* Apr–Sep, *fr* Aug–Nov, Feb–Mar.

Native to Bahamas, Greater Antilles (except Puerto Rico), southeastern United States (Florida). [L2: 303].

Notes: According to Sheila Ward (pers. comm.), the introduced taxa of *Swietenia* mutually hybridize and rehybridize here in Puerto Rico, so that identification of a specimen to a particular name may not be meaningful.

Caoba, Caoba de Santo Domingo, Caoba dominicana.

Toona ciliata M. Roem., Fam. Nat. Syn. Monogr. 1:139. 1846.

Tree planted and persistent in Sierra de Luquillo, Southwestern Coastal Lowlands (Lajas); 10–400 m; *fl* Mar–May (two specimens), *fr* Feb, May, Sep.

Native to Afghanistan, tropical Asia, China, Australia. [L2: 305].

Cedro de Himalaya, Tun.

Trichilia hirta L., Syst. Nat. (ed. 10) 2:1020. 1759.

Tree in Northern Coastal Lowlands, Northern Limestone Hills, Eastern Coast (Ceiba), Sierra de Cayey, Las Tetas de Cayey, Las Piedras Chiquitas, Southeastern Foothills, Eastern Cordillera (Comerío),

Southern Coastal Lowlands, Southern Coastal Limestone (Guayanilla), Guánica Forest Reserve, Southwestern Coastal Lowlands, Sierra Bermeja, Western Coast (Añasco, Rincón), Vieques, Culebra; 1–825 m; *fl* Feb–Nov, *fr* throughout the year.

Antilles, Mexico, Central America, northwestern and west-central South America. [L2: 306].

Anastasio, Cabo de hacha, Guaita, Jobillo, Molinillo, Palo de anastasio, Retamo, Tinacio.

Trichilia pallida Sw., Prodr. 67. 1788.

Tree in Northern Coastal Lowlands, Northern Limestone Hills, Río Abajo Forest Reserve, Sierra de Luquillo, Eastern Coast, Sierra de Cayey, Las Tetas de Cayey, Eastern and Central Cordillera, Susúa Forest Reserve, Maricao Forest Reserve, Guánica Forest Reserve; 1–1050 m; *fl* & *fr* throughout the year.

Greater Antilles (Hispaniola), Lesser Antilles, Trinidad, Mexico, Central America, northern South America to Bolivia, Paraguay. [L2: 308].

Caracolillo, Gaeta, Gaita, Ramoncillo.

Trichilia triacantha Urb., Symb. Antill. 1(2):329. 1899.

Shrub in Southern Coastal Limestone (Peñuelas), Guánica Forest Reserve; 15–150 m; *fl* Dec–Jan, May, Aug–Sep, *fr* Mar, Jul–Sep.

Endemic. [L2: 308].

Notes: This taxon is on the U.S. Fish & Wildlife Service list of threatened and endangered plants.

Bariaco, Guayabacón.

TAXA CITED BY PREVIOUS AUTHORS, BUT NOT INCLUDED

ONLY UNDER CULTIVATION

Azadirachta indica A. Juss. Syn.: *Melia azadirachta* L.
Tree cited by Liogier & Martorell (1982, 2000) and Liogier (1988) as planted and persistent, rare as a shade and ornamental tree. [L2: 295].

Khaya anthotheca (Welw.) C. DC. Syn.: *Khaya nyasica* Stapf ex Baker f.
Tree cited by Liogier (1988) and Liogier & Martorell (2000) as planted as an experimental timber and naturalized in certain wet areas. [L2: 481].

Khaya senegalensis (Desr.) A. Juss.
Tree cited by Liogier (1988) and Liogier & Martorell (2000) as planted as an experimental timber and becoming naturalized in wet areas. [L2: 481].

MENISPERMACEAE

Cissampelos pareira L., Sp. Pl. 2:1031. 1753.

Vine in Northern Coastal Lowlands, Northern Limestone Hills, Río Abajo Forest Reserve, Sierra de Luquillo, Eastern Coast, Sierra de Cayey, Southeastern Foothills, Eastern and Central Cordillera, Susúa Forest Reserve, Maricao Forest Reserve, Western Coast (Hormigueros), Vieques, Culebra; 1–1100 m; *fl* & *fr* throughout the year.

West Indies, Mexico, Central America, South America. [L1: 229; AR: 313].

Bejuco de mona, Oreja de ratón, Pareira.

Hyperbaena domingensis (DC.) Benth., J. Proc. Linn. Soc., Bot. 5(Suppl. 2):50. 1861.
Bas.: *Cocculus domingensis* DC. (1817).

Woody vine in Northern Limestone Hills; 5–100 m; *fl* May–Jul, Nov, *fr* Oct–Dec, Mar–May.

Greater Antilles (Cuba, Hispaniola), Virgin Islands (St. John, St. Thomas), Lesser Antilles, Trinidad and Tobago, northwestern, west-central South America and Venezuela. [L1: 231; AR: 313].

Hyperbaena laurifolia (Poir.) Urb., Symb. Antill. 1(2):304. 1899.
Bas.: *Cissampelos laurifolia* Poir. (1804).

Woody vine in Northern Limestone Hills, Southeastern Foothills (Salinas); 1–100 m; *fl* Mar–May, *fr* Dec–Sep.

Greater Antilles (Hispaniola), Virgin Islands (St. Thomas). [L1: 232; AR: 314].

MENYANTHACEAE

Nymphoides indica (L.) Kuntze, Revis. Gen. Pl. 2:429. 1891 "Nymphodes(oides) indicum."
Bas.: *Menyanthes indica* L. (1753).
Syn.: *Nymphoides humboldtiana* (Kunth) Kuntze (1891) "Nymphodes Humboldtianum."

Aquatic herb in Northern Coastal Lowlands, Northern Limestone Hills (Florida, Arecibo); 1–100 m; *fl* Mar–Nov, *fr* Sep, Feb–Mar.

Pantropical. [L4: 199].

Trébol de agua.

MOLLUGINACEAE

Mollugo nudicaulis Lam., Encycl. 4(1):234. 1797.

Herb in Northern Coastal Lowlands (Fajardo), Southern Coastal Lowlands (Guayama), Susúa Forest Reserve, Southwestern Coastal Lowlands (Cabo Rojo), Vieques; 1–150 m; *fl* & *fr* Sep–May.

Probably native to Africa, Asia. [L1: 196].

Mollugo verticillata L., Sp. Pl. 1:89. 1753.
Syn.: *Mollugo berteroana* Ser. (1824) "Berteriana."

Herb in Northern Coastal Lowlands, Las Piedras Chiquitas, Southern Coastal Lowlands (Caja de Muertos, Ponce), Southwestern Coastal Lowlands, Sierra Bermeja, Vieques; 1–450 m; *fl* & *fr* throughout the year.

Pantropical. [L1: 196].
Alfombra.

MORACEAE

Artocarpus altilis (Parkinson) Fosberg, J. Washington Acad. Sci. 31(3):95. 1941.
Bas.: *Sitodium altile* Parkinson (1773).
Syn.: *Artocarpus communis* J.R. Forst. & G. Forst. (1775).
Tree planted and persistent in Northern Coastal Lowlands, Eastern Coast (Cuchilla de Panduras), Western Cordillera (Maricao); 1–420 m; *fl* Jun, Oct (two specimens), *fr* Oct (one specimen).
Native to Indonesia, Papua New Guinea, various Pacific islands. [L1: 52].
Árbol de pan, Lavapén, Mapén, Palo de pan, Pan, Pana, Pana forastera, Pana de pepitas, Panapén.

Castilla elastica Sessé in Cervantes, Gaz. Lit. México Supl. 2 Jul:7. 1794.
Tree planted and persistent in Northern Limestone Hills, Río Abajo Forest Reserve, Eastern Cordillera (Corozal, Morovis), Western Cordillera (Maricao), Western Coast (Mayagüez); 50–300 m; *fl* (?), *fr* May–Jul, Oct.
Native to Mexico, Central America, northwestern South America (Colombia, Ecuador). [L1: 54].
Cauchera, Caucho, Goma, Palo de goma.

Dorstenia contrajerva L., Sp. Pl. 1:121. 1753.
Herb in Northern Limestone Hills (Quebradillas, Isabela), Río Abajo Forest Reserve, Las Tetas de Cayey, Las Piedras Chiquitas, Southeastern Foothills (Salinas), Eastern Cordillera (Cayey, Cañón San Cristóbal), Central Cordillera (Ponce), Quebrada de Los Cedros, Susúa Forest Reserve; 25–825 m; *fl* & *fr* Apr–Dec.
Antilles, Mexico, Central America, northwestern, west-central South America and Venezuela. [L1: 58].
Contrayerba, Tusilla.

Fatoua villosa (Thunb.) Nakai, Bot. Mag. (Tokyo) 41:516. 1927.
Bas.: *Urtica villosa* Thunb. (1784).
Herb planted and persistent in Northern Coastal Lowlands; 20–80 m; *fl* & *fr* Dec–Apr, Jul–Sep.
Native to Taiwan, Japan. [L5: 389]

Ficus altissima Blume, Bijdr. Fl. Ned. Ind. 444. 1825.
Tree planted and persistent in Northern Coastal Lowlands (Guaynabo), Caguas Valley (Gurabo); 50–100 m; *fl* & *fr* Apr (two specimens). [*Liogier 36278* (UPR); *Axelrod 9762* (UPRRP)]
Native to tropical Asia, China. [B&W6: 533].

Ficus americana Aubl., Hist. Pl. Guiane 2:952. 1775.
Syn.: *Ficus perforata* L. (1775).
Ficus sintenisii Warb. (1903).
Tree in Northern Coastal Lowlands, Northern Limestone Hills, Río Abajo Forest Reserve, Sierra de Luquillo, Eastern Coast (Cuchilla de Panduras), Sierra de Cayey, Las Tetas de Cayey, Eastern and Central Cordillera, Maricao Forest Reserve; 1–1100 m; *fl* & *fr* throughout the year.
West Indies, Mexico, Central America, northern South America to Brazil, Bolivia. [L1: 65].
Higuillo prieto, Jagüey, Jagüey colorado, Jagüey prieto.

Ficus citrifolia Mill., Gard. Dict. (ed. 8): *Ficus* no. 10. 1768.
Syn.: *Ficus laevigata* Vahl (1805).
Tree, sometimes viny, in Northern Coastal Lowlands, Northern Limestone Hills, Río Abajo Forest Reserve, Eastern Coast, Las Tetas de Cayey, Las Piedras Chiquitas, Southeastern Foothills, Eastern and Central Cordillera, Southern Coastal Lowlands, Southern Coastal Limestone, Susúa Forest Reserve, Guánica Forest Reserve, Southwestern Coastal Lowlands, Western Coast (Rincón), Vieques, Culebra, Mona, Desecheo; 1–950 m; *fl* & *fr* throughout the year.
West Indies, southeastern United States (Florida), Mexico, Central America, South America. [L1: 61; AR: 316].
Jagüeíllo, Jagüey, Jagüey blanco, Jagüey macho, Jigüerillo, Lechecillo.

Ficus crassinervia Desf. ex Willd., Sp. Pl. 4(2):1138. 1806.
Syn.: *Ficus stahlii* Warb. (1903).
Ficus mamillifera Warb. (1903).
Tree in Northern Coastal Lowlands, Northern Limestone Hills, Río Abajo Forest Reserve, Monte Guilarte Forest Reserve, Mona; 1–1100 m; *fl* & *fr* Feb–Nov.
Greater Antilles, Virgin Islands (Tortola), Mexico, northern Central America to Costa Rica. [L1: 65].
Notes: Britton & Wilson (1924), Liogier & Martorell (1982) and Liogier (1985) misapplied the above name to, and Liogier & Martorell (2000) wrongly placed it in synonymy with, *Ficus trigonata*.
Jagüey.

Ficus elastica Roxb. ex Hornem., Hort. Bot. Hafn. Suppl. 7. 1819.
Tree planted and persistent in Sierra de Cayey,

Eastern Cordillera (Aibonito), Southern Coastal Lowlands (Aguada); 250–500 m; *fl* & *fr* Feb, May–Jun, Sep.

Native to India, Myanmar, Malesia. [L1: 63].

Palo de goma.

Ficus kurzii King, Ann. Roy. Bot. Gard. (Calcutta) 1:47, *t.* 57. 1887.

Tree planted and persistent in Northern Coastal Lowlands (San Juan), Caguas Valley (Caguas), Eastern Cordillera (Cidra), Cerro Las Mesas, Western Coast (Mayagüez); 1–300 m; *fl* & *fr* Oct–Mar.

Native to tropical Asia, China. [L1: 61].

Notes: This taxon is often confused with *Ficus benjamina*. According to C.C. Berg (pers. comm.), it should more appropriately be designated as *Ficus kurzii vel. aff.*, since he has never seen in collections from Asian natural habitats material similar to the cultivated material in the Neotropics. Whatever it is, it is more commonly planted in Puerto Rico than *Ficus benjamina*. It can be distinguished from the latter by the presence of distinct vs. indistinct basal veins and by elongated vs. globose synconia.

Ficus lutea Vahl, Enum. Pl. 2:185. 1805.

Tree planted and persistent in Northern Coastal Lowlands (Dorado, Arecibo), Caguas Valley (Gurabo), Southern Coastal Lowlands (Aguirre); 1–150 m; *fl* & *fr* Apr–Jan.

Native to Africa. [L1: 64].

Notes: Liogier & Martorell (1982, 2000) and Liogier (1985) misapplied the name *Ficus nekbudu* Warb. to this taxon.

Ficus microcarpa L.f., Suppl. Pl. 442. 1782.

Tree planted and persistent in Río Abajo Forest Reserve, Caguas Valley; 75–325 m; *fl* & *fr* Jul–Nov.

Native to tropical Asia. [L1: 64].

Jagüey, Laurel, Laurel de la India.

Ficus ovalis (Liebm.) Miq., Ann. Mus. Bot. Lugduno-Batavi 3:298. 1868.

Bas.: *Urostigma ovale* Liebm. (1851).

Tree in Río Abajo Forest Reserve; 325 m; *fl* & *fr* Aug (one specimen). [*Axelrod 8244* (UPRRP)].

Native to Mexico, northern Central America to Costa Rica.

Ficus pumila L., Sp. Pl. 2:1060. 1753.

Woody vine/shrub planted and persistent in Northern Coastal Lowlands, Sierra de Luquillo; 10–1065 m; *fl* & *fr* Oct–Mar.

Native to Vietnam, China, Japan. [L1: 62].

Hiedra, Paz y justicia, Yedra.

Ficus trigonata L., Pl. Surin. 17. 1775.

Tree in Northern Coastal Lowlands, Northern Limestone Hills, Sierra de Luquillo, Eastern Coast (Yabucoa), Sierra de Cayey, Southeastern Foothills (Salinas), Central Cordillera (Orocovis), Western Cordillera, Guánica Forest Reserve, Vieques, Culebra; 1–700 m; *fl* & *fr* throughout the year.

Greater Antilles (Cuba, Hispaniola), Virgin Islands (St. Thomas, Tortola), Lesser Antilles (Montserrat to Martinique). [L1: 62].

Jagüey, Jagüey blanco.

Maclura tinctoria (L.) D. Don ex Steud., Nomencl. Bot. (ed. 2) 2(1–2):87. 1841.

Bas.: *Morus tinctoria* L. (1753).

Syn.: *Chlorophora tinctoria* (L.) Gaudich. ex Benth. & Hook.f. (1880).

Tree in Sierra de Cayey, Las Piedras Chiquitas, Southeastern Foothills (Salinas), Guánica Forest Reserve, Vieques, Mona; 40–500 m; *fl* May–Dec, *fr* Oct–Dec (two specimens).

Antilles, Trinidad and Tobago, Mexico, Central America, South America (scattered). [L1: 57].

Escambrón de madera, Mora, Palo de mora.

Pseudolmedia spuria (Sw.) Griseb., Fl. Brit. W.I. 152. 1859.

Bas.: *Brosimum spurium* Sw. (1788).

Tree in Northern Limestone Hills, Río Abajo Forest Reserve, Sierra de Luquillo, Eastern Coast (Naguabo), Maricao Forest Reserve; 50–325 m; *fl* Mar (one specimen), *fr* (?).

Greater Antilles, Mexico, Central America, northern South America. [L1: 68].

Negra lora.

Trophis racemosa (L.) Urb., Symb. Antill. 4(2):195. 1905.

Bas.: *Bucephalon racemosum* L. (1753).

Tree in Northern Coastal Lowlands, Northern Limestone Hills, Río Abajo Forest Reserve, Maricao Forest Reserve; 25–275 m; *fl* Jun–Jan, *fr* Feb–Mar.

Greater Antilles, Mexico, Central America, northwestern, west-central South America and Venezuela. [L1: 69].

Ramón, Ramoncillo.

TAXA CITED BY PREVIOUS AUTHORS, BUT NOT INCLUDED

ONLY UNDER CULTIVATION

Artocarpus heterophyllus Lam.

Tree cited by Liogier & Martorell (1982, 2000) and Liogier (1985) as occasionally planted and naturalized. [L1: 52]

Notes: Britton & Wilson (1924) misapplied the name *Artocarpus integrifolius* L. to this taxon..

Ficus benjamina L.
Tree cited by Liogier & Martorell (1982, 2000) and Liogier (1985) as planted for shade and ornament and persistent. [L1: 61].

Ficus carica L.
Tree cited by Liogier & Martorell (1982, 2000) and Liogier (1985) as planted for its fruits and persistent. [L1: 61].

Ficus drupacea Thunb.
Tree cited by Liogier & Martorell (1982, 2000) and Liogier (1985) as rarely planted for shade and persistent. [L1: 63].

Notes: Britton & Wilson (1926) used *Ficus mysorensis* Roth ex Roem. & Schult. for the name of this taxon.

Ficus lyrata Warb.
Tree cited by Liogier & Martorell (1982, 2000) and Liogier (1985) as sometimes planted for shade and ornament. [L1: 63].

Ficus religiosa L.
Tree cited by Liogier & Martorell (1982, 2000) and Liogier (1985) as uncommon as an ornamental and shade tree. [L1: 65].

Morus nigra L.
Tree cited by Liogier & Martorell (1982, 2000) and Liogier (1985) as rarely planted as a shade and fruit tree. [L1: 66].

MORINGACEAE

Moringa oleifera Lam., Encycl. 1(2):398. 1785.
Syn.: *Moringa moringa* (L.) Millsp. (1902).

Tree planted and persistent in Northern Coastal Lowlands (Manatí, Aguadilla), Southeastern Foothills (Salinas), Eastern Cordillera (Cayey), Southern Coastal Lowlands (Juana Díaz), Guánica Forest Reserve, Southwestern Coastal Lowlands (San Germán), Vieques, Culebra; 20–240 m; *fl* Aug–Nov, Feb–May, *fr* Feb–May, Aug–Nov.

Native to Indian Subcontinent. [L1: 305].

Angela, Ben, Jazmín francés, Resedá, Sen.

MUNTINGIACEAE

Muntingia calabura L., Sp. Pl. 1:509. 1753.

Small tree in Northern Coastal Lowlands, Northern Limestone Hills, Río Abajo Forest Reserve, Sierra de Luquillo, Caguas Valley (Caguas), Southeastern Foothills (Salinas), Southern Coastal Lowlands, Southern Coastal Limestone, Western Coast; 1–150 m; *fl* & *fr* throughout the year.

Bahamas, Greater Antilles, Trinidad and Tobago, Mexico, Central America, northwestern, west-central South America and Venezuela, Argentina. [L3: 104].

Capulín.

MUSACEAE

Musa ×paradisiaca L. (pro sp.), Sp. Pl. 2:1043. 1753.

Cultivated and escaped throughout the island, rarely collected (because of the common and unmistakable occurrence of this plant, I make an exception to vouchering).

Only cultivated, of obscure Paleotropical origin. [AR&S: 173].

Notes: Cultivated bananas and plantains are triploid and/or interspecific hybrids with varying contributions from each parent so that binomial nomenclature cannot be applied. Among the parents contributing are *Musa acuminata* Colla and *Musa balbisiana* Colla. Triploid plants derived from the former include *Musa cavendishii* Lamb.

Britton & Wilson (1924) used the name *Musa paradisiaca* L. (1753) for the plátano; *Musa sapientum* L. for the guineo; and *Musa cavendishii* Lamb. for the niño. Acevedo-Rodríguez in Acevedo-Rodríguez & Strong (2005) has a discussion of the classification of these cultivars, which are designated by letters, and of its relation to the many common names used for these plants in Puerto Rico.

Guineo, Manzano, Niño, Plátano (these are the most common among many others).

TAXA CITED BY PREVIOUS AUTHORS, BUT NOT INCLUDED

ONLY UNDER CULTIVATION

Musa velutina H. Wendl. & Drude.
Herb cited by Liogier & Martorell (2000) as cultivated and naturalized in wet regions.

MYRICACEAE

Morella cerifera (L.) Small, Fl. S.E. U.S. 337, 1329. 1903.
Bas.: *Myrica cerifera* L. (1753).

Tree in Northern Coastal Lowlands (Dorado to Laguna Tortuguero), Northern Limestone Hills (Arecibo), Río Abajo Forest Reserve, Sierra de Luquillo, Central Cordillera (Adjuntas), Western Cordillera, Cerro Las Mesas; 2–700 m; *fl* Feb–Apr, Jul–Oct, *fr* Aug–Nov, Feb.

Bahamas, Greater Antilles, southeastern and south-central United States, Mexico, Central America. [L1: 40].

Arrayán, Cerero.

Morella holdridgeana (Lundell) Kartesz, Synthesis N. Amer. Fl., version 1 Nomencl. Innov.: no. 16. 1999.
Bas.: *Myrica holdridgeana* Lundell (1942) "Holdrigeana."
Tree in Sierra de Luquillo, Toro Negro Forest Reserve, Maricao Forest Reserve, Cerro Las Mesas; 225–1050 m; *fl* Feb–Sep, *fr* Jan (one specimen).
Endemic. [L1: 41].
Palo de cera.

MYRISTICACEAE

TAXA CITED BY PREVIOUS AUTHORS, BUT NOT INCLUDED

ONLY UNDER CULTIVATION
Myristica fragrans Houtt.
Tree cited by Liogier & Martorell (1982, 2000) and Liogier (1985) as planted for its fruit and persistent. [L1: 247].

MYRSINACEAE

Ardisia elliptica Thunb., Nov. Gen. Pl. 8:119. 1798.
Shrub in Northern Coastal Lowlands, Northern Limestone Hills, Eastern Coast (Humacao), Eastern Cordillera (Aguas Buenas), Western Coast (Mayagüez); 25–325 m; *fl* May–Jun, *fr* Sep–Mar.
Native to tropical Asia, Japan. [L4: 119].
Notes: Liogier (1995) and Liogier & Martorell (2000) misapplied the name *Ardisia solanacea* Roxb. to this taxon.

Ardisia glauciflora Urb., Symb. Antill. 1(2):382. 1899.
Syn.: *Icacorea glauciflora* (Urb.) Britton (1925).
Tree in Sierra de Luquillo, Sierra de Cayey; 275–750 m; *fl* Mar–Jun, *fr* May–Oct.
Endemic. [L4: 117].
Ausubón, Mameyuelo.

Ardisia luquillensis (Britton) Alain, Bull. Torrey Bot. Club 90(3):191. 1963.
Bas.: *Icacorea luquillensis* Britton (1925).
Tree in Sierra de Luquillo, Sierra de Cayey; 650–1050 m; *fl* May–Feb, *fr* Aug–May.
Endemic. [L4: 118].
Bádula, Mameyuelo.

Ardisia obovata Ham., Prodr. Pl. Ind. Occid. 26. 1825.
Syn.: *Icacorea guadalupensis* (Duchass. ex Griseb.) Britton & P. Wilson (1917).
Tree in Northern Coastal Lowlands, Northern Limestone Hills, Río Abajo Forest Reserve, Sierra de Luquillo, Eastern Coast, Sierra de Cayey, Las Tetas de Cayey, Las Piedras Chiquitas, Southeastern Foothills, Eastern Cordillera, Southern Coastal Lowlands (Patillas), Susúa Forest Reserve, Maricao Forest Reserve, Guánica Forest Reserve, Southwestern Coastal Lowlands (Cabo Rojo), Sierra Bermeja, Western Coast (Rincón), Vieques; 10–850 m; *fl* throughout the year, *fr* Oct–Mar, Jun–Jul.
Bahamas, Greater Antilles (Hispaniola), Virgin Islands, Lesser Antilles. [L4: 118].
Bádula, Mameyuelo.

Cybianthus sintenisii (Urb.) G. Agostini, Acta Biol. Venez. 10(2):166. 1980.
Bas.: *Ardisia sintenisii* Urb. (1899).
Syn.: *Grammadenia sintenisii* (Urb.) Mez (1901).
Shrub/small tree in Sierra de Luquillo; 350–1000 m; *fl* Jul–Nov, *fr* Dec, Apr (two specimens).
Endemic. [L4: 120].

Myrsine coriacea (Sw.) R. Br. ex Roem. & Schult., Syst. Veg. 4:511. 1819.
Bas.: *Samara coriacea* Sw. (1788).
Syn.: *Rapanea ferruginea* (Ruiz & Pav.) Mez (1901).
Tree in Río Abajo Forest Reserve, Sierra de Luquillo, Caguas Valley (Caguas), Sierra de Cayey, Las Tetas de Cayey, Eastern Cordillera (La Torrecilla), Central Cordillera, Maricao Forest Reserve; 300–1300 m; *fl* Oct–May, *fr* Dec–Aug.
Antilles, Trinidad, Mexico, Central America, South America. [L4: 122].
Arrayán, Arrayán bobo, Bádula, Cucubano, Mameyuelo, Mantecado, Mantequero, Palo blanco.

Myrsine cubana A. DC., Ann. Sci. Nat., Bot., sér. 2 16:86. 1841.
Syn.: *Myrsine floridana* A. DC. (1834).
Rapanea punctata (Lam.) Lundell (1969).
Tree in Northern Coastal Lowlands, Northern Limestone Hills, Río Abajo Forest Reserve, Susúa Forest Reserve, Sierra Bermeja, Cerro Las Mesas; 1–350 m; *fl* Dec–Sep, *fr* throughout the year.
Bahamas, Greater Antilles (Cuba), southeastern United States (Florida), Mexico, Central America. [L4: 122].
Notes: Britton & Wilson (1925) misapplied the name *Rapanea guianensis* Aubl. to this taxon; subsequently Liogier & Martorell (1982) misapplied the name *Myrsine guianensis* (Aubl.) Kuntze to it.
Bádula, Mameyuelo.

Parathesis crenulata (Vent.) Hook. f. ex Hemsl., Biol. Cent.-Amer., Bot. 2(10):291. 1881.
Bas.: *Ardisia crenulata* Vent. (1803).
Syn.: *Parathesis serrulata* (Sw.) Mez (1901).
Shrub in Northern Coastal Lowlands, Northern Limestone Hills, Río Abajo Forest Reserve, Sierra de Cayey, Eastern Cordillera (Aguas Buenas), Southern Coastal Limestone (Guayanilla), vicinity of Susúa Forest Reserve (Yauco), Western Cordillera; 25–850 m; *fl* Apr–Jul, *fr* Apr–Jan.
Greater Antilles (Hispaniola), Lesser Antilles (Martinique). [L4: 124].
Rascagarganta, Secagarganta.

Wallenia lamarckiana (A. DC.) Mez in Urban, Symb. Antill. 2(3):414. 1901.
Bas.: *Badula lamarckiana* A. DC. (1844).
Syn.: *Wallenia pendula* (Urb.) Mez (1901). *Petesioides pendulum* (Urb.) Britton (1925).
Shrub in Northern Limestone Hills, Río Abajo Forest Reserve, Sierra de Luquillo, Sierra de Cayey, Las Tetas de Cayey, Eastern Cordillera (La Torrecilla), Central Cordillera, Maricao Forest Reserve; 150–1200 m; *fl* Oct–Jan, *fr* Nov–Apr.
Lesser Antilles (Martinique). [L4: 127].
Jacanillo, Quiebrahacha, Jácana macho.

Wallenia yunquensis (Urb.) Mez in Urban, Symb. Antill. 2(3):413. 1901.
Bas.: *Ardisia yunquensis* Urb. (1899).
Syn.: *Petesioides yunquense* (Urb.) Britton (1925).
Tree in Sierra de Luquillo, Toro Negro Forest Reserve; 900–1250 m; *fl* throughout the year, *fr* Oct–Apr.
Endemic. [L4: 129].

TAXA CITED BY PREVIOUS AUTHORS, BUT NOT INCLUDED

DOUBTFUL OCCURRENCE

Stylogyne lateriflora (Sw.) Mez.
Shrub cited by Britton & Wilson (1925) as occurring in the Sierra de Luquillo, in forest near Bañadero, at 600 m, collected only by Eggers (according to Mez); subsequently cited by Liogier & Martorell (1982, 2000) and Liogier (1995). [L4: 126].

MYRTACEAE

Calyptranthes acevedoi Alain, Phytologia 61(6):353. 1986.
Shrub/small tree in Northern Limestone Hills, Río Abajo Forest Reserve; 200–395 m; *fl* Jan–Mar, Jun, *fr* May–Jul.
Endemic. [L3: 383].

Calyptranthes estremerae Alain, Phytologia 58(5):325. 1985 "estremenae."
Shrub/small tree in Northern Limestone Hills, Río Abajo Forest Reserve; 250–350 m; *fl* Jun–Jul (two specimens), *fr* (?).
Endemic. [L3: 384].

Calyptranthes kiaerskovii Krug & Urb. in Urban, Symb. Antill. 1(1):43. 1898.
Shrub in Northern Limestone Hills (Camuy); 100–200 m; *fl* Apr (one specimen), *fr* (?). [*Woodbury a.n. 2290* (UPR); *Popenoe 2441* (SJ)].
Virgin Islands (Virgin Gorda, Tortola). [B&W6: 32, cited only for Tortola].

Calyptranthes krugii Kiaersk., Bot. Tidsskr. 17:248, *t. 11C, f. 1.* 1890.
Syn.: *Calyptranthes martorellii* Alain (1985).
Tree in Sierra de Luquillo, Sierra de Cayey, Central Cordillera; 575–1250 m; *fl* throughout the year, *fr* Sep–Mar.
Endemic. [L3: 384].
Hoja menuda, Limoncillo.

Calyptranthes luquillensis Alain, Bull. Torrey Bot. Club 90(3):189. 1963.
Tree in Sierra de Luquillo; 650 m; *fl* Jun–Jul (two specimens), *fr* (?). [*Woodbury a.n. 2302* (UPR); *Liogier 36948* (UPR); *Proctor 39490, 47715* (SJ)].
Endemic. [L3: 384].

Calyptranthes pallens Griseb. var. **pallens,** Abh. Königl. Ges. Wiss. Göttingen 7:215. 1857 (Syst. Veg. Karaiben 67).
Tree in Northern Coastal Lowlands, Northern Limestone Hills, Guánica Forest Reserve, Mona; 5–70 m; *fl* May–Jul, *fr* Jul–Nov.
Bahamas, Greater Antilles, Lesser Antilles (Guadeloupe, Marie Galante), southeastern United States (Florida). [L3: 385].
Tapón blanco.

Calyptranthes peduncularis Alain, Bull. Torrey Bot. Club 90(3):189. 1963.
Syn.: *Calyptranthes dumetorum* Alain (1965).
Shrub/small tree in Susúa Forest Reserve, Maricao Forest Reserve; 150–600 m; *fl* May–Jun, *fr* Jun–Sep.
Endemic. [L3: 385].

Calyptranthes sintenisii Kiaersk., Bot. Tidsskr. 17:250, *t. 11B.* 1890.
Shrub in Northern Coastal Lowlands (Dorado), Northern Limestone Hills, Río Abajo Forest Reserve, Central Cordillera, Guánica Forest Reserve; 1–950 m; *fl* Apr–Jul, *fr* May–Sep.

Greater Antilles (Hispaniola), Virgin Islands (St. Thomas). [L3: 387].

Hoja menuda, Limoncillo, Limoncillo del monte.

Calyptranthes triflora Alain, Bull. Torrey Bot. Club 90(3):189. 1963 "triflorum."

Shrub in Maricao Forest Reserve; 465–920 m; *fl* May–Jul, *fr* Nov (one specimen).

Greater Antilles (Hispaniola). [L3: 388].

Calyptranthes woodburyi Alain, Phytologia 54(2): 107. 1983.

Tree in Sierra de Luquillo, Sierra de Cayey, Maricao Forest Reserve; 400–845 m; *fl* Apr–Aug, *fr* Jun–Nov.

Endemic. [L3: 388].

Calyptranthes zuzygium (L.) Sw., Prodr. 79. 1788.
Bas.: *Myrtus zuzygium* L. (1759).

Shrub in Northern Limestone Hills, Río Abajo Forest Reserve, Sierra de Cayey; 25–800 m; *fl* May–Jun (two specimens), *fr* Aug (one specimen).

Bahamas, Greater Antilles, southeastern United States (Florida). [L3: 388].

Eucalyptus citriodora Hook. in T.L. Mitchell, J. Exped. Trop. Australia 235. 1848.
Syn.: *Corymbia citriodora* (Hook.) K.D. Hill & L.A.S. Johnson (1995).

Tree planted and persistent in Northern Coastal Lowlands (San Juan), Western Coast (Mayagüez); 5–35 m; *fl* Feb, Jun (two specimens), *fr* Jun (one specimen).

Native to Australia. [L3: 390].

Eucalipto, Eucalipto oloroso, Eucalipto de limón, Eucalipto del pantano.

Eucalyptus deglupta Blume, Mus. Bot. 1:83. 1850.

Tree planted and persistent in Northern Coastal Lowlands (San Juan, Fajardo), Western Coast (Aguadilla); 1–50 m; *fl* May–Sep, *fr* (?).

Native to Malesia. [L3: 390].

Eucalipto.

Eucalyptus robusta Sm., Spec. Bot. New Holland 39, *t. 13*. 1795.

Tree planted and persistent in Northern Coastal Lowlands (San Juan), Sierra de Luquillo, Sierra de Cayey, Toro Negro Forest Reserve, Maricao Forest Reserve; 1–1250 m; *fl* Jun–Mar, *fr* Jun–Feb.

Native to Australia. [L3: 391].

Alcanfor, Eucalipto, Eucalipto de alcanfor, Eucalipto de pantano, Eucalipto medicinal.

Eucalyptus robusta ×tereticornis.
Syn.: *Eucalyptus ×kirtoniana* F. Muell (pro sp.) (1879).

Tree planted and persistent in Northern Coastal Lowlands (San Juan, Trujillo Alto), Sierra de Luquillo, Sierra de Cayey, Maricao Forest Reserve; 35–800 m; *fl* Jun–Nov, *fr* Apr, Nov.

Native to Australia.

Eucalyptus saligna Sm., Trans. Linn. Soc. London 3:285. 1797.

Tree planted and persistent in Sierra de Luquillo; 500 m; *fl* (?), *fr* Jun (one specimen).

Native to Australia.

Eucalyptus urophylla S.T. Blake, Austrobaileya 1(1):7. 1977.

Tree planted and persistent in Sierra de Luquillo, Sierra de Cayey, Toro Negro Forest Reserve, Susúa Forest Reserve, Maricao Forest Reserve, Western Coast (Mayagüez); 1–800 m; *fl* Jan (one specimen), *fr* Feb–Aug.

Native to Indonesia.

Eucalipto, Eucalipto medicinal.

Eugenia axillaris (Sw.) Willd., Sp. Pl. 2(2):960. 1799.
Bas.: *Myrtus axillaris* Sw. (1788).

Shrub/small tree in Northern Coastal Lowlands, Northern Limestone Hills, Southeastern Foothills (Salinas), Guánica Forest Reserve, Sierra Bermeja, Western Coast (Rincón), Vieques, Mona, Desecheo; 1–100 m; *fl* May–Aug, *fr* throughout the year.

West Indies, southeastern United States (Florida), Mexico, northern Central America. [L3: 395].

Grajo, Murta.

Eugenia biflora (L.) DC., Prodr. 3:276. 1828.
Bas.: *Myrtus biflora* L. (1759).
Syn.: *Eugenia lancea* Poir. (1813).
Eugenia boqueronensis Britton (1924).
Eugenia underwoodii Britton (1924).

Tree in Northern Coastal Lowlands, Northern Limestone Hills, Río Abajo Forest Reserve, Eastern Coast, Las Tetas de Cayey, Las Piedras Chiquitas, Southeastern Foothills (Coamo), Eastern Cordillera (Cerro La Peña, Corozal), Central Cordillera, Southern Coastal Lowlands, Southern Coastal Limestone (Guayanilla), Susúa Forest Reserve, Western Cordillera, Southwestern Coastal Lowlands, Sierra Bermeja, Cerro Las Mesas, Vieques, Culebra, Mona; 1–875 m; *fl* Jan–Jun, *fr* Apr–Dec.

Greater Antilles (Jamaica, Hispaniola), Virgin Islands, Lesser Antilles, Mexico, Central America, northwestern, west-central South America and Venezuela. [L3: 396, 406].

Notes: Since the leaves of this taxon vary considerably in shape and texture throughout its range, both *Eugenia boqueronensis* and *Eugenia underwoodii*, formerly considered distinct taxa because of their leaf shape and texture, are here treated as conspecific with the above taxon.

Hoja menuda, Pitangueira.

Eugenia borinquensis Britton in Britton & Wilson, Sci. Surv. Porto Rico & Virgin Islands 6(1):38. 1925.

Tree in Sierra de Luquillo, Sierra de Cayey; 775–1050 m; *fl* Mar–Apr, Aug–Oct, *fr* Sep–Jan, Apr.

Endemic. [L3: 396].

Guayabota, Guayabota de sierra, Limoncillo, Oreja.

Eugenia confusa DC., Prodr. 3:279. 1828.

Shrub in Northern Limestone Hills, Susúa Forest Reserve, Maricao Forest Reserve; 60–900 m; *fl* Mar–Oct, *fr* May, Aug–Dec.

West Indies, southeastern United States (Florida). [L3: 398].

Caracolillo, Cieneguillo.

Eugenia cordata (Sw.) DC. var. **cordata**, Prodr. 3:272. 1828.

Bas.: *Myrtus cordata* Sw. (1788).

Shrub/small tree in Northern Coastal Lowlands (islands adjacent to Fajardo), Vieques, Culebra; 1–300 m; *fl* (?), *fr* (?).

Virgin Islands, Lesser Antilles (St. Barts). [L3: 398].

Eugenia cordata (Sw.) DC. var. **sintenisii** (Kiaersk.) Krug & Urb., Bot. Jahrb. Syst. 19(5):656. 1895.

Bas.: *Eugenia sintenisii* Kiaersk. (1890).

Shrub/small tree in Northern Limestone Hills (Arecibo, Quebradillas), Sierra de Luquillo, Las Tetas de Cayey, Eastern Cordillera (Cayey), Susúa Forest Reserve, Maricao Forest Reserve; 175–700 m; *fl* Apr (one specimen), *fr* Apr–Jun (two specimens).

Virgin Islands, Lesser Antilles. [L3: 399].

Murta.

Eugenia domingensis O. Berg, Linnaea 27(2, 3):296. 1856.

Shrub/small tree in Northern Coastal Lowlands, Northern Limestone Hills (Quebradillas, Isabela), Río Abajo Forest Reserve, Sierra de Luquillo, Eastern Coast (Yabucoa), Sierra de Cayey, Central Cordillera (Ponce), Maricao Forest Reserve; 1–500 m; *fl* Jun–Aug, *fr* Mar, Oct–Nov.

Greater Antilles (Cuba, Hispaniola), Lesser Antilles, Trinidad. [L3: 399].

Notes: Britton & Wilson (1924) misapplied the name *Eugenia aeruginea* DC. to, and Liogier & Martorell (1982) wrongly placed it in synonymy with, this taxon.

Guasábara, Guayabacón, Palo de murta.

Eugenia eggersii Kiaersk., Bot. Tidsskr. 17:268, *t. 12, f. 5.* 1890.

Syn.: *Eugenia cacuminis* Alain (1986).

Shrub/small tree in Sierra de Luquillo, Sierra de Cayey, Las Tetas de Cayey, Central Cordillera; 400–1200 m; *fl* & *fr* throughout the year.

Endemic. [L3: 398, 400].

Guasábara, Guayabacón, Palo de murta.

Eugenia fajardensis (Krug & Urb.) Urb., Symb. Antill. 9(1):109, 513. 1923.

Bas.: *Eugenia fragrans* (Sw.) Willd. var. *fajardensis* Krug & Urb. (1895).

Tree in Northern Coastal Lowlands (Fajardo), Vieques, Culebra; 1–275 m; *fl* Aug–Sep (two specimens), *fr* Jan (one specimen).

Endemic. [B&W6: 364].

Notes: Liogier & Martorell (1982) and Liogier (1994) wrongly placed this taxon in synonymy with *Myrcianthes fragrans*.

Eugenia foetida Pers., Syn. Pl. 2(1):29. 1806.

Syn.: *Eugenia buxifolia* (Sw.) Willd. (1799).

Tree in Northern Limestone Hills, Río Abajo Forest Reserve, Southeastern Foothills (Salinas), Southern Coastal Lowlands (Caja de Muertos), Southern Coastal Limestone (Guayanilla), Guánica Forest Reserve, Southwestern Coastal Lowlands (Cabo Rojo), Vieques, Culebra, Mona, Desecheo; 1–325 m; *fl* May–Feb, *fr* May–Dec.

Bahamas, Greater Antilles, Virgin Islands, Lesser Antilles (Anguilla, St. Martin, St Barts), southeastern United States (Florida), southern Mexico (Yucatan), northern Central America (Belize, Guatemala). [L3: 400].

Anguila, Birijí, Hoja menuda, Pico de paloma.

Eugenia glabrata (Sw.) DC., Prodr. 3:274. 1828.

Bas.: *Myrtus glabrata* Sw. (1788).

Shrub/small tree in Northern Coastal Lowlands (Carolina), Northern Limestone Hills, Southwestern Coastal Lowlands (Cabo Rojo); 10–350 m; *fl* Mar–Apr, Sep, *fr* Mar–Apr (two specimens).

Greater Antilles. [L3: 400].

Eugenia haematocarpa Alain, Bull. Torrey Bot. Club 90(3):190. 1963.

Tree in Northern Limestone Hills (Isabela, Moca), Sierra de Luquillo, Sierra de Cayey; 180–700 m; *fl* May–Jul, *fr* May–Jul, Dec–Jan.

Endemic. [L3: 401].

Notes: This taxon is on the U.S. Fish & Wildlife Service list of threatened and endangered plants.

Uvillo.

Eugenia laevis O. Berg, Linnaea 27(2,3):177. 1856.

Shrub/small tree in Northern Limestone Hills, Southeastern Foothills (Salinas); 30–150 m; *fl* Apr–Jun (two specimens), *fr* Apr–Jun, Sep–Jan.

Greater Antilles (Hispaniola), Mexico, Central America. [L3: 401].

Eugenia ligustrina (Sw.) Willd., Sp. Pl. 2(2):962. 1799.

Bas.: *Myrtus ligustrina* Sw. (1788).

Shrub in Northern Coastal Lowlands (Fajardo and adjacent islands), Northern Limestone Hills, Las Tetas de Cayey, Southeastern Foothills (Salinas), Central Cordillera (Ponce), Southern Coastal Limestone, Guánica Forest Reserve, Southwestern Coastal Lowlands, Sierra Bermeja, Western Coast (Mayagüez), Vieques, Culebra; 5–825 m; *fl* Feb–Sep, *fr* May–Jan.

Antilles, Trinidad and Tobago, South America (Suriname, Venezuela, Brazil, Bolivia). [L3: 402].

Granadillo, Hoja menuda, Palo de muleta.

Eugenia monticola (Sw.) DC., Prodr. 3:275. 1828.

Bas.: *Myrtus monticola* Sw. (1788).

Syn.: *Eugenia maleolens* Pers. (1806).

Shrub/small tree in Northern Coastal Lowlands, Northern Limestone Hills, Río Abajo Forest Reserve, Sierra de Cayey, Las Tetas de Cayey, Las Piedras Chiquitas, Southeastern Foothills, Eastern Cordillera (Aibonito), Central Cordillera (Ponce), Southern Coastal Lowlands, Southern Coastal Limestone (Ponce), Susúa Forest Reserve, Maricao Forest Reserve, Guánica Forest Reserve, Southwestern Coastal Lowlands (Lajas), Sierra Bermeja, Vieques, Culebra, Mona; 1–825 m; *fl* throughout the year, *fr* Sep–Dec.

Antilles, Trinidad and Tobago, northern South America to Ecuador. [L3: 402].

Birijí, Espejuelo, Hoja menuda, Hoja menuda colorada.

Eugenia padronii Alain, Phytologia 61(6):358. 1986.

Shrub/small tree in Guajataca Forest Reserve, Río Abajo Forest Reserve, Susúa Forest Reserve, Maricao Forest Reserve; 150–800 m; *fl* Oct–Mar, *fr* Jan–Mar (two specimens).

Endemic. [L3: 402].

Eugenia procera (Sw.) Poir., Encycl. Suppl. 3(1):129. 1813.

Bas.: *Myrtus procera* Sw. (1788).

Shrub/small tree in Northern Limestone Hills (Aguada, Rincón), Southeastern Foothills, Southern Coastal Lowlands (Caja de Muertos), Guánica Forest Reserve, Sierra Bermeja, Western Coast (Mayagüez), Vieques; 1–300 m; *fl* Jan, Feb (two specimens), *fr* Jul–Sep, Jan–Feb.

Greater Antilles (Cuba, Hispaniola), Virgin Islands, Lesser Antilles, northern South America (Venezuela, Colombia). [L3: 403].

Hoja menuda.

Eugenia pseudopsidium Jacq., Enum. Syst. Pl. 23. 1760.

Shrub in Northern Coastal Lowlands, Northern Limestone Hills, Río Abajo Forest Reserve, Eastern Coast (Punta Guayanés), Las Tetas de Cayey, Eastern Cordillera, Susúa Forest Reserve, Maricao Forest Reserve, Sierra Bermeja, Vieques; 5–825 m; *fl* Jul–Feb, *fr* Aug–Apr.

Greater Antilles (Hispaniola), Virgin Islands, Lesser Antilles, northern South America and Brazil. [L3: 403].

Guayaba silvestre, Quiebrahacha, Pitangueira.

Eugenia rhombea (O. Berg) Krug & Urb., Bot. Jahrb. Syst. 19(5):644. 1895.

Bas.: *Eugenia foetida* Pers. var. *rhombea* O. Berg (1856).

Shrub/small tree in Northern Coastal Lowlands (islands adjacent to Fajardo, Quebradillas, Isabela), Southern Coastal Lowlands (Caja de Muertos), Southern Coastal Limestone (Peñuelas), Guánica Forest Reserve, Southwestern Coastal Lowlands (Cabo Rojo), Mona, Desecheo; 5–150 m; *fl* throughout the year, *fr* Jul, Oct–Feb.

West Indies, southeastern United States (Florida), Mexico, Central America, northern South America (Colombia). [L3: 404].

Guayabilla de costa, Hoja menuda.

Eugenia sessiliflora Vahl, Symb. Bot. 3:64. 1794.

Shrub/small tree in Northern Coastal Lowlands (Fajardo), Las Piedras Chiquitas, Southeastern Foothills, Southwestern Coastal Lowlands (Cabo Rojo), Sierra Bermeja, Vieques, Culebra; 1–550 m; *fl* Apr–Dec, *fr* May–Oct.

Virgin Islands. [L3: 405].

Eugenia stahlii (Kiaersk.) Krug & Urb., Bot. Jahrb. Syst. 19(5):650. 1895.
Bas.: *Myrtus stahlii* Kiaersk. (1890).
Shrub/small tree in Sierra de Luquillo, Sierra de Cayey, Central Cordillera, Maricao Forest Reserve; 350–1300 m; *fl* Feb, May–Jul, *fr* Aug–Sep.
Endemic. [L3: 405].
Ausú, Guayabota, Limoncillo.

Eugenia stewardsonii Britton, Bull. Torrey Bot. Club 51(1):11. 1924.
Tree in Sierra de Cayey, Eastern Cordillera (La Torrecilla), Central Cordillera, Susúa Forest Reserve, Western Cordillera; 500–1225 m; *fl* Dec–Apr, *fr* Mar, Jun.
Greater Antilles (Hispaniola). [L3: 405].
Hoja menuda.

Eugenia woodburyana Alain, Phytologia 47(3):185. 1980.
Shrub/small tree in Southern Coastal Limestone (Guayanilla, Yauco), Guánica Forest Reserve, Southwestern Coastal Lowlands (Cabo Rojo), Sierra Bermeja; 25–300 m; *fl* Aug–Feb, May, *fr* throughout the year.
Endemic. [L3: 407].
Notes: This taxon is on the U.S. Fish & Wildlife Service list of threatened and endangered plants.

Gomidesia lindeniana O. Berg, Linnaea 29(2):208. 1858.
Shrub/small tree in Sierra de Cayey, Las Tetas de Cayey, Eastern Cordillera (Guaynabo, Cerro La Peña), Central Cordillera, Maricao Forest Reserve; 475–1250 m; *fl* Jan, May–Aug, *fr* Nov–Aug.
Antilles, South America (southeastern Brazil). [L3: 408].
Cieneguillo.

Marlierea sintenisii Kiaersk., Bot. Tidsskr. 17:252, *t. 13A, f. 2.* 1890.
Syn.: *Plinia sintenisii* (Kiaersk.) Britton (1925).
Shrub in Sierra de Luquillo; 750–1000 m; *fl* Apr–Sep, *fr* Sep, Dec–Mar.
Endemic. [L3: 409].
Beruquillo.

Melaleuca quinquenervia (Cav.) S.T. Blake, Proc. Roy. Soc. Queensland 69:76. 1958.
Bas.: *Metrosideros quinquenervia* Cav. (1797).
Tree in Northern Coastal Lowlands (reported from San Juan and Laguna Tortuguero); 1–5 m; *fl* Sep–Jun, *fr* May–Sep.
Native to Papua New Guinea, Australia, New Caledonia. [L3: 411].
Notes: Since this taxon is an aggressive and easily recognizable invader, I have made an exception for the absence of vouchered specimens.
Aceite de Cayeput, Bálsamo de Cayeput, Cayeputi.

Mosiera longipes (O. Berg) Small, Man. S.E. Fl. 937. 1933.
Bas.: *Eugenia longipes* O. Berg (1856).
Syn.: *Eugenia bellonis* Krug & Urb. (1895). *Myrtus bellonis* (Krug & Urb.) Burret (1941).
Psidium longipes (O. Berg) McVaugh var. *orbiculare* (O. Berg) McVaugh (1973).
Psidium insulanum Alain (1980).
Shrub in Southern Coastal Lowlands (Caja de Muertos, Ponce), Guánica Forest Reserve, Vieques; 1–25 m; *fl* Jun, Dec (two specimens), *fr* May (one specimen).
Bahamas, Greater Antilles (Jamaica, Hispaniola), northern Lesser Antilles, southeastern United States (Florida). [L3: 422, 431].
Notes: Andrew Salywon (pers. comm.) makes *Eugenia bellonis* conspecific with *Mosiera longipes* because of the similarity of the type of the former to juvenile plants of the latter.

Mosiera xerophytica (Britton) Salywon, J. Bot. Res. Inst. Texas 1(2):900. 2007.
Syn.: *Eugenia xerophytica* Britton (1924).
Shrub/small tree in Southern Coastal Lowlands (Caja de Muertos), Southern Coastal Limestone, Guánica Forest Reserve, Southwestern Coastal Lowlands (Cabo Rojo); 1–150 m; *fl* Mar–Oct, *fr* May–Nov.
Endemic. [L3: 407].
Guayabacón, Hierro.

Myrcia citrifolia (Aubl.) Urb. var. **citrifolia**, Repert. Spec. Nov. Regni Veg. 16:150. 1919.
Bas.: *Myrtus citrifolia* Aubl. (1775).
Tree in Northern Coastal Lowlands, Northern Limestone Hills (Quebradillas, Isabela), Eastern Coast (Punta Guayanés), Las Tetas de Cayey, Eastern Cordillera (Cerro La Peña), Susúa Forest Reserve, Western Cordillera, Sierra Bermeja, Cerro Las Mesas; 1–900 m; *fl* Feb–Aug, *fr* Apr–Dec.
Greater Antilles (Hispaniola), Virgin Islands, Lesser Antilles, northern South America. [L3: 413].
Hoja menuda, Limoncillo del monte.

Myrcia citrifolia (Aubl.) Urb. var. **imrayana** (Griseb.) Stehlé & Quentin, Fl. Guadeloupe 2(3):57. 1949.
Bas.: *Myrcia coriacea* (Vahl) DC. var. *imrayana* Griseb. (1860).

Tree in Sierra de Luquillo, Sierra de Cayey, Las Tetas de Cayey, Susúa Forest Reserve, Central Cordillera (Yauco), Maricao Forest Reserve; 250–1000 m; *fl* Mar–Jul, *fr* Jan, Apr–Oct.

Virgin Islands, Lesser Antilles. [L3: 414].

Myrcia deflexa (Poir.) DC., Prodr. 3:244. 1828.

Bas.: *Eugenia deflexa* Poir. (1813).

Tree in Northern Coastal Lowlands (Guaynabo), Sierra de Luquillo, Sierra de Cayey, Eastern and Central Cordillera, Susúa Forest Reserve, Maricao Forest Reserve, Cerro Las Mesas; 275–925 m; *fl* throughout the year, *fr* Sep–Jun.

Greater Antilles (Cuba, Hispaniola), Lesser Antilles, Trinidad, northern South America to Bolivia. [L3: 414].

Cieneguillo.

Myrcia fallax (Rich.) DC., Prodr. 3:244. 1828.

Bas.: *Eugenia fallax* Rich. (1792).

Syn.: *Myrcia berberis* DC. (1828).

Tree in Sierra de Luquillo, Sierra de Cayey, Central Cordillera; 575–1300 m; *fl* Apr–Aug, *fr* Aug–Jan.

Lesser Antilles, Trinidad and Tobago, southern Central America (Panama), northwestern, west-central South America and Venezuela. [L3: 415].

Hoja menuda, Rama menuda.

Myrcia leptoclada DC., Prodr. 3:244. 1828.

Tree in Northern Limestone Hills, Sierra de Luquillo, Sierra de Cayey, Maricao Forest Reserve; 200–1000 m; *fl* May–Jul, *fr* Mar–Apr, Jul.

Greater Antilles (Jamaica, Hispaniola), Lesser Antilles, Trinidad, Central America. [L3: 415].

Guayabacón, Hoja menuda roja.

Myrcia margarettae (Alain) Alain, Phytologia 58(5): 328. 1985.

Bas.: *Eugenia margarettae* Alain (1963).

Shrub in Northern Limestone Hills (Arecibo, Isabela), Río Abajo Forest Reserve, Central Cordillera (Orocovis); 100–350 m; *fl* Jul (one specimen), *fr* (?).

Endemic. [L3: 416].

Myrcia paganii Krug & Urb., Bot. Jahrb. Syst. 19(5): 587. 1895 "Pagani."

Tree in Northern Limestone Hills, Río Abajo Forest Reserve; 150–250 m; *fl* (?), *fr* (?).

Endemic. [L3: 416].

Notes: This taxon is on the U.S. Fish & Wildlife Service list of threatened and endangered plants.

Ausú.

Myrcia portoricensis (Britton) Cedeño-Mald. & Breckon ex F.S. Axelrod, *comb. nov.*

Bas.: *Calyptranthes portoricensis* Britton, Bull. Torrey Bot. Club 51(1):11. 1924.

Syn.: *Myrcia maricaensis* Alain (1985).

Tree in Toro Negro Forest Reserve, Maricao Forest Reserve; 485–1300 m; *fl* Jun–Aug, *fr* Oct, Jan.

Endemic. [L3: 387, 416].

Notes: José Cedeño-Maldonado and Gary Breckon noticed that the type of *Calyptranthes portoricensis* Britton was referable to the genus *Myrcia* and that, furthermore, the type of *Myrcia maricaensis* was referable to this taxon. Based upon their work, which was never published, I am making the above new combination.

Myrcia splendens (Sw.) DC., Prodr. 3:244. 1828.

Bas.: *Myrtus splendens* Sw. (1788).

Tree in Northern Coastal Lowlands, Río Abajo Forest Reserve, Sierra de Luquillo, Caguas Valley, Sierra de Cayey, Las Tetas de Cayey, Eastern and Central Cordillera, Susúa Forest Reserve, Western Cordillera, Western Coast (Mayagüez); 2–1050 m; *fl* Jun–Feb, *fr* Jul–Apr.

Greater Antilles (Cuba, Hispaniola), Virgin Islands, Lesser Antilles, Trinidad and Tobago, Mexico, Central America, northwestern, west-central South America and Venezuela. [L3: 417].

Hoja menuda, Rama menuda.

Myrcianthes fragrans (Sw.) McVaugh, Fieldiana, Bot. 29(8):485. 1963.

Bas.: *Myrtus fragrans* Sw. (1788), *nom. cons.*

Syn.: *Anamomis fragrans* (Sw.) Griseb. (1860).

Tree in Northern Coastal Lowlands (Dorado), Guánica Forest Reserve, Vieques, Culebra, Mona; 40–300 m; *fl* Jan, May–Sep, *fr* Sep–Nov, Feb.

West Indies, southeastern United States (Florida), Mexico, Central America, northwestern, west-central South America and Venezuela. [L3: 418].

Guayabacón.

Myrciaria borinquena Alain, Phytologia 47(3):187. 1980.

Shrub in Southeastern Foothills, Susúa Forest Reserve, Guánica Forest Reserve, Sierra Bermeja; 60–300 m; *fl* Nov Apr, *fr* May (two specimens).

Endemic. [L3: 420].

Myrciaria floribunda (H. West ex Willd.) O. Berg, Linnaea 27(2,3):330. 1856.

Bas.: *Eugenia floribunda* H. West ex Willd. (1799).

Shrub in Northern Coastal Lowlands (Fajardo), Northern Limestone Hills, Southeastern Foothills (Salinas), Sierra Bermeja, Vieques, Culebra; 10–350 m; *fl* Apr, Aug–Oct, *fr* Sep (one specimen).

Greater Antilles (Cuba, Hispaniola), Lesser Antilles, Mexico, northern South America to Bolivia, Paraguay. [L3: 420].

Mirto, Murta.

Myrciaria myrtifolia Alain, Phytologia 54(2):109. 1983.

Shrub in Southeastern Foothills, Southwestern Coastal Lowlands (Cabo Rojo), Sierra Bermeja, Vieques; 40–600 m; *fl* May (one specimen), *fr* Sep (one specimen).

Endemic. [L3: 422].

Pimenta racemosa (Mill.) J.W. Moore var. **racemosa**, Bernice P. Bishop Mus. Bull. 102:33. 1933.
Bas.: *Caryophyllus racemosus* Mill. (1768).
Syn.: *Amomis caryophyllata* (Jacq.) Krug & Urb. (1894).

Tree planted and persistent in Northern Coastal Lowlands (Guaynabo), Northern Limestone Hills, Eastern Coast (Punta Guayanés), Southern Coastal Lowlands (Ponce), Vieques; 15–100 m; *fl* Feb–Apr (two specimens), *fr* (?).

Greater Antilles (Cuba), Virgin Islands, Lesser Antilles, Trinidad and Tobago, northern South America (Suriname). [L3: 426].

Ausú, Guayabita, Limoncillo, Malagueta.

Pimenta racemosa (Mill.) J.W. Moore var. **grisea** (Kiaersk.) Fosberg, Amer. Midl. Naturalist 27(3):762. 1942.
Bas.: *Pimenta acris* (Sw.) Kostel. var. *grisea* Kiaersk. (1890).
Syn.: *Amomis grisea* Britton (1925).

Tree in Northern Limestone Hills (Arecibo), Río Abajo Forest Reserve, Susúa Forest Reserve, Western Cordillera, Cerro Las Mesas; 115–850 m; *fl* Apr–Jul, *fr* May–Feb.

Greater Antilles (Hispaniola), Virgin Islands. [L3: 425].

Ausú, Limoncillo.

Psidium amplexicaule Pers., Syn. Pl. 2(1):27. 1806.
Syn.: *Eugenia corozalensis* Britton (1924).

Tree in Northern Limestone Hills, Río Abajo Forest Reserve; 100–400 m; *fl* May–Nov, *fr* Jun–Nov.

Greater Antilles (Hispaniola), Virgin Islands, Lesser Antilles (Nevis, Guadeloupe). [L3: 428].

Guayaba de monte.

Psidium guajava L., Sp. Pl. 1:470. 1753.

Shrub/small tree in Northern Coastal Lowlands, Northern Limestone Hills, Río Abajo Forest Reserve, Sierra de Luquillo, Las Tetas de Cayey, Central Cordillera, Southern Coastal Limestone (Guayanilla), Susúa Forest Reserve, Maricao Forest Reserve, Southwestern Coastal Lowlands (San Germán), Vieques, Mona; 1–825 m; *fl* Apr–Jan, *fr* May–Feb.

West Indies, Mexico, Central America, South America. [L3: 430].

Guayaba.

Psidium guineense Sw., Prodr. 77. 1788.

Shrub in Northern Coastal Lowlands (Dorado); 5 m; *fl* Aug (one specimen), *fr* (?). [*Woodbury a.n. 2492* (UPR)].

Bahamas, Greater Antilles (Cuba, Jamaica), Lesser Antilles (Guadeloupe, Martinique), Trinidad and Tobago, Mexico, Central America, South America.

Psidium sintenisii (Kiaersk.) Alain, Mem. New York Bot. Gard. 21(2):138. 1971.
Bas.: *Calyptropsidium sintenisii* Kiaersk. (1890).
Syn.: *Psidium calyptranthoides* Alain (1983).

Tree in Sierra de Luquillo, Maricao Forest Reserve; 700–900 m; *fl* Jun–Aug, *fr* Aug–Nov.

Endemic. [L3: 430]

Hoja menuda.

Siphoneugena densiflora O. Berg in Martius, Fl. Bras. 14(1):379. 1857.
Syn.: *Plinia dussii* (Krug & Urb.) Urb. (1919).

Shrub/small tree in Sierra de Cayey, Central Cordillera, Maricao Forest Reserve; 475–1250 m; *fl* Feb–Jun, *fr* Feb–Sep.

Lesser Antilles (Guadeloupe, Martinique, Grenada), South America (Venezuela, southern Brazil). [L3: 432].

Hoja menuda.

Syzygium jambos (L.) Alston in Trimen, Handb. Fl. Ceylon 6:115. 1931.
Bas.: *Eugenia jambos* L. (1753).
Syn.: *Jambos jambos* (L.) Millsp. (1900).

Tree in Northern Coastal Lowlands, Northern Limestone Hills, Río Abajo Forest Reserve, Sierra de Luquillo, Eastern Coast (Naguabo), Caguas Valley (Las Piedras), Sierra de Cayey, Las Tetas de Cayey, Eastern Cordillera (Cañón San Cristóbal), Central Cordillera, Susúa Forest Reserve, Western Cordillera, Vieques; 1–1200 m; *fl* throughout the year, *fr* Apr–Jul.

Probably native to Malesia. [L3: 434].

Pomarrosa.

Syzygium malaccense (L.) Merr. & L.M. Perry, J. Arnold Arbor. 19(3):215. 1938.
Bas.: *Eugenia malaccensis* L. (1753).
Syn.: *Jambosa malaccensis* (L.) DC. (1828).

Tree in Northern Coastal Lowlands (Carolina), Northern Limestone Hills (Florida), Eastern Coast (Naguabo), Eastern Cordillera (Cayey, Aibonito), Western Cordillera (Maricao), Western Coast (Mayagüez); 50–725 m; *fl* Feb–Jul, *fr* Jun–Aug.

Native to Malesia. [L3: 435].

Notes: Britton & Wilson (1925) used the orthographic variant *Jambos malaccensis* (L.) DC. for the name of this taxon.

Manzana malaya, Ohia, Pomarrosa malaya.

TAXA CITED BY PREVIOUS AUTHORS, BUT NOT INCLUDED

DOUBTFUL OCCURRENCE

Eugenia serrasuela Krug & Urb.
Tree cited by Britton & Wilson (1925) as (according to Bello) occurring near Anones. This taxon was known to them from description only, since specimens of it were apparently studied only by Bello, prior to 1881. They also thought that barren specimens from a tree in the Barrio of Maizales, Sierra de Naguabo, might possibly belong to this taxon. It was subsequently cited by Liogier & Martorell (1982, 2000) and Liogier (1994). [L3: 404].

CITED DUE TO MISIDENTIFICATION

Calyptranthes thomasiana O. Berg.
Shrub cited by Liogier & Martorell (1982) as locally common in Vieques and by Liogier (1994) and Liogier & Martorell (2000) as occurring in forest on Monte Pirata. According to Gary Breckon (unpubl. ms.), the only specimen he has seen is referable to *Myrcianthes fragrans*. [L3: 387].

Eucalyptus resinifera Sm.
Tree cited by Liogier & Martorell (1982) and Liogier (1994) as planted in the mountains. This citation was based on the misidentification of specimens of both *Eucalyptus robusta* and *Eucalyptus urophylla*. [L3: 390].

ONLY UNDER CULTIVATION

Callistemon citrinus (Curtis) Skeels.
Tree cited by Liogier & Martorell (1982, 2000) and Liogier (1994) as planted in parks, homes, and along streets, persisent after cultivation. [L3: 380].

Notes: All specimens that I have thus far seen identified as this taxon are referable to *Callistemon viminalis* (Sol. ex Gaertn.) G. Don.

Eucalyptus camaldulensis Dehnh. Syn.: *Eucalyptus rostrata* Schltdl.
Tree cited by Liogier & Martorell (1982, 2000) and Liogier (1994) as planted and persistent. [L3: 389].

Eugenia uniflora L.
Shrub/small tree cited by Liogier & Martorell (1982, 2000) and Liogier (1994) as planted for its fruit and naturalized. [L3: 406].

Pimenta dioica (L.) Merr. Syn.: *Pimenta pimenta* (L.) Cockerell.
Tree cited by Liogier & Martorell (1982, 2000) and Liogier (1994) as occasionally planted and escaped from cultivation. [L3: 423].

Pseudanamomis umbellulifera (Kunth) Kausel. Syn.: *Anamomis umbellulifera* (Kunth) Britton.
Tree cited by Liogier & Martorell (1982, 2000) and Liogier (1994) as occasionally planted and naturalized. [L3: 426].

Psidium cattleianum Sabine.
Shrub/small tree cited by Liogier (1994) and Liogier & Martorell (2000) as occasionally planted for its edible fruit and escaped from cultivation. [L3: 430].

Syzygium grande (Wight) Wall. Syn.: *Jambos grandis* (Wight) Blume.
Tree cited by Liogier & Martorell (1982, 2000) and Liogier (1994) as occurring in forests, Yauco. [L3: 434].

NYCTAGINACEAE

Boerhavia coccinea Mill., Gard. Dict. (ed. 8): *Boerhavia* no. 4. 1768.

Herb in Northern Coastal Lowlands, Las Piedras Chiquitas, Southeastern Foothills (Coamo), Eastern Cordillera (Comerío), Central Cordillera (Villalba), Southern Coastal Lowlands, Southern Coastal Limestone, Guánica Forest Reserve, Southwestern Coastal Lowlands, Western Coast (Rincón), Vieques, Culebra, Desecheo; 1–500 m; *fl* & *fr* throughout the year.

Pantropical. [L1: 170].

Mata pavo, Tostón, Yerba de puerco.

Boerhavia diffusa L., Sp. Pl. 1:3. 1753, *nom. cons.*

Herb in Northern Coastal Lowlands, Northern Limestone Hills (San Sebastián), Eastern Coast (Humacao, Cayo Santiago), Las Piedras Chiquitas, Southeastern Foothills, Southern Coastal Lowlands (Caja de Muertos), Guánica Forest Reserve, Southwestern Coastal Lowlands, Western Coast (Rincón), Vieques, Culebra, Mona, Desecheo; 1–500 m; *fl* & *fr* throughout the year.

Pantropical. [L1: 171].

Mata pavo, Tostón, Yerba de puerco.

Boerhavia erecta L., Sp. Pl. 1:3. 1753.

Herb in Southeastern Foothills, Southern Coastal

Lowlands, Guánica Forest Reserve, Southwestern Coastal Lowlands (Lajas), Sierra Bermeja, Culebra, Mona; 1–425 m; *fl* & *fr* throughout the year.

Pantropical. [L1: 171].

Bougainvillea ×buttiana Holttum & Standl. (pro sp.), Publ. Field Mus. Nat. Hist., Bot. Ser. 23(2):44. 1944.

Woody vine planted and persistent in Northern Coastal Lowlands, Eastern Coast (Punta Guayanés), Southern Coastal Limestone (Peñuelas), Guánica Forest Reserve, Southwestern Coastal Lowlands (Lajas), Vieques; 5–50 m; *fl* Mar–Jun, Nov, *fr* (?).

Only cultivated. [L1: 173; AR: 321].

Notes: Britton & Wilson (1924), Liogier & Martorell (1982, 2000), Liogier (1985) and Acevedo-Rodríguez (2005) misapplied the name *Bougainvillea glabra* Choisy to this taxon.

This name probably represents the hybrid *Bougainvillea glabra ×peruviana*.

Trinitaria.

Commicarpus scandens (L.) Standl., Contr. U.S. Natl. Herb. 12(8):373. 1909.

Bas.: *Boerhavia scandens* L. (1753).

Herb in Northern Coastal Lowlands (Aguadilla), Southeastern Foothills, Southern Coastal Lowlands, Guánica Forest Reserve, Southwestern Coastal Lowlands, Cerro Las Mesas; 1–300 m; *fl* & *fr* throughout the year.

West Indies, southwestern United States, Mexico, Central America, South America. [L1: 173; AR: 319].

Pegapollo.

Guapira discolor (Spreng.) Little, Phytologia 17(5): 368. 1968.

Bas.: *Pisonia discolor* Spreng. (1825).

Syn.: *Torrubia discolor* (Spreng.) Britton (1904).

Tree in Southwestern Coastal Lowlands, Guánica Forest Reserve, Mona, Desecheo; 5–150 m; *fl* Jan, Apr–Sep, *fr* Mar, Jun–Oct.

Bahamas, Greater Antilles, southeastern United States (Florida). [L1: 176].

Barrehorno.

Guapira domingensis (Heimerl) Alain, Mem. New York Bot. Gard. 21(2):108. 1971.

Bas.: *Pisonia obtusata* Jacq. var. *domingensis* Heimerl (1912).

Tree in Northern Limestone Hills, Susúa Forest Reserve, Maricao Forest Reserve, Guánica Forest Reserve; 20–240 m; *fl* Apr–Nov, *fr* Jun–Jul, Oct.

Greater Antilles (Hispaniola). [L1: 178].

Notes: J.C. Trejo alerted me to the fact that *Guapira obtusata* (Jacq.) Little, to which specimens of the above taxon were previously referred, is, *sensu stricto*, confined to the Bahamas and Cuba.

Corcho, Corcho blanco, Corcho prieto.

Guapira fragrans (Dum. Cours.) Little, Phytologia 17(5):368. 1968.

Bas.: *Pisonia fragrans* Dum. Cours. (1814).

Syn.: *Torrubia fragrans* (Dum. Cours.) Standl. (1916).

Tree in Northern Coastal Lowlands, Northern Limestone Hills, Río Abajo Forest Reserve, Eastern Coast, Las Tetas de Cayey, Las Piedras Chiquitas, Southeastern Foothills, Eastern Cordillera, Central Cordillera (Ponce), Quebrada de Los Cedros, Susúa Forest Reserve, Maricao Forest Reserve, Guánica Forest Reserve, Southwestern Coastal Lowlands (Cabo Rojo), Sierra Bermeja, Vieques, Culebra, Mona, Desecheo; 1–825 m; *fl* Mar–Dec, *fr* May–Dec.

Antilles, Southern Caribbean Islands (Curaçao, Bonaire), Trinidad and Tobago, northern South America. [L1: 176].

Corcho, Emajagua, Majagua, Majagua de mona, Majagua quemona, Palo de corcho, Quemadora.

Mirabilis jalapa L., Sp. Pl. 1:177. 1753.

Herb in Northern Coastal Lowlands, Northern Limestone Hills, Caguas Valley (Juncos), Las Tetas de Cayey, Las Piedras Chiquitas, Eastern and Central Cordillera, Southern Coastal Lowlands, Southern Coastal Limestone, Western Cordillera (Maricao), Western Coast (Aguada); 5–500 m; *fl* & *fr* throughout the year.

Possibly native to Mexico. [L1: 178].

Don Diego de noche, Siciliana.

Neea buxifolia (Hook. f.) Heimerl, Bot. Jahrb. Syst. 21(5):633. 1896.

Bas.: *Eggersia buxifolia* Hook. f. (1883).

Shrub in Northern Coastal Lowlands, Northern Limestone Hills, Río Abajo Forest Reserve, Guánica Forest Reserve, Sierra Bermeja, Vieques, Culebra; 1–350 m; *fl* throughout the year, *fr* Jun–Jan.

Virgin Islands (St. Thomas, St. John, Tortola). [L1: 179].

Pisonia aculeata L., Sp. Pl. 2:1026. 1753.

Syn.: *Pisonia helleri* Standl. (1918).

Woody vine in Northern Coastal Lowlands (Guaynabo), Northern Limestone Hills, Río Abajo Forest Reserve, Sierra de Luquillo, Southeastern Foothills, Guánica Forest Reserve, Sierra Bermeja, Vieques, Culebra; 100–500 m; *fl* Jan–Jun, *fr* Feb–Apr.

West Indies, southern United States (Florida, Texas), Mexico, Central America, South America. [L1: 181; AR: 321].

Escambrón, Uña de gato.

Pisonia albida (Heimerl) Britton ex Standl., N. Amer. Fl. 21(3):190. 1918.
Bas.: *Pisonia subcordata* Sw. f. *albida* Heimerl (1896).

Tree in Southern Coastal Lowlands (Caja de Muertos), Southern Coastal Limestone (Guayanilla), Guánica Forest Reserve, Southwestern Coastal Lowlands, Mona; 10–130 m; *fl* Feb–Jun, *fr* Mar–Jun.

Greater Antilles (Hispaniola). [L1: 182].

Corcho, Corcho blanco, Corcho bobo.

Pisonia subcordata Sw., Prodr. 60. 1788.

Tree in Northern Coastal Lowlands (Fajardo and adjacent islands), Northern Limestone Hills (Quebradillas), Eastern Coast, Southeastern Foothills (Salinas), Guánica Forest Reserve, Vieques, Culebra; 2–150 m; *fl* Dec–Apr, *fr* Apr–Jun.

Greater Antilles (Jamaica), Virgin Islands, Lesser Antilles. [L1: 183].

Corcho, Corcho blanco, Palo bobo.

Pisonia taina Trejo, Harvard Pap. Bot. 10(1):118, *t. 1.* 2005.

Tree in Guajataca Forest Reserve, Eastern Cordillera (La Torrecilla), Susúa Forest Reserve; 200–940 m; *fl* Feb–Apr, Aug, *fr* Aug–Sep (one specimen).

Endemic.

Pisonia sp. 1.

Tree in Northern Limestone Hills (Arecibo), Sierra de Cayey, Eastern Cordillera (La Torrecilla), Central Cordillera (Los Tres Picachos), Maricao Forest Reserve, Western Coast (Rincón); 10–930 m; *fl* Mar–Apr, Jul, *fr* Feb–Apr, Jul.

Endemic.

Notes: George Proctor has referred specimens of this taxon to the name *Pisonia borinquena*, which remains unpublished.

Pisonia sp. 2.

Tree in Northern Coastal Lowlands, Northern Limestone Hills; 1–290 m; *fl* Feb–May, *fr* Mar–Jul.

Endemic.

Notes: George Proctor has referred specimens of this taxon to the name *Pisonia woodburyana*, which remains unpublished.

TAXA CITED BY PREVIOUS AUTHORS, BUT NOT INCLUDED

ONLY UNDER CULTIVATION

Bougainvillea spectabilis Willd.

Woody vine cited by Liogier & Martorell (1982, 2000) and Liogier (1985) as commonly planted in gardens. [L1: 173; AR: 321].

NYMPHAEACEAE

Nymphaea amazonum Mart. & Zucc., Abh. Math.-Phys. Cl. Königl. Bayer. Akad. Wiss. 1:363. 1832.
Syn.: *Castalia amazonum* (Mart. & Zucc.) Britton & P. Wilson (1924).

Aquatic herb in Northern Coastal Lowlands, Eastern Coast (Naguabo), Southwestern Coastal Lowlands (Lajas); 1–12 m; *fl* Dec–Feb, *fr* (?).

Antilles, Tobago, Mexico, Central America, South America. [L1: 217].

Yerba de hicotea, Yerba de jicotea.

Nymphaea ampla (Salisb.) DC., Syst. Nat. 2:54. 1821.
Bas.: *Castalia ampla* Salisb. (1805).

Aquatic herb in Northern Coastal Lowlands, Northern Limestone Hills (Lares), Southwestern Coastal Lowlands (Lajas); 1–200 m; *fl* Jan–Jun, Oct, *fr* (?).

West Indies, southern United States (Florida, Texas), Mexico, Central America. [L1: 218].

Calderón, Flor de agua.

Nymphaea caerulea Savigny, Décade égyptienne 1:71(74?). 1798 (Ann. Mus. Natl. Hist. Nat. 1:366, *t. 25.* 1802).

Aquatic herb in Southern Coastal Lowlands (Santa Isabel); 40 m; *fl* Apr (one specimen), *fr* (?). [*Pomales 39* (UPRRP)].

Native to Africa, Yemen.

Nymphaea conardii Wiersema, Brittonia 36(3):213, *f. 1.* 1984.

Aquatic herb in Western Coast (Añasco); 1 m; *fl* May–Jun (one specimen), *fr* (?). [*Monsegur 734* (MAPR)]

Greater Antilles (Cuba, Hispaniola), central Mexico, Central America, northern South America and Brazil. [L1: 218].

Notes: Britton & Wilson (1924) misapplied the name *Castalia jamesoniana* (Planch.) Britton & P. Wilson, and Liogier & Martorell (1982, 2000) and Liogier (1985) misapplied *Nymphaea jamesoniana* Planch., to this taxon.

Nymphaea odorata Aiton, Hort. Kew. 2:227. 1789.
Syn.: *Castalia odorata* (Aiton) Wood (1806).

Aquatic herb in Northern Coastal Lowlands (Vega Alta), Northern Limestone Hills (Arecibo), Southwestern Coastal Lowlands (Lajas); 3–75 m; *fl* Jan–Sep, *fr* (?).

Bahamas, Greater Antilles (Cuba), eastern and central Canada and United States, Mexico, Central America (Honduras, El Salvador, Nicaragua). [L1: 218].

Lirio de agua, Ninfa.

TAXA NOT RECENTLY COLLECTED

Nymphaea pulchella DC., Syst. Nat. 2:51. 1821.
Syn.: *Castalia pulchella* (DC.) Britton (1906).

Aquatic herb collected in Caguas on 9 Sep 1885 [*Sintenis 2577* (GH)], in Guánica on 2 Feb 1886 [*Sintenis 3662* (GH)], in Mayagüez on 14 May 1913 [*Hess 1511* (MAPR)], in San Juan on 5 Nov 1913 [*Hess 4050* (MAPR)], in Algarroba on 26 Jul 1914 [*Stevenson 2139* (US)], in Hatillo on 1 Mar 1915 [*Britton 5093* (US)], near Boquerón on 18 Feb 1927 [*Britton & Britton 8874* (NY), scan seen], between Arroyo and Patillas on 11 Mar 1927 [*Britton & Britton 9017* (NY), scan seen] and in Barrio Pastillo 23 Feb 1933 [*Britton & Britton* 10087 (US)].

West Indies, Trinidad and Tobago, Mexico, Central America (Nicaragua, Costa Rica, Panama), northern South America to Brazil, Ecuador. [L1: 220].

TAXA CITED BY PREVIOUS AUTHORS, BUT NOT INCLUDED

NO SPECIMENS SEEN

Nymphaea rudgeana G. Mey. Syn.: *Castalia rudgeana* (G. Mey.) Britton & P. Wilson.
Aquatic herb cited by Britton & Wilson (1924) as occurring in Puerto Rico (according to Conard); subsequently cited by Liogier & Martorell (1982, 2000) and by Liogier (1985). [L1: 220].

OCHNACEAE

Ouratea litoralis Urb., Symb. Antill. 1(2):363. 1899.

Tree in Northern Coastal Lowlands, Eastern Coast, Susúa Forest Reserve, Southwestern Coastal Lowlands, Sierra Bermeja; 1–300 m; *fl* & *fr* throughout the year.

Virgin Islands (St. Thomas, St. John, Virgin Gorda). [L3: 213].

Abey amarillo, Abeyuelo amarillo, Abeyuelo perfumado, Doncella.

Ouratea striata (Tiegh.) Urb., Symb. Antill. 5(3): 427. 1908.
Bas.: *Camptouratea striata* Tiegh. (1902).

Tree in Sierra de Cayey, Toro Negro Forest Reserve; 800–1000 m; *fl* (?), *fr* Aug–Nov.

Greater Antilles (Cuba). [L3: 215].

Sauvagesia erecta L., Sp. Pl. 1:203. 1753.

Herb in Northern Coastal Lowlands, Sierra de Luquillo, Eastern Coast, Sierra de Cayey, Eastern and Central Cordillera, Maricao Forest Reserve, Western Coast (Añasco); 1–1000 m; *fl* & *fr* throughout the year.

Pantropical. [L3: 215].

Yerba de San Martín.

TAXA CITED BY PREVIOUS AUTHORS, BUT NOT INCLUDED

DOUBTFUL OCCURRENCE

Ouratea ilicifolia (DC.) Baill.
Tree cited by Liogier & Martorell (1982, 2000) and Liogier (1994) as local and uncommon in dry and moist coastal forests near sea level along the northern, northeastern and southeastern coasts. I have been unable to locate any specimens of this taxon and suspect its citation was based upon the misidenification of sterile specimens of *Drypetes ilicifolia*. [L3: 213].

ONLY UNDER CULTIVATION

Ochna jabotapita L.
Shrub cited by Liogier (1994) and Liogier & Martorell (2000) as planted as an ornamental and escaped. [L3: 211].

Ochna mossambicensis Klotzsch.
Shrub cited by Liogier & Martorell (1982, 2000) and Liogier (1994) as cultivated as an ornamental and naturalized. [L3: 211].

OLACACEAE

Ximenia americana L., Sp. Pl. 2:1193. 1753.

Tree in Southern Coastal Limestone, Guánica Forest Reserve, Southwestern Coastal Lowlands (Cabo Rojo), Vieques; 1–125 m; *fl* Apr–Jul, Nov–Dec, *fr* Dec–Jan, Jun–Jul.

Pantropical and -subtropical. [L1: 96].

OLEACEAE

Chionanthus compactus Sw., Prodr. 13. 1788.
Syn.: *Mayepea caribaea* (Jacq.) Kuntze (1891).
Linociera caribaea (Jacq.) Knobl. (1895).

Tree in Northern Coastal Lowlands (Carolina, Trujillo Alto), Northern Limestone Hills, Sierra de Cayey, Eastern Cordillera (Cañón San Cristóbal), Vieques; 10–500 m; *fl* Apr–Nov, *fr* Jan–Aug.

Greater Antilles (Hispaniola), Lesser Antilles,

Virgin Islands, Trinidad, northern South America (Venezuela, Colombia). [L4: 167].

Avispillo, Hueso.

Chionanthus domingensis Lam., Tabl. Encycl. 1:30. 1791.

Syn.: *Mayepea domingensis* (Lam.) Krug & Urb. (1892). *Linociera domingensis* (Lam.) Knobl. (1895).

Tree in Northern Limestone Hills (Camuy), Río Abajo Forest Reserve, Sierra de Luquillo, Sierra de Cayey, Eastern Cordillera (La Torrecilla), Susúa Forest Reserve, Western Cordillera, Sierra Bermeja; 150–900 m; *fl* Sep, Jan–May, *fr* Jun–Oct.

Greater Antilles, northern Central America (Belize, Guatemala). [L4: 168].

Caracolillo, Huesillo, Hueso, Hueso blanco, Palo blanco, Palo de hueso.

Chionanthus holdridgei (Camp & Monach.) Stearn, Ann. Missouri Bot. Gard. 63(2):357. 1977.

Bas.: *Linociera holdridgei* Camp & Monach. (1939) "holdridgii."

Tree in Northern Limestone Hills (Quebradillas), Susúa Forest Reserve, Guánica Forest Reserve; 150–375 m; *fl* Sep–Mar, *fr* Dec, Apr–Jul.

Endemic. [L4: 168].

Notes: According to Pedro González Gutiérrez (2008), this taxon is very similar to *Chionanthus axilliflorus*, a highly variable taxon with respect to leaf size and thickness.

Espejuelo, Hueso, Hueso blanco, Hueso prieto, Palo de hueso.

Chionanthus ligustrinus (Sw.) Pers., Syn. Pl. 1:9. 1805 "ligustrina."

Bas.: *Thouinia ligustrina* Sw. (1788).

Syn.: *Linociera ligustrina* (Sw.) Sw. (1797).

Shrub/small tree in Northern Limestone Hills; 10–100 m; *fl* Jan, May–Sep, *fr* Aug–Jan.

Greater Antilles. [L4: 169].

Hueso.

Forestiera eggersiana Krug & Urb., Bot. Jahrb. Syst. 15:339. 1892.

Shrub/small tree in Northern Limestone Hills (Quebradillas, Isabela), Las Tetas de Cayey, Las Piedras Chiquitas, Southeastern Foothills (Salinas), Southern Coastal Lowlands (Caja de Muertos), Culebra, Mona; 10–825 m; *fl* Oct–Dec (two specimens), *fr* Mar, Jun–Oct.

Virgin Islands, Lesser Antilles (St. Barts). [L4: 170].

Forestiera segregata (Jacq.) Krug & Urb., Bot. Jahrb. Syst. 15:339. 1892.

Bas.: *Myrica segregata* Jacq. (1789).

Shrub/small tree in Southern Coastal Limestone, Guánica Forest Reserve, Southwestern Coastal Lowlands (Cabo Rojo), Sierra Bermeja, Vieques, Mona, Desecheo; 3–300 m; *fl* Sep–Feb, *fr* Sep–Feb.

West Indies, southeastern United States (Georgia, Florida). [L4: 172].

Fraxinus uhdei (Wenz.) Lingelsh., Bot. Jahrb. Syst. 40(2):221. 1907.

Bas.: *Fraxinus americana* L. var. *uhdei* Wenz. (1883).

Tree planted and persistent in Central Cordillera, Maricao Forest Reserve; 700–1000 m; *fl* Nov–Jan, *fr* Nov–Mar.

Native to Mexico, northern Central America (Guatemala, Honduras). [L4: 173].

Fresno.

Haenianthus salicifolius Griseb., Pl. Wright. 2:518. 1862 (preprint of Mem. Amer. Acad. Arts, ser. 2 8(2):518. 1863).

Syn.: *Haenianthus obovatus* Krug & Urb. (1893). *Haenianthus salicifolius* Griseb. var. *obovatus* (Krug & Urb.) Knobl. (1933).

Tree in Sierra de Luquillo, Sierra de Cayey, Central Cordillera; 850–1300 m; *fl* throughout the year, *fr* Apr, Jul–Jan.

Greater Antilles (Cuba, Hispaniola). [L4: 174].

Aceituno blanco, Hueso, Hueso blanco, Hueso prieto, Palo de hueso.

Jasminum fluminense Vell., Fl. Flumin. 10. 1829.

Woody vine in Northern Coastal Lowlands, Northern Limestone Hills (Quebradillas), Las Piedras Chiquitas, Southeastern Foothills (Salinas), Eastern Cordillera (Barranquitas), Southern Coastal Lowlands, Southern Coastal Limestone, Western Cordillera, Guánica Forest Reserve, Southwestern Coastal Lowlands, Sierra Bermeja, Vieques, Culebra; 1–700 m; *fl* & *fr* throughout the year.

Native to Africa, Saudia Arabia, Yemen. [L4: 176; AR: 323].

Notes: Britton & Wilson (1925) misapplied the name *Jasminum azoricum* L. to this taxon.

Jazmín de canario, Jazmín de trapo, Jazmín oloroso.

Jasminum multiflorum (Burm. f.) Andrews, Bot. Repos. 8: *t. 496*. 1807.

Bas.: *Nyctanthes multiflora* Burm. f. (1768).

Syn.: *Jasminum pubescens* (Retz.) Willd. (1797).

Shrub planted and persistent in Northern Coastal Lowlands (Fajardo), Northern Limestone Hills, Río Abajo Forest Reserve, Caguas Valley (Caguas), Sierra de Cayey, Southeastern Foothills

(Salinas), Eastern Cordillera (Barranquitas), Central Cordillera (Yauco), Western Cordillera (Maricao), Cerro Las Mesas; 100–850 m; *fl* Sep–Jun, *fr* (?).

Native to Indian Subcontinent, Myanmar. [L4: 178; AR: 324].

Jazmín, Jazmín de papel, Yerba de tres estrellas.

Ligustrum japonicum Thunb., Nova Acta Regiae Soc. Sci. Upsal. 3:207. 1780.

Shrub planted and persistent in Guajataca Forest Reserve, Eastern Cordillera (Barranquitas); 200–500 m; *fl* Apr (one specimen), *fr* Apr, Oct (two specimens).

Native to eastern temperate Asia (Taiwan, Korea, Japan). [L4: 179].

Privet japonés.

Ligustrum sinense Lour., Fl. Cochinch. 1:19. 1790.

Shrub planted and persistent in Northern Limestone Hills (Lares), Central Cordillera; 300–1330 m; *fl* Jan–Apr, Jul–Sep, *fr* Mar, Aug–Sep.

Native to China, southeastern Asia (Laos, Vietnam). [L4: 181].

Privet, Privet de la China.

TAXA CITED BY PREVIOUS AUTHORS, BUT NOT INCLUDED

CITED DUE TO MISIDENTIFICATION

Chionanthus axilliflorus (Griseb.) Stearn. Syn.: *Mayepea axilliflora* (Griseb.) Krug & Urb.

Shrub/small tree cited by Britton & Wilson (1925) as occurring at Monte Mirabel, near Guánica; subsequently cited by Liogier & Martorell (1982, 2000) and Liogier (1995) as rare and local in moist or dry limestone forests at lower elevations, in Guajataca Gorge and in Guánica. All specimens that I have thus far seen identified as this taxon are referable to *Chionanthus holdridgei*. [L4: 167].

Forestiera rhamnifolia Griseb.

Tree cited by Liogier & Martorell (1982, 2000) and Liogier (1995) as occurring on limestone plateau, Mona Island. According to Gary Breckon (unpubl. ms.) the specimens upon which this citation was based are referable to *Forestiera segregata*. [L4: 171].

ONLY UNDER CULTIVATION

Jasminum grandiflorum L.

Shrub cited by Britton & Wilson (1925) as occasionally spontaneous after planting; subsequently cited by Liogier & Martorell (1982, 2000) and Liogier (1995) as occasionally persistent after cultivation. [L4: 176; AR: 324].

Jasminum sambac (L.) Aiton.

Shrub cited by Britton & Wilson (1925) as spontaneous after planting; subsequently cited by Liogier & Martorell (1982, 2000) and Liogier (1995). [L4: 178; AR: 327].

Ligustrum ovalifolium Hassk.

Shrub/small tree cited by Liogier & Martorell (1982, 2000) and Liogier (1995) as cultivated as an ornamental and persistent. [L4: 179].

ONAGRACEAE

Ludwigia decurrens Walter, Fl. Carol. 89. 1788.

Herb in Northern Coastal Lowlands, Eastern Coast (Naguabo), Caguas Valley; 1–80 m; *fl* & *fr* Aug–Apr.

Native to east-central, southeastern, and south-central United States, Mexico, Central America, South America. [L4: 69].

Ludwigia erecta (L.) H. Hara, J. Jap. Bot. 28(10):292. 1953.

Bas.: *Jussiaea erecta* L. (1753).

Herb in Northern Coastal Lowlands, Eastern Coast, Caguas Valley, Eastern Cordillera (Cayey), Southwestern Coastal Lowlands (Lajas), Western Coast (Añasco), Vieques; 1–300 m; *fl* & *fr* throughout the year.

West Indies, southeastern United States (Florida), Mexico, Central America, northern South America to Bolivia, Paraguay. [L4: 69].

Yerba de clavo acuática, Yerba de hicotea, Yerba de jicotea.

Ludwigia leptocarpa (Nutt.) H. Hara, J. Jap. Bot. 28(10):292. 1953.

Bas.: *Jussiaea leptocarpa* Nutt. (1818).

Herb in Northern Coastal Lowlands, Western Coast (Mayagüez); 1–25 m; *fl* & *fr* Oct–Dec, Mar–Apr.

Antilles, east-central, southeastern, and south-central United States, Mexico, Central America, South America. [L4: 71].

Ludwigia octovalvis (Jacq.) P. H. Raven, Kew Bull. 15(3):476, *ff. 6d-e, 8h*. 1962.

Bas.: *Oenothera octovalvis* Jacq. (1760).

Syn.: *Jussiaea angustifolia* Lam. (1789).

Herb in Northern Coastal Lowlands, Northern Limestone Hills, Río Abajo Forest Reserve, Sierra de Luquillo, Eastern Coast, Caguas Valley, Sierra de Cayey, Eastern and Central Cordillera, Southern Coastal Lowlands, Susúa Forest Reserve, Southwestern Coastal Lowlands, Western Coast; 1–1200 m; *fl* & *fr* throughout the year.

Pantropical. [L4: 71].

Cangá, Yerba cangá, Yerba de clavo.

Ludwigia peploides (Kunth) P. H. Raven, Reinwardtia 6(4):393. 1963.
Bas.: *Jussiaea peploides* Kunth (1823).

Herb in Northern Coastal Lowlands, Northern Limestone Hills (Florida), Eastern Coast, Eastern Cordillera (Cayey), Southern Coastal Lowlands (Santa Isabel), Southwestern Coastal Lowlands (Lajas); 1–300 m; *fl* & *fr* Dec–Sep.

Greater Antilles, east-central and southern United States, Mexico, Central America, South America. [L4: 73].

Notes: Britton & Wilson (1925) misapplied the name *Jussiaea repens* L. to this taxon.

Yerba de clavo acuática.

Ludwigia peruviana (L.) H. Hara, J. Jap. Bot. 28(10): 293. 1953.
Bas.: *Jussiaea peruviana* L. (1753).

Shrub in Northern Coastal Lowlands (Toa Baja), Northern Limestone Hills (Florida), Sierra de Luquillo; 1–400 m; *fl* & *fr* Jun, Sep–Nov.

Greater Antilles, Trinidad, southeastern United States and Texas, Mexico, Central America, South America. [L4: 74].

Yerba de clavo peluda.

Ludwigia repens J.R. Forst., Fl. Amer. Sept. 6. 1771, *nom. cons.*

Herb in Northern Coastal Lowlands (Dorado), Río Abajo Forest Reserve; 5–300 m; *fl* Feb (one specimen), *fr* Feb, Jun (two specimens). [*Acevedo 375* (SJ); *Woodbury a.n.* 15026 (UPR)].

Bahamas, Greater Antilles, southern United States, Mexico, northern Central America (Guatemala).

TAXA NOT RECENTLY COLLECTED

Ludwigia palustris (L.) Elliott, Sketch Bot. S. Carolina 1(3):211. 1817.
Bas.: *Isnardia palustris* L. (1753).

Herb collected in the vicinity of Dorado between 20–22 Mar 1922 [*Britton et al. 6702* (US)] and in La Sardinera (near Dorado) on 26 Mar 1930 [*Britton & Cowell 9522* (US)].

Greater Antilles, North America, Mexico, Central America, northern South America (Colombia), temperate Europe, Africa, western Asia. [L4: 73].

ORCHIDACEAE

Arundina graminifolia (D. Don) Hochr., Bull. New York Bot. Gard. 6:270. 1910.
Bas.: *Bletia graminifolia* D. Don (1825).

Terrestrial herb in Sierra de Luquillo, Sierra de Cayey, Eastern Cordillera (Cerro La Peña), Central Cordillera (Peñuelas); 400–850 m; *fl* throughout the year, *fr* Feb (one specimen).

Native to tropical Asia, China. [A: 12].

Basiphyllaea corallicola (Small) Ames, Schedul. Orchid. 7:1. 1924.
Bas.: *Carteria corallicola* Small (1910).
Syn.: *Basiphyllaea angustifolia* Schltr. (1925).

Terrestrial herb in Northern Limestone Hills (Toa Alta), Maricao Forest Reserve; 100–725 m; *fl* Aug–Sep, Dec–Jan, *fr* Aug–Sep, Dec.

Bahamas, Greater Antilles (Cuba, Hispaniola), southeastern United States (Florida). [A: 13].

Bletia patula Graham, Edinburgh New. Philos. J. 21:155. 1836 Jul; et Bot. Mag. 63: *t. 3518.* 1836 Sep.

Terrestrial herb in Northern Limestone Hills, Río Abajo Forest Reserve, Maricao Forest Reserve; 50–600 m; *fl* Oct–Jun, *fr* Nov–Jun.

Greater Antilles (Cuba, Hispaniola), Lesser Antilles (Guadeloupe, Martinique), southeastern United States (Florida). [A: 15].

Notes: Britton & Wilson (1924) and Liogier & Martorell (1982) used *Bletia patula* Hook. for the name of this taxon; this erroneus author citation is based, I believe, on a misunderstanding of who was the author of the Botanical Magazine article.

Bletia, Flor de pasmo, Mariposilla.

Brachionidium ciliolatum Garay, J. Arnold Arbor. 50:464, *t. 3g–j.* 1969.

Epiphytic herb in Sierra de Luquillo; 1020 m; *fl* Jan (one specimen), *fr* (?). [*Howard 16929* (AMES)].

Endemic. [A: 15].

Notes: This taxon is known from the type specimen only.

Brachionidium parvum Cogn., Repert. Spec. Nov. Regni Veg. 6:307. 1909.

Terrestrial and epiphytic herb in Sierra de Luquillo, Sierra de Cayey; 500–1050 m; *fl* Nov–Jul, *fr* Dec–Feb, May–Jul.

Greater Antilles (Jamaica, Hispaniola), Lesser Antilles. [A: 17].

Notes: Ackerman (1995), and subsequently Liogier & Martorell (2000), misapplied the name *Brachionidium sherringii* Rolfe to this taxon.

Broughtonia domingensis (Lindl.) Rolfe, Gard. Chron., ser. 3 5:491. 1889.
Bas.: *Cattleya domingensis* Lindl. (1831).
Syn.: *Laeliopsis domingensis* (Lindl.) Lindl. (1853).

Epiphytic herb in Mona; 50 m; *fl* Jun–Aug (two

specimens), *fr* (?). [*Sustache 1069* (SJ); *Proyecto Mona 720* (UPRRP)].

Greater Antilles (Hispaniola). [A: 19].

Campylocentrum fasciola (Lindl.) Cogn. in Martius, Fl. Bras. 3(6):520, *t. 106, f. 1.* 1906.
Bas.: *Angraecum fasciola* Lindl. (1840).
Syn.: *Campylocentrum sullivanii* Fawc. & Rendle (1909).

Epiphytic herb in Northern Coastal Lowlands, Northern Limestone Hills (Toa Baja, Vega Baja), Río Abajo Forest Reserve, Eastern Cordillera (Cañón San Cristóbal), Central Cordillera (Utuado); 1–600 m; *fl* Mar–Nov, *fr* Jun–Nov.

Greater Antilles, Central America, South America. [A: 24].

Campylocentrum jamaicense (Rchb. f. & Wullschl.) Benth. ex Fawc., Prov. List Pl. Jamaica 40. 1893.
Bas.: *Aeranthes jamaicensis* Rchb. f. & Wullschl. (1864).

Epiphytic herb in Central Cordillera (Lares, Adjuntas); 450–740 m; *fl* Feb, Sep–Oct, *fr* Feb (one specimen).

Greater Antilles (Cuba, Jamaica), Lesser Antilles (Guadeloupe, Dominica, Martinique), Trinidad (and probably also in Mexico, Central America and South America).

Notes: Ackerman (1995) treated this taxon as conspecific with *Campylocentrum micranthum*, but he now considers it to be distinct (unpubl. ms.).

Campylocentrum micranthum (Lindl.) Maury, J. Bot. (Morot) 3:273. 1889.
Bas.: *Angraecum micranthum* Lindl. (1835).

Epiphytic herb in Northern Limestone Hills (Florida), Río Abajo Forest Reserve, Sierra de Cayey, Las Tetas de Cayey, Eastern Cordillera; 350–900 m; *fl* Sep–Feb, *fr* May–Oct.

Greater Antilles (Jamaica, Hispaniola), Lesser Antilles (Guadeloupe), Trinidad and Tobago, northern South America (Suriname). [A: 21].

Campylocentrum pachyrrhizum (Rchb. f.) Rolfe, Orchid Rev. 11:246. 1903.
Bas.: *Aeranthes pachyrrhizus* Rchb. f. (1865) "Aëranthus."

Epiphytic herb in Northern Limestone Hills (Vega Baja, Quebradillas), Río Abajo Forest Reserve, Sierra de Cayey; 75–600 m; *fl* Feb (one specimen), *fr* Feb, May–Jul.

Greater Antilles, Trinidad and Tobago, southeastern United States (Florida), Mexico, northern South America to Ecuador. [A: 23].

Campylocentrum pygmaeum Cogn. in Urban, Symb. Antill. 4(1):183. 1903.

Epiphytic herb in Sierra de Cayey, Central Cordillera, Maricao Forest Reserve; 600–920 m; *fl* Jan–Feb, *fr* May, Aug (two specimens).

Lesser Antilles (Guadeloupe). [A: 21].

Cochleanthes flabelliformis (Sw.) R.E. Schult. & Garay, Bot. Mus. Leafl. 18(6):324. 1959.
Bas.: *Epidendrum flabelliforme* Sw. (1788).
Syn.: *Warszewiczella flabelliformis* (Sw.) Cogn. (1903).

Epiphytic and epilithic herb in Río Abajo Forest Reserve, Sierra de Luquillo, Central Cordillera (Los Tres Picachos), Maricao Forest Reserve; 225–800 m; *fl* Sep–Nov, *fr* Nov, Mar (two specimens).

Greater Antilles, Trinidad, Mexico, Central America (Belize, Honduras, Nicaragua), South America (Suriname, Venezuela, Ecuador, Brazil). [A: 26].

Comparettia falcata Poepp. & Endl., Nov. Gen. Sp. Pl. 1:42, *t. 73.* 1836.

Epiphytic herb in Sierra de Luquillo, Central Cordillera; 500–1100 m; *fl* Jan–Mar, *fr* Mar, Jul (two specimens).

Greater Antilles, Mexico, Central America, northwestern South America to Bolivia and Venezuela, Brazil. [A: 28].

Corymborkis forcipigera (Rchb. f. & Warsz.) L.O. Williams, Bot. Mus. Leafl. 12(7):237. 1946 "Corymborchis."
Bas.: *Macrostylis forcipigera* Rchb. f. & Warsz. (1854).

Terrestrial herb in Northern Limestone Hills (Arecibo, Camuy), Río Abajo Forest Reserve; 175–350 m; *fl* Aug–Mar, *fr* Feb (one specimen).

Greater Antilles, Mexico, Central America, northern South America (Colombia). [A: 30].

Notes: Britton & Wilson (1924) and Liogier & Martorell (1982) misapplied the name *Corymborkis flava* (Sw.) Kuntze to this taxon.

Cranichis muscosa Sw., Prodr. 120. 1788.

Terrestrial herb in Northern Limestone Hills (Barceloneta), Río Abajo Forest Reserve, Sierra de Luquillo, Sierra de Cayey, Eastern Cordillera (Aguas Buenas), Central Cordillera, Maricao Forest Reserve; 200–1200 m; *fl* Dec–Mar, *fr* Jan–Apr.

West Indies, Trinidad and Tobago, southeastern United States (Florida), Mexico, Central America, northwestern South America to Peru, Brazil. [A: 31].

Cranichis ricartii Ackerman, Lindleyana 4(1):43, *t. 2.* 1989.

Terrestrial herb in Maricao Forest Reserve; 600–630 m; *fl* Jan (one specimen), *fr* Jan–Feb (one specimen). [*Ackerman 2418* (UPRRP), isotype; *Meléndez 216* (UPRRP)].

Greater Antilles (Cuba, Hispaniola), Lesser Antilles (Guadeloupe). [A: 34].

Notes: This taxon is on the U.S. Fish & Wildlife Service list of threatened and endangered plants.

Cranichis tenuis Rchb. f., Flora 48(18):274. 1865.

Terrestrial herb in Maricao Forest Reserve; 700–750 m; *fl* May, Sep–Nov, *fr* Dec (two specimens).

Greater Antilles (Cuba, Hispaniola). [A: 33].

Cyclopogon cranichoides (Griseb.) Schltr., Beih. Bot. Centralbl., Abt. 2 37(2, Heft 3):387. 1920.
Bas.: *Pelexia cranichoides* Griseb. (1866).
Syn.: *Spiranthes cranichoides* (Griseb.) Cogn. (1909). *Beadlea cranichoides* (Griseb.) Small (1913).

Terrestrial herb in Guajataca Forest Reserve, Río Abajo Forest Reserve, Sierra de Luquillo, Sierra de Cayey, Central Cordillera (Ciales, Ponce), Maricao Forest Reserve; 150–725 m; fl Feb–Apr, fr Mar–Apr.

Antilles, Central America, northern and west-central South America. [A: 35].

Cyclopogon elatus (Sw.) Schltr., Repert. Spec. Nov. Regni Veg. Beih. 6:53. 1919 "elatum;" et Beih. Bot. Centralbl., Abt. 2 37(2, Heft 3):387. 1920.
Bas.: *Satyrium elatum* Sw. (1788).
Syn.: *Spiranthes elata* (Sw.) Rich. (1817). *Beadlea elata* (Sw.) Small (1918).

Terrestrial herb in Northern Limestone Hills, Río Abajo Forest Reserve, Sierra de Luquillo, Sierra de Cayey, Las Tetas de Cayey, Eastern and Central Cordillera, Maricao Forest Reserve; 200–1200 m; *fl* Feb–Mar, *fr* Feb–Apr.

Antilles, Trinidad, southeastern United States (Florida), Mexico, Central America, South America. [A: 37].

Juan ascensio.

Cyclopogon miradorensis Schltr., Repert. Spec. Nov. Regni Veg. 21:332. 1925.

Terrestrial herb in Río Abajo Forest Reserve, Maricao Forest Reserve, Central Cordillera (Peñuelas); 350–1000 m; *fl* Feb–Mar, *fr* Mar (one specimen).

Greater Antilles (Hispaniola), Mexico, Central America. [A: 37].

Notes: Nir (2000) misapplied the name *Cyclopogon bicolor* (Ker Gawl.) Schltr. to this taxon.

Cyrtochilum meirax (Rchb. f.) Dalström, Lindleyana 16(2):70, *t. 2F*. 2001.
Bas.: *Oncidium meirax* Rchb. f. (1854).
Syn.: *Oncidium alatum* Cogn. (1910).

Epiphytic herb in Eastern Cordillera (La Torrecilla), Central Cordillera; 900–1250 m; *fl* Jun–Nov, Feb, *fr* Aug–Nov.

Greater Antilles (Hispaniola), Lesser Antilles (Guadeloupe), northwestern, west-central South America and Venezuela. [A: 121].

Dendrobium crumenatum Sw., J. Bot. (Schrader) 1799(2):237. 1800.

Epiphytic herb in Northern Coastal Lowlands (Toa Baja); 3 m; *fl* Mar, Aug (two specimens), *fr* Aug (one specimen). [*Axelrod 13366* (UPRRP); *Ackerman 4249* (UPRRP)].

Native to tropical Asia, China.

Dendrophylax monteverdi (Rchb. f.) Ackerman & Nir, Lankesteriana 4(1):53. 2004.
Bas.: *Aeranthes monteverdi* Rchb. f. (1865) "Aëranthus."
Syn.: *Campylocentrum monteverdi* (Rchb. f.) Rolfe (1903).
Campylocentrum filiforme (Sw.) Cogn. ex Kuntze (1898). *Harrisella filiformis* (Sw.) Cogn. (1910).

Epiphytic herb in Sierra de Cayey, Eastern Cordillera (Cañón San Cristóbal), Central Cordillera (Adjuntas); 450–900 m; *fl* Jan–Mar (two specimens), *fr* Mar, Aug.

Greater Antilles. [A: 24].

Dendrophylax porrectus (Rchb. f.) Carlsward & Whitten, Int. J. Pl. Sci. 164(1):51. 2003.
Bas.: *Aeranthes porrectus* Rchb. f. (1865) "Aëranthus."
Syn.: *Campylocentrum porrectum* (Rchb. f.) Rolfe (1903). *Harrisella porrecta* (Rchb. f.) Fawc. & Rendle (1909).

Epiphytic herb in Susúa Forest Reserve, Southwestern Coastal Lowlands (Cabo Rojo); 5–300 m; *fl* Jan–Feb (one specimen), *fr* Oct, Apr (two specimens).

Greater Antilles, southeastern United States (Florida), southern Mexico, northern Central America (El Salvador). [A: 87].

Dichaea hystricina Rchb. f., Flora 48(18):279. 1865.

Epiphytic herb in Sierra de Luquillo, Sierra de Cayey; 450–1000 m; *fl* Aug–Oct, *fr* throughout the year.

Greater Antilles (Cuba, Hispaniola), Lesser Antilles (Guadeloupe, Dominica, Grenada), Central America, northwestern South America and Venezuela. [A: 41].

Dichaea latifolia Lindl., Gen. Sp. Orchid. Pl. 208. 1833.

Epiphytic and epilithic herb in Sierra de Luquillo, Eastern Cordillera (La Torrecilla), Maricao Forest Reserve; 700–900 m; *fl* Sep–Dec, *fr* Oct–Feb.

Antilles, Trinidad, northern South America (Venezuela, Colombia). [A: 41].

Notes: Britton & Wilson (1924), Liogier & Martorell (1982) and Nir (2000) misapplied the name *Dichaea muricata* (Sw.) Lindl. to this taxon.

Dichaea pendula (Aubl.) Cogn. in Urban, Symb. Antill. 4(1):182. 1903.
Bas.: *Limodorum pendulum* Aubl. (1775).
Syn.: *Dichaea echinocarpa* (Sw.) Lindl. (1833).
Dichaea swartzii (C. Schweinf.) Garay & H.R. Sweet (1972).

Epiphytic herb in Toro Negro Forest Reserve; 1160–1240 m; *fl* Aug (one specimen), *fr* Sep (one specimen). [*Proctor 39444* (SJ); *Mejía 2169* (MAPR)].

Antilles, Central America, northern South America to Brazil, Bolivia. [A: 44].

Dilomilis montana (Sw.) Summerh., Taxon 10(8):253. 1961.
Bas.: *Epidendrum montanum* Sw. (1788).
Syn.: *Octadesmia montana* (Sw.) Benth. & Hook. f. (1881).

Terrestrial herb in Sierra de Luquillo, Central Cordillera; 700–1300 m; *fl* Dec–Apr, *fr* Dec–Jan (two specimens).

Greater Antilles. [A: 44].

Domingoa haematochila (Rchb. f.) Carabia, Mem. Soc. Cub. Hist. Nat. "Felipe Poey" 17:143. 1943.
Bas.: *Epidendrum haematochilum* Rchb. f. (1865).
Syn.: *Domingoa hymenodes* (Rchb. f.) Schltr. (1913).

Epiphytic herb in Mona; 5–50 m; *fl* Jan, May, Aug, *fr* Apr–Jul, Nov–Jan.

Greater Antilles (Cuba, Hispaniola). [A: 46].

Elleanthus cordidactylus Ackerman, Lindleyana 2:122. 1987.

Epiphytic and epilithic herb in Sierra de Luquillo, Maricao Forest Reserve; 425–450 m; *fl* Nov (two specimens), *fr* Jan, Apr.

Greater Antilles. [A: 47].

Notes: Britton & Wilson (1924) and Liogier & Martorell (1982) misapplied the name *Elleanthus linifolius* C. Presl to this taxon.

Eltroplectris calcarata (Sw.) Garay & H.R. Sweet, J. Arnold Arbor. 53(3):390. 1972.
Bas.: *Neottia calcarata* Sw. (1806).
Syn.: *Eltroplectris acuminata* Raf. (1837).

Terrestrial herb in Northern Limestone Hills, Río Abajo Forest Reserve, Eastern Cordillera (Cerro La Peña), Susúa Forest Reserve, Maricao Forest Reserve; 225–825 m; *fl* Dec–Feb, *fr* Feb–Mar.

Bahamas, Greater Antilles, Lesser Antilles (Grenada), Trinidad, southeastern United States (Florida), northwestern, west-central South America and Venezuela, Brazil. [A: 48].

Encyclia gravida (Lindl.) Schltr., Beih. Bot. Centralbl., Abt. 2 36(2):472. 1918.
Bas.: *Epidendrum gravidum* Lindl. (1849).
Syn.: *Epidendrum sintenisii* Rchb. f. (1885).
Encyclia sintenisii (Rchb. f.) Britton (1924).

Epiphytic herb in Toro Negro Forest Reserve, Maricao Forest Reserve; 875–1000 m; *fl* Oct–Dec (two specimens), *fr* Dec–Jun.

Greater Antilles, Mexico, Central America. [A: 53].

Encyclia isochila (Rchb. f.) Dod, Moscosoa 4:193. 1986.
Bas.: *Epidendrum isochilum* Rchb. f. (1856).
Syn.: *Epidendrum isochilum* Rchb. f. var. *tridens* Rchb. f. (1885).
Epidendrum belvederense Fawc. & Rendle (1909).

Epiphytic herb in Northern Limestone Hills (Lares, Moca), Maricao Forest Reserve; 100–750 m; *fl* Oct (one specimen), *fr* Oct–Nov, Mar–May.

Greater Antilles. [A: 53].

Epidendrum ackermanii Hágsater, Icon. Orchid. (Mexico) 7(4): *t. 701*. 2004.

Terrestrial herb in Sierra de Cayey, Las Tetas de Cayey, Eastern Cordillera (Cerro La Peña, La Torrecilla), Central Cordillera, Maricao Forest Reserve; 500–1200 m; *fl* throughout the year, *fr* Feb–Jul.

Endemic. [A: 59].

Notes: Britton & Wilson (1924) misapplied the name *Amphiglottis secunda* (Jacq.) Salisb. to this taxon; Liogier & Martorell (1982) misapplied *Epidendrum elongatum* Jacq., and Ackerman (1995) and subsequently Liogier & Martorell (2000) misapplied *Epidendrum secundum* to it.

Epidendrum anceps Jacq., Select. Stirp. Amer. Hist. 224, *t. 138*. 1763.
Syn.: *Amphiglottis anceps* (Jacq.) Britton (1924).

Epiphytic and epilithic herb in Northern Limestone Hills (Arecibo, Quebradillas), Río Abajo Forest Reserve, Sierra de Luquillo, Eastern Cordillera (Cerro La Peña), Western Cordillera; 200–750 m; *fl* Jan–Jul, *fr* (?).

Greater Antilles. [A: 58].

Notes: Liogier & Martorell (1982), following Garay & Sweet in Howard (1974), misapplied the name *Epidendrum secundum* to this taxon.

Epidendrum angustilobum Fawc. & Rendle, J. Bot. 47:124. 1909.
Syn.: *Epidendrum latifolium* Garay & H.R. Sweet (1972).

Epiphytic herb in Sierra de Luquillo, Sierra de Cayey, Toro Negro Forest Reserve, Maricao Forest Reserve; 470–1150 m; *fl* Oct–Feb, May, *fr* Jun–Jul, Oct–Nov, Feb.

Greater Antilles, Lesser Antilles, southern Central America, northern South America. [A: 65].

Notes: Ackerman (1995), and subsequently Liogier & Martorell (2000), misapplied the name *Epidendrum carpophorum* Barb. Rodr. to this taxon; Nir (2000) misapplied it to *Epidendrum portoricense*.

Epidendrum antillanum Ackerman & Hágsater, Orquídea (Mexico City) 12(2):226. 1992.
Syn.: *Epidendrum ramosum* Jacq. var. *lanceolatum* Griseb. (1864).

Epiphytic herb in Sierra de Luquillo, Sierra de Cayey, Eastern Cordillera (La Torrecilla), Central Cordillera (Adjuntas); 760–1100 m; *fl* Aug–Sep, Dec–Feb, *fr* (?).

Greater Antilles (Hispaniola), Lesser Antilles. [A: 72].

Epidendrum boricuarum Hágsater & L. Sánchez, Icon Orchid. (Mexico) 2(1): *t. 114*. 1993.

Epiphytic herb in Sierra de Cayey, Las Tetas de Cayey, Eastern and Central Cordillera, Maricao Forest Reserve; 300–975 m; *fl* Nov–Aug, *fr* Jan–Mar.

Greater Antilles (Hispaniola). [A: 58].

Notes: According to Ackerman (1995), this taxon is a member of the *Epidendrum difforme* complex, within which the names *Epidendrum difforme* Jacq. used by Liogier & Martorell (1982) and *Amphiglottis difformis* (Jacq.) Britton used by Britton & Wilson (1924) now refer only to Lesser Antillean plants. Nir (2000) used *Epidendrum umbellatum* Sw. for the name of this taxon.

Epidendrum boricuomutelianum Hágsater & L. Sánchez, Icon Orchid. (Mexico) 7(4): *t. 711*. 2004.

Epiphytic and epilithic herb in Sierra de Luquillo, Sierra de Cayey; 900 m; *fl* Oct–Dec (two specimens), *fr* (?). [*Ackerman 2055* (UPRRP); *Axelrod 8355* (UPRRP)].

Endemic. [A: 66].

Notes: Britton & Wilson (1924) misapplied the name *Amphiglottis pallidiflora* (Hook.) Britton to this taxon; Liogier & Martorell (1982) misapplied *Epidendrum pallidiflorum* Hook., and Ackerman (1995) and subsequently Liogier & Martorell (2000) misapplied *Epidendrum mutelianum* Cogn., to it.

Epidendrum ciliare L., Syst. Nat. (ed. 10) 2:1246. 1759.
Syn.: *Auliza ciliaris* (L.) Salisb. (1812).

Epiphytic and epilithic herb in Northern Coastal Lowlands (Río Grande), Northern Limestone Hills (Toa Baja), Las Tetas de Cayey, Las Piedras Chiquitas, Susúa Forest Reserve, Maricao Forest Reserve, Vieques, Culebra; 1–825 m; *fl* Oct–May, *fr* Feb (one specimen).

Lesser Antilles, Trinidad, Mexico, Central America, northern South America to Bolivia. [A: 62].

Epidendrum jamaicense Lindl., Folia Orchid. 4(*Epidendrum*):82. 1853.

Epiphytic herb in Sierra de Luquillo, Toro Negro Forest Reserve; 375–800 m; *fl* Apr–Aug, *fr* (?).

Antilles. [A: 68].

Notes: Britton & Wilson (1924) misapplied the name *Amphiglottis corymbosa* (Lindl.) Britton to this taxon.

Epidendrum miserrimum Rchb. f., Bonplandia 3:220. 1855.

Epiphytic herb in Sierra de Cayey, Toro Negro Forest Reserve; 600–1325 m; *fl* Oct, Mar–Apr, *fr* Apr, Jul.

Greater Antilles (Hispaniola), Lesser Antilles, southern Central America, northern and northwestern South America. [A: 68].

Epidendrum nocturnum Jacq., Enum. Syst. Pl. 29. 1760.
Syn.: *Amphiglottis nocturna* (Jacq.) Britton (1924).

Epiphytic and epilithic herb in Sierra de Luquillo, Eastern Coast (Naguabo), Sierra de Cayey, Eastern Cordillera (Cerro La Peña); 40–800 m; *fl* Apr, Oct–Dec, *fr* May–Jul, Oct–Jan.

West Indies, southeastern United States (Florida), Mexico, Central America, northern South America to Bolivia. [A: 62].

Flor de San Pedro.

Epidendrum portoricense Hágsater & Ackerman, Icon. Orchid. (Mexico) 3(2): *t. 376*. 1999.

Epiphytic herb in Sierra de Luquillo, Sierra de Cayey; 450–600 m; *fl* Jun–Jul (one specimen), *fr* Sep–Oct (two specimens).

Greater Antilles (Cuba, Hispaniola). [A: 65].

Notes: Ackerman (1995), and subsequently

Liogier & Martorell (2000), misapplied the name *Epidendrum tridens* Poepp. & Endl. to this taxon.

Epidendrum radicans Pav. ex Lindl., Gen. Sp. Orchid. Pl. 104. 1831.

Terrestrial herb planted and persistent in Sierra de Luquillo; 400 m; *fl* Mar–Apr (two specimens), *fr* Mar (one specimen).

Native to Mexico, Central America, northern South America (Colombia). [A: 60].

Epidendrum ramosum Jacq., Enum. Syst. Pl. 29. 1760.

Syn.: *Spathiger ramosus* (Jacq.) Britton (1924).

Epiphytic herb in Northern Limestone Hills (Camuy), Sierra de Luquillo, Sierra de Cayey, Las Tetas de Cayey, Eastern Cordillera (La Torrecilla), Toro Negro Forest Reserve; 175–925 m; *fl* Aug–Apr, *fr* Mar–Apr, Jul–Aug.

Antilles, Mexico, Central America, northern South America to Bolivia. [A: 68].

Epidendrum rigidum Jacq., Enum. Syst. Pl. 29. 1760.

Syn.: *Spathiger rigidus* (Jacq.) Small (1913).

Epiphytic herb in Las Tetas de Cayey, Eastern Cordillera (La Torrecilla), Central Cordillera (Adjuntas); 400–1100 m; *fl* Nov, *fr* Nov–Jan (two specimens).

West Indies, southeastern United States (Florida), Mexico, Central America, South America (scattered). [A: 72].

Epidendrum vincentinum Lindl., J. Bot. (Hooker). 3:88, *t. 11*. 1840.

Terrestrial and epiphytic herb in Sierra de Luquillo, Sierra de Cayey, Maricao Forest Reserve; 700–950 m; *fl* Apr–Jun, *fr* Apr–Jul, Oct–Dec.

Antilles, Trinidad, southern Central America, northwestern and west-central South America. [A: 57].

Eulophia alta (L.) Fawc. & Rendle, Fl. Jamaica 1:112, *t. 22, ff. 4–8*. 1910.

Bas.: *Limodorum altum* L. (1767).

Syn.: *Platypus altus* (L.) Small (1913).

Terrestrial herb in Río Abajo Forest Reserve, Sierra de Luquillo, Eastern Coast (Cuchilla de Panduras), Sierra de Cayey; 250–925 m; *fl* Jun–Jan, *fr* Oct–Jan.

West Indies, Trinidad and Tobago, southeastern United States (Florida), Mexico, Central America, northwestern, west-central South America and Argentina, tropical Africa. [A: 76].

Palmita de San Juan.

Govenia utriculata (Sw.) Lindl., Edwards's Bot. Reg. 25: Misc. 47. 1839.

Bas.: *Limodorum utriculatum* Sw. (1788).

Terrestrial herb in Northern Limestone Hills (Arecibo), Central Cordillera, Maricao Forest Reserve; 200–950 m; *fl* Nov–Dec, *fr* Feb (two specimens).

Bahamas, Greater Antilles, Mexico. [A: 79].

Habenaria alata Hook., Exot. Fl. 3: *t. 169*. 1825.

Terrestrial herb in Toro Negro Forest Reserve, Maricao Forest Reserve; 800–900 m; *fl* Sep–Nov (two specimens), *fr* Nov (two specimens).

Bahamas, Greater Antilles (Cuba, Jamaica), Lesser Antilles, Tobago, Mexico, Central America, northern South America (Venezuela, Colombia) and Ecuador. [A: 84].

Habenaria amalfitana Lehm. & Kränzlin, Bot. Jahrb. Syst. 16(1):113. 1892.

Syn.: *Habenaria dussii* Cogn. (1909).

Terrestrial herb in Sierra de Luquillo; 750–900 m; *fl* Jan–Oct, *fr* Oct–Nov, Mar–Apr.

Lesser Antilles (Antigua, Guadeloupe, St. Vincent), northern South America (Venezuela, Colombia) and Ecuador. [A: 82].

Habenaria eustachya Rchb. f., Ber. Deutsch. Bot. Ges. 3:274. 1885.

Terrestrial herb in Sierra de Cayey, Toro Negro Forest Reserve, Maricao Forest Reserve; 650–900 m; *fl* Oct–Jan, *fr* Dec–Jan.

Greater Antilles, Lesser Antilles (Guadeloupe), Mexico, Central America. [A: 84].

Habenaria monorrhiza (Sw.) Rchb. f., Ber. Deutsch. Bot. Ges. 3:274. 1885.

Bas.: *Orchis monorrhiza* Sw. (1788).

Terrestrial herb in Northern Limestone Hills (Ciales), Río Abajo Forest Reserve, Sierra de Luquillo, Sierra de Cayey, Central Cordillera, Maricao Forest Reserve; 250–1250 m; *fl* Sep–Apr, *fr* Nov–Mar.

Antilles, Southern Caribbean Islands (Curaçao), Trinidad and Tobago, Central America, northwestern, west-central South America and Venezuela. [A: 80].

Habenaria repens Nutt., Gen. N. Amer. Pl. 2:190. 1818.

Terrestrial herb in Northern Limestone Hills (Quebradillas), Río Abajo Forest Reserve, Sierra de Luquillo, Sierra de Cayey; 225–1000 m; *fl* throughout the year, *fr* Mar–Dec.

Greater Antilles, Trinidad, southeastern and south-central United States, Mexico, Central America, northern, west-central South America and Paraguay. [A: 82].

Hapalorchis lineatus (Lindl.) Schltr., Beih. Bot. Centralbl., Abt. 2 37(2, Heft 3):363. 1920.
Bas.: *Spiranthes lineata* Lindl. (1840).
Syn.: *Hapalorchis tenuis* (Lindl.) Schltr. (1920).

Terrestrial herb in Sierra de Luquillo, Sierra de Cayey, Central Cordillera; 650–1100 m; *fl* Mar–Apr, *fr* Mar–Apr.

Greater Antilles, Central America (Guatemala, Nicaragua, Costa Rica), South America (scattered). [A: 85].

Ionopsis satyrioides (Sw.) Rchb. f., Ann. Bot. Syst. 6(5):683. 1864.
Bas.: *Epidendrum satyrioides* Sw. (1788).

Epiphytic herb in Northern Limestone Hills, Río Abajo Forest Reserve, Sierra de Cayey, Eastern and Central Cordillera; 50–1000 m; *fl* Feb–Apr, Aug, *fr* Apr–Aug.

Greater Antilles, Lesser Antilles (Guadeloupe, Martinique, St. Lucia), Central America, northern South America to Ecuador. [A: 88].

Ionopsis utricularioides (Sw.) Lindl., Coll. Bot. 8: *t. 39A*. 1826.
Bas.: *Epidendrum utricularioides* Sw. (1788).

Epiphytic herb in Northern Coastal Lowlands, Northern Limestone Hills, Río Abajo Forest Reserve, Caguas Valley (Caguas), Eastern Cordillera (Cayey, Cañón San Cristóbal), Central Cordillera (Ciales); Guánica Forest Reserve; 1–700 m; *fl* Mar–Sep, *fr* Aug–Oct.

Antilles, southeastern United States (Florida), Mexico, Central America, northern South America to Brazil, Bolivia. [A: 88].

Angelitos.

Isochilus linearis (Jacq.) R. Br. in W.T. Aiton, Hort. Kew. (ed. 2) 5:209. 1813.
Bas.: *Epidendrum lineare* Jacq. (1760).

Epiphytic herb in Central Cordillera, Maricao Forest Reserve; 510–1000 m; *fl* May, Oct, *fr* Feb, May–Jun, Oct.

Antilles, Trinidad, Mexico, Central America, South America. [A: 91].

Jacquiniella globosa (Jacq.) Schltr., Repert. Spec. Nov. Regni Veg. Beih. 7:124. 1920.
Bas.: *Epidendrum globosum* Jacq. (1760).

Epiphytic herb in Sierra de Luquillo, Sierra de Cayey, Las Tetas de Cayey, Eastern Cordillera (La Torrecilla); 150–925 m; *fl* Feb–Oct, *fr* Jan–Oct.

Antilles, Trinidad and Tobago, Mexico, Central America, northern South America to Brazil, Ecuador. [A: 94].

Jacquiniella teretifolia (Sw.) Britton & P. Wilson, Sci. Surv. Porto Rico & Virgin Islands 6(3):340. 1926 "Jacquinella."
Bas.: *Epidendrum teretifolium* Sw. (1788).

Epiphytic herb in Toro Negro Forest Reserve; 1150–1300 m; *fl* Aug, Oct (two specimens), *fr* Nov–Jan.

Greater Antilles, Lesser Antilles (Guadeloupe, Dominica), Mexico, Central America, northern South America (Venezuela, Colombia) and Brazil. [A: 91].

Leochilus puertoricensis M.W. Chase, Syst. Bot. 11(1):242. 1986.

Epiphytic herb in Northern Limestone Hills, Río Abajo Forest Reserve, Sierra de Luquillo, Sierra de Cayey, Eastern Cordillera (Aibonito), Central Cordillera (Ponce); 5–700 m; *fl* Mar–Jun, *fr* May, Aug (two specimens).

Lesser Antilles (Guadeloupe to St. Vincent). [A: 95].

Notes: Britton & Wilson (1924) and Liogier & Martorell (1982) misapplied the name *Leochilus labiatus* (Sw.) Kuntze to this taxon.

Lepanthes caritensis Tremblay & Ackerman, Brittonia 45(4):339, *t. 1*. 1993.

Epiphytic herb in Sierra de Luquillo, Sierra de Cayey; 600–900 m; *fl* Mar–Nov, *fr* (?).

Endemic. [A: 98, 100].

Notes: This taxon contains the Puerto Rican plants previously referred to as *Lepanthes sanguinea* Hook., a taxon found only in Jamaica.

Lepanthes dodiana Stimson, Brittonia 21:338, *t. 4*. 1970.

Epiphytic herb in Sierra de Luquillo, Sierra de Cayey, Eastern Cordillera (La Torrecilla), Toro Negro Forest Reserve; 325–1150 m; *fl* Aug, Nov–Apr, *fr* Mar–Apr, Aug–Oct.

Endemic. [A: 100].

Lepanthes eltoroensis Stimson, Brittonia 21:337, *t. 3*. 1970.

Epiphytic herb in Sierra de Luquillo; 750–900 m; *fl* Jan–Feb (one specimen), *fr* (?). [*Woodbury 20978* (SJ)].

Endemic. [A: 98].

Notes: This taxon is on the U.S. Fish & Wildlife Service list of threatened and endangered plants.

Lepanthes rubripetala Stimson, Brittonia 21:340, *t. 6*. 1970.

Epiphytic herb in Sierra de Luquillo, Sierra de Cayey; 250–750 m; *fl* Aug–Nov, Feb, May, *fr* Aug–Sep.

Endemic. [A: 104].

Lepanthes rupestris Stimson, Brittonia 21:339, *t.* 5. 1970.

Epiphytic herb in Sierra de Luquillo, Sierra de Cayey, Eastern Cordillera (La Torrecilla); 200–925 m; *fl* throughout the year, *fr* Apr–Aug.

Endemic. [A: 102].

Lepanthes selenitepala Rchb. f. subsp. **selenitepala**, Ber. Deutsch. Bot. Ges. 3:279. 1885.

Terrestrial and epiphytic herb in Maricao Forest Reserve; 480–860 m; *fl* Jan–Sep, *fr* (?).

Endemic. [A: 98].

Lepanthes selenitepala Rchb. f. subsp. **ackermanii** Luer, Selbyana 23(1):19, *t. 15.* 2002.

Terrestrial and epiphytic herb in Sierra de Luquillo; 215–450 m; *fl* Aug–Oct (two specimens), *fr* Mar (one specimen).

Endemic. [A: 98].

Lepanthes stimsonii Luer, Selbyana 23(1):22, *f. 17.* 2002.

Syn.: *Lepanthes veleziana* Stimson var. *retusicolumna* Stimson (1970).

Epiphytic herb in Sierra de Luquillo, Sierra de Cayey, Toro Negro Forest Reserve; 700–1260 m; *fl* Jan–Feb (two specimens), *fr* (?).

Endemic. [A: 104].

Lepanthes veleziana Stimson, Brittonia 21:343, *f. 8.* 1970.

Epiphytic herb in Sierra de Luquillo, Sierra de Cayey, Eastern Cordillera (La Torrecilla), Toro Negro Forest Reserve, Maricao Forest Reserve; 250–1300 m; *fl* throughout the year, *fr* Jan–May.

Endemic. [A: 104].

Lepanthes woodburyana Stimson, Brittonia 21:341, *f. 7.* 1970.

Epiphytic herb in Río Abajo Forest Reserve, Sierra de Luquillo, Sierra de Cayey, Eastern Cordillera (La Torrecilla), Central Cordillera, Maricao Forest Reserve; 300–1300 m; *fl* throughout the year, *fr* Aug–Dec.

Endemic. [A: 104].

Liparis nervosa (Thunb.) Lindl., Gen. Sp. Orchid. Pl. 26. 1830.

Bas.: *Ophrys nervosa* Thunb. (1784).

Syn.: *Liparis elata* Lindl. (1828).

Liparis eggersii Rchb. f. (1885).

Epiphytic herb in Northern Limestone Hills, Río Abajo Forest Reserve, Sierra de Luquillo, Sierra de Cayey, Toro Negro Forest Reserve, Western Cordillera; 30–1000 m; *fl* Jul–Jan, *fr* Sep–Mar.

Pantropical. [A: 106].

Liparis saundersiana Rchb. f., Gard. Chron. 1872: 1003. 1872.

Epiphytic herb in Northern Limestone Hills (Arecibo), Maricao Forest Reserve; 400–750 m; *fl* Oct–Nov, *fr* Nov–Jan (two specimens).

Greater Antilles (Jamaica, Hispaniola). [A: 106].

Liparis vexillifera (La Llave & Lex.) Cogn. in Martius, Fl. Bras. 3(4):289. 1895.

Bas.: *Cymbidium vexilliferum* La Llave & Lex. (1825).

Syn.: *Liparis jamaicensis* (Rchb. f.) Lindl. ex Griseb. (1866).

Terrestrial herb in Toro Negro Forest Reserve, Maricao Forest Reserve; 800–1200 m; *fl* & *fr* Nov (one specimen). [*Rodríguez* photograph, *a.n. 35707* (UPRRP)].

Greater Antilles, Trinidad, Mexico, Central America, west-central South America (Ecuador, Peru). [A: 106].

Notes: Liogier & Martorell (1982) misapplied the name *Liparis harrisii* Fawc. & Rendle to this taxon.

Malaxis massonii (Ridl.) Kuntze, Revis. Gen. Pl. 2:673. 1891.

Bas.: *Microstylis massonii* Ridl. (1888).

Terrestrial herb in Northern Limestone Hills, Río Abajo Forest Reserve, Sierra de Luquillo, Sierra de Cayey, Las Tetas de Cayey, Central Cordillera (Utuado); 100–800 m; *fl* Sep–Feb, *fr* Jan–Mar.

Greater Antilles (Hispaniola), Lesser Antilles, Trinidad and Tobago, northern South America (Venezuela). [A: 110].

Mesadenus lucayanus (Britton) Schltr., Beih. Bot. Centralbl., Abt. 2 37(2, Heft 3):368. 1920.

Bas.: *Ibidium lucayanum* Britton (1907).

Syn.: *Spiranthes stahlii* Cogn. (1910).

Terrestrial herb in Las Piedras Chiquitas, Guánica Forest Reserve; 25–550 m; *fl* Feb–Mar, *fr* Feb (one specimen).

Bahamas, Greater Antilles (Cuba, Hispaniola), Lesser Antilles (Antigua), southeastern United States (Florida), Mexico, northern Central America (Guatemala). [A: 116].

Notes: Ackerman (1995), and subsequently Liogier & Martorell (2000) and Nir (2000), misapplied the name *Mesadenus polyanthus* (Rchb. f.) Schltr. to this taxon; Garay & Sweet in Howard (1974), and subsequently Liogier & Martorell (1982), misapplied *Spiranthes polyantha* Rchb. f. to it.

Microchilus familiaris Ormerod, Harvard Pap. Bot. 14(2):121. 2009.

Terrestrial herb in Northern Limestone Hills

(Isabela), Río Abajo Forest Reserve, Sierra de Luquillo, Sierra de Cayey, Central Cordillera, Maricao Forest Reserve; 250–1000 m; *fl* & *fr* Jan–May.

Greater Antilles (Cuba, Hispaniola).

Notes: Specimens of this taxon were previously identified as *Microchilus hirtellus*, but can be distinguished from the latter by, among other things, a rounded leaf base and a laxer inflorescence.

Microchilus hirtellus (Sw.) D. Dietr., Syn. Pl. 5:166. 1852.
Bas.: *Satyrium hirtellum* Sw. (1788).
Syn.: *Physurus hirtellus* (Sw.) Lindl. (1840). *Erythrodes hirtella* (Sw.) Fawc. & Rendle (1910).

Terrestrial herb in Northern Limestone Hills, Río Abajo Forest Reserve, Sierra de Luquillo, Eastern Cordillera (La Torrecilla), Central Cordillera; 100–1000 m; *fl* & *fr* Jan–May.

Antilles, Trinidad and Tobago, northern South America (Suriname, French Guiana, Guyana). [A: 74].

Microchilus plantagineus (L.) D. Dietr., Syn. Pl. 5:166. 1852.
Bas.: *Satyrium plantagineum* L. (1759).
Syn.: *Physurus plantagineus* (L.) Lindl. (1840). *Erythrodes plantaginea* (L.) Fawc. & Rendle (1910).

Terrestrial herb in Sierra de Luquillo, Caguas Valley (Caguas), Sierra de Cayey, Central Cordillera; 300–1250 m; *fl* Nov–Apr, *fr* Feb–Apr.

Antilles. [A: 74].

Mormolyca pudica (Carnevali & J.L. Tapia) M.A. Blanco, Lankesteriana 7(3):531. 2007.
Bas.: *Maxillaria pudica* Carnevali & J.L. Tapia (2001).

Epiphytic herb in Sierra de Luquillo; 200–650 m; *fl* Jan, May, Aug, *fr* Jan, Apr–May (two specimens).

Greater Antilles, Lesser Antilles (Guadeloupe). [A: 114].

Notes: Liogier & Martorell (1982) used *Maxillaria rufescens* Lindl., *sensu lato*, for the name of this taxon; Ackerman (1995), and subsequenty Nir (2000) and Liogier & Martorell (2000), misapplied the name *Maxillaria acutifolia* Lindl. to it.

Oeceoclades maculata (Lindl.) Lindl., Gen. Sp. Orchid. Pl. 237. 1833.
Bas.: *Angraecum maculatum* Lindl. (1821).

Terrestrial herb in Northern Coastal Lowlands (Guaynabo, Laguna Tortuguero), Northern Limestone Hills, Río Abajo Forest Reserve, Sierra de Luquillo, Eastern Coast, Eastern Cordillera (Cayey, Comerío), Central Cordillera, Southern Coastal Lowlands (Patillas), Susúa Forest Reserve, Western Cordillera, Guánica Forest Reserve, Southwestern Coastal Lowlands (Cabo Rojo), Sierra Bermeja, Western Coast, Vieques, Mona, Desecheo; 10–750 m; *fl* Aug–Dec, *fr* Oct–Apr.

Native to tropical Africa. [A: 119].

Oncidium altissimum (Jacq.) Sw., Kongl. Vetensk. Acad. Nya Handl. 21:240. 1800.
Bas.: *Epidendrum altissimum* Jacq. (1760).
Syn.: *Oncidium wydleri* Rchb. f. (1885).

Epiphytic and epilithic herb in Northern Limestone Hills, Río Abajo Forest Reserve, Sierra de Luquillo, Las Tetas de Cayey, Eastern Cordillera (Cañón San Cristóbal), Central Cordillera, Susúa Forest Reserve, Maricao Forest Reserve; 50–920 m; *fl* Mar–Nov, *fr* Mar, Jun (two specimens).

Virgin Islands (St. Thomas), Lesser Antilles. [A: 121].

Ornithidium coccineum (Jacq.) Salisb. ex R. Br. in W.T. Aiton, Hort. Kew. (ed. 2) 5:210. 1813.
Bas.: *Epidendrum coccineum* Jacq. (1760).
Syn.: *Maxillaria coccinea* (Jacq.) L.O. Williams (1954).

Epiphytic herb in Sierra de Luquillo, Sierra de Cayey, Eastern Cordillera (La Torrecilla), Toro Negro Forest Reserve; 200–1250 m; *fl* Jan–May, *fr* Feb, May–Jun.

Greater Antilles (Hispaniola), Lesser Antilles, Trinidad. [A: 112].

Pelexia adnata (Sw.) Spreng., Syst. Veg. 3:704. 1826.
Bas.: *Satyrium adnatum* Sw. (1788).
Syn.: *Spiranthes adnata* (Sw.) Benth. ex Fawc. (1893).

Terrestrial herb in Río Abajo Forest Reserve, Sierra de Luquillo, Central Cordillera (Ciales, Ponce), Western Cordillera; 200–825 m; *fl* Apr–May, *fr* Apr–May.

West Indies, southeastern United States (Florida), Mexico, Central America, northern South America (Venezuela). [A: 124].

Hachuela.

Pleurothallis angustifolia Lindl., Edwards's Bot. Reg. 21: *sub t. 1797*. 1835.
Syn.: *Pleurothallis wilsonii* Lindl. (1858).

Epiphytic herb in Northern Limestone Hills (Florida, Moca), Sierra de Cayey, Las Tetas de Cayey, Southeastern Foothills (Salinas), Central Cordillera, Maricao Forest Reserve; 100–800 m; *fl* Aug–Jan, *fr* Oct–May.

Greater Antilles, Lesser Antilles (Dominica, Guadeloupe), Mexico, Central America, northern South America (Venezuela, Colombia). [A: 127].

Pleurothallis appendiculata Cogn. in Urban, Symb. Antill. 7(2):174. 1912.

Epiphytic herb in Maricao Forest Reserve; 800 m; *fl* (?), *fr* Jul (one specimen). [*Woodbury a.n. 1839* (SJ)].

Greater Antilles (Cuba, Hispaniola). [A: 135].

Pleurothallis aristata Hook., Ann. Nat. Hist. 2(11):329, *t. 15*. 1839.

Syn.: *Pleurothallis urbaniana* Rchb. f. (1885).

Epiphytic herb in Sierra de Luquillo, Central Cordillera (Adjuntas), Maricao Forest Reserve; 600–1200 m; *fl* Apr–Oct, *fr* Sep (one specimen).

Greater Antilles, Lesser Antilles, southern Central America, northern South America to Peru. [A: 127].

Pleurothallis domingensis Cogn. in Urban, Symb. Antill. 6(2):402. 1909.

Epiphytic herb in Sierra de Luquillo, Sierra de Cayey, Eastern Cordillera (La Torrecilla), Central Cordillera, Maricao Forest Reserve; 475–1300 m; *fl* Aug–Mar, *fr* Aug–Apr.

Greater Antilles (Cuba, Hispaniola). [A: 132].

Notes: Britton & Wilson (1924) misapplied the name *Pleurothallis crassipes* Lindl. to this taxon.

Pleurothallis gelida Lindl., Edwards's Bot. Reg. 27: Misc. 91. 1841.

Epiphytic herb in Northern Limestone Hills (Quebradillas), Toro Negro Forest Reserve, Maricao Forest Reserve; 200–960 m; *fl* Jan–Mar, *fr* Jan–Mar.

Greater Antilles, southeastern United States (Florida), Mexico, Central America, northwestern, west-central South America and Venezuela. [A: 127].

Pleurothallis oblongifolia Lindl., Companion Bot. Mag. 2(24):355. 1836.

Epiphytic herb in Sierra de Luquillo; 950 m; *fl* Feb (one specimen), *fr* (?). [*Tremblay a.n. 28293* (UPRRP)].

Greater Antilles, northern South America (Venezuela).

Pleurothallis pruinosa Lindl., Edwards's Bot. Reg. 28: Misc. 75. 1842.

Epiphytic and epilithic herb in Sierra de Luquillo, Maricao Forest Reserve; 215–625 m; *fl* Dec, Mar–Apr, *fr* Jan–Jul.

Antilles, Trinidad, Central America, northern, west-central South America and Brazil. [A: 127].

Pleurothallis pubescens Lindl., Companion Bot. Mag. 2(24):355. 1836.

Syn.: *Pleurothallis coriacea* Bello (1883).

Epiphytic herb in Las Tetas de Cayey, Eastern Cordillera (Cayey, Aibonito), Central Cordillera, Maricao Forest Reserve; 450–950 m; *fl* Sep–Feb, *fr* Nov–Apr, Jul.

Greater Antilles (Hispaniola), Mexico, Central America, South America (scattered). [A: 135].

Pleurothallis quadrifida (La Llave & Lex.) Lindl., Edwards's Bot. Reg. 28: Misc. 70. 1842.

Bas.: *Dendrobium quadrifidum* La Llave & Lex. (1825).

Syn.: *Pleurothallis racemiflora* Lindl. (1825).

Pleurothallis ghiesbreghtiana A. Rich. & Galeotti (1845).

Pleurothallis longissima Lindl. (1859).

Epiphytic herb in Maricao Forest Reserve; 525–650 m; *fl* Mar (one specimen), *fr* Mar–May.

Greater Antilles, Mexico, Central America, northern South America (Venezuela, Colombia). [A: 132].

Notes: Ackerman (1995), and subsequently Liogier & Martorell (2000), misapplied the name *Pleurothallis racemiflora* Lindl. ex. Hook. to this taxon; that name is based upon *Epidendrum racemiflorum* Sw., which is a synonym of *Pleurothallis oblongifolia* Lindl. The homonym *Pleurothallis racemiflora* Lindl., which indeed is a synonym of this taxon, was published a year later in Loddiges's "Botanical Cabinet."

Pleurothallis ruscifolia (Jacq.) R. Br. in W.T. Aiton, Hort. Kew. (ed. 2) 5:211. 1813.

Bas.: *Epidendrum ruscifolium* Jacq. (1760).

Epiphytic and epilithic herb in Sierra de Luquillo, Eastern Cordillera (La Torrecilla), Central Cordillera, Maricao Forest Reserve; 250–1300 m; *fl* Nov–Jun, *fr* Jul–Nov.

Antilles, Trinidad, southern Central America, northern South America to Brazil, Bolivia. [A: 132].

Polystachya concreta (Jacq.) Garay & H.R. Sweet, Orquideologia 9(3):206. 1974 Nov; et in Howard, Fl. Lesser Antilles 1:178. 1974 Dec.

Bas.: *Epidendrum concretum* Jacq. (1760).

Syn.: *Polystachya minuta* (Aubl.) Britton (1903), *nom. illeg.*

Epiphytic herb in Northern Limestone Hills, Sierra de Luquillo, Eastern Coast, Sierra de Cayey, Las Tetas de Cayey, Eastern Cordillera (Cerro La Peña, Cidra), Central Cordillera (Peñuelas), Maricao Forest Reserve; 300–825 m; *fl* Sep–May, *fr* Oct–Nov, Feb, May.

West Indies, Trinidad and Tobago, southeastern United States (Florida), Central America (Nicaragua, Costa Rica), northwestern, west-central South America and Argentina. [A: 137].

Parásita.

Polystachya foliosa (Hook.) Rchb. f., Ann. Bot. Syst. 6(4):640. 1863.

Bas.: *Stelis foliosa* Hook. (1839).

Syn.: *Polystachya cerea* Lindl. (1840).
Polystachya minor Fawc. & Rendle (1910).

Epiphytic herb in Northern Limestone Hills (Arecibo), Río Abajo Forest Reserve, Sierra de Luquillo, Eastern Cordillera (Cerro La Peña), Maricao Forest Reserve; 300–550 m; *fl* May–Jun, Sep–Feb, *fr* Feb–Nov.

West Indies, Southern Caribbean Islands (Curaçao), Trinidad, Mexico, Central America, northwestern, west-central South America and Venezuela, Argentina. [A: 137].

Parásita.

Ponthieva racemosa (Walter) C. Mohr, Contr. U.S. Natl. Herb. 6:460. 1901.
Bas.: *Arethusa racemosa* Walter (1788).

Terrestrial herb in Northern Limestone Hills, Río Abajo Forest Reserve, Central Cordillera, Maricao Forest Reserve; 200–800 m; *fl* Jan–Mar, *fr* Jan (one specimen).

West Indies, southeastern and south-central United States, Mexico, Central America, northern South America (Venezuela). [A: 140].

Ponthieva ventricosa (Griseb.) Fawc. & Rendle, Fl. Jamaica 1:39. 1910.
Bas.: *Cranichis ventricosa* Griseb. (1864).

Terrestrial herb in Northern Limestone Hills, Río Abajo Forest Reserve, Eastern Cordillera (Corozal), Maricao Forest Reserve; 100–800 m; *fl* Nov–Mar, *fr* Jan–May.

Greater Antilles. [A: 140].

Prescottia oligantha (Sw.) Lindl., Gen. Sp. Orchid. Pl. 454. 1840.
Bas.: *Cranichis oligantha* Sw. (1788).

Terrestrial herb in Northern Coastal Lowlands (Dorado to Laguna Tortuguero), Río Abajo Forest Reserve, Sierra de Luquillo, Sierra de Cayey, Las Tetas de Cayey, Eastern Cordillera (Cidra, Corozal), Central Cordillera, Maricao Forest Reserve; 1–1000 m; *fl* Jan–Mar, *fr* Feb–Apr.

West Indies, Central America, northwestern, west-central South America and Venezuela, Argentina. [A: 142].

Prescottia stachyodes (Sw.) Lindl., Edwards's Bot. Reg. 22: *sub t. 1915* ("*t. 1916*"). 1836.
Bas.: *Cranichis stachyodes* Sw. (1788).
Syn.: *Prescottia pellucida* Lindl. (1858).

Terrestrial herb in Sierra de Luquillo, Sierra de Cayey, Central Cordillera (Adjuntas), Maricao Forest Reserve; 200–1050 m; *fl* Jan–Apr, *fr* Jan–Apr.

Antilles, Mexico, Central America, South America (scattered). [A: 142].

Prosthechea cochleata (L.) W. E. Higgins, Phytologia 82(5):377. 1998.
Bas.: *Epidendrum cochleatum* L. (1763).
Syn.: *Anacheilium cochleatum* (L.) Hoffmanns. (1842). *Encyclia cochleata* (L.) Dressler (1961).

Epiphytic and epilithic herb in Northern Limestone Hills (Arecibo), Río Abajo Forest Reserve, Sierra de Luquillo, Las Tetas de Cayey, Las Piedras Chiquitas, Eastern Cordillera (Cerro La Peña), Central Cordillera, Western Cordillera, Sierra Bermeja; 100–900 m; *fl* throughout the year, *fr* Jan–Mar, Jun–Aug.

West Indies, southeastern United States (Florida), Mexico, Central America, northern South America. [A: 50].

Cañuela, Orquídea negra.

Prosthechea pygmaea (Hook.) W. E. Higgins, Phytologia 82(5):380. 1998.
Bas.: *Epidendrum pygmaeum* Hook. (1833).
Syn.: *Encyclia pygmaea* (Hook.) Dressler (1961).
Hormidium tripterum (Brongn.) Cogn. (1898).

Terrestrial, epiphytic and epilithic herb in Sierra de Luquillo, Las Tetas de Cayey, Central Cordillera, Maricao Forest Reserve; 200–1025 m; *fl* Nov–Mar, *fr* Jan–May.

Greater Antilles, Lesser Antilles (Dominica), southeastern United States (Florida), Mexico, Central America, northern South America to Brazil, Bolivia. [A: 50].

Psilochilus macrophyllus (Lindl.) Ames, Orchidaceae 7:45, *t. 110*. 1922.
Bas.: *Pogonia macrophylla* Lindl. (1858).

Terrestrial herb in Eastern Cordillera (Cerro La Peña), Central Cordillera, Maricao Forest Reserve; 500–1100 m; *fl* Jan–Sep, *fr* Mar, Jul–Nov.

Antilles, Mexico, Central America, northern South America. [A: 147].

Psychilis kraenzlinii (Bello) Sauleda, Phytologia 65(1):15. 1988.
Bas.: *Epidendrum kraenzlinii* Bello (1883).

Epiphytic herb in Northern Coastal Lowlands (Río Grande, Laguna Tortuguero), Northern Limestone Hills, Río Abajo Forest Reserve, Las Piedras Chiquitas, Susúa Forest Reserve, Maricao Forest Reserve; 1–650 m; *fl* throughout the year, *fr* Jun (one specimen).

Endemic. [A: 149].

Psychilis krugii (Bello) Sauleda, Phytologia 65(1):17. 1988.

Bas.: *Epidendrum krugii* Bello (1883).
Syn.: *Encyclia krugii* (Bello) Britton & P. Wilson (1930).

Epiphytic herb in Southern Coastal Limestone, Susúa Forest Reserve, Guánica Forest Reserve, Sierra Bermeja; 10–300 m; *fl* throughout the year, *fr* Dec–Feb, Jun–Sep.

Endemic. [A: 149].

Psychilis macconnelliae Sauleda, Phytologia 65(1):18. 1988.

Epiphytic herb in Eastern Coast (Punta Guayanés), Vieques, Culebra; 1–100 m; *fl* Jan–Feb, Jun–Aug, *fr* Feb, Jun–Aug.

Virgin Islands, Lesser Antilles (St. Kitts). [A: 151].

Notes: Britton & Wilson (1924) misapplied the name *Encyclia papilionacea* (Vahl) Schltr., and Liogier & Martorell (1982) misapplied *Epidendrum bifidum* Aubl., to this taxon.

Psychilis monensis Sauleda, Phytologia 65(1):20. 1988.

Epiphytic herb in Mona; 40–100 m; *fl* Jan–Feb, May–Oct, *fr* Feb, Aug (two specimens).

Endemic. [A: 151].

Sacoila lanceolata (Aubl.) Garay, Bot. Mus. Leafl. 28(4):352. 1982.
Bas.: *Limodorum lanceolatum* Aubl. (1775).
Syn.: *Stenorrhynchos lanceolatum* (Aubl.) Rich. ex Spreng. (1826). *Spiranthes lanceolata* (Aubl.) León (1946).

Terrestrial herb in Northern Coastal Lowlands (San Juan, Vega Baja), Sierra de Luquillo, Sierra de Cayey, Las Tetas de Cayey, Central Cordillera (Ponce), Susúa Forest Reserve, Maricao Forest Reserve, Western Coast (Mayagüez); 1–700 m; *fl* Mar–Jun, *fr* Apr–Jun.

Neotropical. [A: 156].

Scaphyglottis modesta (Rchb. f.) Schltr., Repert. Spec. Nov. Regni Veg. 23:46. 1926.
Bas.: *Tetragamestus modestus* Rchb. f. (1854).

Epiphytic herb in Sierra de Luquillo, Maricao Forest Reserve; 450–550 m; *fl* Jan (one specimen), *fr* Jun–Jul.

Greater Antilles (Cuba, Hispaniola), Lesser Antilles, northern South America to Brazil, Ecuador. [A: 152].

Scaphyglottis punctulata (Rchb. f.) C. Schweinf., Bot. Mus. Leafl. 17(2):47. 1955.
Bas.: *Ponera punctulata* Rchb. f. (1855).
Syn.: *Helleriella punctulata* (Rchb. f.) Garay & H.R. Sweet (1974).

Epiphytic herb in Sierra de Luquillo, Toro Negro Forest Reserve; 500–1250 m; *fl* Sep, Jan–Mar, *fr* Oct, Jan–Jun.

Lesser Antilles, southern Central America (Panama), northwestern and west-central South America. [A: 87].

Spathoglottis plicata Blume, Bijdr. Fl. Ned. Ind. 401, *t. 76.* 1825.

Terrestrial herb in Río Abajo Forest Reserve, Sierra de Luquillo, Sierra de Cayey, Eastern Cordillera (La Torrecilla); 250–900 m; *fl* Sep–Jun, *fr* Sep–Apr.

Native to tropical Asia. [A: 152].

Spiranthes torta (Thunb.) Garay & H.R. Sweet in Howard, Fl. Lesser Antilles 1:77, *t.* 22. 1974.
Bas.: *Ophrys torta* Thunb. (1791).
Syn.: *Ibidium tortile* (Sw.) House (1906).

Terrestrial herb in Northern Coastal Lowlands, Northern Limestone Hills (Moca), Río Abajo Forest Reserve, Sierra de Luquillo, Sierra de Cayey, Eastern and Central Cordillera, Susúa Forest Reserve, Western Cordillera, Western Coast (Añasco); 1–950 m; *fl* Sep–May, *fr* Mar–May.

West Indies, Trinidad, southern United States (Florida, Louisiana), Mexico, Central America (except Panama). [A: 156].

Ibis.

Stelis perpusilliflora Cogn. in Urban, Symb. Antill. 6(2):386. 1909.

Epiphytic herb in Sierra de Luquillo, Sierra de Cayey, Toro Negro Forest Reserve; 550–1250 m; *fl* May, Sep–Nov, *fr* Feb–May, Aug–Nov.

Lesser Antilles (Guadeloupe, Dominica). [A: 156].

Stelis pygmaea Cogn. in Urban, Symb. Antill. 6(2):390. 1909.

Epiphytic herb in Sierra de Luquillo; 450–600 m; *fl* Jul (one speciman), *fr* Apr–Aug.

Greater Antilles (Hispaniola), Lesser Antilles (Guadeloupe, Dominica, Martinique, St. Vincent). [A: 155].

Tetramicra canaliculata (Aubl.) Urb., Repert. Spec. Nov. Regni Veg. 15:306. 1918.
Bas.: *Limodorum canaliculatum* Aubl. (1775).
Syn.: *Tetramicra elegans* (Ham.) Cogn. (1910).

Terrestrial herb in Susúa Forest Reserve, Sierra Bermeja; 150–300 m; *fl* Dec–May, *fr* (?).

Greater Antilles (Hispaniola), Virgin Islands, Lesser Antilles. [A: 159].

Tolumnia prionochila (Kränzlin) Braem, Orchidee (Hamburg) 37(2):58. 1986.
Bas.: *Oncidium prionochilum* Kränzlin (1922).

Terrestrial and epiphytic herb in Culebra; 1–100 m; *fl* May–Jul, *fr* (?).

Virgin Islands. [A: 161].

Tolumnia variegata (Sw.) Braem, Orchidee (Hamburg) 37(2):59. 1986.
Bas.: *Epidendrum variegatum* Sw. (1788).
Syn.: *Oncidium variegatum* (Sw.) Sw. (1800).

Epiphytic herb in Northern Coastal Lowlands, Northern Limestone Hills, Río Abajo Forest Reserve, Las Tetas de Cayey, Southeastern Foothills (Coamo), Eastern and Central Cordillera, Susúa Forest Reserve, Maricao Forest Reserve, Guánica Forest Reserve, Southwestern Coastal Lowlands (Cabo Rojo), Vieques, Culebra, Mona; 1–800 m; *fl* throughout the year, *fr* Oct–Feb, Jun.

Greater Antilles (Cuba, Hispaniola), Virgin Islands. [A: 161].

Notes: Britton & Wilson (1924) and Liogier & Martorell (1982) misapplied the name *Oncidium leiboldii* Rchb. f. to this taxon.

Angelitos, Angelitos del cafetal.

Trichosalpinx dura (Lindl.) Luer, Phytologia 54(5): 395. 1983.
Bas.: *Pleurothallis dura* Lindl. (1859).
Syn.: *Pleurothallis foliata* Griseb. (1864).

Epiphytic herb in Sierra de Luquillo; 350–1000 m; *fl* (?), *fr* (?). [*Stenzel a.n.* 33723 (UPRRP)].

Antilles, southern Central America, northwestern, west-central South America and Venezuela. [A: 164].

Triphora amazonica Schltr., Beih. Bot. Centralbl., Abt. 2 42(2):75. 1925.
Syn.: *Triphora latifolia* G.M. Luer (1969).

Terrestrial herb in Sierra de Luquillo; 250–650 m; *fl* Dec, Mar (two specimens), *fr* May–Jun (two specimens).

Greater Antilles (Hispaniola), Trinidad, southeastern United States (Florida), South America (French Guiana, Guyana, Brazil). [A: 168].

Triphora hassleriana (Cogn. ex Chodat & Hassl.) Schltr., Beih. Bot. Centralbl., Abt. 2 42(2):76. 1925.
Bas.: *Pogonia hassleriana* Cogn. ex Chodat & Hassl. (1907).

Terrestrial herb in Sierra de Cayey; 250–400 m; *fl* Jun–Jul (two specimens), *fr* Jul (one specimen). [*Ackerman 2436, 2611* (UPRRP)]

Greater Antilles (Hispaniola), Mexico, South America (Colombia, Ecuador, Paraguay, Argentina). [A: 165].

Triphora surinamensis (Lindl.) Britton in Britton & Wilson, Sci. Surv. Porto Rico & Virgin Islands 5(2):184. 1924.
Bas.: *Pogonia surinamensis* Lindl. (1843).

Terrestrial herb in Sierra de Luquillo, Sierra de Cayey, Central Cordillera (Ciales), Maricao Forest Reserve; 400–900 m; *fl* Apr–Aug, *fr* Jul (one specimen).

Greater Antilles, Lesser Antilles (Guadeloupe, Dominica, St. Lucia), Trinidad, northern South America to Brazil. [A: 165].

Vanilla barbellata Rchb. f., Flora 48(18):274. 1865.

Vine in Northern Limestone Hills, Guajataca Forest Reserve, Southern Coastal Limestone, Susúa Forest Reserve, Guánica Forest Reserve, Sierra Bermeja, Vieques; 10–300 m; *fl* Apr–Jul, *fr* Jan, Apr–Sep.

Bahamas, Greater Antilles (Cuba, Hispaniola), Virgin Islands, southeastern United States (Florida). [A: 171; AR: 430].

Vanilla claviculata Sw., Nova Acta Regiae Soc. Sci. Upsal. 6:66, *t. 5, f. 1.* 1799.

Vine in Susúa Forest Reserve, Maricao Forest Reserve, Sierra Bermeja; 150–900 m; *fl* Apr–Jun, *fr* Jan, May–Jun.

Greater Antilles. [A: 174; AR: 430].

Notes: Liogier & Martorell (1982, 2000) and Ackerman (1995) used *Vanilla claviculata* (W. Wright) Sw. for the name of this taxon; but, the cited basionym, *Epidendrum claviculatum* Sw., turns out to be a *nom. illeg. hom.*

Vanilla dilloniana Correll, Amer. Orchid Soc. Bull. 15(7):331, *tt. 1–7.* 1946.

Vine in Southern Coastal Limestone (Yauco), Susúa Forest Reserve, Maricao Forest Reserve; 75–800 m; *fl* Apr–Jun, *fr* Apr–Jun, Oct.

Greater Antilles (Cuba, Hispaniola), southeastern United States (Florida). [A: 171; AR: 430].

Gungulén.

Vanilla mexicana Mill., Gard. Dict. (ed. 8): *Vanilla* no. 1. 1768.
Syn.: *Epidendrum vanilla* L. (1753). *Vanilla vanilla* (L.) Britton (1924).
Vanilla inodora Schiede (1829).

Vine in Guajataca Forest Reserve, Sierra de Luquillo; 150–550 m; *fl* (?), *fr* (?).

Greater Antilles, Lesser Antilles (Guadeloupe to St. Lucia), Trinidad and Tobago, southeastern United States (Florida), Mexico, Central America, northern South America to Ecuador. [A: 174; AR: 432].

Notes: Britton & Wilson (1924) wrongly treated *Vanilla planifolia* as a synonym of *Vanilla vanilla*.

Lombricera.

Vanilla planifolia Andrews, Bot. Repos. 8: *t. 538.* 1808.

Vine in Northern Limestone Hills (Morovis), Sierra de Luquillo, Central Cordillera (Ponce), Maricao Forest Reserve; 150–825 m; *fl* Mar–Jul, *fr* (?).

Native to Mexico, Central America (probably). [A: 176; AR: 432].

Vainilla.

Vanilla poitaei Rchb. f., Linnaea 41(1):66. 1876.

Syn.: *Vanilla eggersii* Rolfe (1896).

Vine in Northern Limestone Hills, Río Abajo Forest Reserve, Sierra de Luquillo, Eastern Coast (Punta Guayanés), Susúa Forest Reserve, Maricao Forest Reserve, Southwestern Coastal Lowlands (Cabo Rojo); 10–900 m; *fl* Apr–Jul, *fr* throughout the year.

Bahamas, Greater Antilles (Cuba, Hispaniola). [A: 169; AR: 434].

Notes: Liogier & Martorell (1982) wrongly placed *Vanilla eggersii* in synonymy with *Vanilla dilloniana*.

Vanilla pompona Schiede, Linnaea 4(4):573. 1829.

Vine in Western Coast (Mayagüez), Maricao Forest Reserve; 5–800 m; *fl* (?), *fr* (?).

Lesser Antilles (Guadeloupe, Dominica, Martinique), Trinidad, Mexico, Central America, northern South America to Paraguay. [A: 174; AR: 434].

Wullschlaegelia calcarata Benth., J. Linn. Soc., Bot. 18:342. 1880.

Terrestrial herb in Sierra de Luquillo; 200–650 m; *fl* Mar–Aug, *fr* Mar–Aug.

Greater Antilles (Hispaniola), Lesser Antilles (Guadeloupe to St. Lucia), Trinidad, Central America, northern South America to Bolivia, Brazil. [A: 176].

Notes: Britton & Wilson (1926), Liogier & Martorell (1982, 2000) and Ackerman (1995) misapplied the name *Wullschlaegelia aphylla* (Sw.) Rchb. f. to this taxon.

Zeuxine strateumatica (L.) Schltr., Bot. Jahrb. Syst. 45(3):394. 1911.

Bas.: *Orchis strateumatica* L. (1753).

Terrestrial herb in Northern Limestone Hills (Lares); 300 m; *fl* Oct (one specimen), *fr* (?). [*Muñiz a.n. 14518* (UPRRP), photograph].

Native to tropical Asia, China, Japan. [A: 178].

TAXA NOT RECENTLY COLLECTED

Brassavola nodosa (L.) Lindl., Gen. Sp. Orchid. Pl. 114. 1831.

Bas.: *Epidendrum nodosum* L. (1753).

Epiphytic herb collected in Puerto Rico (no locality given) in 1874 [*Kuntze a.n. 168087* (NY), scan seen].

Greater Antilles (Jamaica), Mexico, Central America, northern South America (Venezuela, Colombia). [A: 19].

Bulbophyllum pachyrachis (A. Rich.) Griseb., Fl. Brit. W.I. 613. 1864 "*Bolbophyllum pachyrrachis.*"

Bas.: *Pleurothallis pachyrachis* A. Rich. (1850).

Syn.: *Bolbophyllaria pachyrachis* (A. Rich.) Rchb. f. (1861).

Epiphytic and epilithic herb collected in Utuado on 6 Mar 1887 [*Sintenis 6471* (NY), scan seen].

Greater Antilles, Trinidad, Mexico, Central America, northern and northwestern South America to Bolivia. [A: 20].

Notes: Ackerman (1995) claimed that it had more recently been observed by Donald Dod in the Maricao Forest Reserve, but no voucher was made.

Camaridium grisebachianum (Nir & Dod) M.A. Blanco, Lankesteriana 7(3):520. 2007.

Bas.: *Maxillaria grisebachiana* Nir & Dod (2000).

Epiphytic herb collected near Utuado on 25 Feb 1887 [*Sintenis 6300* (NY), scan seen].

Greater Antilles (Cuba, Hispaniola). [A: 114].

Notes: Britton & Wilson (1924) used *Ornithidium confertum* Griseb., and Liogier & Martorell (1982) used *Maxillaria conferta* (Griseb.) C. Schweinf. ex León, both *sensu lato*, for the name of this taxon; Ackerman (1995), and subsequently Liogier & Martorell (2000), used *Maxillaria parviflora* (Poepp. & Endl.) Garay, *sensu lato*, for it.

Cyrtopodium punctatum (L.) Lindl., Gen. Sp. Orchid. Pl. 188. 1833.

Bas.: *Epidendrum punctatum* L. (1759).

Epiphytic herb collected in Bayamón on 17 Jan 1945 [*Dávila Olivo 1846* (UPR)].

Greater Antilles (Cuba, Hispaniola), southeastern United States (Florida), northwestern, west-central South America and Venezuela, Argentina. [A: 40].

Cañuela, Piñuela.

Malaxis major (Rchb. f.) León ex A.D. Hawkes, Phytologia 3:260. 1950.

Bas.: *Microstylis major* Rchb. f. (1888).

Terrestrial herb known from a specimen in the Reichenbach Orchid Herbarium from an unknown collector [type specimen, locality unknown, not seen].

Lesser Antilles (Guadeloupe, Dominica, Martinique). [A: 110].

Notes: This taxon is known to occur in Puerto Rico from the type specimen only.

Nidema ottonis (Rchb. f.) Britton & Millsp., Bahama Fl. 94. 1920.
Bas.: *Epidendrum ottonis* Rchb. f. (1858).

Epiphytic herb collected on palm trunks near Utuado on 5 Apr 1887 [*Sintenis 6608* (NY), scan seen].

Greater Antilles, Trinidad, Central America (Nicaragua, Costa Rica, Panama), northwestern South America to Bolivia and Venezuela. [A: 116].

Pleurothallis obovata (Lindl.) Lindl., Edwards's Bot. Reg. 28: Misc. 75. 1842.
Bas.: *Specklinia obovata* Lindl. (1839).

Epiphytic herb collected in the Luquillo Mountains on 19 Feb 1935 [*Horne* 4 (NY), scan seen].

Greater Antilles (Cuba, Dominican Republic), Central America (Nicaragua), South America (scattered). [A: 130].

TAXA CITED BY PREVIOUS AUTHORS, BUT NOT INCLUDED

NO SPECIMENS SEEN

Eurystyles ananassocomos (Rchb. f.) Schltr.
Epiphytic herb cited by Liogier & Martorell (1982, 2000) as occurring on trunks of palms in Carite Forest, collected by Roy Woodbury, not collected again; subsequently cited by Ackerman (1995), who states that the plant was not likely to have been misidentified at the generic level. [A: 78].

Galeandra beyrichii Rchb. f.
Terrestrial herb cited by Ackerman (1995) as having been observed in the Maricao Forest Reserve by R. Padrón and J. Ricart, between 1980 and 1984. A single photograph was made whose whereabouts is now unknown. [A: 79].

Habenaria distans Griseb.
Terrestrial herb cited by Britton & Wilson (1924) as occurring in Cerro Las Mesas; subsequently cited by Liogier & Martorell (1982) as occurring in the same locality and by Liogier & Martorell (2000) as occurring in moist or wet regions at middle to high elevations; cited by Ackerman (1995) also as occurring in the Maricao Forest and in El Yunque, as reported by Donald Dod. [A: 82].

Ida barringtoniae (Sm.) A. Ryan & Oakeley. Syn.: *Lycaste barringtoniae* (Sm.) Lindl.
Epiphytic herb cited by Ackerman (1995) as, according to Donald Dod, discovered in Maricao by H. Winters. [A: 109].

Lepanthopsis melanantha (Rchb. f.) Ames.
Epiphytic herb cited by Ackerman (1995) as recorded by Winters & Schweinfurth from San Germán, but he was unable to locate a specimen purportedly made of it; he remarked that Donald Dod claimed to have seen plants in the same geographical area. [A: 105].

DOUBTFUL OCCURRENCE

Koellensteinia graminea (Lindl.) Rchb. f.
Terrestrial herb cited by Britton & Wilson (1924) as occurring on trees near Utuado; subsequently cited by Liogier & Martorell (1982) and by Ackerman (1995), who was unable to locate any specimen and suspected that the identification was mistaken and then by Liogier & Martorell (2000) as occurring in moist regions at middle elevations, very rare. [A: 95].

Tropidia polystachya (Sw.) Ames.
Terrestrial herb cited by Liogier & Martorell (1982) as collected both by Winters and by Dod in western Puerto Rico; however, Ackerman (1995), stating that the specimen was immature and the identification doubtful, excluded the taxon. [A: 180].

CITED DUE TO MISIDENTIFICATION

Brassavola cucullata (L.) R. Br. Syn.: *Epidendrum cucullatum* L.
Epiphytic herb cited by Liogier & Martorell (1982) as occurring on 'rocks and trees, Puerto Rico,' a misreading of on 'rocks and trees, St. Thomas' as stated by Britton and Wilson (1924). They later (2000) corrected the error. [A: 18].

Habenaria odontopetala Rchb. f.
Terrestrial herb cited by Liogier & Martorell (1982, 2000) and by Ackerman (1995). Ackerman (pers. comm.) now believes that all the specimens previously so identified are referable to *Habenaria eustachya*. [A: 84].

Malaxis spicata Sw.
Terrestrial herb cited by Britton & Wilson (1924) as occurring in wet woods and river banks at lower and middle elevations; subsequently cited by Liogier & Martorell (1982, 2000). Ackerman (1995) found most, if not all, specimens so identified to be referable to *Malaxis massonii*. [A: 110].

Stenorrhynchos speciosum (Jacq.) Rich. ex Spreng.
Syn.: *Spiranthes speciosa* (Jacq.) A. Rich.
Terrestrial herb cited by Britton & Wilson (1924) as occurring in primeval forest near Lares; subsequently cited by Liogier & Martorell (1982, 2000) and by

Ackerman (1995). Ackerman (pers. comm.) now believes that red-flowered specimens of *Sacoila lanceolata* may have been confused with it and that, therefore, it does not occur in Puerto Rico. [A: 158].

Triphora trianthophora (Sw.) Rydb.
Terrestrial herb cited by Liogier & Martorell (1982) as occurring in wet or moist western districts. Ackerman (1995) found that the specimens upon which the citation was based are referable to *Triphora amazonica*. [A: 180].

ONLY UNDER CULTIVATION
Phaius tancarvilleae (Banks ex L'Hér.) Blume.
Terrestrial herb cited by Ackerman (1995) as known from a naturalized population in Luquillo; subsequently cited by Liogier & Martorell (2000). [A: 124].

OROBANCHACEAE

Agalinis fasciculata (Elliott) Raf., New Fl. 2:63. 1837.
Bas.: *Gerardia fasciculata* Elliott (1822).

Herb in Northern Coastal Lowlands (Laguna Tortuguero), Sierra de Cayey, Eastern Cordillera (Cerro La Peña, Aguas Buenas, Aibonito), Central Cordillera; 300–1075 m; *fl* Apr–Oct, *fr* May–Oct.

Greater Antilles (Hispaniola), eastern and central United States. [L4: 480].

Yerba Verónica.

Alectra aspera (Cham. & Schltdl.) L.O. Williams, Fieldiana, Bot. 34(8):118. 1972.
Bas.: *Glossostylis aspera* Cham. & Schltdl. (1828).
Syn.: *Alectra fluminensis* (Vell.) Stearn (1971). *Melasma melampyroides* (Rich.) Pennell ex Britton & P. Wilson (1925).

Herb in Río Abajo Forest Reserve; 325 m; *fl* Sep (one specimen), *fr* (?). [*Liogier 30847* (UPR)].

Greater Antilles (Jamaica, Hispaniola), Lesser Antilles, Central America, northwestern, west-central South America and Venezuela. [L4: 481].

Notes: Liogier & Martorell (1972) used the homonym *Alectra aspera* (Cham. & Schltdl.) Stearn for the name of this taxon.

Yerba de hierro.

Buchnera longifolia Kunth in Humboldt, Bonpland & Kunth, Nov. Gen. Sp. 2:274 [folio]; 2:340 [quarto]. 1818.
Syn.: *Buchnera elongata* Sw. (1788), *nom. illeg. superfl.*

Herb in Northern Coastal Lowlands, Eastern Cordillera (Aguas Buenas); 1–550 m; *fl* Apr–Jun, Sep–Nov, *fr* Jun, Sep–Nov.

Bahamas, Greater Antilles, Trinidad, southern United States (Georgia, Alabama, Mississippi), Central America, southern South America. [L4: 490].

Espiga de San Antonio.

TAXA CITED BY PREVIOUS AUTHORS, BUT NOT INCLUDED

CITED DUE TO MISIDENTIFICATION
Buchnera americana L. Syn.: *Buchnera floridana* Gand.
Cited by Liogier (1995) and Liogier & Martorell (2000) as rare at middle elevations. All specimens that I have seen so identified are referable to *Buchnera longifolia*. [L4: 489].

OXALIDACEAE

Averrhoa carambola L., Sp. Pl. 1:428. 1753.

Tree planted and persistent in Southern Coastal Lowlands (Ponce), Western Coast (Mayagüez); 15–100 m; *fl* Oct, Feb, May, *fr* Feb (one specimen).

Native to Indonesia. [L2: 233].

Carambola, Jalea.

Oxalis barrelieri L., Sp. Pl. (ed. 2) 1:624. 1762.

Herb in Northern Coastal Lowlands, Sierra de Luquillo, Eastern Coast, Caguas Valley, Southeastern Foothills (Patillas, Arroyo), Eastern Cordillera (Corozal, Cayey), Western Coast (Mayagüez), Vieques; 1–650 m; *fl* & *fr* throughout the year.

Greater Antilles (Hispaniola), Lesser Antilles, Trinidad, southern Central America, northern South America to Bolivia. [L2: 235].

Oxalis corniculata L., Sp. Pl. 1:435. 1753.
Syn.: *Xanthoxalis corniculata* (L.) Small (1903).

Herb in Northern Coastal Lowlands, Eastern Coast (Ceiba), Sierra de Cayey, Las Piedras Chiquitas, Eastern Cordillera (Corozal), Central Cordillera, Southern Coastal Lowlands (Santa Isabel); 5–1335 m; *fl* & *fr* throughout the year.

Worldwide weed of obscure origin. [L2: 235].

Trebolillo, Trebolillo amarillo, Trebolillo de jardín, Vinagrillo.

Oxalis debilis Kunth var. **corymbosa** (DC.) Lourteig, Ann. Missouri Bot. Gard. 67(4):840. 1981.
Bas.: *Oxalis corymbosa* DC. (1824).
Syn.: *Ionoxalis martiana* (Zucc.) Small (1903).

Herb in Northern Coastal Lowlands (San Juan), Northern Limestone Hills, Río Abajo Forest Reserve, Sierra de Luquillo, Caguas Valley (Caguas), Eastern Cordillera (Corozal), Central Cordillera, Susúa Forest Reserve, Western Cordillera; 5–1300 m; *fl* Jan–Jul, *fr* Jan–May.

Native to central and southern South America. [L2: 235].

Vinagrillo morado.

Oxalis eggersii Urb., Symb. Antill. 5(3):375. 1908.

Herb in Las Tetas de Cayey, Las Piedras Chiquitas; 300–800 m; *fl* & *fr* Oct–Dec. [*Proctor 42490, 42703* (SJ); *Axelrod 645, 3014* (UPRRP)].

Bahamas, Greater Antilles (Cuba, Hispaniola). [L2: 236].

Oxalis latifolia Kunth in Humboldt, Bonpland & Kunth, Nov. Gen. Sp. 5:184 [folio]; 5:237 [quarto], *t. 467.* 1822.

Syn.: *Ionoxalis intermedia* (A. Rich.) Small (1907).

Herb in Northern Coastal Lowlands (San Juan, Isabela), Eastern Cordillera (Cañón San Cristóbal); 10–600 m; *fl* May–Jul, Oct, *fr* May–Oct.

Greater Antilles, Lesser Antilles (Guadeloupe, Martinique), southwestern United States (New Mexico, California), Mexico, Central America, northern South America to Bolivia. [L2: 237].

Lóbulo, Trébol de jardín, Violeta de monte.

Oxalis rugeliana Urb., Symb. Antill. 7(2):234. 1912.

Herb in Southeastern Foothills (Coamo); 250–350 m; *fl* Nov (one specimen), *fr* (?). [*Woodbury a.n. 8502* (SJ)].

Greater Antilles (Cuba, Hispaniola). [L2: 237].

TAXA CITED BY PREVIOUS AUTHORS, BUT NOT INCLUDED

CITED DUE TO MISIDENTIFICATION

Oxalis violacea L.

Herb cited by Liogier & Martorell (1982, 2000) and Liogier (1988) as cultivated and escaped, based on the misidentification of a specimen of *Oxalis latifolia* Kunth. [L2: 237].

PANDANACEAE

TAXA CITED BY PREVIOUS AUTHORS, BUT NOT INCLUDED

ONLY UNDER CULTIVATION

Pandanus dubius Spreng.

Suffrutescent plant cited by Liogier & Martorell (1982, 2000) as cultivated and probably persistent.

Notes: Liogier & Martorell (1982, 2000) used *Pandanus pacificus* J.H. Veitch ex Mast., a *nom. nud.*, for the name of this taxon.

Pandanus odorifer (Forssk.) Kuntze. Syn.: *Pandanus odoratissimus* L.f. *Pandanus fascicularis* Lam.

Suffrutescent plant cited by Liogier & Martorell (1982, 2000) as introduced and persistent.

Pandanus tectorius Parkinson ex Du Roi. Syn.: *Pandanus veitchii* Mast. *Pandanus baptistii* Misonne.

Suffrutescent plant cited by Liogier & Martorell (1982, 2000) as introduced and persistent.

Pandanus utilis Bory.

Suffrutescent plant cited by Liogier & Martorell (1982, 2000) as planted as an ornamental and persistent.

PAPAVERACEAE

Argemone mexicana L., Sp. Pl. 1:508. 1753.

Herb in Northern Coastal Lowlands, Eastern Coast (Naguabo), Las Piedras Chiquitas, Southeastern Foothills, Eastern Cordillera (Aibonito), Southern Coastal Lowlands, Southwestern Coastal Lowlands, Sierra Bermeja, Vieques; 10–650 m; *fl* & *fr* throughout the year.

West Indies, eastern and south-central United States, Mexico, Central America, South America; exact native range uncertain. [L1: 272].

Cardo santo.

Bocconia frutescens L., Sp. Pl. 1:505. 1753.

Shrub in Northern Limestone Hills, Río Abajo Forest Reserve, Sierra de Cayey, Las Piedras Chiquitas, Southeastern Foothills (Salinas), Central Cordillera, Maricao Forest Reserve, Western Coast (Añasco); 225–1300 m; *fl* Sep–Apr, *fr* Jul–Mar.

West Indies, Mexico, Central America, northwestern, west-central South America and Venezuela. [L1: 275].

Palo de pan cimarrón, Pan cimarrón, Panapén cimarrón, Panilla.

PASSIFLORACEAE

Passiflora berteroana Balb. ex DC., Prodr. 3:325. 1828 "Berteriana."

Vine in Guánica Forest Reserve; 1–50 m; *fl* Sep, Dec (two specimens), *fr* (?). [*Liogier 33732, 34436*].

Greater Antilles (Cuba, Hispaniola). [L3: 291; AR: 328].

Passiflora bilobata Juss., Ann. Mus. Natl. Hist. Nat. 6:107, *t. 37. f. 2.* 1805.

Syn.: *Passiflora anadenia* Urb. (1902).

Vine in Las Piedras Chiquitas, Southeastern Foothills (Coamo), Southern Coastal Limestone, Susúa Forest Reserve, Maricao Forest Reserve, Guánica Forest Reserve; 50–650 m; *fl* May–Jun, Oct, Jan, *fr* Jan, Apr–Jun.

Greater Antilles (Hispaniola). [L3: 290, 291; AR: 328].

Notes: The linear leaves purported to be the distinguishing characteristic of *Passiflora anadenia* are found on young plants of this taxon.

Passiflora edulis Sims, Bot. Mag. 45: *t. 1989*. 1818.

Vine in Northern Coastal Lowlands, Northern Limestone Hills, Sierra de Luquillo, Eastern Coast (Naguabo), Sierra de Cayey, Las Tetas de Cayey, Central Cordillera (Adjuntas), Susúa Forest Reserve, Guánica Forest Reserve, Western Coast (Aguada), Culebra; 1–975 m; *fl* Mar–Nov, *fr* Mar–May, Aug–Oct.

Native to southern South America (Paraguay, Argentina) and Brazil. [L3: 291; AR: 328].

Parcha.

Passiflora foetida L., Sp. Pl. 2:959. 1753.

Syn.: *Passiflora foetida* L. var. *hispida* (DC. ex Triana & Planch.) Killip ex Gleason (1931).

Passiflora foetida L. var. *riparia* (C. Wright ex Griseb.) Killip (1938).

Vine in Northern Coastal Lowlands, Sierra de Luquillo, Eastern Coast (Ceiba), Southeastern Foothills (Cayey), Southern Coastal Lowlands, Maricao Forest Reserve, Western Coast (Mayagüez), Vieques; 1–660 m; *fl* & *fr* throughout the year.

West Indies, Trinidad and Tobago, southern United States (Florida, Texas, Arizona), Mexico, Central America, South America. [L3: 292; AR: 331].

Flor de pasión silvestre, Tagua-tagua.

Passiflora laurifolia L., Sp. Pl. 2:956. 1753.

Vine in Northern Coastal Lowlands, Northern Limestone Hills, Southern Coastal Lowlands (Santa Isabel), Cerro Las Mesas; 15–300 m; *fl* Nov, Feb, Aug, *fr* Aug (two specimens).

Antilles, northern South America to Bolivia. [L3: 292; AR: 331].

Parcha.

Passiflora maliformis L., Sp. Pl. 2:956. 1753.

Vine in Northern Coastal Lowlands (Dorado); 5 m; *fl* Apr, Sep (two specimens), *fr* Sep (one specimen). [*Woodbury a.n. 2187* (UPR)].

Native to northwestern, west-central South America and Venezuela. [L3: 292; AR: 333].

Parcha cimarrona.

Passiflora multiflora L., Sp. Pl. 2:956. 1753.

Vine in Northern Limestone Hills, Southeastern Foothills (Coamo), Cerro de Las Cuevas; 10–525 m; *fl* Dec–Mar, *fr* Nov–Apr, Jul.

Bahamas, Greater Antilles (Cuba, Hispaniola), Virgin Islands (St. John, Tortola), southeastern United States (Florida). [L3: 292; AR: 333].

Passiflora murucuja L., Sp. Pl. 2:957. 1753.

Vine in Northern Limestone Hills, Río Abajo Forest Reserve; 20–300 m; *fl* Apr (one specimen), *fr* Apr–May (two specimens).

Greater Antilles (Hispaniola). [L3: 292; AR: 334].

Passiflora quadrangularis L., Syst. Nat. (ed. 10) 2:1248. 1759.

Vine in Río Abajo Forest Reserve; 300 m; *fl* (?), *fr* (?). [*Acevedo-Rodríguez 10558* (US)].

Probably originating in the Neotropics. [L3: 292; AR: 334].

Granadilla.

Passiflora rubra L., Sp. Pl. 2:956. 1753.

Vine in Northern Coastal Lowlands (Trujillo Alto, Guaynabo), Northern Limestone Hills, Río Abajo Forest Reserve, Sierra de Luquillo, Caguas Valley (Caguas), Sierra de Cayey, Las Piedras Chiquitas, Southeastern Foothills (Cayey), Eastern Cordillera (Cerro La Peña, Cayey), Central Cordillera, Southern Coastal Limestone (Yauco), Susúa Forest Reserve, Maricao Forest Reserve; 50–750 m; *fl* & *fr* throughout the year.

West Indies, Trinidad, Mexico, northern South America to Bolivia. [L3: 294; AR: 334].

Flor de pasión, Parcha colorada.

Passiflora serratodigitata L., Sp. Pl. 2:960. 1753 "serrato-digitata."

Vine in Sierra de Luquillo; 80–300 m; *fl* Aug, Jan, Oct, *fr* Aug, Feb (two specimens).

Greater Antilles (Hispaniola), Lesser Antilles, Trinidad and Tobago, southern Central America (Panama), northern South America to Brazil, Bolivia. [L3: 294; AR: 337].

Tagua-tagua.

Passiflora sexflora Juss., Ann. Mus. Natl. Hist. Nat. 6:110, *t. 37*, *f. 1*. 1805.

Vine in Northern Limestone Hills (Florida), Río Abajo Forest Reserve, Sierra de Luquillo, Sierra de Cayey, Southeastern Foothills (Coamo), Eastern Cordillera (Cerro La Peña, La Torrecilla), Central Cordillera, Maricao Forest Reserve; 150–1300 m; *fl* Sep–May, *fr* Jul–Apr.

Greater Antilles, Lesser Antilles (Saba, St. Kitts), southeastern United States (Florida), Mexico, Central America, northwestern South America. [L3: 295; AR: 337].

Parchita.

Passiflora suberosa L., Sp. Pl. 2:958. 1753.

Vine in Northern Coastal Lowlands, Northern Limestone Hills, Río Abajo Forest Reserve, Eastern Coast, Las Tetas de Cayey, Las Piedras Chiquitas, Southeastern Foothills, Eastern Cordillera (Cidra), Central Cordillera, Southern Coastal Lowlands, Southern Coastal Limestone, Susúa Forest Reserve, Western Cordillera, Guánica Forest Reserve, Southwestern Coastal Lowlands, Sierra Bermeja, Cerro Las Mesas, Western Coast (Rincón), Vieques, Culebra, Mona, Desecheo; 1–850 m; *fl* & *fr* throughout the year.

West Indies, southern United States (Texas, Florida), Mexico, Central America, South America. [L3: 295; AR: 339].

Flor de pasión, Parcha yedra.

Passiflora tulae Urb., Symb. Antill. 1(2):374. 1899.

Vine in Guajataca Forest Reserve, Río Abajo Forest Reserve, Caguas Valley (Caguas), Sierra de Cayey, Las Tetas de Cayey, Eastern Cordillera (Cerro La Peña), Monte Guilarte Forest Reserve, Maricao Forest Reserve; 300–1000 m; *fl* Jan–Apr, *fr* Feb–Jun.

Endemic. [L3: 296; AR: 339].

Flor de pasión, Tagua-tagua serrana.

PEDALIACEAE

Sesamum indicum L., Sp. Pl. 2:634. 1753, *nom. cons.*

Syn.: *Sesamum orientale* L. (1753), *nom. rej.*

Herb planted and persistent in Northern Coastal Lowlands (Carolina); 5–25 m; *fl* Jun (one specimen), *fr* (?). [*Liogier 34406* (UPR)].

Native to India. [L4: 558].

Ajonjolí.

PENTAPHYLACACEAE

Cleyera albopunctata (Griseb.) Krug & Urb., Bot. Jahrb. Syst. 21(5):537. 1896.

Bas.: *Ternstroemia albopunctata* Griseb. (1866).

Syn.: *Eroteum albopunctatum* (Griseb.) Britton (1924).

Tree in Sierra de Luquillo, Sierra de Cayey; 750–1000 m; *fl* Jul–Aug, Nov, *fr* (?).

Greater Antilles (Cuba, Hispaniola). [L3: 221].

Teta prieta.

Ternstroemia heptasepala Krug & Urb., Bot. Jahrb. Syst. 21(5):530. 1896.

Syn.: *Taonabo heptasepala* (Krug & Urb.) Britton (1924).

Tree in Sierra de Luquillo; 775–1000 m; *fl* Apr–Aug, *fr* Jul–Aug.

Endemic. [L3: 224].

Ternstroemia luquillensis Krug & Urb., Bot. Jahrb. Syst. 21(5):531. 1896.

Syn.: *Taonabo luquillensis* (Krug & Urb.) Britton (1924).

Tree in Sierra de Luquillo, Maricao Forest Reserve; 375–750 m; *fl* Sep (two specimens), *fr* (?).

Endemic. [L3: 224].

Notes: This taxon is on the U.S. Fish & Wildlife Service list of threatened and endangered plants.

Palo colorado.

Ternstroemia peduncularis DC., Mém. Soc. Phys. Genève 1:409. 1822.

Syn.: *Taonabo peduncularis* (DC.) Britton (1918).

Tree in Eastern Coast (Punta Guayanés); 50–100 m; *fl* Mar–May (two specimens), *fr* Mar, Jun, Oct.

Greater Antilles (Cuba, Hispaniola), Virgin Islands (St. John, St. Croix), Lesser Antilles. [L3: 226].

Ternstroemia stahlii Krug & Urb., Bot. Jahrb. Syst. 21(5):527. 1896.

Syn.: *Taonabo stahlii* (Krug & Urb.) Britton (1924).

Taonabo pachyphylla (Krug & Urb.) Britton (1924).

Tree in Northern Coastal Lowlands (Dorado to Laguna Tortuguero), Northern Limestone Hills, Río Abajo Forest Reserve, Sierra de Luquillo, Eastern Cordillera (La Torrecilla), Susúa Forest Reserve, Maricao Forest Reserve, Cerro Las Mesas; 1–750 m; *fl* throughout the year, *fr* Jan–Oct.

Endemic. [L3: 226].

Cupeyillo, Mamey del cura, Palo de buey.

Ternstroemia subsessilis (Britton) Kobuski, J. Arnold Arbor. 24:66. 1943.

Bas.: *Taonabo subsessilis* Britton (1924).

Shrub/small tree in Sierra de Luquillo, Maricao Forest Reserve; 675–900 m; *fl* (?), *fr* Jun (one specimen).

Endemic. [L3: 226].

Notes: This taxon is on the U.S. Fish & Wildlife Service list of threatened and endangered plants.

PHYLLANTHACEAE

Breynia disticha J.R. Forst. & G. Forst., Char. Gen. Pl. 146, *t. 73.* 1775.

Shrub planted and persistent in Piñones Forest Reserve, Central Cordillera (Ciales); 1–550 m; *fl* Feb (one specimen), *fr* (?).

Native to the southwestern Pacific (New Caledonia, Vanuatu). [L2: 361].

Carnaval, Nevado.

Flueggea acidoton (L.) G.L. Webster, Allertonia 3(4):299. 1984.
Bas.: *Adelia acidoton* L. (1759).
Syn.: *Securinega acidoton* (L.) Fawc. & Rendle (1919).

Shrub in Southern Coastal Lowlands, Guánica Forest Reserve, Southwestern Coastal Lowlands (Cabo Rojo), Sierra Bermeja, Vieques, Culebra, Mona, Desecheo; 2–50 m; *fl* Jun–Nov, *fr* Mar, Jul–Oct.

Bahamas, Greater Antilles, Virgin Islands. [L2: 430].

Hieronyma clusioides (Tul.) Griseb., Pl. Wright. 1:157. 1860 (preprint of Mem. Amer. Acad. Arts, n.s. 8(1):157. 1861).
Bas.: *Stilaginella clusioides* Tul. (1851).

Tree in Northern Limestone Hills (Arecibo), Río Abajo Forest Reserve, Monte Guilarte Forest Reserve, Maricao Forest Reserve; 250–900 m; *fl* Jun–Jul, *fr* Jun–Oct.

Endemic. [L2: 401].

Notes: The spelling of the genus name *Hieronyma* has been conserved against such orthographic variants as *Hyeronima* and *Hieronima*.

Cedro macho.

Margaritaria nobilis L.f., Suppl. Pl. 428. 1782.
Syn.: *Phyllanthus nobilis* (L.f.) Müll. Arg. (1866).

Tree in Northern Coastal Lowlands (Fajardo), Northern Limestone Hills, Sierra de Luquillo, Eastern Coast, Southeastern Foothills (Cayey), Southern Coastal Lowlands (Patillas), Susúa Forest Reserve, Maricao Forest Reserve, Southwestern Coastal Lowlands (Cabo Rojo), Vieques; 1–375 m; *fl* Feb, May–Jul, *fr* May–Nov.

Antilles, Mexico, Central America, South America. [L2: 407].

Amortiguado, Avispillo, Higuillo, Millo, Palo de millo, Siete cueros, Yuquillo.

Phyllanthus acidus (L.) Skeels, U.S.D.A. Bur. Pl. Industr. Bull. 148:17. 1909.
Bas.: *Averrhoa acida* L. (1753).
Syn.: *Cicca disticha* L. (1767).

Tree planted and persistent in Northern Coastal Lowlands, Northern Limestone Hills (San Sebastián), Mona; 1–300 m; *fl* Apr (one specimen), *fr* Mar–Oct.

Native to Brazil. [L2: 416].

Cereza amarilla, Cerezo agrio, Grosella, Grosella blanca.

Phyllanthus amarus Schumach., Beskr. Guin. Pl. 421. 1827.

Herb in Northern Coastal Lowlands, Eastern Coast (Maunabo), Caguas Valley (Caguas), Southern Coastal Lowlands (Coamo, Guayama), Southwestern Coastal Lowlands (Cabo Rojo), Sierra Bermeja, Vieques, Mona, Desecheo; 1–250 m; *fl* & *fr* throughout the year.

West Indies, Mexico, Central America, South America. [L2: 417].

Phyllanthus cuneifolius (Britton) Croizat, J. Washington Acad. Sci. 33:12. 1943.
Bas.: *Andrachne cuneifolia* Britton (1920).

Shrub in Las Piedras Chiquitas, Susúa Forest Reserve, Guánica Forest Reserve, Southwestern Coastal Lowlands (Cabo Rojo); 1–500 m; *fl* throughout the year, *fr* Jan–May, Aug.

Greater Antilles (Cuba, Hispaniola). [L2: 418].

Phyllanthus debilis Klein ex Willd., Sp. Pl. 4(1):582. 1805.

Herb in Northern Coastal Lowlands (San Juan); 5–15 m; *fl* & *fr* May (one specimen). [*Liogier 28890* (UPR)].

Native to Indian Subcontinent. [L2: 418].

Notes: Since this taxon has been collected only on the grounds of the University of Puerto Rico Botanical Garden, it was probably an inadvertent and only temporary introduction.

Phyllanthus epiphyllanthus L. subsp. **epiphyllanthus**, Sp. Pl. 2:981. 1753.
Syn.: *Xylophylla epiphyllanthus* (L.) Britton (1913).

Shrub in Northern Limestone Hills, Río Abajo Forest Reserve, Southern Coastal Limestone (Ponce), Susúa Forest Reserve, Guánica Forest Reserve, Vieques, Mona; 1–350 m; *fl* Mar–Dec, *fr* Mar–Nov.

West Indies. [L2: 419].

Bayoneta, Lengua de vaca.

Phyllanthus fraternus G.L. Webster, Contr. Gray Herb. 176:53. 1955.

Herb in Northern Coastal Lowlands (Guaynabo), Western Coast (Mayagüez); 1–50 m; *fl* & *fr* Aug–Oct (two specimens). [*Liogier 32323* (UPR); *Unander 1449* (UPR)].

Native to Indian Subcontinent. [L5: 407].

Phyllanthus juglandifolius Willd. subsp. **juglandifolius**, Enum. Pl. Suppl. 64. 1813.
Syn.: *Asterandra grandifolia* (L.) Britton (1918).

Shrub/small tree in Northern Coastal Lowlands (Laguna Tortuguero), Northern Limestone Hills, Southeastern Foothills (Salinas), Eastern Cordillera (Corozal), Cerro de Las Cuevas; 10–450 m; *fl* Mar–Jul, Oct, *fr* Mar–Nov.

Greater Antilles (Cuba, Hispaniola), Virgin Islands (St. Thomas). [L2: 419].

Gamo de costa, Jagüerillo, Tobillo.

Phyllanthus niruri L., Sp. Pl. 2:981. 1753.

Syn.: *Phyllanthus lathyroides* Kunth (1817).

Herb in Northern Limestone Hills, Río Abajo Forest Reserve, Eastern Coast, Sierra de Cayey (San Lorenzo), Las Tetas de Cayey, Maricao Forest Reserve, Western Coast (Mayagüez), Vieques; 5–825 m; *fl* & *fr* Apr–Nov.

Greater Antilles (Hispaniola), Lesser Antilles, Mexico, Central America, South America. [L2: 420].

Quinino de pobre, Viernes Santo.

Phyllanthus pentaphyllus C. Wright ex Griseb. subsp. **pentaphyllus**, Nachr. Königl. Ges. Wiss. Georg-Augusts-Univ. 1:167. 1865.

Suffrutescent herb in Las Tetas de Cayey, Southern Coastal Limestone (Guayanilla), Susúa Forest Reserve, Maricao Forest Reserve, Guánica Forest Reserve; 2–825 m; *fl* & *fr* Jun, Oct–Apr.

West Indies. [L2: 421].

Phyllanthus pentaphyllus C. Wright ex Griseb. subsp. **polycladus** (Urb.) G.L. Webster, Contr. Gray Herb. 176:56. 1955.

Bas.: *Phyllanthus polycladus* Urb. (1899).

Suffrutescent herb in Northern Coastal Lowlands (islands adjacent to Fajardo), Southern Coastal Lowlands (Caja de Muertos), Southern Coastal Limestone (Ponce), Guánica Forest Reserve; 5–150 m; *fl* Apr–Sep, Dec–Jan, *fr* May–Sep, Dec–Jan.

Virgin Islands (Anegada), Lesser Antilles (Guadeloupe). [L2: 421].

Phyllanthus stipulatus (Raf.) G.L. Webster, Contr. Gray Herb. 176:53. 1955.

Bas.: *Moeroris stipulata* Raf. (1838).

Syn.: *Phyllanthus diffusus* Klotzsch (1853).

Herb in Northern Coastal Lowlands, Sierra de Luquillo, Eastern Coast (Cuchilla de Panduras), Caguas Valley, Central Cordillera (Utuado, Adjuntas), Maricao Forest Reserve; 1–800 m; *fl* & *fr* Feb–Sep.

Widespread in the Neotropics. [L2: 421].

Phyllanthus tenellus Roxb., Fl. Ind. (ed. 1832) 3:668. 1832

Herb in Caguas Valley (Caguas); 100 m; *fl* & *fr* Apr.

Native to eastern tropical Africa, Yemen. [L2: 422].

Phyllanthus urinaria L., Sp. Pl. 2:982. 1753.

Herb in Northern Coastal Lowlands, Río Abajo Forest Reserve, Sierra de Luquillo, Eastern Coast, Caguas Valley (Caguas), Sierra de Cayey, Southeastern Foothills (Arroyo), Eastern Cordillera (Cañón San Cristóbal), Central Cordillera, Western Cordillera (Maricao), Cerro Las Mesas; 5–925 m; *fl* & *fr* throughout the year.

Native to tropical Asia, China, Japan. [L2: 422].

Savia sessiliflora (Sw.) Willd., Sp. Pl. 4(2):771. 1806.

Bas.: *Croton sessiliflorus* Sw. (1788) "sessiliflorum."

Shrub in Northern Coastal Lowlands (Loíza), Northern Limestone Hills (Isabela), Eastern Coast (Ceiba), Las Piedras Chiquitas, Southeastern Foothills, Eastern Cordillera (Cayey, Aibonito), Southern Coastal Limestone, Susúa Forest Reserve, Guánica Forest Reserve, Southwestern Coastal Lowlands (Cabo Rojo), Sierra Bermeja, Vieques, Desecheo; 10–550 m; *fl* May–Nov, *fr* throughout the year.

Greater Antilles, Virgin Islands. [L2: 427].

Amansa guapo, Carbonero de costa, Garrote.

TAXA CITED BY PREVIOUS AUTHORS, BUT NOT INCLUDED

CITED DUE TO MISIDENTIFICATION

Phyllanthus acuminatus Vahl. Syn.: *Conami conami* (Sw.) Britton.

Shrub cited by Liogier & Martorell (1982, 2000) and Liogier (1988) as collected by Roy Woodbury in Guajataca, probably an escape. This citation is based upon the midisdentification of a sterile specimen of *Phyllanthus juglandifolius* subsp. *juglandifolius*. [L2: 416].

ONLY UNDER CULTIVATION

Flueggea virosa (Roxb. ex Willd.) Voight. Syn.: *Securinega virosa* (Roxb. ex Willd.) Baill. *Conami portoricensis* (Kuntze) Britton.

Shrub cited by Britton & Wilson (1924) as collected by O. Kuntze between Cayey and Guayama, probably an escape from cultivation; subsequently cited by Liogier & Martorell (1982, 2000) and Liogier (1988). [L2: 395].

Phyllanthus emblica L.

Tree cited by Liogier & Martorell (1982, 2000) and Liogier (1988) as cultivated and persistent in Mayagüez. [L2: 418].

PHYTOLACCACEAE

Microtea debilis Sw., Prodr. 53. 1788.

Herb in Northern Coastal Lowlands, Eastern Coast, Caguas Valley (Caguas), Sierra de Cayey, Vieques; 1–500 m; *fl* & *fr* throughout the year.

Antilles, Trinidad and Tobago, Central America, South America. [L1: 185].

Petiveria alliacea L., Sp. Pl. 1:342. 1753.

Herb in Northern Coastal Lowlands, Northern Limestone Hills, Eastern Coast, Caguas Valley, Southeastern Foothills, Eastern and Central Cordillera, Southern Coastal Limestone, Guánica Forest Reserve, Southwestern Coastal Lowlands, Sierra Bermeja, Western Coast, Vieques, Culebra, Mona, Desecheo; 1–1000 m; *fl* & *fr* throughout the year.

West Indies, southern United States (Florida, Texas), Central America, South America. [L1: 186].

Anamú.

Phytolacca icosandra L., Syst. Nat. (ed. 10) 2:1040. 1759.

Herb in Northern Coastal Lowlands (Vega Alta to Laguna Tortuguero); 1–4 m; *fl* & *fr* Aug–Nov.

Bahamas, Greater Antilles, Trinidad, Mexico, Central America, northwestern, west-central South America and Venezuela. [L1: 188].

Bella sombra.

Phytolacca rivinoides Kunth & C.D. Bouché, Sp. Nov. Hort. Berol. 15; et Ann. Sci. Nat., Bot., sér. 3 11:231. 1849.

Scandent herb in Northern Coastal Lowlands (Guaynabo), Río Abajo Forest Reserve, Sierra de Luquillo, Caguas Valley (Caguas), Sierra de Cayey, Eastern Cordillera (Cerro La Peña), Central Cordillera, Western Cordillera; 1–1300 m; *fl* & *fr* throughout the year.

Antilles, Mexico, Central America, South America. [L1: 189].

Juan de Vargas.

Rivina humilis L., Sp. Pl. 1:121. 1753.

Herb in Northern Coastal Lowlands, Northern Limestone Hills (Morovis), Eastern Coast, Las Piedras Chiquitas, Southeastern Foothills, Southern Coastal Lowlands, Southern Coastal Limestone (Guayanilla), Guánica Forest Reserve, Southwestern Coastal Lowlands, Sierra Bermeja, Vieques, Culebra, Mona, Desecheo; 1–550 m; *fl* throughout the year, *fr* Jun–Mar.

West Indies, southern United States, Mexico, Central America, South America. [L1: 190].

Carmín.

Trichostigma octandrum (L.) H. Walter in Engler, Pflanzenr. IV. 83(Heft 39):109. 1909.
Bas.: *Rivina octandra* L. (1756).

Suffrutescent herb in Northern Coastal Lowlands, Northern Limestone Hills, Río Abajo Forest Reserve, Las Piedras Chiquitas, Southeastern Foothills, Eastern and Central Cordillera, Southern Coastal Lowlands, Susúa Forest Reserve, Southwestern Coastal Lowlands, Sierra Bermeja, Western Coast (Rincón), Vieques, Culebra, Mona; 1–1050 m; *fl* Oct–May, *fr* Dec–Jul.

West Indies, southeastern United States (Florida), Mexico, Central America, South America. [L1: 193; AR: 342].

Bejuco de nasa, Bejuco de palma, Bejuco de paloma.

TAXA NOT RECENTLY COLLECTED

Microtea portoricensis Urb., Ber. Deutsch Bot. Ges. 3(8):324. 1885.

Herb collected in Cabo Rojo on 20 Jan 1885 [*Sintenis 717* (GOET), type specimen, scan seen] and also reported by Britton & Wilson (1924) as occurring in Trujillo Alto.

Greater Antilles (Cuba, Hispaniola). [L1: 185].

Stegnosperma cubense A. Rich. in Sagra, Hist. Fis. Cuba, Bot. 10:309, 12: *t. 44*. 1845.

Shrub collected in Aibonito on 11 Jan 1931 [*Britton & Horne 9628* (NY)].

Greater Antilles, Mexico, northern Central America to Nicaragua. [L1: 191; AR: 341].

TAXA CITED BY PREVIOUS AUTHORS, BUT NOT INCLUDED

ONLY UNDER CULTIVATION

Agdestis clematidea Moc. & Sessé ex DC.

Vine cited by Liogier & Martorell (1982, 2000) and Liogier (1985) as cultivated and escaped; cited by Acevedo-Rodríguez (2005) as only occasional in gardens. [L1: 184; AR: 341].

PICRAMNIACEAE

Picramnia pentandra Sw., Fl. Ind. Occid. 1:220. 1797.

Tree in Northern Coastal Lowlands, Northern Limestone Hills, Río Abajo Forest Reserve, Caguas Valley (Caguas), Las Tetas de Cayey, Las Piedras Chiquitas, Eastern Cordillera, Quebrada de Los Cedros, Guánica Forest Reserve, Southwestern Coastal Lowlands (Cabo Rojo), Sierra Bermeja; 5–825 m; *fl* Mar–Jul, Oct–Nov, *fr* throughout the year.

West Indies, southeastern United States (Florida), northern South America (Venezuela, Colombia). [L2: 280].

Guarema, Hueso.

PIPERACEAE

Peperomia alata Ruiz & Pav., Fl. Peruv. 1:31, *t.* 48, *f. b.* 1798.

Herb in Northern Limestone Hills (Hatillo), Sierra de Luquillo, Eastern Cordillera (La Torrecilla), Central Cordillera, Maricao Forest Reserve; 225–1150 m; *fl* & *fr* throughout the year.

Greater Antilles, Mexico, Central America, South America. [L1: 20].

Peperomia cogniauxii Urb., Symb. Antill. 1(2):292. 1899.

Creeping herb in Sierra de Cayey, Eastern Cordillera (La Torrecilla), Toro Negro Forest Reserve, Maricao Forest Reserve; 470–940 m; *fl* & *fr* Feb–Apr, Jul–Aug, Nov.

Greater Antilles (Hispaniola). [L1: 22].

Yerba de guabá.

Peperomia dendrophila Schltdl. & Cham., Linnaea 5(1):74. 1830.

Syn.: *Peperomia maxonii* C. DC. (1912).

Epiphytic herb in Eastern Cordillera (La Torrecilla), Central Cordillera (Adjuntas, Peñuelas); 900–1200 m; *fl* & *fr* Sep, Dec–Mar.

Greater Antilles (Hispaniola), Mexico, Central America, western South America (Colombia, Ecuador, Peru). [L5: 388].

Peperomia distachya (L.) A. Dietr., Sp. Pl. 1:156. 1831.

Bas.: *Piper distachyon* L. (1753).

Creeping herb in Sierra de Luquillo, Sierra de Cayey, Central Cordillera, Maricao Forest Reserve; 350–1100 m; *fl* & *fr* Jan–Oct.

Greater Antilles, southern Central America, northwestern, west-central South America and Venezuela. [L1: 22].

Peperomia dominicana C. DC., J. Bot. 4:146. 1866.

Herb in Toro Negro Forest Reserve; 900 m; *fl* & *fr* Mar (one specimen). [*Breckon 3121* (MAPR)].

Greater Antilles (Cuba, Hispaniola).

Peperomia emarginella (Sw. ex Wikstr.) C. DC. in A. de Candolle, Prodr. 16(1):437. 1869.

Bas.: *Piper emarginellum* Sw. ex Wikstr. (1828).

Delicate creeping herb in Sierra de Luquillo, Eastern Coast (Cuchilla de Panduras), Caguas Valley (Caguas), Sierra de Cayey; Central Cordillera (Adjuntas, Peñuelas); 375–1050 m; *fl* & *fr* Jun–Nov.

Antilles, southern Central America, northwestern, west-central South America and Venezuela. [L1: 22].

Peperomia glabella (Sw.) A. Dietr., Sp. Pl. 1:156. 1831.

Bas.: *Piper glabellum* Sw. (1788).

Syn.: *Peperomia sintenisii* C. DC. (1902).

Creeping herb in Northern Coastal Lowlands (Río Grande), Northern Limestone Hills (Quebradillas), Río Abajo Forest Reserve, Sierra de Luquillo, Eastern Coast, Sierra de Cayey, Las Tetas de Cayey, Las Piedras Chiquitas, Southeastern Foothills, Eastern and Central Cordillera, Maricao Forest Reserve, Vieques; 1–1000 m; *fl* & *fr* throughout the year.

West Indies, Trinidad, southeastern United States (Florida), Mexico, Central America, northwestern, west-central South America and Venezuela. [L1: 23, 27].

Notes: Britton & Wilson (1924) wrongly placed *Peperomia sintenisii* in synonymy with *Peperomia dendrophila*.

Peperomia hernandiifolia (Vahl) A. Dietr., Sp. Pl. 1:157. 1831 "hernandiaefolia."

Bas.: *Piper hernandiifolium* Vahl (1804) "hernandifolium."

Creeping herb in Sierra de Luquillo, Central Cordillera; 400–1300 m; *fl* & *fr* throughout the year.

Antilles, Mexico, Central America, northwestern South America and Venezuela, Brazil. [L1: 23].

Peperomia humilis A. Dietr., Sp. Pl. 1:168. 1831.

Syn.: *Peperomia questeliana* Stehlé & Trel. (1940).

Herb in Northern Coastal Lowlands (Río Grande, Manatí), Northern Limestone Hills, Las Piedras Chiquitas, Southeastern Foothills (Coamo), Eastern Cordillera (Cayey, Aibonito), Central Cordillera (Yauco), Susúa Forest Reserve, Guánica Forest Reserve, Sierra Bermeja, Vieques, Mona, Desecheo; 1–875 m; *fl* & *fr* throughout the year.

Antilles, southeastern United States (Florida), Central America. [L1: 26].

Peperomia maculosa (L.) Hook., Exot. Fl. 2: *ad t.* 92. 1824.

Bas.: *Piper maculosum* L. (1753).

Fleshy herb in Central Cordillera, Maricao Forest Reserve; 650–1100 m; *fl* & *fr* Dec–Jul.

Greater Antilles, Central America, northwestern, west-central South America and Venezuela. [L1: 23].

Peperomia magnoliifolia (Jacq.) A. Dietr., Sp. Pl. 1:153. 1831 "magnoliaefolia."

Bas.: *Piper magnoliifolium* Jacq. (1791) "magnoliaefolium."

Fleshy herb in Northern Coastal Lowlands (Río Grande), Northern Limestone Hills, Río Abajo Forest

Reserve, Eastern Cordillera (Cayey), Susúa Forest Reserve, Maricao Forest Reserve, Southwestern Coastal Lowlands (Cabo Rojo), Vieques; 1–900 m; *fl* & *fr* throughout the year.

West Indies, southeastern United States (Florida), Mexico, Central America, northwestern South America and Venezuela. [L1: 24].

Yerba de medio real.

Peperomia myrtifolia (Vahl) A. Dietr., Sp. Pl. 1:147. 1831.

Bas.: *Piper myrtifolium* Vahl (1804).

Syn.: *Peperomia wheeleri* Britton (1924).

Herb in Northern Limestone Hills (Florida, Isabela), Río Abajo Forest Reserve, Las Tetas de Cayey, Eastern Cordillera (Cañón San Cristóbal), Vieques, Culebra; 25–800 m; *fl* & *fr* May–Dec.

Virgin Islands (St. John, St. Croix, Virgin Gorda), Lesser Antilles. [L1: 24, 29].

Notes: *Peperomia myrtifolia* is a highly variable taxon throughout the Lesser Antilles and the Virgin Islands. I have thus far been unable to discriminate between exsiccatae of it and of what, in Puerto Rico, has been called *Peperomia wheeleri*. Until such time as marked differences among populations of *Peperomia myrtifolia* are demonstrated, I am treating *Peperomia wheeleri* as encompassed within it.

Peperomia wheeleri is on the U.S. Fish & Wildlife Service list of threatened and endangered plants.

Peperomia obtusifolia (L.) A. Dietr., Sp. Pl. 1:154. 1831.

Bas.: *Piper obtusifolium* L. (1753).

Fleshy herb in Northern Limestone Hills (Quebradillas, San Sebastián), Río Abajo Forest Reserve, Sierra de Luquillo, Sierra de Cayey, Eastern Cordillera (Aibonito), Maricao Forest Reserve; 200–800 m; *fl* & *fr* throughout the year.

West Indies, southeastern United States (Florida), Mexico, Central America, northern South America to Bolivia. [L1: 25].

Peperomia parvulifolia Trel., Repert. Spec. Nov. Regni Veg. 23:325, 332. 1927.

Herb in Northern Limestone Hills (Camuy); 300 m; *fl* & *fr* Oct (one specimen). [*Proctor 48910* (SJ)].

Greater Antilles (Hispaniola). [L5: 388].

Peperomia pellucida (L.) Kunth in Humboldt, Bonpland & Kunth, Nov. Gen. Sp. 1:53 [folio]; 1:64 [quarto]. 1816.

Bas.: *Piper pellucidum* L. (1753).

Herb in Northern Coastal Lowlands, Eastern Coast (Naguabo), Caguas Valley (San Lorenzo), Southern Coastal Lowlands, Western Coast (Mayagüez), Vieques; 5–250 m; *fl* & *fr* throughout the year.

Originating in the Neotropics. [L1: 25].

Alumbre, Coclaria, Frescura, Peletaria.

Peperomia portoricensis Urb., Symb. Antill. 1(2):292. 1899.

Creeping herb in Central Cordillera, Western Cordillera (Maricao); 470–1150 m; *fl* & *fr* Aug–Nov, Feb–May.

Greater Antilles (Cuba, Hispaniola). [L5: 388, 389].

Notes: Liogier (1985) wrongly placed this taxon in synonymy with *Peperomia cogniauxii*, but subsequently (1997) corrected the error.

Peperomia quadrifolia (L.) Kunth in Humboldt, Bonpland & Kunth, Nov. Gen. Sp. 1:57 [folio]; 1:69 [quarto]. 1816.

Bas.: *Piper quadrifolium* L. (1762).

Epiphytic herb in Toro Negro Forest Reserve; 1200 m; *fl* & *fr* Jun (one specimen). [*Woodbury a.n. 27447* (UPR)].

Greater Antilles, Mexico, Central America, northwestern and west-central South America. [L1: 25].

Peperomia rhombea Ruiz & Pav., Fl. Peruv. 1:31, *t. 46, f. c.* 1798.

Syn.: *Peperomia myrtillus* Miq. (1843).

Herb in Central Cordillera (Ponce, Yauco); 500–850 m; *fl* & *fr* Jan, Jul two specimens). [*Proctor 40067* (SJ); *Caraballo 2665* (UPR)].

Greater Antilles, Central America, northwestern, west-central South America and Venezuela. [L1: 26].

Peperomia robustior (Dahlst.) Urb., Symb. Antill. 4(1):191. 1903.

Bas.: *Peperomia tenerrima* Schltdl. & Cham. f. *robustior* Dahlst. (1900).

Creeping herb in Sierra de Luquillo, Sierra de Cayey; 400–800 m; *fl* & *fr* (?).

Greater Antilles, northern South America. [L1: 26].

Peperomia rotundifolia (L.) Kunth in Humboldt, Bonpland & Kunth, Nov. Gen. Sp. 1:54 [folio]; 1:65 [quarto]. 1816.

Bas.: *Piper rotundifolium* L. (1753).

Creeping herb in Northern Limestone Hills, Sierra de Luquillo, Eastern Cordillera (Corozal, Cañón San Cristóbal), Western Cordillera (Maricao); 90–600 m; *fl* & *fr* Mar–Nov.

Antilles, Trinidad, southeastern United States (Florida), Mexico, Central America, South America. [L1: 26; AR: 344].

Bejuco de alcanfor, Peperomia, Yerba linda, Yerba de medio real, Yerba de ratón.

Peperomia serpens (Sw.) Loudon, Hort. Brit. (ed. 2) 13. 1830.
Bas.: *Piper serpens* Sw. (1788).
Creeping herb in Northern Limestone Hills, Río Abajo Forest Reserve, Sierra de Luquillo, Eastern Coast (Cuchilla de Panduras), Caguas Valley (Las Piedras), Sierra de Cayey, Eastern Cordillera (Cañón San Cristóbal), Maricao Forest Reserve; 100–600 m; *fl* & *fr* throughout the year.
Antilles, Trinidad, Mexico, Central America, South America. [L1: 27].
Vinagre.

Peperomia tenella (Sw.) A. Dietr., Sp. Pl. 1:153. 1831.
Bas.: *Piper tenellum* Sw. (1788).
Creeping herb in Toro Negro Forest Reserve; 1125–1300 m; *fl* & *fr* Aug–Oct, Jan–May.
Antilles, Central America, northern South America (Venezuela, Colombia) and Brazil. [L1: 27].

Peperomia urocarpa Fisch. & C.A. Mey., Index Seminum (St. Petersburg) 4:42. 1838.
Creeping herb in Northern Limestone Hills, Río Abajo Forest Reserve, Central Cordillera, Maricao Forest Reserve; 200–950 m; *fl* & *fr* throughout the year.
Greater Antilles (Cuba, Hispaniola), Lesser Antilles (Saba, St. Kitts, Guadeloupe), Mexico, Central America, South America. [L1: 28].

Piper aduncum L., Sp. Pl. 1:29. 1753.
Shrub in Northern Coastal Lowlands, Northern Limestone Hills, Río Abajo Forest Reserve, Sierra de Luquillo, Eastern Coast, Caguas Valley (Las Piedras), Sierra de Cayey, Las Tetas de Cayey, Eastern Cordillera (Comerío), Central Cordillera, Susúa Forest Reserve, Western Cordillera, Western Coast (Aguada), Vieques; 1–1150 m; *fl* & *fr* throughout the year.
Antilles, Mexico, Central America, South America. [L1: 31].
Higuillo, Higuillo de hoja menuda.

Piper amalago L., Sp. Pl. 1:29. 1753.
Shrub in Northern Coastal Lowlands, Northern Limestone Hills, Río Abajo Forest Reserve, Eastern Coast (Naguabo), Caguas Valley (Juncos), Sierra de Cayey, Las Tetas de Cayey, Las Piedras Chiquitas, Southeastern Foothills, Eastern Cordillera (Aibonito), Central Cordillera, Southern Coastal Limestone, Susúa Forest Reserve, Western Cordillera, Guánica Forest Reserve, Cerro Las Mesas, Western Coast, Vieques; 1–850 m; *fl* & *fr* throughout the year.
Antilles, Mexico, Central America, South America. [L1: 30].
Higuillo de limón, Higuillo oloroso.

Piper blattarum Spreng., Neue Entdeck. Pflanzenk. 2:100. 1821.
Shrub in Northern Limestone Hills, Río Abajo Forest Reserve, Sierra de Luquillo, Sierra de Cayey, Central Cordillera, Susúa Forest Reserve, Maricao Forest Reserve; 225–1330 m; *fl* & *fr* throughout the year.
Virgin Islands (St. Thomas, recorded by Eggers). [L1: 32].
Higuillo, Higuillo oloroso.

Piper glabrescens (Miq.) C. DC. in A. de Candolle, Prodr. 16(1):271. 1869.
Bas.: *Artanthe glabrescens* Miq. (1845).
Syn.: *Piper treleaseanum* Britton & P. Wilson (1924).
Shrub in Northern Coastal Lowlands (Bayamón), Northern Limestone Hills, Río Abajo Forest Reserve, Sierra de Luquillo, Eastern Coast (Maunabo), Caguas Valley (Caguas), Sierra de Cayey, Eastern Cordillera (La Torrecilla), Central Cordillera, Maricao Forest Reserve; 100–900 m; *fl* & *fr* throughout the year.
Antilles, Trinidad, Central America, South America. [L1: 32].

Piper hispidum Sw., Prodr. 15. 1788.
Syn.: *Piper scabrum* Sw. (1797).
Shrub in Northern Coastal Lowlands (Bayamón), Northern Limestone Hills, Río Abajo Forest Reserve, Sierra de Luquillo, Eastern Coast, Sierra de Cayey, Central Cordillera, Western Cordillera; 125–1100 m; *fl* & *fr* Feb–Nov.
Antilles, Trinidad and Tobago, Mexico, Central America, South America. [L1: 33].

Piper jacquemontianum Kunth, Linnaea 13(6):631. 1840.
Syn.: *Piper wydlerianum* C. DC. (1869).
Shrub in Northern Coastal Lowlands, Northern Limestone Hills, Río Abajo Forest Reserve, Sierra de Luquillo, Eastern Coast, Sierra de Cayey, Southern Coastal Lowlands (Patillas), Susúa Forest Reserve, Southwestern Coastal Lowlands (San Germán), Western Coast (Mayagüez); 1–500 m; *fl* & *fr* throughout the year.
Greater Antilles (Hispaniola), Mexico, Central America. [L1: 33].
Notes: Britton & Wilson (1924) misapplied

the name *Piper citrifolium* Lam. to, and Liogier & Martorell (1982) wrongly made it a synonym of, this taxon; Liogier (1985) invented the combination *Piper jacquemontianum* (Kunth) DC. for the name of this taxon.

Piper marginatum Jacq., Collectanea 4:128. 1791.

Shrub in Northern Coastal Lowlands, Northern Limestone Hills, Río Abajo Forest Reserve, Southeastern Foothills, Eastern Cordillera (Corozal); 1–350 m; *fl* & *fr* throughout the year.

Greater Antilles (Cuba, Hispaniola), Trinidad, Central America, northern South America to Peru. [L1: 34].

Higuillo oloroso, Oloroso.

Piper peltatum L., Sp. Pl. 1:30. 1753 "pelatum."

Syn.: *Pothomorphe peltata* (L.) Miq. (1840).

Herb in Northern Limestone Hills, Río Abajo Forest Reserve, Eastern Coast, Caguas Valley (Caguas), Sierra de Cayey, Las Piedras Chiquitas, Eastern Cordillera (Cerro La Peña), Central Cordillera, Southern Coastal Limestone (Guayanilla), Western Cordillera, Cerro Las Mesas, Western Coast; 5–750 m; *fl* & *fr* throughout the year.

Antilles, Mexico, Central America, northern South America to Bolivia. [L1: 17].

Notes: Liogier & Martorell (1982, 2000) and Liogier (1985) used *Lepianthes peltata* (L.) Raf., a *nom. inval.*, for the name of this taxon.

Baquiña.

Piper swartzianum (Miq.) C. DC. in A. de Candolle, Prodr. 16(1):265. 1869.

Bas.: *Artanthe swartziana* Miq. (1844).

Shrub in Northern Limestone Hills, Río Abajo Forest Reserve; 200–325 m; *fl* & *fr* Jan–Mar, Jun–Aug.

Greater Antilles (Hispaniola). [L1: 34].

Notes: Little, Woodbury, & Wadsworth (1974) misapplied the name *Piper tuberculatum* Jacq. to this taxon.

Higuillo.

Piper umbellatum L., Sp. Pl. 1:30. 1753.

Syn.: *Pothomorphe umbellata* (L.) Miq. (1840).

Herb in Northern Coastal Lowlands (Guaynabo), Northern Limestone Hills, Río Abajo Forest Reserve, Sierra de Luquillo, Eastern Coast (Maunabo), Sierra de Cayey, Las Tetas de Cayey, Eastern Cordillera (Cerro La Peña), Central Cordillera, Western Cordillera, Cerro Las Mesas; 150–1320 m; *fl* & *fr* throughout the year.

Antilles, Mexico, Central America, northern South America to Brazil, Bolivia. [L1: 18].

Notes: Britton & Wilson (1924) placed *Pothomorphe umbellata* in synonymy with *Pothomorphe peltata*. Liogier & Martorell (1982) and Liogier (1985) used *Lepianthes umbellata* (L.) Raf., a *nom. inval.*, for the name of this taxon; Liogier & Martorell (2000), following Howard (1988), placed it in synonymy with *Lepianthes peltata*.

Baquiña.

Piper sp.

Shrub in Northern Limestone Hills (Arecibo), Río Abajo Forest Reserve, Toro Negro Forest Reserve, 225–900 m; *fl* & *fr* Jun–Oct, Jan–Feb.

Notes: This plant may well be a hybrid between *Piper glabrescens* and *Piper jacquemontianum*.

TAXA NOT RECENTLY COLLECTED

Peperomia megalopoda Trel. in Britton & Wilson, Sci. Surv. Porto Rico & Virgin Islands 6(4):533. 1930.

Herb collected between Guayama and Cayey on 5 Dec 1929 [by *E.G. Britton 9352* (NY), type specimen].

Endemic. [L1: 24].

Notes: Known from the type specimen only.

Peperomia yabucoana Urb. & C. DC. in Urban, Symb. Antill. 3(2):232. 1902.

Herb collected in Yabucoa [*Sintenis 5127*, type specimen, not seen], between Cayey and Guayama on 22 Jul 1901 [*Underwood & Griggs 320* (NY)] and in the Sierra de Yabucoa on 26 Feb 1922 [*Britton et al. 6251* (US)].

Endemic. [L1: 29].

TAXA CITED BY PREVIOUS AUTHORS, BUT NOT INCLUDED

NO SPECIMENS SEEN

Peperomia tetraphylla (G. Forst.) Hook. & Arn.
Herb cited by Liogier & Martorell (1982, 2000) and Liogier (1985) as occurring in wet forests, Luquillo Mountains, reported by Roy Woodbury. [L1: 28].

CITED DUE TO MISIDENTIFICATION

Piper dilatatum Rich.
Herb cited by Liogier & Martorell (1982, 2000) and Liogier (1985) as occurring in shaded valleys, Río Abajo Forest and Cerro de Punta. This citation was based on the misidentification of specimens of *Piper hispidum* and *Piper sp.* [L1: 32].

PLANTAGINACEAE

Angelonia salicariifolia Humb. & Bonpl., Pl. Aequinoct. 2:92, *t. 108*. 1812 "salicariaefolia."

Herb in Northern Limestone Hills (Bayamón, Morovis), Caguas Valley (Caguas), Sierra de Cayey, Western Cordillera (Maricao); 100–600 m; *fl* Apr–Jul, Oct, *fr* Oct (two specimens).

Possibly native to Brazil. [L4: 483].

Angelón.

Bacopa innominata (M. Gómez) Alain, Revista Soc. Cub. Bot. 13:61. 1957.
Bas.: *Conobea innominata* M. Gómez (1894).
Syn.: *Herpestis rotundifolia* C.F. Gaertn. (1807).

Trailing herb in Piñones Forest Reserve; 1 m; *fl* & *fr* Jun (one specimen). [*Woodbury a.n. 3355* (UPR)].

Greater Antilles, Lesser Antilles (Guadeloupe), southeastern United States. [L4: 485].

Bacopa monnieri (L.) Pennell, Proc. Acad. Nat. Sci. Philadelphia 98:94. 1946.
Bas.: *Lysimachia monnieri* L. (1756).
Syn.: *Braimia monnieri* (L.) Drake (1893) "Monnieria."

Low creeping herb in Northern Coastal Lowlands, Northern Limestone Hills, Río Abajo Forest Reserve, Sierra de Luquillo, Eastern Coast (Humacao including Cayo Santiago), Southeastern Foothills, Susúa Forest Reserve, Guánica Forest Reserve, Southwestern Coastal Lowlands (Cabo Rojo), Western Coast (Rincón), Vieques; 1–300 m; *fl* & *fr* throughout the year.

Pantropical. [L4: 485].

Yerba de culebra.

Bacopa repens (Sw.) Wettst. in Engler & Prantl, Nat. Pflanzenfam. 4(3b):76. 1891.
Bas.: *Gratiola repens* Sw. (1788).
Syn.: *Macuillamia repens* (Sw.) Pennell (1923).

Aquatic herb in Northern Coastal Lowlands, Eastern Coast (Maunabo); 1–100 m; *fl* & *fr* Mar, Jun, Sep, Dec.

Greater Antilles, Trinidad, Mexico, Central America, northwestern, west-central South America and Venezuela. [L4: 487].

Bacopa stricta (Schrad.) Wettst. ex Edwall, Bol. Commiss. Geogr. Estado São Paulo 13:176, 180. 1897.
Bas.: *Herpestis stricta* Schrad (1822).
Syn.: *Caconapea stricta* (Schrad.) Britton (1925).

Herb in Northern Coastal Lowlands (Bayamón, Laguna Tortuguero), Northern Limestone Hills (Florida), Sierra de Luquillo, Sierra de Cayey, Toro Negro Forest Reserve, Southwestern Coastal Lowlands (Lajas); 250–1025 m; *fl* & *fr* throughout the year.

Greater Antilles (Hispaniola), northern South America to Bolivia. [L4: 487].

Notes: Liogier & Martorell (1982, 2000) and Liogier (1995) used the homonym *Bacopa stricta* (Schrad.) Rob. for the name of this taxon.

Yerba de culebra.

Mecardonia procumbens (Mill.) Small, Fl. S.E. U.S. 1065, 1338. 1903.
Bas.: *Erinus procumbens* Mill. (1768).

Herb in Northern Coastal Lowlands, Río Abajo Forest Reserve, Sierra de Luquillo, Caguas Valley (Caguas), Sierra de Cayey, Eastern Cordillera (Cayey), Central Cordillera, Western Coast (Mayagüez); 5–1200 m; *fl* & *fr* throughout the year.

Greater Antilles (Cuba, Hispaniola), Lesser Antilles, southeastern and south central United States, Mexico, Central America, South America. [L4: 498].

Yerba de culebra amarilla.

Micranthemum umbrosum (J.F. Gmel.) S.F. Blake, Rhodora 17(199):131. 1915.
Bas.: *Globifera umbrosa* J.F. Gmel. (1791).
Syn.: *Hemianthus callitrichoides* Griseb. (1862).

Creeping herb in Northern Coastal Lowlands (Dorado), Sierra de Luquillo, Eastern Cordillera (Cañón San Cristóbal), Central Cordillera (Utuado), Maricao Forest Reserve; 1–900 m; *fl* & *fr* Mar–May, Aug–Sep, Dec.

Bahamas, Greater Antilles, Lesser Antilles (Guadeloupe, Dominica, Grenada), Trinidad, southeastern and south-central United States, Mexico, South America. [L4: 491, 500].

Plantago major L., Sp. Pl. 1:112. 1753.

Herb in Northern Coastal Lowlands, Northern Limestone Hills, Sierra de Luquillo, Sierra de Cayey, Eastern Cordillera (Barranquitas including La Torrecilla), Central Cordillera, Maricao Forest Reserve, Southwestern Coastal Lowlands (Lajas), Western Coast (Mayagüez); 1–1200 m; *fl* & *fr* Dec–Sep.

Native to Eurasia. [L5: 63].

Llantén, Plantago.

Russelia equisetiformis Schltdl. & Cham., Linnaea 6(2):377. 1831.

Herb planted and persistent in Northern Limestone Hills; 60–100 m; *fl* Feb, May–Jul, Oct–Nov, *fr* Feb (one specimen).

Probably native to Mexico. [L4: 502].

Coral de Italia, Lluvia de coral.

Scoparia dulcis L., Sp. Pl. 1:116. 1753.

Suffrutescent herb in Northern Coastal

Lowlands, Eastern Coast, Caguas Valley (Caguas), Eastern Cordillera (Aibonito), Central Cordillera (Utuado), Southwestern Coastal Lowlands (Cabo Rojo), Western Coast, Vieques; 1–260 m; *fl* & *fr* throughout the year.

Neotropical. [L4: 503].

Escoba amarga, Mastuerzo, Orozús.

Stemodia durantifolia (L.) Sw., Obs. Bot. 240. 1791.
Bas.: *Capraria durantifolia* L. (1759).

Herb in Northern Coastal Lowlands (Cataño); 1 m; *fl* & *fr* Feb (one specimen). [*Woodbury a.n. 10734* (SJ)].

Greater Antilles, Lesser Antilles (Antigua, Guadeloupe, Grenada), Trinidad, Mexico, Central America, South America (scattered). [L4: 507].

Verónica.

Stemodia maritima L., Syst. Nat. (ed. 10) 2:1118. 1759.

Suffrutescent herb in Northern Coastal Lowlands (San Juan), Southern Coastal Lowlands (Salinas); 1 m; *fl* May, Aug (two specimens), *fr* Jan (one specimen).

Bahamas, Greater Antilles, Southern Caribbean Islands (Curaçao), Mexico, northern Central America (Belize). [L4: 507].

Orozús, Verónica del mar.

Stemodia verticillata (Mill.) Hassl., Trab. Mus. Farmacol. 21:110. 1909.
Bas.: *Erinus verticillatus* Mill. (1768).
Syn.: *Lendneria verticillata* (Mill.) Britton (1925).

Herb in Northern Coastal Lowlands, Caguas Valley (Caguas); 1–55 m; *fl* & *fr* Jan–Apr, Jul.

Antilles, Mexico, Central America, South America. [L4: 507].

Veronica peregrina L., Sp. Pl. 1:14. 1753.

Herb in Northern Limestone Hills (Florida), Toro Negro Forest Reserve; 150–1000 m; *fl* & *fr* Apr–May (two specimens). [*Liogier 28915, 33132* (UPR)].

Bahamas, Greater Antilles (Jamaica, Hispaniola), Canada, United States, South America (scattered). [L4: 509].

TAXA CITED BY PREVIOUS AUTHORS, BUT NOT INCLUDED

ONLY UNDER CULTIVATION

Lophospermum erubescens D. Don. Syn.: *Maurandya erubescens* (D. Don) A. Gray.
Vine cited by Liogier & Martorell (1982, 2000) and Liogier (1995) as occasionally grown and escaped and by Acevedo-Rodríguez (2005) as cultivated and naturalized, uncommon. [L4: 497; AR: 379].

Plantago lanceolata L.
Herb cited by Britton & Wilson (1925) as occurring as a garden weed, El Condado; subsequently cited by Liogier & Martorell (1982, 2000) and Liogier (1997) as probably not persistent. [L5: 63].

PLUMBAGINACEAE

Plumbago scandens L., Sp. Pl. (ed. 2) 1:215. 1762.

Herb in Northern Coastal Lowlands (Manatí), Las Piedras Chiquitas, Southeastern Foothills, Eastern Cordillera (Aibonito), Southern Coastal Lowlands (Salinas), Quebrada de Los Cedros, Guánica Forest Reserve, Sierra Bermeja, Mona, Desecheo; 10–650 m; *fl* throughout the year, *fr* Nov–Feb, May–Aug.

West Indies, Trinidad, southern United States (Florida, Texas, Arizona), Mexico, Central America, South America. [L4: 134; AR: 345].

Higuillo, Meladillo, Meladillo silvestre.

TAXA CITED BY PREVIOUS AUTHORS, BUT NOT INCLUDED

ONLY UNDER CULTIVATION

Plumbago auriculata Lam. Syn.: *Plumbago capensis* Thunb.
Herb cited by Liogier & Martorell (1982, 2000) and Liogier (1994) as widely cultivated and persistent. [L4: 134].

Plumbago indica L.
Herb cited by Liogier & Martorell (1982, 2000) and Liogier (1994) as cultivated and persistent. [L4: 134].

POACEAE (GRAMINEAE)

Agrostis hyemalis (Walter) Britton, Sterns & Poggenb., Prelim. Cat. 68. 1888.
Bas.: *Cornucopiae hyemale* Walter (1788).

Herb in Toro Negro Forest Reserve; 1000–1340 m; *fl* & *fr* Jan–Oct.

Native to Canada, United States, Mexico, eastern Russia, Japan, Korea.

Andropogon bicornis L., Sp. Pl. 2:1046. 1753 "bicorne," *nom. cons.*

Herb in Northern Coastal Lowlands, Sierra de Luquillo, Eastern Coast (Punta Guayanés), Caguas Valley (Caguas), Sierra de Cayey, Central Cordillera, Western Cordillera; 1–925 m; *fl* & *fr* throughout the year.

Antilles, Mexico, Central America, South America.

Barba de indio, Matojo de techar, Rabo de gato, Zacatón.

Andropogon fastigiatus Sw., Prodr. 26. 1788.
Syn.: *Diectomis fastigiata* (Sw.) P. Beauv. (1812).

Herb in Northern Coastal Lowlands (Laguna Tortuguero), Susúa Forest Reserve, Maricao Forest Reserve, Southwestern Coastal Lowlands (Cabo Rojo), Sierra Bermeja; 1–900 m; *fl* & *fr* Nov–Feb.

Greater Antilles, Lesser Antilles (Dominica, Grenada), Mexico, Central America, northern South America to Bolivia, tropical Africa, southeastern Asia.

Notes: Liogier & Martorell (1982, 2000) used the homonym *Diectomis fastigiata* (Sw.) Kunth for the name of this taxon.

Andropogon glomeratus (Walter) Britton, Sterns & Poggenb., Prelim. Cat. 67. 1888.
Bas.: *Cinna glomerata* Walter (1788).

Herb in Northern Coastal Lowlands, Northern Limestone Hills, Río Abajo Forest Reserve, Sierra de Cayey, Toro Negro Forest Reserve, Western Cordillera; 2–800 m; *fl* & *fr* Mar–Sep.

West Indies, eastern and southern United States, Mexico, Central America, northern South America (Venezuela, Colombia).

Matojilla, Matojo de escoba del fogón, Yerba barbuda.

Andropogon gracilis Spreng., Syst. Veg. 1:284. 1824.
Syn.: *Schizachyrium gracile* (Spreng.) Nash (1903).

Herb in Northern Coastal Lowlands (Laguna Tortuguero), Northern Limestone Hills, Río Abajo Forest Reserve, Eastern Cordillera (Corozal), Central Cordillera (Utuado), Cerro Las Mesas, Western Coast (Rincón); 1–500 m; *fl* & *fr* Jun–Mar.

Greater Antilles, Lesser Antilles (Guadeloupe), southeastern United States (Florida).

Pajón hembra.

Andropogon leucostachyus Kunth in Humboldt, Bonpland & Kunth, Nov. Gen. Sp. 1:150 [folio]; 1:187 [quarto]. 1816.

Herb in Northern Coastal Lowlands, Sierra de Luquillo, Eastern Coast, Sierra de Cayey, Central Cordillera, Cerro Las Mesas, Western Coast; 1–1300 m; *fl* & *fr* Jan–Oct.

Greater Antilles, Trinidad and Tobago, Mexico, Central America, South America.

Matojillo, Plumilla.

Andropogon virgatus Ham., Prodr. Pl. Ind. Occid. 9. 1825.
Syn.: *Hypogynium virgatum* (Ham.) Dandy (1931).

Herb in Northern Coastal Lowlands (Laguna Tortuguero), Sierra de Luquillo, Eastern Cordillera (Aibonito); 1–600 m; *fl* & *fr* Nov–Aug.

Greater Antilles (Cuba, Hispaniola), Trinidad, Mexico, Central America, South America.

Rabo de ratón.

Andropogon virginicus L., Sp. Pl. 2:1046. 1753 "virginicum."

Herb in Northern Coastal Lowlands (Vega Baja to Laguna Tortuguero), Sierra de Luquillo, Eastern Coast (Cuchilla de Panduras), Western Cordillera (Maricao); 1–1050 m; *fl* & *fr* May–Oct.

Bahamas, Greater Antilles, Trinidad and Tobago, eastern Canada, eastern, central United States and California, Mexico, Central America, northern South America (Colombia).

Matojo de escoba.

Anthenantia lanata (Kunth) Benth., J. Linn. Soc., Bot. 19:39. 1881.
Bas.: *Paspalum lanatum* Kunth (1816).
Syn.: *Leptocoryphium lanatum* (Kunth) Nees (1829).

Herb in Cerro Las Mesas; 330 m; *fl* & *fr* Mar (one specimen). [*Gould 15834* (UPR)].

Greater Antilles (Cuba, Hispaniola), Trinidad, Mexico, Central America, South America.

Lanilla.

Anthephora hermaphrodita (L.) Kuntze, Revis. Gen. Pl. 2:759. 1891.
Bas.: *Tripsacum hermaphroditum* L. (1759).

Herb in Northern Coastal Lowlands, Southern Coastal Lowlands (Caja de Muertos, Santa Isabel), Southwestern Coastal Lowlands (Cabo Rojo); 1–100 m; *fl* & *fr* Mar–Aug, Nov–Dec.

West Indies, Mexico, Central America, northern South America to Peru.

Aristida adscensionis L., Sp. Pl. 1:82. 1753.

Herb in Southern Coastal Limestone (Ponce), Guánica Forest Reserve, Southwestern Coastal Lowlands, Mona, Desecheo; 5–50 m; *fl* & *fr* throughout the year.

West Indies, southwestern and south-central United States, Mexico, Central America, South America (scattered).

Aristida calcicola Hitchc. & Ekman, N. Amer. Flora 17(5):405. 1935.

Herb in Southern Coastal Limestone; 20–80 m; *fl* & *fr* Mar, Jun, Sep–Nov.

Greater Antilles (Cuba, Hispaniola).

Aristida chaseae Hitchc., Contr. U.S. Natl. Herb. 22(7):575. 1924.

Herb in Southwestern Coastal Lowlands (Cabo Rojo), Sierra Bermeja; 40–300 m; *fl* & *fr* Jul–Apr.

Endemic.

Notes: Britton & Wilson (1923) misapplied the name *Aristida cognata* Trin. & Rupr. to this taxon.

This taxon is on the U.S. Fish & Wildlife Service list of threatened and endangered plants.

Aristida portoricensis Pilg. in Urban, Symb. Antill. 4(1):100. 1903.

Herb in Sierra Bermeja, Cerro Las Mesas; 260–380 m; *fl* & *fr* Nov–Jun.

Endemic.

Notes: This taxon is on the U.S. Fish & Wildlife Service list of threatened and endangered plants.

Aristida refracta Griseb., Cat. Pl. Cub. 228. 1866.

Herb in Eastern Coast (Punta Guayanés), Southwestern Coastal Lowlands (Cabo Rojo); 40 m; *fl* & *fr* Oct, Jan (two specimens). [*Woodbury a.n. 5566, 15170* (UPR); *Axelrod 9564* (UPRRP)].

Greater Antilles.

Aristida spiciformis Elliott, Sketch Bot. S. Carolina 1(2):141. 1816.

Herb in Northern Coastal Lowlands (Laguna Tortuguero); 1 m; *fl* & *fr* Sep–Nov, Jan–Mar.

Greater Antilles (Cuba), southeastern and south-central United States.

Aristida suringarii Henrard, Meded. Rijks-Herb. 54(B):616, *t.* 1928.

Herb in Mona; 1–40 m; *fl* & *fr* Mar, Aug (two specimens). [*Proctor 45848* (SJ); *González Más 8a* (MAPR)].

Virgin Islands, Lesser Antilles, northern South America (Venezuela).

Arthraxon castratus (Griff.) V. Naray. ex Bor, Fl. Assam 5:376. 1940.

Bas.: *Andropogon castratus* Griff. (1851).

Herb in Central Cordillera (Adjuntas); 400 m; *fl* & *fr* Feb (one specimen). [*Liogier 31714* (UPR)].

Native to Indian Subcontinent, Vietnam, Java.

Arthrostylidium farctum (Aubl.) Soderstr. & Lourteig, Phytologia 64(2):163. 1987.

Bas.: *Arundo farcta* Aubl. (1775).

Syn.: *Arthrostylidium capillifolium* Griseb. (1862).

Viny herb in Northern Limestone Hills, Río Abajo Forest Reserve, Las Tetas de Cayey, Eastern Cordillera (Cañón San Cristóbal), Central Cordillera, Southern Coastal Limestone (Guayanilla), Susúa Forest Reserve, Maricao Forest Reserve, Vieques; 100–1300 m; *fl* & *fr* (?).

Bahamas, Greater Antilles (Cuba, Hispaniola), Virgin Islands (St. Thomas, St. John). [AR: 436].

Arthrostylidium multispicatum Pilg. in Urban, Symb. Antill. 2(3):341. 1901.

Viny herb in Central Cordillera, Maricao Forest Reserve; 700–1100 m; *fl* & *fr* Feb–Mar.

Greater Antilles (Cuba, Hispaniola), Trinidad and Tobago. [AR: 437].

Millo trepador.

Arthrostylidium sarmentosum Pilg. in Urban, Symb. Antill. 4(1):108. 1903.

Viny herb in Sierra de Luquillo, Sierra de Cayey, Eastern Cordillera (La Torrecilla), Central Cordillera, Maricao Forest Reserve; 350–1330 m; *fl* & *fr* Jan–Aug.

Greater Antilles (Cuba, Hispaniola), Trinidad and Tobago, northern South America (Venezuela). [AR: 437].

Bambú trepador.

Arundinaria viridistriata (Regel) Makino ex Nakai, J. Jap. Bot. 10:568. 1934.

Bas.: *Bambusa viridistriata* Regel (1867) "viridi-striata."

Herb planted and persistent in Cerro Las Mesas; 250–300 m; *fl* & *fr* (?). [*Proctor 43919* (SJ)].

Cultivated and naturalized in Japan; origin uncertain.

Arundinella hispida (Humb. & Bonpl. ex Willd.) Kuntze, Revis. Gen. Pl. 2:761. 1891.

Bas.: *Andropogon hispidus* Humb. & Bonpl. ex Willd. (1806).

Syn.: *Arundinella confinis* (Schult.) Hitchc. & Chase (1917).

Herb in Sierra de Luquillo, Sierra de Cayey, Eastern Cordillera (Cerro La Peña), Central Cordillera, Western Cordillera, Cerro Las Mesas, Western Coast (Mayagüez); 250–1000 m; *fl* & *fr* Dec–Mar.

Antilles, Trinidad and Tobago, Mexico, Central America, South America.

Rabo de gato.

Arundo donax L., Sp. Pl. 1:81. 1753.

Suffrutescent herb in Southern Coastal Lowlands, Southwestern Coastal Lowlands (Guánica), Western Coast (Añasco); 1–30 m; *fl* & *fr* Jul–Dec, Mar.

Native to eastern Europe (Ukraine), northern Africa, temperate and tropical Asia.

Caña de castilla, Guajana.

Axonopus compressus (Sw.) P. Beauv., Ess. Agrostogr. 12, 154. 1812.
Bas.: *Milium compressum* Sw. (1788).

Herb in Northern Coastal Lowlands, Northern Limestone Hills, Río Abajo Forest Reserve, Sierra de Luquillo, Eastern Coast, Caguas Valley (Gurabo), Sierra de Cayey, Eastern Cordillera (Aibonito, Cañón San Cristóbal), Susúa Forest Reserve, Southwestern Coastal Lowlands (Cabo Rojo), Vieques; 1–925 m; *fl* & *fr* throughout the year.

West Indies, southeastern and south-central United States, Mexico, Central America, South America.

Grama colorada, Pasto alfombra.

Axonopus fissifolius (Raddi) Kuhlm., Relat. Commiss. Linhas Telegr. Estratég. Matto Grosso Amazonas 11:87. 1922.
Bas.: *Paspalum fissifolium* Raddi (1823).
Syn.: *Axonopus affinis* Chase (1938).

Herb in Northern Coastal Lowlands (Laguna Tortuguero), Guajataca Forest Reserve, Sierra de Luquillo, Sierra de Cayey; 1–975 m; *fl* & *fr* Feb–Apr, Sep–Oct.

Greater Antilles, southeastern and south-central United States and California, Mexico, Central America, South America.

Grama colorada fina, Pasto alfombra fino.

Bambusa longispiculata Gamble ex Brandis, Indian Trees 668. 1906.

Suffrutescent herb planted and persistent in Sierra de Luquillo, Toro Negro Forest Reserve; 150–1000 m; *fl* & *fr* Feb–Jun.

Native to India, Bangladesh, Myanmar.

Bambusa multiplex (Lour.) Raeusch. ex Schult. & Schult. f., Syst. Veg. 7(2):1350. 1830.
Bas.: *Arundo multiplex* Lour. (1790).
Syn.: *Bambusa glaucescens* (Willd.) Holttum (1956).

Suffrutescent herb planted and persistent in Susúa Forest Reserve; 110–185 m; *fl* & *fr* Feb, May (two specimens).

Probably native to southeastern Asia.

Bambusa tulda Roxb., Fl. Ind. (ed. 1832) 2:193. 1832.

Suffrutescent herb planted and persistent in Sierra de Luquillo, Toro Negro Forest Reserve; 600–1000 m; *fl* & *fr* Feb (one specimen).

Native to India, Bangladesh, Myanmar, Thailand, China.

Notes: Britton & Wilson (1930) used the orthographic variant *Bambos tulda* Roxb. for the name of this taxon.

Bambusa tuldoides Munro, Trans. Linn. Soc. London 26(1):93. 1868.

Suffrutescent herb planted and persistent in Toro Negro Forest Reserve, Maricao Forest Reserve; 850–1000 m; *fl* & *fr* Feb–Mar, Aug.

Native to Vietnam, China.

Bambusa vulgaris Schrad. ex J.C. Wendl., Coll. Pl. 2:26, *t. 47.* 1808.

Arborescent herb planted and persistent in Northern Coastal Lowlands, Sierra de Luquillo, Caguas Valley (San Lorenzo), Sierra de Cayey, Susúa Forest Reserve, Maricao Forest Reserve, Southwestern Coastal Lowlands (Cabo Rojo), Vieques; 5–900 m; *fl* & *fr* Mar–May, Aug–Oct.

Probably native to southeastern Asia.

Notes: Britton & Wilson (1923) used *Bambos vulgaris* Schrad. a *nom. nud.*, for the name of this taxon.

Bambú, Bambúa, Caña India.

Bothriochloa pertusa (L.) Camus, Ann. Soc. Linn. Lyon, sér 2 76:164. 1931.
Bas.: *Holcus pertusus* L. (1771).

Herb in Northern Coastal Lowlands, Las Piedras Chiquitas, Southeastern Foothills, Southern Coastal Lowlands, Southern Coastal Limestone, Susúa Forest Reserve, Guánica Forest Reserve, Southwestern Coastal Lowlands, Vieques, Culebra, Mona, Desecheo; 1–550 m; *fl* & *fr* throughout the year.

Native to Indian Subcontinent, southeastern Asia.

Huracán, Yerba agria de Barbados, Yerba amarga, Yerba huracán.

Bouteloua americana (L.) Scribn., Proc. Acad. Nat. Sci. Philadelphia 43(2):306. 1891.
Bas.: *Aristida americana* L. (1759).

Herb in Northern Coastal Lowlands, Eastern Coast (Ceiba), Las Piedras Chiquitas, Southern Coastal Lowlands (Caja de Muertos), Southern Coastal Limestone (Ponce), Guánica Forest Reserve, Southwestern Coastal Lowlands (Cabo Rojo), Sierra Bermeja, Vieques, Mona; 1–550 m; *fl* & *fr* Oct–Jan.

West Indies, southern Mexico, Central America, northwestern South America and Venezuela.

Yerba mesquite.

Bouteloua juncea (Desv. ex P. Beauv.) Hitchc., Contr. U.S. Natl. Herb. 17(3):343. 1913.
Bas.: *Triathera juncea* Desv. ex P. Beauv. (1812).

Herb in Guánica Forest Reserve; 5–20 m; *fl* & *fr* Mar, Nov–Dec.

Greater Antilles (Cuba, Hispaniola).

Bouteloua repens (Kunth) Scribn., Bull. Div. Agrostol. U.S.D.A. 4:9. 1897.

Bas.: *Dinebra repens* Kunth (1816).
Syn.: *Bouteloua heterostega* (Trin.) Griffiths (1912).

Herb in Susúa Forest Reserve, Guánica Forest Reserve, Southwestern Coastal Lowlands (Cabo Rojo), Sierra Bermeja, Western Coast (Rincón); 1–300 m; *fl* & *fr* throughout the year.

Greater Antilles (Cuba, Hispaniola), Lesser Antilles (St. Martin), southwestern United States, Mexico, Central America, northern South America (Venezuela, Colombia).

Lamilla, Yerba mesquite.

Cenchrus brownii Roem. & Schult., Syst. Veg. 2:258. 1817.
Syn.: *Cenchrus viridis* Spreng. (1824).

Herb in Northern Coastal Lowlands, Northern Limestone Hills (Quebradillas), Eastern Coast (Ceiba), Caguas Valley (Caguas), Southern Coastal Lowlands, Guánica Forest Reserve, Southwestern Coastal Lowlands (Guánica, Sabana Grande), Vieques, Mona; 1–200 m; *fl* & *fr* Jun–Feb.

West Indies, southeastern United States and Texas, Mexico, Central America, northern South America to Bolivia.

Abrojito.

Cenchrus ciliaris L., Mant. Pl. 2:302. 1771.
Syn.: *Pennisetum ciliare* (L.) Link (1827).

Herb in Las Piedras Chiquitas, Southeastern Foothills (Salinas), Southern Coastal Lowlands, Guánica Forest Reserve, Southwestern Coastal Lowlands, Sierra Bermeja, Vieques, Mona; 1–550 m; *fl* & *fr* throughout the year.

Native to southern Europe (Sicily), Africa, Saudia Arabia, western Asia, Indian Subcontinent.

Yerba buffel, Yerba de salinas.

Cenchrus clandestinus (Hochst. ex Chiov.) Morrone, Ann. Bot. (Oxford), n.s. 106(1):127. 2010.
Bas.: *Pennisetum clandestinum* Hochst. ex Chiov. (1903).

Herb in Southwestern Coastal Lowlands (Cabo Rojo); (?) m; *fl* & *fr* (?). [*Proctor 44954* (SJ)].

Native to Africa.

Cenchrus echinatus L., Sp. Pl. 2:1050. 1753.

Herb in Northern Coastal Lowlands, Northern Limestone Hills, Río Abajo Forest Reserve, Eastern Coast (Humacao), Caguas Valley (Caguas), Southeastern Foothills, Eastern Cordillera (Cayey, Cañón San Cristóbal), Southern Coastal Lowlands, Susúa Forest Reserve, Maricao Forest Reserve, Guánica Forest Reserve, Southwestern Coastal Lowlands, Western Coast, Vieques, Culebra, Mona, Desecheo; 1–720 m; *fl* & *fr* throughout the year.

Pantropical.

Abrojo, Cadillo.

Cenchrus incertus M.A. Curtis, Boston J. Nat. Hist. 1(2):135. 1835.

Herb in Northern Coastal Lowlands, Southern Coastal Limestone (Guayanilla), Guánica Forest Reserve, Southwestern Coastal Lowlands (Cabo Rojo), Vieques, Mona, Desecheo; 1–5 m; *fl* & *fr* throughout the year.

West Indies, United States, Mexico, Central America, South America.

Notes: The above name may be a synonym of *Cenchrus spinifex* Cav. (Icon. 5:38, *t. 461.* 1799), an earlier name. For the time being, however, Zuloaga in Soreng et al. (online) is rejecting that name "until further studies can be made on the isotype and holotype."

Britton & Wilson (1923) misapplied the name *Cenchrus carolinianus* Walter to this taxon; Liogier & Martorell (1982) wrongly treated the former, as well as *Cenchrus tribuloides*, as synonyms of it.

Abrojo de dunas, Abrojo de playa.

Cenchrus myosuroides Kunth in Humboldt, Bonpland & Kunth, Nov. Gen. Sp. 1:94 [folio]; 1:115 [quarto], *t. 35.* 1816.
Syn.: *Cenchropsis myosuroides* (Kunth) Nash (1903).

Herb in Mona, Desecheo; 1–100 m; *fl* & *fr* Feb–Jul.

Bahamas, Greater Antilles (Cuba, Hispaniola), southeastern and south-central United States, Mexico, southern Central America (Costa Rica), western, southern South America and Brazil.

Abrojo de espiga, Abrojo gigante.

Cenchrus purpureus (Schumach.) Morrone, Ann. Bot. (Oxford), n.s. 106(1):129. 2010.
Bas.: *Pennisetum purpureum* Schumach. (1827).

Herb in Northern Coastal Lowlands, Río Abajo Forest Reserve, Sierra de Luquillo, Caguas Valley (Caguas), Sierra de Cayey, Eastern Cordillera (Corozal, Barranquitas), Central Cordillera, Southern Coastal Lowlands (Ponce), Susúa Forest Reserve, Maricao Forest Reserve, Southwestern Coastal Lowlands (Lajas), Western Coast; 2–950 m; *fl* & *fr* Sep–Jun.

Native to tropical Africa.

Yerba elefante.

Cenchrus setaceus (Forssk.) Morrone, Ann. Bot. (Oxford), n.s. 106(1):129. 2010.
Bas.: *Phalaris setacea* Forssk. (1775).

Syn.: *Pennisetum setaceum* (Forssk.) Chiov. (1923).

Herb in Central Cordillera (Juana Díaz); 250 m; *fl* & *fr* Nov (one specimen). [*Axelrod 14043* (UPRRP)].

Native to northern and tropical Africa, Arabian Peninsula, Middle East (Lebanon, Syria).

Cenchrus tribuloides L., Sp. Pl. 2:1050. 1753.

Herb in Northern Coastal Lowlands (Loíza), Southwestern Coastal Lowlands (Cabo Rojo), Mona; 1 m; *fl* & *fr* Mar–May, Oct.

Bahamas, Greater Antilles, eastern United States, South America (Brazil).

Chloris barbata Sw., Fl. Ind. Occid. 1:200. 1797.

Syn.: *Chloris inflata* Link (1821).
Chloris paraguaiensis Steud. (1854).

Herb in Northern Coastal Lowlands, Southeastern Foothills, Eastern Cordillera (Cayey), Southern Coastal Lowlands, Susúa Forest Reserve, Guánica Forest Reserve, Southwestern Coastal Lowlands, Sierra Bermeja, Western Coast, Vieques, Culebra, Mona, Desecheo; 1–425 m; *fl* & *fr* throughout the year.

Widespread in tropics and subtropics, origin uncertain.

Horquetilla morada, Paragüita morada.

Chloris ciliata Sw., Prodr. 25. 1788.

Herb in Northern Coastal Lowlands, Southern Coastal Lowlands (Caja de Muertos), Guánica Forest Reserve, Southwestern Coastal Lowlands (Cabo Rojo), Western Coast (Añasco), Culebra; 1–30 m; *fl* & *fr* Jun–Oct, Jan.

West Indies, south-central United States (Texas), Mexico, northern Central America (Belize), South America (scattered).

Cola de caballo.

Chloris radiata (L.) Sw., Prodr. 26. 1788.

Bas.: *Agrostis radiata* L. (1759).

Herb in Northern Coastal Lowlands, Eastern Coast (Naguabo), Caguas Valley, Sierra de Cayey, Las Tetas de Cayey, Eastern Cordillera (Corozal), Central Cordillera (Utuado, Adjuntas); 1–900 m; *fl* & *fr* Jul–Apr.

West Indies, southern Mexico, Central America, northern South America to Bolivia.

Grama de costa, Horquetilla.

Chloris sagraeana A. Rich. in Sagra, Hist. Fis. Cuba, Bot. 11:315. 1850.

Herb in Northern Coastal Lowlands, Sierra de Luquillo, Sierra de Cayey, Eastern Cordillera (Corozal, La Torrecilla), Central Cordillera (Adjuntas); 5–925 m; *fl* & *fr* throughout the year.

West Indies.

Chrysopogon zizanioides (L.) Roberty, Bull. Inst. Franç. Afrique Noire 22:106. 1960.

Bas.: *Phalaris zizanioides* L. (1771).
Syn.: *Vetiveria zizanioides* (L.) Nash (1903).
Anatherum zizanioides (L.) Hitchc. & Chase (1917).

Herb in Northern Coastal Lowlands (Manatí), Eastern Coast (Cuchilla de Panduras), Sierra de Cayey (San Lorenzo); 1–500 m; *fl* & *fr* Sep–Oct (two specimens).

Native to Indian Subcontinent, Indo-China.

Baúl del pobre, Pacholí.

Chusquea abietifolia Griseb., Fl. Brit. W.I. 529. 1864.

Woody vine in Central Cordillera (Los Tres Picachos), Maricao Forest Reserve 450–1200 m; *fl* & *fr* (?).

Greater Antilles. [AR: 439].

Coix lacryma-jobi L., Sp. Pl. 2:972. 1753.

Herb in Northern Coastal Lowlands, Sierra de Luquillo, Eastern Coast (Cuchilla de Panduras), Sierra de Cayey, Eastern and Central Cordillera, Susúa Forest Reserve, Western Cordillera (Maricao), Southwestern Coastal Lowlands (Cabo Rojo), Western Coast; 1–1000 m; *fl* & *fr* Jul–Apr.

Native to Indian Subcontinent, southeastern Asia.

Camándulas.

Cymbopogon citratus (DC.) Stapf, Bull. Misc. Inform. Kew 1906:322, 357. 1906.

Bas.: *Andropogon citratus* DC. (1813) "citratum."

Herb planted and persistent in Eastern Coast (Cuchilla de Panduras), Sierra de Cayey; 500 m; *fl* & *fr* Oct–Nov (two specimens). [*Liogier 31011, 34545* (UPR)].

Native to the Old World.

Limoncillo, Zorra de limón.

Cynodon dactylon (L.) Pers., Syn. Pl. 1:85. 1805.

Bas.: *Panicum dactylon* L. (1753).
Syn.: *Capriola dactylon* (L.) Kuntze (1891).

Herb in Northern Coastal Lowlands, Northern Limestone Hills (San Sebastián), Eastern Coast, Caguas Valley (Gurabo), Southeastern Foothills (Salinas), Eastern Cordillera (Barranquitas), Southern Coastal Lowlands, Guánica Forest Reserve, Southwestern Coastal Lowlands, Cerro Las Mesas, Western Coast, Vieques, Mona; 1–600 m; *fl* & *fr* throughout the year.

Native to the Old World.

Hala que te quedas, Pelo de brujas, Pepe Ortiz, Yerba Bermuda.

Cynodon nlemfuensis Vanderyst, Bull. Agric. Congo Belge 13:342. 1922.

Herb in Northern Coastal Lowlands, Northern Limestone Hills (Quebradillas), Caguas Valley, Las Tetas de Cayey, Southeastern Foothills (Salinas), Eastern Cordillera (Corozal, Cayey), Central Cordillera (Orocovis), Southwestern Coastal Lowlands (Cabo Rojo); 1–830 m; *fl* & *fr* Sep–Jun.

Native to tropical Africa.

Yerba de estrella.

Dactylis glomerata L., Sp. Pl. 1:71. 1753 "glomeratus."

Herb in Toro Negro Forest Reserve; 1340 m; *fl* & *fr* May (one specimen). [*Liogier 28928* (UPR)].

Native to Europe, Northern Africa, temperate Asia, Indian Subcontinent.

Dactyloctenium aegyptium (L.) Willd., Enum. Pl. 2:1029. 1809 "aegyptiacum."

Bas.: *Cynosurus aegyptius* L. (1753).

Herb in Northern Coastal Lowlands, Southeastern Foothills, Southern Coastal Lowlands (Caja de Muertos, Guayanilla), Susúa Forest Reserve, Guánica Forest Reserve, Southwestern Coastal Lowlands, Western Coast (Rincón), Vieques, Culebra, Mona; 1–425 m; *fl* & *fr* throughout the year.

Native to Africa, temperate and tropical Asia.

Yerba egipcia.

Dichanthelium aciculare (Desv. ex Poir.) Gould & C.A. Clark var. **aciculare**, Ann. Missouri Bot. Gard. 65(4):1116. 1979.

Bas.: *Panicum aciculare* Desv. ex Poir. (1816).

Syn.: *Panicum chrysopsidifolium* Nash (1903).

Herb in Northern Coastal Lowlands (Laguna Tortuguero), Sierra de Luquillo, Maricao Forest Reserve, Cerro Las Mesas; 1–1000 m; *fl* & *fr* Jun–Mar.

Greater Antilles, eastern United States, Mexico, Central America, northern South America (Venezuela, Colombia).

Notes: Liogier & Martorell (1982) wrongly placed *Panicum chrysopsidifolium* in synonymy with *Dichanthelium acuminatum* var. *acuminatum*.

Dichanthelium aciculare (Desv. ex Poir.) Gould & C.A. Clark var. **ramosum** (Griseb.) Davidse, Novon 2(2):104. 1992.

Bas.: *Panicum neuranthum* Griseb. var. *ramosum* Griseb. (1866).

Syn.: *Panicum fusiforme* Hitchc. (1909).

Herb in Sierra de Luquillo; 1000 m; *fl* & *fr* Aug–Oct (two specimens).

West Indies, southeastern United States, Mexico, northern Central America to Nicaragua, northern South America (Venezuela).

Notes: Liogier & Martorell (1982) placed *Panicum fusiforme* in synonymy with *Dichanthelium aciculare, sensu lato*.

Dichanthelium acuminatum (Sw.) Gould & C.A. Clark var. **acuminatum**, Ann. Missouri Bot. Gard. 65(4):1121. 1979.

Bas.: *Panicum acuminatum* Sw. (1788).

Syn.: *Dichanthelium acuminatum* (Sw.) Gould & C.A. Clark var. *implicatum* (Scribn.) Gould & C.A. Clark (1979).

Dichanthelium ovale (Elliott) Gould & C.A. Clark var. *addisonii* (Nash) Gould & C.A. Clark (1979).

Herb in Northern Coastal Lowlands (Laguna Tortuguero), Sierra de Luquillo, Sierra de Cayey, Eastern Cordillera (Cerro La Peña), Central Cordillera, Maricao Forest Reserve, Cerro Las Mesas; 2–1200 m; *fl* & *fr* Jan–Sep.

Bahamas, Greater Antilles, Canada, United States, Mexico, Central America, northwestern South America.

Yerba de ramillete.

Dichanthelium acuminatum (Sw.) Gould & C.A. Clark var. **longiligulatum** (Nash) Gould & C.A. Clark, Ann. Missouri Bot. Gard. 65(4):1127. 1979.

Bas.: *Panicum longiligulatum* Nash (1899).

Syn.: *Panicum leucothrix* Nash (1897).

Herb in Sierra de Luquillo; 650–1050 m; *fl* & *fr* Dec–Feb, May–Sep.

West Indies, eastern and central United States, Mexico, northern Central America to Nicaragua, northern South America (Venezuela, Colombia).

Notes: Liogier & Martorell (1982), following Gould & Clark (1978), wrongly placed *Panicum leucothrix* in synonymy with *Dichanthelium acuminatum* var. *implicatum*.

Dichanthelium dichotomum (L.) Gould var. **roanokense** (Ashe) LeBlond, Sida 19(4):831. 2001.

Bas.: *Panicum roanokense* Ashe (1898).

Herb in Northern Coastal Lowlands (Laguna Tortuguero); 1–2 m; *fl* & *fr* Mar (one specimen). [*Woodbury a.n. 4171* (UPR)].

Bahamas, Greater Antilles, eastern and central United States.

Notes: Liogier & Martorell (1982, 2000) used *Dichanthelium dichotomum, sensu lato*, for the name of this taxon.

Dichanthelium portoricense (Ham.) B.F. Hansen

& Wunderlin, Ann. Missouri Bot. Gard. 75(4):1649. 1989.
Bas.: *Panicum portoricense* Ham. (1825).
Syn.: *Dichanthelium sabulorum* (Lam.) Gould & C.A. Clark var. *thinium* (Hitchc. & Chase) Gould & C.A. Clark (1979).

Herb in Northern Coastal Lowlands (Dorado to Laguna Tortuguero), Sierra de Luquillo, Eastern Cordillera (Cerro La Peña); 1–1050 m; *fl* & *fr* throughout the year.

West Indies, eastern and central United States, Mexico, northern Central America (Belize, Honduras).

Dichanthelium scoparium (Lam.) Gould, Brittonia 26(1):60. 1974.
Bas.: *Panicum scoparium* Lam. (1798).

Herb in Sierra de Luquillo, Central Cordillera (Utuado, Adjuntas); 425–935 m; *fl* & *fr* Dec–Apr, Jul–Aug.

Greater Antilles (Cuba, Hispaniola), eastern and central United States.

Dichanthelium strigosum (Muhl. ex Elliott) Freckmann, Brittonia 33(3):457. 1981.
Bas.: *Panicum strigosum* Muhl. ex Elliott (1816).
Syn.: *Dichanthelium strigosum* (Muhl. ex Elliott) Freckmann var. *glabrescens* (Griseb.) Freckmann (1981).
Panicum polycaulon Nash (1897).

Herb in Northern Coastal Lowlands (Laguna Tortuguero), Sierra de Luquillo, Sierra de Cayey, Maricao Forest Reserve, Cerro Las Mesas; 1–900 m; *fl* & *fr* Jun–Nov, Feb–Mar.

Greater Antilles, southern United States (Florida to Louisiana).

Dichanthium annulatum (Forssk.) Stapf in Prain, Fl. Trop. Afr. 9(1):178. 1917.
Bas.: *Andropogon annulatus* Forssk. (1775).

Herb in Northern Coastal Lowlands, Sierra de Luquillo, Eastern Coast (Ceiba), Sierra de Cayey, Southeastern Foothills, Central Cordillera (Orocovis, Adjuntas), Southern Coastal Lowlands, Maricao Forest Reserve, Guánica Forest Reserve, Southwestern Coastal Lowlands, Western Coast; 1–900 m; *fl* & *fr* throughout the year.

Native to Africa, Saudia Arabia, western temperate Asia, tropical Asia.

Pajón, Yerba de vías.

Dichanthium aristatum (Poir.) C.E. Hubb., Bull. Misc. Inform. Kew 1939:654. 1939.
Bas.: *Andropogon aristatus* Poir. (1811).

Herb in Northern Coastal Lowlands (islands adjacent to Fajardo, San Juan), Eastern Cordillera (Cayey), Western Coast (Mayagüez); 1–400 m; *fl* & *fr* Oct–Mar.

Native to India.

Dichanthium caricosum (L.) A. Camus, Bull. Mus. Natl. Hist. Nat. 27:549. 1921.
Bas.: *Andropogon caricosus* L. (1763) "caricosum."

Herb in Southwestern Coastal Lowlands (Cabo Rojo); 1 m; *fl* & *fr* May (one specimen). [*Proctor 47879* (SJ)].

Native to tropical Asia, China.

Digitaria bicornis (Lam.) Roem. & Schult., Syst. Veg. 2:470. 1817.
Bas.: *Paspalum bicorne* Lam. (1791).

Herb in Northern Coastal Lowlands, Las Piedras Chiquitas, Southeastern Foothills, Eastern Cordillera (Corozal), Southern Coastal Lowlands, Susúa Forest Reserve, Guánica Forest Reserve, Southwestern Coastal Lowlands, Western Coast (Añasco), Vieques, Culebra, Mona; 1–550 m; *fl* & *fr* Jul–Apr.

Native to tropical Asia, Australia.

Digitaria ciliaris (Retz.) Koeler, Descr. Gram. 27. 1802.
Bas.: *Panicum ciliare* Retz. (1786–1787).

Herb in Northern Coastal Lowlands, Río Abajo Forest Reserve, Sierra de Luquillo, Caguas Valley (Caguas), Sierra de Cayey, Southeastern Foothills (Coamo), Eastern Cordillera (Corozal, Barranquitas), Central Cordillera, Southwestern Coastal Lowlands, Western Coast (Mayagüez), Vieques, Culebra, Desecheo; 1–975 m; *fl* & *fr* Mar–Dec.

Tropics and subtropics; native range uncertain.

Pata de gallina, Pendejuelo, Yerba de juey.

Digitaria eggersii (Hack.) Henrard, Blumea 1(1):97. 1934.
Bas.: *Panicum eggersii* Hack. (1901).

Herb in Sierra Bermeja; 250–300 m; *fl* & *fr* Oct–Nov (one specimen). [*Proctor 44125* (SJ), not seen; *McKenzie 808* (MO), not seen].

Virgin Islands.

Digitaria eriantha Steud., Flora 12(2):468. 1829.
Syn.: *Digitaria decumbens* Stent (1930).
Digitaria pentzii Stent (1930).

Herb in Northern Coastal Lowlands, Eastern Cordillera (Corozal); 1–400 m; *fl* & *fr* Jun–Jan.

Native to southern Africa.

Pangola.

Digitaria filiformis (L.) Koeler var. **dolichophylla** (Henrard) Wipff, Phytologia 80(5):348. 1997.
Bas.: *Digitaria dolichophylla* Henrard (1753).

Herb in Northern Coastal Lowlands (Laguna Tortuguero); 1–2 m; *fl* & *fr* Aug–Sep (two specimens). [*Woodbury a.n. 1051067* (NY); *Woodbury a.n. P00527038* (P), not seen].

Greater Antilles (Cuba), southeastern United States (Florida).

Notes: Britton & Wilson (1923) used *Syntherisma filiforme* (L.) Nash, *sensu lato*, for the name of this taxon; Liogier & Martorell (1982) wrongly placed the former in synonymy with *Digitaria panicea*.

Digitaria fragilis (Steud.) Luces, J. Washington Acad. Sci. 32(6):160. 1942.
Bas.: *Paspalum fragile* Steud. (1853).

Herb in Cerro Las Mesas; 370 m; *fl* & *fr* (?). [*Proctor 43911* (SJ), not seen].

Native to northwestern, west-central South America and Brazil.

Digitaria horizontalis Willd., Enum. Pl. 1:92. 1809.
Syn.: *Syntherisma digitatum* (Sw.) Hitchc. (1908) "digitata."

Herb in Northern Coastal Lowlands, Eastern Coast, Las Piedras Chiquitas, Southeastern Foothills (Coamo), Central Cordillera, Southwestern Coastal Lowlands, Vieques, Mona; 1–1200 m; *fl* & *fr* throughout the year.

West Indies, southeastern United States, Mexico, Central America, South America.

Pata de gallina, Pendejuelo.

Digitaria insularis (L.) Fedde, Just's Bot. Jahresber. 31:778. 1904.
Bas.: *Andropogon insularis* L. (1759) "insulare."
Syn.: *Valota insularis* (L.) Chase (1906).

Herb in Northern Coastal Lowlands, Eastern Coast, Southeastern Foothills (Salinas), Eastern Cordillera, Southern Coastal Lowlands, Southern Coastal Limestone (Guayanilla), Southwestern Coastal Lowlands, Western Coast (Rincón), Vieques, Culebra, Mona, Desecheo; 1–925 m; *fl* & *fr* throughout the year.

West Indies, southern United States (scattered), Central America, South America.

Notes: Liogier & Martorell (1982, 2000) used *Digitaria insularis* (L.) Mez, a *nom. nud.*, for the name of this taxon.

Rabo de zorra, Zorra.

Digitaria ischaemum (Schreb.) Schreb. ex Muhl., Descr. Gram. 131. 1817.
Bas.: *Panicum ischaemum* Schreb. (1804).
Syn.: *Syntherisma ischaemum* (Schreb.) Nash (1912).

Herb in Sierra de Luquillo; 1065 m; *fl* & *fr* Oct (one specimen). [*Proctor 42279* (SJ)].

Native to Europe and temperate Asia.

Digitaria longiflora (Retz.) Pers., Syn. Pl. 1:85. 1805.
Bas.: *Paspalum longiflorum* Retz. (1786–1787).

Herb in Northern Coastal Lowlands (Laguna Tortuguero), Toro Negro Forest Reserve; 1–1200 m; *fl* & *fr* Feb, May–Oct.

Native to tropical and southern Africa, tropical Asia, China, Australia.

Digitaria nuda Schumach., Beskr. Guin. Pl. 45. 1827.

Herb in Northern Coastal Lowlands (San Juan), Sierra de Cayey, Eastern Cordillera (Aibonito); 10–500 m; *fl* & *fr* Feb–May.

Native to Africa.

Digitaria setigera Roth in Roemer & Schultes, Syst. Veg. 2:474. 1817.

Herb in Northern Coastal Lowlands, Sierra de Luquillo, Eastern Coast, Caguas Valley (Gurabo), Sierra de Cayey, Southeastern Foothills, Eastern Cordillera (Cayey), Central Cordillera, Southern Coastal Lowlands (Patillas); 1–900 m; *fl* & *fr* throughout the year.

Native to eastern tropical Africa (Tanzania), western Indian Ocean (Mauritius and the Seychelles), tropical Asia, Australia, Hawaii.

Pendejuelo.

Digitaria violascens Link, Hort. Berol. 1:229. 1827.
Syn.: *Syntherisma chinense* (Nees) Hitchc. (1922) "chinensis."

Herb in Northern Coastal Lowlands, Sierra de Luquillo, Eastern Coast (Punta Guayanés), Sierra de Cayey, Central Cordillera, Maricao Forest Reserve; 1–1300 m; *fl* & *fr* throughout the year.

Native to Afghanistan, tropical Asia, China, Japan.

Echinochloa colona (L.) Link, Hort. Berol. 2:209. 1833.
Bas.: *Panicum colonum* L. (1759).

Herb in Northern Coastal Lowlands, Northern Limestone Hills, Río Abajo Forest Reserve, Eastern Coast, Caguas Valley, Southeastern Foothills, Eastern Cordillera (Aibonito), Central Cordillera (Adjuntas), Southern Coastal Lowlands, Guánica Forest Reserve, Southwestern Coastal Lowlands (Lajas), Western Coast, Vieques, Culebra, Mona; 1–740 m; *fl* & *fr* throughout the year.

Throughout tropic, subtropic and warm temperate regions.

Arrocillo, Arroz de monte, Grama pintada.

Echinochloa crus-pavonis (Kunth) Schult., Mant. 2:269. 1824 "Crus pavonis."
Bas.: *Oplismenus crus-pavonis* Kunth (1816).

Herb in Northern Coastal Lowlands, Eastern Coast, Eastern Cordillera (Aibonito), Southern Coastal Lowlands, Southwestern Coastal Lowlands; 1–500 m; *fl* & *fr* throughout the year.

Greater Antilles, Lesser Antilles (Guadeloupe, Martinique), Trinidad and Tobago, southern United States, Mexico, Central America, South America, tropical and southern Africa.

Arrocillo.

Echinochloa polystachya (Kunth) Hitchc., Contr. U.S. Natl Herb. 22(3):135, *t. 26.* 1920.
Bas.: *Oplismenus polystachyus* (1816).

Herb in Northern Coastal Lowlands, Eastern Coast (Maunabo), Eastern Cordillera (Corozal), Southern Coastal Lowlands (Guayanilla), Southwestern Coastal Lowlands (Lajas); 1–400 m; *fl* & *fr* throughout the year.

Greater Antilles, Lesser Antilles (Antigua, Martinique, St. Lucia), southern United States (Florida, Louisiana, Texas), Mexico, Central America, South America.

Yerba de río, Pasto alemán.

Eleusine indica (L.) Gaertn., Fruct. Sem. Pl. 1:8. 1788.
Bas.: *Cynosurus indicus* L. (1753).

Herb in Northern Coastal Lowlands, Río Abajo Forest Reserve, Sierra de Luquillo, Eastern Coast, Caguas Valley (Juncos), Sierra de Cayey (San Lorenzo), Las Tetas de Cayey, Las Piedras Chiquitas, Southeastern Foothills (Coamo), Eastern Cordillera, Central Cordillera (Adjuntas), Southern Coastal Lowlands, Susúa Forest Reserve, Guánica Forest Reserve, Western Coast, Vieques, Culebra, Mona; 1–925 m; *fl* & *fr* throughout the year.

Native to the Paleotropics.

Pata de gallina.

Eragrostis barrelieri Daveau, J. Bot. (Morot) 8:289. 1894.

Herb in Northern Coastal Lowlands (San Juan); 5–15 m; *fl* & *fr* Aug (one specimen). [*Liogier 34426* (UPR)].

Native to southern Europe, northern Africa, western Asia, Pakistan.

Eragrostis ciliaris (L.) R. Br. var. **laxa** Kuntze, Revis. Gen. Pl. 2:774. 1891.

Herb in Northern Coastal Lowlands, Eastern Coast (Cuchilla de Panduras), Caguas Valley (Caguas), Sierra de Cayey, Las Piedras Chiquitas, Southeastern Foothills (Coamo), Southern Coastal Lowlands (Caja de Muertos), Guánica Forest Reserve, Southwestern Coastal Lowlands, Western Coast (Mayagüez), Vieques, Mona; 5–600 m; *fl* & *fr* throughout the year.

Native to Africa, Arabian Peninsula, Indian Subcontinent, Indo-China, Philippines.

Notes: Britton & Wilson (1923) used the homonym *Eragrostis ciliaris* (L.) Link for the name of this taxon.

Eragrostis curvula (Schrad.) Nees, Fl. Afr. Austral. Ill. 397. 1841.
Bas.: *Poa curvula* Schrad. (1821).

Herb in Northern Coastal Lowlands (San Juan); 10–15 m; *fl* & *fr* Oct–Nov (two specimens). [*Liogier 33717* (UPR)].

Native to eastern tropical and southern Africa.

Melena.

Eragrostis elliottii S. Watson, Proc. Amer. Acad. Arts 25:140. 1890.

Herb in Northern Coastal Lowlands (Vega Alta to Manatí); 1–2 m; *fl* & *fr* Jun–Mar.

Bahamas, Greater Antilles, southeastern and south-central United States, Mexico, northern Central America (Belize).

Eragrostis hypnoides (Lam.) Britton, Sterns & Poggenb., Prelim. Cat. 69. 1888.
Bas.: *Poa hypnoides* Lam. (1791).

Herb in Northern Coastal Lowlands (Laguna Tortuguero), Northern Limestone Hills (San Sebastián), Southwestern Coastal Lowlands (Lajas); 1–10 m; *fl* & *fr* Feb–Aug.

Greater Antilles, Trinidad, Canada, United States, Mexico, Central America, South America.

Eragrostis pectinacea (Michx.) Nees ex Jedwabn. var. **miserrima** (E. Fourn.) Reeder, Phytologia 60(2):154. 1986.
Bas.: *Eragrostis purshii* Schrad. var. *miserrima* E. Fourn. (1886).
Syn.: *Eragrostis tephrosanthos* Schult. (1824).

Herb in Northern Coastal Lowlands (San Juan, Aguadilla), Eastern Coast, Central Cordillera (Adjuntas), Southern Coastal Lowlands (Ponce), Susúa Forest Reserve, Southwestern Coastal Lowlands (Cabo Rojo), Vieques, Culebra; 1–550 m; *fl* & *fr* May–Feb.

West Indies, Trinidad, southern United States and New York to Missouri, Mexico, Central America, northern South America to Peru.

Eragrostis pilosa (L.) P. Beauv., Ess. Agrostogr. 71, 162, 175. 1812.
Bas.: *Poa pilosa* L. (1753).

Herb in Northern Coastal Lowlands, Eastern

Coast (Maunabo), Southern Coastal Limestone (Sabana Grande), Susúa Forest Reserve, Southwestern Coastal Lowlands (Cabo Rojo), Vieques, Culebra; 1–200 m; *fl* & *fr* throughout the year.

Native to Europe, Africa, temperate and tropical Asia.

Eragrostis tenella (L.) P. Beauv. ex Roem. & Schult., Syst. Veg. 2:576. 1817.
Bas.: *Poa tenella* L. (1753).
Syn.: *Eragrostis amabilis* (L.) Wight & Arn. (1834).

Herb in Northern Coastal Lowlands, Eastern Coast (Cayo Santiago), Caguas Valley (Caguas), Southeastern Foothills (Guayama), Eastern Cordillera (La Torrecilla), Southern Coastal Lowlands (Caja de Muertos), Guánica Forest Reserve, Southwestern Coastal Lowlands (Cabo Rojo), Vieques, Culebra, Mona; 1–925 m; *fl* & *fr* throughout the year.

Native to Africa, temperate and tropical Asia.

Amor, Yerba de amor, Yerba romántica.

Eremochloa ophiuroides (Munro) Hack. in A. & C. de Candolle, Monogr. Phan. 6:261. 1889.
Bas.: *Ischaemum ophiuroides* Munro (1860).

Herb in Northern Coastal Lowlands (Loíza to San Juan), Toro Negro Forest Reserve; 2–1300 m; *fl* & *fr* May–Oct.

Native to Vietnam, southern China, Taiwan.

Ciempiés.

Eriochloa polystachya Kunth in Humboldt, Bonpland & Kunth, Nov. Gen. Sp. 1:79 [folio]; 1:95 [quarto], *t. 31*. 1816.
Syn.: *Eriochloa subglabra* (Nash) Hitchc. (1909).

Herb in Northern Coastal Lowlands, Northern Limestone Hills (Lares, San Sebastián), Sierra de Luquillo, Eastern Coast, Caguas Valley (Caguas), Sierra de Cayey, Eastern Cordillera, Central Cordillera (Adjuntas), Southwestern Coastal Lowlands (Lajas), Western Coast (Mayagüez); 1–850 m; *fl* & *fr* Aug–May.

Greater Antilles (Cuba, Jamaica), Lesser Antilles (Guadeloupe), Trinidad, Mexico, Central America (Honduras, Costa Rica), northern South America to Peru.

Malojilla.

Eriochloa punctata (L.) Ham., Prodr. Pl. Ind. Occ. 5. 1825.
Bas.: *Milium punctatum* L. (1759).

Herb in Northern Coastal Lowlands, Northern Limestone Hills (Lares), Sierra de Luquillo, Eastern Coast, Southeastern Foothills (Salinas), Central Cordillera (Adjuntas), Southern Coastal Lowlands (Guayama); 1–800 m; *fl* & *fr* throughout the year.

Antilles, south-central United States, Mexico, Central America, South America.

Eriochrysis cayennensis P. Beauv., Ess. Agrostogr. 8, *t. 4*, *f. 11*. 1812 "cayanensis."

Herb in Northern Limestone Hills (Moca), Eastern Coast (Yabucoa), Sierra de Cayey, Eastern Cordillera (Cerro La Peña), Central Cordillera (Ciales, Adjuntas), Maricao Forest Reserve; 50–1000 m; *fl* & *fr* throughout the year.

Greater Antilles (Hispaniola), Trinidad, Mexico, Central America, South America.

Moco de pavo, Uña de gato.

Eustachys petraea (Sw.) Desv., Nouv. Bull. Sci. Soc. Philom. Paris 2:189. 1810.
Bas.: *Chloris petraea* Sw. (1788).

Herb in Northern Coastal Lowlands, Southwestern Coastal Lowlands (Cabo Rojo), Western Coast, Mona; 1–2 m; *fl* & *fr* throughout the year.

Bahamas, Greater Antilles, Lesser Antilles (Barbados), southeastern and south-central United States, Mexico, Central America, South America.

Yerba de deo.

Gymnopogon foliosus (Willd.) Nees in Martius et al., Fl. Bras. Enum. Pl. 2(1):426. 1829 (Agrost. Bras.).
Bas.: *Chloris foliosa* Willd. (1806).

Herb in Northern Coastal Lowlands (Laguna Tortuguero); 1–2 m; *fl* & *fr* Aug, Dec–Mar.

Greater Antilles (Hispaniola), Virgin Islands (St. Thomas), northern South America to Brazil.

Gynerium sagittatum (Aubl.) P. Beauv., Ess. Agrostogr. 138, 153, *t. 24*, *f. 6*. 1812.
Bas.: *Saccharum sagittatum* Aubl. (1775).

Suffrutescent herb in Northern Coastal Lowlands (Isabela), Eastern Coast (Maunabo), Caguas Valley (Gurabo), Western Coast (Mayagüez), Vieques; 1–50 m; *fl* & *fr* Jun–Jul, Oct.

Antilles, Mexico, Central America, South America.

Caña brava, Caña cimarrona, Caña India, Guajana.

Heteropogon contortus (L.) P. Beauv. ex Roem. & Schult., Syst. Veg. 2:836. 1817.
Bas.: *Andropogon contortus* L. (1753) "contortum."

Herb in Sierra de Luquillo, Southeastern Foothills (Coamo), Southern Coastal Lowlands (Peñuelas, Caja de Muertos), Southern Coastal Limestone, Guánica Forest Reserve, Southwestern

Coastal Lowlands, Sierra Bermeja, Cerro Las Mesas, Mona; 1–600 m; *fl* & *fr* throughout the year.

Tropical and warm temperate regions.

Yerba torcida.

Holcus lanatus L., Sp. Pl. 2:1048. 1753.

Herb in Toro Negro Forest Reserve; 1300 m; *fl* & *fr* May–Jul (two specimens). [*Vives a.n. 2705* (SJ); *Liogier 36840* (UPR)].

Native to Europe, northern Africa, west-central Asia.

Homolepis aturensis (Kunth) Chase, Proc. Biol. Soc. Wash. 24:146, *f. 12.* 1911.
Bas.: *Panicum aturense* Kunth (1816).

Herb in Eastern Coast (Ceiba); 15–20 m; *fl* & *fr* Sep–Oct (one specimen). [*Caraballo 2723* (UPR)].

Native to Trinidad, southern Mexico, Central America, northern South America to Peru.

Homolepis glutinosa (Sw.) Zuloaga & Soderstr., Smithsonian Contr. Bot. 59:19. 1985.
Bas.: *Panicum glutinosum* Sw. (1788).

Herb in Northern Coastal Lowlands (Laguna Tortuguero), Sierra de Luquillo, Sierra de Cayey, Central Cordillera, Maricao Forest Reserve; 625–1300 m; *fl* & *fr* May–Jan.

Greater Antilles (Cuba, Jamaica), Mexico, Central America, South America.

Yerba pegajosa.

Hymenachne amplexicaulis (Rudge) Nees in Martius et al., Fl. Bras. Enum. Pl. 2(1):276. 1829 (Agrost. Bras.).
Bas.: *Panicum amplexicaule* Rudge (1805).

Herb in Northern Coastal Lowlands, Eastern Coast, Caguas Valley, Eastern Cordillera (Aibonito), Southern Coastal Lowlands (Guayama), Southwestern Coastal Lowlands, Western Coast; 1–250 m; *fl* & *fr* Oct–Jul.

Antilles, Trinidad, Mexico, Central America, South America.

Trompetilla.

Hymenachne donacifolia (Raddi) Chase, J. Washington Acad. Sci. 13(9):177. 1923.
Bas.: *Panicum donacifolium* Raddi (1823).

Herb in Eastern Coast (Naguabo, Humacao); 10–40 m; *fl* & *fr* Jan, Jun (two specimens). [*Axelrod 11040, 12598* (UPRRP)].

Native to Central America, South America.

Hyparrhenia rufa (Nees) Stapf in Prain, Fl. Trop. Afr. 9(2):304. 1919.
Bas.: *Trachypogon rufus* Nees (1829).

Herb in Northern Coastal Lowlands, Eastern Cordillera (Corozal), Southwestern Coastal Lowlands, Cerro Las Mesas, Western Coast; 5–400 m; *fl* & *fr* Nov–Apr.

Native to tropical and southern Africa.

Cola de caballo, Jaragua falsa.

Ichnanthus nemorosus (Sw.) Döll in Martius, Fl. Bras. 2(2):289. 1877.
Bas.: *Panicum nemorosum* Sw. (1788).

Herb in Northern Limestone Hills (Isabela), Río Abajo Forest Reserve, Eastern Cordillera (Cerro La Peña), Central Cordillera; 100–1000 m; *fl* & *fr* Jan–Apr, Aug–Oct.

Antilles, Trinidad and Tobago, Mexico, Central America.

Ichnanthus pallens (Sw.) Munro ex Benth., Fl. Hongk. 414. 1861.
Bas.: *Panicum pallens* Sw. (1788).
Syn.: *Ichnanthus axillaris* (Nees) Hitchc. & Chase (1917).

Herb in Northern Coastal Lowlands, Northern Limestone Hills, Río Abajo Forest Reserve, Sierra de Luquillo, Eastern Coast, Sierra de Cayey, Las Tetas de Cayey, Eastern Cordillera (Aibonito), Central Cordillera, Susúa Forest Reserve, Western Cordillera, Western Coast (Mayagüez); 50–1225 m; *fl* & *fr* throughout the year.

Antilles, Trinidad, Central America, South America.

Carrucillo, Carruzo, Cohítre falso.

Ichnanthus tenuis (J. Presl & C. Presl) Hitchc. & Chase, Contr. U.S. Natl. Herb. 18(7):334. 1917.
Bas.: *Oplismenus tenuis* J. Presl & C. Presl (1830).

Herb in Eastern Coast (Cuchilla de Panduras); 500 m; *fl* & *fr* Sep (one specimen). [*Liogier 30977* (UPR)].

Trinidad, Mexico, Central America, South America.

Imperata brasiliensis Trin., Mém. Acad. Imp. Sci. St.-Pétersbourg, Sér. 6, Sci. Math. 2(3):331. 1832.

Herb in Maricao Forest Reserve, 500–700 m; *fl* & *fr* Apr (one specimen). [*Mckenzie 601* (SJ)].

Bahamas, Greater Antilles, Trinidad and Tobago, Mexico, Central America, South America.

Guajanilla.

Isachne angustifolia Nash, Bull. Torrey Bot. Club 30(7):377. 1903 "angustifolium."

Herb in Sierra de Luquillo, Sierra de Cayey, Central Cordillera; 475–1340 m; *fl* & *fr* throughout the year.

Lesser Antilles (St. Kitts, Guadeloupe, Martinique).

Yerba de alambre.

Ischaemum rugosum Salisb., Icon. Stirp. Rar. 1, *t. 1*. 1791.

Herb in Western Coast (Añasco); 1–50 m; *fl* & *fr* (?). [*Cruz Pérez a.n. 2595063* (US)].

Native to tropical Asia, China, Australia.

Lasiacis divaricata (L.) Hitchc., Contr. U.S. Natl. Herb. 15:16. 1910.

Bas.: *Panicum divaricatum* L. (1759).

Syn.: *Lasiacis harrisii* Nash (1913).

Viny herb in Northern Coastal Lowlands, Northern Limestone Hills, Río Abajo Forest Reserve, Sierra de Luquillo, Eastern Coast, Caguas Valley (Caguas), Sierra de Cayey, Las Tetas de Cayey, Southeastern Foothills, Eastern Cordillera (Aibonito), Central Cordillera (Orocovis), Southern Coastal Lowlands (Caja de Muertos), Southern Coastal Limestone, Susúa Forest Reserve, Maricao Forest Reserve, Guánica Forest Reserve, Southwestern Coastal Lowlands (Cabo Rojo), Sierra Bermeja, Cerro Las Mesas, Vieques, Culebra, Mona; 1–800 m; *fl* & *fr* throughout the year.

West Indies, southeastern United States (Florida), Mexico, Central America, South America. [AR: 439].

Pito, Yerba de caña.

Lasiacis ligulata Hitchc. & Chase, Contr. U.S. Natl. Herb. 18(7):337. 1917.

Viny herb in Sierra de Luquillo, Sierra de Cayey, Western Cordillera; 400–850 m; *fl* & *fr* Jan–Mar, Jul.

Greater Antilles, Lesser Antilles (Dominica, St. Vincent), Trinidad and Tobago, northern South America to Bolivia. [AR: 440].

Lasiacis sloanei (Griseb.) Hitchc., Bot. Gaz. 51:302. 1911.

Bas.: *Panicum sloanei* Griseb. (1864).

Herb in Northern Limestone Hills (Toa Alta); 50–100 m; *fl* & *fr* Nov (one specimen). [*Woodbury s.n.* (SJ), not seen].

Greater Antilles, Lesser Antilles (Antigua, Martinique, Grenada), Mexico, Central America, northwestern South America and Venezuela.

Notes: According to Liogier & Martorell (1982), because of misidentification by Hitchcock, Britton & Wilson (1923) had misapplied the above name to specimens of *Lasiacis divaricata*.

Lasiacis sorghoidea (Ham.) Hitchc. & Chase var. **sorghoidea**, Contr. U.S. Natl. Herb. 18(7):338. 1917.

Bas.: *Panicum sorghoideum* Ham. (1825).

Syn.: *Lasiacis maculata* (Aubl.) Urb. (1921).

Viny herb in Sierra de Luquillo, Caguas Valley (Caguas), Sierra de Cayey, Eastern Cordillera (Cañón San Cristóbal), Central Cordillera, Susúa Forest Reserve, Western Cordillera, Vieques; 150–825 m; *fl* & *fr* Nov–Jul.

West Indies, Trinidad, Mexico, Central America, South America. [AR: 440].

Carrucillo, Pito de alfombra.

Leersia hexandra Sw., Prodr. 21. 1788.

Syn.: *Homalocenchrus hexandrus* (Sw.) Kuntze (1891).

Herb in Northern Coastal Lowlands, Sierra de Luquillo, Sierra de Cayey; 1–500 m; *fl* & *fr* throughout the year.

Pantropical.

Arrocillo rosado.

Leersia monandra Sw., Prodr. 21. 1788.

Syn.: *Homalocenchrus monandrus* (Sw.) Kuntze (1891).

Herb in Southeastern Foothills (Coamo), Southern Coastal Limestone (Guayanilla), Sierra Bermeja, Culebra (Cayo Luis Peña); 1–300 m; *fl* & *fr* Oct–Dec.

Greater Antilles, Lesser Antilles (Martinique), southern United States (Florida, Texas), Mexico, South America (Brazil).

Leptochloa fusca (L.) Kunth subsp. **uninervia** (J. Presl) N. Snow, Novon 8(1):79. 1998.

Bas.: *Megastachya uninervia* J. Presl (1830).

Syn.: *Leptochloa uninervia* (J. Presl) Hitchc. & Chase (1917).

Herb in Northern Coastal Lowlands, Southern Coastal Lowlands, Western Coast (Añasco); 1–5 m; *fl* & *fr* Jan–Apr, Aug.

Bahamas, Greater Antilles (Cuba, Jamaica), southern half of United States, Mexico, Central America, South America.

Leptochloa panicea (Retz.) Ohwi subsp. **brachiata** (Steud.) N. Snow, Novon 8(1):79. 1998.

Bas.: *Leptochloa brachiata* Steud. (1854).

Herb in Northern Coastal Lowlands, Cambalache Forest Reserve, Caguas Valley (Gurabo), Southeastern Foothills (Salinas), Southern Coastal Lowlands, Guánica Forest Reserve, Southwestern Coastal Lowlands (Cabo Rojo), Western Coast, Culebra, Desecheo; 15–150 m; *fl* & *fr* throughout the year.

Antilles, Tobago, southern half of United States, Mexico, Central America, South America (scattered).

Notes: Britton & Wilson (1923) and Liogier & Martorell (1982, 2000) used *Leptochloa filiformis* (Lam.) P. Beauv., a *nom. inval.*, for the name of this taxon; Liogier & Martorell (2000) misapplied *Leptochloa mucronata* (Michx.) Kunth to it.

Yerba de hilo.

Leptochloa scabra Nees in Martius et al., Fl. Bras. Enum. Pl. 2(1):435. 1829 (Agrost. Bras.).

Herb in Northern Coastal Lowlands, Sierra de Cayey, Eastern Coast, Caguas Valley, Southern Coastal Lowlands (Guayama, Coamo), Guánica Forest Reserve, Southwestern Coastal Lowlands (Cabo Rojo), Western Coast, Vieques; 1–150 m; *fl* & *fr* Aug–May.

Southern United States (Alabama, Louisiana), Mexico, Central America, South America (scattered).

Notes: Specimens of this taxon have commonly been misidentified as *Leptochloa virgata*.

Leptochloa virgata (L.) P. Beauv., Ess. Agrostogr. 71, 161, 166, *t. 15. f. 1.* 1812.

Bas.: *Cynosurus virgatus* L. (1759).

Herb in Eastern Coast (Ceiba), Sierra de Cayey, Vieques; 1–5 m; *fl* & *fr* Jun–Aug, Dec.

West Indies, southern United States (South Carolina, Florida, Texas), Mexico, Central America, South America (scattered).

Lithachne pauciflora (Sw.) P. Beauv., Ess. Agrostogr. 135, 168. 1812 "Lithacne."

Bas.: *Olyra pauciflora* Sw. (1788).

Herb in Northern Coastal Lowlands, Northern Limestone Hills, Río Abajo Forest Reserve, Sierra de Cayey, Las Tetas de Cayey, Eastern Cordillera (Cañón San Cristóbal), Central Cordillera, Susúa Forest Reserve, Western Cordillera (Maricao), Guánica Forest Reserve; 10–750 m; *fl* & *fr* throughout the year.

Greater Antilles (Cuba, Hispaniola), Lesser Antilles (Antigua, Guadeloupe, Martinique), Mexico, Central America, South America (scattered).

Diente de perro, Perla.

Lolium perenne L., Sp. Pl. 1:83. 1753.

Herb in Sierra de Luquillo, Toro Negro Forest Reserve; 1050–1330 m; *fl* & *fr* Mar–Sep.

Native to Europe, northern Africa, western Asia, Indian Subcontinent.

Luziola subintegra Swallen, Ann. Missouri Bot. Gard. 30(2):165. 1943.

Aquatic herb in Northern Coastal Lowlands (Toa Baja), Western Coast; 1–25 m; *fl* & *fr* Jul, Oct. [*Caudales a.n. 9582, 11086* (UPR)].

Greater Antilles (Cuba), Mexico, Central America, South America (scattered).

Megathyrsus maximus (Jacq.) B.K. Simon & S.W.L. Jacobs, Austrobaileya 6(3):572. 2003.

Bas.: *Panicum maximum* Jacq. (1786).

Herb in Northern Coastal Lowlands, Northern Limestone Hills, Eastern Coast (Humacao), Caguas Valley, Southeastern Foothills, Eastern Cordillera (Cayey), Central Cordillera, Southern Coastal Lowlands, Southern Coastal Limestone (Guayanilla), Susúa Forest Reserve, Guánica Forest Reserve, Southwestern Coastal Lowlands, Sierra Bermeja, Western Coast, Vieques, Culebra, Mona, Desecheo; 1–1200 m; *fl* & *fr* throughout the year.

Native to Africa, Yemen.

Guinea, Yerba de Guinea.

Melinis minutiflora P. Beauv., Ess. Agrostogr. 54, *t. 11, f. 4.* 1812.

Herb in Northern Limestone Hills (Isabela), Sierra de Luquillo, Eastern Coast (Punta Guayanés), Sierra de Cayey, Eastern and Central Cordillera, Susúa Forest Reserve, Western Cordillera, Cerro Las Mesas; 25–1200 m; *fl* & *fr* Nov–May.

Native to Africa.

Yaraguá, Yerba de yaraguá, Yerba melao, Zacate gordura.

Melinis repens (Willd.) Zizka, Biblioth. Bot. 138:55. 1988.

Bas.: *Saccharum repens* Willd. (1797).

Syn.: *Rhynchelytrum repens* (Willd.) C.E. Hubb. (1934). *Tricholaena repens* (Willd.) Hitchc. (1936).

Tricholaena rosea Nees (1835).

Herb in Northern Coastal Lowlands, Río Abajo Forest Reserve, Eastern Coast (Ceiba, Punta Guayanés), Las Piedras Chiquitas, Southeastern Foothills (Coamo), Central Cordillera (Ponce), Southern Coastal Lowlands, Southern Coastal Limestone (Ponce), Susúa Forest Reserve, Maricao Forest Reserve, Guánica Forest Reserve, Southwestern Coastal Lowlands (Cabo Rojo), Cerro Las Mesas, Vieques, Mona, Desecheo; 2–600 m; *fl* & *fr* Mar–Nov.

Native to Africa, Arabian peninsula, India.

Arrocillo colorado, Pasto Natal, Yerba del Natal, Yerba rosada.

Mnesithea granularis (L.) de Koning & Sosef, Blumea 31(2):295. 1986.

Bas.: *Cenchrus granularis* L. (1771).

Syn.: *Hackelochloa granularis* (L.) Kuntze (1891). *Rytilix granularis* (L.) Skeels (1913).

Herb in Western Coast (Rincón); 1–25 m; *fl* & *fr* Sep (one specimen). [*Woodbury a.n. 24640* (UPR)].

Native to Africa, tropical Asia, China, Japan, Australia.

Yerba de grano.

Muhlenbergia capillaris (Lam.) Trin., Gram. Unifl. Sesquifl. 191, 296, *t. 5, f. 15.* 1824 "Mühlenbergia."
Bas.: *Stipa capillaris* Lam. (1791).

Herb in Northern Coastal Lowlands (Laguna Tortuguero); 1 m; *fl* & *fr* May, Sep. [*Woodbury a.n. 4300* (UPR); *Woodbury a.n. 826054* (NY), scan seen; *Proctor 45020* (SJ)].

Bahamas, Greater Antilles (Cuba), eastern and south-central United States, Mexico, northern Central America (Guatemala).

Rabo de pavo real, Yerba de pelos.

Olyra latifolia L., Syst. Nat. (ed. 10) 2:1261. 1759.

Herb in Northern Coastal Lowlands (Fajardo, Toa Baja), Northern Limestone Hills, Río Abajo Forest Reserve, Sierra de Luquillo, Eastern Coast, Sierra de Cayey, Southeastern Foothills, Eastern Cordillera (Cerro La Peña, Cañón San Cristóbal), Central Cordillera (Ponce), Western Cordillera, Southwestern Coastal Lowlands, Sierra Bermeja, Western Coast, Vieques; 10–600 m; *fl* & *fr* throughout the year.

Antilles, Mexico, Central America, South America. [AR: 440].

Carrucillo, Cortadera, Lambedora, Lamedora, Sonadora.

Oplismenus hirtellus (L.) P. Beauv. subsp. **hirtellus**, Ess. Agrostogr. 54, 168, 170. 1812.
Bas.: *Panicum hirtellum* L. (1759).

Herb in Río Abajo Forest Reserve, Sierra de Luquillo, Eastern Coast (Naguabo), Sierra de Cayey, Southeastern Foothills (Coamo), Monte Guilarte Forest Reserve, Maricao Forest Reserve; 125–1050 m; *fl* & *fr* Oct–Mar.

West Indies, Trinidad and Tobago, Mexico, Central America, South America.

Carruzo, Yerba de madera, Yerba de maleza.

Oplismenus hirtellus (L.) P. Beauv. subsp. **setarius** (Lam.) Mez ex Ekman, Ark. Bot. 11(4):26. 1912.
Bas.: *Panicum setarium* Lam. (1791).

Herb in Northern Coastal Lowlands, Northern Limestone Hills, Eastern Coast (Cayo Santiago), Southeastern Foothills (Guayama), Toro Negro Forest Reserve, Susúa Forest Reserve, Western Coast (Mayagüez), Vieques, Mona; 1–500 m; *fl* & *fr* Jun–Mar.

West Indies, southeastern and south-central United States, Mexico, Central America, South America.

Carruzo, Yerba de madera, Yerba de maleza.

Oryza latifolia Desv., J. Bot. Agric. 1:77. 1813.

Herb in Eastern Coast (Humacao); 1–2 m; *fl* & *fr* May (one specimen). [*Woodbury a.n. 4306* (UPR)].

Native to Greater Antilles (Cuba), Trinidad and Tobago, Mexico, Central America, South America.

Oryza sativa L., Sp. Pl. 1:333. 1753.

Herb planted and persistent in Northern Coastal Lowlands; 1–5 m; *fl* & *fr* Oct–Dec, Mar.

Probably native to tropical Asia, China.

Arroz.

Panicum aquaticum Poir., Encycl. Suppl. 4(1):281. 1816.

Herb in Northern Coastal Lowlands (Vega Baja, Laguna Tortuguero), Sierra de Luquillo; 1–1040 m; *fl* & *fr* Apr–May, Aug.

Greater Antilles (Cuba), Lesser Antilles (Guadeloupe), Trinidad, Mexico, Central America, southern and central South America.

Yerba acuática.

Panicum dichotomiflorum Michx., Fl. Bor.-Amer. 1:48. 1803.

Herb in Sierra de Luquillo, Sierra de Cayey, Toro Negro Forest Reserve; 700–1000 m; *fl* & *fr* May–Jan.

Bahamas, Greater Antilles (Cuba), Lesser Antilles (Guadeloupe), eastern half of Canada and United States, southern Central America (Panama), South America.

Panicum diffusum Sw., Prodr. 23. 1788.

Herb in Southwestern Coastal Lowlands (Cabo Rojo), Western Coast (Mayagüez, Moca); 10–100 m; *fl* & *fr* Feb (one specimen). [*González Más 4014* (MAPR); *Proctor 43932* (SJ)].

West Indies, south-central United States (Texas), Mexico, Central America, northern South America and Brazil.

Panicum elephantipes Nees ex Trin., Gram. Panic. 206. 1826.

Herb in Northern Coastal Lowlands (Río Grande, Loíza), Eastern Coast (Naguabo); 1–10 m; *fl* & *fr* Dec–May.

Greater Antilles, Mexico, Central America, South America (scattered).

Panicum ghiesbreghtii E. Fourn., Mexic. Pl. 2:29. 1886.

Herb in Northern Coastal Lowlands (Fajardo), Eastern Coast (Maunabo), Southwestern Coastal Lowlands (Cabo Rojo); 1–10 m; *fl* & *fr* Sep–Nov.

Bahamas, Greater Antilles, Lesser Antilles (St. Martin, Antigua, Guadeloupe), south-central United

States (Texas), Mexico, Central America, northwestern, west-central South America and Venezuela.

Panicum hirsutum Sw., Fl. Ind. Occid. 1:173. 1797.
Herb in Eastern Coast (Humacao); 1–3 m; *fl* & *fr* May (one specimen). [*Woodbury a.n. 13762* (SJ)].
Greater Antilles, Lesser Antilles (Antigua, Barbuda), south-central United States (Texas), Mexico, Central America, South America (scattered).

Panicum miliaceum L., Sp. Pl. 1:58. 1753.
Herb in Northern Coastal Lowlands (San Juan); 10 m; *fl* & *fr* Apr (one specimen). [*Proctor 46345* (US)].
Native to temperate Asia.

Panicum parvifolium Lam., Tabl. Encycl. 1:173. 1791.
Herb in Northern Coastal Lowlands (Dorado to Laguna Tortuguero), Sierra de Luquillo; 1–800 m; *fl* & *fr* Jun–Mar.
Greater Antilles, Trinidad, Central America, northern South America to Paraguay.
Notes: Liogier & Martorell (2000) wrongly used *Panicum parviflorum* Lam., probably an orthographic error, for the name of this taxon; the earlier edition (1982) had the correct name.
Yerba de piedra.

Panicum rigidulum Bosc ex Nees in Martius et al., Fl. Bras. Enum Pl. 2(1):163. 1829 (Agrost. Bras.).
Syn.: *Panicum condensum* Nash (1903).
Herb in Northern Coastal Lowlands (Laguna Tortuguero); 1–2 m; *fl* & *fr* Mar–May, Oct.
Bahamas, Greater Antilles, Lesser Antilles (Guadeloupe), eastern half of United States, Mexico, northern Central America (Belize, Guatemala).

Panicum stenodes Griseb., Fl. Brit. W.I. 547. 1864.
Herb in Northern Coastal Lowlands (Laguna Tortuguero); 1–2 m; *fl* & *fr* May (one specimen).
Greater Antilles, Trinidad, Central America, northern South America to Bolivia.

Panicum stevensianum Hitchc. & Chase, Contr. U.S. Natl. Herb. 17(6):498, *t.* 76. 1915.
Herb in Northern Limestone Hills (Florida); 100–250 m; *fl* & *fr* Jun–Aug (two specimens). [*Caudales a.n. 11011, 11387* (UPR)].
Greater Antilles (Cuba).

Panicum tenerum Beyr. ex Trin., Mém. Acad. Imp. Sci. St.-Pétersbourg, Sér. 6, Sci. Math., Seconde Pt. Sci. Nat. 3,1(2–3):341. 1834.
Herb in Northern Coastal Lowlands (Laguna Tortuguero); 1–2 m; *fl* & *fr* Oct–Nov, Feb–Jun.
Bahamas, Greater Antilles (Cuba, Hispaniola), southeastern and south-central United States, Mexico, Central America.

Panicum trichanthum Nees in Martius et al., Fl. Bras. Enum. Pl. 2(1):210. 1829 (Agrost. Bras.).
Herb in Northern Limestone Hills, Sierra de Luquillo, Sierra de Cayey, Eastern Cordillera (Cayey), Central Cordillera (Ponce, Peñuelas), Maricao Forest Reserve; 150–650 m; *fl* & *fr* Nov–Jul.
Greater Antilles, Mexico, Central America, South America.

Panicum trichidiachne Döll in Martius, Fl. Bras. 2(2):339, *t.* 49. 1877.
Syn.: *Panicum schiffneri* Hack. (1906).
Herb in Sierra de Cayey, Las Tetas de Cayey, Eastern Cordillera (La Torrecilla), Central Cordillera; 600–1100 m; *fl* & *fr* Oct–Mar.
Lesser Antilles (Martinique), Mexico, southern Central America (Costa Rica), South America (Venezuela, Brazil).

Panicum trichoides Sw., Prodr. 24. 1788.
Herb in Northern Coastal Lowlands (Carolina, Guaynabo), Northern Limestone Hills (Florida, Arecibo), Río Abajo Forest Reserve, Sierra de Luquillo, Eastern Coast (Cuchilla de Panduras), Caguas Valley (Caguas), Sierra de Cayey (San Lorenzo), Eastern and Central Cordillera, Western Cordillera (Maricao), Vieques; 3–625 m; *fl* & *fr* throughout the year.
West Indies, Trinidad, Mexico, Central America, South America.
Ilusión, Cohitrillo fino.

Panicum venezuelae Hack., Oesterr. Bot. Z. 51:368. 1901.
Herb in Las Tetas de Cayey; 830 m; *fl* & *fr* Feb (one specimen). [*Proctor 42997* (SJ)].
Native to Greater Antilles (Cuba), northern Central America (Guatemala, Honduras), South America (Venezuela, Brazil).

Pappophorum pappiferum (Lam.) Kuntze, Revis. Gen. Pl. 3(3):365. 1898.
Bas.: *Saccharum pappiferum* Lam. (1791).
Herb in Guánica Forest Reserve, Southwestern Coastal Lowlands (Cabo Rojo), Mona, Desecheo; 1–200 m; *fl* & *fr* Mar–Nov.
Greater Antilles (Cuba, Hispaniola), Lesser Antilles, Mexico, South America (scattered).
Plumacho.

Paspalidium geminatum (Forssk.) Stapf in Prain, Fl. Trop. Afr. 9(4):583. 1920.
Bas.: *Panicum geminatum* Forssk. (1775).

Herb in Northern Coastal Lowlands, Southern Coastal Lowlands (Guayama), Guánica Forest Reserve, Southwestern Coastal Lowlands, Vieques; 1–50 m; *fl* & *fr* throughout the year.

Throughout tropical and warm-temperate regions of the world.

Yerba de agua, Yerba de pantano.

Paspalum arundinaceum Poir., Encycl. Suppl. 4(1): 310. 1816.

Syn.: *Paspalum secans* Hitchc. & Chase (1917).

Herb in Northern Coastal Lowlands (Laguna Tortuguero), Sierra de Luquillo, Eastern Coast (Ceiba, Punta Guayanés), Western Cordillera (Maricao); 1–670 m; *fl* & *fr* Mar, Jul.

Bahamas, Greater Antilles, Lesser Antilles (Antigua, Guadeloupe, Grenadines), Central America, northern South America (Guyana).

Notes: Liogier & Martorell (1982) wrongly placed *Paspalum secans* in synonymy with *Paspalum pleostachyum*.

Paspalum blodgettii Chapm., Fl. South. U.S. 571. 1860.

Syn.: *Paspalum simpsonii* Nash (1897).

Herb in Northern Coastal Lowlands (San Juan), Northern Limestone Hills (Quebradillas), Susúa Forest Reserve, Mona; 25–100 m; *fl* & *fr* Mar–Nov.

Bahamas, Greater Antilles, southeastern United States (Florida), Mexico, northern Central America.

Paspalum boscianum Flüggé, Gram. Monogr., Paspalum 170. 1810 "Paspalus boscianus."

Herb in Northern Coastal Lowlands (Guaynabo, Laguna Tortuguero), Eastern Cordillera (Corozal), Central Cordillera (Utuado), Maricao Forest Reserve; 1–900 m; *fl* & *fr* Feb–Apr, Jul–Oct.

Greater Antilles (Hispaniola), Virgin Islands, southeastern and south-central United States, Mexico, Central America, northern South America to Bolivia.

Notes: Britton & Wilson (1923) misapplied the name *Paspalum melansopermum* Desv. to, and Liogier & Martorell (1982) wrongly placed it in synonymy with, this taxon.

Paspalum caespitosum Flüggé, Gram. Monogr., Paspalum 161. 1810.

Syn.: *Paspalum poiretii* Roem. & Schult. (1817).

Herb in Southern Coastal Limestone, Guánica Forest Reserve, Western Coast, Vieques, Mona; 1–200 m; *fl* & *fr* throughout the year.

Bahamas, Greater Antilles, southeastern United States (Florida, Alabama), Mexico, northern Central America.

Notes: Liogier & Martorell (1982) wrongly placed *Paspalum poiretii* in synonymy with *Paspalum saugetii*.

Paspalum clavuliferum C. Wright in Sauvalle, Anales Acad. Ci. Méd. Habana 8:203. 1871.

Herb in Eastern Coast (Punta Guayanés); 25 m; *fl* & *fr* Jun (one specimen). [*Axelrod 13547* (UPRRP)].

Greater Antilles (Cuba, Hispaniola), Mexico, Central America, northern South America to Brazil.

Paspalum conjugatum P. J. Bergius, Acta Helv. Phys.-Math. 7:129, *t. 8*. 1772.

Herb in Northern Coastal Lowlands, Northern Limestone Hills (Isabela), Río Abajo Forest Reserve, Sierra de Luquillo, Eastern Coast (Humacao), Caguas Valley (Caguas), Sierra de Cayey, Eastern Cordillera (Corozal, La Torrecilla), Central Cordillera (Adjuntas), Cerro de Las Cuevas, Susúa Forest Reserve, Western Coast; 1–1050 m; *fl* & *fr* throughout the year.

Pantropical and -subtropical.

Horquetilla, Horquetilla blanca, Grama amarga.

Paspalum decumbens Sw., Prodr. 22. 1788.

Herb in Northern Coastal Lowlands (Fajardo), Northern Limestone Hills (San Sebastián), Sierra de Luquillo, Eastern Coast, Caguas Valley (Caguas), Sierra de Cayey, Eastern Cordillera (Cerro La Peña), Central Cordillera, Maricao Forest Reserve, Cerro Las Mesas, Western Coast (Añasco); 50–1000 m; *fl* & *fr* throughout the year.

Greater Antilles, Lesser Antilles (Guadeloupe), Trinidad, Mexico, Central America, northwestern, west-central South America and Venezuela.

Pendejuelo.

Paspalum dilatatum Poir., Encycl. 5:35. 1804.

Herb in Northern Coastal Lowlands, Northern Limestone Hills (Lares), Sierra de Luquillo, Caguas Valley (Gurabo), Central Cordillera (Utuado); 1–700 m; *fl* & *fr* Feb–Apr, Jul–Oct.

Native to southern South America and Brazil.

Dalis, Yerba dalis.

Paspalum distichum L., Syst. Nat. (ed. 10) 2:855. 1759.

Herb in Northern Coastal Lowlands, Susúa Forest Reserve, Guánica Forest Reserve, Southwestern Coastal Lowlands (Cabo Rojo); 1–200 m; *fl* & *fr* Feb, Jun–Oct.

Probably originating in the Neotropics.

Saladillo, Salaíllo.

Paspalum fasciculatum Willd. ex Flüggé, Gram. Monogr., Paspalum 69. 1810.

Herb in Northern Coastal Lowlands, Sierra de Luquillo, Eastern Cordillera (Aguas Buenas); 10–1000 m; *fl* & *fr* Jul–Nov.

Greater Antilles, Lesser Antilles (Guadeloupe, St. Lucia), Trinidad, Mexico, Central America, South America (scattered).

Venezolana.

Paspalum fimbriatum Kunth in Humboldt, Bonpland & Kunth, Nov. Gen. Sp. 1:77 [folio]; 1:93 [quarto], *t. 28*. 1816.

Herb in Northern Coastal Lowlands (islands adjacent to Fajardo, San Juan), Northern Limestone Hills, Río Abajo Forest Reserve, Sierra de Cayey, Las Piedras Chiquitas, Southeastern Foothills, Eastern Cordillera, Toro Negro Forest Reserve, Cerro de Las Cuevas, Southern Coastal Limestone (Guayanilla), Susúa Forest Reserve, Western Cordillera, Guánica Forest Reserve, Southwestern Coastal Lowlands (Lajas), Sierra Bermeja, Cerro Las Mesas, Vieques; 10–775 m; *fl* & *fr* Jul–Apr.

West Indies, Trinidad and Tobago, southeastern United States (Florida), Mexico, Central America (Belize, Guatemala, Panama), northern South America (Venezuela, Colombia) and Brazil.

Pata de conejo.

Paspalum laxum Lam., Tabl. Encycl. 1:176. 1791.

Syn.: *Paspalum glabrum* Poir. (1804).

Herb in Northern Coastal Lowlands, Northern Limestone Hills, Río Abajo Forest Reserve, Eastern Coast, Eastern Cordillera (Cañón San Cristóbal), Southern Coastal Lowlands (Caja de Muertos), Southern Coastal Limestone (Ponce), Susúa Forest Reserve, Western Cordillera (Las Marías), Guánica Forest Reserve, Southwestern Coastal Lowlands, Cerro Las Mesas, Western Coast, Vieques, Mona, Desecheo; 1–450 m; *fl* & *fr* throughout the year.

West Indies, southeastern United States (Georgia, Florida), northern Central America (Belize), northern South America (Guyana, Colombia).

Matojo de arena.

Paspalum macrophyllum Kunth in Humboldt, Bonpland & Kunth, Nov. Gen. Sp. 1:76 [folio]; 1:92 [quarto]. 1816.

Herb in Northern Coastal Lowlands (San Juan), Western Cordillera (Maricao), Western Coast (Mayagüez); 10–700 m; *fl* & *fr* Mar–Apr, Jun–Sep.

Native to northwestern South America and Venezuela.

Matojo de choroní.

Paspalum millegrana Schrad. in Schultes, Mant. 2:175. 1824 "Paspalus."

Herb in Northern Coastal Lowlands, Northern Limestone Hills (Quebradillas), Sierra de Luquillo, Eastern Coast, Caguas Valley (Juncos), Sierra de Cayey, Eastern Cordillera (Aibonito), Southern Coastal Lowlands (Guayama), Western Cordillera (Maricao), Southwestern Coastal Lowlands (Lajas), Western Coast (Rincón); 1–500 m; *fl* & *fr* throughout the year.

West Indies, Trinidad and Tobago, Central America, South America (scattered).

Cortadera, Paja brava.

Paspalum minus E. Fourn., Mexic. Pl. 2:6. 1886.

Herb in Northern Coastal Lowlands, Sierra de Luquillo, Sierra de Cayey, Central Cordillera (Orocovis), Western Coast (Moca); 1–1050 m; *fl* & *fr* Feb, May–Nov.

Greater Antilles, southern United States (Florida to Texas), Mexico, Central America, South America (scattered).

Paspalum molle Poir., Encycl. 5:34. 1804.

Syn.: *Paspalum portoricense* Nash (1903).

Herb in Northern Limestone Hills, Las Tetas de Cayey, Southeastern Foothills (Coamo), Susúa Forest Reserve, Maricao Forest Reserve; 50–850 m; *fl* & *fr* Feb–Jul, Oct.

Bahamas, Lesser Antilles (Antigua, Guadeloupe, Dominica), northern South America (Venezuela, Brazil).

Yerba menuda.

Paspalum notatum Flüggé, Gram. Monogr., Paspalum 106. 1810.

Herb in Northern Coastal Lowlands (San Juan, Aguadilla), Vieques; 1–10 m; *fl* & *fr* Mar, Aug–Oct.

West Indies, Mexico, Central America, southern South America to Peru.

Bahía, Yerba bahía.

Paspalum orbiculatum Poir., Encycl. 5:32. 1804.

Herb in Sierra de Luquillo, Eastern Coast (Cuchilla de Panduras), Sierra de Cayey (San Lorenzo); 350–600 m; *fl* & *fr* Oct, Feb–Mar.

Antilles, Trinidad, Mexico, Central America, South America.

Gramilla.

Paspalum paniculatum L., Syst. Nat. (ed. 10) 2:855. 1759.

Herb in Northern Coastal Lowlands, Northern Limestone Hills, Sierra de Luquillo, Eastern Coast (Ceiba), Caguas Valley, Sierra de Cayey, Eastern and Central Cordillera, Western Cordillera (Maricao), Western Coast; 1–1050 m; *fl* & *fr* throughout the year.

West Indies, Mexico, Central America, South America, tropical Africa.

Arrocillo, Arrocillo de altura, Grasilla, Yerba peluda.

Paspalum pleostachyum Döll in Martius, Fl. Bras. 2(2):58. 1877.

Herb in Northern Coastal Lowlands, Río Abajo Forest Reserve, Sierra de Luquillo, Eastern Coast, Caguas Valley (Gurabo), Sierra de Cayey, Eastern Cordillera, Central Cordillera (Orocovis), Maricao Forest Reserve, Guánica Forest Reserve, Cerro Las Mesas, Western Coast (Rincón), Vieques; 1–1000 m; *fl* & *fr* throughout the year.

Bahamas, Greater Antilles (Hispaniola), Lesser Antilles, South America (Brazil).

Paspalum plicatulum Michx., Fl. Bor.-Amer. 1:45. 1803.

Herb in Northern Coastal Lowlands, Northern Limestone Hills, Sierra de Luquillo, Eastern Coast, Caguas Valley (Caguas), Sierra de Cayey, Las Tetas de Cayey, Eastern Cordillera, Susúa Forest Reserve, Southwestern Coastal Lowlands (Cabo Rojo), Sierra Bermeja, Western Coast (Mayagüez); 1–1050 m; *fl* & *fr* Mar–Dec.

Antilles, Trinidad and Tobago, southeastern and south-central United States, Mexico, Central America, South America.

Cortadera dulce, Gramalotillo, Manga larga.

Paspalum pulchellum Kunth, Mém. Mus. Hist. Nat. 2:68. 1815.

Herb in Cerro Las Mesas; 340 m; *fl* & *fr* Aug–Sep (two specimens). [*McKenzie 711* (SJ); *Proctor 44998* (SJ)].

Native to Greater Antilles (Cuba, Hispaniola), Trinidad, Central America, northern South America to Brazil.

Paspalum rupestre Trin., Linnaea 10(3):293. 1836.
Syn.: *Paspalum leoninum* Chase (1911).

Herb in Susúa Forest Reserve, Western Cordillera, Cerro Las Mesas; 300–800 m; *fl* & *fr* Sep–Dec, Mar.

Greater Antilles (Cuba, Hispaniola).

Paspalum saugetii Chase, Contr. U.S. Natl. Herb. 28(1):147, *t. 90*. 1929.

Herb in Northern Coastal Lowlands (Laguna Tortuguero, Isabela), Northern Limestone Hills, Río Abajo Forest Reserve, Las Tetas de Cayey, Southern Coastal Lowlands (Caja de Muertos), Susúa Forest Reserve, Vieques, Mona; 5–830 m; *fl* & *fr* Feb–Nov.

Bahamas, Greater Antilles, Lesser Antilles (Martinique).

Paspalum setaceum Michx. var. **stramineum** (Nash) D.J. Banks, Sida 2(4):276, *f. 6*. 1966.
Bas.: *Paspalum stramineum* Nash (1901).

Herb in Northern Coastal Lowlands, Northern Limestone Hills (Rincón), Western Cordillera (Las Marías), Southwestern Coastal Lowlands (Cabo Rojo), Cerro Las Mesas, Vieques; 1–400 m; *fl* & *fr* throughout the year.

West Indies, eastern half of United States, Mexico.

Notes: Britton & Wilson (1923) misapplied the name *Paspalum propinquum* Nash, and Liogier & Martorell (1982, 2000) misapplied *Paspalum setaceum* Michx. var. *ciliatifolium* (Michx.) Vasey, to this taxon.

Paspalum urvillei Steud., Syn. Pl. Glumac. 1:24. 1853.

Herb in Northern Coastal Lowlands (Fajardo, San Juan), Sierra de Luquillo, Sierra de Cayey (San Lorenzo), Central Cordillera; 5–1330 m; *fl* & *fr* Dec–Sep.

Native from southern South America to Bolivia.

Vasey, Yerba vasey.

Paspalum vaginatum Sw., Prodr. 21. 1788.

Herb in Northern Coastal Lowlands, Eastern Coast (Cayo Santiago), Caguas Valley (San Lorenzo), Southern Coastal Lowlands (Caja de Muertos), Guánica Forest Reserve, Southwestern Coastal Lowlands (Cabo Rojo), Vieques, Mona; 1–200 m; *fl* & *fr* throughout the year.

Tropics and subtropics.

Paspalum virgatum L., Syst. Nat. (ed. 10) 2:855. 1759.

Herb in Northern Coastal Lowlands, Northern Limestone Hills (San Sebastián), Eastern Coast (Humacao), Caguas Valley, Sierra de Cayey, Eastern Cordillera (Cayey), Western Cordillera (Las Marías), Western Coast (Añasco), Vieques; 1–500 m; *fl* & *fr* throughout the year.

Antilles, southern United States (Alabama, Texas), Mexico, Central America, South America.

Cortadero, Matojo blanco.

Pharus lappulaceus Aubl., Hist. Pl. Guiane 2:859. 1775.
Syn.: *Pharus glaber* Kunth (1816).

Herb in Northern Coastal Lowlands, Northern Limestone Hills, Río Abajo Forest Reserve, Eastern Coast (Yabucoa), Sierra de Cayey (San Lorenzo), Las Tetas de Cayey, Southeastern Foothills (Cayey), Quebrada de Los Cedros, Western Cordillera, Vieques; 2–830 m; *fl* & *fr* throughout the year.

Antilles, Mexico, Central America, South America.

Pegadora, Prenda de oro.

Pharus latifolius L., Syst. Nat. (ed. 10) 2:1269. 1759.

Herb in Northern Limestone Hills (Toa Baja, Dorado), Río Abajo Forest Reserve, Sierra de Luquillo, Sierra de Cayey, Central Cordillera (Ponce), Maricao Forest Reserve; 350–700 m; *fl* & *fr* throughout the year.

Greater Antilles, Lesser Antilles (Guadeloupe, Martinique, Grenada), Trinidad, Mexico, Central America, northwestern, west-central South America and Venezuela, Brazil.

Cortadora, Lamedora.

Pharus parvifolius Nash, Bull. Torrey Bot. Club 35(6):301. 1908.

Herb in Río Abajo Forest Reserve, Southern Coastal Lowlands (Patillas); 50–300 m; *fl* & *fr* Oct, Jan–Feb.

Greater Antilles, Trinidad, Mexico, Central America, northern South America to Bolivia.

Phragmites australis (Cav.) Trin. ex Steud., Nomencl. Bot. (ed. 2) 1:143. 1840; et 2:324. 1841.
Bas.: *Arundo australis* Cav. (1799).
Syn.: *Phragmites phragmites* (L.) H. Karst. (1880).

Suffrutescent herb in Northern Coastal Lowlands, Eastern Coast (Naguabo); 1 m; *fl* & *fr* Jan–Mar, Jun–Jul.

Probably native to north temperate regions.

Caña de indio, Caña de pantano.

Phyllostachys meyeri McClure, J. Washington Acad. Sci. 35 (10):286, *t. 1.* 1945.

Suffrutescent herb planted and persistent in Maricao Forest Reserve; 800 m; *fl* (?), *fr* (?). [*Breckon 8269* (MAPR)].

Native to China.

Poa annua L., Sp. Pl. 1:68. 1753.

Herb in Sierra de Luquillo, Toro Negro Forest Reserve; 800–1300 m; *fl* & *fr* Jan–Jun, Sep–Oct.

Probably native to Europe.

Polypogon viridis (Gouan) Breistr., Bull. Soc. Bot. France 110(89 Sess. Extraord.):56. 1966.
Bas.: *Agrostis viridis* Gouan (1762).

Herb in Toro Negro Forest Reserve; 1185–1300 m; *fl* & *fr* Apr–Jun. [*Vives a.n. 3311* (SJ)].

Native to Europe, northern Africa, western temperate Asia, Indian Subcontinent.

Rottboellia cochinchinensis (Lour.) Clayton, Kew Bull. 35(4):817. 1981.
Bas.: *Stegosia cochinchinensis* Lour. (1790).
Syn.: *Rottboellia exaltata* L.f. (1782).

Herb in Northern Coastal Lowlands, Southern Coastal Lowlands (Ponce), Western Cordillera, Southwestern Coastal Lowlands (Lajas); 1–500 m; *fl* & *fr* Feb–Oct.

Native to tropical Africa, tropical Asia, China, Japan, Australia.

Yerba caminadora, Picante.

Saccharum officinarum L., Sp. Pl. 1:54. 1753.

Herb in Northern Coastal Lowlands (Fajardo), Southern Coastal Lowlands (Guayama), Sierra de Luquillo, Eastern Coast (Naguabo), Eastern Cordillera (Barranquitas); 2–700 m; *fl* Nov–Jan, *fr* (?).

Cultivated in the tropics and subtropics of the world; probably originating in New Guinea.

Caña, Caña de azúcar.

Saccharum spontaneum L., Mant. Pl. 2:183. 1771.

Herb in Northern Coastal Lowlands, Northern Limestone Hills (Bayamón), Río Abajo Forest Reserve, Caguas Valley (Caguas), Sierra de Cayey, Southeastern Foothills (Cayey), Eastern Cordillera (Aguas Buenas); 4–750 m; *fl* & *fr* Aug–Dec, Mar.

Native to southern Europe (Sicily), northern and tropical Africa, temperate and tropical Asia.

Caña de azúcar silvestre, Caña silvestre.

Sacciolepis indica (L.) Chase, Proc. Biol. Soc. Wash. 21:8. 1908.
Bas.: *Aira indica* L. (1753).

Herb in Sierra de Luquillo; 150–900 m; *fl* & *fr* Sep–Oct, Jan–Jun.

Native to tropical and southern Africa, tropical Asia, China, Japan, Australia.

Sacciolepis striata (L.) Nash, Bull. Torrey Bot. Club 30(7):383. 1903.
Bas.: *Holcus striatus* L. (1753).

Herb in Northern Coastal Lowlands; 1–8 m; *fl* & *fr* Mar–Oct.

Greater Antilles, southeastern and south-central United States, Mexico, Central America, northern South America and Brazil.

Yerba de pantano.

Schizachyrium brevifolium (Sw.) Nees ex Büse in Miq., Pl. Jungh. 359. 1854.
Bas.: *Andropogon brevifolius* Sw. (1788).

Herb in Northern Coastal Lowlands, Sierra de Luquillo, Eastern Coast (Humacao), Sierra de Cayey, Central Cordillera (Orocovis); 1–700 m; *fl* & *fr* Oct–Jan.

Widespread in the tropics and subtropics.

Cerrillo dulce, Paja de colchón, Serrillo dulce.

Schizachyrium sanguineum (Retz.) Alston in Trimen, Handb. Fl. Ceylon 6:334. 1931.
Bas.: *Rottboellia sanguinea* Retz. (1783).
Syn.: *Schizachyrium sanguineum* (Retz.) Alston var. *brevipedicellatum* (Beal) S.L. Hatch (1978).
Schizachyrium semiberbe Nees (1829).
Schizachyrium domingense (Spreng. ex Schult.) Nash (1912).

Herb in Northern Limestone Hills (Moca), Sierra de Luquillo, Susúa Forest Reserve, Maricao Forest Reserve, Cerro Las Mesas; 250–700 m; *fl* & *fr* Sep–Dec, Mar.

Bahamas, Greater Antilles, Trinidad, southern United States, Mexico, Central America, South America.

Schizachyrium tenerum Nees in Martius et al., Fl. Bras. Enum. Pl. 2(1):336. 1829 (Agrost. Bras.).

Herb in Cerro Las Mesas; 400 m; *fl* & *fr* Aug (one specimen). [*Liogier 30912* (UPR)].

Greater Antilles, southeastern and south-central United States, Mexico, Central America, South America.

Setaria barbata (Lam.) Kunth, Révis. Gramin. 1:47. 1829.
Bas.: *Panicum barbatum* Lam. (1791).
Syn.: *Chaetochloa barbata* (Lam.) Hitchc. & Chase (1917).

Herb in Northern Coastal Lowlands, Northern Limestone Hills (Quebradillas, Lares), Caguas Valley (Caguas), Central Cordillera, Susúa Forest Reserve, Western Cordillera (Maricao), Southwestern Coastal Lowlands (Cabo Rojo); 1–900 m; *fl* & *fr* Oct–May.

Native to tropical Africa, tropical Asia.

Palmilla, Rabo de foca, Rabo de zorra, Yerba María.

Setaria macrostachya Kunth in Humboldt, Bonpland & Kunth, Nov. Gen. Sp. 1:91 [folio]; 1:110 [quarto]. 1816.

Herb in Southwestern Coastal Lowlands (Cabo Rojo), Culebra; 10–25 m; *fl* & *fr* Oct (one specimen). [*Proctor 44951, 45068* (US), the latter not seen].

Greater Antilles (Cuba), south-central and southwestern United States, Mexico, Central America (Nicaragua).

Setaria magna Griseb., Fl. Brit. W.I. 554. 1864.
Syn.: *Chaetochloa magna* (Griseb.) Scribn. (1897).

Herb in Northern Coastal Lowlands (Laguna Tortuguero); 1–2 m; *fl* & *fr* Oct (one specimen). [*Gayá González a.n. 5879* (MAPR)].

Greater Antilles (Cuba, Jamaica), Lesser Antilles (Guadeloupe, Martinique), southeastern, east-central and south-central United States, Mexico, southern Central America (Costa Rica), South America (scattered).

Setaria parviflora (Poir.) Kerguélen, Lejeunia, n.s. 120:161. 1987.
Bas.: *Cenchrus parviflorus* Poir. (1804).
Syn.: *Setaria geniculata* P. Beauv. (1812).
Chaetochloa geniculata (Lam.) Millsp. & Chase (1903).

Herb in Northern Coastal Lowlands, Río Abajo Forest Reserve, Sierra de Luquillo, Eastern Coast (Humacao), Caguas Valley (Caguas), Las Piedras Chiquitas, Eastern Cordillera (Corozal, Cayey), Central Cordillera, Southern Coastal Lowlands (Guayama), Susúa Forest Reserve, Maricao Forest Reserve, Western Coast, Vieques, Mona; 2–1200 m; *fl* & *fr* throughout the year.

West Indies, United States (except north-central states), Mexico, Central America, South America.

Notes: Liogier & Martorell (1982, 2000) used the name *Setaria geniculata* (Lam.) P. Beauv., a combination that was never made, for this taxon.

Arrocillo, Cepillo de fregar botellas, Cepillo de dientes, Deshollinador.

Setaria pradana (León ex Hitchc.) León, Contr. Ocas. Mus. Hist. Nat. Colegio "De La Salle" 8:164. 1946 (Fl. Cuba 1:164. 1946).
Bas.: *Panicum pradanum* León ex Hitchc. (1936).

Herb in Guánica Forest Reserve; 10–100 m; *fl* & *fr* Nov (one specimen). [*Liogier 33664* (UPR)].

Greater Antilles (Cuba, Hispaniola).

Setaria pumila (Poir.) Roem. & Schult., Syst. Veg. 2:891. 1817.
Bas.: *Panicum pumilum* Poir. (1816).

Herb in Río Abajo Forest Reserve; 275–325 m; *fl* & *fr* Jan (one specimen).

Possibly native to the Paleotropics.

Setaria setosa (Sw.) P. Beauv. var. **setosa,** Ess. Agrostogr. 51, 171, 178. 1812.
Bas.: *Panicum setosum* Sw. (1788).
Syn.: *Chaetochloa setosa* (Sw.) Scribn. (1897).
Setaria rariflora J.C. Mikan ex Trin. (1821).
Chaetochloa rariflora (J.C. Mikan ex Trin.) Hitchc. & Chase (1917).

Herb in Northern Coastal Lowlands (islands adjacent to Fajardo), Northern Limestone Hills (Quebradillas), Sierra de Luquillo, Las Tetas de Cayey, Las Piedras Chiquitas, Southeastern Foothills, Eastern Cordillera (Cañón San Cristóbal), Southern Coastal Lowlands (Coamo, Caja de Muertos), Southern Coastal Limestone (Guayanilla), Susúa Forest Reserve, Guánica

Forest Reserve, Southwestern Coastal Lowlands, Sierra Bermeja, Vieques, Culebra, Mona, Desecheo; 3–800 m; *fl* & *fr* throughout the year.

West Indies, Trinidad, South America (scattered).

Yerba suave.

Setaria setosa (Sw.) P. Beauv. var. **leiophylla** (Nees) Arechav., Anales Mus. Nac. Montevideo 1:171. 1894.

Bas.: *Panicum leiophyllum* Nees (1829).

Herb in Southern Coastal Lowlands (Ponce), Susúa Forest Reserve; 60 m; *fl* & *fr* Jan, Jun (two specimens). [*Axelrod 7377* (UPRRP); *Breckon 8173* (MAPR)].

Greater Antilles, Lesser Antilles (Antigua, Dominica, Martinique), South America (Colombia, Brazil).

Setaria tenacissima Schrad. in Schultes, Mant. 2:279. 1824.

Syn.: *Chaetochloa tenacissima* (Schrad.) Hitchc. & Chase (1917).

Herb in Central Cordillera; 500–900 m; *fl* & *fr* Apr, Aug, Nov. [*Liogier 32142, 35688* (UPR); *Caraballo 2536* (UPR)].

Greater Antilles (Hispaniola), Trinidad, Central America, northwestern, north-central South America and Venezuela.

Setaria utowanaea (Scribn.) Pilg. in Engler & Prantl, Nat. Pflanzenfam. (ed. 2) 14e:72. 1940.

Bas.: *Panicum utowanaeum* Scribn. (1900).

Herb in Southern Coastal Limestone, Guánica Forest Reserve, Southwestern Coastal Lowlands (Cabo Rojo), Vieques, Culebra, Mona, Desecheo; 1–100 m; *fl* & *fr* throughout the year.

Greater Antilles (Cuba), Virgin Islands (St. John, Anegada), Lesser Antilles, northern South America (Venezuela, Colombia).

Setaria vulpiseta (Lam.) Roem. & Schult., Syst. Veg. 2:495. 1817.

Bas.: *Panicum vulpisetum* Lam. (1798).

Syn.: *Chaetochloa vulpiseta* (Lam.) Hitchc. & Chase (1917).

Herb in Eastern Cordillera (Aguas Buenas), Central Cordillera; 200–1000 m; *fl* & *fr* Dec–Apr, Jul–Sep.

Greater Antilles (Cuba, Hispaniola), Trinidad, Mexico, Central America, South America.

Rabo de ardilla.

Sorghastrum stipoides (Kunth) Nash, N. Amer. Fl. 17:129. 1912.

Bas.: *Andropogon stipoides* Kunth (1816).

Herb in Northern Coastal Lowlands (Laguna Tortuguero); 1 m; *fl* & *fr* Sep (one specimen). [*Proctor 45036* (SJ), not seen].

Greater Antilles (Cuba), Mexico, South America.

Sorghum bicolor subsp. **bicolor** (L.) Moench, Methodus. 207. 1794.

Bas.: *Holcus bicolor* L. (1771).

Syn.: *Holcus sorghum* L. (1753).

Sorghum saccharatum (L.) Moench (1794).

Herb in Northern Limestone Hills (Arecibo), Caguas Valley (Gurabo), Southeastern Foothills (Coamo); 100–200 m; *fl* & *fr* Jan–Mar, Sep–Oct.

Native to Africa.

Millo, Sorgo, Yerba de Sudán.

Sorghum bicolor (L.) Moench subsp. **verticilliflorum** (Steud.) de Wet ex Wiersema & J. Dahlb., Taxon 56(3):944. 2007.

Bas.: *Andropogon verticilliflorus* Steud. (1854).

Syn.: *Sorghum verticilliflorum* (Steud.) Stapf (1917).

Herb in Southwestern Coastal Lowlands (Lajas), Western Coast; 2–20 m; *fl* & *fr* May–Feb.

Native to Africa.

Millito.

Sorghum halepense (L.) Pers., Syn. Pl. 1:101. 1805.

Bas.: *Holcus halepensis* L. (1753).

Herb in Northern Coastal Lowlands, Eastern Cordillera (Corozal), Central Cordillera (Yauco), Southern Coastal Lowlands, Guánica Forest Reserve, Southwestern Coastal Lowlands (Lajas), Western Coast; 1–220 m; *fl* & *fr* throughout the year.

Native to northern Africa, western and central temperate Asia, Indian Subcontinent.

Yerba Johnson.

Spartina patens (Aiton) Muhl., Cat. Pl. Amer. Sept. 8. 1813.

Bas.: *Dactylis patens* Aiton (1789).

Herb in Northern Coastal Lowlands, Southern Coastal Lowlands (Ponce), Western Coast (Rincón), Vieques; 0–1 m; *fl* & *fr* (?).

Coasts of Bahamas, Greater Antilles, Virgin Islands (St. Croix, St. John), Lesser Antilles (Antigua, Guadeloupe), eastern Canada, eastern and south-central United States.

Yerba de sal.

Sporobolus cubensis Hitchc., Contr. U.S. Natl. Herb. 12(6):237. 1909.

Herb in Cerro Las Mesas; 350 m; *fl* & *fr* Apr (one specimen). [*Proctor 43320* (SJ)].

Greater Antilles (Cuba), Central America, northern South America to Bolivia.

Matojo cubano.

Sporobolus domingensis (Trin.) Kunth, Enum. Pl. 1:214. 1833.
Bas.: *Vilfa domingensis* Trin. (1821).

Herb in Northern Coastal Lowlands (Arecibo), Eastern Coast (Humacao), Southern Coastal Lowlands (Guayama); 1–3 m; *fl* & *fr* May, Aug, Nov–Dec.

Bahamas, Greater Antilles, southern United States (Georgia, Florida, Texas), Mexico, northern South America (Venezuela).

Notes: Britton & Wilson (1923) misapplied the above name to *Sporobolus pyramidatus*.

Sporobolus indicus (L.) R. Br., Prodr. 170. 1810.
Bas.: *Agrostis indica* L. (1753).
Syn.: *Sporobolus berteroanus* (Trin.) Hitchc. & Chase (1917).

Herb in Sierra de Luquillo, Eastern Coast (Yabucoa), Sierra de Cayey, Las Tetas de Cayey, Eastern and Central Cordillera, Western Cordillera (Maricao); 250–1200 m; *fl* & *fr* throughout the year.

Tropic and warm temperate regions.

Notes: Britton & Wilson (1923) misapplied the above name to *Sporobolus jacquemontii*.

Cerrillo, Matojillo, Matojo de burro, Serrillo.

Sporobolus jacquemontii Kunth, Révis. Gramin. 2:427, *t. 127*. 1831.

Herb in Northern Coastal Lowlands, Cambalache Forest Reserve, Río Abajo Forest Reserve, Sierra de Luquillo, Eastern Coast (Ceiba), Caguas Valley, Sierra de Cayey (San Lorenzo), Las Piedras Chiquitas, Southeastern Foothills, Eastern Cordillera (Cañón San Cristóbal), Central Cordillera (Ponce), Southern Coastal Lowlands (Caja de Muertos), Southern Coastal Limestone, Susúa Forest Reserve, Guánica Forest Reserve, Southwestern Coastal Lowlands, Cerro Las Mesas, Western Coast (Aguada), Vieques, Mona, Desecheo; 1–600 m; *fl* & *fr* throughout the year.

West Indies, Trinidad and Tobago, Mexico, Central America, northern South America to Bolivia, tropical Africa.

Cerrillo, Matojo de burro.

Sporobolus pyramidatus (Lam.) Hitchc., Man. Grasses W. Ind. 84. 1936 (U.S. Dept. Agric. Misc. Publ. 243).
Bas.: *Agrostis pyramidata* Lam. (1791).
Syn.: *Sporobolus argutus* (Nees) Kunth (1833).

Herb in Northern Coastal Lowlands (Arecibo), Southern Coastal Lowlands, Guánica Forest Reserve, Southwestern Coastal Lowlands, Vieques, Culebra, Mona; 1–65 m; *fl* & *fr* throughout the year.

West Indies, central and south-central United States, Mexico, Central America, South America (scattered).

Sporobolus tenuissimus (Mart. ex Schrank) Kuntze, Revis. Gen. Pl. 3(3):369. 1898.
Bas.: *Panicum tenuissimum* Mart. ex Schrank (1822).

Herb in Northern Coastal Lowlands, Caguas Valley, Sierra de Cayey, Eastern and Central Cordillera, Southern Coastal Lowlands (Coamo), Maricao Forest Reserve; 1–800 m; *fl* & *fr* throughout the year.

Bahamas, Greater Antilles (Jamaica, Hispaniola), Lesser Antilles, Mexico, Central America, northwestern, west-central South America and Venezuela.

Sporobolus virginicus (L.) Kunth, Révis. Gramin. 1:67. 1829.
Bas.: *Agrostis virginica* L. (1753).

Herb in Northern Coastal Lowlands, Eastern Coast (Ceiba), Southern Coastal Lowlands, Guánica Forest Reserve, Southwestern Coastal Lowlands (Cabo Rojo), Western Coast (Rincón), Vieques, Culebra, Mona; 1–50 m; *fl* & *fr* Feb, May–Nov.

Widespread in the tropics and subtropics.

Matojo de burro, Matojo de playa.

Steinchisma laxa (Sw.) Zuloaga, Amer J. Bot. 90(5):817. 2003.
Bas.: *Panicum laxum* Sw. (1788).

Herb in Northern Coastal Lowlands, Northern Limestone Hills, Río Abajo Forest Reserve, Sierra de Luquillo, Eastern Coast, Caguas Valley (Caguas), Sierra de Cayey, Eastern and Central Cordillera, Susúa Forest Reserve, Western Cordillera, Southwestern Coastal Lowlands (Lajas), Western Coast, Vieques; 1–1050 m; *fl* & *fr* Mar–Dec.

West Indies, Trinidad and Tobago, Mexico, Central America, South America.

Malojillo del monte.

Stenotaphrum secundatum (Walter) Kuntze, Revis. Gen. Pl. 2:794. 1891 "secundum."
Bas.: *Ischaemum secundatum* Walter (1788).

Herb in Northern Coastal Lowlands, Northern Limestone Hills (Toa Alta, Isabela), Sierra de Luquillo, Las Tetas de Cayey, Maricao Forest Reserve, Western Coast (Rincón); 2–800 m; *fl* & *fr* throughout the year.

West Indies, southern and south-central United States, Mexico, Central America, South America, western tropical Africa.

Cinta, Cintillo, Grama, Grama blanca, Grama dulce, San Agustín.

Tarigidia axelrodii A.S. Vega & Rúgolo, Syst. Bot. 35(1):96, *ff.* 1–3. 2010.

Herb in Southwestern Coastal Lowlands (Cabo Rojo); 1 m; *fl* & *fr* Jun, Nov–Dec. [*Axelrod 14076* (US, holotype); *Axelrod 13906, 14227* (UPRRP)].

Endemic.

Themeda quadrivalvis (L.) Kuntze, Revis. Gen. Pl. 2:794. 1891.

Bas.: *Andropogon quadrivalvis* L. (1774).

Herb in Southern Coastal Limestone, Southwestern Coastal Lowlands (Lajas); 10–100 m; *fl* & *fr* Oct–Feb.

Native to India, Nepal.

Tragus berteronianus Schult., Mant. 2:205. 1824.

Syn.: *Nazia aliena* (Spreng.) Scribn. (1899).

Herb in Southern Coastal Limestone, Guánica Forest Reserve, Southwestern Coastal Lowlands (Cabo Rojo), Vieques, Culebra; 1–100 m; *fl* & *fr* Jun–Dec.

Native to Africa, Saudi Arabia, Iran, Afghanistan, Pakistan, China.

Hincadora.

Tripsacum andersonii J.R. Gray, Phytologia 33(3): 204, *t. 1.* 1976.

Herb in Western Cordillera (Maricao); 300–840 m; *fl* & *fr* Apr, Nov. [*Proctor 45216, 46807* (SJ); *Atha 779* (UPRRP)].

Probably native to Mesoamerica.

Uniola virgata (Poir.) Griseb., Fl. Brit. W.I. 531. 1864.

Bas.: *Poa virgata* Poir. (1804).

Syn.: *Leptochloopsis virgata* (Poir.) H.O. Yates (1966).

Suffrutescent herb in Southern Coastal Lowlands (Caja de Muertos), Southern Coastal Limestone, Guánica Forest Reserve, Southwestern Coastal Lowlands (Cabo Rojo), Vieques, Mona, Desecheo; 1–75 m; *fl* & *fr* throughout the year.

Bahamas, Greater Antilles, Virgin Islands.

Yerba de alambre.

Urochloa adspersa (Trin.) R.D. Webster, Syst. Bot. 13(4):607. 1988.

Bas.: *Panicum adspersum* Trin. (1826).

Syn.: *Brachiaria adspersa* (Trin.) Parodi (1969). *Brachiaria echinulata* (Mez) Parodi (1969).

Herb in Northern Coastal Lowlands, Southeastern Foothills, Southern Coastal Lowlands, Southwestern Coastal Lowlands, Vieques, Culebra, Mona, Desecheo; 1–625 m; *fl* & *fr* throughout the year.

West Indies, southeastern United States (Alabama, Florida), southern South America.

Millito, Cohitrillo.

Urochloa arrecta (Hack. ex T. Durand & Schinz) Morrone & Zuloaga, Darwiniana 31(1–4):69. 1992.

Bas.: *Panicum arrectum* Hack. ex T. Durand & Schinz (1894).

Herb in Northern Coastal Lowlands (Toa Baja), Sierra de Cayey, Central Cordillera; 1–900 m; *fl* & *fr* Mar, Jul–Nov.

Native to eastern tropical and southern Africa.

Taner, Gola.

Urochloa brizantha (Hochst. ex A. Rich.) R.D. Webster, Austral. Paniceae 233. 1987.

Bas.: *Panicum brizanthum* Hochst. ex A. Rich. (1850).

Syn.: *Brachiaria brizantha* (Hochst. ex A. Rich.) Stapf (1919).

Herb in Central Cordillera (Utuado), Maricao Forest Reserve, Cerro Las Mesas; 260–900 m; *fl* & *fr* Mar, Jun–Oct.

Native to tropical and southern Africa.

Yerba signal.

Urochloa decumbens (Stapf) R.D. Webster, Austral. Paniceae 234. 1987.

Bas.: *Brachiaria decumbens* Stapf (1919).

Herb in Northern Coastal Lowlands (San Juan), Caguas Valley (Gurabo), Eastern Cordillera (Corozal), Central Cordillera (Orocovis), Maricao Forest Reserve, Western Coast (Mayagüez); 5–825 m; *fl* & *fr* throughout the year.

Native to eastern and west-central tropical Africa.

Urochloa distachya (L.) T.Q. Nguyen, Novosti Sist. Vyssh. Rast. 1966:13. 1966.

Bas.: *Panicum distachyon* L. (1771).

Syn.: *Brachiaria subquadripara* (Trin.) Hitchc. (1931). *Urochloa subquadripara* (Trin.) R.D. Webster (1987).

Herb in Northern Coastal Lowlands, Northern Limestone Hills (Isabela), Eastern Coast, Caguas Valley (Caguas), Southeastern Foothills, Eastern Cordillera (Cayey), Southern Coastal Lowlands, Southern Coastal Limestone, Southwestern Coastal Lowlands, Western Coast, Vieques, Culebra, Mona; 1–600 m; *fl* & *fr* throughout the year.

Native to tropical Asia, China, Japan, Australia.

Gramita.

Urochloa fusca (Sw.) B.F. Hansen & Wunderlin, Novon 11(3):368. 2001.

Bas.: *Panicum fuscum* Sw. (1788).

Syn.: *Panicum fasciculatum* Sw. (1788). *Brachiaria fasciculata* (Sw.) Parodi (1969). *Urochloa fasciculata* (Sw.) R.D. Webster ex Morrone & Zuloaga (1992).

Herb in Northern Coastal Lowlands, Northern Limestone Hills, Sierra de Luquillo, Sierra de Cayey (San Lorenzo), Las Piedras Chiquitas, Southeastern Foothills, Central Cordillera (Ponce), Southern Coastal Lowlands, Southern Coastal Limestone (Peñuelas), Susúa Forest Reserve, Western Cordillera (San Germán), Southwestern Coastal Lowlands, Sierra Bermeja, Western Coast, Culebra, Mona, Desecheo; 2–800 m; *fl* & *fr* throughout the year.

West Indies, southern United States, Mexico, Central America, northern South America.

Notes: Liogier & Martorell (1982) used the homonym *Brachiaria fasciculata* (Sw.) S.T. Blake for the name of this taxon.

Urochloa mosambicensis (Hack.) Dandy, J. Bot. 69(2):54. 1931.

Bas.: *Panicum mosambicense* Hack. (1888).

Herb in Southern Coastal Lowlands (Santa Isabel); 15 m; *fl* & *fr* Jan (one specimen). [*Axelrod 12422* (UPRRP)].

Native to southern and eastern tropical Africa.

Urochloa mutica (Forssk.) T.Q. Nguyen, Novosti Sist. Vyssh. Rast. 3:13. 1966.

Bas.: *Panicum muticum* Forssk. (1775).

Syn.: *Panicum barbinode* Trin. (1832).

Brachiaria purpurascens (Raddi) Henrard (1940).

Herb in Northern Coastal Lowlands, Northern Limestone Hills (Bayamón), Sierra de Luquillo, Eastern Coast, Sierra de Cayey, Southeastern Foothills (Cayey), Eastern Cordillera (Aguas Buenas, Corozal), Central Cordillera (Orocovis, Utuado), Southern Coastal Lowlands, Guánica Forest Reserve, Southwestern Coastal Lowlands (Lajas), Western Coast (Mayagüez); 1–650 m; *fl* & *fr* Oct–Jul.

Probably originating in tropical Africa.

Malojillo.

Urochloa plantaginea (Link) R.D. Webster, Syst. Bot. 13(4):607. 1988.

Bas.: *Panicum plantagineum* Link (1827).

Syn.: *Brachiaria plantaginea* (Link) Hitchc. (1909).

Herb in Sierra de Cayey (Guayama), Cerro Las Mesas, Western Coast (Mayagüez); 1–500 m; *fl* & *fr* Mar (one specimen). [*Gould 15752, 15821* (US), both not seen; *González Más 4261* (MAPR)].

Greater Antilles (Cuba, Hispaniola), Lesser Antilles (St. Lucia), Trinidad, Mexico, Central America, South America (scattered).

Urochloa reptans (L.) Stapf in Prain, Fl. Trop. Afr. 9(4):601. 1920.

Bas.: *Panicum reptans* L. (1759).

Syn.: *Brachiaria reptans* (L.) C.A. Gardner & C.E. Hubb (1938).

Herb in Southeastern Foothills, Southern Coastal Lowlands (Caja de Muertos), Susúa Forest Reserve, Southwestern Coastal Lowlands, Vieques, Culebra; 5–150 m; *fl* & *fr* Jun–Feb.

Probably originating in tropical Asia.

Alpiste cimarrón.

Vulpia bromoides (L.) Gray, Nat. Arr. Brit. Pl. 2:124. 1821.

Bas.: *Festuca bromoides* L. (1753).

Herb in Toro Negro Forest Reserve; 1200 m; *fl* & *fr* Mar (one specimen). [*Gould 15853* (UPR)].

Native to Europe, eastern and west-central tropical and northeastern Africa, western temperate Asia.

Vulpia myuros (L.) C.C. Gmel., Fl. Bad. 1:8. 1805.

Bas.: *Festuca myuros* L. (1753).

Herb in Toro Negro Forest Reserve; 1200 m; *fl* & *fr* May (one specimen). [*Liogier 28929* (UPR)].

Native to Europe, northern Africa, western temperate Asia, Indian Subcontinent.

Zoysia matrella (L.) Merr. var. **matrella**, Philipp. J. Sci., C 7(4):230. 1912.

Bas.: *Agrostis matrella* L. (1771).

Herb in Toro Negro Forest Reserve, Cerro Las Mesas; 280–1340 m; *fl* & *fr* Mar (one specimen). [*Gould 15836, 15857* (US), the latter not seen].

Native to southeastern Asia, China, Japan, Australia.

Yerba de Manila, Zoisia.

Zoysia matrella (L.) Merr. var. **pacifica** Goudswaard, Blumea 26(1):172. 1980.

Herb in Northern Coastal Lowlands, Sierra de Luquillo, Toro Negro Forest Reserve, Maricao Forest Reserve, Southwestern Coastal Lowlands (Cabo Rojo), Vieques; 1–1300 m; *fl* & *fr* Feb–Nov.

Native to Asia, Pacific islands.

Notes: Liogier & Martorell (1982, 2000) misapplied the name *Zoysia matrella* (L.) Merr. var. *tenuifolia* (Thiele) Sasaki to this taxon.

Zoisia.

TAXA NOT RECENTLY COLLECTED

Axonopus aureus P. Beauv., Ess. Agrostogr. 12, 154. 1812.

Herb collected between Río Piedras and Trujillo Alto in Sep 1913 [*Hioram 8838* (NY)] and on 9 Dec

1913 [*Chase 560* (NY)], in Utuado [*Sargent 3279* (US)], in Río Piedras on 24 Nov 1916 [*Stevenson 3282* (UPR)] and on 1 Dec 1938 [*García Molinari 163* (UPR)].

Trinidad, tropical continental America.

Bothriochloa saccharoides (Sw.) Rydb., Brittonia 1(2):81. 1931.
Bas.: *Andropogon saccharoides* Sw. (1788).
Syn.: *Amphilophis saccharoides* (Sw.) Nash (1912).

Herb collected in Maricao [*Chase 6226*, not seen], in Mayagüez on 26 Oct 1913 [*Chase 6270*] and in Adjuntas on 25 Jul 1938 [*Alberts 223* (US)].

. Greater Antilles, southern United States, Mexico, Central America, northern South America (Venezuela, Colombia).

Digitaria argillacea (Hitchc. & Chase) Fernald, Rhodora 22(258):104. 1920.
Bas.: *Syntherisma argillaceum* Hitchc. & Chase (1917) "argillacea."

Herb collected in Monte Mesa (Cerro Las Mesas) on 26 Oct 1913 [*Chase 6271, 6277* (US)]; also collected, according to Britton & Wilson (1923), at Monte Alegrillo near Maricao.

Greater Antilles, Central America.

Digitaria sacchariflora (Nees) Henrard, Blumea 1:99. 1934.
Bas.: *Trichachne sacchariflora* Nees (1829).

Herb collected on Mona on cultivated land near coast [*Hess 433* (US, not seen)].

Native to Brazil and southern South America.

Digitaria villosa (Walter) Pers., Syn. Pl. 1:85. 1805.
Bas.: *Syntherisma villosum* Walter (1788) "villosa."

Herb collected in Trujillo Alto [*Chase 6271* (US), not seen] and in Campo Alegre, Manatí [*Chase 6806* (US), not seen], at Lake Loisa, vicinity of San Juan, [*Chase 6784* (US), not seen], at Laguna Tortuguero on 5 Feb 1915 [*Britton et al. 3858* (NY)] and in Campo Seco, Cataño [*Molinari*, not seen].

Greater Antilles (Cuba, Hispaniola), southeastern and-central United States, Mexico, Central America.

Imperata contracta (Kunth) Hitchc., Rep. (Annual) Missouri Bot. Gard. 4:146. 1893.
Bas.: *Saccharum contractum* Kunth (1816).

Herb collected in Hato Grande, Monte Gregorio (Bayamón) in 1885 [*Sintenis 2720* (US)], west of Barceloneta on 7 Nov 1913 [*Chase 6438* (US)], on the road to Guayama on 3 Dec 1913 [*Chase 6746* (US)] and at Alto de la Bandera on 9 Jan 1943 [*Sargent 3054* (US)].

Greater Antilles, Lesser Antilles (Guadeloupe, Dominica, Martinique), Mexico, Central America, northern South America to Paraguay.

Guajanilla.

Isachne polygonoides (Lam.) Döll, in Martius, Fl. Bras. 2(2):273. 1877.
Bas.: *Panicum polygonoides* Lam. (1798).

Herb collected in a marsh in Utuado on 11 Nov 1943 [*Sargent 3276* (US)].

Greater Antilles (Cuba, Hispaniola), Trinidad, northern South America to Bolivia.

Lasiacis sorghoidea (Ham.) Hitchc. & Chase var. **patentiflora** (Hitchock & Chase) Davidse, Ann. Missouri Bot. Gard. 64(2):375. 1978.
Bas.: *Lasiacis patentiflora* Hitchc. & Chase (1917).

Herb collected between Bayamón and Toa Alta on 6 Nov 1913 [*Chase 6412* (US)] and in Cerro Las Mesas on 13 Dec 1913 [*Chase 6824* (US)].

Greater Antilles (Hispaniola), Lesser Antilles (Domenica, St. Vincent, Grenada), northern South America (Venezuela, Colombia), Ecuador and Paraguay.

Moorochloa eruciformis (Sm.) Veldkamp, Reinwardtia 12(2):139. 2004.
Bas.: *Panicum eruciforme* Sm. (1806).

Herb collected in Juana Díaz [*Sargent 3194* (US), not seen].

Native to Paleotropics and -subtropics.

Oplismenus compositus (L.) P. Beauv., Ess. Agrostogr. 54, 169. 1812.
Bas.: *Panicum compositum* L. (1753).

Herb collected in Jayuya on 19 Sep 1943 [*Sargent 3152* (US)].

Native to southern Mexico, Central America (scattered), northern South America (Venezuela, Colombia) and Ecuador, tropical Africa, temperate and tropical Asia, Australia.

Panicum cayennense Lam., Tabl. Encycl. 1:173. 1791.

Herb collected in Moca on 7 Sep 1938 [*Sargent 555* (MO), not seen].

Greater Antilles (Cuba), Lesser Antilles (Dominica), Mexico, Central America, South America.

Panicum hirtum Lam., Encycl. 4(2):741. 1798.

Herb collected in Puerto Rico, without specific locality or date, [*LeDru s.n.* (US)].

Native to Trinidad, Mexico, Central America, northern, northwestern South America and Brazil.

Panicum hylaeicum Mez, Notizbl. Bot. Gart. Berlin-Dahlem 7:75. 1917.

Herb collected in El Yunque on 7 Sep 1941 [*Blomquist 11748* (US), not seen].

Greater Antilles (Cuba), Lesser Antilles (Dominica), Mexico, Central America, South America.

Paspalum densum Poir., Encycl. 5:32. 1804.

Herb collected in Cataño in Nov 1883 [*Stahl 42* (US)] and west of Vega Baja on 10 Dec 1913 [*Chase 6791* (US)].

Greater Antilles, Lesser Antilles (Guadeloupe, St. Martin, St. Vincent), southern Central America (Panama), South America to Brazil.

Paspalum parviflorum Rhode ex Flüggé, Gram. Monogr., Paspalum 98. 1810.

Herb collected in Puerto Rico, without specific locality or date, [*Desontaines a.n. 2855752* (US), fragment of type specimen, not seen].

South America.

Notes: Known in Puerto Rico from the type specimen only.

Sorghastrum setosum (Griseb.) Hitchc., Contr. U.S. Natl. Herb. 12(6):195. 1909.
Bas.: *Andropogon setosus* Griseb. (1866).
Syn.: *Sorghastrum parviflorum* Hitchc. & Chase (1917).

Herb collected between Río Piedras and Trujillo Alto in Sep 1913 [*Hioram a.n. 991100* (NY)], in Vega Baja on 7 Nov 1913 [*Chase 6421* (NY)] and in Río Piedras on 26 Jun 1938 [*García Molinari 204* (NY)].

Greater Antilles, Trinidad, Mexico, Central America, South America.

TAXA CITED BY PREVIOUS AUTHORS, BUT NOT INCLUDED

NO SPECIMENS SEEN

Cenchrus americanus (L.) Morrone. Syn.: *Setaria glauca* (L.) P. Beauv.
Herb cited by Liogier & Martorell (1982, 2000) as occurring in forests in Maricao and Guilarte.

Cenchrus polystachios (L.) Morrone. Syn.: *Pennisetum setosum* (Sw.) Rich.
Herb cited by Liogier & Martorell (1982, 2000) as common in southern districts.

Digitaria similis Beetle ex Gould.
Herb cited by Liogier & Martorell (1982, 2000) as collected by Hess, fairly common in the southern regions of the island.

Eragrostis glutinosa (Sw.) Trin.
Herb cited by Liogier & Martorell (1982, 2000) as occurring on sandy soil, northern Puerto Rico.

Saccharum bengalense Retz.
Herb cited by Liogier & Martorell (2000) as introduced and naturalized in the Cataño area, according to Roy Woodbury.

Urochloa platyphylla (Munro ex C. Wright) R.D. Webster. Syn.: *Brachiaria extensa* Chase.
Herb cited by Liogier & Martorell (2000) as occurring in moist places.

DOUBTFUL OCCURRENCE

Aristida swartziana Steud.
Herb cited by Liogier & Martorell (2000) as occurring in dry southern districts.

Notes: Britton & Wilson (1923) wrongly placed this taxon in synonymy with *Aristida cognata*.

Leptochloa fusca (L.) Kunth subsp. ***fascicularis*** (Lam.) N. Snow. Syn.: *Diplachne fascicularis* (Lam.) P. Beauv. *Leptochloa fascicularis* (Lam.) A. Gray.
Herb cited by Liogier & Martorell (1982, 2000) as occurring in ditches and moist grounds, listed by J. Otero.

CITED DUE TO MISIDENTIFICATION

Bothriochloa ischaemum (L.) Keng. Syn.: *Amphilophis ischaemum* (L.) Nash.
Herb cited by Liogier & Martorell (1982, 2000) as common on roadsides eastern Puerto Rico. All specimens that I have seen so identified are referable to other taxa.

Chloris cubensis Hitchc. & Ekman.
Herb cited by Liogier & Martorell (1982, 2000) as occurring in dry open grassy areas. Specimens of *Chloris sagraeana* were misidentified as this taxon.

Digitaria panicea (Sw.) Urb.
Herb cited by Liogier & Martorell (1982, 2000) as occurring on sandy soil, northern districts. Specimens of *Digitaria filiformis* var. *dolichophylla* were misidentified as this taxon.

Digitaria sanguinalis (L.) Scop. Syn: *Syntherisma sanguinale* (L.) Dulac.
Herb cited by Liogier & Martorell (1982, 2000) as occurring in waste and cultivated grounds. Specimens of *Digitaria setigera* were misidentified as this taxon.

Echinochloa crus-galli (L.) P. Beauv. "crus galli."
Herb cited by Britton & Wilson (1923) as occurring on roadsides and wet grounds; subsequently cited by Liogier & Martorell (1982, 2000). All specimens that

I have seen so identified are referable to *Echinochloa crus-pavonis*.

Eragrostis urbaniana Hitchc.
Herb cited by Liogier & Martorell (1982, 2000) as occurring on limestone rocks, Vieques; however, no such specimen has been found. The only specimen so identified that I have seen [*Britton et al. 3854* (NY)] was later referred to *Eragrostis ciliaris*.

Lasiacis grisebachii (Nash) Hitchc. var. ***grisebachii***.
Herb cited by Liogier & Martorell (1982, 2000) as occurring in Cerro Las Mesas, based on the misidentification by F. Gould of a specimen of *Lasiacis divaricata*.

Leptochloa nealleyi Vasey.
Herb cited, as *Leptochloa neallei*, by Liogier & Martorell (1982, 2000) as occurring in wet places, northern Puerto Rico, based on the misidentification of specimens of *Leptochloa scabra*.

Leptochloa panicoides (J. Presl) Hitchc.
Herb cited by Liogier & Martorell (1982, 2000) as occurring as a weed in Río Piedras based on the misidentification of specimens of *Leptochloa scabra*.

Paspalidium chapmanii (Vasey) R. W. Pohl. Syn.: *Setaria chapmanii* (Vasey) Pilg.
Herb cited by Liogier & Martorell (1982, 2000) as occurring on white sand, Laguna Tortuguero, based on the misidentification of a specimen of *Paspalidium geminatum*.

Paspalum bakeri Hack.
Herb cited by Liogier & Martorell (1982, 2000) as occurring in Cañón San Cristóbal. The specimen that I saw identified as this taxon is referable to *Paspalum laxum*. I have been unable to locate a specimen collected in Cabo Rojo [*Proctor 43754* (SJ)], but I suspect that it too was misidentified.

Paspalum convexum Humb. & Bonpl. ex Flüggé.
Herb cited by Liogier & Martorell (1982, 2000) as occurring in grassy place, Laguna Tortuguero, according to Roy Woodbury, but based on a misidentification.

Paspalum dispar Chase.
Herb cited by Liogier & Martorell (1982, 2000) as occurring in open places, central mountains. The specimen that I saw so identified is referable to *Paspalum decumbens*.

Paspalum maritimum Trin.
Herb cited by Liogier & Martorell (1982, 2000) based on the misidentification of specimens of *Paspalum macrophyllum*.

Poa pratensis L.
Herb cited by Liogier & Martorell (1982, 2000) as occurring on the summit of Luquillo Mountains based on the misidentification of specimens of *Poa annua*.

Themeda arguens (L.) Hack.
Herb cited by Liogier & Martorell (1982, 2000) based on the misidentification of specimens of *Andropogon virginicum*.

Tripsacum laxum Nash.
Herb cited by Liogier & Martorell (1982, 2000) based on the misidentification of specimens of *Tripsacum andersonii*.

ONLY UNDER CULTIVATION

Avena sativa L.
Herb cited by Britton & Wilson (1923) as occurring on waste grounds, San Juan; subsequently cited by Liogier & Martorell (1982, 2000) as perhaps not persistent.

Cymbopogon nardus (L.) Rendle.
Herb cited by Britton & Wilson (1923) as locally spontaneous after cultivation; subsequently cited by Liogier & Martorell (1982, 2000).

Polytrias indica (Houtt.) Veldkamp. Syn.: *Polytrias praemorsa* (Nees) Hack. *Polytrias amaura* (Büse) Kuntze.
Herb cited by Liogier & Martorell (2000) as cultivated as a ground cover and escaped.

Tripsacum latifolium Hitchc.
Herb cited by Liogier & Martorell (1982, 2000) as introduced and escaped in western areas, though their citation was probably based on the misidentification of specimens of *Tripsacum andersonii*. It had, however, been collected by Britton & Britton under cultivation at the Agricultural Station in Mayagüez.

Triticum aestivum L.
Herb cited by Liogier & Martorell (1982, 2000) as cultivated, perhaps not permanent.

Urochloa ruziziensis (R. Germ. & C. Evrard) Crins. Syn.: *Brachiaria ruziziensis* R. Germ. & C. Evrard.
Herb cited by Liogier & Martorell (1982, 2000) as planted as fodder and persistent.

Zea mays L.
Herb cited by Britton & Wilson (1923) as widely cultivated and occasionally spontaneous after cultivation; subsequently cited by Liogier & Martorell (1982, 2000).

POLYGALACEAE

Badiera penaea (L.) DC., Prodr. 1:335. 1824.
Bas.: *Polygala penaea* L. (1753).
Syn.: *Badiera portoricensis* Britton (1915).

Tree in Southern Coastal Limestone (Peñuelas), Susúa Forest Reserve, Maricao Forest Reserve, Guánica Forest Reserve, Southwestern Coastal Lowlands (Cabo Rojo), Sierra Bermeja; 1–775 m; *fl* Dec–Jan, Apr–Sep, *fr* Jul–Feb.

Bahamas, Greater Antilles (Hispaniola), Virgin Islands (Virgin Gorda). [L2: 335].

Phlebotaenia cowellii Britton, Torreya 7:38. 1907.
Syn.: *Polygala cowellii* (Britton) S.F. Blake (1916).

Tree in Northern Limestone Hills, Río Abajo Forest Reserve, Southeastern Foothills (Guayama), Southern Coastal Lowlands (Juana Díaz), Southern Coastal Limestone (Guayanilla, Yauco), Susúa Forest Reserve, Maricao Forest Reserve, Guánica Forest Reserve, Southwestern Coastal Lowlands (Cabo Rojo); 5–600 m; *fl* Feb–Mar, Jun–Oct, *fr* Mar (one specimen).

Endemic. [L2: 332].

Árbol de violeta, Caracolillo, Palo de hueso, Palo de tortuga, Palo de violeta, Tortuguero, Violeta.

Polygala crucianelloides DC., Prodr. 1:329. 1824.

Herb in Río Abajo Forest Reserve; 325 m; *fl* Apr–Jun (two specimens), *fr* Jun (one specimen). [*Acevedo 362* (SJ); *Woodbury a.n. 8598, 8599* (SJ)].

Greater Antilles (Hispaniola). [L2: 334].

Polygala hecatantha Urb., Symb. Antill. 1(2):331. 1899.

Herb in Southern Coastal Limestone, Guánica Forest Reserve, Southwestern Coastal Lowlands (Cabo Rojo), Western Cordillera (Maricao); 5–600 m; *fl* Oct–Jul, *fr* Oct–Dec, Mar, Jul.

Greater Antilles (Hispaniola), Virgin Islands (Anegada). [L2: 334].

Polygala longicaulis Kunth in Humboldt, Bonpland & Kunth, Nov. Gen. Sp. 5:310 [folio]; 5:396 [quarto]. 1823.

Herb in Northern Coastal Lowlands (Laguna Tortuguero), Eastern Coast (Punta Guayanés), Cerro Las Mesas; 1–300 m; *fl* & *fr* Jun–Feb.

Greater Antilles (Cuba, Hispaniola), Trinidad, Central America, South America. [L2: 334].

Polygala paniculata L., Syst. Nat. (ed. 10) 2:1154. 1759.

Herb in Northern Coastal Lowlands, Northern Limestone Hills (San Sebastián), Sierra de Luquillo, Sierra de Cayey, Las Tetas de Cayey, Southeastern Foothills (Salinas), Eastern and Central Cordillera, Maricao Forest Reserve, Cerro Las Mesas; 1–1300 m; *fl* & *fr* throughout the year.

Antilles, Mexico, Central America, South America. [L2: 335].

Orosne.

Polygala planellasii Molinet & M. Gómez, Anales Soc. Esp. Hist. Nat. 19:233. 1890.

Herb in Cerro Las Mesas; 300 m; *fl* Apr–Jun, *fr* May–Jun (one specimen). [*Vives a.n. 11837* (SJ)].

Lesser Antilles (Guadeloupe). [L5: 406].

Securidaca diversifolia (L.) S.F. Blake, Contr. U.S. Natl. Herb. 23(3):594. 1923.
Bas.: Polygala diversifolia L. (1753).
Syn.: *Elsota diversifolia* (L.) S.F. Blake (1924).

Vine in Cerro Las Mesas; 50-75 m; *fl* Feb-Mar (one specimen), *fr* (?). [*Breckon 9391* (MAPR)].

Native to Lesser Antilles, Mexico, Central America, northern and western South America, Brazil and Argentina. [L2: 336; AR: 347].

Securidaca virgata Sw., Prodr. 104. 1788.
Syn.: *Elsota virgata* (Sw.) Kuntze (1891).

Vine in Northern Coastal Lowlands, Northern Limestone Hills, Sierra de Luquillo, Sierra de Cayey, Las Piedras Chiquitas, Southeastern Foothills, Central Cordillera, Susúa Forest Reserve, Western Cordillera, Southwestern Coastal Lowlands (Cabo Rojo), Sierra Bermeja, Cerro Las Mesas, Western Coast, Vieques; 1–800 m; *fl* & *fr* throughout the year.

Greater Antilles (Cuba, Hispaniola). [L2: 336; AR: 347].

Bejuco de sopla, Jaboncillo.

POLYGONACEAE

Antigonon leptopus Hook. & Arn., Bot. Beechey Voy. 308, *t. 69*. 1838.

Vine in Northern Coastal Lowlands, Northern Limestone Hills (Camuy, San Sebastián), Eastern Coast (Naguabo), Caguas Valley (Caguas), Sierra de Cayey, Las Piedras Chiquitas, Southeastern Foothills (Salinas), Eastern Cordillera (Corozal, Comerío), Southern Coastal Lowlands, Southern Coastal Limestone, Guánica Forest Reserve, Southwestern Coastal Lowlands, Western Coast (Mayagüez), Vieques; 1–550 m; *fl* throughout the year, *fr* Apr–Dec.

Native to Mexico, northern Central America (Guatemala). [L1: 117; AR: 350].

Bellísima, Bellosinia, Coral, Coralillo, Coralina.

Coccoloba buchii O.C. Schmidt, Ark. Bot. 20A(15): 32. 1926.

Tree in Northern Limestone Hills, Río Abajo Forest Reserve; 50–400 m; *fl* Feb–Jun, *fr* Jun (one specimen).

Greater Antilles (Hispaniola).

Coccoloba costata C. Wright in Sauvalle, Anales Acad. Ci. Méd. Habana 7:343. 1870.
Syn.: *Coccoloba rupicola* Urb. (1909) "rupicolae."

Tree in Northern Limestone Hills, Río Abajo Forest Reserve, Central Cordillera (Utuado), Susúa Forest Reserve; 40–500 m; *fl* Feb–Aug, Nov, *fr* Jun–Nov, Feb.

Greater Antilles (Cuba, Hispaniola). [L1: 121].

Notes: Britton & Wilson (1924) used the orthographic variant *Coccolobis* for the name of the genus of the above synonym.

Uvilla.

Coccoloba diversifolia Jacq., Enum. Syst. Pl. 19. 1760.
Syn.: *Coccoloba laurifolia* Jacq. (1798).

Tree in Northern Coastal Lowlands, Northern Limestone Hills, Eastern Coast (Maunabo), Southeastern Foothills (Coamo), Southern Coastal Limestone, Susúa Forest Reserve, Guánica Forest Reserve, Southwestern Coastal Lowlands (Lajas), Sierra Bermeja, Western Coast (Mayagüez), Vieques, Mona, Desecheo; 1–325 m; *fl* Apr–Jan, *fr* throughout the year.

Bahamas, Greater Antilles, Lesser Antilles (St. Kitts, Antigua, Barbuda), southeastern United States (Florida), Mexico, northern Central America. [L1: 121].

Notes: Britton & Wilson (1924) used the orthographic variant *Coccolobis* for the name of the genus of the above synonym.

Cucubano, Gateado, Palo bobo, Uverillo, Uvilla.

Coccoloba krugii Lindau, Bot. Jahrb. Syst. 13:145, *t. 5, f. 14.* 1890.

Tree in Southern Coastal Lowlands, Southern Coastal Limestone, Guánica Forest Reserve, Southwestern Coastal Lowlands, Vieques, Mona; 1–150 m; *fl* Jun–Dec, *fr* Jul–Feb.

Bahamas, Greater Antilles, Virgin Islands, Lesser Antilles (Anguilla, Antigua, Barbuda). [L1: 121].

Notes: Britton & Wilson (1924) used the orthographic variant *Coccolobis* for the name of the genus of this taxon.

Coccoloba krugii ×uvifera.

Tree in Northern Coastal Lowlands, Vieques, Culebra, Mona, Desecheo; 1–100 m; *fl* Jun–Aug, *fr* Apr (one specimen).

Virgin Islands (St. John, Virgin Gorda). [L1: 122].

Coccoloba microstachya Willd., Sp. Pl. 2(1):459. 1799.

Tree in Northern Coastal Lowlands (Fajardo and adjacent islands), Northern Limestone Hills, Río Abajo Forest Reserve, Eastern Coast, Las Piedras Chiquitas, Southern Coastal Lowlands, Southern Coastal Limestone, Susúa Forest Reserve, Maricao Forest Reserve, Guánica Forest Reserve, Southwestern Coastal Lowlands (Cabo Rojo), Sierra Bermeja, Cerro Las Mesas, Vieques, Mona; 1–750 m; *fl* May–Feb, *fr* throughout the year.

Greater Antilles (Hispaniola), Virgin Islands. [L1: 123].

Notes: Britton & Wilson (1924) misapplied the name *Coccoloba obtusifolia* Jacq. (as "*Coccolobis*") to this taxon.

Negra loca, Uverilla, Uverillo, Uvillo.

Coccoloba pubescens L., Syst. Nat. (ed. 10) 2:1007. 1759.
Syn.: *Coccoloba grandifolia* Jacq. (1760).

Tree in Northern Limestone Hills, Río Abajo Forest Reserve, Maricao Forest Reserve, Mona; 25–600 m; *fl* Feb–Jun, *fr* Jan, Jul (two specimens).

Greater Antilles (Hispaniola), Lesser Antilles. [L1: 123].

Notes: Britton & Wilson (1924) used the orthographic variant *Coccolobis* for the name of the genus of the above synonym.

Moralón.

Coccoloba pyrifolia Desf., Cat. Pl. Horti Paris (ed. 3) 389. 1829.

Tree in Northern Limestone Hills, Río Abajo Forest Reserve, Sierra de Luquillo, Sierra de Cayey, Las Tetas de Cayey, Southeastern Foothills (Salinas), Eastern and Central Cordillera, Susúa Forest Reserve, Maricao Forest Reserve, Cerro Las Mesas; 15–1000 m; *fl* Nov–Aug, *fr* Nov–Jul.

Endemic. [L1: 124].

Notes: Britton & Wilson (1924) used the orthographic variant *Coccolobis pirifolia* for the name of this taxon.

Uvera.

Coccoloba rugosa Desf., Cat. Pl. Horti Paris (ed. 3) 389. 1829.

Tree in Northern Coastal Lowlands (Fajardo), Sierra de Luquillo, Eastern Coast; 10–200 m; *fl* Jan–Sep, *fr* Jan–Jul.

Virgin Islands (St. Thomas, but only historically). [L1: 124].

Notes: Britton & Wilson (1924) used the orthographic variant *Coccolobis* for the name of the genus of this taxon.

Ortegón.

Coccoloba sintenisii Urb. in Lindau, Bot. Jahrb. Syst. 13:157. 1890.

Tree in Northern Coastal Lowlands (Fajardo), Guajataca Forest Reserve, Las Tetas de Cayey, Susúa Forest Reserve, Maricao Forest Reserve, Vieques; 150–800 m; *fl* Sep–Jun, *fr* May–Jun, Nov.

Endemic. [L1: 125].

Notes: Britton & Wilson (1924) used the orthographic variant *Coccolobis* for the name of the genus of this taxon.

Uvero de monte.

Coccoloba swartzii Meisn. in A. de Candolle, Prodr. 14(1):159. 1856.

Tree in Northern Coastal Lowlands (Fajardo and adjacent islands to Carolina), Northern Limestone Hills, Sierra de Luquillo, Eastern Coast, Sierra de Cayey, Las Tetas de Cayey, Eastern and Central Cordillera, Southern Coastal Lowlands (Caja de Muertos), Maricao Forest Reserve, Guánica Forest Reserve, Southwestern Coastal Lowlands, Vieques, Culebra, Mona, Desecheo; 1–925 m; *fl* throughout the year, *fr* Apr–Sep, Dec–Jan.

West Indies, Central America, northern South America. [L1: 125].

Notes: According to Liogier & Martorell (1982), the specimen Britton & Wilson (1924) cited as *Coccolobis swartzii* Meisn. var. (?) *portoricensis* Meisn. is referable to this taxon.

Ortegón, Uvilla.

Coccoloba swartzii Meisn. f. **urbaniana** (Lindau) R.A. Howard, J. Arnold Arbor. 37:328. 1956.

Bas.: *Coccoloba urbaniana* Lindau (1890).

Syn.: *Coccoloba borinquensis* Britton (1924).

Tree in Río Abajo Forest Reserve, Sierra de Cayey, Central Cordillera, Maricao Forest Reserve; 300–1300 m; *fl* Feb–Nov, *fr* Jun–Oct, Feb.

Endemic. [L1: 125].

Notes: Britton & Wilson (1924) used the orthographic variant *Coccolobis* for the name of the genus of the above synonym.

Coccoloba tenuifolia L., Syst. Nat. (ed. 10) 2:1007. 1759.

Tree in Northern Limestone Hills (Quebradillas, Aguadilla); 5–85 m; *fl* Aug–Sep (two specimens), *fr* Nov (one specimen).

Bahamas, Greater Antilles (Cuba, Jamaica). [L1: 125].

Coccoloba uvifera (L.) L., Syst. Nat. (ed. 10) 2:1007. 1759.

Bas.: *Polygonum uviferum* L. (1753) "Uvifera."

Shrub/small tree in Northern Coastal Lowlands, Eastern Coast (Cayo Santiago), Southern Coastal Lowlands, Guánica Forest Reserve, Southwestern Coastal Lowlands (Cabo Rojo), Western Coast, Vieques, Culebra, Mona, Desecheo; 1–150 m; *fl* throughout the year, *fr* Jun–Jan.

West Indies, southeastern United States (Florida), Mexico, Central America, northwestern South America. [L1: 126].

Notes: Britton & Wilson (1924) used the homonym, and orthographic variant, *Coccolobis uvifera* (L.) Jacq. for the name of this taxon.

Uva de mar, Uva de playa, Uvas, Uvero.

Coccoloba venosa L., Syst. Nat. (ed. 10) 2:1007. 1759.

Tree in Northern Coastal Lowlands, Northern Limestone Hills (Quebradillas), Las Tetas de Cayey, Las Piedras Chiquitas, Southeastern Foothills, Eastern Cordillera (Cañón San Cristóbal), Southern Coastal Lowlands (Ponce), Southwestern Coastal Lowlands (Lajas), Sierra Bermeja, Vieques, Culebra, Mona; 1–800 m; *fl* May–Jan, *fr* Jul–Oct.

Greater Antilles (Hispaniola), Lesser Antilles, Trinidad and Tobago, Mexico, northern South America (Venezuela). [L1: 126].

Notes: Britton & Wilson (1924) used the orthographic variant *Coccolobis* for the name of the genus of this taxon.

Calambreña.

Fagopyrum esculentum Moench, Methodus 290. 1794.

Syn.: *Fagopyrum fagopyrum* (L.) H. Karst. (1881).

Herb in Northern Coastal Lowlands (San Juan), Sierra de Luquillo; 1–500 m; *fl* Feb, May (two specimens), *fr* (?). [*Vives a.n. 28156* (MAPR); *Axelrod 12950* (UPRRP)].

Probably native to China.

Polygonum acuminatum Kunth in Humboldt, Bonpland & Kunth, Nov. Gen. Sp. 2:142 [folio]; 2:178 [quarto]. 1818.

Syn.: *Persicaria acuminata* (Kunth) M. Gómez (1896).

Herb in Northern Coastal Lowlands, Eastern Coast (Naguabo); 1–25 m; *fl* Oct–Jan, May–Jun, *fr* May, Sep (two specimens).

Greater Antilles, Lesser Antilles (Guadeloupe,

Marie Galante), Trinidad, Central America, South America. [L1: 129].

Polygonum ferrugineum Wedd., Ann. Sci. Nat. Bot., sér. 3 13:252. 1850.

Herb in Northern Coastal Lowlands (Laguna Tortuguero), Northern Limestone Hills (San Sebastián), Central Cordillera (Villalba), Southern Coastal Lowlands, Guánica Forest Reserve, Southwestern Coastal Lowlands; 1–200 m; *fl* May–Feb, *fr* Feb, May–Sep.

Greater Antilles (Cuba, Hispaniola), Virgin Islands (St. Croix), Mexico, Central America, South America.

Polygonum glabrum Willd., Sp. Pl. 2(1):447. 1799.
Syn.: *Persicaria portoricensis* (Bertero ex Small) Small (1903).

Herb in Northern Coastal Lowlands, Northern Limestone Hills (San Sebastián), Caguas Valley (Caguas), Eastern Cordillera (Aibonito), Central Cordillera (Villalba), Southwestern Coastal Lowlands, Western Coast (Aguada); 1–250 m; *fl* throughout the year, *fr* Jan–May, Aug–Oct.

Widely naturalized thoughout tropics and subtropics; native range uncertain. [L1: 130].

Yerba de hicotea, Yerba de jicotea.

Polygonum hydropiperoides Michx., Fl. Bor.-Amer. 1:239. 1803.

Herb in Northern Coastal Lowlands (Laguna Tortuguero), Northern Limestone Hills (Florida); 1–150 m; *fl* Dec–Jan, May, *fr* Dec–Jan (one specimen).

Bahamas, Greater Antilles (Cuba, Hispaniola), Canada, United States, Central America, northern South America to Bolivia, Paraguay. [L1: 130].

Polygonum punctatum Elliott, Sketch Bot. S. Carolina 1(5):455. 1817.
Syn.: *Persicaria punctata* (Elliott) Small (1903).

Herb in Northern Coastal Lowlands, Northern Limestone Hills, Río Abajo Forest Reserve, Sierra de Luquillo, Eastern Coast, Caguas Valley, Sierra de Cayey, Southeastern Foothills (Cayey), Eastern and Central Cordillera, Southern Coastal Lowlands (Santa Isabel), Western Coast (Rincón); 1–1200 m; *fl* & *fr* throughout the year.

West Indies, Canada, United States, Mexico, Central America, South America. [L1: 130].

Yerba de hicotea, Yerba de jicotea.

Polygonum segetum Kunth in Humboldt, Bonpland & Kunth, Nov. Gen. Sp. 2:142 [folio]; 2:177 [quarto]. 1818.
Syn.: *Persicaria segetum* (Kunth) Small (1903) "segeta."

Herb in Northern Coastal Lowlands (Laguna Tortuguero), Guánica Forest Reserve; 1–5 m; *fl* Aug (one specimen), *fr* Aug, Nov (two specimens). [*Caudales a.n. 11401* (UPR); *Liogier 34573* (UPR)].

Greater Antilles, Mexico, Central America, northwestern and west-central South America. [L1: 131].

Rumex crispus L., Sp. Pl. 1:335. 1753.

Herb in Sierra de Cayey, Eastern Cordillera (Cayey), Central Cordillera, Maricao Forest Reserve; 400–1330 m; *fl* Jan–May, *fr* Oct, Jan–Feb, May.

Native to Europe, northern Africa, temperate Asia. [L1: 131].

Col agria, Vinagrillo.

Triplaris cumingiana Fisch. & C.A. Mey. ex C.A. Mey., Mém. Acad. Imp. Sci. St.-Pétersbourg, Sér. 6, Sci. Math., Seconde Pt. Sci. Nat. 4:148. 1845.

Tree planted and persistent in Northern Coastal Lowlands (Río Grande), Central Cordillera (Juana Díaz); 5–270 m; *fl* Mar–Apr (two specimens), *fr* Mar (one specimen). [*Axelrod 6026* (UPRRP)].

Native to southern Central America, northwestern, west-central South America and Venezuela. [L1: 133].

Triplaria.

TAXA CITED BY PREVIOUS AUTHORS, BUT NOT INCLUDED

CITED DUE TO MISIDENTIFICATION

Coccoloba pallida C. Wright ex Griseb.
Tree cited by Liogier & Martorell (1982, 2000) and Liogier (1985) as occurring in coastal forest on limestone, near Quebradillas. This citation was based on the misidentification of a specimen of *Coccoloba swartzii*. [L1: 123].

ONLY UNDER CULTIVATION

Antigonon guatimalense Meisn. Syn.: *Antigonon macrocarpum* Britton & Small.
Vine cited by Liogier & Martorell (1982, 2000) and Liogier (1985) as spontaneous after cultivation and by Acevedo-Rodríguez (2005) as cultivated and spontaneous. [L1: 117; AR: 349].

Homalocladium platycladum (F. Muell.) L.H. Bailey.
Syn.: *Muehlenbeckia platyclados* (F. Muell.) Meisn.
Suffrutescent herb cited by Liogier & Martorell (1982, 2000) and Liogier (1985) as introduced and naturalized. [L1: 127].

PONTEDERIACEAE

Eichhornia azurea (Sw.) Kunth, Enum. Pl. 4:129. 1843.

Bas.: *Pontederia azurea* Sw. (1788).

Aquatic herb in Southern Coastal Lowlands, Southwestern Coastal Lowlands (Cabo Rojo); 1–10 m; *fl* & *fr* Jul–Aug (two specimens). [*Liogier 33264* (NY); *Caudales Aug 1982* (NY)].

Greater Antilles (Jamaica), Mexico, Central America, South America. [AR&S: 171].

Eichhornia crassipes (Mart.) Solms in A. & C. de Candolle, Monogr. Phan. 4:527. 1883.

Bas.: *Pontederia crassipes* Mart. (1823).

Syn.: *Piaropus crassipes* (Mart.) Britton (1893).

Aquatic herb in Northern Coastal Lowlands, Southern Coastal Lowlands, Guánica Forest Reserve, Southwestern Coastal Lowlands (Lajas), Western Coast; 1–50 m; *fl* Apr–Jan, *fr* (?).

Native to northern South America and Brazil. [AR&S: 170].

Flor de agua, Jacinto de agua.

Eichhornia diversifolia (Vahl) Urb., Symb. Antill. 4:147. 1903.

Bas.: *Heteranthera diversifolia* Vahl (1805).

Syn.: *Piaropus diversifolius* (Vahl) P. Wilson (1923).

Aquatic herb known from a single collection by Roy Woodbury without date, without locality and without number. [*Woodbury a.n. 4833* (UPR)].

Greater Antilles (Hispaniola), southern Central America, northern South America and Bolivia. [AR&S: 170].

Flor de agua.

Heteranthera limosa (Sw.) Willd., Neue Schriften Ges. Naturf. Freunde Berlin 3:439. 1801.

Bas.: *Pontederia limosa* Sw. (1788).

Aquatic herb in Western Coast (Mayagüez); 1 m; *fl* Jul, Dec (two specimens), *fr* (?). [*Caudales 33256* (UPR)].

Greater Antilles, Mexico, Central America, South America (scattered). [AR&S: 172].

Heteranthera reniformis Ruiz & Pav., Fl. Peruv. 1:43, *t. 71a*. 1798.

Herb in Northern Coastal Lowlands, Sierra de Luquillo, Eastern Coast, Caguas Valley, Central Cordillera (Utuado); 5–700 m; *fl* Apr–Jan, *fr* Oct (one specimen).

Greater Antilles (Jamaica, Hispaniola), Lesser Antilles (Antigua, Guadeloupe), Mexico, Central America, South America (scattered). [AR&S: 172].

PORTULACACEAE

Portulaca caulerpoides Britton & P. Wilson, Bull. Torrey Bot. Club 42(9):515. 1915.

Herb in Southern Coastal Lowlands (Caja de Muertos), Guánica Forest Reserve, Mona; 20–70 m; *fl* Mar–Nov, *fr* Feb–Jul, Oct–Nov.

Virgin Islands (St. Thomas). [L1: 203].

Portulaca grandiflora Hook., Bot. Mag. 56: *t. 2885*. 1829.

Herb in Southwestern Coastal Lowlands; 15–70 m; *fl* & *fr* Mar, Oct (two specimens). [*Breckon 5325, 6064* (MAPR)].

Native to southern South America (Uruguay, Argentina). [L1: 203].

Don Diego americano, Don Diego gigante.

Portulaca halimoides L., Sp. Pl. (ed. 2) 1:639. 1762.

Herb in Southern Coastal Lowlands, Guánica Forest Reserve, Southwestern Coastal Lowlands; 1–100 m; *fl* Aug–Nov, *fr* Sep–Mar.

West Indies, Mexico, Central America, northwestern, west-central South America and Venezuela. [L1: 203].

Portulaca oleracea L., Sp. Pl. 1:445. 1753.

Herb in Northern Coastal Lowlands, Eastern Coast (Ceiba), Southeastern Foothills, Central Cordillera (Utuado), Southern Coastal Lowlands, Guánica Forest Reserve, Southwestern Coastal Lowlands (Cabo Rojo), Western Coast (Mayagüez), Vieques, Culebra, Mona, Desecheo; 1–300 m; *fl* & *fr* throughout the year.

Worldwide weed of unknown origin. [L1: 204].

Verdolaga.

Portulaca pilosa L., Sp. Pl. 1:445. 1753.

Herb in Northern Coastal Lowlands, Northern Limestone Hills (Quebradillas), Eastern Coast (Cayo Santiago), Las Piedras Chiquitas, Southeastern Foothills (Coamo), Eastern Cordillera (Cayey), Southwestern Coastal Lowlands (Cabo Rojo), Sierra Bermeja, Vieques, Culebra; 1–500 m; *fl* & *fr* throughout the year.

West Indies, Trinidad, southern United States, Mexico, Central America, northwestern, west-central South America and Venezuela, Paraguay. [L1: 204].

Don Diego.

Portulaca quadrifida L., Mant. Pl. 1:73. 1767.

Herb in Southern Coastal Lowlands, Guánica Forest Reserve, Southwestern Coastal Lowlands,

Sierra Bermeja, Culebra, Mona; 1–70 m; *fl* & *fr* Nov–Feb.

Pantropical (except Australia). [L1: 204].

Yerba de pollo.

Portulaca rubricaulis Kunth in Humboldt, Bonpland & Kunth, Nov. Gen. Sp. 6:58 [folio]; 6:73 [quarto]. 1823.

Syn.: *Portulaca phaeosperma* Urb. (1905).

Herb in Guánica Forest Reserve, Vieques, Mona; 1–25 m; *fl* Feb–Nov, *fr* throughout the year.

Bahamas, Greater Antilles, Lesser Antilles (St. Barts, La Desirade, Les Saintes), southeastern United States (Florida), Mexico, Central America, South America (Venezuela, Ecuador). [L1: 205].

Portulaca teretifolia Kunth in Humboldt, Bonpland & Kunth, Nov. Gen. Sp. 6:59 [folio]; 6:73 [quarto]. 1823.

Syn.: *Portulaca poliosperma* Urb. (1905).

Herb in Northern Coastal Lowlands (Vega Baja to Laguna Tortuguero), Susúa Forest Reserve, Southwestern Coastal Lowlands (Cabo Rojo), Vieques; 2–150 m; *fl* Feb–Nov, *fr* Feb–Jun, Sep–Nov.

Greater Antilles (Cuba), Lesser Antilles, Central America, northern and northwestern South America. [L1: 206].

POTAMOGETONACEAE

Potamogeton foliosus Raf., Med. Repos., ser. 2 5:354. 1808 "foliosum."

Aquatic herb in Northern Coastal Lowlands (Quebradillas), Western Coast (Mayagüez); 1–10 m; *fl* (?), *fr* (?). [*Liogier 33432* (UPR); *Proctor 44117* (US)].

Greater Antilles (Cuba, Jamaica), Canada, United States, Central America (Guatemala, Costa Rica). [AR&S: 77].

Potamogeton illinoensis Morong, Bot. Gaz. 5(5):50. 1880.

Aquatic herb in Northern Coastal Lowlands (Toa Baja); 1 15 m; *fl* (?), *fr* (?). [*Liogier 33773* (UPR)].

Bahamas, Greater Antilles (Cuba, Jamaica), Canada, United States, Mexico, Central America, northwestern and west-central South America. [AR&S: 77].

Potamogeton nodosus Poir., Encycl. Suppl. 4(2):535. 1816.

Syn.: *Potamogeton insulanus* Hagstr. (1916).

Aquatic herb in Northern Coastal Lowlands; 1–25 m; *fl* Jan–Feb, May, *fr* (?).

Native to north temperate regions. [AR&S: 77].

Notes: Britton & Wilson (1923) misapplied the names *Potamogeton epihydrus* Raf. and *Potamogeton fluitans* Roth to this taxon.

Lino de agua.

PRIMULACEAE

Anagallis arvensis L., Sp. Pl. 1:148. 1753.

Herb in Sierra de Cayey; 450–765 m; *fl* Feb–Apr, Sep, *fr* Feb (two specimens).

Native to western Europe. [L4: 130].

Anagallis pumila Sw., Prodr. 40. 1788.

Herb in Sierra de Cayey; 730 m; *fl* & *fr* Jun (one specimen). [*Axelrod 7917* (UPRRP)].

Native to Greater Antilles (Jamaica), southeastern United States (Florida), Mexico, Central America, South America (scattered).

Samolus valerandi L., Sp. Pl. 1:171. 1753.

Syn.: *Samolus parviflorus* Raf. (1818).

Herb in Eastern Cordillera (Cañón San Cristóbal) 450 m; *fl* & *fr* Feb (one specimen). [*Liogier 37202* (UPR)].

Bahamas, Greater Antilles (Cuba, Hispaniola), Canada, United States, Mexico, South America. [L4: 132].

PROTEACEAE

TAXA CITED BY PREVIOUS AUTHORS, BUT NOT INCLUDED

ONLY UNDER CULTIVATION

Grevillea robusta A. Cunn. ex R. Br.

Tree cited by Liogier & Martorell (1982, 2000) and Liogier (1985) as widely planted for shade and ornament. [L1: 90].

Macadamia integrifolia Maiden & Betche.

Tree cited by Liogier & Martorell (1982, 2000) and Liogier (1985) as planted and naturalized. [L1: 93].

PUTRANJIVACEAE

Drypetes alba Poit., Mém. Hist. Nat. Paris 1:157, *t.* 7. 1815.

Tree in Northern Limestone Hills, Río Abajo Forest Reserve, Southeastern Foothills (Salinas), Susúa Forest Reserve, Maricao Forest Reserve; 50–800 m; *fl* Dec–Mar, *fr* Mar–Sep.

Greater Antilles, Lesser Antilles (Guadeloupe). [L2: 386].

Cafeíllo, Hueso, Palo de vaca, Palo de vaca blanco.

Drypetes glauca Vahl, Eclog. Amer. 3:49. 1807.

Tree in Sierra de Luquillo, Sierra de Cayey,

Central Cordillera, Maricao Forest Reserve; 200–1100 m; *fl* Jul–Nov, *fr* throughout the year.

Greater Antilles (Hispaniola), Lesser Antilles. [L2: 387].

Cafeíllo, Palo blanco, Palo de aceituna, Varital.

Drypetes ilicifolia Krug & Urb., Bot. Jahrb. Syst. 15:352. 1892.

Tree in Northern Coastal Lowlands (Manatí), Northern Limestone Hills, Río Abajo Forest Reserve; 5–300 m; *fl* Feb (one specimen), *fr* Feb–Jun.

Greater Antilles (Jamaica). [L2: 387].

Encinillo.

Drypetes lateriflora (Sw.) Krug & Urb., Bot. Jahrb. Syst. 15:357. 1892.

Bas.: *Schaefferia lateriflora* Sw. (1788).

Shrub/small tree in Northern Limestone Hills; 50–250 m; *fl* Feb (one specimen), *fr* Apr–Aug.

Bahamas, Greater Antilles, southeastern United States (Florida), Mexico, Central America. [L2: 387].

Ciruela de Guayana, Cueriduro.

RANUNCULACEAE

Clematis dioica L., Syst. Nat. (ed. 10) 2:1084. 1759.

Vine in Northern Limestone Hills (Aguadilla); 150–200 m; *fl* (?) & *fr* (?). [*Quigley 515* (SJ)].

Greater Antilles, Mexico, northern Central America (Belize). [L1: 226; AR: 352].

Notes: In floristic treatments of Puerto Rico prior to that of Acevedo-Rodríguez (2003, 2005) the three Puerto Rican taxa of *Clematis* had been treated as a single one under the above name.

Cabellos de ángel.

Clematis flammulastrum Griseb., Pl. Wright. 1:153. 1860 (preprint of Mem. Amer. Acad. Arts, n.s. 8(1):153. 1861).

Vine in Northern Limestone Hills, Sierra de Luquillo, Sierra de Cayey, Southeastern Foothills, Central Cordillera, Maricao Forest Reserve; 50–720 m; *fl* Oct–Nov, *fr* Oct–Mar.

Greater Antilles. [L1: 226; AR: 352].

Cabellos de ángel.

Clematis polygama Jacq., Enum. Syst. Pl. 24. 1760.

Vine in Northern Limestone Hills (Ciales, Lares), Río Abajo Forest Reserve, Las Tetas de Cayey, Las Piedras Chiquitas, Southeastern Foothills (Coamo), Maricao Forest Reserve, Sierra Bermeja; 250–720 m; *fl* Oct–Dec, Apr, *fr* Nov–Dec (two specimens).

Greater Antilles, Mexico, Central America. [L1: 226; AR: 352].

Cabellos de ángel.

Ranunculus recurvatus Poir. var. **tropicus** (Griseb.) Fawc. & Rendle, Fl. Jamaica 3:188. 1914.

Bas.: *Ranunculus repens* L. var. *tropicus* Griseb. (1859).

Herb in Toro Negro Forest Reserve; 1000–1150 m; *fl* Jun (two specimens), *fr* Sep (one specimen). [*Acevedo 6* (SJ)].

Greater Antilles. [L1: 227].

RHAMNACEAE

Auerodendron pauciflorum Alain, Phytologia 50(3):164, *f. s.n.* 1982.

Shrub/small tree in Northern Limestone Hills (Arecibo, Isabela); 35 m; *fl* Feb (one specimen), *fr* (?).

Endemic. [L3: 77].

Notes: This taxon is on the U.S. Fish & Wildlife Service list of threatened and endangered plants.

Colubrina arborescens (Mill.) Sarg., Trees & Shrubs 2(3):167. 1911.

Bas.: *Ceanothus arborescens* Mill. (1768).

Syn.: *Colubrina colubrina* (Jacq.) Millsp. (1900).

Tree in Northern Coastal Lowlands, Northern Limestone Hills, Southern Coastal Lowlands, Southern Coastal Limestone, Susúa Forest Reserve, Guánica Forest Reserve, Southwestern Coastal Lowlands, Western Coast (Rincón), Vieques, Culebra, Mona, Desecheo; 1–250 m; *fl* & *fr* throughout the year.

West Indies, southeastern United States (Florida), southern Mexico, Central America, northwestern and west-central South America. [L3: 79].

Abejuelo, Abeyuelo, Achiotillo, Aguacatillo, Aguaitarán, Catire, Guitarán, Mabí, Ratón, Sanguinaria.

Colubrina elliptica (Sw.) Brizicky & W.L. Stern, Trop. Woods 109:95. 1958.

Bas.: *Rhamnus elliptica* Sw. (1788) "ellipticus."

Syn.: *Colubrina reclinata* (L'Hér.) Brogn. (1826).

Tree in Northern Coastal Lowlands (islands adjacent to Fajardo), Southeastern Foothills (Cayey), Southern Coastal Lowlands (Caja de Muertos), Southern Coastal Limestone, Susúa Forest Reserve, Guánica Forest Reserve, Southwestern Coastal Lowlands (Cabo Rojo), Sierra Bermeja, Culebra, Mona, Desecheo; 5–425 m; *fl* May–Dec, *fr* Sep–May.

West Indies, southeastern United States (Florida), Mexico, Central America, northwestern South America and Venezuela. [L3: 79].

Catire, Guitarán, Mabí, Palo de mabí.

Colubrina glandulosa Perkins var. **antillana** (M.C. Johnst.) M.C. Johnst., Brittonia 23(1):17. 1971.
Bas.: *Colubrina rufa* (Vell.) Reissek var. *antillana* M.C. Johnst. (1963).

Tree in Northern Limestone Hills, Río Abajo Forest Reserve, Sierra de Luquillo, Central Cordillera (Orocovis); 120–625 m; *fl* Oct–Nov, Apr–May, *fr* Jun–Dec.

Greater Antilles (Cuba, Hispaniola). [L3: 81].

Colubrina verrucosa (Urb.) M.C. Johnst., Wrightia 3(5):93. 1963.
Bas.: *Hybosperma verrucosum* Urb. (1926).
Syn.: *Hybosperma spinosum* Urb. (1899).

Shrub in Susúa Forest Reserve, Guánica Forest Reserve, Sierra Bermeja; 145–300 m; *fl* Jul–Jan, *fr* Apr, Jul–Jan.

Greater Antilles (Hispaniola). [L3: 81].

Gouania lupuloides (L.) Urb., Symb. Antill. 4(3): 378. 1910.
Bas.: *Banisteria lupuloides* L. (1753).

Vine in Northern Coastal Lowlands, Northern Limestone Hills, Eastern Coast (Yabucoa), Sierra de Cayey, Las Tetas de Cayey, Las Piedras Chiquitas, Southeastern Foothills, Eastern Cordillera (Cañón San Cristóbal), Central Cordillera (Orocovis), Southern Coastal Lowlands, Southern Coastal Limestone, Susúa Forest Reserve, Western Cordillera (Maricao), Guánica Forest Reserve, Southwestern Coastal Lowlands, Western Coast, Sierra Bermeja, Vieques, Culebra; 1–925 m; *fl* Oct–Nov, *fr* Nov–Apr.

West Indies, Trinidad and Tobago, southeastern United States (Florida), Mexico, Central America, northwestern, west-central South America and Venezuela. [L3: 82; AR: 354].

Bejuco indio, Bejuco de sopla, Mascapelo, Sopla.

Gouania polygama (Jacq.) Urb., Symb. Antill. 4(3): 378. 1910.
Bas.: *Rhamnus polygama* Jacq. (1760) "polygamus."

Vine in Northern Limestone Hills, Eastern Coast (Cuchilla de Panduras), Central Cordillera (Adjuntas), Southern Coastal Limestone (Peñuelas), Susúa Forest Reserve, Western Cordillera (Maricao), Cerro Las Mesas; 75–225 m; *fl* Nov, *fr* Jan (two specimens).

Antilles, southeastern United States (Florida), Mexico, Central America, South America. [L3: 84; AR: 355].

Bejuco indio, Bejuco de sopla, Mascapelo, Sopla.

Krugiodendron ferreum (Vahl) Urb., Symb. Antill. 3(2):314. 1902.
Bas.: *Rhamnus ferrea* Vahl (1794).

Tree in Northern Coastal Lowlands (islands adjacent to Fajardo), Northern Limestone Hills, Las Piedras Chiquitas, Southeastern Foothills (Coamo), Southern Coastal Lowlands, Southern Coastal Limestone, Guánica Forest Reserve, Southwestern Coastal Lowlands (Cabo Rojo), Sierra Bermeja, Western Coast (Mayagüez), Vieques, Culebra, Mona, Desecheo; 1–500 m; *fl* Apr–Oct, *fr* May–Nov.

West Indies, southeastern United States (Florida), Mexico, Central America. [L3: 84].

Bariaco, Espejuelo, Palo de hierro, Quiebrahacha.

Maesopsis eminii Engl., Pflanzenw. Ost-Afrikas C:255. 1895.

Tree planted and persistent in Sierra de Luquillo, Eastern Cordillera (Cayey); 500–800 m; *fl* (?), *fr* Jan (one specimen). [*Liogier 27351, 36224* (UPR); *Axelrod 12501* (UPRRP)].

Native to tropical Africa. [L3: 86].

Reynosia krugii Urb., Symb. Antill. 1(2):354. 1899.

Tree in Northern Limestone Hills, Río Abajo Forest Reserve, Eastern Cordillera (Cayey), Southern Coastal Limestone (Guayanilla), Susúa Forest Reserve, Western Cordillera, Guánica Forest Reserve; 60–800 m; *fl* Jan–Jun, Oct, *fr* Feb–Sep.

Endemic. [L3: 88].

Reynosia uncinata Urb., Symb. Antill. 1(2):355. 1899.

Tree in Northern Limestone Hills (Isabela), Southern Coastal Lowlands, Southern Coastal Limestone, Guánica Forest Reserve, Southwestern Coastal Lowlands (Cabo Rojo), Sierra Bermeja, Vieques, Mona; 1–100 m; *fl* Feb–Jun, Sep–Oct, *fr* Feb, May–Nov.

Greater Antilles (Hispaniola), Virgin Islands, Lesser Antilles (Anguilla). [L3: 89].

Cascarroya, Chicharrón.

Reynosia vivesiana Trejo, Novon 18(3):390, *t. 1.* 2008.

Shrub/small tree in Guánica Forest Reserve; 50–150 m; *fl* Jun–Sep, Dec, *fr* Jun.

Endemic. [L3: 88].

Notes: Specimens of this taxon were previously referred to *Reynosia guama* Urb., which is confined to the Virgin Islands.

Guamá.

Rhamnus sphaerosperma Sw. var. **longipes** M.C. Johnst. & L.A. Johnst., Fl. Neotrop. Monogr. 20:62, *f. 8, 22.* 1978.

Shrub/small tree in Río Abajo Forest Reserve,

Toro Negro Forest Reserve, Maricao Forest Reserve; 300–1335 m; *fl* Jan–Jul, *fr* Nov–Feb, Jun–Jul.

Greater Antilles (Hispaniola). [L3: 90].

Ziziphus mauritiana Lam., Encycl. 3(1):319. 1789.
Syn.: *Ziziphus jujuba* (L.) Lam. (1789), *nom. illeg.*

Tree in Northern Coastal Lowlands (Fajardo and adjacent islands), Guánica Forest Reserve, Southwestern Coastal Lowlands (Cabo Rojo), Vieques; 1–300 m; *fl* Jul–Jan, *fr* Jan–Mar.

Possibly native to India, Myanmar. [L3: 92].

Aprín, Yuyubí.

Ziziphus reticulata (Vahl) DC., Prodr. 2:20. 1825.
Bas.: *Paliurus reticulatus* Vahl (1807).
Syn.: *Sarcomphalus reticulatus* (Vahl) Urb. (1899).

Shrub/small tree in Southern Coastal Lowlands (Salinas, Santa Isabel), Southern Coastal Limestone (Guayanilla), Guánica Forest Reserve, Southwestern Coastal Lowlands, Sierra Bermeja, Vieques, Desecheo; 1–200 m; *fl* May–Feb, *fr* Jul–Mar.

Greater Antilles (Hispaniola), Virgin Islands, Lesser Antilles (Antigua, Barbuda). [L3: 93].

Azufaifo, Cacao rojo, Cascarilla, Cascarroya, Espejuelo.

Ziziphus rignonii Delponte, Mem. Reale Accad. Sci. Torino, ser. 2 14:404, *t. 7*. 1854.
Syn.: *Sarcomphalus domingensis* (Vahl) Urb. (1899).

Tree in Northern Coastal Lowlands, Vieques; 1–15 m; *fl* May–Aug, Dec, *fr* Sep–Dec.

Greater Antilles (Hispaniola), Virgin Islands, Lesser Antilles (Anguilla, St. Martin). [L3: 93].

Ziziphus taylorii (Britton) M.C. Johnst., Amer. J. Bot. 51:1118. 1964.
Bas.: *Sarcomphalus taylorii* Britton (1905).

Shrub in Mona; 1–50 m; *fl* Apr–Jul, Oct–Nov, *fr* Apr–Jan.

Bahamas, Turks and Caicos. [L3: 93].

RHIZOPHORACEAE

Cassipourea guianensis Aubl., Hist. Pl. Guiane 1:529, *t. 211*. 1775 "Guyannensis."
Syn.: *Cassipourea elliptica* (Sw.) Poir. (1811).
Cassipourea alba Griseb. (1857).

Shrub/small tree in Northern Coastal Lowlands, Northern Limestone Hills, Río Abajo Forest Reserve, Sierra de Luquillo, Sierra de Cayey, Las Tetas de Cayey, Eastern Cordillera (Cerro La Peña, La Torrecilla), Central Cordillera, Susúa Forest Reserve, Maricao Forest Reserve; 150–950 m; *fl* throughout the year, *fr* Aug–Apr.

Antilles, Mexico, Central America, northern South America to Brazil, Peru. [L3: 362].

Aceitillo, Murta, Palo de gongolí, Palo de orejas, Palo de toro.

Rhizophora mangle L., Sp. Pl. 1:443. 1753.

Tree in coastal swamps of Northern Coastal Lowlands, Eastern Coast (Cayo Santiago), Southern Coastal Lowlands, Guánica Forest Reserve, Southwestern Coastal Lowlands, Vieques, Culebra, Mona; 0 m; *fl* & *fr* throughout the year.

Coastal swamps of West Indies, Mexico, Central America, northern South America to Peru, tropical West Africa. [L3: 364].

Mangle, Mangle colorado, Mangle de chifle, Mangle zapatero.

ROSACEAE

Prunus myrtifolia (L.) Urb., Symb. Antill. 5(1):93. 1904.
Bas.: *Celastrus myrtifolius* L. (1753).
Syn.: *Laurocerasus myrtifolia* (L.) Britton (1908).

Tree in Northern Limestone Hills, Río Abajo Forest Reserve, Central Cordillera (Adjuntas), Maricao Forest Reserve; 10–1050 m; *fl* Jun–Aug, Nov–Feb, *fr* Jun, Dec (two specimens).

Bahamas, Greater Antilles, Trinidad and Tobago, southeastern United States (Florida), Mexico, Central America, northern South America to Bolivia, Paraguay. [L1: 319].

Prunus occidentalis Sw., Prodr. 80. 1788.
Syn.: *Laurocerasus occidentalis* (Sw.) M. Roem. (1847).

Tree in Sierra de Luquillo, Central Cordillera, Maricao Forest Reserve; 600–1060 m; *fl* Feb (one specimen), *fr* Mar, Jun (two specimens).

Greater Antilles, Lesser Antilles (Guadeloupe), Mexico, Central America, South America. [L1: 320].

Almendrillo, Almendrón.

Rubus florulentus Focke, Abh. Naturwiss. Vereins Bremen 11:411. 1890.

Shrub in Central Cordillera, Maricao Forest Reserve; 500–1300 m; *fl* & *fr* Mar–May.

Endemic. [L1: 322; AR: 357].

Fresa de montaña.

Rubus probus L.H. Bailey, Gentes Herb. 1:150. 1923.

Shrub in Central Cordillera; 800–1250 m; *fl* & *fr* Nov–May.

Native to Australia (Queensland). [L1: 323].

Notes: The type specimen was taken from a plant grown at the Puerto Rico Agricultural Experimental Station in Río Piedras.

Rubus rosifolius Sm., Pl. Icon. Ined. 3: *t. 60*. 1791 "rosaefolius."
Syn.: *Rubus coronarius* (Sims) Sweet (1826).

Shrub in Northern Coastal Lowlands (Carolina), Río Abajo Forest Reserve, Sierra de Luquillo, Sierra de Cayey, Eastern and Central Cordillera, Maricao Forest Reserve; 200–925 m; *fl* & *fr* throughout the year.

Native to western Indian Ocean (Mauritius), Malesia, China, Australia, southwestern Pacific islands. [L1: 322].

Frambuesa común, Fresa, Rosa minadora, Rosa de novia, Zarza.

TAXA CITED BY PREVIOUS AUTHORS, BUT NOT INCLUDED

ONLY UNDER CULTIVATION

Eriobotrya japonica (Thunb.) Lindl.
Tree cited by Liogier & Martorell (1982, 2000) and Liogier (1985) as planted for its fruits and becoming naturalized at higher altitudes. [L1: 317].

Rosa indica L.
Shrub cited by Britton & Wilson (1924) as locally subspontaneous after cultivation; subsequently cited by Liogier & Martorell (1982, 2000) and Liogier (1985). [L1: 321].

Rosa sempervirens L.
Shrub cited by Britton & Wilson (1924) as locally subspontaneous after cultivation at middle elevations; subsequently cited by Liogier & Martorell (1982, 2000) and Liogier (1985). [L1: 321].

RUBIACEAE

Catesbaea melanocarpa Krug & Urb. in Urban, Symb. Antill. 1(3):427. 1899.

Shrub in Guánica Forest Reserve, Southwestern Coastal Lowlands (Cabo Rojo); 75–85 m; *fl* Sep (one specimen), *fr* Feb–Mar.

Virgin Islands (St. Croix), Lesser Antilles (Antigua, Barbuda, Guadeloupe). [L5: 75].

Notes: This taxon is on the U.S. Fish & Wildlife Service list of threatened and endangered plants.

Catesbaea parviflora Sw., Prodr. 30. 1788.

Shrub in Southern Coastal Limestone, Susúa Forest Reserve, Guánica Forest Reserve, Vieques; 5–200 m; *fl* Jun–Dec, *fr* Oct–Jun.

Bahamas, Greater Antilles, southeastern United States (Florida). [L5: 76].

Chiococca alba (L.) Hitchc., Rep. (Annual) Missouri Bot. Gard. 4:94. 1893.
Bas.: *Lonicera alba* L. (1753).
Syn.: *Chiococca parvifolia* Wullschl. ex Griseb. (1861).
Chiococca micrantha J.R. Johnst. (1905).

Vine in Northern Coastal Lowlands, Northern Limestone Hills, Río Abajo Forest Reserve, Sierra de Luquillo, Eastern Coast, Sierra de Cayey, Las Tetas de Cayey, Las Piedras Chiquitas, Southeastern Foothills, Eastern Cordillera (Comerío), Central Cordillera (Orocovis), Southern Coastal Lowlands (Patillas), Southern Coastal Limestone, Susúa Forest Reserve, Maricao Forest Reserve, Guánica Forest Reserve, Southwestern Coastal Lowlands, Sierra Bermeja, Cerro Las Mesas, Western Coast, Vieques, Culebra, Mona; 1–850 m; *fl* & *fr* throughout the year.

West Indies, southern United States (Florida, Texas), Mexico, Central America, South America. [L5: 77; AR: 359].

Bejuco de berac.

Chione venosa (Sw.) Urb. var. **venosa,** Symb. Antill. 4(4):594. 1911.
Bas.: *Jacquinia venosa* Sw. (1788).

Tree in Northern Limestone Hills, Río Abajo Forest Reserve, Sierra de Luquillo; 20–350 m; *fl* Apr–Jul, *fr* Mar, Jun–Oct.

Greater Antilles (Hispaniola), Lesser Antilles, Trinidad, southern Mexico, Central America (Nicaragua, Costa Rica), west-central South America. [L5: 79].

Martín Ávila, Palo blanco, Santa Olalla.

Coccocypselum herbaceum Aubl., Hist. Pl. Guiane 1:68. 1775 "Coccocipsilum."
Syn.: *Tontanea herbacea* (Aubl.) Standl. (1921).

Low creeping herb in Northern Coastal Lowlands (Laguna Tortuguero), Sierra de Cayey, Eastern Cordillera (Cerro La Peña), Central Cordillera, Maricao Forest Reserve; 1–1225 m; *fl* Aug–May, *fr* throughout the year.

Greater Antilles, Central America, northern South America. [L5: 81].

Notes: Britton & Wilson (1925) used *Tontanea herbacea* (Lam.) Standl., based on the homonym *Coccocypselum herbaceum* Lam., for the name of this taxon; Liogier (1997) and Liogier & Martorell (2000) used *Coccocypselum herbaceum* P. Browne, which was not validly published.

Balsamilla, Fruta de paloma, Yerba de guabá.

Coffea arabica L., Sp. Pl. 1:172. 1753.

Shrub planted and persistent in Northern Coastal Lowlands (Guaynabo), Northern Limestone Hills, Río Abajo Forest Reserve, Sierra de Cayey,

Las Tetas de Cayey, Southeastern Foothills, Eastern Cordillera (Cayey), Central Cordillera, Western Cordillera; 175–1000 m; *fl* Feb–May, *fr* throughout the year.

Native to Eastern Africa (Ethiopia, Sudan, Kenya). [L5: 83].

Café.

Coffea liberica W. Bull ex Hiern, Trans. Linn. Soc. London, Bot. 1(4):171, *t. 24*. 1876.

Syn.: *Coffea dewevrei* De Wild. & T. Durand (1899). *Coffea liberica* W. Bull ex Hiern var. *dewevrei* (De Wild. & T. Durand) Lebrun (1941).

Shrub planted and persistent in Northern Limestone Hills, Río Abajo Forest Reserve, Caguas Valley (Caguas), Sierra de Cayey, Central Cordillera, Western Cordillera (Maricao, San Germán), Western Coast (Mayagüez); 70–600 m; *fl* Nov, Feb, May–Jun, *fr* Feb–Nov.

Native to tropical Africa. [L5: 83].

Café excelsa, Café de Liberia, Café libérico.

Colleteria seminervis (Urb. & Ekman) D.W. Taylor, Syst. & Geogr. Pl. 73(2):206. 2003.

Bas.: *Chione seminervis* Urb. & Ekman (1926).

Shrub/small tree in Toro Negro Forest Reserve, Maricao Forest Reserve; 850–1300 m; *fl* Aug, Oct, *fr* Oct, Feb–Mar.

Greater Antilles (Hispaniola). [L5: 79].

Diodia apiculata (Willd. ex Roem. & Schult.) K. Schum., Bot. Jahrb. Syst. 10:313. 1889.

Bas.: *Spermacoce apiculata* Willd. ex Roem. & Schult. (1818).

Suffrutescent herb in Northern Coastal Lowlands, Eastern Coast, Southern Coastal Lowlands (Coamo), Susúa Forest Reserve, Guánica Forest Reserve, Southwestern Coastal Lowlands (Cabo Rojo), Sierra Bermeja, Cerro Las Mesas; 1–300 m; *fl* & *fr* throughout the year.

Greater Antilles, Lesser Antilles (St. Barts, Antigua, St. Vincent), Mexico, Central America, northern South America to Bolivia, Paraguay. [L5: 86].

Notes: Britton & Wilson (1925) used *Diodia rigida* Cham. & Schltdl., a *nom. inval.*, for the name of this taxon.

Diodia sarmentosa Sw., Prodr. 30. 1788.

Scandent shrub in Northern Coastal Lowlands, Northern Limestone Hills, Sierra de Luquillo, Eastern Coast, Caguas Valley (Caguas), Sierra de Cayey, Eastern Cordillera (Cerro La Peña, Cidra, Corozal), Central Cordillera (Villalba, Utuado), Susúa Forest Reserve; 1–825 m; *fl* & *fr* Jan–Oct.

Greater Antilles, Lesser Antilles (Montserrat, Guadeloupe, Martinique), Mexico, Central America, northern South America to Brazil, Peru, tropical Africa. [L5: 87; AR: 360].

Diodia serrulata (P. Beauv.) G. Taylor in Exell, Cat. Vasc. Pl. S. Tomé 220. 1944.

Bas.: *Spermacoce serrulata* P. Beauv. (1805).

Syn.: *Diodia maritima* Thonn. (1827).

Trailing suffrutescent herb in Northern Coastal Lowlands, Northern Limestone Hills; 1–100 m; *fl* Oct–Dec, Mar, *fr* Sep–Dec, Mar–May.

Greater Antilles (Cuba, Hispaniola), Virgin Islands (St. Thomas), Atlantic and Caribbean coasts of Central America, northern South America (Colombia), western tropical Africa. [L5: 87].

Erithalis fruticosa L., Syst. Nat. (ed. 10) 2:930. 1759.

Syn.: *Erithalis revoluta* Urb. (1903).

Shrub/small tree in Northern Coastal Lowlands, Northern Limestone Hills, Río Abajo Forest Reserve, Eastern Coast, Southern Coastal Lowlands, Southern Coastal Limestone, Guánica Forest Reserve, Southwestern Coastal Lowlands, Western Coast, Vieques, Culebra, Mona; 1–300 m; *fl* & *fr* throughout the year.

West Indies, southeastern United States (Florida), southern Mexico (Yucatan), northern Central America (Belize). [L5: 89].

Jayajabico, Manglillo, Tea.

Erithalis odorifera Jacq., Select. Stirp. Amer. Hist. 72, *t. 173. f. 23*. 1763.

Shrub/small tree in Northern Coastal Lowlands (Río Grande), Río Abajo Forest Reserve, Southwestern Coastal Lowlands (Cabo Rojo), Vieques; 1–360 m; *fl* & *fr* Jun–Jul, Nov.

West Indies. [L5: 90].

Notes: This taxon is difficult to distinguish from *Erithalis fruticosa* especially in exsiccatae [see Negrón-Ortiz (2005)]; accordingly, its distribution and range of flowering and fruiting may be more extensive than given here.

Jayajabico, Manglillo, Tea.

Ernodea littoralis Sw., Prodr. 29. 1788.

Shrub in Northern Coastal Lowlands, Northern Limestone Hills (Arecibo), Southern Coastal Lowlands (Salinas), Southern Coastal Limestone, Susúa Forest Reserve, Maricao Forest Reserve, Guánica Forest Reserve, Southwestern Coastal Lowlands, Vieques, Mona, Desecheo; 1–450 m; *fl* Sep–Jun, *fr* throughout the year.

West Indies, southeastern United States (Florida), Mexico, northern Central America. [L5: 91].

Exostema caribaeum (Jacq.) Roem. & Schult., Syst. Veg. 5:18. 1819.
Bas.: *Cinchona caribaea* Jacq. (1760).

Tree in Northern Limestone Hills, Eastern Coast (Ceiba), Southeastern Foothills, Southern Coastal Lowlands (Caja de Muertos), Southern Coastal Limestone, Susúa Forest Reserve, Guánica Forest Reserve, Southwestern Coastal Lowlands (Cabo Rojo), Sierra Bermeja, Western Coast (Rincón), Vieques, Culebra, Mona; 5–600 m; *fl* May–Feb, *fr* throughout the year.

West Indies, Mexico, Central America (except Panama). [L5: 92].

Albarillo, Cuero de sapo, Guacarán, Lechecillo, Palo de jazmín, Palo de quina, Quina, Teílla.

Exostema ellipticum Griseb., Pl. Wright. 2:504. 1862 (preprint of Mem. Amer. Acad. Arts, ser. 2 8(2):504. 1863).

Tree in Maricao Forest Reserve; 465–750 m; *fl* Mar–Jul, *fr* Feb–Mar, Jun–Aug.

Greater Antilles (Cuba, Hispaniola). [L5: 94].

Plateado.

Faramea occidentalis (L.) A. Rich., Mém. Soc. Hist. Nat. Paris 5:176. 1834.
Bas.: *Ixora occidentalis* L. (1759).

Shrub in Northern Coastal Lowlands, Northern Limestone Hills, Río Abajo Forest Reserve, Sierra de Luquillo, Las Tetas de Cayey, Eastern Cordillera (Aibonito, Cañón San Cristóbal), Southern Coastal Lowlands (Patillas), Sierra Bermeja, Vieques; 1–825 m; *fl* Apr–Nov, *fr* throughout the year.

Antilles, Trinidad and Tobago, Mexico, Central America, northern South America to Bolivia. [L5: 98].

Notes: The above literature citation was preprinted in 1830, as Mém. Rubiac. 96; however, since according to TL-2 “Richard’s preprints were prepared in very limited numbers for a few friends,” this preprint does not, according to the ICBN (Vienna Code 29.1), constitute publication.

Café cimarrón, Cafeíllo, Guacarán, Palo de toro.

Genipa americana L., Syst. Nat. (ed. 10) 2:931. 1759.

Tree in Northern Coastal Lowlands, Northern Limestone Hills (San Sebastián), Río Abajo Forest Reserve, Sierra de Luquillo, Southeastern Foothills, Eastern Cordillera (Cerro La Peña), Vieques; 1–425 m; *fl* Apr–Jul, Oct–Jan, *fr* Jul–Feb.

Greater Antilles (Cuba, Hispaniola), Virgin Islands, Lesser Antilles, Trinidad, Mexico, Central America, South America. [L5: 91].

Jagua.

Geophila repens (L.) I.M. Johnst., Sargentia 8:281. 1949.
Bas.: *Rondeletia repens* L. (1759).
Syn.: *Geophila herbacea* (Jacq.) K. Schum. (1891).

Low creeping herb in Northern Coastal Lowlands (Dorado, Vega Alta), Northern Limestone Hills, Western Coast (Mayagüez); 5–340 m; *fl* Jul, Oct–Dec, *fr* Mar, Aug–Dec.

Pantropical. [L5: 99].

Yerba de guabá.

Gonzalagunia spicata (Lam.) M. Gómez, Dicc. Bot. Nombres Vulg. Cub. Puerto-Riq. 106. 1889.
Bas.: *Lygistum spicatum* Lam. (1792).

Shrub in Northern Coastal Lowlands, Northern Limestone Hills, Río Abajo Forest Reserve, Sierra de Luquillo, Eastern Coast, Caguas Valley, Sierra de Cayey, Las Tetas de Cayey, Southeastern Foothills (Arroyo), Eastern and Central Cordillera, Maricao Forest Reserve, Cerro Las Mesas, Western Coast (Mayagüez); 1–1000 m; *fl* & *fr* throughout the year.

Greater Antilles (Hispaniola), Virgin Islands, Lesser Antilles, Trinidad and Tobago, northern, northwestern South America and Brazil. [L5: 100].

Notes: Britton & Wilson (1925) misapplied the name *Duggenia hirsuta* (Jacq.) Britton to this taxon; Howard (1989), Acevedo-Rodríguez (1996), Liogier (1997) and Liogier & Martorell (2000) misapplied the name *Gonzalagunia hirsuta* (Jacq.) K. Schum. to it. According to Charlotte Taylor (pers. comm.), the type of the basionym *Justicia hirsuta* Jacq. is a mixed collection and the part first assigned (by Urban) to that name actually belongs in the Acanthaceae, as do all the names based upon it.

Mata de mariposa, Palo pelado, Rabo de ratón, Yerba pelada.

Guettarda elliptica Sw., Prodr. 59. 1788.

Shrub in Southern Coastal Lowlands (Caja de Muertos), Southern Coastal Limestone, Guánica Forest Reserve, Southwestern Coastal Lowlands (Cabo Rojo), Sierra Bermeja, Vieques, Mona, Desecheo; 5–150 m; *fl* Apr–Oct, *fr* Jun–Feb.

Bahamas, Greater Antilles, Virgin Islands, southeastern United States (Florida), Mexico, Central America, northern South America (Venezuela, Colombia). [L5: 103].

Cucubano liso.

Guettarda krugii Urb., Symb. Antill. 1(3):431. 1899.

Shrub/small tree in Southern Coastal Lowlands (Guayanilla), Southern Coastal Limestone, Guánica

Forest Reserve, Southwestern Coastal Lowlands (Cabo Rojo), Mona; 3–150 m; *fl* May–Jan, *fr* Jun–Dec.
Bahamas, Turks and Caicos. [L5: 103].
Cucubano.

Guettarda odorata (Jacq.) Lam., Tabl. Encycl. 2:219. 1819.
Bas.: *Laugieria odorata* Jacq. (1760).
Syn.: *Guettarda parviflora* Vahl (1798).
Shrub/small tree in Northern Coastal Lowlands (islands adjacent to Fajardo), Eastern Coast, Southern Coastal Lowlands (Patillas), Sierra Bermeja, Vieques, Culebra; 50–300 m; *fl* May–Nov, *fr* Sep–Dec.
Greater Antilles (Cuba), Virgin Islands, Lesser Antilles, Tobago, northern South America (Venezuela). [L5: 104].
Cucubano de Vieques.

Guettarda ovalifolia Urb., Symb. Antill. 1(3):432. 1899.
Tree in Northern Limestone Hills, Río Abajo Forest Reserve, Sierra de Cayey, Las Tetas de Cayey, Eastern Cordillera (La Torrecilla), Central Cordillera, Susúa Forest Reserve, Maricao Forest Reserve; 50–1100 m; *fl* throughout the year, *fr* Sep–Mar.
Greater Antilles (Hispaniola). [L5: 104].
Cucubano.

Guettarda pungens Urb., Symb. Antill. 1(3):434. 1899.
Shrub/small tree in Northern Limestone Hills (Arecibo), Río Abajo Forest Reserve, Susúa Forest Reserve, Maricao Forest Reserve; 150–900 m; *fl* Nov–Jan, May–Jul, *fr* Aug–May.
Greater Antilles (Hispaniola). [L5: 105].
Cerote, Roseta.

Guettarda scabra (L.) Vent., Choix Pl. *t. 1*. 1803.
Bas.: *Matthiola scabra* L. (1753).
Tree in Northern Coastal Lowlands, Northern Limestone Hills, Río Abajo Forest Reserve, Eastern Coast (Punta Guayanés), Sierra de Luquillo, Las Tetas de Cayey, Southeastern Foothills (Coamo), Southern Coastal Lowlands (Patillas), Susúa Forest Reserve, Western Cordillera, Guánica Forest Reserve, Southwestern Coastal Lowlands (Cabo Rojo), Sierra Bermeja, Cerro Las Mesas, Western Coast (Rincón), Vieques; 1–900 m; *fl* & *fr* throughout the year.
West Indies, Trinidad, southeastern United States (Florida). [L5: 106].
Notes: Britton & Wilson (1925) used *Guettarda scabra* (L.) Lam., a *comb. superfl.* for the name of this taxon.
Cucubano, Palo de cucubano, Palo de dajao, Serrazuela.

Guettarda valenzuelana A. Rich. in Sagra, Hist. Fis. Cuba, Bot. 11:20. 1850.
Syn.: *Guettarda laevis* Urb. (1899).
Tree in Sierra de Luquillo, Sierra de Cayey (San Lorenzo), Central Cordillera, Maricao Forest Reserve; 275–700 m; *fl* Jun–Nov, *fr* Jun–Jan.
Greater Antilles (Cuba, Hispaniola). [L5: 106].
Cucubano de monte.

Hamelia axillaris Sw., Prodr. 46. 1788.
Shrub in Northern Coastal Lowlands (Dorado), Northern Limestone Hills, Río Abajo Forest Reserve, Sierra de Luquillo, Sierra de Cayey, Eastern Cordillera (Cañón San Cristóbal), Central Cordillera (Coamo), Quebrada de Los Cedros, Maricao Forest Reserve; 10–500 m; *fl* May–Nov, *fr* Jul–Feb.
Greater Antilles, Virgin Islands (St. Croix), Lesser Antilles (St. Kitts), Trinidad, southern Mexico, Central America, northwestern, west-central South America and Venezuela. [L5: 108].
Balsamillo, Bálsamo amarillo.

Hamelia patens Jacq., Enum. Syst. Pl. 16. 1760.
Syn.: *Hamelia erecta* Jacq. (1760).
Shrub in Northern Coastal Lowlands, Northern Limestone Hills, Río Abajo Forest Reserve, Sierra de Luquillo, Sierra de Cayey (San Lorenzo), Las Piedras Chiquitas, Southeastern Foothills, Eastern Cordillera (Corozal, Aibonito), Central Cordillera, Southern Coastal Limestone, Susúa Forest Reserve, Western Cordillera, Southwestern Coastal Lowlands, Western Coast; 1–750 m; *fl* & *fr* throughout the year.
West Indies, southeastern United States (Florida), Mexico, Central America, South America. [L5: 109].
Bálsamo, Bálsamo colorado, Doña Julia, Pata de pájaro.

Hillia parasitica Jacq., Enum. Syst. Pl. 18. 1760.
Shrub in Sierra de Luquillo, Sierra de Cayey, Las Tetas de Cayey, Eastern Cordillera (Cerro La Peña, La Torrecilla), Central Cordillera, Maricao Forest Reserve; 500–1325 m; *fl* throughout the year, *fr* Aug–Mar.
Antilles, Trinidad, northern South America to Brazil, Bolivia. [L5: 113; AR: 360].
Flor de sierra, Hilia, Tibey trepador.

Ixora ferrea (J.F. Gmel.) Benth., Linnaea 23(4):447. 1850.
Bas.: *Siderodendrum ferreum* J.F. Gmel. (1791).
Tree in Northern Coastal Lowlands (Río Grande), Northern Limestone Hills, Sierra de Luquillo, Sierra de Cayey, Las Tetas de Cayey, Eastern Cordillera

(Cañón San Cristóbal), Susúa Forest Reserve, Maricao Forest Reserve, Cerro Las Mesas; 1–825 m; *fl* throughout the year, *fr* Oct–Apr.

Greater Antilles (Cuba, Hispaniola), Virgin Islands, Lesser Antilles, northern South America (Venezuela). [L5: 116].

Notes: Britton & Wilson (1925), Liogier & Martorell (1982,2000) and Liogier(1997) used *Ixora ferrea* (Jacq.) Benth., a *nom. inval.*, for the name of this taxon.

Cafeíllo, Dajao, Palo de dajao, Palo de hierro.

Ixora finlaysoniana Wall. ex G. Don, Gen. Hist. 3:572. 1834.

Shrub/small tree planted and persistent in Northern Coastal Lowlands (Río Grande), Eastern Coast (Ceiba), Southern Coastal Lowlands (Ponce), Vieques; 1–125 m; *fl* Dec–Jan, Apr–Aug, *fr* (?).

Probably native to Thailand. [L5: 116].

Notes: Liogier & Martorell (1982, 2000) and Liogier (1997) misapplied the name *Ixora thwaitesii* Hook. f. to this taxon.

Bola de nieve, Nevado.

Lasianthus lanceolatus (Griseb.) Urb., Symb. Antill. 1(3):449. 1899.
Bas.: *Hoffmannia lanceolata* Griseb. (1862).
Syn.: *Lasianthus moralesii* (Griseb.) C. Wright (1869).

Shrub, sometimes scandent, in Sierra de Luquillo, Sierra de Cayey, Eastern Cordillera (La Torrecilla), Central Cordillera, Maricao Forest Reserve; 425–1300 m; *fl* Jul–Apr, *fr* Aug–May.

Greater Antilles. [L5: 118; AR: 362].

Aroma, Mata de peo.

Lucya tetrandra (L.) K. Schum. in Engler & Prantl, Nat. Planzenfam. 4(4):27. 1891.
Bas.: *Peplis tetrandra* L. (1760).
Syn.: *Clavenna tetrandra* (L.) Standl. (1918).

Low herb in Northern Limestone Hills, Río Abajo Forest Reserve, Eastern Cordillera (Corozal, Cañón San Cristóbal), Southern Coastal Limestone (Guayanilla); 50–500 m; *fl* Oct–Jul, *fr* Oct–Feb, May–Jun.

Greater Antilles. [L5: 119].

Machaonia portoricensis Baill., Bull. Mens. Soc. Linn. Paris 1:204. 1879.

Shrub in Southern Coastal Limestone, Susúa Forest Reserve, Guánica Forest Reserve, Southwestern Coastal Lowlands (Cabo Rojo), Sierra Bermeja; 5–300 m; *fl* throughout the year, *fr* Mar–Dec.

Endemic. [L5: 121].

Alfilerillo, Roseta.

Margaritopsis microdon (DC.) C.M. Taylor, Syst. & Geogr. Pl. 75(2):169. 2005.
Bas.: *Rondeletia microdon* DC. (1830).
Syn.: *Psychotria microdon* (DC.) Urb. (1928).
Psychotria pinularis Sessé & Moc. (1894).

Shrub/small tree in Northern Coastal Lowlands, Northern Limestone Hills (Quebradillas), Eastern Coast, Las Tetas de Cayey, Las Piedras Chiquitas, Southeastern Foothills, Southern Coastal Lowlands (Patillas), Guánica Forest Reserve, Southwestern Coastal Lowlands, Vieques, Mona; 5–585 m; *fl* Jun–Nov, Feb, *fr* Jul–Apr.

Greater Antilles (Cuba, Hispaniola), Virgin Islands, Lesser Antilles, Mexico, Central America, northwestern, west-central South America and Venezuela. [L5: 147; AR: 362].

Margaritopsis nutans (Sw.) C.M. Taylor, Syst. & Geogr. Pl. 75(2):169. 2005.
Bas.: *Psychotria nutans* Sw. (1788).

Shrub/small tree in Mona; 40–75 m; *fl* Jun–Aug (two specimens), *fr* Mar, Jun, Oct.

Greater Antilles (Cuba, Hispaniola). [L5: 148].

Cachimbo de Mona.

Mitracarpus hirtus (L.) DC., Prodr. 4:572. 1830.
Bas.: *Spermacoce hirta* L. (1762).
Syn.: *Mitracarpus villosus* (Sw.) Cham. & Schltdl. ex DC. (1830).

Suffrutescent herb in Northern Coastal Lowlands, Eastern Coast (Naguabo), Caguas Valley, Vieques; 1–125 m; *fl* & *fr* Oct–May.

Antilles, southeastern and south-central United States, Mexico, Central America, northern South America to Bolivia, Paraguay. [L5: 123].

Mitracarpus maxwelliae Britton & P. Wilson, Sci. Surv. Porto Rico & Virgin Islands 6(2):257. 1925.

Suffrutescent herb in Guánica Forest Reserve; 5–150 m; *fl* & *fr* Oct–Jun.

Endemic. [L5: 123].

Notes: This taxon is on the U.S. Fish & Wildlife Service list of threatened and endangered plants.

Mitracarpus polycladus Urb., Symb. Antill. 3(3):389. 1903.

Suffrutescent herb in Guánica Forest Reserve; 1–50 m; *fl* & *fr* Jul–Jan.

Virgin Islands (Anegada), Lesser Antilles (Saba). [L5: 124].

Notes: This taxon is on the U.S. Fish & Wildlife Service list of threatened and endangered plants.

Mitracarpus portoricensis (Urb.) Urb., Symb. Antill. 4(4):609. 1911.

Bas.: *Mitracarpus frigidus* (Willd. ex Roem. & Schult.) K. Schum. var. *portoricensis* Urb. (1903).

Suffrutescent herb in Northern Coastal Lowlands (Vega Alta, Laguna Tortuguero), Susúa Forest Reserve, Maricao Forest Reserve; 1–700 m; *fl* & *fr* throughout the year.

Endemic. [L5: 125].

Morinda citrifolia L., Sp. Pl. 1:176. 1753.

Shrub/small tree in Northern Coastal Lowlands, Eastern Coast (Punta Guayanés), Western Coast, Vieques, Culebra, Mona; 1–25 m; *fl* & *fr* throughout the year.

Native to tropical Asia, China, Japan, Australia; exact native range uncertain. [L5: 126].

Gardenia hedionda, Noni.

Neolamarckia cadamba (Roxb.) F. Bosser, Bull. Mus. Natl. Hist. Nat., B, Adansonia 6(3):247. 1985.

Bas.: *Nauclea cadamba* Roxb. (1824).

Syn.: *Anthocephalus cadamba* (Roxb.) Miq. (1856).

Tree planted and persistent in Río Abajo Forest Reserve, Sierra de Luquillo, Monte Guilarte Forest Reserve; 250–1000 m; *fl* Jun (two specimens), *fr* Nov–Dec.

Native to tropical Asia, China. [L5: 127].

Kadam.

Neolaugeria resinosa (Vahl) Nicolson, Brittonia 31(1):121. 1979.

Bas.: *Laugeria resinosa* Vahl (1797).

Syn.: *Terebraria resinosa* (Vahl) Sprague (1932).

Tree in Northern Coastal Lowlands (Dorado), Northern Limestone Hills, Río Abajo Forest Reserve, Sierra de Luquillo, Eastern Cordillera (Cerro La Peña), Susúa Forest Reserve, Western Cordillera; 25–800 m; *fl* Jul–Oct, *fr* Jun–Jan.

Bahamas, Greater Antilles (Hispaniola), Lesser Antilles, Trinidad. [L5: 129].

Aquilón, Colorado, Espejuelo.

Notopleura guadalupensis (DC.) C.M. Taylor, Ann. Missouri Bot. Gard. 88(3):512. 2001.

Bas.: *Loranthus guadalupensis* DC. (1830).

Syn.: *Psychotria guadalupensis* (DC.) R.A. Howard (1966).

Psychotria grosourdyana (Baill.) Urb. (1911).

Epiphytic herb in Sierra de Luquillo, Sierra de Cayey, Eastern Cordillera (La Torrecilla), Central Cordillera, Maricao Forest Reserve; 475–1325 m; *fl* & *fr* throughout the year.

Antilles, Trinidad, Mexico, Central America, northern South America (Venezuela, Colombia). [L5: 145].

Notes: Liogier & Martorell (1982, 2000) and Liogier (1997) used *Psychotria guadaloupensis* (DC.) R.A. Howard, an orthographic variant of the above synonym, for the name of this taxon.

Notopleura uliginosa (Sw.) Bremek., Recueil Trav. Bot. Néerl. 31:290. 1934.

Bas.: *Psychotria uliginosa* Sw. (1788).

Suffrutescent herb in Sierra de Luquillo, Sierra de Cayey, Eastern Cordillera (La Torrecilla), Central Cordillera, Maricao Forest Reserve; 100–1300 m; *fl* Sep–Jan, *fr* throughout the year.

Antilles, Trinidad, southern Mexico, Central America, northern South America. [L5: 150].

Tres cabezas.

Oldenlandia callitrichoides Griseb., Pl. Wright. 2:506. 1862 (preprint of Mem. Amer. Acad. Arts, n.s. 8(2):506. 1863).

Syn.: *Hedyotis callitrichoides* (Griseb.) W.H. Lewis (1961). *Oldenlandiopsis callitrichoides* (Griseb.) Terrell & W.H. Lewis (1990).

Herb in Northern Coastal Lowlands (Toa Alta), Northern Limestone Hills (San Sebastián), Sierra de Luquillo; 5–600 m; *fl* & *fr* Apr–Jun.

Bahamas, Greater Antilles (Cuba, Hispaniola), Lesser Antilles (St. Barts, Guadeloupe, Dominica), Mexico, Central America (Nicaragua, Panama). [L5: 130].

Oldenlandia corymbosa L., Sp. Pl. 1:119. 1753.

Syn.: *Hedyotis corymbosa* (L.) Lam. (1792).

Herb in Northern Coastal Lowlands, Eastern Coast (Punta Guayanés), Caguas Valley (Caguas), Southern Coastal Lowlands (Patillas), Western Coast (Hormigueros), Vieques, Mona; 1–100 m; *fl* & *fr* throughout the year.

Native to tropical Africa, India. [L5: 111].

Oldenlandia lancifolia (Schumach.) DC., Prodr. 4:425. 1830.

Bas.: *Hedyotis lancifolia* Schumach. (1827).

Syn.: *Hedyotis commutata* Schult. & Schult. f. (1827).

Herb in Northern Coastal Lowlands, Río Abajo Forest Reserve, Sierra de Luquillo, Eastern Coast, Caguas Valley, Sierra de Cayey, Western Coast; 1–600 m; *fl* & *fr* Sep–Jun.

Native to tropical Africa. [L5: 111].

Notes: Britton & Wilson (1925) misapplied the name *Oldenlandia herbacea* (L.) DC. to this taxon.

Oldenlandia uniflora L., Sp. Pl. 1:119. 1753.
Syn.: *Hedyotis uniflora* (L.) Lam. (1792).
Herb in Northern Coastal Lowlands (Vega Alta to Manatí); 1–5 m; *fl* & *fr* Mar–Apr, Sep–Oct.
Greater Antilles (Cuba, Jamaica), southeastern and south-central United States. [L5: 111].
Graciosa.

Palicourea alpina (Sw.) DC., Prodr. 4:528. 1830.
Bas.: *Psychotria alpina* Sw. (1788).
Syn.: *Palicourea brevithyrsa* Britton & Standl. (1925).
Shrub in Sierra de Luquillo, Central Cordillera; 750–1330 m; *fl* Dec–Sep, *fr* Jul–Mar.
Greater Antilles. [L5: 132].

Palicourea crocea (Sw.) Schult. in Roemer & Schultes, Syst. Veg. 5:193. 1819.
Bas.: *Psychotria crocea* Sw. (1788).
Shrub in Northern Coastal Lowlands, Northern Limestone Hills, Río Abajo Forest Reserve, Sierra de Luquillo, Eastern Coast, Caguas Valley (Gurabo), Sierra de Cayey, Southeastern Foothills (Arroyo), Eastern Cordillera, Southern Coastal Lowlands, Western Cordillera, Cerro Las Mesas, Western Coast (Mayagüez); 1–850 m; *fl* & *fr* throughout the year.
Antilles, Mexico, Central America, South America. [L5: 132].
Cachimbo.

Palicourea croceoides Ham., Prodr. Pl. Ind. Occid. 29. 1825.
Syn.: *Palicourea riparia* Benth. (1841). *Palicourea crocea* (Sw.) Schult. var. *riparia* (Benth.) Griseb. (1864).
Shrub in Northern Coastal Lowlands, Northern Limestone Hills, Sierra de Luquillo, Eastern Coast, Caguas Valley (Juncos), Sierra de Cayey, Eastern Cordillera (Cerro La Peña, Corozal), Western Cordillera, Cerro Las Mesas; 1–850 m; *fl* & *fr* throughout the year.
Greater Antilles (Hispaniola), Lesser Antilles, northern South America to Brazil, Bolivia. [L5: 134].
Cachimbo, Palo de cachimbo.

Palicourea guianensis Aubl., Hist. Pl. Guiane 1:173, *t. 66.* 1775.
Syn.: *Palicourea barbinervia* DC. (1830).
Shrub in Northern Limestone Hills (Florida, Isabela), Sierra de Cayey, Central Cordillera, Maricao Forest Reserve; 300–800 m; *fl* Oct, Jan–Apr, *fr* Aug, Nov–Jan.
Greater Antilles (Cuba, Hispaniola), Trinidad, Central America, northern South America to Brazil, Bolivia. [L5: 134].
Bálsamo real, Tafetán.

Phialanthus myrtilloides Griseb., Fl. Brit. W.I. 335. 1861.
Syn.: *Phialanthus grandifolius* Alain (1965).
Shrub/small tree in Northern Limestone Hills (Camuy, Isabela), Central Cordillera (Utuado), Susúa Forest Reserve, Maricao Forest Reserve; 150–875 m; *fl* throughout the year, *fr* Sep–Jun.
Bahamas, Greater Antilles (Cuba, Jamaica). [L5: 137].
Notes: I find that the purported differences between *Phialanthus myrtilloides* and *Phialanthus grandifolius* are not so discrete as to merit the recognition of distinct taxa.
Aquilón, Aquilón prieto.

Psychotria berteroana DC., Prodr. 4:515. 1830 "Berteriana."
Shrub/small tree in Río Abajo Forest Reserve, Sierra de Luquillo, Caguas Valley (Caguas), Sierra de Cayey, Central Cordillera, Maricao Forest Reserve; 200–1300 m; *fl* Dec–Aug, *fr* throughout the year.
Antilles, Trinidad, Mexico, Central America, northern South America to Bolivia. [L5: 141].
Cachimbo común, Cotorra.

Psychotria brachiata Sw., Prodr. 45. 1788.
Shrub in Northern Coastal Lowlands (Fajardo, Bayamón), Northern Limestone Hills (San Sebastián), Sierra de Luquillo, Eastern Coast, Caguas Valley (Gurabo), Sierra de Cayey, Central Cordillera, Western Coast (Mayagüez); 50–800 m; *fl* May–Nov, *fr* Aug–Mar.
Greater Antilles, Trinidad, Central America, northwestern, west-central South America and Venezuela. [L5: 141].
Palo de cachimbo.

Psychotria brownei Spreng., Syst. Veg. 1:742. 1824.
Shrub/small tree in Northern Coastal Lowlands, Southeastern Foothills, Sierra Bermeja, Vieques, Culebra; 1–300 m; *fl* Aug, Nov, Mar, *fr* Aug–Mar.
Greater Antilles (Cuba, Hispaniola), Virgin Islands. [L5: 143].

Psychotria deflexa DC., Prodr. 4:510. 1830.
Shrub in Guajataca Forest Reserve, Sierra de Luquillo, Sierra de Cayey, Central Cordillera (Adjuntas), Susúa Forest Reserve, Maricao Forest Reserve; 220–950 m; *fl* Apr, Jul–Nov, *fr* Aug–Nov, Mar.
Greater Antilles (Cuba, Hispaniola), Trinidad, Mexico, Central America, northern South America to Bolivia, Paraguay. [L5: 143].

Notes: Britton & Wilson (1925) misapplied the name *Psychotria patens* Sw. to this taxon.

Psychotria domingensis Jacq., Enum. Syst. Pl. 16. 1760.
Syn.: *Palicourea domingensis* (Jacq.) DC. (1830).
Shrub in Northern Coastal Lowlands, Northern Limestone Hills, Río Abajo Forest Reserve, Southeastern Foothills (Salinas), Eastern Cordillera (Cayey), Quebrada de Los Cedros, Vieques; 5–375 m; *fl* & *fr* Jun–Nov.
Greater Antilles, Lesser Antilles (Antigua, St. Eustatius, Guadeloupe), Mexico, Central America, northwestern South America. [L5: 144].

Psychotria grandis Sw., Prodr. 43. 1788.
Shrub/small tree in Northern Limestone Hills, Río Abajo Forest Reserve, Sierra de Luquillo, Central Cordillera, Maricao Forest Reserve, Western Coast (Mayagüez); 50–1300 m; *fl* May–Jun, *fr* Oct–Mar.
Greater Antilles, Central America, northwestern South America and Venezuela. [L5: 144].
Cachimbo grande, Palo moro.

Psychotria hoffmannseggiana (Schult.) Müll. Arg. var. **tribracteata** (C. Wright ex Griseb.) Steyerm., Mem. New York Bot. Gard. 23:608. 1972.
Bas.: *Psychotria tribracteata* C. Wright ex Griseb. (1866).
Shrub in Maricao Forest Reserve; 800–820 m; *fl* May (one specimen), *fr* Feb–May (two specimens). [*Cedeño-Maldonado 426, 494, 564* (MAPR)].
Greater Antilles (Cuba, Hispaniola). [L5: 145].
Notes: Britton & Wilson (1925) misapplied the name *Psychotria involucrata* Sw. to this taxon.
Cachimbo verde.

Psychotria ligustrifolia (Northr.) Millsp., Publ. Field Columb. Mus., Bot. Ser. 2:172. 1906.
Bas.: *Myrstiphyllum ligustrifolium* Northr. (1902).
Shrub in Northern Coastal Lowlands (San Juan), Cerro Las Mesas, Western Coast; 3–300 m; *fl* Nov–Jul, *fr* Mar–Jul, Oct–Nov.
Bahamas, Greater Antilles (Cuba, Hispaniola), southeastern United States (Florida). [L5: 146].

Psychotria maleolens Urb., Symb. Antill. 1(3):444. 1899.
Shrub in Sierra de Luquillo, Sierra de Cayey, Maricao Forest Reserve; 350–1000 m; *fl* Nov–Aug, *fr* throughout the year.
Endemic. [L5: 146].
Aroma, Cachimbo de gato, Mata de peo.

Psychotria maricaensis Urb., Symb. Antill. 1(3): 447. 1899.
Shrub in Río Abajo Forest Reserve, Central Cordillera (Adjuntas), Susúa Forest Reserve, Western Cordillera, Cerro Las Mesas; 150–1150 m; *fl* Nov–Jan, Mar–Aug, *fr* May, Aug–Jan.
Endemic. [L5: 147].
Cachimbo de Maricao.

Psychotria nervosa Sw., Prodr. 43. 1788.
Syn.: *Psychotria undata* Jacq. (1798).
Shrub in Northern Coastal Lowlands, Northern Limestone Hills, Río Abajo Forest Reserve, Eastern Coast (Yabucoa), Las Tetas de Cayey, Las Piedras Chiquitas, Eastern Cordillera (Cidra, Cañón San Cristóbal), Cerro de Las Cuevas, Quebrada de Los Cedros, Susúa Forest Reserve, Guánica Forest Reserve, Sierra Bermeja, Western Coast (Rincón), Vieques, Mona; 1–825 m; *fl* Feb–Oct, *fr* throughout the year.
West Indies, Mexico, Central America, northwestern South America. [L5: 147].
Bálsamo, Palo moro.

Psychotria pubescens Sw., Prodr. 44. 1788.
Shrub in Northern Limestone Hills, Río Abajo Forest Reserve, Eastern Coast (Punta Guayanés), Central Cordillera (Ponce), Quebrada de Los Cedros, Susúa Forest Reserve, Guánica Forest Reserve; 5–350 m; *fl* & *fr* throughout the year.
Bahamas, Greater Antilles, Lesser Antilles (St. Kitts), Mexico, Central America, northern South America (Venezuela). [L5: 148].
Cachimbo.

Psychotria revoluta DC., Prodr. 4:517. 1830.
Shrub in Southwestern Coastal Lowlands (Cabo Rojo); 20–85 m; *fl* Sep (one specimen), *fr* Sep, Dec. [*Trejo 2770, 2771* (UPR)].
Greater Antilles (Cuba, Hispaniola). [L5: 149].

Psychotria tenuifolia Sw., Prodr. 43. 1788.
Shrub in Sierra de Cayey (San Lorenzo), Las Tetas de Cayey, Eastern Cordillera (Cayey, La Torrecilla), Central Cordillera (Yauco), Maricao Forest Reserve; 600–975 m; *fl* Apr–Jul, *fr* Oct–Jan.
Antilles, Mexico, Central America, northwestern and north-central South America. [L5: 149].

Psychotria sp.
Shrub in Sierra de Cayey; 625–875 m; *fl* Jul, Dec (two specimens), *fr* Nov–Dec (one specimen). [*Taylor 6345* (UPRRP); *Axelrod 6633, 13716* (UPRRP)].

Notes: Specimens of this taxon have previously been misidentified as *Psychotria hoffmannseggiana* var. *tribracteata*.

Randia aculeata L., Sp. Pl. 2:1192. 1753.
Syn.: *Randia mitis* L. (1753).
Shrub/small tree in Northern Coastal Lowlands, Northern Limestone Hills, Río Abajo Forest Reserve, Eastern Coast, Las Piedras Chiquitas, Southeastern Foothills, Eastern Cordillera (Comerío), Quebrada de Los Cedros, Susúa Forest Reserve, Maricao Forest Reserve, Guánica Forest Reserve, Southwestern Coastal Lowlands, Sierra Bermeja, Vieques, Culebra, Mona; 1–550 m; *fl* Feb–Nov, *fr* throughout the year.
West Indies, Mexico, Central America, northwestern, west-central South America and Venezuela. [L5: 152].
Árbol de Navidad, Cabai nacte, Cabai nagte, Cambrón, Escambrón, Palo de cotorra, Palo de Navidad, Sota caballo, Tintillo.

Randia portoricensis (Urb.) Standl., Contr. U.S. Natl. Herb. 20:201. 1919.
Bas.: *Basanacantha portoricensis* Urb. (1908).
Shrub in Southern Coastal Limestone (Peñuelas, Yauco), Guaníca Forest Reserve, Southwestern Coastal Lowlands (Cabo Rojo); 10–140 m; *fl* Sep–Oct, Jan, *fr* Sep–Oct, Jan.
Endemic. [L5: 154].

Richardia scabra L., Sp. Pl. 1:330. 1753.
Herb in Northern Coastal Lowlands, Eastern Coast (Humacao), Eastern Cordillera (Corozal), Southern Coastal Lowlands (Arroyo); 1–400 m; *fl* & *fr* throughout the year.
Greater Antilles (Cuba), southeastern and south-central United States, Mexico, Central America, northwestern, west-central South America and Brazil, Paraguay. [L5: 155].

Rondeletia inermis (Spreng.) Krug & Urb. in Urban, Symb. Antill. 1(3):416. 1899.
Bas.: *Catesbaea inermis* Spreng. (1824).
Shrub/small tree in Northern Coastal Lowlands, Northern Limestone Hills, Río Abajo Forest Reserve, Las Piedras Chiquitas, Southeastern Foothills, Eastern Cordillera (Cañón San Cristóbal), Cerro de Las Cuevas, Southern Coastal Lowlands, Southern Coastal Limestone, Susúa Forest Reserve, Western Cordillera, Guánica Forest Reserve, Southwestern Coastal Lowlands, Sierra Bermeja, Cerro Las Mesas; 1–650 m; *fl* & *fr* throughout the year.
Endemic. [L5: 157].
Cordobancillo.

Rondeletia pilosa Sw., Prodr. 41. 1788.
Shrub/small tree in Northern Coastal Lowlands (Fajardo to Río Grande), Eastern Coast (Ceiba); 1–40 m; *fl* Aug–Nov, Feb, *fr* Aug–Oct, Feb.
Virgin Islands. [L5: 157].
Cachimbo peludo, Cordobancillo peludo, Lechecillo.

Rondeletia portoricensis Krug & Urb. in Urban, Symb. Antill. 1(3):414. 1899.
Shrub/small tree in Sierra de Luquillo, Eastern Cordillera (Cayey, La Torrecilla), Central Cordillera, Maricao Forest Reserve; 400–1300 m; *fl* Apr–Aug, *fr* Aug–Apr.
Endemic. [L5: 158].
Juan Tomás.

Rosenbergiodendron formosum (Jacq.) Fagerl., Svensk Bot. Tidskr. 42:152. 1948.
Bas.: *Mussaenda formosa* Jacq. (1760).
Syn.: *Randia formosa* (Jacq.) K. Schum. (1889).
Shrub planted and persistent in Northern Coastal Lowlands (Carolina), Western Coast (Mayagüez); 25–125 m; *fl* May–Jun, Sep, *fr* Mar, Jun.
Native to southern Central America (Panama), northern South America to Ecuador. [L5: 152].

Sabicea villosa Schult. in Roemer & Schultes, Syst. Veg. 5:265. 1819.
Syn.: *Sabicea hirsuta* Kunth (1820).
Vine in Sierra de Luquillo, Sierra de Cayey, Central Cordillera (Adjuntas), Western Cordillera (San Germán); 100–740 m; *fl* Sep–Jan, Apr, *fr* Sep–Apr.
Trinidad, Mexico, Central America, northern South America to Bolivia. [L5: 160; AR: 364].

Schradera exotica (J.F. Gmel.) Standl., Publ. Field Mus. Nat. Hist., Bot. Ser. 4(8):286. 1929.
Bas.: *Urceolaria exotica* J.F. Gmel. (1791).
Syn.: *Schradera vahlii* Steyerm. (1964).
Woody vine in Northern Limestone Hills, Sierra de Luquillo, Sierra de Cayey, Central Cordillera, Maricao Forest Reserve; 350–1300 m; *fl* Nov–Apr, *fr* Jun–Sep.
Greater Antilles (Hispaniola), Lesser Antilles. [L5: 161; AR: 366].

Scolosanthus grandifolius Krug & Urb., Ber. Deutsch. Bot. Ges. 15(heft 4):267, *t. 9*, *ff. 23, 24*. 1897.
Shrub in Maricao Forest Reserve; 700–850 m; *fl* Apr (one specimen), *fr* Apr (one specimen).
Greater Antilles (Hispaniola). [L5: 163].
Espuela de galán, Picha de gato.

Scolosanthus portoricensis Borhidi, Acta Bot. Hung. 29(1–4):202. 1983.

Shrub in Susúa Forest Reserve, Maricao Forest Reserve; 200–650 m; *fl* Apr–Aug, Nov, *fr* (?).

Endemic. [L5: 163].

Notes: Liogier & Martorell (1982) misapplied the name *Scolosanthus densiflorus* Urb. to this taxon.

Alfiler, Tintillo.

Scolosanthus versicolor Vahl, Eclog. Amer. 1:11. 1797.

Shrub in Northern Coastal Lowlands (islands adjacent to Fajardo), Eastern Coast (Punta Guayanés), Southeastern Foothills (Guayama), Southern Coastal Limestone, Susúa Forest Reserve, Guánica Forest Reserve, Sierra Bermeja; 50–300 m; *fl* Nov–Dec, May, *fr* May–Jun, Oct–Dec.

Virgin Islands. [L5: 165].

Alfiler, Tintillo.

Spermacoce confusa Rendle, J. Bot. 74:12, *ff. d-f.* 1936.

Herb in Northern Coastal Lowlands, Northern Limestone Hills, Río Abajo Forest Reserve, Eastern Coast (Ceiba), Las Piedras Chiquitas, Southeastern Foothills, Eastern Cordillera (Aibonito), Cerro de Las Cuevas, Southern Coastal Lowlands, Southern Coastal Limestone, Susúa Forest Reserve, Guánica Forest Reserve, Southwestern Coastal Lowlands, Sierra Bermeja, Western Coast (Rincón), Vieques, Mona; 1–600 m; *fl* & *fr* throughout the year.

West Indies, Mexico, Central America, northwestern, west-central South America and Venezuela. [L5: 168].

Yerba de hierro.

Spermacoce densiflora (DC.) Alain, Phytologia 54(2):113. 1983.

Bas.: *Borreria densiflora* DC. (1830).

Suffrutescent herb in Northern Coastal Lowlands (Loíza), Vieques; 1–10 m; *fl* & *fr* Oct–Nov, Feb.

Greater Antilles (Jamaica), Lesser Antilles, Central America, South America. [L5: 168].

Spermacoce ocymifolia Willd. ex Roem. & Schult., Syst. Veg. 3:530. 1818.

Syn.: *Hemidiodia ocymifolia* (Willd. ex Roem. & Schult.) K. Schum. (1888). *Diodia ocymifolia* (Willd. ex Roem. & Schult.) Bremek. (1934) "ocimifolia."

Suffrutescent herb in Northern Coastal Lowlands, Northern Limestone Hills, Río Abajo Forest Reserve, Sierra de Luquillo, Eastern Coast, Sierra de Cayey, Eastern Cordillera (Barranquitas including La Torrecilla), Central Cordillera, Western Cordillera, Sierra Bermeja; 1–900 m; *fl* & *fr* throughout the year.

Antilles, Mexico, Central America, northern South America to Brazil, Bolivia. [L5: 85].

Poaya.

Spermacoce ovalifolia (M. Martens & Galeotti) Hemsl., Biol. Cent.-Amer., Bot. 2(7):59. 1881.

Bas.: *Borreria ovalifolia* M. Martens & Galeotti (1844).

Herb in Northern Coastal Lowlands, Northern Limestone Hills (Camuy), Sierra de Luquillo, Eastern Coast, Caguas Valley, Sierra de Cayey, Southeastern Foothills (Cayey), Eastern Cordillera (Comerío), Vieques; 1–550 m; *fl* & *fr* throughout the year.

Bahamas, Greater Antilles, Central America (Nicaragua, Costa Rica), northwestern, west-central South America and Venezuela. [L5: 169].

Notes: Britton & Wilson misapplied the name *Borreria ocymoides* (Burm. f.) DC., first (1925) to this taxon and later (1926) to *Borreria prostrata*, a synonym of *Spermacoce prostrata*; Liogier & Martorell (2000) misapplied *Spermacoce eryngioides* (Cham. & Schltdl.) Kuntze to this taxon.

Botoncillo.

Spermacoce prostrata Aubl., Hist. Pl. Guiane 1:58, *t. 20, f.* 3. 1775.

Syn.: *Borreria prostrata* (Aubl.) Miq. (1851).

Herb in Northern Coastal Lowlands, Northern Limestone Hills (Moca), Sierra de Luquillo, Eastern Coast, Caguas Valley (Juncos), Sierra de Cayey, Eastern Cordillera (Corozal), Western Coast (Mayagüez), Mona; 3–900 m; *fl* & *fr* throughout the year.

West Indies, southeastern United States (Florida), Central America, northern South America to Brazil, Bolivia. [L5: 169].

Notes: Britton & Wilson (1926) misapplied the name *Borreria ocymoides* (Burm. f.) DC. to this taxon; Liogier & Martorell (1982) misapplied *Spermacoce repens* (DC.) Fosberg & Powell to it and Liogier (1997) wrongly placed that name in synonymy with it.

Botoncillo.

Spermacoce remota Lam., Tabl. Encycl. 1:273. 1792.

Syn.: *Spermacoce assurgens* Ruiz & Pav. (1798). *Borreria assurgens* (Ruiz & Pav.) Griseb. (1874) "Borrera."

Herb in Northern Coastal Lowlands, Northern Limestone Hills, Río Abajo Forest Reserve, Sierra de Luquillo, Eastern Coast, Caguas Valley (Las Piedras), Sierra de Cayey, Las Tetas de Cayey, Southeastern Foothills (Guayama), Eastern and Central Cordillera, Southern Coastal Lowlands (Patillas), Western Cordillera, Guánica Forest Reserve, Southwestern Coastal Lowlands (Cabo Rojo), Western Coast

(Aguada), Vieques, Culebra; 1–1300 m; *fl* & *fr* throughout the year.

West Indies, southeastern United States (Georgia, Florida), Mexico, Central America, South America. [L5: 166, 170].

Notes: Britton & Wilson (1925) misapplied the name *Borreria laevis* (Lam.) Griseb., Liogier & Martorell (1982) misapplied *Borreria brachysepala* Urb., and Liogier (1997) and Liogier & Martorell (2000) misapplied *Spermacoce sintenisii* (Urb.) Alain to this taxon.

Yerba de garro.

Spermacoce verticillata L., Sp. Pl. 1:102. 1753.
Syn.: *Borreria verticillata* (L.) G. Mey. (1818).

Suffrutescent herb in Northern Coastal Lowlands, Northern Limestone Hills, Sierra de Luquillo, Eastern Coast, Sierra de Cayey, Las Tetas de Cayey, Las Piedras Chiquitas, Southeastern Foothills (Cayey), Eastern and Central Cordillera, Susúa Forest Reserve, Maricao Forest Reserve, Southwestern Coastal Lowlands (Lajas), Sierra Bermeja, Western Coast, Vieques; 1–1175 m; *fl* & *fr* throughout the year.

West Indies, southern United States (Florida, Texas), Mexico, Central America, South America. [L5: 170].

Botón blanco.

Stenostomum acutatum DC., Prodr. 4:460. 1830.
Syn.: *Antirhea acutata* (DC.) Urb. (1899).

Tree in Southern Coastal Lowlands, Southern Coastal Limestone, Guánica Forest Reserve, Southwestern Coastal Lowlands, Vieques, Mona; 1–100 m; *fl* & *fr* throughout the year.

Greater Antilles (Hispaniola), Lesser Antilles, Southern Caribbean Islands. [L5: 70].

Boje, Manglillo, Quina.

Stenostomum coriaceum (Vahl) Griseb., Fl. Brit. W.I. 334. 1861.
Bas.: *Laugeria coriacea* Vahl (1797).
Syn.: *Antirhea coriacea* (Vahl) Urb. (1899).

Shrub/small tree in Northern Limestone Hills; 25–100 m; *fl* May–Jun (two specimens), *fr* Jul (one specimen). [*Gentry 66112* (UPRRP)].

Greater Antilles (Jamaica), Lesser Antilles. [L5: 70].

Boje, Quina.

Stenostomum lucidum (Sw.) C.F. Gaertn., Suppl. Carp. 1:69. 1806.
Bas.: *Laugeria lucida* Sw. (1788).
Syn.: *Antirhea lucida* (Sw.) Hook. f. (1873).

Tree in Northern Limestone Hills, Southern Coastal Limestone (Peñuelas), Guánica Forest Reserve, Southwestern Coastal Lowlands (Isla Magueyes); 20–150 m; *fl* May–Dec, *fr* Jun–Jan.

Bahamas, Greater Antilles, Virgin Islands (St. Croix), southern Mexico (Yucatan), northern Central America (Guatemala). [L5: 71].

Palo llorón.

Stenostomum obtusifolium (Urb.) Britton & P. Wilson, Sci. Surv. Porto Rico & Virgin Islands 6(2):237. 1925.
Bas.: *Antirhea obtusifolia* Urb. (1899).

Tree in Sierra de Luquillo, Sierra de Cayey, Maricao Forest Reserve; 350–850 m; *fl* May–Jul, Oct, *fr* Jan–Jun.

Endemic. [L5: 71].

Cucubano, Quina, Quina del monte, Quina roja, Tortuguillo.

Stenostomum sintenisii (Urb.) Britton & P. Wilson, Sci. Surv. Porto Rico & Virgin Islands 6(2):238. 1925.
Bas.: *Antirhea sintenisii* Urb. (1899).
Syn.: *Stenostomum portoricense* Britton & P. Wilson (1930). *Antirhea portoricensis* (Britton & P. Wilson) Standl. (1934).

Tree in Northern Limestone Hills, Río Abajo Forest Reserve; 25–325 m; *fl* May–Dec, *fr* Jul–Sep, Jan–Feb.

Greater Antilles (Hispaniola). [L5: 73].

Notes: The type specimens and other exsiccatae of *Stenostomum portoricense* and *Stenostomum sintenisi* do not present to my eye such differences as to allow a clear-cut separation of two taxa.

Quina.

Strumpfia maritima Jacq., Enum. Syst. Pl. 28. 1760.

Shrub in Northern Coastal Lowlands (islands adjacent to Fajardo), Southern Coastal Lowlands, Guánica Forest Reserve, Southwestern Coastal Lowlands (Guánica), Vieques, Mona; 1–50 m; *fl* & *fr* throughout the year.

West Indies, southeastern United States (Florida), southern Mexico (Yucatan). [L5: 172].

Lirio.

Vangueria madagascariensis J.F. Gmel., Syst. Nat. 2(1):367. 1791.

Tree planted and persistent in Northern Coastal Lowlands (Carolina, Dorado), Sierra de Cayey (Patillas), Southeastern Foothills, Central Cordillera (Ciales, Orocovis); 10–600 m; *fl* (?), *fr* Sep–Nov, Feb–Apr.

Native to tropical Africa. [L5: 174].

Tamarindo americano, Tamarindo forastero.

TAXA NOT RECENTLY COLLECTED

Spermacoce tenuior L., Sp. Pl. 1:102. 1753.
Syn.: *Spermacoce riparia* Cham. & Schltdl. (1828).

Herb collected near Cabo Rojo on 22 Jan 1885 [*Sintenis 890* (GH)].

Bahamas, Greater Antilles, southeastern and southern United States, Mexico, Central America, South America (scattered). [L5: 170].

Notes: Britton & Wilson (1925) misapplied the name *Spermacoce confusa* to this taxon.

TAXA CITED BY PREVIOUS AUTHORS, BUT NOT INCLUDED

CITED DUE TO MISIDENTIFICATION

Exostema sanctae-luciae (Kentish) Britten.
Tree cited by Britton & Wilson (1925) as occurring in a forest on Monte Alegrillo, near Maricao, collected only by Sintenis; subsequently cited by Liogier & Martorell (1982), who claimed it as also occurring in Bayamón and the Luquillo Mountains. The specimens upon which these citations were based were later referred to *Exostema ellipticum* Griseb. [B&W6: 228].

Randia parvifolia Lam.
Shrub cited by Liogier & Martorell (1982, 2000) and Liogier (1997) as occurring on dry limestone, Mona. According to Gary Breckon (pers. comm.), all specimens identified as this taxon are referable to *Randia aculeata*. [L5: 154].

Sabicea cinerea Aubl.
Vine cited by Liogier & Martorell (1982) as having been collected by Woodbury in Dorado Forest; subsequently cited by Liogier (1997) and Liogier & Martorell (2000), who claimed it as also occurring in Las Piedras Chiquitas and Vieques. Specimens previously identified as this taxon, all sterile, are referable to *Tetrapterys inaequalis* in the Malpighiaceae. [L5: 159; A-R 364].

ONLY UNDER CULTIVATION

Calycophyllum candidissimum (Vahl) DC.
Tree cited by Liogier & Martorell (1982, 2000) and Liogier (1997) as uncommonly planted and persistent as an ornamental. [L5: 74].

Gardenia jasminoides J. Ellis. Syn.: *Gardenia augusta* (L.) Merr.
Shrub cited by Britton & Wilson (1925) as occasionally escaped from cultivation in mountain gardens; subsequently cited by Liogier & Martorell (1982, 2000) and Liogier (1997). [L5: 96].

Ixora acuminata Roxb.
Shrub cited by Britton & Wilson (1925) as persistent in gardens; subsequently cited by Liogier & Martorell (1982, 2000) and Liogier (1997) as much planted as an ornamental, persistent after cultivation. [L5: 115].

Ixora coccinea L.
Shrub cited by Liogier & Martorell (1982, 2000) and Liogier (1997) as much planted as an ornamental and naturalized. [L5: 115].

Pentas lanceolata (Forssk.) Defl.
Suffrutescent herb/subshrub cited by Liogier & Martorell (1982, 2000) and Liogier (1997) as cultivated and persistent in humid places. [L5: 135].

RUPPIACEAE

Ruppia maritima L., Sp. Pl. 1:127. 1753.

Aquatic herb in Northern Coastal Lowlands (Manatí, Arecibo), Guánica Forest Reserve, Vieques; 1 m; *fl* Jan (one specimen), *fr* Jan, Apr–Jun.

Nearly cosmopolitan. [AR&S: 75].

Yerba de zanja.

TAXA NOT RECENTLY COLLECTED

Ruppia didyma Sw. ex Wikstr., Kongl. Svenska Vetensk. Acad. Handl. 1825:427. 1826.
Syn.: *Ruppia anomala* Ostenf. (1915).

Aquatic herb collected in Guánica on 11–12 Mar 1913 [*Britton & Shafer 1870* (NY)].

West Indies (scattered), Mexico. [AR&S: 75].

Notes: Liogier & Martorell (1982) wrongly placed *Ruppia anomala* in synonymy with *Ruppia maritima*.

RUSCACEAE (including DRACAENACEAE)

Dracaena fragrans (L.) Ker Gawl., Bot. Mag. 27: *t. 1081*. 1808.
Bas.: *Aletris fragrans* L. (1762).

Shrub/small tree planted and persistent in Northern Limestone Hills (Lares), Río Abajo Forest Reserve, Sierra de Luquillo, Eastern Cordillera (Aibonito), Central Cordillera (Adjuntas), Western Cordillera (Maricao); 280–600 m; *fl* Nov–Feb, May, *fr* Feb–Apr, Jul–Aug.

Native to tropical Africa. [AR&S: 130].

Cocomacaco, Drecina.

Sansevieria concinna N.E. Br., Bull. Misc. Inform. Kew 1915:233. 1915.

Herb planted and persistent in Guánica Forest

Reserve; 10 m; *fl* May, Dec (two specimens), *fr* (?). [*Breckon 5573* (MAPR); *Axelrod 13918* (UPRRP)]

Native to Mozambique.

Sansevieria cylindrica Bojer, Bot. Mag. 85: *t. 5093*. 1859.

Herb planted and persistent in Guánica Forest Reserve; 10 m; *fl* Dec–Feb, *fr* (?). [*Knudsen 68* (UPRRP); *Axelrod 13919* (UPRRP); *Monsegur 666* (MAPR)].

Native to tropical Africa. [AR&S: 131].

Sansiviera redonda.

Sansevieria hyacinthoides (L.) Druce, Rep. Bot. Exch. Club Soc. Brit. Isles 3(5):423. 1914.

Bas.: *Aloe hyacinthoides* L. (1753).

Syn.: *Cordyline guineensis* (L.) Britton (1918).

Herb planted and persistent in Northern Coastal Lowlands, Northern Limestone Hills (Ciales, Arecibo), Eastern Cordillera (Cayey, Barranquitas), Central Cordillera (Villalba), Southern Coastal Lowlands (Guayama), Southern Coastal Limestone (Guayanilla), vicinity of Susúa Forest Reserve (Sabana Grande), Western Cordillera (San Germán), Guánica Forest Reserve, Cerro Las Mesas, Western Coast (San Germán, Rincón), Vieques; 1–700 m; *fl* Aug–Apr, *fr* (?).

Native to southern Africa. [AR&S: 131].

Chucho, Cocuisa, Lengua de chucho, Lengua de vaca, Sansiviera.

Sansevieria trifasciata Prain, Bengal Pl. 2:1054. 1903.

Herb planted and persistent in Northern Coastal Lowlands, Cambalache Forest Reserve, Vieques; 1–50 m; *fl* Feb–Apr, Aug–Oct, *fr* (?).

Native to western Africa (Zaire, Nigeria). [AR&S: 133].

RUTACEAE

Amyris elemifera L., Syst. Nat. (ed. 10) 2:1000. 1759.

Tree in Northern Limestone Hills, Río Abajo Forest Reserve, Las Piedras Chiquitas, Eastern Cordillera (Cidra), Southern Coastal Lowlands (Caja de Muertos), Southern Coastal Limestone, Susúa Forest Reserve, Maricao Forest Reserve, Guánica Forest Reserve, Southwestern Coastal Lowlands (Cabo Rojo), Sierra Bermeja, Vieques, Culebra, Mona, Desecheo; 1–700 m; *fl* throughout the year, *fr* Jan–Oct.

West Indies, Trinidad and Tobago, southeastern United States (Florida), Mexico, northern Central America. [L2: 255].

Cuabilla, Tea.

Citrus ×aurantiifolia (Christm.) Swingle (pro sp.), J. Washington Acad. Sci. 3(18):465. 1913 "aurantifolia."

Bas.: *Limonia ×aurantiifolia* Christm. (1777) "aurantifolia."

Tree planted and persistent in Northern Coastal Lowlands (Loíza, Carolina), Northern Limestone Hills (Quebradillas), Southern Coastal Lowlands, Susúa Forest Reserve, Guánica Forest Reserve, Vieques, Mona; 1–200 m; *fl* Jan–Jun, *fr* Mar–Oct.

Probably native to tropical Asia. [L2: 257].

Lima, Limón, Limón agrio.

Citrus ×aurantium L. (pro sp.), Syst. Pl. 2:782. 1753.

Tree planted and persistent in Northern Limestone Hills, Sierra de Luquillo, Eastern and Central Cordillera, Southern Coastal Lowlands (Ponce), Guánica Forest Reserve, Southwestern Coastal Lowlands (Lajas), Vieques; 10–950 m; *fl* Feb–Apr, *fr* Feb–Sep.

Native to Indo-China. [L2: 258].

Naranja, Naranja agria.

Citrus ×jambhiri Lush. (pro sp.), Indian Forester 36:342. 1910.

Tree planted and persistent in Sierra de Luquillo, Monte Guilarte Forest Reserve; 450–1025 m; *fl* (?), *fr* Mar, Jun–Aug.

Possibly native to India. [L2: 260].

Notes: Britton & Wilson first (1924) misapplied the name *Citrus limonum* Risso and, later (1930), *Citrus limonia* Osbeck to this hybrid; Little & Wadsworth (1964), Liogier & Martorell (1982, 2000) and Liogier (1988) misapplied *Citrus limon* (L.) Burm. f. to it.

This name probably represents the hybid *Citrus medica ×reticulata*.

Limón de cabro.

Citrus medica L., Sp. Pl. 2:782. 1753.

Tree planted and persistent in Central Cordillera (Adjuntas); 600 m; *fl* Sep, Feb (two specimens), *fr* Feb–Mar (two specimens).

Probably native to northeastern India. [L2: 261].

Cidra.

Citrus ×paradisi Macfad. (pro sp.), Bot. Misc. 1:304. 1830.

Tree planted and persistent in Eastern Cordillera (Aibonito), Central Cordillera (Ciales), Mona; 5–500 m; *fl* Feb–Apr (two specimens), *fr* Mar, Aug (two specimens).

Widely cultivated in tropics and subtropics. [L2: 261].

Notes: This name probably represents the hybid *Citrus maxima ×sinensis*.

Toronja.

Citrus reticulata Blanco, Fl. Filip. 610. 1837.

Tree planted and persistent in Sierra de Luquillo, Central Cordillera; 400–700 m; *fl* Mar–Apr, Sep, *fr* Mar (one specimen).

Probably native to southeastern Asia. [L2: 262].

Notes: Britton & Wilson (1924) misapplied the name *Citrus nobilis* Lour. to this taxon.

Mandarina, Tangerina.

Citrus ×sinensis (L.) Osbeck (pro sp.), Reise Ostindien 250. 1765.

Bas.: *Citrus aurantium* L. var. *sinensis* L. (1753).

Tree planted and persistent in Río Abajo Forest Reserve, Sierra de Cayey, Central Cordillera (Juana Díaz), Susúa Forest Reserve, Maricao Forest Reserve; 225–1025 m; *fl* Feb–Apr, *fr* Mar–May, Aug–Sep.

Probably native to southeastern Asia. [L2: 262].

China, China dulce.

Murraya paniculata (L.) Jack, Malayan Misc. 1:31. 1820.

Bas.: *Chalcas paniculata* L. (1767).

Syn.: *Murraya exotica* L. (1771) "Murraea." *Chalcas exotica* (L.) Millsp. (1895).

Shrub/small tree in Northern Coastal Lowlands, Northern Limestone Hills (Morovis, Quebradillas), Eastern Coast (Naguabo), Caguas Valley (Caguas), Central Cordillera (Ponce), Western Coast (Rincón); 1–250 m; *fl* Nov–Jul, *fr* Sep–Nov, Feb–Apr.

Native to tropical Asia, China, Australia. [L2: 264].

Café de la India, Mirto.

Pilocarpus racemosus Vahl, Eclog. Amer. 1:29, *t. 10*. 1797.

Tree in Northern Limestone Hills, Río Abajo Forest Reserve, Susúa Forest Reserve, Maricao Forest Reserve, Cerro Las Mesas, Vieques, Culebra; 150–700 m; *fl* Jul, Oct–Feb, *fr* Jul–Nov, Feb–Apr.

Greater Antilles (Cuba, Hispaniola), Virgin Islands, Lesser Antilles (Montserrat, Guadeloupe, Martinique), Mexico, Central America, northern South America (Venezuela, Colombia). [L2: 266].

Aceitillo.

Ravenia urbanii Engl. in Urban, Bot. Jahrb. Syst. 21(5):552. 1896.

Tree in Sierra de Luquillo, Sierra de Cayey; 325–750 m; *fl* Oct–Jul, *fr* May (one specimen).

Endemic. [L2: 266].

Tortugo prieto.

Triphasia trifolia (Burm. f.) P. Wilson, Torreya 9(2): 33. 1909.

Bas.: *Limonia trifolia* Burm. f. (1768).

Shrub in Northern Coastal Lowlands (Río Grande, Laguna Tortuguero), Eastern Coast (Naguabo), Vieques, Mona; 1–90 m; *fl* Mar–Apr (one specimen), *fr* Feb–Nov.

Possibly native to southeastern Asia. [L2: 270].

China de Bakón, China forastera, Chinita, Limoncito, Limón de Jerusalén.

Zanthoxylum bifoliolatum Leonard, J. Washington Acad. Sci. 14:413. 1924.

Shrub/small tree in Maricao Forest Reserve; 575–850 m; *fl* May–Jun, Sep–Oct, *fr* May, Aug–Jan.

Bahamas, Greater Antilles (Hispaniola). [L2: 272].

Zanthoxylum caribaeum Lam., Encycl. 2(1):39. 1786.

Tree in Northern Coastal Lowlands, Northern Limestone Hills; 5–150 m; *fl* Sep, Dec (two specimens), *fr* May–Aug.

Lesser Antilles, Mexico, Central America, northwestern, west-central South America and Venezuela, Paraguay. [L2: 272].

Espino rubial.

Zanthoxylum flavum Vahl, Eclog. Amer. 3:48. 1807.

Syn.: *Fagara flava* (Vahl) Krug & Urb. (1896).

Tree in Northern Coastal Lowlands, Northern Limestone Hills (Quebradillas), Southern Coastal Lowlands (Guayanilla), Maricao Forest Reserve, Guánica Forest Reserve, Southwestern Coastal Lowlands (Cabo Rojo), Sierra Bermeja; 5–800 m; *fl* Jun–Jul, Nov, Feb, *fr* Jul–Feb.

West Indies, southeastern United States (Florida). [L2: 273].

Aceitillo.

Zanthoxylum martinicense (Lam.) DC., Prodr. 1:726. 1824.

Bas.: *Fagara martinicensis* Lam. (1792).

Tree in Northern Coastal Lowlands, Northern Limestone Hills, Sierra de Cayey, Las Tetas de Cayey, Las Piedras Chiquitas, Southeastern Foothills (Coamo), Eastern Cordillera, Central Cordillera (Ponce), Susúa Forest Reserve, Maricao Forest Reserve, Vieques; 25–825 m; *fl* Oct–Nov, Feb–Jul, *fr* Oct–Jul.

Antilles, Trinidad and Tobago, northwestern, west-central South America and Venezuela. [L2: 273].

Ayúa, Cenizo, Espinosa, Espino, Espino rubial.

Zanthoxylum monophyllum (Lam.) P. Wilson, Bull. Torrey Bot. Club 37(2):86. 1910.

Bas.: *Fagara monophylla* Lam. (1792).

Tree in Northern Coastal Lowlands, Eastern

Coast (Ceiba), Las Piedras Chiquitas, Southeastern Foothills, Southern Coastal Lowlands (Patillas, Ponce), Susúa Forest Reserve, Guánica Forest Reserve, Southwestern Coastal Lowlands, Sierra Bermeja, Vieques, Culebra; 25–600 m; *fl* May–Nov, *fr* Jul–Jan.

Greater Antilles (Hispaniola), Lesser Antilles, southern Central America, northern South America (Venezuela, Colombia). [L2: 275].

Carubio, Enrubio, Espino, Espino blanco, Espino rubial, Mapurito, Palo rubio, Rubia.

Zanthoxylum punctatum Vahl in West, Bidr. Beskr. Ste. Croix 310. 1793.
Syn.: *Fagara trifoliata* Sw. (1788).

Tree in Northern Limestone Hills, Río Abajo Forest Reserve, Eastern Cordillera (Cayey), Susúa Forest Reserve, Maricao Forest Reserve, Sierra Bermeja, Mona; 3–750 m; *fl* Sep–Oct, Jan–Apr, *fr* throughout the year.

Antilles. [L2: 275].

Alfiler.

Zanthoxylum spinifex (Jacq.) DC., Prodr. 1:728. 1824.
Bas.: *Fagara spinifex* Jacq. (1809).

Shrub in Southeastern Foothills (Salinas, Coamo); 150 m; *fl* May (one specimen), *fr* (?). [*Axelrod 1421* (UPRRP); *Ross SAN674* (scan) (UPRRP)].

Antilles, northern South America (Venezuela). [L2: 276].

Niaragato.

Zanthoxylum thomasianum (Krug & Urb.) Krug & Urb. ex P. Wilson, N. Amer. Fl. 25:182. 1911.
Bas.: *Fagara thomasiana* Krug & Urb. (1896).

Shrub in Northern Limestone Hills, Las Piedras Chiquitas, Southeastern Foothills (Salinas); 30–550 m; *fl* Mar (one specimen), *fr* May (one specimen).

Virgin Islands (St. Thomas, St. John). [L2: 276].

Notes: This taxon is on the U.S. Fish & Wildlife Service list of threatened and endangered plants.

TAXA CITED BY PREVIOUS AUTHORS, BUT NOT INCLUDED

NO SPECIMENS SEEN

Amyris balsamifera L.
Shrub/small tree cited by Britton & Wilson (1924) as occurring in woodlands near Guánica; subsequently cited by Liogier & Martorell (1982, 2000) and Liogier (1988). [L2: 255].

ONLY UNDER CULTIVATION

Citrus ×limettioides Tanaka (pro sp.).
Tree cited by Liogier (1988) and Liogier & Martorell (2000) as cultivated and escaped. [L2: 260].

Notes: Little & Wadsworth (1964) did not consider this taxon distinct from *Citrus ×aurantiifolia*; Liogier (1988) and Liogier & Martorell (2000) misapplied the name *Citrus limetta* Risso to it.

Citrus maxima (Burm.) Merr. Syn.: *Citrus grandis* (L.) Osbeck.
Tree cited by Liogier & Martorell (1982, 2000) and Liogier (1988) as cultivated for its fruit and often escaped from cultivation. [L2: 260].

Fortunella margarita (Lour.) Swingle.
Tree cited by Liogier & Martorell (1982, 2000) and Liogier (1988) as occasionally grown for its fruit and persistent. [L2: 263].

Ruta chalepensis L.
Herb cited by Britton & Wilson (1925) as locally spontaneous after cultivation; subsequently cited by Liogier & Martorell (1982, 2000) and Liogier (1988). [L2: 269].

SABIACEAE

Meliosma herbertii Rolfe, Bull. Misc. Inform. Kew 1893:244. 1893.

Tree in Sierra de Luquillo, Central Cordillera, Maricao Forest Reserve; 400–800 m; *fl* Jan, Aug (two specimens), *fr* Feb, Jun.

Greater Antilles (Hispaniola), Lesser Antilles, northwestern, west-central South America and Venezuela. [L3: 71].

Aguacatillo, Arroyo, Cacaíllo, Cacao bobo.

Meliosma obtusifolia (Bello) Krug & Urb., Ber. Deutsch. Bot. Ges. 13:216. 1895.
Bas.: *Atelandra obtusifolia* Bello (1881).

Tree in Eastern Cordillera (Barranquitas including La Torrecilla), Central Cordillera (Adjuntas), Maricao Forest Reserve; 500–925 m; *fl* Jul, Oct–Nov, *fr* Apr, Jul–Aug, Nov.

Endemic. [L3: 71].

Arroyo, Cacaíllo, Cacao bobo, Cerrillo, Ciralillo, Guayarote.

SALICACEAE (including FLACOURTIACEAE)

Banara portoricensis Krug & Urb. in Urban, Symb. Antill. 1(2):370. 1899.

Tree in Sierra de Luquillo, Central Cordillera, Maricao Forest Reserve; 700–1300 m; *fl* Feb–Apr (two specimens), *fr* May, Aug.

Endemic. [L3: 258].

Caracolillo, Palo de ramón, Tostado.

Banara vanderbiltii Urb., Symb. Antill. 3(2):320. 1902.
Tree in Northern Limestone Hills (Dorado), Las Tetas de Cayey 10–840 m; *fl* May–Jun, Dec, *fr* Jun–Oct.
Endemic. [L3: 258].
Notes: This taxon is on the U.S. Fish & Wildlife Service list of threatened and endangered plants.

Casearia aculeata Jacq., Enum. Syst. Pl. 21. 1760.
Tree in Northern Coastal Lowlands, Northern Limestone Hills, Guánica Forest Reserve, Southwestern Coastal Lowlands, Sierra Bermeja, Western Coast (Mayagüez); 90–500 m; *fl* Jan–Oct, *fr* Apr–May, Aug–Oct.
Greater Antilles, Central America, South America. [L3: 261].
Cambrón, Rabo de ratón.

Casearia arborea (Rich.) Urb., Symb. Antill. 4(3):421. 1910.
Bas.: *Samyda arborea* Rich. (1792).
Tree in Northern Coastal Lowlands, Sierra de Luquillo, Eastern Coast, Caguas Valley (Caguas), Sierra de Cayey, Eastern Cordillera (Cerro La Peña), Central Cordillera, Maricao Forest Reserve, Western Coast (Mayagüez); 15–900 m; *fl* & *fr* throughout the year.
Greater Antilles, Central America, northern South America to Brazil, Bolivia. [L3: 261].
Cabrilla, Guía verde, Palo de queresas, Rabo de ratón, Rabojunco.

Casearia decandra Jacq., Enum. Syst. Pl. 21. 1760.
Tree in Northern Coastal Lowlands, Northern Limestone Hills, Río Abajo Forest Reserve, Eastern Coast, Sierra de Cayey, Las Tetas de Cayey, Las Piedras Chiquitas, Southeastern Foothills, Eastern Cordillera (Corozal, Aibonito), Central Cordillera, Susúa Forest Reserve, Maricao Forest Reserve, Western Coast, Vieques; 1–1050 m; *fl* Dec–Aug, *fr* Nov–Aug.
Greater Antilles (Hispaniola), Virgin Islands, Lesser Antilles, Central America, South America. [L3: 262].
Caracolillo, Cerezo, Corcho blanco, Cotorrerillo, Duro-duro, Guía mansa, Palo blanco, Tostado.

Casearia guianensis (Aubl.) Urb., Symb. Antill. 3(2):322. 1902.
Bas.: *Iroucana guianensis* Aubl. (1775).
Tree in Northern Coastal Lowlands, Northern Limestone Hills, Río Abajo Forest Reserve, Eastern Coast, Southeastern Foothills, Eastern Cordillera (Comerío, Corozal), Central Cordillera (Ponce), Cerro de Las Cuevas, Quebrada de Los Cedros, Maricao Forest Reserve, Southwestern Coastal Lowlands (Cabo Rojo), Western Coast (Mayagüez), Vieques; 1–800 m; *fl* Sep–May, *fr* throughout the year.
Antilles, southern Central America, northern South America (Venezuela, Colombia) and Brazil. [L3: 262].
Cafeíllo, Cafetillo, Palo blanco.

Casearia sylvestris Sw., Fl. Ind. Occid. 2:752. 1798.
Tree in Northern Coastal Lowlands, Northern Limestone Hills, Río Abajo Forest Reserve, Sierra de Luquillo, Eastern Coast, Caguas Valley, Sierra de Cayey (San Lorenzo), Las Piedras Chiquitas, Southeastern Foothills, Eastern and Central Cordillera, Cerro de Las Cuevas, Southern Coastal Limestone (Guayanilla), Susúa Forest Reserve, Maricao Forest Reserve, Southwestern Coastal Lowlands, Cerro Las Mesas, Western Coast (Mayagüez), Vieques; 1–900 m; *fl* throughout the year, *fr* May–Jan.
Antilles, Trinidad and Tobago, Mexico, Central America, South America. [L3: 264].
Cafeíllo, Cafeíllo cimarrón, Café silvestre, Laurel espada, Palo blanco, Palo de queresas, Sarna de perro.

Homalium racemosum Jacq., Enum. Syst. Pl. 24. 1760.
Syn.: *Homalium leiogynum* S.F. Blake (1919).
Homalium pleiandrum S.F. Blake (1919).
Tree in Northern Coastal Lowlands (Fajardo), Northern Limestone Hills, Río Abajo Forest Reserve, Sierra de Luquillo, Eastern Coast (Punta Guayanés), Sierra de Cayey, Las Tetas de Cayey, Southeastern Foothills (Cayey), Eastern and Central Cordillera, Susúa Forest Reserve, Maricao Forest Reserve; 50–900 m; *fl* Dec–Aug, *fr* Jan–Aug.
Antilles, Mexico, Central America, northwestern, west-central South America and Venezuela. [L3: 268].
Caracolillo, Cereza, Guajanilla, Palo de ramón, Tostado.

Laetia procera (Poepp.) Eichler in Martius, Fl. Bras. 13(1):453. 1871.
Bas.: *Samyda procera* Poepp. (1845).
Syn.: *Casearia bicolor* Urb. (1899).
Tree in Sierra de Luquillo; 350–500 m; *fl* (?), *fr* Jan–Mar, Aug.
Greater Antilles (Hispaniola), Trinidad, Central America, northwestern, west-central South America and Venezuela. [L3: 270].
Almendrillo, Cotorrerillo, Cuero de sapo, Talantrón.

Lunania ekmanii Urb., Ark. Bot. 17(7):42. 1922.
Tree in Sierra de Luquillo, Maricao Forest Reserve; 350–750 m; *fl* Feb–Aug, *fr* Apr–Jul.

Greater Antilles (Hispaniola). [L3: 272].

Prockia crucis L., Syst. Nat. (ed. 10) 2:1074. 1759.

Shrub/small tree in Las Tetas de Cayey, Sierra Bermeja, Vieques; 25–825 m; *fl* (?), *fr* Jun, Dec (two specimens).

Greater Antilles (Cuba, Hispaniola), Lesser Antilles, Mexico, Central America, South America. [L3: 274].

Salix humboldtiana Willd., Sp. Pl. 4(2):657. 1806.

Tree in Northern Coastal Lowlands (Río Grande); 10 m; *fl* (?), *fr* (?). [*Caraballo 1649* (UPR)].

Native to southern Mexico, Central America, western and southern South America and Brazil. {L1: 37].

Notes: Britton & Wilson (1924) used *Salix chilensis* Molina for the name of this taxon.

Samyda dodecandra Jacq., Enum. Syst. Pl. 21. 1760.

Shrub/small tree in Northern Coastal Lowlands (Fajardo and adjacent islands), Northern Limestone Hills (Isabela), Eastern Coast, Las Piedras Chiquitas, Southeastern Foothills, Southern Coastal Limestone (Guayanilla, Yauco), Susúa Forest Reserve, Guánica Forest Reserve, Southwestern Coastal Lowlands (Cabo Rojo), Sierra Bermeja, Vieques, Mona; 5–500 m; *fl* & *fr* throughout the year.

Antilles, Southern Caribbean Islands (Curaçao), Trinidad. [L3: 276].

Guayabilla.

Samyda spinulosa Vent., Choix Pl. *t. 43*. 1808.

Shrub/small tree in Northern Coastal Lowlands (Carolina), Northern Limestone Hills, Río Abajo Forest Reserve, Sierra de Luquillo, Susúa Forest Reserve, Maricao Forest Reserve, Guánica Forest Reserve; 25–700 m; *fl* & *fr* throughout the year.

Virgin Islands (St. Thomas). [L3: 277].

Gaita, Guía, Níspero cimarrón.

Xylosma buxifolia A. Gray, Pl. Wright. 1:155. 1860 (preprint of Mem. Amer. Acad. Arts, n.s. 8(1):155. 1861) "buxifolium."

Tree in Northern Limestone Hills, Southeastern Foothills (Coamo), Guánica Forest Reserve, Sierra Bermeja; 25–300 m; *fl* Dec, May, *fr* Jul–Aug (two specimens).

Bahamas, Greater Antilles (Cuba, Hispaniola), Virgin Islands, Lesser Antilles. [L3: 279].

Roseta.

Xylosma pachyphylla (Krug & Urb.) Urb., Symb. Antill. 1(2):371. 1899 "pachyphyllum."

Bas.: *Myroxylon pachyphyllum* Krug & Urb. (1892).

Tree in Susúa Forest Reserve, Maricao Forest Reserve; 250–900 m; *fl* Apr–May (two specimens), *fr* May–Aug.

Endemic. [L3: 279].

Xylosma schaefferioides A. Gray, Pl. Wright. 1:155. 1860 (preprint of Mem. Amer. Acad. Arts, n.s. 8(1):155. 1861).

Tree in Northern Limestone Hills (Arecibo), Río Abajo Forest Reserve, Eastern Cordillera (Cañón San Cristóbal), Maricao Forest Reserve; 275–850 m; *fl* Oct–Dec (two specimens), *fr* Jan, Jul.

Greater Antilles. [L3: 281].

Xylosma schwaneckeana (Krug & Urb.) Urb., Symb. Antill. 1(2):371. 1899 "Schwaneckeanum."

Bas.: *Myroxylon schwaneckeanum* Krug & Urb. (1892).

Shrub/small tree in Sierra de Luquillo, Sierra de Cayey; 400–800 m; *fl* May–Oct, *fr* Apr, Aug–Oct.

Endemic. [L3: 281].

Candela, Palo colorado, Palo de candela.

TAXA CITED BY PREVIOUS AUTHORS, BUT NOT INCLUDED

ONLY UNDER CULTIVATION

Dovyalis hebecarpa (Gardner) Warb.

Shrub/small tree cited by Liogier & Martorell (1982, 2000) and Liogier (1994) as occasionally planted for its edible fruits. [L3: 265].

Flacourtia indica (Burm. f.) Merr.

Shrub/small tree cited by Liogier & Martorell (1982, 2000) and Liogier (1994) as occasionally planted for its edible fruits. [L3: 266].

Flacourtia inermis Roxb.

Shrub/small tree cited by Liogier & Martorell (1982, 2000) and Liogier (1994) as occasionally planted for its edible fruits. [L3: 267].

Oncoba echinata Oliv.

Shrub/small tree cited by Liogier & Martorell (1982, 2000) and Liogier (1994) as introduced as an ornamental and medicinal plant. [L3: 273].

Salix babylonica L.

Tree cited by Liogier & Martorell (1982, 2000) and Liogier (1985) as planted and persistent. [L1: 37].

SANTALACEAE (including EREMOLEPIDACEAE AND VISCACEAE)

Dendrophthora bermejae Kuijt, Carlo, & Aukema, Contr. Univ. Michigan Herb. 24:115, *t. 1.* 2005.

Parasitic shrub in Southwestern Coastal Lowlands (Cabo Rojo); 12–100 m; *fl* & *fr* Jan, Jun–Sep.

Endemic.

Notes: This plant has thus far been found to parasitize only *Guaiacum officinale*.

Dendrophthora brachylepis Urb., Ark. Bot. 17(7): 31. 1922.

Parasitic shrub in Northern Coastal Lowlands (Fajardo), Northern Limestone Hills, Sierra de Luquillo, Las Piedras Chiquitas, Susúa Forest Reserve, Western Cordillera, Guánica Forest Reserve; 15–550 m; *fl* & *fr* throughout the year.

Greater Antilles (Hispaniola). [L5: 393].

Notes: Liogier (1997) and Liogier & Martorell (2000) misapplied the name *Dendrophthora serpyllifolia* (Griseb.) Krug & Urb. to this taxon.

Dendrophthora domingensis (Spreng.) Eichler in Martius, Fl. Bras. 5(2):103. 1868.
Bas.: *Viscum domingense* Spreng. (1824).

Parasitic shrub in Northern Limestone Hills, Río Abajo Forest Reserve, Susúa Forest Reserve, Western Cordillera; 150–850 m; *fl* & *fr* throughout the year.

Greater Antilles (Cuba, Hispaniola). [L1: 102].

Dendrophthora flagelliformis (Lam.) Krug & Urb., Ber. Deutsch. Bot. Ges. 14:287. 1896.
Bas.: *Viscum flagelliforme* Lam. (1789).
Syn.: *Dendrophthora wrightii* Eichler (1868).

Parasitic shrub in Northern Coastal Lowlands (San Juan, Guaynabo), Northern Limestone Hills, Southeastern Foothills (Coamo), Eastern Cordillera (Cerro La Peña, Aibonito), Central Cordillera (Orocovis), Susúa Forest Reserve, Western Cordillera, Guánica Forest Reserve; 100–700 m; *fl* & *fr* throughout the year.

Greater Antilles (Cuba, Hispaniola). [L1: 103].

Eubrachion ambiguum (Hook. & Arn.) Engl. in Engler & Prantl, Nat. Pflanzenfam. 3(1):192. 1889.
Bas.: *Viscum ambiguum* Hook. & Arn. (1833).
Syn.: *Eubrachion ambiguum* (Hook. & Arn.) Engl. var. *jamaicense* Krug & Urb. (1897).

Parasitic shrub in Central Cordillera (Adjuntas, Yauco), Maricao Forest Reserve; 650–1100 m; *fl* & *fr* May–Jan.

Greater Antilles (Jamaica, Hispaniola), southeastern and south-central South America. [L1: 104; L5: 393].

Phoradendron anceps (Spreng.) M. Gómez, Anales Inst. Segunda Enseñ. 2:170. 1895.
Bas.: *Viscum anceps* Spreng. (1824).
Syn.: *Phoradendron chrysocarpum* Krug & Urb. (1897).
Phoradendron barahonae Urb. & Trel. (1914).
Phoradendron helleri Trel. (1916).

Parasitic shrub in Northern Coastal Lowlands, Northern Limestone Hills, Sierra de Luquillo, Las Tetas de Cayey, Las Piedras Chiquitas, Southeastern Foothills, Eastern and Central Cordillera, Southern Coastal Lowlands, Susúa Forest Reserve, Maricao Forest Reserve, Guánica Forest Reserve, Southwestern Coastal Lowlands; 1–1000 m; *fl* & *fr* throughout the year.

Antilles, Trinidad, northern South America (Venezuela). [L1: 107; L5: 394].

Pata de gallina.

Phoradendron berteroanum (DC.) Griseb., Fl. Brit. W.I. 313. 1860 "Berterianum."
Bas.: *Viscum berteroanum* DC. (1830) "Berterianum."
Syn.: *Phoradendron dichotomum* (Bertero ex Spreng.) Krug & Urb. (1897).

Parasitic shrub in Northern Limestone Hills, Central Cordillera (Utuado), Maricao Forest Reserve; 150–875 m; *fl* & *fr* Jun (one specimen).

Greater Antilles, Central America (Nicaragua, Panama), northwestern and west-central South America. [L1: 107].

Yerba capitana.

Phoradendron dipterum Eichler in Martius, Fl. Bras. 5(2):109. 1868.
Syn.: *Phoradendron tetrapterum* Krug & Urb. (1897).

Parasitic shrub in Las Piedras Chiquitas, Southeastern Foothills, Central Cordillera (Utuado), Susúa Forest Reserve, Maricao Forest Reserve, Guánica Forest Reserve; 150–850 m; *fl* & *fr* throughout the year.

Greater Antilles, Lesser Antilles (Martinique, St. Lucia, St. Vincent), Trinidad, Central America (Nicaragua, Costa Rica), northwestern, west-central South America and Venezuela, Argentina. [L1: 110].

Notes: This plant has thus far been found to parasitize only other taxa of *Phoradendron*.

Phoradendron hexastichum (DC.) Griseb., Fl. Brit. W.I. 313. 1860.
Bas.: *Viscum hexastichum* DC. (1830).

Parasitic shrub in Sierra de Luquillo, Sierra de Cayey, Central Cordillera, Maricao Forest Reserve; 350–1150 m; *fl* & *fr* throughout the year.

Antilles, Mexico, Central America (Belize, Panama), northwestern, west-central South America and Venezuela. [L1: 107].

Phoradendron piperoides (Kunth) Trel., *Phoradendron* 145, *tt. 217–222.* 1916.
Bas.: *Loranthus piperoides* Kunth (1820).

Parasitic shrub in Northern Limestone Hills, Río Abajo Forest Reserve, Sierra de Luquillo, Sierra de Cayey, Southeastern Foothills, Eastern Cordillera (Corozal), Central Cordillera, Southern Coastal Limestone (Guayanilla), Susúa Forest Reserve, Western Cordillera; 100–1100 m; *fl* & *fr* throughout the year.

Antilles, Trinidad and Tobago, Mexico, Central America, South America (scattered). [L1: 108].

Conde, Golondrina.

Phoradendron quadrangulare (Kunth) Griseb., Fl. Brit. W.I. 711. 1864.
Bas.: *Loranthus quadrangularis* Kunth (1820).
Syn.: *Phoradendron randiae* (Bello) Britton (1924).

Parasitic shrub in Northern Coastal Lowlands (Laguna Tortuguero), Northern Limestone Hills (Quebradillas), Las Piedras Chiquitas, Southeastern Foothills, Central Cordillera (Ponce), Southern Coastal Lowlands, Susúa Forest Reserve, Maricao Forest Reserve, Southwestern Coastal Lowlands, Sierra Bermeja; 5–720 m; *fl* & *fr* throughout the year.

Greater Antilles, Lesser Antilles (Antigua, Guadeloupe, St. Vincent), Central America, northwestern, west-central South America and Venezuela, Argentina. [L1: 110].

Golondrina de canario.

Phoradendron racemosum (Aubl.) Krug & Urb., Bot. Jahrb. Syst. 24(1):46. 1897.
Bas.: *Viscum racemosum* Aubl. (1775).

Parasitic shrub in Northern Coastal Lowlands (Dorado), Sierra de Luquillo, Sierra de Cayey, Western Cordillera, Cerro Las Mesas; 1–700 m; *fl* & *fr* Dec–Mar, Jun, Sep.

Bahamas, Greater Antilles (Cuba, Hispaniola), Virgin Islands (St. Thomas), northern South America to Brazil, Bolivia. [L1: 108].

Phoradendron trinervium (Lam.) Griseb., Fl. Brit. W.I. 314. 1860.
Bas.: *Viscum trinervium* Lam. (1789).

Parasitic shrub in Northern Limestone Hills (Lares), Las Piedras Chiquitas, Southeastern Foothills (Salinas), Eastern Cordillera (Aibonito), Guánica Forest Reserve, Southwestern Coastal Lowlands, Sierra Bermeja; 1–550 m; *fl* & *fr* throughout the year.

West Indies (except Cuba), Central America (Guatemala, Belize, Panama), northern South America (Venezuela, Colombia). [L1: 110].

TAXA NOT RECENTLY COLLECTED

Antidaphne wrightii (Griseb.) Kuijt, Syst. Bot. Monogr. 18:40. 1988.
Bas.: *Eremolepis wrightii* Griseb. (1860).
Syn.: *Ixidium wrightii* (Griseb.) Eichler (1868).

Parasitic shrub collected in the Sierra de Naguabo on 30 Oct 1886 [*Sintenis 5347* (US)].

Greater Antilles (Cuba, Hispaniola), northern South America (Venezuela). [L1: 104; L5: 393].

Phoradendron rubrum (L.) Griseb., Fl. Brit. W.I. 314. 1860.
Bas.: *Viscum rubrum* L. (1753).

Parasitic shrub collected near Manatí on 9 Apr 1887 [*Sintenis 6624* (L), scan seen].

Bahamas, Cayman Islands, Greater Antilles (Cuba, Hispaniola).

TAXA CITED BY PREVIOUS AUTHORS, BUT NOT INCLUDED

CITED DUE TO MISIDENTIFICATION

Phoradendron mucronatum (DC.) Krug & Urb. Parasitic shrub cited by Liogier & Martorell (1982) and Liogier (1985) as occurring in thickets in Guánica. [L1: 108].

SAPINDACEAE

Allophylus crassinervis Radlk. in Engler & Prantl, Nat. Pflanzenfam. 3(5):312. 1895.

Tree in Northern Limestone Hills, Río Abajo Forest Reserve, Central Cordillera (Adjuntas), Susúa Forest Reserve, Maricao Forest Reserve; 150–1000 m; *fl* Oct–Mar, *fr* Dec–May.

Greater Antilles (Cuba, Hispaniola). [L3: 45].

Palo blanco.

Allophylus racemosus Sw., Prodr. 62. 1788.
Syn.: *Allophylus occidentalis* (Sw.) Radlk. (1890).

Tree in Northern Coastal Lowlands (Dorado), Northern Limestone Hills, Las Tetas de Cayey, Southeastern Foothills (Salinas), Central Cordillera

(Adjuntas, Peñuelas), Maricao Forest Reserve, Guánica Forest Reserve; 25–825 m; *fl* Nov–May, *fr* Jun–Jul, Oct–Nov.

Antilles, Mexico, Central America, northwestern South America. [L3: 45].

Palo blanco, Quiebrahacha.

Cardiospermum corindum L., Sp. Pl. (ed. 2) 1:526. 1762.

Vine in Northern Coastal Lowlands (islands adjacent to Fajardo: Icacos), Northern Limestone Hills (San Sebastián), Southwestern Coastal Lowlands (Cabo Rojo), Western Coast (Mayagüez), Vieques, Culebra, Mona, Desecheo; 1–250 m; *fl* throughout the year, *fr* Nov–Jul.

Antilles, Mexico, Central America, South America, tropical Africa, Yemen. [L3: 49; AR: 367].

Farolito.

Cardiospermum halicacabum L. var. **halicacabum**, Sp. Pl. 1:366. 1753.

Vine in Northern Limestone Hills (San Sebastián), Southwestern Coastal Lowlands, Mona; 5–200 m; *fl* Feb, Jul–Aug, *fr* Feb, Aug.

Pantropical. [L3: 49; AR: 368].

Bombilla, Bombilla menor.

Cardiospermum halicacabum L. var. **microcarpum** (Kunth) Blume, Rumphia 3:185. 1847.

Bas.: *Cardiospermum microcarpum* Kunth (1821).

Vine in Northern Coastal Lowlands, Northern Limestone Hills (Camuy), Eastern Coast (Ceiba), Southeastern Foothills (Salinas), Southern Coastal Lowlands (Caja de Muertos, Ponce), Southwestern Coastal Lowlands (Cabo Rojo), Western Coast (Aguadilla), Vieques, Mona; 1–300 m; *fl* Jan–Oct, *fr* Dec–Sep.

Pantropical. [L3: 49; AR: 370].

Farolito.

Cupania americana L., Sp. Pl. 1:200. 1753.

Tree in Northern Coastal Lowlands, Northern Limestone Hills, Río Abajo Forest Reserve, Eastern Coast (Punta Guayanés), Caguas Valley (Caguas), Sierra de Cayey, Southeastern Foothills, Eastern Cordillera, Central Cordillera (Adjuntas), Susúa Forest Reserve, Western Cordillera, Southwestern Coastal Lowlands (Cabo Rojo), Sierra Bermeja, Western Coast; 1–675 m; *fl* Nov–May, *fr* Mar–Sep.

Antilles, northwestern, west-central South America and Venezuela. [L3: 51, 52].

Guara, Guara blanca.

Cupania triquetra A. Rich. in Sagra, Hist. Fis. Cuba, Bot. 10:119. 1845.

Tree in Northern Coastal Lowlands, Northern Limestone Hills, Caguas Valley (Caguas), Vieques; 1–400 m; *fl* Feb–Mar (two specimens), *fr* Feb–Mar.

Greater Antilles (Hispaniola), Virgin Islands (St. Thomas, St. John), Lesser Antilles. [L3: 52].

Guara, Guara blanca.

Dodonaea elaeagnoides Rudolphi ex Ledeb. in Ledebour & Alderstam, Diss. Bot. Pl. Doming. 18 (no. VII). 1805.

Syn.: *Dodonaea ehrenbergii* Schltdl. (1844).

Shrub in Southwestern Coastal Lowlands (Cabo Rojo), Mona; 1–5 m; *fl* Nov, Feb, Jul, *fr* Jan–Jul.

Bahamas, Greater Antilles, Lesser Antilles (La Désirade, St. Eustatius, Barbuda), southeastern United States (Florida). [L3: 53].

Guitarán.

Dodonaea viscosa Jacq., Enum Syst. Pl. 19. 1760.

Shrub in Northern Coastal Lowlands, Southern Coastal Lowlands (Arroyo), Southwestern Coastal Lowlands, Western Coast (Añasco), Vieques, Mona; 1–50 m; *fl* Aug–May, *fr* throughout the year.

Pantropical. [L3: 54].

Notes: Liogier & Martorell (1982, 2000) and Liogier (1994) used *Dodonaea viscosa* (L.) Jacq., a *nom. inval.*, for the name of this taxon.

Castaño, Chamiso, Gitarán, Guitarán, Quitarán.

Exothea paniculata (Juss.) Radlk. in Durand, Index Gen. Phan. 81. 1888.

Bas.: *Melicocca paniculata* Juss. (1817).

Tree in Northern Limestone Hills, Río Abajo Forest Reserve, Las Piedras Chiquitas, Southwestern Coastal Lowlands (Cabo Rojo), Mona; 30–550 m; *fl* Jan–Apr, Aug, *fr* Mar–Aug.

West Indies, southeastern United States (Florida), northern Central America (Guatemala, Belize), northwestern South America. [L3: 55].

Gaita, Guacarán.

Hypelate trifoliata Sw., Prodr. 61. 1788.

Tree in Southern Coastal Limestone, Guánica Forest Reserve, Vieques, Mona; 25–100 m; *fl* Jun–Jul, Oct, *fr* Jul–Dec.

Bahamas, Greater Antilles, Lesser Antilles (Anguilla, St. Martin), southeastern United States (Florida). [L3: 57].

Cigua, Melcocha, Melocha.

Matayba domingensis (DC.) Radlk., Sitzungsber. Math.-Phys. Cl. Königl. Bayer. Akad. Wiss. München 9:624. 1879.

Bas.: *Ratonia domingensis* DC. (1824).

Tree in Sierra de Luquillo, Sierra de Cayey, Eastern Cordillera (Cerro La Peña), Central Cordillera, Maricao Forest Reserve; 500–1300 m; *fl* Oct–Mar, *fr* Oct, Jan–Apr.

Greater Antilles (Cuba, Hispaniola), Virgin Islands (Tortola). [L3: 59].

Doncella, Escoba, Negra lora, Ratón, Tea cimarrona.

Melicoccus bijugatus Jacq., Enum. Syst. Pl. 19. 1760.
Syn.: *Melicocca bijuga* L. (1762), *nom. illeg. superfl.*

Tree in Northern Coastal Lowlands, Guajataca Forest Reserve, Las Piedras Chiquitas, Southeastern Foothills, Southern Coastal Lowlands, Southern Coastal Limestone (Guayanilla), Susúa Forest Reserve, Guánica Forest Reserve, Southwestern Coastal Lowlands, Sierra Bermeja, Vieques, Mona; 1–500 m; *fl* Apr–Nov, *fr* Jun–Oct.

Native to northern South America (Venezuela, Colombia). [L3: 61].

Quenepa.

Paullinia pinnata L., Sp. Pl. 1:366. 1753.

Vine in Northern Coastal Lowlands, Northern Limestone Hills, Río Abajo Forest Reserve, Sierra de Luquillo, Eastern Coast, Sierra de Cayey, Las Tetas de Cayey, Southeastern Foothills (Cayey), Eastern Cordillera, Toro Negro Forest Reserve, Southern Coastal Lowlands (Coamo), Maricao Forest Reserve, Southwestern Coastal Lowlands, Sierra Bermeja, Vieques; 1–825 m; *fl* May–Nov, *fr* throughout the year.

Greater Antilles (Cuba, Hispaniola), Lesser Antilles, Trinidad, Central America, South America, tropical Africa. [L3: 63; AR: 371].

Bejuco de costilla, Bejuco de guajanilla, Bejuco de paloma, Juanillo.

Sapindus saponaria L., Sp. Pl. 1:367. 1753.

Tree in Southern Coastal Lowlands, Southwestern Coastal Lowlands (Lajas), Western Coast (Mayagüez), Vieques; 15–45 m; *fl* May, Sep–Nov, *fr* Feb–Apr, Jul.

West Indies, southeastern and south-central United States, Mexico, Central America, South America. [L3: 64].

Jaboncillo, Saponaria.

Serjania diversifolia (Jacq.) Radlk., Monogr. *Serjania* 136. 1875.
Bas.: *Paullinia diversifolia* Jacq. (1760).

Vine in Northern Limestone Hills (Bayamón, Isabela); 230–300 m; *fl* (?), *fr* (?). [*Acevedo-Rodríguez 13462, 13751* (US)].

Bahamas, Greater Antilles (Cuba, Hispaniola), northern South America (Venezuela). [L3: 66; AR: 373].

Serjania lucida Schumach., Skr. Naturhist.-Selsk. 3(2):128. 1794.

Vine in Northern Coastal Lowlands (islands adjacent to Fajardo: Icacos; Fajardo to Carolina), Eastern Coast (Ceiba), Vieques, Culebra; 1–200 m; *fl* Sep–Mar, *fr* Jan–Mar.

Virgin Islands. [AR: 374].

Notes: Britton & Wilson (1924) and Liogier (1994) placed this taxon in synonymy with *Serjania polyphylla*.

Bejuco de canastas, Bejuco de corrales, Bejuco de costillas.

Serjania polyphylla (L.) Radlk., Monogr. *Serjania* 179. 1875.
Bas.: *Paullinia polyphylla* L. (1753).

Vine in Northern Coastal Lowlands, Eastern Coast (Ceiba), Northern Limestone Hills, Río Abajo Forest Reserve, Las Tetas de Cayey, Las Piedras Chiquitas, Southeastern Foothills, Eastern Cordillera (Comerío), Southern Coastal Lowlands (Caja de Muertos), Susúa Forest Reserve, Guánica Forest Reserve, Southwestern Coastal Lowlands, Sierra Bermeja, Western Coast, Culebra, Mona; 1–825 m; *fl* Aug–Apr, *fr* Nov–Mar.

Greater Antilles (Hispaniola), Virgin Islands. [L3: 66; AR: 374].

Bejuco de canastas, Bejuco de corrales, Bejuco de costillas.

Thouinia striata Radlk. var. **striata** in Engler & Prantl, Nat. Pflanzenfam. 3(5):311. 1895.
Syn.: *Thyana striata* (Radlk.) Britton (1924).

Shrub/small tree in Northern Limestone Hills, Río Abajo Forest Reserve, Las Piedras Chiquitas, Southeastern Foothills (Coamo), Eastern Cordillera, Southern Coastal Limestone, Susúa Forest Reserve, Western Cordillera, Southwestern Coastal Lowlands (Cabo Rojo), Sierra Bermeja; 25–600 m; *fl* Aug–Apr, *fr* Aug–Jan, May.

Endemic. [L3: 69].

Ceboruquillo, Guabá, Quebracha, Quiebrahacha, Serrasuela.

Thouinia striata Radlk. var. **portoricensis** (Radlk.) Votava & Alain, Phytologia 50(3):166. 1982.
Bas.: *Thouinia portoricensis* Radlk. (1895).
Syn.: *Thyana portoricensis* (Radlk.) Britton (1924).

Shrub/small tree in Southern Coastal Limestone, Guánica Forest Reserve, Southwestern Coastal

Lowlands (Lajas), Sierra Bermeja; 1–300 m; *fl* Mar–Oct, *fr* Sep–Jun.

Endemic. [L3: 68].

Ceboruquillo, Quebracho, Serrasuela.

TAXA CITED BY PREVIOUS AUTHORS, BUT NOT INCLUDED

CITED DUE TO MISIDENTIFICATION

Matayba oppositifolia (A. Rich.) Britton.

Tree cited by Britton & Wilson (1925) as occurring in mountainous forests. According to Little et al. (1974), the barren specimens upon which this citation was based were referable to *Matayba domingensis*. [B&W5: 528].

ONLY UNDER CULTIVATION

Blighia sapida K.D. Koenig.

Tree cited by Liogier & Martorell (1982, 2000) and Liogier (1994) as rarely planted and persistent. [L3: 47].

Cardiospermum grandiflorum Sw.

Vine cited by Acevedo-Rodríguez (2005) as naturalized at the Experimental Station in Mayagüez. [AR: 368].

Paullinia fuscescens Kunth.

Vine cited by Acevedo-Rodríguez (2005) as naturalized at the Experimental Station in Mayagüez. [AR: 370].

SAPOTACEAE

Chrysophyllum argenteum Jacq., Enum. Syst. Pl. 15. 1760.

Tree in Northern Coastal Lowlands, Northern Limestone Hills, Río Abajo Forest Reserve, Sierra de Luquillo, Eastern Cordillera (La Torrecilla), Central Cordillera; 150–1000 m; *fl* Jun–Nov, *fr* Jan–Jun.

Greater Antilles (Cuba, Hispaniola), Lesser Antilles, Trinidad and Tobago, Central America, northern South America to Bolivia, Paraguay. [L4: 137].

Caimitillo, Caimito verde, Lechecillo.

Chrysophyllum bicolor Poir., Encycl. Suppl. 2(1): 15, *t. 185*. 1811.

Tree in Northern Limestone Hills (Quebradillas), Vieques; 25–225 m; *fl* Jan (one specimen), *fr* Nov, Jan (two specimens). [*Breckon 9108* (MAPR); *Vives a.n. 46652* (UPRRP)].

Greater Antilles (Hispaniola), Virgin Islands. [L4: 139].

Notes: This taxon was treated by Pennington (1990) as conspecific with *Chrysophyllum cainito*; Acevedo-Rodríguez (1996) maintains it as distinct taxon because of striking differences in the fruits of the two (that of *Chrysophyllum bicolor* is asymmetrically ellipsoid rather than globose). Gary Breckon (pers. comm.) thinks the *Chrysophylum cainito* may be a cultivar of *Chrysophyllum bicolor*.

Caimitillo, Caimito cimarrón, Lechecillo.

Chrysophyllum cainito L., Sp. Pl. 1:192. 1753.

Tree planted and persistent in Northern Coastal Lowlands, Guajataca Forest Reserve, Southwestern Coastal Lowlands (San Germán); 1–15 m; *fl* Jul–Nov, *fr* Apr (one specimen).

Probably native to Greater Antilles. [L4: 139].

Caimito.

Chrysophyllum oliviforme L., Syst. Nat. (ed. 10) 2:937. 1759.

Tree in Northern Limestone Hills, Central Cordillera (Adjuntas), Southern Coastal Limestone (Yauco), Maricao Forest Reserve, Guánica Forest Reserve, Southwestern Coastal Lowlands (Cabo Rojo), Western Coast (Rincón), Mona; 3–900 m; *fl* Jul–Nov, *fr* Oct–Mar.

Bahamas, Greater Antilles, southeastern United States (Florida). [L4: 140].

Caimitillo, Caimitillo de perro, Teta de burra.

Chrysophyllum pauciflorum Lam., Tabl. Encycl. 2:44. 1794.

Tree in Cambalache Forest Reserve, Eastern Coast, Las Tetas de Cayey, Las Piedras Chiquitas, Southeastern Foothills, Eastern Cordillera (Cayey), Susúa Forest Reserve, Maricao Forest Reserve, Vieques; 25–825 m; *fl* May–Jan, *fr* Sep, Dec–Jan.

Virgin Islands. [L4: 140].

Caimito de perro.

Manilkara bidentata (A. DC.) A. Chev., Rev. Bot. Appl. Agric. Trop. 12(128):270. 1932.
Bas.: *Mimusops bidentata* A. DC. (1844).
Syn.: *Manilkara nitida* (Sessé & Moc.) Dubard (1915).

Tree in Northern Coastal Lowlands, Northern Limestone Hills (Arecibo), Río Abajo Forest Reserve, Sierra de Luquillo, Eastern Coast (Maunabo); 1–450 m; *fl* May, Sep (two specimens), *fr* Nov–Mar.

Greater Antilles (Hispaniola), Virgin Islands (St. John, Tortola), Lesser Antilles, Trinidad, southern Central America (Panama), northern South America to Brazil, Bolivia. [L4: 141].

Acana, Ausubo, Balatá.

Manilkara pleeana (Pierre ex Baill.) Cronquist, Bull. Torrey Bot. Club 72(6):554. 1945.
Bas.: *Mimusops pleeana* Pierre ex Baill. (1891).
Syn.: *Manilkara duplicata* (Sessé & Moc.) Dubard (1915).

Tree in Northern Coastal Lowlands (Fajardo, Luquillo), Northern Limestone Hills (Quebradillas, Isabela), Eastern Coast (Cuchilla de Panduras), Southern Coastal Limestone (Guayanilla), Susúa Forest Reserve, Guánica Forest Reserve, Southwestern Coastal Lowlands (Lajas), Sierra Bermeja; 1–200 m; *fl* throughout the year, *fr* Jun–Sep, Dec–Jan.

Endemic. [L4: 143].

Acana, Ausuba, Ausubo machuelo, Mameyuelo, Zapote de costa, Zapote negro.

Manilkara valenzuelana (A. Rich.) T.D. Penn., Fl. Neotrop. Monogr. 52:84, *ff. 12a, 13.* 1990.
Bas.: *Lucuma valenzuelana* A. Rich. (1850).
Syn.: *Manilkara albescens* (Griseb.) Cronquist (1945).

Tree in Northern Limestone Hills (Quebradillas, Isabela), Eastern Cordillera (Cañón San Cristóbal); 125–300 m; *fl* Jun (two specimens), *fr* Jan–Mar.

Greater Antilles (Cuba, Hispaniola). [L4: 143].

Ausubo de costa, Nisperillo.

Manilkara zapota (L.) P. Royen, Blumea 7(2):410. 1953.
Bas.: *Achras zapota* L. (1753).
Syn.: *Sapota achras* Mill. (1768).

Tree planted and persistent in Northern Coastal Lowlands, Southern Coastal Lowlands (Ponce), Southwestern Coastal Lowlands (Cabo Rojo), Sierra Bermeja; 15–200 m; *fl* May–Nov, *fr* Jul, Oct, Jan–Feb.

Native to Mexico, Central America. [L4: 144].

Notes: In Britton & Wilson (1925) the homonym *Achras zapota* L. (1759) is treated as the basionym of *Sapota achras*, while *Achras zapota* L. (1753) is treated as a separate taxon with an accompanying description that fits *Pouteria sapota*.

Níspero.

Micropholis garciniifolia Pierre, Not. Bot. 38. 1891.

Tree in Sierra de Luquillo; 450–1025 m; *fl* Jun–Sep, Dec, *fr* throughout the year.

Endemic. [L4: 145].

Caimitillo, Caimitillo blanco, Caimitillo cimarrón, Caimitillo verde, Guayabota.

Micropholis guyanensis (A. DC.) Pierre, Not. Bot. 40. 1891.
Bas.: *Sideroxylon guyanense* A. DC. (1844).
Syn.: *Micropholis chrysophylloides* Pierre (1891). *Micropholis curvata* (Pierre) Pierre & Urb. (1904) "curvatae."

Tree in Sierra de Luquillo, Sierra de Cayey, Central Cordillera, Maricao Forest Reserve; 350–1250 m; *fl* throughout the year, *fr* Oct–Jun.

Lesser Antilles, southern Central America, northwestern, west-central South America and Venezuela, Brazil. [L4: 145].

Caimitillo, Leche prieta, Mesa.

Pouteria dictyoneura (Griseb.) Radlk. subsp. **fuertesii** (Urb.) Cronquist, Lloydia 9(4):267. 1946.
Bas.: *Paralabatia fuertesii* Urb. (1912).
Syn.: *Paralabatia portoricensis* Britton & P. Wilson (1926).

Tree in Northern Coastal Lowlands (Carolina), Northern Limestone Hills, Río Abajo Forest Reserve, Eastern Cordillera (Cañón San Cristóbal), Susúa Forest Reserve; 10–375 m; *fl* May–Jun, *fr* Jul–Aug.

Greater Antilles (Hispaniola). [L4: 148].

Pouteria hotteana (Urb. & Ekman) Baehni, Candollea 9:423, *ff. 94, 95a*. 1942.
Bas.: *Labatia hotteana* Urb. & Ekman (1929).

Tree in Maricao Forest Reserve; 425–800 m; *fl* Aug (one specimen), *fr* Sep–Nov.

Greater Antilles (Hispaniola). [L4: 148].

Almendrón.

Pouteria multiflora (A. DC.) Eyma, Meded. Bot. Mus. Herb. Rijks Univ. Utrecht 27:164; et Recueil Trav. Bot. Néerl. 33:164. 1936.
Bas.: *Lucuma multiflora* A. DC. (1844).

Tree in Northern Coastal Lowlands (Río Grande), Northern Limestone Hills, Sierra de Luquillo, Eastern Coast (Yabucoa), Sierra de Cayey, Eastern Cordillera (Aibonito), Central Cordillera, Susúa Forest Reserve, Maricao Forest Reserve; 1–800 m; *fl* Feb–Oct, *fr* throughout the year.

Greater Antilles (Jamaica), Virgin Islands (St. John), Lesser Antilles, Trinidad, southern Central America (Panama), northwestern, west-central South America and Venezuela. [L4: 149].

Jácana.

Sideroxylon cubense (Griseb.) T.D. Penn., Fl. Neotrop. Monogr. 52:129. 1990.
Bas.: *Bumelia cubensis* Griseb. (1866).
Syn.: *Dipholis sintenisiana* Pierre (1904).

Tree in Northern Limestone Hills, Río Abajo Forest Reserve, Toro Negro Forest Reserve, Susúa Forest Reserve, Maricao Forest Reserve, Cerro Las Mesas; 250–1300 m; *fl* May–Nov, *fr* Sep–Mar.

Greater Antilles (Cuba, Hispaniola), Lesser Antilles (St. Kitts). [L4: 152].

Boje, Espejuelo, Quiebrahacha.

Sideroxylon foetidissimum Jacq. subsp. **foetidissimum**, Enum. Syst. Pl. 15. 1760.
Syn.: *Masticodendron foetidissimum* (Jacq.) H.J. Lam. (1939).

Tree in Northern Coastal Lowlands, Northern Limestone Hills, Southern Coastal Lowlands (Guayama), Guánica Forest Reserve, Southwestern Coastal Lowlands (Lajas), Sierra Bermeja, Vieques, Culebra, Mona; 3–150 m; *fl* Jul–Sep, *fr* Aug–Mar.

West Indies, Trinidad and Tobago, southeastern United States (Florida), Mexico, northern Central America (Guatemala, Belize). [L4: 152].

Tabloncillo, Tortugo, Tortugo amarillo, Tortugo colorado.

Sideroxylon obovatum Lam., Tabl. Encycl. 2:42. 1794.
Syn.: *Bumelia obovata* (Lam.) A. DC. (1844). *Bumelia krugii* Pierre (1904).

Tree in Northern Coastal Lowlands (Fajardo), Northern Limestone Hills (Arecibo), Río Abajo Forest Reserve, Southern Coastal Lowlands (Santa Isabel), Southern Coastal Limestone, Guánica Forest Reserve, Southwestern Coastal Lowlands, Sierra Bermeja, Vieques, Mona, Desecheo; 1–325 m; *fl* & *fr* throughout the year.

Greater Antilles (Hispaniola), Virgin Islands, Lesser Antilles, northern South America (Venezuelan islands). [L4: 153].

Alquitrán, Araña gato, Lechecillo, Peine, Quiebrahacha.

Sideroxylon portoricense Urb. subsp. **portoricense**, Symb. Antill. 5(1):134. 1904.
Syn.: *Mastichodendron portoricense* (Urb.) Alain (1985).
Dipholis bellonis Urb. (1904). *Bumelia bellonis* (Urb.) Stearn (1968).

Tree in Northern Limestone Hills, Río Abajo Forest Reserve, Central Cordillera; 130–800 m; *fl* Jul–Aug, *fr* Feb–Mar.

Greater Antilles. [L4: 155].

Tabloncillo, Varital.

Sideroxylon salicifolium (L.) Lam., Tabl. Encycl. 2:42. 1794.
Bas.: *Achras salicifolia* L. (1762).
Syn.: *Bumelia salicifolia* (L.) Sw. (1788). *Dipholis salicifolia* (L.) A. DC. (1844).

Tree in Northern Limestone Hills, Río Abajo Forest Reserve, Sierra de Luquillo, Las Piedras Chiquitas, Southeastern Foothills (Coamo), Southern Coastal Limestone (Guayanilla), Guánica Forest Reserve, Southwestern Coastal Lowlands (Cabo Rojo), Sierra Bermeja, Vieques, Mona; 25–700 m; *fl* Feb–Aug, *fr* May–Sep.

West Indies, southeastern United States (Florida), Mexico, northern Central America (Guatemala, Belize). [L4: 155].

Almendrón, Sabina, Sanguinaria.

TAXA CITED BY PREVIOUS AUTHORS, BUT NOT INCLUDED

DOUBTFUL OCCURRENCE

Manilkara jaimiqui (C. Wright ex Griseb.) Dubard subsp. ***emarginata*** (L.) Cronquist.

Tree cited by Britton & Wilson (1926) as occurring between Lajas and La Parguera, southwestern Puerto Rico; subsequently cited by Little et al. (1974) and by Liogier & Martorell (1982) as rare, but not cited thereafter. [B&W6: 366].

Notes: Britton & Wilson (1926) used the homonym *Manilkara emarginata* Britton & P. Wilson for the name of this taxon; Liogier (1982) used *Manilkara jaimiqui* (C. Wright ex Griseb.) Dubard, *sensu lato*, for it.

ONLY UNDER CULTIVATION

Pouteria campechiana (Kunth) Baehni.

Tree cited by Liogier & Martorell (1982, 2000) and Liogier (1995) as planted for its fruit and escaped. [L4: 147].

Pouteria sapota (Jacq.) H.E. Moore & Stearn.

Tree cited by Britton & Wilson (1924) as occurring in woodlands and hillsides at lower elevations and planted for its fruit and for shade; subsequently cited by Liogier & Martorell (1982, 2000) and Liogier (1995) as rarely planted at lower elevations. [L4: 149].

Notes: Britton & Wilson (1925) used *Achras zapota* L. (1753) for the name of this taxon.

SCHLEGELIACEAE

Schlegelia brachyantha Griseb., Cat. Pl. Cub. 191. 1866.
Syn.: *Schlegelia portoricensis* (Urb.) Britton (1925). *Schlegelia brachyantha* Griseb. subsp. *portoricensis* (Urb.) Borhidi (1975).

Woody vine in Northern Limestone Hills (Arecibo), Río Abajo Forest Reserve, Sierra de Luquillo, Sierra de Cayey, Las Tetas de Cayey, Eastern Cordillera (Cerro La Peña, La Torrecilla), Central Cordillera, Maricao Forest Reserve; 300–1250 m; *fl* & *fr* throughout the year.

Greater Antilles (Cuba, Hispaniola), southern Central America, northern South America (Venezuela, Colombia). [L4: 543; AR: 377].

Higüerito de sierra, Tulipa.

SCHOEPFIACEAE

Schoepfia arenaria Britton in Urban, Symb. Antill. 5(2):181. 1907.

Shrub in Northern Coastal Lowlands (Fajardo),

Northern Limestone Hills, Río Abajo Forest Reserve; 5–395 m; *fl* Dec–Mar, Jun, Sep, *fr* Jan–Jul.

Endemic. [L1: 94].

Notes: This taxon is on the U.S. Fish & Wildlife Service list of threatened and endangered plants.

Schoepfia obovata C. Wright in Sauvalle, Anales Acad. Ci. Méd. Habana 5:289. 1868.

Shrub/small tree in Southern Coastal Lowlands (Caja de Muertos), Southern Coastal Limestone (Peñuelas), Susúa Forest Reserve, Maricao Forest Reserve, Guánica Forest Reserve, Southwestern Coastal Lowlands (Guánica), Vieques; 150–275 m; *fl* Aug–May, *fr* throughout the year.

Bahamas, Greater Antilles (Cuba, Hispaniola), Virgin Islands (St. John, Anegada). [L1: 95].

Araña.

Schoepfia schreberi J.F. Gmel., Syst. Nat. 2(1):376. 1791.

Syn.: *Schoepfia chrysophylloides* (A. Rich.) Planch. (1854).

Tree in Northern Limestone Hills (coastal); 15–35 m; *fl* Feb–May, *fr* Feb–Mar.

West Indies, Trinidad and Tobago, Mexico, Central America, northern South America (Venezuela, Colombia). [L1: 96].

SCROPHULARIACEAE (including BUDDLEJACEAE AND MYOPORACEAE)

Bontia daphnoides L., Sp. Pl. 2:638. 1753.

Shrub/small tree in Northern Coastal Lowlands, Eastern Coast (Cayo Santiago), Southern Coastal Lowlands, Southwestern Coastal Lowlands (Cabo Rojo), Vieques; 1–3 m; *fl* Oct–Apr, Jul, *fr* throughout the year.

West Indies, northern South America (Guyana, Venezuela). [L5: 60].

Mangle, Mangle bobo, Manzanilla.

Buddleja davidii Franch., Nouv. Arch. Mus. Hist. Nat., sér. 2 10:65. 1887.

Shrub in Toro Negro Forest Reserve; 800 m; *fl* Mar–May (two specimens), *fr* (?). [*Liogier 35080* (UPR)].

Native to China. [L4: 182].

Heliotropo arbóreo.

Buddleja madagascariensis Lam., Encycl. 1(2):513. 1785.

Scandent shrub in Toro Negro Forest Reserve; 800 m; *fl* Apr, Sep (two specimens), *fr* (?). [*Liogier 33103* (UPR); *Acevedo-Rodríguez 11454* (NY), scan seen].

Native to Madagascar. [L4: 184; AR: 141].

Capraria biflora L., Sp. Pl. 2:628. 1753.

Suffrutescent herb in Northern Coastal Lowlands, Eastern Coast (Cayo Santiago), Las Piedras Chiquitas, Southeastern Foothills, Southern Coastal Lowlands, Guánica Forest Reserve, Southwestern Coastal Lowlands, Sierra Bermeja, Western Coast (Rincón), Vieques, Culebra, Mona; 1–500 m; *fl* & *fr* throughout the year.

West Indies, southern United States (Florida, Texas), Mexico, Central America, South America. [L4: 491].

Té del país.

Scrophularia minutiflora Pennell, Proc. Acad. Nat. Sci. Philadelphia 75:18. 1923.

Herb in Central Cordillera; 600–1100 m; *fl* Dec–Feb, *fr* Dec–Feb, Jun.

Greater Antilles. [L4: 505].

Perejil cimarrón.

SIMAROUBACEAE

Castela erecta Turpin, Ann. Mus. Nat. Hist. 7:80, *t. 5b*. 1806.

Syn.: *Castelaria nicholsonii* (Hook.) Small (1911).

Shrub in Southwestern Coastal Lowlands (Cabo Rojo); 1–25 m; *fl* Oct–Mar, *fr* Oct (one specimen).

Virgin Islands (St. Croix), Lesser Antilles (Anguilla, Antigua, Barbuda), northern South America (Venezuela, Colombia). [L2: 278].

Picrasma excelsa (Sw.) Planch., London J. Bot. 5:574. 1846.

Bas.: *Quassia excelsa* Sw. (1788).

Syn.: *Aeschrion excelsa* (Sw.) Kuntze (1891).

Tree in Northern Limestone Hills, Las Piedras Chiquitas, Central Cordillera (Adjuntas); 10–600 m; *fl* Sep–Oct (two specimens), *fr* Jul (one specimen).

Antilles, Central America (Nicaragua, Costa Rica), northwestern, west-central South America and Venezuela. [L2: 281].

Leña amarga, Palo amargo.

Simarouba tulae Urb., Jahrb. Königl. Bot. Gart. Berlin 4:245. 1886.

Shrub/small tree in Northern Limestone Hills (Arecibo, Isabela), Río Abajo Forest Reserve, Sierra de Luquillo, Sierra de Cayey, Susúa Forest Reserve, Maricao Forest Reserve; 150–825 m; *fl* Feb–Oct, *fr* Aug–Mar.

Endemic. [L2: 285].

Aceitillo cimarrón, Aceitillo falso.

TAXA CITED BY PREVIOUS AUTHORS, BUT NOT INCLUDED

ONLY UNDER CULTIVATION

Quassia amara L.

Small tree cited by Britton & Wilson (1924) as planted for ornament and interest, occasionally spontaneous after cultivation; subsequently cited by Liogier & Martorell (1982, 2000) and Liogier (1995). [L2: 283].

SMILACACEAE

Smilax coriacea Spreng., Syst. Veg. 2:103. 1825.

Vine in Northern Coastal Lowlands, Northern Limestone Hills, Río Abajo Forest Reserve, Sierra de Luquillo, Sierra de Cayey, Las Piedras Chiquitas, Southeastern Foothills (Cayey), Susúa Forest Reserve, Maricao Forest Reserve, Guánica Forest Reserve, Southwestern Coastal Lowlands, Western Coast (Rincón), Vieques; 1–1000 m; *fl* Feb–Sep, *fr* Feb–Nov.

Bahamas, Greater Antilles (Hispaniola), Virgin Islands, Lesser Antilles. [AR&S: 79; AR: 443].

Notes: According to Acevedo-Rodríguez (2005), Liogier & Martorell (1982) misapplied the name *Smilax havanensis* Jacq. to this taxon and wrongly treated *Smilax coriacea* as a synonym of it; Liogier & Martorell (2000), used the correct name, but wrongly maintained the synonymy.

Dunguey, Dunguey blanco.

Smilax domingensis Willd., Sp. Pl. 4(2):783. 1806.

Vine in Northern Coastal Lowlands (San Juan, Laguna Tortuguero), Northern Limestone Hills, Río Abajo Forest Reserve, Sierra de Cayey, Central Cordillera, Susúa Forest Reserve, Maricao Forest Reserve; 1–1000 m; *fl* Mar–Sep, *fr* Jul–Mar.

Bahamas, Greater Antilles, Virgin Islands (St. John), Mexico, Central America, northwestern, west-central South America and Venezuela. [AR&S: 80; AR: 444].

Bejuco de membrillo, Raíz de zarzaparilla, Zarzaparilla.

SOLANACEAE

Acnistus arborescens (L.) Schltdl., Linnaea 7(1):67. 1832.

Bas.: *Atropa arborescens* L. (1756).

Syn.: *Dunalia arborescens* (L.) Sleumer (1950).

Shrub in Sierra de Cayey, Las Tetas de Cayey, Southeastern Foothills, Eastern Cordillera, Central Cordillera (Peñuelas); 400–1025 m; *fl* Jan–Mar, Jun, Sep–Oct, *fr* Oct, Jan, Apr–May.

Antilles, Mexico, southern Central America, northern South America to Brazil, Peru. [L4: 424].

Galán arbóreo, Gallinero, Palo de gallina.

Browallia americana L., Sp. Pl. 2:631. 1753.

Herb in Central Cordillera, Western Cordillera; 400–950 m; *fl* & *fr* throughout the year.

Antilles, Mexico, Central America, northwestern, west-central South America and Venezuela, Brazil. [L4: 426].

Catalina, Margarita morada, Pensamiento de pobre, Teresita.

Brugmansia ×candida Pers. (pro sp.), Syn. Pl. 1:216. 1805.

Syn.: *Datura candida* (Pers.) Saff. (1921).

Shrub in Northern Coastal Lowlands (Trujillo Alto), Eastern Cordillera (Comerío), Central Cordillera, Western Cordillera (Maricao); 200–600 m; *fl* Sep–Nov, Feb–Mar, *fr* (?).

Possibly native to Peru. [L4: 428].

Notes: This name represents the hybid *Brugmansia aurea ×versicolor.*

Cornucopia, Floripondio.

Brugmansia suaveolens (Humb. & Bonpl. ex Willd.) Bercht. & J. Presl, Přir. Rostlin 1(Solanaceae):45. 1823.

Bas.: *Datura suaveolens* Humb. & Bonpl. ex Willd. (1809).

Shrub in Sierra de Luquillo, Sierra de Cayey, Central Cordillera, Western Cordillera (Maricao); 300–875 m; *fl* Nov, Feb–Jun, *fr* (?).

Native to central South America (Brazil, Peru, Bolivia). [L4: 430].

Aguacero, Campana, Campana blanca, Campana de Paris, Campana del río, Floripondio.

Brunfelsia americana L., Sp. Pl. 1:191. 1753.

Shrub in Northern Limestone Hills, Southeastern Foothills (Coamo), Susúa Forest Reserve, Southwestern Coastal Lowlands (Lajas), Vieques; 85–225 m; *fl* Mar–Jun, Nov, *fr* Apr–Jul.

Greater Antilles (Hispaniola), Virgin Islands, Lesser Antilles, Trinidad. [L4: 431].

Aguacero, Alhelí falso, Trompeta de ángel, Tulipán sencillo.

Brunfelsia densifolia Krug & Urb., Notizbl. Königl. Bot. Gart. Berlin 1(10):324. 1897.

Shrub in Maricao Forest Reserve; 600–900 m; *fl* Mar–Jun, Sep, *fr* Jan, May–Sep.

Endemic. [L4: 431].

Brunfelsia lactea Krug & Urb., Notizbl. Königl. Bot. Gart. Berlin 1(10):323. 1897.

Shrub in Sierra de Luquillo, Sierra de Cayey, Eastern Cordillera (La Torrecilla), Central Cordillera (Adjuntas); 600–1150 m; *fl* Jan–Oct, *fr* throughout the year.

Endemic. [L4: 432].

Aguacero, Jazmín de monte, Vega blanca.

Brunfelsia nitida Benth. in A. de Candolle, Prodr. 10:201. 1846.

Shrub/small tree planted and persistent in Northern Coastal Lowlands (Dorado); 50 m; *fl* Mar–Sep, *fr* (?).

Native to Greater Antilles (Cuba). [L4: 432].

Brunfelsia portoricensis Krug & Urb., Notizbl. Königl. Bot. Gart. Berlin 1(10):322. 1897.

Shrub in Northern Coastal Lowlands (Quebradillas), Northern Limestone Hills (Arecibo, Isabela), Sierra de Luquillo, Maricao Forest Reserve; 1–725 m; *fl* Apr–Jun, Nov, *fr* Apr–Jun (two specimens).

Endemic. [L4: 432].

Brunfelsia sp.

Shrub in Guajataca Forest Reserve; 250 m; *fl* May–Jun (one specimen), *fr* (?). [*Vives a.n. 45704* (UPRRP)].

Notes: Specimens of this taxon show affinity to *Brunfelsia abbottii* Leonard, which is endemic to Hispaniola, but more material is yet needed to determine its identity.

Capsicum annuum L. var. **glabriusculum** (Dunal) Heiser & Pickersgill, Baileya 19(4):156. 1975. Bas.: *Capsicum hispidum* Dunal var. *glabriusculum* Dunal (1852).

Suffrutescent herb/shrub in Northern Coastal Lowlands, Northern Limestone Hills, Sierra de Cayey, Las Tetas de Cayey, Las Piedras Chiquitas, Southeastern Foothills, Susúa Forest Reserve, Southwestern Coastal Lowlands, Sierra Bermeja, Vieques, Culebra; 1–800 m; *fl* & *fr* throughout the year.

West Indies, southeastern and south-central United States, Mexico, Central America, northern South America (Colombia). [L4: 436].

Ají, Ají picante, Ají caballero, Ají pico de paloma.

Capsicum frutescens L., Sp. Pl. 1:189. 1753.

Suffrutescent herb/shrub in Northern Coastal Lowlands, Northern Limestone Hills (Toa Baja), Río Abajo Forest Reserve, Eastern Coast (Ceiba), Las Tetas de Cayey, Southeastern Foothills (Salinas), Mona; 1–825 m; *fl* & *fr* throughout the year.

Native to Central America, northern South America to Brazil, Peru. [L4: 436].

Ají, Ají picante, Ají caballero.

Cestrum citrifolium Retz. in Hoffmann, Phytogr. Bl. 1:36. 1803.

Shrub in Northern Coastal Lowlands, Northern Limestone Hills, Eastern Coast, Las Piedras Chiquitas, Southeastern Foothills, Central Cordillera, Susúa Forest Reserve, Maricao Forest Reserve, Southwestern Coastal Lowlands (Lajas), Western Coast (Rincón), Vieques; 1–650 m; *fl* Sep–Jun, *fr* Dec–Jun.

Greater Antilles (Cuba, Hispaniola), Lesser Antilles. [L4: 438].

Notes: Britton & Wilson (1925), Liogier & Martorell (1982, 2000) and Liogier (1995), among many others, misapplied the name *Cestrum laurifolium* L'Hér. to this taxon, since, according to Michael Nee (unpubl. ms.), the type of the former is the same as that for *Cestrum macrophyllum*.

Dama de noche, Galán del monte.

Cestrum diurnum L., Sp. Pl. 1:191. 1753.

Shrub in Northern Coastal Lowlands, Northern Limestone Hills, Caguas Valley (Caguas), Eastern Cordillera (Comerío), Western Coast; 1–400 m; *fl* & *fr* throughout the year.

Bahamas, Greater Antilles. [L4: 438].

Dama de día, Galán de día.

Cestrum macrophyllum Vent., Choix Pl. *t. 18*. 1803.

Shrub in Northern Limestone Hills, Río Abajo Forest Reserve, Sierra de Luquillo, Eastern Coast (Cuchilla de Panduras), Caguas Valley (Caguas), Sierra de Cayey, Las Tetas de Cayey, Eastern and Central Cordillera, Maricao Forest Reserve; 200–1250 m; *fl* Aug–Apr, *fr* Nov–Jun.

Greater Antilles (Hispaniola). [L4: 440].

Notes: According to Michael Nee (unpubl. ms.), although the type for the name *Cestrum laurifolium* is the same as that for *Cestrum macrophyllum* and the former name has priority (1788), it is a *nom. illeg. superfl.*, as L'Héritier cited *Cestrum venenatum* Mill. (1768) in synonymy.

Galán de monte.

Cestrum nocturnum L., Sp. Pl. 1:191. 1753.

Scandent shrub in Northern Limestone Hills (Isabela, Moca), Río Abajo Forest Reserve, Caguas Valley (Caguas); 200–400 m; *fl* Oct–Jun, *fr* Feb–May.

Native to Greater Antilles (Cuba, Jamaica), Mexico, Central America. [L4: 440].

Dama de noche.

Datura inoxia Mill., Gard. Dict. (ed. 8): *Datura* no. 5. 1768.

Herb in Northern Coastal Lowlands, Southern Coastal Lowlands, Guánica Forest Reserve, Southwestern Coastal Lowlands, Sierra Bermeja, Vieques, Culebra; 1–200 m; *fl* & *fr* throughout the year.

Antilles (possibly naturalized), south-central United States (Texas), Mexico, Central America, northwestern, west-central South America and Venezuela, Paraguay. [L4: 442].

Notes: Britton & Wilson (1925) misapplied the name *Datura metel* to this taxon.

Campana blanca, Campana del pasto, Chamisco blanco, Chamisco doble.

Datura metel L., Sp. Pl. 1:179. 1753.

Syn.: *Datura fastuosa* L. (1759).

Herb in Northern Coastal Lowlands, Southern Coastal Lowlands (Salinas); 1–75 m; *fl* & *fr* Sep–Apr.

Possibly originating in the West Indies. [L4: 442].

Chamisco amarillo, Chamisco morado.

Datura stramonium L., Sp. Pl. 1:179. 1753.

Herb in Caguas Valley (Gurabo), Southern Coastal Lowlands (Coamo); 50–100 m; *fl* & *fr* Aug, Nov (two specimens). [*Woodbury a.n. 28507* (UPR); Axelrod 8500 (UPRRP)].

Probably native to Mexico. [L4: 442].

Belladona de pobre, Chamisco, Estramonio, Peo de fraile.

Goetzea elegans Wydler, Linnaea 5(3):423, *t. 8*. 1830.

Tree in Northern Limestone Hills (Arecibo to Isabela), Vieques; 20–200 m; *fl* Apr–May, Aug–Nov, *fr* Apr, Aug–Sep.

Endemic. [L4: 444].

Notes: This taxon is on the U.S. Fish & Wildlife Service list of threatened and endangered plants.

Manzanilla, Mata buey.

Jaltomata antillana (Krug & Urb.) D'Arcy, Ann. Missouri Bot. Gard. 65(2):780. 1979.

Bas.: *Saracha antillana* Krug & Urb. (1895).

Herb in Toro Negro Forest Reserve; 1300 m; *fl* & *fr* Jul (one specimen). [*Liogier 9533* (NY)].

Greater Antilles. [L4: 447].

Lycianthes virgata (Lam.) Bitter, Abh. Naturwiss. Vereins Bremen 24(2):370. 1920.

Bas.: *Solanum virgatum* Lam. (1794).

Vine in Northern Limestone Hills (Florida, Quebradillas), Río Abajo Forest Reserve, Sierra de Cayey, Las Tetas de Cayey, Eastern Cordillera (La Torrecilla), Central Cordillera; 85–1100 m; *fl* Apr–Nov, *fr* Apr, Jul–Dec.

Greater Antilles (Cuba, Hispaniola). [L4: 440; AR: 380].

Berenjena de paloma.

Lycium americanum Jacq., Select. Stirp. Amer. Hist. 50. 1763.

Shrub in Mona (Monito); 1–5 m; *fl* Jun (one specimen), *fr* Jun–Aug (two specimens). [*Woodbury Mt-27* (UPR); *Breckon 5160* (MAPR)].

Greater Antilles (Cuba, Hispaniola), Lesser Antilles, Trinidad, South America (scattered). [L4: 449].

Physalis angulata L., Sp. Pl. 1:183. 1753.

Herb in Northern Coastal Lowlands, Caguas Valley (Caguas), Eastern Cordillera (Aibonito), Southern Coastal Lowlands, Southwestern Coastal Lowlands, Western Coast, Vieques, Culebra; 1–600 m; *fl* & *fr* throughout the year.

Throughout tropical, subtropical, and warm temperate regions. [L4: 456].

Alquequenje, Sacabuche.

Physalis cordata Mill., Gard. Dict. (ed. 8): *Physalis* no. 14. 1768.

Herb in Southeastern Foothills (Coamo); 300 m; *fl* & *fr* Aug–Sep (one specimen). [*del Llano a.n. 1008447* (NY), scan seen].

West Indies, Mexico, Central America, northwestern, west-central South America and Venezuela, Brazil. [L4: 457].

Alquequenje, Sacabuche.

Physalis ignota Britton, Mem. Torrey Bot. Club 16(2):100. 1920.

Herb in Southwestern Coastal Lowlands (Guánica); 1–5 m; *fl* & *fr* Dec (one specimen). [*Liogier 35745* (NY), scan seen].

Greater Antilles, southern Mexico, Central America, northern South America (Guyana).

Physalis pubescens L., Sp. Pl. 1:183. 1753.

Syn.: *Physalis turbinata* Medik. (1780).

Herb in Northern Coastal Lowlands, Northern Limestone Hills (Arecibo), Río Abajo Forest Reserve, Las Piedras Chiquitas, Southwestern Coastal Lowlands (Guánica), Vieques; 1–550 m; *fl* & *fr* Sep–Jun.

West Indies, United States, Mexico, Central America, South America. [L4: 457].

Alquequenje, Sacabuche peludo.

Solandra grandiflora Sw., Kongl. Vetensk. Acad. Nya Handl. 8:300, *t. 11*. 1787.

Syn.: *Swartzia grandiflora* J.F. Gmel. (1791).

Woody vine in Northern Limestone Hills (Quebradillas), Sierra de Luquillo, Central Cordillera (Juana Díaz), Maricao Forest Reserve; 600–750 m; *fl* Jul–Nov, *fr* (?).

Antilles, southern Mexico, Central America, northwestern, west-central South America and Venezuela. [L4: 458; AR: 380].

Bejuco de peo.

Solanum americanum Mill., Gard. Dict. (ed. 8): *Solanum* no. 5. 1768.

Syn.: *Solanum caribaeum* Dunal (1852).

Suffrutescent herb in Northern Coastal Lowlands, Northern Limestone Hills, Río Abajo Forest Reserve, Eastern Coast, Sierra de Cayey, Las Tetas de Cayey, Las Piedras Chiquitas, Eastern and Central Cordillera, Southern Coastal Lowlands, Southwestern Coastal Lowlands (Lajas), Western Coast (Aguada), Vieques, Culebra; 1–950 m; *fl* & *fr* throughout the year.

West Indies, southern and western United States, Mexico, Central America, South America. [L4: 461].

Notes: Britton & Wilson (1925) used *Solanum nigrum* L., *sensu lato*, for the name of this taxon.

Mata gallina, Yerba mora.

Solanum bahamense L., Sp. Pl. 1:188. 1753.

Syn.: *Solanum racemosum* Jacq. (1760).

Solanum persicifolium Dunal (1813).

Shrub in Northern Coastal Lowlands, Southeastern Foothills, Southern Coastal Lowlands, Southern Coastal Limestone, Guánica Forest Reserve, Southwestern Coastal Lowlands, Sierra Bermeja, Western Coast (Aguada), Vieques, Culebra, Mona, Desecheo; 1–425 m; *fl* & *fr* throughout the year.

West Indies, southeastern United States (Florida). [L4: 470, 472].

Berenjena de playa, Rubia.

Solanum campechiense L., Sp. Pl. 1:187. 1753.

Syn.: *Solanum guanicense* Urb. (1899).

Herb in Northern Limestone Hills (San Sebastián), Southwestern Coastal Lowlands (Lajas); 5–200 m; *fl* Feb (one specimen), *fr* Feb, May–Jul.

Greater Antilles, Mexico, Central America, west-central South America. [L4: 463].

Solanum capsicoides All., Auct. Syn. Meth. Stirp. Hort. Regii Taur. 12. 1773 (preprint of Mélanges Philos. Math. Soc. Roy. Turin 5:64. 1774).

Suffrutescent herb in Northern Limestone Hills, Río Abajo Forest Reserve, Eastern Cordillera, Central Cordillera (Ciales), Maricao Forest Reserve; 150–700 m; *fl* Apr–May, Aug–Jan, *fr* throughout the year.

Native to Brazil. [L4: 463].

Solanum elaeagnifolium Cav., Icon. 3(1):22, *t. 243.* 1795–1796.

Suffrutescent herb in Northern Coastal Lowlands (Vega Baja, Manatí), Southern Coastal Lowlands, Guánica Forest Reserve, Southwestern Coastal Lowlands, Sierra Bermeja; 2–150 m; *fl* Mar–Nov, *fr* Jun–Nov.

Bahamas, Greater Antilles (Jamaica), Virgin Islands (Anegada), Southern Caribbean Islands (Curaçao), United States (except northern part), Mexico, southern South America. [L4: 464].

Solanum ensifolium Dunal in A. de Candolle, Prodr. 13(1):186. 1852.

Syn.: *Solanum drymophilum* O.E. Schulz (1909).

Suffrutescent herb/shrub in Northern Limestone Hills (Florida, Arecibo), Río Abajo Forest Reserve, Las Tetas de Cayey, Southeastern Foothills (Cayey), Southern Coastal Lowlands (Coamo); 70–825 m; *fl* & *fr* throughout the year.

Endemic. [L4: 464].

Notes: This taxon is on the U.S. Fish & Wildlife Service list of threatened and endangered plants.

Erubia.

Solanum erianthum D. Don, Prodr. Fl. Nepal. 96. 1825.

Shrub/small tree in Northern Limestone Hills, Southern Coastal Limestone (Peñuelas), Guánica Forest Reserve, Mona; 1–200 m; *fl* Mar–Nov, *fr* May–Dec.

Bahamas, Greater Antilles, Virgin Islands, southern United States (Florida, Texas), Mexico, Central America, west-central South America. [L4: 465].

Notes: Britton & Wilson (1925) misapplied the name *Solanum verbascifolium* L. to this taxon.

Berenjena de paloma, Tabacón afelpado.

Solanum jamaicense Mill., Gard. Dict. (ed. 8): *Solanum* no. 17. 1768.

Straggling shrub in Sierra de Luquillo, Eastern Coast (Cuchilla de Panduras), Sierra de Cayey; 400–500 m; *fl* & *fr* May–Sep.

Greater Antilles, Lesser Antilles (Martinique, St. Lucia, Grenada), Tobago, Mexico, Central America, northwestern, west-central South America and Venezuela. [L4: 465].

Solanum lycopersicum L., Sp. Pl. 1:185. 1753.

Syn.: *Lycopersicon lycopersicum* (L.) H. Karst. (1882).

Lycopersicon esculentum Mill. (1768), *nom. cons.*

Herb planted and persistent in Northern Coastal Lowlands, Eastern Coast (Naguabo), Eastern Cordillera (Aibonito); 10–300 m; *fl* Feb–Jun, Oct, *fr* Jun, Oct (two specimens).

Possibly native to Mexico, Central America, western South America (Ecuador, Peru). [L4: 451].

Tomate.

Solanum mammosum L., Sp. Pl. 1:187. 1753.

Suffrutescent herb/shrub in Northern Coastal Lowlands, Caguas Valley, Eastern Cordillera (Aibonito), Western Coast (Mayagüez); 15–500 m; *fl* Sep–Jan, Apr, *fr* Apr, Jul–Jan.

Antilles, Mexico, Central America, northwestern, west-central South America and Venezuela. [L4: 467].

Berenjena de cucarachas, Berenjena de marimbo.

Solanum nigrescens M. Martens & Galeotti, Bull. Acad. Roy. Sci. Bruxelles 12(1):140. 1845.

Herb in Sierra de Luquillo; 300–600 m; *fl* Apr (one specimen), *fr* (?). [*McClain 358* (NY)].

Native to Mexico, Central America, South America. [L4: 468].

Solanum nudum Dunal, Solan. Syn. 20. 1816.

Syn.: *Solanum antillarum* O.E. Schulz (1909).

Tree in Northern Limestone Hills (Arecibo), Río Abajo Forest Reserve, Sierra de Luquillo, Sierra de Cayey, Eastern Cordillera (La Torrecilla), Central Cordillera; 225–1100 m; *fl* May–Sep, *fr* Jul–Dec.

Greater Antilles (Cuba, Hispaniola), Lesser Antilles (St. Vincent, Grenada), Mexico, Central America, northwestern, west-central South America and Venezuela. [L4: 468].

Solanum polygamum Vahl, Symb. Bot. 3:39, *t. 55*. 1794.

Shrub in Vieques, Culebra; 100–140 m; *fl* Jun (one specimen), *fr* Jun (one specimen). [*Griffis a.n. 744863* (NY); *Woodbury a.n. 17407, 174111, 17413* (SJ)].

Greater Antilles (Hispaniola), Virgin Islands. [L4: 470].

Solanum rugosum Dunal in A. de Candolle, Prodr. 13(1):108. 1852.

Shrub/small tree in Northern Limestone Hills (Florida), Sierra de Luquillo, Sierra de Cayey, Las Tetas de Cayey, Eastern Cordillera (Corozal, Barranquitas including La Torrecilla), Central Cordillera, Maricao Forest Reserve, Western Coast (Mayagüez); 200–1000 m; *fl* Jan–Oct, *fr* throughout the year.

Greater Antilles (Hispaniola), Lesser Antilles (Guadeloupe, Dominica, Martinique), Central America, northern South America to Brazil, Peru. [L4: 472].

Sepí, Tabacón, Tabacón áspero.

Solanum seaforthianum Andrews, Bot. Repos. 8: *t. 504*. 1808.

Vine in Northern Coastal Lowlands, Sierra de Cayey, Eastern Cordillera (Barranquitas); 1–600 m; *fl* Aug–May, *fr* Aug–Feb, May.

Native to (though possibly naturalized in some) Trinidad, Mexico, Central America, northern South America (Venezuela, Colombia). [L4: 473; AR: 383].

Falsa belladonna, Jazmín de Italia.

Solanum torvum Sw., Prodr. 47. 1788.

Shrub in Northern Coastal Lowlands, Northern Limestone Hills, Río Abajo Forest Reserve, Sierra de Luquillo, Eastern Coast, Caguas Valley, Sierra de Cayey, Las Tetas de Cayey, Southeastern Foothills, Eastern and Central Cordillera, Susúa Forest Reserve, Maricao Forest Reserve, Southwestern Coastal Lowlands, Sierra Bermeja, Western Coast, Vieques; 1–1050 m; *fl* & *fr* throughout the year.

West Indies, Mexico, Central America, northern South America to Brazil, Ecuador. [L4: 473].

Berenjena cimarrona.

Solanum viarum Dunal in A. de Candolle, Prodr. 13(1):240. 1852.

Suffrutescent herb/shrub in Northern Limestone Hills, Western Coast (Mayagüez); 2–250 m; *fl* & *fr* Mar–Jun.

Native to southern South America and Brazil.

Solanum wendlandii Hook. f., Bot. Mag. 113: *t. 6914*. 1887.

Herb planted and persistent in Central Cordillera; 700 m; *fl* May (one specimen), *fr* (?). [*Liogier 28969* (UPR)].

Native to southern Mexico, Central America, northwestern and west-central South America. [L4: 474; AR: 383].

Campana de Pascua, Flor de la mañana, Papa aérea.

Solanum woodburyi R.A. Howard, J. Arnold Arbor. 47(2):138, *t. 2*. 1966.

Shrub in Sierra de Luquillo; 775–800 m; *fl* May–Jul, *fr* Jun–Jul (one specimen).

Endemic. [L4: 475].

TAXA CITED BY PREVIOUS AUTHORS, BUT NOT INCLUDED

NO SPECIMENS SEEN

Cestrum alternifolium (Jacq.) O.E. Schulz.

Shrub cited by Britton & Wilson (1925) as recorded as found by Eggers in the Luquillo Mountains; subsequently cited by Liogier & Martorell (1982, 2000) and Liogier (1995). [L4: 437].

Cestrum salicifolium Jacq.

Shrub cited by Liogier & Martorell (1982, 2000) and Liogier (1995), according to Francey, collected by Sintenis in woods in Bayamón. [L4: 441].

Solanum lanceifolium Jacq. "lanceaefolium."
Woody vine cited by Liogier & Martorell (1982, 2000) and Liogier (1995) as occurring in a forest, Vieques, and by Acevedo-Rodríguez (2005), who refers only to the former citation. [L4: 466; AR: 382].

Solanum pterocaulon Dunal.
Herb cited by Britton & Wilson (1930) and Liogier & Martorell (1982) as found by Otto Kuntze at Caguas in 1874. [L4: 468].

Notes: Liogier (1995) wrongly considered this taxon to be a synonym of *Solanum nigrescens*.

DOUBTFUL OCCURRENCE
Solanum riedlei Dunal.
Shrub cited by Britton & Wilson (1925) as a species doubtfully attributed to Porto Rico, collected by Riedlé, and not known from elsewhere; subsequently cited by Liogier & Martorell (1982, 2000) and Liogier (1995). [L4: 472].

CITED DUE TO MISIDENTIFICATION
Solanum mucronatum O.E. Schulz.
Shrub cited by Britton & Wilson (1925) as recorded as occurring in mountain woodlands in Calabazas, near Coamo; subsequently cited by Liogier & Martorell (1982, 2000) and Liogier (1995). According to Acevedo-Rodríguez (1996), this taxon was known only from three sterile specimens, which, he now reports (pers. comm.), have been referred by Sandra Knapp to *Cordia alliodora*. [L4: 468].

Solanum pyrifolium Lam.
Vine cited by Liogier (1995) and Liogier & Martorell (2000) as collected at Cabezas de San Juan, Fajardo. According to Acevedo-Rodríguez (2005), this citation was based upon the misidentification of a specimen of *Solanum seaforthianum*. [L4: 471].

ONLY UNDER CULTIVATION
Capsicum annuum L. var. ***annuum***.
Herb cited by and Liogier (1995) and Liogier & Martorell (2000) as cultivated and escaped. [L4: 434].

Nicotiana tabacum L.
Herb cited by Britton & Wilson (1925) as occasionally spontaneous after cultivation; subsequently cited by Liogier & Martorell (1982, 2000) and Liogier (1995). [L4: 452].

Petunia integrifolia (Hook.) Schinz & Thell. Syn.: *Petunia violacea* Lindl.
Herb cited by Liogier & Martorell (1982, 2000) and Liogier (1995) as cultivated as an ornamental and subspontaneous. [L4: 454].

Solanum melongena L.
Herb cited by Liogier & Martorell (1982, 2000) and Liogier (1995) as widely grown for its fruit and sometimes escaped. [L4: 467].

Solanum quitoense Lam.
Shrub cited by Liogier & Martorell (1982, 2000) and Liogier (1995) as introduced for its fruit, planted in different localities, and collected as an escape in the Toro Negro Mountains. [L4: 471].

Solanum tuberosum L.
Herb cited by Liogier & Martorell (1982, 2000) and Liogier (1995) as cultivated and persistent in the mountains. [L4: 474].

SPHENOCLEACEAE

Sphenoclea zeylanica Gaertn., Fruct. Sem. Pl. 1:113, *t. 24, f. 5.* 1788.
Syn.: *Pongatium zeylanicum* (Gaertn.) Kuntze (1891).

Herb in Northern Coastal Lowlands, Eastern Coast, Caguas Valley, Southwestern Coastal Lowlands (Lajas); 1–100 m; *fl* & *fr* throughout the year.
Native to tropical Africa. [L5: 216].
Campanilla.

STAPHYLEACEAE

Turpinia occidentalis (Sw.) G. Don, Gen. Hist. 2:3. 1832.
Bas.: *Staphylea occidentalis* Sw. (1788).
Syn.: *Turpinia paniculata* Vent. (1807).

Tree in Northern Limestone Hills, Río Abajo Forest Reserve, Eastern Coast (Cuchilla de Panduras), Sierra de Cayey, Las Tetas de Cayey, Eastern Cordillera (La Torrecilla), Central Cordillera, Maricao Forest Reserve; 300–1300 m; *fl* Mar–Aug, *fr* Jun–Nov.

Antilles, Mexico, Central America, northwestern, west-central South America and Venezuela, Brazil. [L3: 37].

Avispillo, Cedro hembra, Eugenio, Lilayo, Palo mingo, Saúco cimarrón, Sauquillo.

STRELITZIACEAE

TAXA CITED BY PREVIOUS AUTHORS, BUT NOT INCLUDED

ONLY UNDER CULTIVATION
Ravenala madagascariensis Sonn.
Tree cited by Liogier & Martorell (1982, 2000) as occasionally planted and persistent.

STYRACACEAE

Styrax portoricensis Krug & Urb., Bot. Jahrb. Syst. 15:337. 1892.

Tree in Sierra de Luquillo, Sierra de Cayey; 300–400 m; *fl* (?), *fr* (?).

Endemic. [L4: 164].

Notes: This taxon is on the U.S. Fish & Wildlife Service list of threatened and endangered plants.

Palo de jazmín.

SURIANACEAE

Suriana maritima L., Sp. Pl. 1:284. 1753.

Shrub along coast in Northern Coastal Lowlands, Eastern Coast (Punta Guayanés), Southern Coastal Lowlands, Guánica Forest Reserve, Southwestern Coastal Lowlands, Vieques, Mona; 1–5 m; *fl* & *fr* throughout the year.

West Indies, southeastern United States (Florida), Central America, northern South America, Madagascar, Polynesia, New Guinea, Australia. [L2: 285].

Guitarán, Temporana.

SYMPLOCACEAE

Symplocos lanata Krug & Urb., Bot. Jahrb. Syst. 15:335. 1892.

Tree in Sierra de Luquillo, Toro Negro Forest Reserve; 800–1300 m; *fl* Mar–Jun, *fr* Jul–Aug.

Endemic. [L4: 161].

Níspero cimarrón.

Symplocos martinicensis Jacq., Enum. Syst. Pl. 24. 1760.

Syn.: *Symplocos polyantha* Krug & Urb. (1892).

Symplocos latifolia Krug & Urb. (1892).

Tree in Northern Coastal Lowlands, Guajataca Forest Reserve, Sierra de Luquillo, Caguas Valley (Gurabo), Eastern Cordillera (Aguas Buenas), Central Cordillera (Utuado), Maricao Forest Reserve, Cerro Las Mesas; 1–900 m; *fl* Feb–Jul, *fr* Feb–Sep.

Greater Antilles (Cuba, Jamaica), Virgin Islands, Lesser Antilles, Trinidad, northern Central America (Belize), northern South America (Guyana, Colombia). [L4: 161, L5: 410].

Notes: Liogier (1997) and Liogier & Martorell (2000) treated *Symplocos latifolia* as a separate taxon, but I do not believe that the character differences merit such a separation.

Aceituna, Aceituna blanca, Aceituna cimarrona, Cuero de sapo.

Symplocos micrantha Krug & Urb., Bot. Jahrb. Syst. 15:336. 1892.

Tree in Sierra de Luquillo, Toro Negro Forest Reserve; 500–1300 m; *fl* Dec–May, *fr* Apr–May (two specimens).

Endemic. [L4: 163].

Aceitunillo.

TALINACEAE

Talinum fruticosum (L.) Juss., Gen. Pl. 312. 1789.

Bas.: *Portulaca fruticosa* L. (1759).

Syn.: *Talinum triangulare* (Jacq.) Willd. (1799).

Herb in Northern Coastal Lowlands (San Juan), Northern Limestone Hills, Caguas Valley (Caguas), Central Cordillera (Ponce), Southern Coastal Lowlands (Ponce), Guánica Forest Reserve, Southwestern Coastal Lowlands, Sierra Bermeja, Culebra; 1–300 m; *fl* throughout the year, *fr* Apr–Jan.

Bahamas, Antilles, Mexico, Central America, South America. [L1: 207].

Verdolaga francesa, Verdolaguilla.

Talinum paniculatum (Jacq.) Gaertn., Fruct. Sem. Pl. 2(2):219, *t. 128, f. 13.* 1791.

Bas.: *Portulaca paniculata* Jacq. (1760).

Herb in Northern Coastal Lowlands (Cataño), Las Tetas de Cayey, Las Piedras Chiquitas, Southeastern Foothills, Eastern Cordillera, Central Cordillera (Juana Díaz), Cerro de Las Cuevas, Southern Coastal Lowlands (Yauco), Guánica Forest Reserve, Southwestern Coastal Lowlands (Cabo Rojo), Sierra Bermeja, Vieques, Culebra, Mona, Desecheo; 30–825 m; *fl* & *fr* Mar–Dec.

Antilles, Trinidad, southern United States, Mexico, Central America, South America (scattered). [L1: 206].

TAMARICACEAE

TAXA CITED BY PREVIOUS AUTHORS, BUT NOT INCLUDED

ONLY UNDER CULTIVATION

Tamarix aphylla (L.) H. Karst. Syn.: *Tamarix articulata* Vahl.

Tree cited by Britton & Wilson (1930) as grown at the Forest Station, Río Piedras and by Liogier & Martorell (1982, 2000) and Liogier (1994) as rarely planted and persistent around houses in dry areas. [L3: 242].

Tamarix gallica L.

Tree cited by Britton & Wilson (1930) as seen in a yard in Juana Díaz and by Liogier & Martorell (1982, 2000) and Liogier (1994) as rarely planted for ornament and persistent. [L3: 242].

TETRACHONDRACEAE

Polypremum procumbens L., Sp. Pl. 1:111. 1753.

Low herb in Northern Coastal Lowlands (Vega Alta to Manatí); 1–5 m; *fl* & *fr* Sep–Nov, Feb–Jun.

Bahamas, Greater Antilles (Cuba, Jamaica), southern United States, Mexico, Central America, northern South America to Brazil, Peru. [L4: 186].

THEACEAE

Laplacea portoricensis (Krug & Urb.) Dyer, Index Kew. Suppl. 2:86. 1904.

Bas.: *Haemocharis portoricensis* Krug & Urb. (1896).

Tree in Sierra de Luquillo, Sierra de Cayey; 250–800 m; *fl* Dec–Jan (two specimens), *fr* (?). [*Proctor 46380, 51518* (SJ); *Flores Colón a.n. 34984* (UPR)].

Greater Antilles (Hispaniola). [L3: 222].

Maricao, Maricao verde, Niño de cota.

THEOPHRASTACEAE

Bonellia pauciflora (B. Ståhl & F.S. Axelrod) B. Ståhl & Källersjö, Novon 14(1):117. 2004.

Bas.: *Jacquinia pauciflora* B. Ståhl & F.S. Axelrod (1998).

Shrub in Northern Limestone Hills (Quebradillas, Isabela); 250–350 m; *fl* Apr–Jun, *fr* Jul (one specimen).

Endemic. [L4: 114].

Bonellia umbellata (A. DC.) B. Ståhl & Källersjö, Novon 14(1):118. 2004.

Bas.: *Jacquinia umbellata* A. DC. (1844).

Shrub in Northern Limestone Hills (Camuy, Quebradillas), Southeastern Foothills (Coamo), Southern Coastal Limestone (Yauco), Susúa Forest Reserve, Western Cordillera, Guánica Forest Reserve, Sierra Bermeja; 150–750 m; *fl* Mar–Jun, *fr* throughout the year.

Greater Antilles (Hispaniola). [L4: 115].

Chirre, Chirriador.

Jacquinia arborea Vahl, Eclog. Amer. 1:26. 1797.

Syn.: *Jacquinia barbasco* Mez (1903).

Shrub/small tree in Northern Coastal Lowlands, Southern Coastal Lowlands, Southern Coastal Limestone (Guayanilla), Guánica Forest Reserve, Southwestern Coastal Lowlands, Vieques, Culebra, Mona; 1–30 m; *fl* Nov–May, *fr* Mar–Oct.

Antilles. [L4: 113].

Notes: Howard (1989) and Liogier (1997) misapplied the name *Jacquinia armillaris* Jacq. to this taxon.

Azúcares, Barbasco, Bizcocho.

Jacquinia berteroi Spreng., Syst. Veg. 1:668. 1824 "Berterii."

Shrub/small tree in Northern Coastal Lowlands (islands adjacent to Fajardo), Northern Limestone Hills (Quebradillas), Southern Coastal Limestone, Guánica Forest Reserve, Sierra Bermeja, Vieques, Culebra; 10–300 m; *fl* May–Nov, *fr* throughout the year.

Bahamas, Greater Antilles (Cuba, Hispaniola), Virgin Islands, Lesser Antilles. [L4: 113].

Mercocha.

TAXA CITED BY PREVIOUS AUTHORS, BUT NOT INCLUDED

CITED DUE TO MISIDENTIFICATION

Bonellia stenophylla (Urb.) B. Ståhl & Källersjö.

Syn.: *Jacquinia stenophylla* Urb.

Shrub cited by Liogier & Martorell (1982) and Liogier (1995) as occurring in dry thickets, Guánica. There is at UPR a Liogier specimen of this taxon that was collected in Guantanamo, Cuba, but which bears a label with a map of Puerto Rico. That specimen is probably the source of this citation. [L4: 114].

THYMELAEACEAE

Daphnopsis americana (Mill.) J.R. Johnst. subsp. **caribaea** (Griseb.) Nevling, Ann. Missouri Bot. Gard. 46(4):315. 1960.

Bas.: *Daphnopsis caribaea* Griseb. (1860).

Tree in Northern Coastal Lowlands (Fajardo), Sierra de Luquillo, Eastern Coast, Sierra de Cayey, Las Tetas de Cayey, Southeastern Foothills (Coamo), Eastern Cordillera (Cerro La Peña), Central Cordillera (Adjuntas, Yauco), Vieques; 25–975 m; *fl* Jun–Mar, *fr* Sep–Feb.

Lesser Antilles, Tobago, southern Central America, northern South America (Venezuela, Colombia). [L3: 340].

Emajagua de sierra, Majagua de sierra, Maho, Mahout.

Daphnopsis helleriana Urb., Symb. Antill. 2(3):453. 1901.

Shrub/small tree in Northern Limestone Hills, Río Abajo Forest Reserve; 50–350 m; *fl* Feb–Mar, Jun–Oct, *fr* Oct, Jan–Mar, Jul.

Endemic. [L3: 340].

Notes: This taxon is on the U.S. Fish & Wildlife Service list of threatened and endangered plants.

Daphnopsis philippiana Krug & Urb., Bot. Jahrb. Syst. 15:348. 1892.

Shrub/small tree in Sierra de Luquillo, Sierra de Cayey, Eastern Cordillera (La Torrecilla), Central Cordillera, Maricao Forest Reserve; 350–1300 m; *fl* throughout the year, *fr* Jan–Oct.

Endemic. [L3: 340].

Cieneguillo, Emajagua brava, Emajagua de sierra, Majagua brava, Majagua quemadora, Quemadora.

TROPAEOLACEAE

Tropaeolum majus L., Sp. Pl. 1:345 "minus;" 2: errata. 1753.

Vine planted and escaped in Las Tetas de Cayey, Central Cordillera (Adjuntas); 500–1000 m; *fl* Nov, Feb–Mar, *fr* (?).

Only cultivated; probably of hybrid origin in South America. [L2: 241; AR: 385].

Jacinto.

TURNERACEAE

Piriqueta cistoides (L.) Griseb., Fl. Brit. W.I. 298. 1860.

Bas.: *Turnera cistoides* L. (1762).

Herb in Northern Coastal Lowlands, Northern Limestone Hills, Eastern Coast, Caguas Valley (Gurabo), Southern Coastal Lowlands (Guayama), Susúa Forest Reserve, Western Coast (Mayagüez); 1–200 m; *fl* & *fr* throughout the year.

West Indies, southeastern United States, Mexico, Central America, South America. [L3: 283].

Notes: Britton & Wilson (1924) used *Piriqueta cistoides* (L.) G. Mey. ex Steud., a *nom. nud.*, for the name of this taxon.

Piriqueta racemosa (Jacq.) Sweet, Hort. Brit. (ed. 1) 154. 1826.

Bas.: *Turnera racemosa* Jacq. (1777).
Syn.: *Piriqueta ovata* (Bello) Urb. (1883).

Herb in Northern Limestone Hills (Quebradillas), Southwestern Coastal Lowlands, Sierra Bermeja, Western Coast (Rincón), Mona; 1–200 m; *fl* & *fr* throughout the year.

Greater Antilles (Hispaniola), Southern Caribbean Islands (Curaçao, Bonaire), South America (scattered). [L3: 283].

Piriqueta viscosa Griseb., Cat. Pl. Cub. 114. 1866.

Herb in Northern Coastal Lowlands (Laguna Tortuguero), Eastern Coast (Punta Guayanés); 25 m; *fl* & *fr* Feb, Aug (two specimens). [*Woodbury a.n. 5276* (UPR); *Axelrod 2015, 2744* (UPRRP)].

Greater Antilles (Cuba), Virgin Islands (St. Thomas), southern Central America, South America (Venezuela, Brazil, Bolivia). [L3: 284].

Turnera diffusa Willd. in Schultes, Syst. Veg. 6:679. 1820.

Suffrutescent herb in Southern Coastal Limestone, Susúa Forest Reserve, Guánica Forest Reserve, Southwestern Coastal Lowlands, Sierra Bermeja, Vieques, Mona; 2–300 m; *fl* throughout the year, *fr* Feb–Jun, Sep–Nov.

Bahamas, Greater Antilles, Virgin Islands, south-central United States (Texas), Mexico, Central America, South America (Brazil). [L3: 285].

Damiana.

Turnera ulmifolia L., Sp. Pl. 1:271. 1753.

Suffrutescent herb in Northern Coastal Lowlands, Eastern Coast, Eastern Cordillera (Aibonito), Southwestern Coastal Lowlands (Cabo Rojo), Western Coast, Vieques, Culebra; 1–50 m; *fl* & *fr* throughout the year.

West Indies, southeastern United States (Florida), Mexico, Central America, northwestern, west-central South America and Venezuela. [L3: 286].

Marilópez.

TAXA NOT RECENTLY COLLECTED

Turnera pumilea L., Syst. Nat. (ed. 10) 2:965. 1759.

Herb collected in Coamo on 9 Dec 1885 [*Sintenis 2995* (US)] and on 9 Feb 1922 [*Sintenis 5959* (NY)] and at Punta Casabe (Cabo Rojo) on 26 Feb 1913 [*Britton et al. 4659* (NY)].

Greater Antilles, Southern Caribbean Islands (Curaçao), tropical South America. [L3: 286].

TYPHACEAE

Typha domingensis Pers., Syn. Pl. 2(2):532. 1807.

Herb in Northern Coastal Lowlands, Río Abajo Forest Reserve, Eastern Coast, Central Cordillera, Southern Coastal Lowlands (Yauco), Southwestern Coastal Lowlands (Lajas), Western Coast (Mayagüez), Vieques; 1–1000 m; *fl* throughout the year, *fr* Mar–May.

Bahamas, Greater Antilles, Virgin Islands, southern half of United States, Mexico, Central America, South America (scattered). [AR&S: 393].

Notes: Britton & Wilson (1923) misapplied the name *Typha angustifolia* L. to this taxon.

Aneas, Eneas, Yerba de eneas.

ULMACEAE

Celtis iguanaea (Jacq.) Sarg., Silva 7:64. 1895 "iguanaeus."

Bas.: *Rhamnus iguanaea* Jacq. (1760) "iguanaeus."
Syn.: *Momisia iguanaea* (Jacq.) Rose & Standl. (1912).

Scandent shrub in Northern Coastal Lowlands,

Northern Limestone Hills (Hatillo, Quebradillas), Eastern Coast (Cayo Santiago), Las Piedras Chiquitas, Southeastern Foothills, Southern Coastal Lowlands, Guánica Forest Reserve, Southwestern Coastal Lowlands (Lajas), Vieques, Culebra, Desecheo; 1–500 m; *fl* Jun–Jul, *fr* Oct–Jul.

West Indies, Trinidad, southeastern United States (Florida, Alabama), Mexico, Central America, South America. [L1: 46; AR: 387].

Azufaifo.

Celtis trinervia Lam., Encycl. 4(1):140. 1797.

Tree in Northern Coastal Lowlands (Fajardo), Northern Limestone Hills, Las Piedras Chiquitas, Southeastern Foothills (Guayama), Southern Coastal Limestone (Guayanilla), Guánica Forest Reserve, Southwestern Coastal Lowlands, Vieques, Mona, Desecheo; 100–550 m; *fl* Jun–Jul, Nov, *fr* Nov–Jan.

Greater Antilles, Virgin Islands, Mexico, Central America (scattered). [L1: 46].

Almez, Guacimilla, Raspador, Ruiseñor.

Trema lamarckianum (Schult.) Blume, Mus. Bot. 2:58. 1856 "lamarckiana."

Bas.: *Celtis lamarckiana* Schult. (1820).

Tree in Río Abajo Forest Reserve, Southeastern Foothills (Coamo), Eastern Cordillera (Cañón San Cristóbal), Susúa Forest Reserve, Western Cordillera; 150–850 m; *fl* Feb–Oct, *fr* throughout the year.

West Indies, southeastern United States (Florida). [L1: 48].

Cabrilla, Palo de cabrilla.

Trema micranthum (L.) Blume, Mus. Bot. 2:58. 1856 "micrantha."

Bas.: *Rhamnus micrantha* L. (1759).

Tree in Northern Coastal Lowlands, Río Abajo Forest Reserve, Sierra de Luquillo, Eastern Coast (Punta Guayanés), Sierra de Cayey (San Lorenzo), Las Piedras Chiquitas, Southeastern Foothills, Eastern and Central Cordillera, Maricao Forest Reserve, Vieques; 1–1100 m; *fl* Dec–Sep, *fr* throughout the year.

Antilles, southeastern United States (Florida), Mexico, Central America, South America. [L1: 49].

Cabra, Guacimilla, Palo de cabra.

URTICACEAE (including CECROPIACEAE)

Boehmeria cylindrica (L.) Sw., Prodr. 34. 1788.

Bas.: *Urtica cylindrica* L. (1753).

Shrub in Northern Limestone Hills (Ciales), Río Abajo Forest Reserve, Toro Negro Forest Reserve; 150–1330 m; *fl* & *fr* throughout the year.

Greater Antilles, Trinidad, Canada, eastern and central United States, Central America, South America. [L1: 72].

Boehmeria repens (Griseb.) Wedd. in A. de Candolle, Prodr. 16(1):201. 1869.

Bas.: *Pouzolzia repens* Griseb. (1860).

Herb in Northern Limestone Hills (Ciales), Río Abajo Forest Reserve, Sierra de Cayey, Eastern Cordillera (Cayey); 200–640 m; *fl* & *fr* Oct, Jan–Feb.

Greater Antilles (Cuba, Hispaniola), Lesser Antilles (Dominica, Grenada). [L1: 74].

Cecropia schreberiana Miq. subsp. **antillarum** (Snethl.) C.C. Berg & P. Franco, Fl. Neotrop. Monogr. 94:167, *t. 43.7.* 2005.

Bas.: *Cecropia antillarum* Snethl. (1923).

Tree in Northern Limestone Hills (Isabela), Sierra de Luquillo, Eastern Coast (Punta Guayanés), Vieques; 25–400 m; *fl* & *fr* May–Aug.

Greater Antilles (Hispaniola), Virgin Islands.

Cecropia schreberiana Miq. subsp. **schreberiana** in Martius, Fl. Bras. 4(1):150. 1853.

Tree in Northern Coastal Lowlands (San Juan), Northern Limestone Hills, Sierra de Luquillo, Eastern Coast (Cuchilla de Panduras), Caguas Valley (Las Piedras), Sierra de Cayey, Las Tetas de Cayey, Eastern Cordillera (Cerro La Peña, Aibonito), Central Cordillera, Maricao Forest Reserve, Cerro Las Mesas; 15–1300 m; *fl* & *fr* Jul–Apr.

Antilles. [L1: 55].

Notes: Britton & Wilson (1924) and Liogier & Martorell (1982) misapplied the name *Cecropia peltata* L. to this taxon.

Grayumo hembra, Llagrumo, Llagrumo hembra, Yagrumo, Yagrumo hembra.

Laportea aestuans (L.) Chew, Gard. Bull. Singapore 21(2):200. 1965.

Bas.: *Urtica aestuans* L. (1759).

Herb in Northern Coastal Lowlands, Río Abajo Forest Reserve, Caguas Valley, Eastern Cordillera (Cayey, Aibonito), Southwestern Coastal Lowlands (Cabo Rojo), Western Coast (San Germán), Vieques; 5–250 m; *fl* & *fr* Jul–Apr.

Bahamas, Greater Antilles (Jamaica, Hispaniola), Mexico, Central America, northern South America to Peru, southern tropical Africa, western Indian Ocean (Madagascar, Mauritius, Reunion). [L1: 75].

Notes: Britton & Wilson (1924) used *Fleurya aestuans* (L.) Gaudich., a *nom. inval.*, for the name of this taxon.

Picapica.

Phenax sonneratii (Poir.) Wedd. in A. de Candolle, Prodr. 16(1):235(37). 1869.
Bas.: *Parietaria sonneratii* Poir. (1804).

Herb in Río Abajo Forest Reserve, Eastern Cordillera (Cañón San Cristóbal), Central Cordillera (Ciales); 225–700 m; *fl* & *fr* May–Aug.

Native to tropical Asia.

Pilea depressa (Sw.) Blume, Mus. Bot. 2(3):46. 1856.
Bas.: *Urtica depressa* Sw. (1787).

Herb planted and persistent in Eastern Coast (Cuchilla de Panduras), Eastern Cordillera (Barranquitas); 450–700 m; *fl* & *fr* Oct (one specimen). [*Liogier 30014, 31022* (UPR)].

Greater Antilles. [L1: 78].

Pilea herniarioides (Sw.) Wedd., Ann. Sci. Nat., Bot., sér. 3 18:207. 1852.
Bas.: *Urtica herniarioides* Sw. (1787).

Prostrate herb in Eastern Cordillera (Cañón San Cristóbal), Toro Negro Forest Reserve; 450–1260 m; *fl* & *fr* Dec–Jan.

Bahamas, Greater Antilles, Virgin Islands (Anegada), Lesser Antilles (Antigua, Montserrat), southeastern United States (Florida, Alabama), Mexico, Central America. [L1: 78].

Notes: Britton & Wilson (1924), Liogier (1985) and Liogier & Martorell (2000) used *Pilea herniarioides* (Sw.) Lindl., a *nom. inval.*, for the name of this taxon.

Pilea hyalina Fenzl, Nov. Gen. Sp. Vasc. 4. 1849 (preprint of Denkschr. Kaiserl. Akad. Wiss., Wien. Math.-Naturwiss. Kl. 1:256. 1850).

Herb in Western Coast (San Germán); 120 m; *fl* & *fr* Nov (one specimen). [*Breckon 8101* (MAPR)].

Native to Mexico, Central America, South America.

Pilea inaequalis (Juss. ex Poir.) Wedd., Ann. Sci. Nat. Bot., sér 3 18:229. 1852.
Bas.: *Urtica inaequalis* Juss. ex Poir. (1816).

Herb in Northern Limestone Hills, Río Abajo Forest Reserve, Sierra de Luquillo, Sierra de Cayey, Central Cordillera; 55–1300 m; *fl* & *fr* throughout the year.

Antilles, Trinidad and Tobago. [L1: 79].

Pilea krugii Urb., Symb. Antill. 1(2):298. 1899.

Herb in Sierra de Luquillo, Sierra de Cayey; 350–1050 m; *fl* & *fr* Mar–Nov.

Endemic. [L1: 79].

Pilea leptophylla Urb., Symb. Antill. 1(2):297. 1899.

Herb in Eastern Coast; 100–360 m; *fl* & *fr* Mar, Jun, Sep–Oct.

Endemic. [L1: 79].

Pilea margarettae Britton in Britton & Wilson, Sci. Surv. Porto Rico & Virgin Islands 5(2):246. 1924.

Herb in Northern Limestone Hills, Mona, Las Piedras Chiquitas; 5–600 m; *fl* & *fr* Oct–Nov, Feb–May.

Greater Antilles (Cuba), Virgin Islands (St. Croix, Anegada). [L1: 80].

Pilea microphylla (L.) Liebm., Kongel. Danske Vidensk. Selsk. Skr., Naturvidensk. Math. Afd., ser. 5 2:296. 1851 (Mexic. Neldeagt. Pl.).
Bas.: *Parietaria microphylla* L. (1759).

Herb in Northern Coastal Lowlands, Northern Limestone Hills, Río Abajo Forest Reserve, Sierra de Luquillo, Las Tetas de Cayey, Las Piedras Chiquitas, Southeastern Foothills (Coamo), Eastern Cordillera (Corozal), Central Cordillera, Southern Coastal Lowlands (Salinas), Susúa Forest Reserve, Western Cordillera, Vieques, Mona, Desecheo; 10–825 m; *fl* & *fr* throughout the year.

West Indies, southeastern and south-central United States, Mexico, Central America, northwestern, west-central South America and Venezuela, Argentina. [L1: 81].

Madreperla.

Pilea multicaulis Urb., Symb. Antill. 1(2):295. 1899.

Herb in Sierra de Luquillo; 650–750 m; *fl* (?), *fr* (?). [*Liogier 28534* (UPR)].

Endemic. [L1: 81].

Pilea nummulariifolia (Sw.) Wedd., Ann. Sci. Nat., Bot., sér. 3 18:225. 1852 "nummulariaefolia."
Bas.: *Urtica nummulariifolia* Sw. (1787) "nummularifolia."

Creeping herb in Northern Coastal Lowlands (Trujillo Alto), Northern Limestone Hills, Río Abajo Forest Reserve, Sierra de Cayey, Las Tetas de Cayey, Las Piedras Chiquitas, Central Cordillera (Ponce), Maricao Forest Reserve, Vieques; 10–825 m; *fl* & *fr* Jun–Jul.

Antilles, Mexico, Central America. [L1: 81].

Yerba de culebra.

Pilea obtusata Liebm., Kongel. Danske Vidensk. Selsk. Skr., Naturvidensk. Math. Afd., ser. 5 2:300. 1851.

Herb in Northern Limestone Hills (Florida, Arecibo), Río Abajo Forest Reserve, Sierra de Luquillo, Sierra de Cayey, Central Cordillera, Maricao Forest Reserve; 150–1250 m; *fl* & *fr* throughout the year.

Greater Antilles (Jamaica). [L1: 82].

Pilea parietaria (L.) Blume, Mus. Bot. 2(3):48. 1856.
Bas.: *Urtica parietaria* L. (1753).

Herb in Northern Limestone Hills, Río Abajo Forest Reserve, Caguas Valley (Caguas), Sierra de Cayey, Las Tetas de Cayey, Southeastern Foothills, Eastern Cordillera (Corozal, Aibonito), Central Cordillera, Western Cordillera; 75–1300 m; *fl* & *fr* throughout the year.

Antilles, Mexico, Central America, northern South America (Venezuela, Colombia). [L1: 82].

Verdolaguilla.

Pilea repens (Sw.) Wedd., Ann. Sci. Nat., Bot., sér. 3 18:225. 1852.
Bas.: *Urtica repens* Sw. (1787).

Herb in Northern Limestone Hills (Ciales), Río Abajo Forest Reserve, Quebrada de Los Cedros; 100–350 m; *fl* & *fr* Feb, Jun–Aug.

Greater Antilles. [L1: 83].

Pilea semidentata (Juss. ex Poir.) Wedd., Ann. Sci. Nat., Bot., sér. 3 18:215. 1852.
Bas.: *Urtica semidentata* Juss. ex Poir. (1816).

Herb in Northern Limestone Hills, Río Abajo Forest Reserve, Sierra de Luquillo, Caguas Valley (Caguas), Sierra de Cayey, Las Tetas de Cayey, Southeastern Foothills (Coamo), Eastern Cordillera (Cayey, La Torrecilla), Central Cordillera, Susúa Forest Reserve, Western Cordillera; 100–1200 m; *fl* & *fr* Feb–Nov.

Greater Antilles (Hispaniola), Lesser Antilles. [L1: 83].

Pilea tenerrima Miq., Linnaea 26(2):219. 1854.

Herb in Vieques, Mona; 40–70 m; *fl* & *fr* Mar, Jun, Nov. [*Woodbury M-321* (UPR); *Woodbury a.n. 17409* (SJ); *Sustache 1091* (SJ)].

Bahamas, Greater Antilles (Cuba), Virgin Islands. [L1: 84].

Musgo.

Pilea yunquensis (Urb.) Britton & P. Wilson, Sci. Surv. Porto Rico & Virgin Islands 5(2):247. 1924.
Bas.: *Pilea parietaria* (L.) Blume var. *yunquensis* Urb. (1899).

Herb in Sierra de Luquillo, Central Cordillera; 900–1300 m; *fl* & *fr* throughout the year.

Endemic. [L1: 84].

Pouzolzia occidentalis (Liebm.) Wedd., Arch. Mus. Hist. Nat. 9(1–2):410. 1857.
Bas.: *Leucococcus occidentalis* Liebm. (1851).

Shrub/small tree in Southeastern Foothills (Salinas), Central Cordillera (Ponce); 300–600 m; *fl* & *fr* Jun (one specimen). [*Liogier 34631* (UPR)].

Native to Central America, northwestern, west-central South America and Venezuela. [L1: 85].

Rousselia humilis (Sw.) Urb., Symb. Antill. 4(2):205. 1905.
Bas.: *Urtica humilis* Sw. (1785).

Herb in Northern Limestone Hills; 150–250 m; *fl* & *fr* Mar–Apr, Aug–Oct.

Bahamas, Greater Antilles, Lesser Antilles (St. Martin, Saba, Dominica), Mexico, northern Central America (Guatemala, Belize). [L1: 86].

Urera baccifera (L.) Gaudich. ex Wedd., Ann. Sci. Nat., Bot., sér. 3 18:199. 1852.
Bas.: *Urtica baccifera* L. (1763).

Shrub/small tree in Northern Coastal Lowlands (San Juan, Camuy), Northern Limestone Hills, Río Abajo Forest Reserve, Sierra de Luquillo, Eastern Coast (Cuchilla de Panduras), Sierra de Cayey (San Lorenzo), Las Tetas de Cayey, Eastern Cordillera (Corozal), Central Cordillera, Maricao Forest Reserve, Western Coast (Mayagüez); 2–825 m; *fl* Aug–Apr, *fr* Sep–Jan.

Greater Antilles, Trinidad and Tobago, Mexico, Central America, South America. [L1: 87].

Notes: Britton & Wilson (1924) used *Urera baccifera* (L.) Gaudich., a *nom. inval.*, for the name of this taxon.

Ortiga, Ortiga brava.

Urera chlorocarpa Urb., Symb. Antill. 1:293. 1899.

Shrub in Sierra de Luquillo, Sierra de Cayey, Central Cordillera; 500–1000 m; *fl* Apr, Jul–Aug, *fr* Feb–Mar (two specimens).

Endemic. [L1: 89].

Ortiga.

TAXA NOT RECENTLY COLLECTED

Pilea rivoriae Wedd., Ann. Sci. Nat., Bot., sér 3 18:210. 1852.

Herb collected in Adjuntas on 25 Mar 1886 [*Sintenis 3979* (GH)].

Greater Antilles (Cuba, Hispaniola), Lesser Antilles. [L5: 391].

Urera caracasana (Jacq.) Griseb., Fl. Brit. W.I. 154. 1859.
Bas.: *Urtica caracasana* Jacq. (1798).

Shrub/small tree collected in the Sierra de Naguabo on 27 Oct 1886 [*Sintenis 5411* (US)].

Greater Antilles, Mexico, southern Central America, northwestern, west-central South America and Venezuela. [L1: 89].

Mal hombre, Ortiga colorada.

TAXA CITED BY PREVIOUS AUTHORS, BUT NOT INCLUDED

NO SPECIMENS SEEN

Pilea sanctae-crucis Liebm.

Herb cited by Britton & Wilson (1924) as, according to Urban, occurring in the Sierra de Luquillo; subsequently cited by Liogier & Martorell (1982, 2000) and Liogier (1985). [L1: 83].

CITED DUE TO MISIDENTIFICATION

Laportea cuneata (A. Rich.) Chew.

Herb cited by Liogier & Martorell (1982) and Liogier (1985) as growing in a nursery, Río Piedras, but subsequently reported by Liogier (1997) and Liogier & Martorell (2000) to be based on the misidentification of a specimen of *Fatoua villosa*. [L1: 76, L5: 390].

ONLY UNDER CULTIVATION

Boehmeria nivea (L.) Gaudich.

Herb cited by Liogier & Martorell (1982, 2000) and Liogier (1985) as cultivated. [L1: 74].

VALERIANACEAE

Valeriana scandens L., Sp. Pl. (ed. 2) 1:47. 1762.

Vine in Northern Limestone Hills (Florida), Río Abajo Forest Reserve, Eastern Cordillera (La Torrecilla), Central Cordillera; 325–1100 m; *fl* & *fr* Jan–Jul.

Greater Antilles (Cuba, Hispaniola), southeastern United States (Florida), Mexico, Central America, northwestern, west-central South America and Venezuela, Argentina. [L5: 181; AR: 388].

Valeriana.

VERBENACEAE

Bouchea prismatica (L.) Kuntze, Revis. Gen. Pl. 2:502. 1891.

Bas.: *Verbena prismatica* L. (1753).

Suffrutescent herb in Northern Coastal Lowlands, Eastern Coast (Ceiba), Southeastern Foothills, Cerro de Las Cuevas, Southern Coastal Lowlands, Southern Coastal Limestone (Ponce), Guánica Forest Reserve, Southwestern Coastal Lowlands, Sierra Bermeja, Cerro Las Mesas, Vieques, Culebra; 5–550 m; *fl* & *fr* Jun–Jan.

West Indies, southern United States (Alabama, Texas, Arizona), Central America, northwestern, west-central South America and Venezuela, Brazil. [L4: 341].

Citharexylum caudatum L., Fl. Jamaic. 18. 1759 "Citharexylon."

Tree in Sierra de Luquillo, Sierra de Cayey, Central Cordillera, Maricao Forest Reserve; 600–1300 m; *fl* & *fr* throughout the year.

Bahamas, Greater Antilles, Lesser Antilles (Martinique), Central America, northern South America (Colombia). [L4: 344].

Higüerillo, Péndula de sierra.

Citharexylum spinosum L., Sp. Pl. 2:625. 1753.

Syn.: *Citharexylum fruticosum* L. (1759) "Citharexylon."

Citharexylum pentandrum Vent. (1801).

Tree in Northern Coastal Lowlands, Northern Limestone Hills, Sierra de Luquillo, Eastern Coast, Caguas Valley (Caguas), Las Piedras Chiquitas, Southeastern Foothills, Eastern Cordillera, Southern Coastal Lowlands, Southern Coastal Limestone, Susúa Forest Reserve, Maricao Forest Reserve, Guánica Forest Reserve, Southwestern Coastal Lowlands, Sierra Bermeja, Vieques, Culebra, Desecheo; 1–850 m; *fl* & *fr* throughout the year.

West Indies, Trinidad and Tobago, southeastern United States (Florida), southern Central America (Panama), northern South America. [L4: 344].

Bálsamo, Higüerillo, Palo de guitarra, Péndula.

Duranta erecta L., Sp. Pl. 2:637. 1753.

Syn.: *Duranta repens* L. (1753).

Shrub in Southeastern Foothills, Southern Coastal Lowlands (Caja de Muertos), Guánica Forest Reserve, Vieques, Mona, Desecheo; 50–350 m; *fl* Feb–Nov, *fr* Jul–Apr.

West Indies, southwestern United States, Mexico, Central America, South America. [L4: 355].

Azotacaballo, Cuentas de oro, Lila, Lluvia.

Lantana camara L., Sp. Pl. 2:627. 1753, *nom. cons.*

Shrub in Northern Coastal Lowlands, Northern Limestone Hills (Isabela), Río Abajo Forest Reserve, Sierra de Luquillo, Eastern Coast, Caguas Valley, Sierra de Cayey, Las Piedras Chiquitas, Southeastern Foothills, Eastern Cordillera (Aguas Buenas, La Torrecilla), Central Cordillera, Southern Coastal Lowlands, Maricao Forest Reserve, Guánica Forest Reserve, Southwestern Coastal Lowlands, Sierra Bermeja, Western Coast, Vieques, Mona; 1–1150 m; *fl* & *fr* throughout the year.

West Indies. [L4: 361, 363, 366].

Notes: Based on a reexamination of the type specimen, Roger Sanders (2006) changed the concept of *Lantana camara* such that the name now applies to what would formerly have been considered in Puerto Rico to be either *Lantana urticifolia* Mill., *Lantana urticifolia* Mill. subsp. *moldenkei* R. W. Sanders or *Lantana arida* Britton. The name *Lantana strigoca-*

mara now applies to what would formerly have been considered to be the above taxon.

Cariaquillo, Cariaquillo espinoso.

Lantana exarata Urb. & Ekman, Ark. Bot. 22A(17): 103. 1929.

Shrub in Southern Coastal Limestone, Guánica Forest Reserve, Southwestern Coastal Lowlands (Cabo Rojo); 10–100 m; *fl* throughout the year, *fr* Dec, Mar, Jun–Jul.

Greater Antilles (Cuba, Hispaniola). [L4: 363].

Lantana horrida Kunth in Humboldt, Bonpland & Kunth, Nov. Gen. Sp. 2:211 [folio]; 2:261 [quarto]. 1818.

Shrub in Southwestern Coastal Lowlands (Isla Magueyes); 5–10 m; *fl* Nov (one specimen), *fr* (?).

Native to southwestern United States (Arizona), Mexico. [L4: 363].

Notes: According to Roger Sanders (2006), the name *Lantana glandulosissima* Hayek has been misapplied to this taxon.

Lantana involucrata L., Cent. Pl. II 22. 1756.

Shrub in Northern Coastal Lowlands, Northern Limestone Hills, Eastern Coast (Humacao), Las Tetas de Cayey, Las Piedras Chiquitas, Southeastern Foothills, Southern Coastal Lowlands, Southern Coastal Limestone, Guánica Forest Reserve, Southwestern Coastal Lowlands, Sierra Bermeja, Western Coast (Rincón), Vieques, Culebra, Mona, Desecheo; 1–550 m; *fl* & *fr* throughout the year.

West Indies, southeastern United States (Florida), Mexico, Central America, northern South America. [L4: 363].

Cariaquillo Santa María, Santa María.

Lantana reticulata Pers., Syn. Pl. 2(1):141. 1806.

Shrub in Las Tetas de Cayey, Eastern Cordillera (Cayey, Aibonito), Southern Coastal Limestone (Peñuelas), Guánica Forest Reserve, Southwestern Coastal Lowlands (Cabo Rojo); 1–825 m; *fl* Aug–Mar, *fr* Oct–Feb.

Antilles, Trinidad, Central America, northern South America. [L4: 364].

Lantana scabrida Aiton, Hort. Kew. 2:352. 1789.

Shrub in Sierra de Luquillo, Eastern Coast (Cuchilla de Panduras), Sierra de Cayey, Eastern Cordillera (La Torrecilla), Central Cordillera, Southern Coastal Lowlands (Patillas); 100–1150 m; *fl* Feb–Nov, *fr* Feb–Apr, Jul–Aug.

Neotropical.

Lantana strigocamara R. W. Sanders, Sida 22(1):392, *t.* 4. 2006.

Shrub in Northern Coastal Lowlands, Cambalache Forest Reserve, Sierra de Luquillo, Eastern Cordillera (Cayey), Southern Coastal Lowlands (Guayama, Caja de Muertos), Guánica Forest Reserve, Southwestern Coastal Lowlands (Cabo Rojo), Western Coast (Aguadilla); 2–800 m; *fl* May–Jan, *fr* Jun–Oct.

Pantropical.

Lantana strigosa (Griseb.) Urb., Symb. Antill. 7(3):352. 1912.

Bas.: *Lantana reticulata* Pers. var. *strigosa* Griseb. (1866).

Shrub in Guánica Forest Reserve; 50–100 m; *fl* Oct (one specimen), *fr* (?).

Greater Antilles (Cuba). [L4: 365].

Lippia alba (Mill.) N.E. Br. ex Britton & P. Wilson, Sci. Surv. Porto Rico & Virgin Islands 6(1):141. 1925.

Bas.: *Lantana alba* Mill. (1768).

Shrub in Northern Coastal Lowlands, Northern Limestone Hills, Eastern Coast (Ceiba), Caguas Valley, Southern Coastal Lowlands (Coamo), Vieques; 1–90 m; *fl* & *fr* throughout the year.

West Indies, south-central United States (Texas), Mexico, Central America, South America. [L4: 367].

Poleo.

Lippia micromera Schauer in A. de Candolle, Prodr. 11:587. 1847.

Syn.: *Lippia helleri* Britton (1903). *Lippia micromera* Schauer var. *helleri* (Britton) Moldenke (1937).

Shrub in Northern Coastal Lowlands (Loíza), Las Piedras Chiquitas, Central Cordillera (Yauco), Sierra Bermeja; 2–900 m; *fl* & *fr* Jun–Mar.

Greater Antilles (Cuba, Hispaniola), Lesser Antilles (Dominica), Trinidad, Central America (Honduras, Nicaragua), northern South America. [L4: 369].

Majorana, Orégano chiquito, Orégano del país.

Nashia inaguensis Millsp., Publ. Field Columb. Mus., Bot. Ser. 2:177. 1906.

Shrub in Southern Coastal Limestone (Peñuelas); 50–100 m; *fl* Mar–May (two specimens), *fr* Jan–May.

Bahamas.

Phyla dulcis (Trevir.) Moldenke, Torreya 34:9. 1934.

Bas.: *Lippia dulcis* Trevir. (1826).

Suffrutescent herb in Northern Coastal Lowlands (Arecibo), Northern Limestone Hills (San Sebastián), Sierra de Cayey, Southeastern Foothills (Coamo), Eastern Cordillera (Barranquitas), Central

Cordillera (Villalba, Ponce), Quebrada de Los Cedros; 15–600 m; *fl* throughout the year, *fr* Feb–Jul, Oct.

Greater Antilles (Hispaniola), Mexico, Central America, northern South America (Venezuela, Colombia). [L4: 369].

Notes: Liogier & Martorell (1982) and Liogier (1995) wrongly placed *Phyla scaberrima* (Juss. ex Pers.) Moldenke in synonymy with this taxon.

Phyla fruticosa (Mill.) K. Kenn. ex Wunderlin & B.F. Hansen, Bot. Explor. (Florida) 3:38. 2003.
Bas.: *Verbena fruticosa* Mill. (1768).
Syn.: *Lippia strigulosa* M. Martens & Galeotti (1844).

Herb in Northern Limestone Hills (San Sebastián), Southern Coastal Lowlands, Guánica Forest Reserve, Southwestern Coastal Lowlands; 1–100 m; *fl* & *fr* Feb–Oct.

Antilles, southeastern United States (Florida), Mexico, Central America, northern South America. [L4: 371].

Notes: Britton & Wilson (1925) misapplied the name *Lippia reptans* Kunth, and Liogier & Martorell (1982) misapplied *Lippia nodiflora* (L.) Michx. var. *reptans* (Kunth) Kuntze, to this taxon.

Cidrón, Yerba de sapo.

Phyla nodiflora (L.) Greene, Pittonia 4(20E):46. 1899.
Bas.: *Verbena nodiflora* L. (1753).
Syn.: *Lippia nodiflora* (L.) Michx. (1803).

Herb in Northern Coastal Lowlands, Eastern Coast (Humacao including Cayo Santiago), Western Coast, Vieques; 1–200 m; *fl* & *fr* throughout the year.

West Indies, southern United States, Mexico, Central America, South America. [L4: 369].

Cidrón, Yerba de sapo.

Phyla stoechadifolia (L.) Small, Bull. Torrey Bot. Club 36(3):162. 1909.
Bas.: *Verbena stoechadifolia* L. (1753).
Syn.: *Lippia stoechadifolia* (L.) Kunth (1818).

Suffrutescent herb in Northern Limestone Hills (San Sebastián); 200 m; *fl* & *fr* Feb, May (two specimens). [*Proctor 49283* (SJ); *Axelrod 437* (UPRRP)].

Bahamas, Greater Antilles, Lesser Antilles (Antigua, Guadeloupe), southeastern United States (Florida), Central America (scattered), northern South America. [L4: 370].

Priva lappulacea (L.) Pers., Syn. Pl. 2(1):139. 1806.
Bas.: *Verbena lappulacea* L. (1753).

Herb in Northern Coastal Lowlands, Northern Limestone Hills, Río Abajo Forest Reserve, Eastern Coast (Cayo Santiago), Las Piedras Chiquitas, Southeastern Foothills, Southern Coastal Lowlands, Southern Coastal Limestone (Peñuelas), Sierra Bermeja, Western Coast (Rincón), Vieques; 1–500 m; *fl* & *fr* throughout the year.

West Indies, southern United States (Florida, Texas), Mexico, Central America, northern South America to Bolivia. [L4: 375].

Abrojo, Pegapega.

Stachytarpheta cayennensis (Rich.) Vahl, Enum. Pl. 1:208. 1804 "cajanensis."
Bas.: *Verbena cayennensis* Rich. (1792).
Syn.: *Valerianoides cayennense* (Rich.) Kuntze (1891).
Stachytarpheta urticifolia Sims (1816).

Suffrutescent herb in Northern Coastal Lowlands, Northern Limestone Hills (Arecibo), Sierra de Luquillo, Eastern Coast, Caguas Valley, Sierra de Cayey, Southeastern Foothills (Arroyo), Eastern and Central Cordillera, Maricao Forest Reserve, Western Coast (Mayagüez); 50–900 m; *fl* & *fr* throughout the year.

Antilles, Mexico, Central America, South America. [L4: 378].

Stachytarpheta jamaicensis (L.) Vahl, Enum. Pl. 1:206. 1804.
Bas.: *Verbena jamaicensis* L. (1753).
Syn.: *Valerianoides jamaicense* (L.) Kuntze (1891).

Herb in Northern Coastal Lowlands, Northern Limestone Hills, Río Abajo Forest Reserve, Sierra de Luquillo, Eastern Coast, Caguas Valley (Caguas), Sierra de Cayey, Las Tetas de Cayey, Eastern Cordillera, Cerro de Las Cuevas, Southern Coastal Limestone (Guayanilla), Western Cordillera, Guánica Forest Reserve, Southwestern Coastal Lowlands, Sierra Bermeja, Western Coast, Vieques, Culebra, Mona; 1–925 m; *fl* & *fr* throughout the year.

West Indies, southeastern United States (Florida, Alabama), Mexico, Central America, northwestern, west-central South America and Venezuela, Brazil. [L4: 378].

Bretónica, Verbena.

Stachytarpheta strigosa Vahl, Enum. Pl. 1:207. 1804.
Syn.: *Valerianoides strigosum* (Vahl) Britton (1925).

Suffrutescent herb in Northern Coastal Lowlands (islands adjacent to Fajardo), Las Piedras Chiquitas, Southeastern Foothills, Southern Coastal Lowlands, Susúa Forest Reserve, Guánica Forest Reserve, Southwestern Coastal Lowlands (Cabo Rojo), Sierra Bermeja, Vieques, Mona; 20–600 m; *fl* & *fr* Sep–Jun.

Greater Antilles (Hispaniola), Virgin Islands. [L4: 380].

Notes: This taxon differs from *Stachytarpheta*

jamaicensis only in having its parts covered with varying amounts of pluricellular hairs, which are also occasionally present to a small extent on specimens identified as *Stachytarpheta jamaicensis*. It may thus be conspecific with the latter.

Tamonea boxiana (Moldenke) R.A. Howard, Phytologia 65(4):288. 1988.
Bas.: *Ghinia boxiana* Moldenke (1942).

Suffrutescent herb/shrub in Southern Coastal Limestone, Guánica Forest Reserve, Southwestern Coastal Lowlands, Sierra Bermeja; 3–300 m; *fl* & *fr* throughout the year.

Virgin Islands, Lesser Antilles (Antigua, Barbuda). [L4: 381].

Notes: Britton & Wilson (1925) misapplied the name *Ghinia spinosa* (Sw.) Britton & P. Wilson, and Liogier & Martorell (1982) misapplied the name *Tamonea spinosa* Sw., to this taxon.

Cardero.

Verbena bonariensis L., Sp. Pl. 1:20. 1753.

Herb in Sierra de Cayey, Las Tetas de Cayey, Eastern Cordillera, Toro Negro Forest Reserve; 400–1300 m; *fl* & *fr* Feb–Nov.

Native to southern, central South America and Colombia. [L4: 385].

Verbena scabra Vahl, Eclog. Amer. 2:2. 1798.

Herb in Northern Limestone Hills, Río Abajo Forest Reserve, Southeastern Foothills, Eastern and Central Cordillera; 100–1250 m; *fl* & *fr* throughout the year.

Bahamas, Greater Antilles, southern United States, Mexico. [L4: 385].

Verbena ortiga.

TAXA NOT RECENTLY COLLECTED

Lantana trifolia L., Sp. Pl. 2:626. 1753.

Shrub collected in Guayama on 16 Mar 1874 [*O. Kuntze Herbarium 562a*, not seen].

Greater Antilles, (St. Vincent, Barbados), Central America, northern South America. [L4: 365].

Priva portoricensis Urb., Symb. Antill. 4(4):534. 1911.

Herb collected near Guánica on 2 Feb 1886 [*Sintenis 3597* (US), type specimen].

Endemic. [L4: 377].

TAXA CITED BY PREVIOUS AUTHORS, BUT NOT INCLUDED

DOUBTFUL OCCURRENCE

Stachytarpheta* ×*hybrida Moldenke (pro sp.).

Herb cited by Liogier & Martorell (1982, 2000) and Liogier (1995) as occurring in wet places.

Notes: According to Liogier (1995), this taxon is a cross between *Stachytarpheta jamaicensis* and *Stachytarpheta strigosa*. [L4: 380].

ONLY UNDER CULTIVATION

Aloysia citriodora Palau. Syn.: *Aloysia triphylla* (L'Hér.) Britton.

Shrub cited by Britton & Wilson (1925) as sparingly spontaneous after cultivation; subsequently cited by Liogier & Martorell (1982, 2000) and Liogier (1995). [L4: 337].

Glandularia aristigera (S. Moore) Tronc. Syn.: *Verbena tenuisecta* Briq.

Herb cited by Liogier & Martorell (1982, 2000) and Liogier (1995) as cultivated and escaped. [L4: 387].

Lantana montevidensis (Spreng.) Briq.

Shrub cited by Liogier & Martorell (1982, 2000) and Liogier (1995) as cultivated as an ornamental and escaped. [L4: 364].

Petrea volubilis L. Syn.: *Petrea kohautiana* C. Presl.

Woody vine cited by Liogier & Martorell (1982, 2000) and Liogier (1995) as planted for ornament and persistent and by Acevedo-Rodríguez (2005) as cultivated. [L4: 373, 374; AR: 396].

VIOLACEAE

Hybanthus linearifolius (Vahl) Urb., Symb. Antill. 5(3):436. 1908.
Bas.: *Viola linearifolia* Vahl (1793).
Syn.: *Ionidium linearifolium* (Vahl) Britton (1924).
Ionidium portoricense Krug & Urb. (1908).

Herb in Northern Coastal Lowlands (islands adjacent to Fajardo: Isla Ramos; Laguna Tortuguero), Northern Limestone Hills, Southeastern Foothills, Susúa Forest Reserve, Southwestern Coastal Lowlands (Cabo Rojo); 1–250 m; *fl* & *fr* May–Oct.

Greater Antilles (Cuba, Hispaniola), Virgin Islands, Lesser Antilles, Trinidad. [L3: 252].

Chancleta.

TAXA CITED BY PREVIOUS AUTHORS, BUT NOT INCLUDED

ONLY UNDER CULTIVATION

Viola odorata L.

Herb cited by Liogier & Martorell (1982, 2000) and Liogier (1994) as much grown and subspontaneous. [L3: 254].

VITACEAE

Cissus erosa Rich., Actes Soc. Hist. Nat. Paris 1:106. 1792.

Vine in Northern Coastal Lowlands (Vega Alta, Laguna Tortuguero), Northern Limestone Hills (Lares), Sierra de Luquillo, Sierra de Cayey, Eastern Cordillera (Cerro La Peña), Central Cordillera, Western Cordillera; 1–900 m; *fl* Mar–Dec, *fr* Jul–Jan.

Greater Antilles (Hispaniola), Trinidad and Tobago, northern South America to Bolivia, Paraguay. [L3: 98; AR: 397].

Caro de tres hojas.

Cissus obovata Vahl, Symb. Bot. 3:19. 1794.
Syn.: *Cissus tuberculata* Jacq. (1797).
Cissus caustica Tussac (1808).

Vine in Northern Limestone Hills, Río Abajo Forest Reserve, Sierra de Luquillo, Las Tetas de Cayey, Cerro de Las Cuevas, Southwestern Coastal Lowlands (Cabo Rojo), Mona; 1–825 m; *fl* & *fr* Jul–Aug, Nov.

Bahamas, Greater Antilles (Hispaniola), Virgin Islands, northern Lesser Antilles. [L3: 99; AR: 397].

Bejuco de gongolí.

Cissus trifoliata (L.) L., Syst. Nat. (ed. 10) 2:897. 1759.
Bas.: *Sicyos trifoliatus* L. (1753) "trifoliata."

Vine in Northern Coastal Lowlands (Fajardo and adjacent islands, Laguna Tortuguero), Southeastern Foothills (Coamo), Southern Coastal Lowlands, Southern Coastal Limestone (Guayanilla), Guánica Forest Reserve, Southwestern Coastal Lowlands, Sierra Bermeja, Vieques, Mona; 1–150 m; *fl* Apr–Sep, *fr* May–Nov.

Bahamas, Greater Antilles, Virgin Islands, southern United States, Mexico. [L3: 99; AR: 399].

Bejuco de caro, Caro.

Cissus verticillata (L.) Nicolson & C.E. Jarvis subsp. **verticillata**, Taxon 33(4):727. 1984.
Bas.: *Viscum verticillatum* L. (1753).
Syn.: *Cissus sicyoides* L. (1759).

Vine in Northern Coastal Lowlands, Northern Limestone Hills, Río Abajo Forest Reserve, Sierra de Luquillo, Eastern Coast, Caguas Valley (Caguas), Sierra de Cayey, Las Tetas de Cayey, Southeastern Foothills, Eastern Cordillera (Aguas Buenas, Cayey), Central Cordillera, Cerro de Las Cuevas, Southern Coastal Lowlands, Southern Coastal Limestone, Susúa Forest Reserve, Maricao Forest Reserve, Guánica Forest Reserve, Southwestern Coastal Lowlands (Lajas), Western Coast, Vieques, Culebra, Mona, Desecheo; 1–825 m; *fl* throughout the year, *fr* Apr–Jan.

West Indies, southeastern United States (Florida), Mexico, Central America, South America. [L3: 100; AR: 401].

Bejuco de caro, Caro.

Vitis tiliifolia Humb. & Bonpl. ex Schult. in Roemer & Schultes, Syst. Veg. 5:320. 1819 "tiliaefolia."

Vine in Northern Limestone Hills, Río Abajo Forest Reserve, Sierra de Luquillo, Sierra de Cayey, Southeastern Foothills (Cayey), Central Cordillera (Juana Díaz); 35–750 m; *fl* Apr–May, *fr* May–Aug.

Antilles, Mexico, Central America, northwestern, west-central South America and Venezuela. [L3: 101; AR: 403].

Bejuco de agua, Parra cimarrona, Uva cimarrona.

TAXA CITED BY PREVIOUS AUTHORS, BUT NOT INCLUDED

DOUBTFUL OCCURRENCE

Ampelopsis arborea (L.) Koehne.
Vine mentioned by Howard (1989) as occurring in Puerto Rico; subsequently cited by Liogier (1994) and Liogier & Martorell (2000). [L3: 96; AR: 403].

Cissus intermedia A. Rich.
Vine cited by Britton & Wilson (1924) as occurring near Peñuelas and Manatí (according to Urban); subsequently cited by Liogier & Martorell (1982, 2000) and Liogier (1994). Acevedo-Rodríguez (2005) remarks that since, of the two specimens known of this species, one was misidentified and the other lost, he considers that the taxon is not found in Puerto Rico. [L3: 98; AR: 401].

ONLY UNDER CULTIVATION

Vitis vinifera L.
Vine cited by Liogier & Martorell (1982, 2000) and Liogier (1994) as commonly grown, becoming permanent mostly in the southern dry districts at lower elevations. [L3: 102].

XYRIDACEAE

Xyris difformis Chapm. var. **curtissii** (Malme) Kral, Sida 2(3):255. 1966.
Bas.: *Xyris curtissii* Malme (1913).

Herb in Northern Coastal Lowlands (Laguna Tortuguero); 1–2 m; *fl* & *fr* Mar (one specimen). [*Vélez-Gavilán 374* (MAPR)].

Native to southeastern and south-central United States, Mexico, northern Central America (Belize). [AR&S: 232].

Xyris elliottii Chapm., Fl. South. U.S. 500. 1860.

Herb in Northern Coastal Lowlands (Vega Alta

to Laguna Tortuguero); 1–2 m; *fl* & *fr* throughout the year.

Greater Antilles (Cuba), southeastern United States, northern Central America (Belize). [AR&S: 233].

Xyris jupicai Rich., Actes Soc. Hist. Nat. Paris 1:106. 1792.

Herb in Northern Coastal Lowlands (Dorado to Laguna Tortuguero), Sierra de Luquillo; 1–1000 m; *fl* & *fr* Feb–Nov.

Greater Antilles, Trinidad, east-central, southeastern and south-central United States, Mexico, Central America, South America (scattered). [AR&S: 233].

Notes: Liogier & Martorell (1982, 2000) misapplied the name *Xyris caroliniana* Walter to this taxon.

ZINGIBERACEAE

Alpinia purpurata (Vieill.) K. Schum. in Engler, Pflanzenr. IV 46(Heft 20):323, *t. 40A*. 1904.
Bas.: *Guillainia purpurata* Vieill. (1866).

Herb planted and persistent in Sierra de Luquillo, Central Cordillera (Adjuntas); 700–950 m; *fl* Feb–May, *fr* (?).

Native to Papua New Guinea, Southwestern Pacific (New Caledonia, Solomon Islands, Vanuatu). [AR&S: 180].

Notes: Liogier & Martorell (2000) used *Alpinia purpurea* Vieill. ex K. Schum. for the name of this taxon.

Jengibre rojo.

Alpinia zerumbet (Pers.) B.L. Burtt & R.M. Sm., Notes Roy. Bot. Gard. Edinburgh 31(2):204. 1972.
Bas.: *Costus zerumbet* Pers. (1805).
Syn.: *Languas speciosa* (J.C. Wendl.) Small (1913).

Herb planted and persistent in Río Abajo Forest Reserve, Sierra de Luquillo, Sierra de Cayey, Central Cordillera, Maricao Forest Reserve; 300–680 m; *fl* Nov–May, *fr* (?).

Native to tropical Asia, China, Japan. [AR&S: 179].

Boca de dragón, Pimienta Angola.

Etlingera elatior (Jack) R.M. Sm., Notes Roy. Bot. Gard. Edinburgh 43(2):244. 1986.
Bas.: *Alpinia elatior* Jack (1822).
Syn.: *Nicolaia elatior* (Jack) Horan. (1862).

Robust herb planted and persistent in Sierra de Luquillo, Toro Negro Forest Reserve; 225–925 m; *fl* May, Aug–Sep (two specimens), *fr* Nov (one specimen).

Native to Malesia. [AR&S: 180].

Antorcha.

Hedychium coronarium J. Koenig in Retz., Observ. Bot. 3:73. 1783.

Herb planted and persistent in Río Abajo Forest Reserve, Sierra de Luquillo, Central Cordillera, Maricao Forest Reserve; 100–1280 m; *fl* Jul–Nov, *fr* (?).

Native to India, Nepal, Myanmar, China. [AR&S: 181].

Dulce nieve, Jazmín cimarrón, Jazmín de río, Mariposa blanca, Narciso, Nardo.

Renealmia alpinia (Rottb.) Maas, Acta Bot. Neerl. 24(5–6):474. 1976.
Bas.: *Amomum alpinia* Rottb. (1775).
Syn.: *Alpinia exaltata* Roem. & Schult. (1817).

Herb planted and persistent in Sierra de Luquillo; 250–500 m; *fl* Oct, Mar–Apr, *fr* Oct, Jan–Apr.

Native to Lesser Antilles, Trinidad and Tobago, Mexico, Central America, northwestern and west-central South America and Venezuela. [AR&S: 182].

Bihao, Bijao, Jengibre de jardín.

Renealmia aromatica (Aubl.) Griseb., Abh. Königl. Ges. Wiss. Göttingen 7:275. 1857.
Bas.: *Alpinia aromatica* Aubl. (1775).
Syn.: *Renealmia occidentalis* (Sw.) Sweet (1830).

Herb in Northern Coastal Lowlands, Northern Limestone Hills, Río Abajo Forest Reserve, Sierra de Luquillo, Caguas Valley (Caguas), Sierra de Cayey, Maricao Forest Reserve, Western Coast (Mayagüez); 50–700 m; *fl* & *fr* Jul–Jan.

Antilles, Central America, northern South America to Bolivia. [AR&S: 183].

Bihao, Bijao, Narciso.

Renealmia jamaicensis (Gaertn.) Horan. var. **puberula** (Gagnep.) Maas, Acta Bot. Neerl. 24(5–6):477. 1976.
Bas.: *Renealmia antillarum* (Roem. & Schult.) Gagnep. var. *puberula* Gagnep. (1903).

Herb in Sierra de Luquillo, Sierra de Cayey, Central Cordillera, Maricao Forest Reserve; 550–1250 m; *fl* throughout the year, *fr* Aug–Mar.

Greater Antilles (Hispaniola). [AR&S: 184].

Notes: Britton & Wilson (1923) used *Alpinia antillarum* Roem. & Schult., which is now considered to be a synonym of *Renealmia jamaicensis* (Gaertn.) Horan. var. *jamaicensis*, for the name of this taxon.

Narciso colorado.

Zingiber montanum (J. Koenig) Link ex A. Dietr., Sp. Pl. 1:52. 1831.
Bas.: *Amomum montanum* J. Koenig (1783).
Syn.: *Zingiber cassumunar* Roxb. (1810).

Herb in Northern Limestone Hills (Isabela); 200–250 m; *fl* Oct (one specimen), *fr* (?). [*Axelrod 11949* (UPRRP)]

Native to India. [AR&S: 185].

Jengibre colorado.

Zingiber zerumbet (L.) Roscoe ex Sm., Exot. Bot. 2:105, *t. 112*. 1806.
Bas.: *Amomum zerumbet* L. (1753).

Herb in Northern Coastal Lowlands (Fajardo), Northern Limestone Hills, Río Abajo Forest Reserve, Eastern Coast (Naguabo), Sierra de Cayey, Central Cordillera, Susúa Forest Reserve, Western Cordillera, Cerro Las Mesas, Western Coast (Mayagüez); 15–720 m; *fl* Jul–Oct, *fr* (?).

Native to India, Malay Penisula. [AR&S: 185].

Jengibre amargo.

TAXA CITED BY PREVIOUS AUTHORS, BUT NOT INCLUDED

CITED DUE TO MISIDENTIFICATION

Hedychium flavum Roxb.
Herb cited by Liogier & Martorell (2000) as naturalized in the Luquillo Mountains. According to Acevedo-Rodríguez & Strong (2005), the specimens upon which this citation was based are referable to *Hedychium coronarium*. [AR&S: 181].

ONLY UNDER CULTIVATION

Curcuma longa L.
Herb cited by Britton & Wilson (1924) as persistent after cultivation; subsequently cited by Liogier & Martorell (1982, 2000).

Kaempferia rotunda L.
Herb cited by Liogier & Martorell (1982, 2000) as cultivated and persistent.

Zingiber officinale Roscoe. Syn.: *Zingiber zingiber* H. Karst.
Herb cited by Liogier & Martorell (1982, 2000) as spontaneous after cultivation for centuries.

ZYGOPHYLLACEAE

Guaiacum officinale L., Sp. Pl. 1:381. 1753 "Guajacum."

Tree in Southern Coastal Limestone, Guánica Forest Reserve, Southwestern Coastal Lowlands, Vieques, Culebra; 1–150 m; *fl* Feb–Nov, *fr* throughout the year.

West Indies, northern South America. [L2: 247].

Guayacán, Guayaco.

Guaiacum sanctum L., Sp. Pl. 1:382. 1753 "Guajacum."

Tree in Southern Coastal Lowlands (Ponce), Southern Coastal Limestone, Guánica Forest Reserve, Southwestern Coastal Lowlands (Cabo Rojo), Vieques, Mona; 5–150 m; *fl* Mar–Sep, *fr* Feb–Nov.

Bahamas, Greater Antilles (Cuba, Hispaniola), southeastern United States (Florida), eastern Mexico, Central America. [L2: 247].

Guayacán blanco, Guayacancillo, Guayacán de vera.

Kallstroemia maxima (L.) Hook. & Arn., Bot. Beechey Voy. 282. 1838 Jul; et Torr. & A. Gray, Fl. N. Amer. 1:213. 1838 Oct.
Bas.: *Tribulus maximus* L. (1753).

Prostrate suffrutescent herb in Northern Coastal Lowlands, Eastern Coast, Las Piedras Chiquitas, Southeastern Foothills, Eastern Cordillera (Aibonito), Southern Coastal Lowlands, Southern Coastal Limestone, Guánica Forest Reserve, Southwestern Coastal Lowlands, Western Coast, Vieques, Mona; 1–650 m; *fl* & *fr* Jan–Oct.

West Indies, southeastern and south-central United States, Mexico, Central America, northern South America (Venezuela, Colombia). [L2: 249].

Abrojo, Caldero, Verdolaga de abrojo.

Kallstroemia pubescens (G. Don) Dandy in Keay, Kew Bull. 10(1):138. 1955.
Bas.: *Tribulus pubescens* G. Don (1831).
Syn.: *Kallstroemia caribaea* Rydb. (1910).

Prostrate herb in Southeastern Foothills (Coamo), Southern Coastal Lowlands (Caja de Muertos), Guánica Forest Reserve, Vieques, Culebra; 10–250 m; *fl* & *fr* May–Jul, Oct.

Greater Antilles (Jamaica), Lesser Antilles, Southern Caribbean Islands (Curaçao), Tobago, Mexico, Central America, northwestern, west-central South America and Venezuela. [L2: 249].

Tribulus cistoides L., Sp. Pl. 1:387. 1753.

Prostrate spreading herb in Northern Coastal Lowlands (Fajardo, San Juan), Southwestern Coastal Lowlands (Isla Magueyes), Mona; 1–25 m; *fl* throughout the year, *fr* Feb–Nov.

Native to tropical Africa, Malesia, Australia. [L2: 251].

Abrojo, Cabeza de arriero, Yerba de araña.

ACKNOWLEDGMENTS

Many people have made this work possible, but one person is paramount. James D. Ackerman of the University of Puerto Rico at Río Piedras, both boss and colleague, was responsible for setting up the myriad of conditions whereby a philosopher of science converted himself into a practicing botanist. I hope that this book goes some way toward repaying my debt to him.

Two friends stand out in their foolish devotion to my labors. Mark Nir has been tireless in his search for references and material at the New York Botanical Garden (NY)—no task, no matter how menial it may have appeared, was refused. Gary Breckon (now retired and much missed) of the University of Puerto Rico at Mayagüez (MAPR) was indefatigable in his use of and comments on this work as it progressed—I am sempiternally grateful to him for pushing me to include many of the little things that make it more useful

In Puerto Rico, the curators of the various herbaria, Gary Breckon (MAPR), Eugenio Santiago and his assistant Marcus Caraballo at the University of Puerto Rico Botanical Garden (UPR), and José Sustache at the Puerto Rico Department of Natural Resources and the Environment (SJ), have allowed me unfettered use of their facilities. Outside of Puerto Rico, Pedro Acevedo-Rodríguez and Mark Strong have helped out, especially in my use of the herbarium at the Smithsonian Institution (US); Barbara Thiers has given me access to the herbarium and library at New York Botanical Garden (NY); Emily Wood has given me access to the herbaria at Harvard University (GH, AMES) and Judy Warnement has helped me out at the incomparable library there.

In taxonomic matters, Kanchi Gandhi (GH) and Rosemary Davies of Kew Botanic Gardens (K) have patiently answered numerous nomenclatural questions and I owe much of my nomenclatural education to both. I have also learned much on both nomenclatural and more general taxonomic matters from give-and-take with John Wiersema, who directs the Germplasm Resources Information Network online (GRIN), and with both Richard Wunderlin and Bruce Hansen of the University of South Florida (USF), who are responsible for the Atlas of Florida Vascular Plants online (I am also especially indebted to Bruce Hansen who read through my finished manuscript and corrected numerous nomenclatural slipups and typographical errors). James Solomon, who is in charge of the Missouri Botanical Garden's VAST nomenclatural database online (TROPICOS), was helpful in the early stages of this work, as was Dan Nicolson (US).

Various experts have vetted whole families of this work. Not only were the taxonomic comments of William Anderson of the University of Michigan (MICH) on the Malpighiaceae helpful, but also, unwittingly, his stylistic comments set the goals for what I was to do in the work. Many years ago my friend Charlotte Taylor, then at UPRRP, gave me much needed encouragement in my early botanical struggles; more recently, Charlotte, who has been for many years at the Missouri Botanical Garden (MO), vetted my treatment of the Rubiaceae and also gave me encouragement in what proved to be a long undertaking. Paul Fryxell, formerly of the University of Texas at Austin (TEX), helped me, unstintingly as he has done with many others, with the Malvaceae. Mark Strong (US) answered my pointed questions on the Cyperaceae with aplomb and to both him and Pedro Acevedo-Rodríguez I am indebted, aside from many monocot identifications, for their comments both personal and published on that group. In the case of Pedro, the debt extends to the vines of Puerto Rico and to the Sapindaceae. Both Dieter Wasshausen (US) and Thomas Daniel of the California Academy of Sciences (CAS) looked over my treatment of the Acanthaceae. Frank Almeda (CAS) and Fabian Michelangeli (NY) helped with the Melastomataceae. Job Kuijt of the University of Victoria (UVIC) gave me useful advice on the Loranthaceae and the Santalaceae. Pedro A. González Gutiérrez of the Centro de Investigaciones y Servicios Ambientales y Tecnológicos de Holguín (Cuba) was responsible for several major changes in the Aquifoliaceae and the Oleaceae. For the Orchidaceae, I am indebted to James Ackerman for access to his soon to be published monograph on the Orchidaceae of the Greater Antilles and to both him and Mark Nir for numerous helpful suggestions on this difficult family and on this manuscript in general. Yeoman work was done by Alan R. Smith of the University of California at Berkeley (UC), who annotated a whole raft of specimens that enabled me to clarify many problems of identification and nomenclature. Furthermore, through his comments on the large section on ferns and fern allies, I was able to bring their ever-changing taxonomy up-to-date, at least at the time of publication and to correct errors in their distribution—he was truly my fern guru.

Other experts have vetted difficult genera: David Conant of Lyndon State College, Vermont (LSC) and Marcus Lehnert of the Staatliches Museum für Naturkunde, Stuttgart (STU), *Alsophila* and *Cyathea*; John

Strother (UC), *Ambrosia* and *Wedelia* (Asteraceae); Guy Nesom, formerly at the Botanical Research Institute of Texas (BRIT), *Gamochaeta*, *Gnaphalium*, *Pseudognaphalium*, and *Symphyotrichum* (Asteraceae); Roger Sanders, formerly at BRIT, now at Bryan College, *Lantana* (Verbenaceae); Mihal Costea of Wilfrid Laurier University (WLU), *Cuscuta* (Convolvulaceae); Anthony Bean of the Queensland Herbarium (BRI), *Eucalyptus* (Myrtaceae) and *Rubus* (Rosaceae); Melissa Luckow of Cornell University (BH), *Desmanthus* (Fabaceae-Mimosoideae); Guido Mathieu (BR), *Peperomia* (Piperaceae); Andrea Vega of the Universidad de Buenos Aires (BAA) and Zulma Rúgolo of the Instituto de Botánica Darwinion (SI), *Digitaria* (Poaceae); and David Seigler of the University of Illinois at Urbana-Champaign (ILL), *Vachelia* and *Senegalia* (Fabaceae-Mimosoideae). C.C. Berg of both the University of Bergen (BG) and the Leiden Herbarium (L) sent me his treatment of *Ficus* (Moraceae) in the Antilles and it became the basis for my treatment. Gerrit Davidse (MO) helped out with various grass determinations and Paul Peterson (US) and Fernando Zuloaga (SI) showed me their manuscripts for a forthcoming grass flora of Puerto Rico.

There were also many botanists, amateur and professional, too numerous to mention here, who, having helped out with an issue here and there, get their deserving, but unfortunately anonymous, thanks here.

The physiographic maps that accompany the text were adapted from Gould et al. (2009) by Olga Ramos of the GIS and Remote Sensing Lab, USDA Forest Service, International Institute of Tropical Forestry. I am grateful to its Director, Ariel Lugo, for the loan of her incomparable services.

I am beholden to Elvira Cuevas, Director of CREST-CATEC, for considering this book of sufficient value to merit defrayal of its publication costs with funds from her agency.

Many thanks to Barney Lipscomb, director of the Botanical Research Institute of Texas Press, who, in patiently educating me in and leading me through the rituals of publication, has produced the wonderful design of this book.

Finally, I thank those friends and relations who put me up and put up with me while visiting various herbaria: Emma Duchane and Jayant Shah of Cambridge, Massachusetts; Stephen and Cynthia Weidman, my brother and my sister-in-law, of Fairfax, Virginia; and, once again, Mark and Hannelore Nir of Dobbs Ferry, New York.

Who could be more worthy of acknowledgement than those who have patiently suffered through years of the enthusiasms and compulsions of a cranky botanist? To my still patient and encouraging wife, Jean Turnquist, a 'partner' indeed in this enterprise, and to my children, Ysaaca, Amittai, and Barzilai, who have put much distance in years and space from forced botanical marches, I dedicate this book.

METALEGOMENA

As soon as this book has been sent to press, I can assure the reader that there will have been name changes in some of the taxa and families listed herein. This is a consequence of the attention, perhaps undue, taxonomy has recently been paying to developing a classification that reflects the relative origins of species.

I have not in fact made all the changes that I have been aware of, preferring to wait on some till adoption was more general. Thus, while I may have jumped the gun on changes in the Polypodiaceae (if only in the Notes), I have preferred to wait to make changes in *Heliotropium* and *Tournefortia* (both in the Boraginaceae). I have also delayed sinking the Turneraceae in the Passifloraceae. Since I am sure that at some point I will accept changes such as these, I want to be able to supply readers with updates to my book—a "Vademecum 1.n" if you will. In such updates I would also both correct errors that eluded proofreading and add new species and changes in the knowledge of the distribution and phenology of known species that came about through continuing plant collection.

For such updates, I refer readers to the website of the Herbarium of the Biology Department of the University of Puerto Rico (http://herbario.uprr.pr/) and also to my email (vademecum.upr@gmail.com).

REFERENCES

Acevedo-Rodríguez, P. 2005. Vines and climbing plants of Puerto Rico and the Virgin Islands. Contr. U.S. Natl. Herb. 51:1–483.

Acevedo-Rodríguez, P. and Collaborators. 1996. Flora of St. John, U. S. Virgin Islands. Mem. New York Bot. Gard. 78:1–581.

Acevedo-Rodríguez, P. and M.T. Strong. 2005. Monocotyledons and Gymnosperms of Puerto Rico and the Virgin Islands. Contr. U.S. Natl. Herb. 52:1–415.

Acevedo-Rodríguez, P. and M.T. Strong. 2007 onwards. Catalogue of the seed plants of the West Indies Website. (http://botany.si.edu/antilles/WestIndies/results.cfm).

Ackerman, J.D. 1995. An orchid flora of Puerto Rico and the Virgin Islands. Mem. New York Bot. Gard. 73:1–203.

Adams, C.D. with G.R. Proctor, R.W. Read, and Others. 1972. Flowering plants of Jamaica. University of the Antilles, Mona, Jamaica.

Barneby, R.C. 1991. Sensitivae censitae: a description of the genus *Mimosa* Linnaeus (Mimosacaeae) in the New World. Mem. New York Bot. Gard. 65:1–835.

Barneby, R.C. 1996. Silk tree, guanacaste, monkey's earring: a generic system for the synandrous Mimosaceae of the Americas. Part III: Calliandra. Mem. New York Bot. Gard. 74:1–223.

Barneby, R.C. and J.W. Grimes. 1996. Silk tree, guanacaste, monkey's earring: a generic system for the synandrous Mimosaceae of the Americas. Part I: Abarema, Albizia, and allies. Mem. New York Bot. Gard. 74:1–292.

Barneby, R.C. and J.W. Grimes. 1997. Silk tree, guanacaste, monkey's earring: a generic system for the synandrous Mimosaceae of the Americas. Part II: Pithecellobium, Cojoba, and Zygia. Mem. New York Bot. Gard. 74:1–161.

Barneby, R.C. and H.S. Irwin. 1982. The American Cassiinae: a synoptical revision of Leguminosae tribe Cassieae subtribe Cassiinae in the New World. 2 parts. Mem. New York Bot. Gard. 35:1–918.

Berry, P.E., K. Yatskievych, and B.K. Holst (eds.). 1995–2005. Flora of the Venezuelan Guayana. 9 vols. Missouri Botanical Garden Press, St. Louis.

Berry, P.E., A.L. Hipp, K.J. Wurdack, B. van Ee, and R. Riina. 2005. Molecular phylogenetics of the giant genus *Croton* and tribe Crotoneae (Euphorbiaceae sensu stricto) using ITS and trnL-trnF DNA sequence data. Amer. J. Bot. 92:1520–1534.

Berry, P.E. and B.W. van Ee. 2009. A phylogenetic and taxonomic review of *Croton* (Euphorbiaceae *s. s.*) on Jamaica including the description of *Croton jamaicensis*, a new species of section *Eluteria*. Syst. Bot. 34:129–140.

La Biblioteca Digital del Real Jardín Botánico (CSIC). http://bibdigital.rjb.csic.es/spa/presentacion.php/.

Biodiversity Heritage Library. ttp://www.biodiversitylibrary.org/.

Botanicus. http://www.botanicus.org/.

Britton, N.L. and P. Wilson. 1923–1930. Botany of Porto Rico and the Virgin Islands: descriptive flora - Spermatophyta in Scientific survey of Puerto Rico and the Virgin Islands, Vols. 5–6. New York Academy of Sciences, New York.

Brummitt, R.K. and C.E. Powell. 1992. Authors of plant names. Royal Botanic Gardens, Kew.

Calonje, M., A.W. Meerow, and D.W. Stevenson. 2010. The typification of *Zamia erosa* and the priority of that name over *Z. amblyphyllidia*. Taxon 59:637–642.

Cornejo, X. and H.H. Iltis. 2008. The reinstatement of *Capparidastrum* (Capparaceae). Harvard Pap. Bot. 13: 229–236.

Dassanayake, M.D. et al. (eds.). 1980–1995. A revised handbook to the flora of Ceylon. 9 vols. Smithsonian Institution, Washington, D.C.

Ebihara, A., J.-Y. Dubuisson, K. Iwatsuki, S. Hennequin, and M. Ito. 2006. A taxonomic revision of Hymenophyllaceae. Blumea 51:221–280.

Field Museum. Online. Neotropical herbarium specimens. (http://fm1.fieldmuseum.org/vrrc/).

Fairchild Tropical Garden. Online. Virtual herbarium. (http://www.virtualherbarium.org/).

Flora of North America Editorial Committee. 1993 onwards. Flora of North America, all published volumes. Oxford University Press, New York and Oxford.

González Gutiérrez, P.A. 2007. The genus *Ilex* (Aquifoliaceae) in the Antilles. Harvard Pap. Bot. 12:435–482.

González Gutiérrez, P.A. 2008. Oleaceae. Flora de la República de Cuba 14(2). A.R. Gantner Verlag, Ruggell, Liechtenstein.

González-Más, A. 1964. Cyperaceae of Puerto Rico. Ph. D. Thesis, Louisiana State University and Agricultural and Mechanical College, Baton Rouge, Louisiana.

Gould, F.W. and C.A. Clark. 1978. *Dichanthelium* (Poaceae) in the United States and Canada. Ann. Missouri Bot. Gard. 65(4):1088–1132. The actual year of publication was 1979.

Gould, W.A., B. Edwards, M.E. Jiménez, S. Martinuzzi, and M. Quiñones. 2009. Physiography of the Puerto Rican Archipelago. Scale 1:260 000. IITF-RMAP-07. US Department of Agriculture Forest Service, International Institute of Tropical Forestry, Río Piedras, Puerto Rico.

Henderson, A., G. Galeano, and R. Bernal. 1995. Field guide to the palms of the Americas. Princeton Univeristy Press, Princeton, New Jersey.

Hitchcock, A.S. 1936. Manual of the grasses of the West Indies. Misc. Publ. No. 243, United States Department of Agriculture. United States Government Printing Office, Washington, D.C.

Howard, R.A. 1974–1989. Flora of the Lesser Antilles. 6 vols. Arnold Arboretum of Jamaica Plain, Massachusetts.

Iltis, H.H. and T.S. Cochrane. 2007. Studies in the Cleomaceae V: a new genus and ten new combinations for the Flora of North America. Novon 17:447–451.

Iltis, H.H. and X. Cornejo. 2010. Studies in the Capparaceae XXIX: synopsis of Quadrella, a Mesoamerican and West Indian genus. J. Bot. Res. Inst. Texas 4(1)117–132.

The International Plant Names Index (IPNI). Online. http://www.ipni.org.

Judd, W.S. and C.A. Kabat. 2005. *Miconia* sects. *Amblyarrhena*, *Miconia*, and *Tamonea* (Melastomataceae) in Hispaniola, with a note on the recognition of *Miconia pyramidalis*. Moscosoa 14:83–99.

Kuijt, J. 2003. Monograph of *Phoradendron* (Viscaceae). Syst. Bot. Mongr. 66:1–643.

Lawrence, G.H.M., A.F.G. Buchheim, G.S. Daniels, and H. Dolezal (eds.). 1968. B-P-H: Botanico-Periodicum-Huntianum. Hunt Botanical Library, Pittsburgh, Pennsylvania.

Liogier, H.A. 1985–1997. Descriptive flora of Puerto Rico and adjacent islands. 5 vols. Editorial de la Universidad de Puerto Rico, Río Piedras, Puerto Rico.

Liogier, H.A. and L.F. Martorell. 1982. Flora of Puerto Rico and the adjacent islands: a systematic synopsis. Editorial de la Universidad de Puerto Rico, Río Piedras, Puerto Rico.

Liogier, H.A. and L.F. Martorell. 2000. Flora of Puerto Rico and the adjacent islands: a systematic synopsis. 2nd ed. Editorial de la Universidad de Puerto Rico, Río Piedras, Puerto Rico.

Little, E.L., Jr. and F.H. Wadsworth. 1964. Common trees of Puerto Rico and the Virgin Islands. U. S. Department of Agriculture, Washington, D.C.

Little, E.L., Jr. and F.H. Wadsworth. 1989. Common trees of Puerto Rico and the Virgin Islands. 2nd printing with supplement and added names. Privately printed by the authors, Washington, D.C.

Little, E.L., Jr., R.O. Woodbury, and F.H. Wadsworth. 1974. Common trees of Puerto Rico and the Virgin Islands, vol. 2. U. S. Department of Agriculture, Washington, D.C.

Martorell, L.F., H.A. Liogier, and R.O. Woodbury. 1981. Catálogo de los nombres vulgares y científicos de las plantas de Puerto Rico. Boletin 263, Universidad de Puerto Rico, Recinto de Mayagüez, Estación Experimental Agrícola, Río Piedras, Puerto Rico

Maxon, W.R. 1926. Botany of Porto Rico and the Virgin Islands: descriptive flora – Pteridophyta. In: Scientific survey of Puerto Rico and the Virgin Islands, Vol. 6. New York Academy of Sciences, New York.

McNeill, J. et al. (eds.). 2006. International code of botanical nomenclature (Vienna code). Regnum Vegetabile 146. A.R. G. Gantner Verlag KG.

McVaugh, R. 1984. Flora Novo-Galiciana: a descriptive account of the vascular plants of Western Mexico. Vol. 12: Compositae. University of Michigan Press, Ann Arbor.

Mickel, J.T. and A.R. Smith. 2004. The pteridophytes of Mexico. Mem. New York Bot. Gard. 88:1–1054.

Missouri Botanical Garden. Online. VAST (VAScular Tropicos) nomenclatural database and associated authority files. (http://mobot.mobot.org/W3T/Search/vast.html).

Moran, R.C., J. Prado, and P.H. Labiak. 2009. *Megalastrum* (Dryopteridaceae) in the West Indies. Brittonia, 61(3): 273–292.

Negrón-Ortiz, V. 2005. Taxonomic revision of the Neotropical genus *Erithalis* (Rubiaceae: Chiococcaceae). Sida 21(3):1565–1598.

New York Botanical Garden. Online. The C.V. Starr virtual herbarium: plants and fungi of Puerto Rico. (http://sciweb.nybg.org/Science2/hcol/pr/index.asp).

NICOLSON, D.H. with R.A. DEFELIPPS, A.C. NICOLSON, AND OTHERS. 1991. Flora of Dominica, part 2: Dicotyledonae. Smithsonian Contr. Bot. 77:1–274.

NIR, M.A. 2000. Orchidaceae antillanae. DAG Media Publishing, New York.

ØLLGAARD, B. 1983. Lycopodiaceae. In: R. Stolze, ed., Fern and fern allies of Guatemala, Part III. Fieldiana, Bot., n.s. 12:20–44.

PICÓ, R. 1974. The geography of Puerto Rico. Aldine Publishing, Chicago, Illinois.

PROCTOR, G.R. 1989. Ferns of Puerto Rico and the Virgin Islands. Mem. New York Bot. Gard. 53:1–387.

SANDERS, R.W. 2006. Taxonomy of *Lantana* sect. *Lantana* (Verbenaceae): I. Correct application of *Lantana camara* and associated names. Sida 22:381–421.

SEIGLER, D.S. AND J.E. EBBINGER. 2005. New contributions in the genus *Vachellia* (Fabaceae: Mimosoideae) from the New World. Phytologia 87:139–178.

SEIGLER, D.S., J.E. EBBINGER, AND J.T. MILLER. 2006. The genus *Senegalia* (Fabaceae: Mimosoideae) from the New World. Phytologia 88:38–93.

SMITH, A.R., K.M. PRYER, E. SCHUETTPELZ, P. KORALL, H. SCHNEIDER, AND P.G. WOLF. 2006. A classification for extant ferns. Taxon 55(3):705–731.

SMITH, A.R., H.-P. KREIER, C.H. HAUFLER, T.A. RANKER, AND H. SCHNEIDER. 2006. *Serpocaulon* (Polypodiaceae), a new genus segregated from *Polypodium*. Taxon 55:919–930.

SORENG, R.J. ET AL. Online. Catalogue of New World grasses (Poaceae). (http://mobot.mobot.org/Pick/Search/nwgc.html).

STAFLEU, F.A. AND R.S. COWAN. 1976. Taxonomic literature: a selective guide to botanical publications and collections with dates, commentaries and types. 2nd ed. 6 vols. Bohn, Scheltema, & Holkema, Utrecht.

STEVENS, P.F. 2001 onwards. Angiosperm Phylogeny Website. Version 7, May 2006 [and more or less continuously updated since]. http://www.mobot.org/MOBOT/research/APweb/.

STEVENS, W.D., C. ULLOA ULLOA, A. POOL, AND O.M. MONTIEL (EDS.). 2001. Flora de Nicaragua. 3 vols. Missouri Botanical Garden Press, St. Louis, Missouri.

USDA, ARS, NATIONAL GENETIC RESOURCES PROGRAM. Online. Germplasm Resources Information Network - (GRIN). National Germplasm Resources Laboratory, Beltsville, Maryland. (http://www.ars-grin.gov/cgi-bin/npgs/html/tax_search.pl).

WORLD CHECKLIST OF SELECTED PLANT FAMILIES. 2009. The Board of Trustees of the Royal Botanic Gardens, Kew. (apps.kew.org/wcsp/home.do).

WUNDERLIN, R.P. 1998. A guide to the vascular plants of Florida. University Press of Florida, Gainesville.

WUNDERLIN, R.P AND B.F. HANSEN. 2000. Flora of Florida. Vol 1: Pteridophytes and Gymnosperms. University Press of Florida, Gainesville.

WUNDERLIN, R.P. AND B.F. HANSEN. Online. Atlas of Florida vascular plants (http://www.plantatlas.usf.edu/). Institute for Systematic Botany, University of South Florida, Tampa.

INDEX OF SCIENTIFIC NAMES

Accepted names are in **bold,** *excluded names* in ***bold italics***, *synonyms* in *italics*; families which contain only excluded names are in *CAPITAL ITALICS*.

B

E

M

O

Q

R

S

T

W

INDEX OF SPANISH COMMON NAMES

Note that here may be more than one taxon with the same common Spanish name on a given page.

C

Q

R

CORRIGENDA

Page 17

Bolbitis nicotianifolia =

Mickelia nicotianifolia (Sw.) R.C. Moran, Labiak & Sundue, Brittonia 62(4):347. 2010.
Bas.: *Acrostichum nicotianifolium* Sw. (1806).
Syn.: *Leptochilus nicotianifolius* (Sw.) C. Chr. (1904). *Bolbitis nicotianifolia* (Sw.) Alston (1932).

Bolbitis pergamentacea =

Mickelia pergamentacea (Maxon) R.C. Moran, Labiak & Sundue, Brittonia 62(4):350. 2010.
Bas.: *Leptochilus pergamentaceus* Maxon (1924).
Syn.: *Bolbitis pergamentacea* (Maxon) Ching (1934).

Page 32

Terpsichore anfractuosa =

Ascogrammitis anfractuosa (Kunze ex Klotzsch) Sundue, Brittonia 62(4):365–370, 3, 13D, 17H–K. 2010.
Bas.: *Polypodium anfractuosum* Kunze ex Klotzsch (1847).
Syn.: *Grammitis anfractuosa* (Kunze ex Klotzsch) Proctor (1961). *Terpsichore anfractuosa* (Kunze ex Klotzsch) B. León & A.R. Sm. (2003).

Aguadilla
Quebradillas
Hatillo
Barceloneta
Dorado
Isabela
Camuy
Arecibo
Manatí
Vega Baja
Vega Alta
Toa Baja
Cataño
Loíza
San Juan
Carolina
Aguada
Moca
Florida
Toa Alta
Bayamón
Río Grande
Luquillo
Culebra
Rincón
San Sebastián
Morovis
Corozal
Guaynabo
Trujillo Alto
Canóvanas
Fajardo
Añasco
Ciales
Naranjito
Lares
Utuado
Aguas Buenas
Gurabo
Ceiba
Las Marías
Comerío
Juncos
Naguabo
Orocovis
Mayagüez
Jayuya
Barranquitas
Caguas
Las Piedras
Adjuntas
Maricao
Cidra
San Lorenzo
Humacao
Hormigueros
Villalba
Aibonito
Vieques
Cayey
San Germán
Yauco
Coamo
Sabana Grande
Peñuelas
Yabucoa
Cabo Rojo
Ponce
Patillas
Juana Díaz
Guayanilla
Salinas
Santa Isabel
Maunabo
Lajas
Arroyo
Guayama
Guánica
0
20
40 km
N

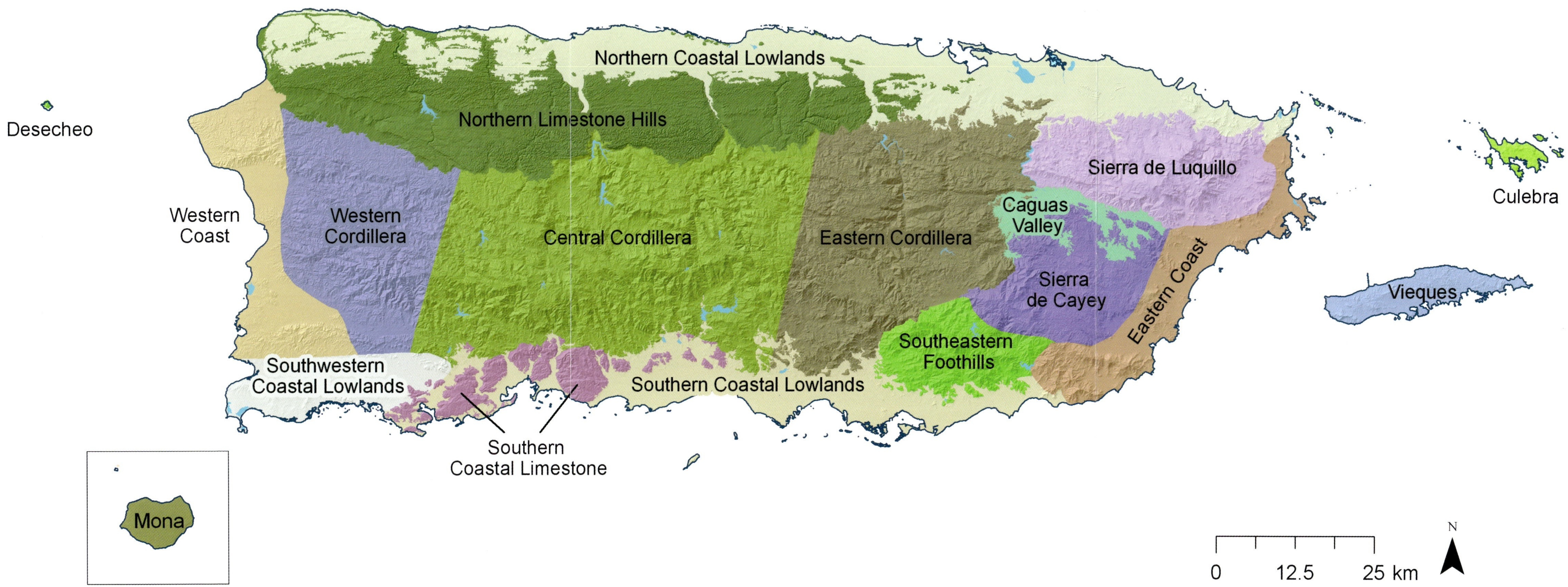

Desecheo
Northern Coastal Lowlands
Northern Limestone Hills
Western Coast
Western Cordillera
Central Cordillera
Eastern Cordillera
Sierra de Luquillo
Caguas Valley
Sierra de Cayey
Eastern Coast
Culebra
Vieques
Southeastern Foothills
Southwestern Coastal Lowlands
Southern Coastal Lowlands
Southern Coastal Limestone
Mona
0
12.5
25 km
N

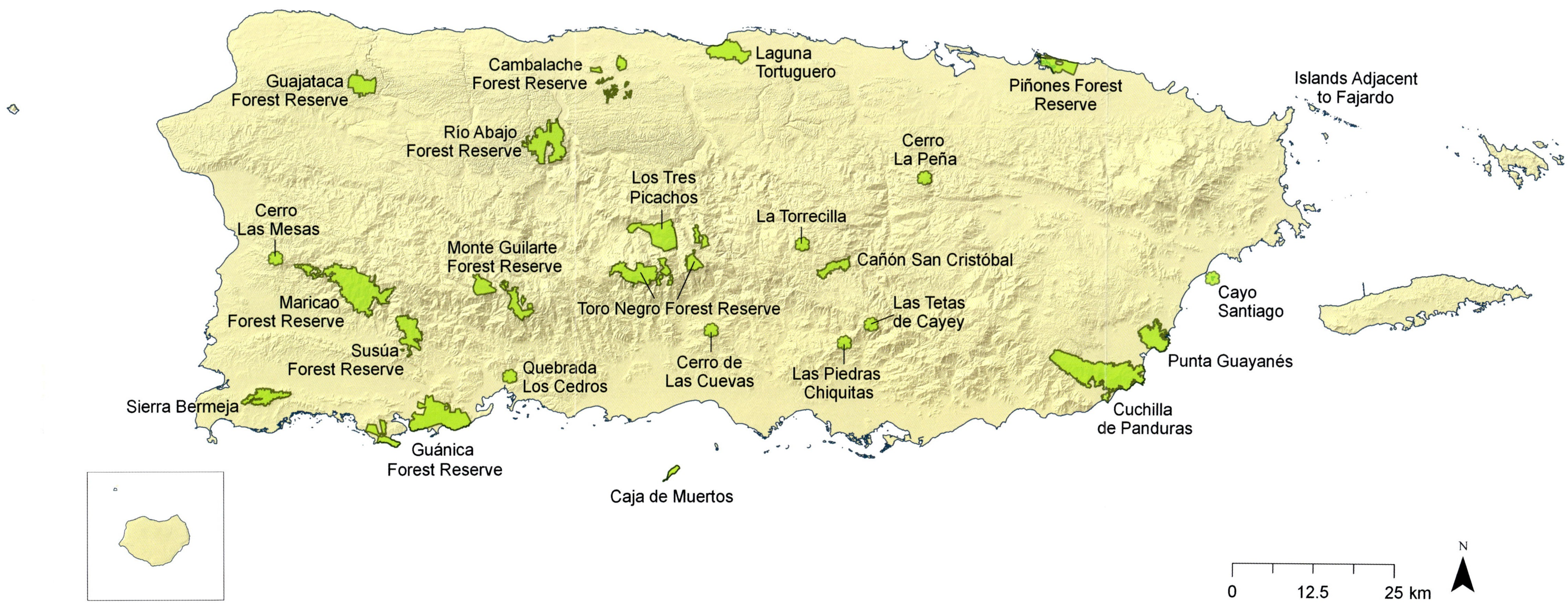

Guajataca
Forest Reserve
Cambalache
Forest Reserve
Laguna
Tortuguero
Piñones Forest
Reserve
Islands Adjacent
to Fajardo
Río Abajo
Forest Reserve
Cerro
La Peña
Los Tres
Picachos
Cerro
Las Mesas
La Torrecilla
Monte Guilarte
Forest Reserve
Cañón San Cristóbal
Maricao
Forest Reserve
Toro Negro Forest Reserve
Las Tetas
de Cayey
Cayo
Santiago
Susúa
Forest Reserve
Quebrada
Los Cedros
Cerro de
Las Cuevas
Las Piedras
Chiquitas
Punta Guayanés
Sierra Bermeja
Cuchilla
de Panduras
Guánica
Forest Reserve
Caja de Muertos
0
12.5
25 km
N